STUDENT'S SOLUTIONS MANUAL

JEFFERY A. COLE
Anoka-Ramsey Community College

BEGINNING ALGEBRA
ELEVENTH EDITION

Margaret L. Lial
American River College

John Hornsby
University of New Orleans

Terry McGinnis

Addison-Wesley
is an imprint of

Addison-Wesley
is an imprint of

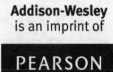

www.pearsonhighered.com

Preface

This *Student's Solutions Manual* contains solutions to selected exercises in the text *Beginning Algebra, Eleventh Edition* by Margaret L. Lial, John Hornsby, and Terry McGinnis. It contains solutions to the Now Try Exercises, the odd-numbered exercises in each section, all Relating Concepts exercises, as well as solutions to all the exercises in the review sections, the chapter tests, and the cumulative review sections.

This manual is a text supplement and should be read along *with* the text. You should read all exercise solutions in this manual because many concept explanations are given and then used in subsequent solutions. All concepts necessary to solve a particular problem are not reviewed for every exercise. If you are having difficulty with a previously covered concept, refer back to the section where it was covered for more complete help.

A significant number of today's students are involved in various outside activities, and find it difficult, if not impossible, to attend all class sessions; this manual should help meet the needs of these students. In addition, it is my hope that this manual's solutions will enhance the understanding of all readers of the material and provide insights to solving other exercises.

I appreciate feedback concerning errors, solution correctness or style, and manual style. Any comments may be sent directly to me at the address below, at jeff.cole@anokaramsey.edu, or in care of the publisher, Pearson Addison-Wesley.

I would like to thank Mary Johnson and Marv Riedesel, formerly of Inver Hills Community College, for their careful accuracy checking and valuable suggestions; Karen Hartpence, for creating the new art pieces; and the authors and Maureen O'Connor and Mary St. Thomas, of Pearson Addison-Wesley, for entrusting me with this project.

Jeffery A. Cole
Anoka-Ramsey Community College
11200 Mississippi Blvd. NW
Coon Rapids, MN 55433

Table of Contents

CHAPTER 1   THE REAL NUMBER SYSTEM

1.1   Fractions

1.1   Now Try Exercises

N1.

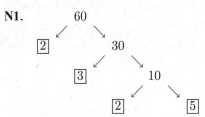

Writing 60 as the product of primes gives us

$$60 = 2 \cdot 2 \cdot 3 \cdot 5.$$

N2. $\dfrac{30}{42} = \dfrac{5 \cdot 6}{7 \cdot 6} = \dfrac{5 \cdot 1}{7 \cdot 1} = \dfrac{5}{7}$

N3. (a) $\dfrac{4}{7} \cdot \dfrac{5}{8} = \dfrac{4 \cdot 5}{7 \cdot 8}$ *Multiply numerators.*
Multiply denominators.

$= \dfrac{4 \cdot 5}{7 \cdot 2 \cdot 4}$ *Factor.*

$= \dfrac{5}{14}$ *Write in lowest terms.*

(b)

$3\dfrac{2}{5} \cdot 6\dfrac{2}{3} = \dfrac{17}{5} \cdot \dfrac{20}{3}$ *Change both mixed numbers to improper fractions.*

$= \dfrac{17 \cdot 20}{5 \cdot 3}$ *Multiply numerators.*
Multiply denominators.

$= \dfrac{17 \cdot 5 \cdot 4}{5 \cdot 3}$ *Factor.*

$= \dfrac{68}{3}, \text{ or } 22\dfrac{2}{3}$ *Write as a mixed number.*

N4. (a) $\dfrac{2}{7} \div \dfrac{8}{9} = \dfrac{2}{7} \cdot \dfrac{9}{8}$ *Multiply by the reciprocal of the second fraction.*

$= \dfrac{2 \cdot 3 \cdot 3}{7 \cdot 2 \cdot 4}$

$= \dfrac{9}{28}$

(b) $3\dfrac{3}{4} \div 4\dfrac{2}{7} = \dfrac{15}{4} \div \dfrac{30}{7}$ *Change both mixed numbers to improper fractions.*

$= \dfrac{15}{4} \cdot \dfrac{7}{30}$ *Multiply by the reciprocal of the second fraction.*

$= \dfrac{15 \cdot 7}{4 \cdot 2 \cdot 15}$

$= \dfrac{7}{8}$

N5. $\dfrac{1}{8} + \dfrac{3}{8} = \dfrac{1+3}{8}$ *Add numerators; denominator does not change.*

$= \dfrac{4}{8}$

$= \dfrac{1 \cdot 4}{2 \cdot 4}$ *Factor.*

$= \dfrac{1}{2}$

N6. (a) $\dfrac{5}{12} + \dfrac{3}{8}$

Since $12 = 2 \cdot 2 \cdot 3$ and $8 = 2 \cdot 2 \cdot 2$, the least common denominator must have three factors of 2 (from 8) and one factor of 3 (from 12), so it is $2 \cdot 2 \cdot 2 \cdot 3 = 24$.

Write each fraction with a denominator of 24.

$\dfrac{5}{12} = \dfrac{5 \cdot 2}{12 \cdot 2} = \dfrac{10}{24}$ and $\dfrac{3}{8} = \dfrac{3 \cdot 3}{8 \cdot 3} = \dfrac{9}{24}$

Now add.

$\dfrac{5}{12} + \dfrac{3}{8} = \dfrac{10}{24} + \dfrac{9}{24} = \dfrac{10+9}{24} = \dfrac{19}{24}$

(b)

$3\dfrac{1}{4} + 5\dfrac{5}{8} = \dfrac{13}{4} + \dfrac{45}{8}$ *Change both mixed numbers to improper fractions.*

The least common denominator is 8, so write each fraction with a denominator of 8.

$\dfrac{45}{8}$ and $\dfrac{13}{4} = \dfrac{13 \cdot 2}{4 \cdot 2} = \dfrac{26}{8}$

Now add.

$\dfrac{13}{4} + \dfrac{45}{8} = \dfrac{26}{8} + \dfrac{45}{8} = \dfrac{26+45}{8}$

$= \dfrac{71}{8}, \text{ or } 8\dfrac{7}{8}$

N7. (a) $\dfrac{5}{11} - \dfrac{2}{9}$

Since $11 = 11$ and $9 = 3 \cdot 3$, the least common denominator is $3 \cdot 3 \cdot 11 = 99$. Write each fraction with a denominator of 99.

$\dfrac{5}{11} = \dfrac{5 \cdot 9}{11 \cdot 9} = \dfrac{45}{99}$ and $\dfrac{2}{9} = \dfrac{2 \cdot 11}{9 \cdot 11} = \dfrac{22}{99}$

Now subtract.

$\dfrac{5}{11} - \dfrac{2}{9} = \dfrac{45}{99} - \dfrac{22}{99} = \dfrac{23}{99}$

(b) $4\frac{1}{3} - 2\frac{5}{6} = \frac{13}{3} - \frac{17}{6}$ *Change each mixed number into an improper fraction.*

The least common denominator is 6. Write each fraction with a denominator of 6. $\frac{17}{6}$ remains unchanged, and

$$\frac{13}{3} = \frac{13 \cdot 2}{3 \cdot 2} = \frac{26}{6}.$$

Now subtract.

$$\frac{13}{3} - \frac{17}{6} = \frac{26}{6} - \frac{17}{6} = \frac{26 - 17}{6} = \frac{9}{6}$$

Now reduce.

$$\frac{9}{6} = \frac{3 \cdot 3}{2 \cdot 3} = \frac{3}{2}, \text{ or } 1\frac{1}{2}$$

N8. To find out how long each piece must be, divide the total length by the number of pieces.

$$10\frac{1}{2} \div 4 = \frac{21}{2} \div \frac{4}{1} = \frac{21}{2} \cdot \frac{1}{4} = \frac{21}{8}, \text{ or } 2\frac{5}{8}$$

Each piece should be $2\frac{5}{8}$ feet long.

N9. **(a)** In the circle graph, the sector for Other is the smallest, so Other had the least number of Internet users.

(b) As in Example 9(b) (using $\frac{1}{3}$ for $\frac{7}{20}$),

$$\frac{1}{3}(1000) \approx 333 \text{ million}.$$

(c) As in Example 9(c),

$$\frac{7}{20}(970) = \frac{7}{20} \cdot \frac{970}{1} = \frac{679}{2}$$
$$= 339\frac{1}{2} \text{ million, or } 339{,}500{,}000.$$

1.1 Section Exercises

1. True; the number above the fraction bar is called the numerator and the number below the fraction bar is called the denominator.

3. False; this is an improper fraction. Its value is 1.

5. False; the fraction $\frac{13}{39}$ can be written in lowest terms as $\frac{1}{3}$ since $\frac{13}{39} = \frac{13 \cdot 1}{13 \cdot 3} = \frac{1}{3}$.

7. False; *product* refers to multiplication, so the product of 10 and 2 is 20. The *sum* of 10 and 2 is 12.

9. Since 19 has only itself and 1 as factors, it is a prime number.

11. $30 = 2 \cdot 15$
$= 2 \cdot 3 \cdot 5$

Since 30 has factors other than itself and 1, it is a composite number.

13. $64 = 2 \cdot 32$
$= 2 \cdot 2 \cdot 16$
$= 2 \cdot 2 \cdot 2 \cdot 8$
$= 2 \cdot 2 \cdot 2 \cdot 2 \cdot 4$
$= 2 \cdot 2 \cdot 2 \cdot 2 \cdot 2 \cdot 2$

Since 64 has factors other than itself and 1, it is a composite number.

15. As stated in the text, the number 1 is neither prime nor composite, by agreement.

17. $57 = 3 \cdot 19$, so 57 is a composite number.

19. Since 79 has only itself and 1 as factors, it is a prime number.

21. $124 = 2 \cdot 62$
$= 2 \cdot 2 \cdot 31$,

so 124 is a composite number.

23. $500 = 2 \cdot 250$
$= 2 \cdot 2 \cdot 125$
$= 2 \cdot 2 \cdot 5 \cdot 25$
$= 2 \cdot 2 \cdot 5 \cdot 5 \cdot 5$,

so 500 is a composite number.

25. $3458 = 2 \cdot 1729$
$= 2 \cdot 7 \cdot 247$
$= 2 \cdot 7 \cdot 13 \cdot 19$

Since 3458 has factors other than itself and 1, it is a composite number.

27. $\frac{8}{16} = \frac{1 \cdot 8}{2 \cdot 8} = \frac{1}{2}$

29. $\frac{15}{18} = \frac{3 \cdot 5}{3 \cdot 6} = \frac{5}{6}$

31. $\frac{64}{100} = \frac{4 \cdot 16}{4 \cdot 25} = \frac{16}{25}$

33. $\frac{18}{90} = \frac{1 \cdot 18}{5 \cdot 18} = \frac{1}{5}$

35. $\frac{144}{120} = \frac{6 \cdot 24}{5 \cdot 24} = \frac{6}{5}$

37. $\frac{16}{24} = \frac{2 \cdot 8}{3 \cdot 8} = \frac{2}{3}$

Therefore, **C** is correct.

39. $\frac{4}{5} \cdot \frac{6}{7} = \frac{4 \cdot 6}{5 \cdot 7} = \frac{24}{35}$

41. $\frac{2}{3} \cdot \frac{15}{16} = \frac{2 \cdot 15}{3 \cdot 16} = \frac{2 \cdot 3 \cdot 5}{3 \cdot 2 \cdot 8} = \frac{5}{8}$

43. $\frac{1}{10} \cdot \frac{12}{5} = \frac{1 \cdot 12}{10 \cdot 5} = \frac{1 \cdot 2 \cdot 6}{2 \cdot 5 \cdot 5} = \frac{6}{25}$

45. $\dfrac{15}{4} \cdot \dfrac{8}{25} = \dfrac{15 \cdot 8}{4 \cdot 25}$

$\qquad = \dfrac{3 \cdot 5 \cdot 4 \cdot 2}{4 \cdot 5 \cdot 5}$

$\qquad = \dfrac{3 \cdot 2}{5}$

$\qquad = \dfrac{6}{5}, \text{ or } 1\dfrac{1}{5}$

47. $21 \cdot \dfrac{3}{7} = \dfrac{21 \cdot 3}{1 \cdot 7}$

$\qquad = \dfrac{3 \cdot 7 \cdot 3}{1 \cdot 7}$

$\qquad = \dfrac{3 \cdot 3}{1} = 9$

49. $3\dfrac{1}{4} \cdot 1\dfrac{2}{3}$

Change both mixed numbers to improper fractions.

$3\dfrac{1}{4} = 3 + \dfrac{1}{4} = \dfrac{12}{4} + \dfrac{1}{4} = \dfrac{13}{4}$

$1\dfrac{2}{3} = 1 + \dfrac{2}{3} = \dfrac{3}{3} + \dfrac{2}{3} = \dfrac{5}{3}$

$3\dfrac{1}{4} \cdot 1\dfrac{2}{3} = \dfrac{13}{4} \cdot \dfrac{5}{3}$

$\qquad = \dfrac{13 \cdot 5}{4 \cdot 3}$

$\qquad = \dfrac{65}{12}, \text{ or } 5\dfrac{5}{12}$

51. $2\dfrac{3}{8} \cdot 3\dfrac{1}{5}$

Change both mixed numbers to improper fractions.

$2\dfrac{3}{8} = 2 + \dfrac{3}{8} = \dfrac{16}{8} + \dfrac{3}{8} = \dfrac{19}{8}$

$3\dfrac{1}{5} = 3 + \dfrac{1}{5} = \dfrac{15}{5} + \dfrac{1}{5} = \dfrac{16}{5}$

$2\dfrac{3}{8} \cdot 3\dfrac{1}{5} = \dfrac{19}{8} \cdot \dfrac{16}{5}$

$\qquad = \dfrac{19 \cdot 16}{8 \cdot 5}$

$\qquad = \dfrac{19 \cdot 2 \cdot 8}{8 \cdot 5}$

$\qquad = \dfrac{38}{5}, \text{ or } 7\dfrac{3}{5}$

53. $\dfrac{5}{4} \div \dfrac{3}{8} = \dfrac{5}{4} \cdot \dfrac{8}{3}$ *Multiply by the reciprocal of the second fraction.*

$\qquad = \dfrac{5 \cdot 8}{4 \cdot 3}$

$\qquad = \dfrac{5 \cdot 4 \cdot 2}{4 \cdot 3}$

$\qquad = \dfrac{5 \cdot 2}{3}$

$\qquad = \dfrac{10}{3}, \text{ or } 3\dfrac{1}{3}$

55. $\dfrac{32}{5} \div \dfrac{8}{15} = \dfrac{32}{5} \cdot \dfrac{15}{8}$ *Multiply by the reciprocal of the second fraction.*

$\qquad = \dfrac{32 \cdot 15}{5 \cdot 8}$

$\qquad = \dfrac{8 \cdot 4 \cdot 3 \cdot 5}{1 \cdot 5 \cdot 8}$

$\qquad = \dfrac{4 \cdot 3}{1} = 12$

57. $\dfrac{3}{4} \div 12 = \dfrac{3}{4} \cdot \dfrac{1}{12}$ *Multiply by the reciprocal of 12.*

$\qquad = \dfrac{3 \cdot 1}{4 \cdot 12}$

$\qquad = \dfrac{3 \cdot 1}{4 \cdot 3 \cdot 4}$

$\qquad = \dfrac{1}{4 \cdot 4} = \dfrac{1}{16}$

59. $6 \div \dfrac{3}{5} = \dfrac{6}{1} \cdot \dfrac{5}{3}$ *Multiply by the reciprocal of the second fraction.*

$\qquad = \dfrac{6 \cdot 5}{1 \cdot 3}$

$\qquad = \dfrac{2 \cdot 3 \cdot 5}{1 \cdot 3}$

$\qquad = \dfrac{2 \cdot 5}{1} = 10$

61. $6\dfrac{3}{4} \div \dfrac{3}{8}$

Change the first number to an improper fraction.

$6\dfrac{3}{4} = 6 + \dfrac{3}{4} = \dfrac{24}{4} + \dfrac{3}{4} = \dfrac{27}{4}$

$6\dfrac{3}{4} \div \dfrac{3}{8} = \dfrac{27}{4} \cdot \dfrac{8}{3}$ *Multiply by the reciprocal of the second fraction.*

$\qquad = \dfrac{27 \cdot 8}{4 \cdot 3}$

$\qquad = \dfrac{3 \cdot 9 \cdot 2 \cdot 4}{4 \cdot 3}$

$\qquad = \dfrac{9 \cdot 2}{1} = 18$

63. $2\dfrac{1}{2} \div 1\dfrac{5}{7}$

Change both mixed numbers to improper fractions.

$$2\dfrac{1}{2} = 2 + \dfrac{1}{2} = \dfrac{4}{2} + \dfrac{1}{2} = \dfrac{5}{2}$$

$$1\dfrac{5}{7} = 1 + \dfrac{5}{7} = \dfrac{7}{7} + \dfrac{5}{7} = \dfrac{12}{7}$$

$$2\dfrac{1}{2} \div 1\dfrac{5}{7} = \dfrac{5}{2} \div \dfrac{12}{7}$$

$$= \dfrac{5}{2} \cdot \dfrac{7}{12} \quad \textit{Multiply by the reciprocal of the second fraction.}$$

$$= \dfrac{5 \cdot 7}{2 \cdot 12}$$

$$= \dfrac{35}{24}, \text{ or } 1\dfrac{11}{24}$$

65. $2\dfrac{5}{8} \div 1\dfrac{15}{32}$

Change both mixed numbers to improper fractions.

$$2\dfrac{5}{8} = 2 + \dfrac{5}{8} = \dfrac{16}{8} + \dfrac{5}{8} = \dfrac{21}{8}$$

$$1\dfrac{15}{32} = 1 + \dfrac{15}{32} = \dfrac{32}{32} + \dfrac{15}{32} = \dfrac{47}{32}$$

$$2\dfrac{5}{8} \div 1\dfrac{15}{32} = \dfrac{21}{8} \div \dfrac{47}{32}$$

$$= \dfrac{21}{8} \cdot \dfrac{32}{47}$$

$$= \dfrac{21 \cdot 32}{8 \cdot 47}$$

$$= \dfrac{21 \cdot 8 \cdot 4}{8 \cdot 47}$$

$$= \dfrac{21 \cdot 4}{47}$$

$$= \dfrac{84}{47}, \text{ or } 1\dfrac{37}{47}$$

67. A common denominator for $\dfrac{p}{q}$ and $\dfrac{r}{s}$ must be a multiple of both denominators, q and s. Such a number is $q \cdot s$. Therefore, **A** is correct.

69. $\dfrac{7}{15} + \dfrac{4}{15} = \dfrac{7 + 4}{15} = \dfrac{11}{15}$

71. $\dfrac{7}{12} + \dfrac{1}{12} = \dfrac{7 + 1}{12}$

$$= \dfrac{8}{12}$$

$$= \dfrac{2 \cdot 4}{3 \cdot 4} = \dfrac{2}{3}$$

73. $\dfrac{5}{9} + \dfrac{1}{3}$

Since $9 = 3 \cdot 3$, and 3 is prime, the LCD (least common denominator) is $3 \cdot 3 = 9$.

$$\dfrac{1}{3} = \dfrac{1}{3} \cdot \dfrac{3}{3} = \dfrac{3}{9}$$

Now add the two fractions with the same denominator.

$$\dfrac{5}{9} + \dfrac{1}{3} = \dfrac{5}{9} + \dfrac{3}{9} = \dfrac{8}{9}$$

75. $\dfrac{3}{8} + \dfrac{5}{6}$

Since $8 = 2 \cdot 2 \cdot 2$ and $6 = 2 \cdot 3$, the LCD is $2 \cdot 2 \cdot 2 \cdot 3 = 24$.

$$\dfrac{3}{8} = \dfrac{3}{8} \cdot \dfrac{3}{3} = \dfrac{9}{24} \text{ and } \dfrac{5}{6} \cdot \dfrac{4}{4} = \dfrac{20}{24}$$

Now add fractions with the same denominator.

$$\dfrac{3}{8} + \dfrac{5}{6} = \dfrac{9}{24} + \dfrac{20}{24} = \dfrac{29}{24}, \text{ or } 1\dfrac{5}{24}$$

77. $3\dfrac{1}{8} + 2\dfrac{1}{4}$

$$3\dfrac{1}{8} = 3 + \dfrac{1}{8} = \dfrac{24}{8} + \dfrac{1}{8} = \dfrac{25}{8}$$

$$2\dfrac{1}{4} = 2 + \dfrac{1}{4} = \dfrac{8}{4} + \dfrac{1}{4} = \dfrac{9}{4}$$

$$3\dfrac{1}{8} + 2\dfrac{1}{4} = \dfrac{25}{8} + \dfrac{9}{4}$$

Since $8 = 2 \cdot 2 \cdot 2$ and $4 = 2 \cdot 2$, the LCD is $2 \cdot 2 \cdot 2$ or 8.

$$3\dfrac{1}{8} + 2\dfrac{1}{4} = \dfrac{25}{8} + \dfrac{9 \cdot 2}{4 \cdot 2}$$

$$= \dfrac{25}{8} + \dfrac{18}{8}$$

$$= \dfrac{43}{8}, \text{ or } 5\dfrac{3}{8}$$

79. $3\dfrac{1}{4} + 1\dfrac{4}{5}$

$$3\dfrac{1}{4} = 3 + \dfrac{1}{4} = \dfrac{12}{4} + \dfrac{1}{4} = \dfrac{13}{4}$$

$$1\dfrac{4}{5} = 1 + \dfrac{4}{5} = \dfrac{5}{5} + \dfrac{4}{5} = \dfrac{9}{5}$$

Since $4 = 2 \cdot 2$, and 5 is prime, the LCD is $2 \cdot 2 \cdot 5 = 20$.

$$3\dfrac{1}{4} + 1\dfrac{4}{5} = \dfrac{13 \cdot 5}{4 \cdot 5} + \dfrac{9 \cdot 4}{5 \cdot 4}$$

$$= \dfrac{65}{20} + \dfrac{36}{20}$$

$$= \dfrac{101}{20}, \text{ or } 5\dfrac{1}{20}$$

81. $\dfrac{7}{9} - \dfrac{2}{9} = \dfrac{7-2}{9} = \dfrac{5}{9}$

83. $\dfrac{13}{15} - \dfrac{3}{15} = \dfrac{13-3}{15}$

$= \dfrac{10}{15}$

$= \dfrac{2 \cdot 5}{3 \cdot 5} = \dfrac{2}{3}$

85. $\dfrac{7}{12} - \dfrac{1}{3}$

Since $12 = 4 \cdot 3$ (12 is a multiple of 3), the LCD is 12.

$$\dfrac{1}{3} \cdot \dfrac{4}{4} = \dfrac{4}{12}$$

Now subtract fractions with the same denominator.

$$\dfrac{7}{12} - \dfrac{1}{3} = \dfrac{7}{12} - \dfrac{4}{12} = \dfrac{3}{12} = \dfrac{1 \cdot 3}{4 \cdot 3} = \dfrac{1}{4}$$

87. $\dfrac{7}{12} - \dfrac{1}{9}$

Since $12 = 2 \cdot 2 \cdot 3$ and $9 = 3 \cdot 3$, the LCD is $2 \cdot 2 \cdot 3 \cdot 3 = 36$.

$$\dfrac{7}{12} = \dfrac{7}{12} \cdot \dfrac{3}{3} = \dfrac{21}{36} \text{ and } \dfrac{1}{9} \cdot \dfrac{4}{4} = \dfrac{4}{36}$$

Now subtract fractions with the same denominator.

$$\dfrac{7}{12} - \dfrac{1}{9} = \dfrac{21}{36} - \dfrac{4}{36} = \dfrac{17}{36}$$

89. $4\dfrac{3}{4} - 1\dfrac{2}{5}$

$4\dfrac{3}{4} = 4 + \dfrac{3}{4} = \dfrac{16}{4} + \dfrac{3}{4} = \dfrac{19}{4}$

$1\dfrac{2}{5} = 1 + \dfrac{2}{5} = \dfrac{5}{5} + \dfrac{2}{5} = \dfrac{7}{5}$

Since $4 = 2 \cdot 2$, and 5 is prime, the LCD is $2 \cdot 2 \cdot 5 = 20$.

$4\dfrac{3}{4} - 1\dfrac{2}{5} = \dfrac{19 \cdot 5}{4 \cdot 5} - \dfrac{7 \cdot 4}{5 \cdot 4}$

$= \dfrac{95}{20} - \dfrac{28}{20}$

$= \dfrac{67}{20}, \text{ or } 3\dfrac{7}{20}$

91. $6\dfrac{1}{4} - 5\dfrac{1}{3}$

$6\dfrac{1}{4} = 6 + \dfrac{1}{4} = \dfrac{24}{4} + \dfrac{1}{4} = \dfrac{25}{4}$

$5\dfrac{1}{3} = 5 + \dfrac{1}{3} = \dfrac{15}{3} + \dfrac{1}{3} = \dfrac{16}{3}$

Since $4 = 2 \cdot 2$, and 3 is prime, the LCD is $2 \cdot 2 \cdot 3 = 12$.

$6\dfrac{1}{4} - 5\dfrac{1}{3} = \dfrac{25}{4} - \dfrac{16}{3}$

$= \dfrac{25 \cdot 3}{4 \cdot 3} - \dfrac{16 \cdot 4}{3 \cdot 4}$

$= \dfrac{75}{12} - \dfrac{64}{12}$

$= \dfrac{11}{12}$

93. Multiply the number of cups of water per serving by the number of servings.

$\dfrac{3}{4} \cdot 8 = \dfrac{3}{4} \cdot \dfrac{8}{1}$

$= \dfrac{3 \cdot 8}{4 \cdot 1}$

$= \dfrac{3 \cdot 2 \cdot 4}{4 \cdot 1}$

$= \dfrac{3 \cdot 2}{1} = 6 \text{ cups}$

For 8 microwave servings, 6 cups of water will be needed.

95. The difference in length is found by subtracting.

$3\dfrac{1}{4} - 2\dfrac{1}{8} = \dfrac{13}{4} - \dfrac{17}{8}$

$= \dfrac{13 \cdot 2}{4 \cdot 2} - \dfrac{17}{8} \quad LCD = 8$

$= \dfrac{26}{8} - \dfrac{17}{8}$

$= \dfrac{9}{8}, \text{ or } 1\dfrac{1}{8}$

The difference is $1\dfrac{1}{8}$ inches.

97. The difference between the two measures is found by subtracting, using 16 as the LCD.

$\dfrac{3}{4} - \dfrac{3}{16} = \dfrac{3 \cdot 4}{4 \cdot 4} - \dfrac{3}{16}$

$= \dfrac{12}{16} - \dfrac{3}{16}$

$= \dfrac{12-3}{16} = \dfrac{9}{16}$

The difference is $\dfrac{9}{16}$ inch.

99. The perimeter is the sum of the measures of the 5 sides.

$$196 + 98\frac{3}{4} + 146\frac{1}{2} + 100\frac{7}{8} + 76\frac{5}{8}$$
$$= 196 + 98\frac{6}{8} + 146\frac{4}{8} + 100\frac{7}{8} + 76\frac{5}{8}$$
$$= 196 + 98 + 146 + 100 + 76 + \frac{6+4+7+5}{8}$$
$$= 616 + \frac{22}{8} \quad \left(\frac{22}{8} = 2\frac{6}{8} = 2\frac{3}{4}\right)$$
$$= 618\frac{3}{4} \text{ feet}$$

The perimeter is $618\frac{3}{4}$ feet.

101. Divide the total board length by 3.

$$15\frac{5}{8} \div 3 = \frac{125}{8} \div \frac{3}{1}$$
$$= \frac{125}{8} \cdot \frac{1}{3}$$
$$= \frac{125 \cdot 1}{8 \cdot 3}$$
$$= \frac{125}{24}, \text{ or } 5\frac{5}{24}$$

The length of each of the three pieces must be $5\frac{5}{24}$ inches.

103. To find the number of cakes the caterer can make, divide $15\frac{1}{2}$ by $1\frac{3}{4}$.

$$15\frac{1}{2} \div 1\frac{3}{4} = \frac{31}{2} \div \frac{7}{4}$$
$$= \frac{31}{2} \cdot \frac{4}{7}$$
$$= \frac{31 \cdot 2 \cdot 2}{2 \cdot 7}$$
$$= \frac{62}{7}, \text{ or } 8\frac{6}{7}$$

There is not quite enough sugar for 9 cakes. The caterer can make 8 cakes with some sugar left over.

105. Multiply the amount of fabric it takes to make one costume by the number of costumes.

$$2\frac{3}{8} \cdot 7 = \frac{19}{8} \cdot \frac{7}{1}$$
$$= \frac{19 \cdot 7}{8 \cdot 1}$$
$$= \frac{133}{8}, \text{ or } 16\frac{5}{8} \text{ yd}$$

For 7 costumes, $16\frac{5}{8}$ yards of fabric would be needed.

107. Subtract the heights to find the difference.

$$10\frac{1}{2} - 7\frac{1}{8} = \frac{21}{2} - \frac{57}{8}$$
$$= \frac{21 \cdot 4}{2 \cdot 4} - \frac{57}{8} \quad LCD = 8$$
$$= \frac{84}{8} - \frac{57}{8}$$
$$= \frac{27}{8}, \text{ or } 3\frac{3}{8}$$

The difference in heights is $3\frac{3}{8}$ inches.

109. The sum of the fractions representing the U.S. foreign-born population from Latin America, Asia, or Europe is

$$\frac{27}{50} + \frac{27}{100} + \frac{7}{50} = \frac{27 \cdot 2}{50 \cdot 2} + \frac{27}{100} + \frac{7 \cdot 2}{50 \cdot 2}$$
$$= \frac{54 + 27 + 14}{100}$$
$$= \frac{95}{100}.$$

So the fraction representing the U.S. foreign-born population from other regions is

$$1 - \frac{95}{100} = \frac{100}{100} - \frac{95}{100}$$
$$= \frac{5}{100} = \frac{1}{20}.$$

111. Multiply the fraction representing the U.S. foreign-born population from Europe, $\frac{7}{50}$, by the total number of foreign-born people in the U.S., approximately 38 million.

$$\frac{7}{50} \cdot 38 = \frac{7}{50} \cdot \frac{38}{1} = \frac{7 \cdot 2 \cdot 19}{2 \cdot 25} = \frac{133}{25}, \text{ or } 5\frac{8}{25}$$

There were approximately $5\frac{8}{25}$ million (or 5,320,000) foreign-born people in the U.S. in 2006 who were born in Europe.

113. Observe that there are 24 dots in the entire figure, 6 dots in the triangle, 12 dots in the rectangle, and 2 dots in the overlapping region.

(a) $\frac{12}{24} = \frac{1}{2}$ of all the dots are in the rectangle.

(b) $\frac{6}{24} = \frac{1}{4}$ of all the dots are in the triangle.

(c) $\frac{2}{6} = \frac{1}{3}$ of the dots in the triangle are in the overlapping region.

(d) $\frac{2}{12} = \frac{1}{6}$ of the dots in the rectangle are in the overlapping region.

1.2 Exponents, Order of Operations, and Inequality

1.2 Now Try Exercises

N1. (a) $6^2 = 6 \cdot 6 = 36$

  (b) $\left(\dfrac{4}{5}\right)^3 = \underbrace{\dfrac{4}{5} \cdot \dfrac{4}{5} \cdot \dfrac{4}{5}}_{} = \dfrac{64}{125}$

    $\dfrac{4}{5}$ is used as a factor 3 times.

N2. (a) $15 - 2 \cdot 6 = 15 - 12$ *Multiply.*
              $= 3$ *Subtract.*

  (b) $6(2 + 4) - 7 \cdot 5$
    $= 6(6) - 7 \cdot 5$ *Add inside parentheses.*
    $= 36 - 35$ *Multiply.*
    $= 1$ *Subtract.*

  (c) $8 \cdot 10 \div 4 - 2^3 + 3 \cdot 4^2$
    $= 8 \cdot 10 \div 4 - 8 + 3 \cdot 16$ *Use exponents.*
    $= 80 \div 4 - 8 + 48$ *Multiply.*
    $= 20 - 8 + 48$ *Divide.*
    $= 12 + 48$ *Subtract.*
    $= 60$ *Add.*

N3. (a) $7[(3^2 - 1) + 4]$
    $= 7[(9 - 1) + 4]$ *Use the exponent.*
    $= 7[8 + 4]$ *Subtract inside parens.*
    $= 7(12)$ *Add inside parentheses.*
    $= 84$ *Multiply.*

  (b) $\dfrac{9(14 - 4) - 2}{4 + 3 \cdot 6} = \dfrac{9(10) - 2}{4 + 3 \cdot 6}$ *Subt. inside parentheses.*

    $= \dfrac{90 - 2}{4 + 18}$ *Multiply.*

    $= \dfrac{88}{22}$ *Subtract and add.*

    $= 4$ *Divide.*

N4. (a) The statement $12 \neq 10 - 2$ is *true* because 12 *is not equal to* 8.

  (b) The statement $5 > 4 \cdot 2$ is *false* because 5 *is less than* 8.

  (c) The statement $7 \leq 7$ is *true* since $7 = 7$.

  (d) Write the fractions with a common denominator. The statement $\frac{5}{9} > \frac{7}{11}$ is equivalent to the statement $\frac{55}{99} > \frac{63}{99}$. Since 55 is *less* than 63, the original statement is *false*.

N5. (a) "Ten is not equal to eight minus two" is written $10 \neq 8 - 2$.

  (b) "Fifty is greater than fifteen" is written $50 > 15$.

  (c) "Eleven is less than or equal to twenty" is written $11 \leq 20$.

N6. $8 < 9$ may be written as $9 > 8$.

1.2 Section Exercises

1. False; 6^2 means that 6 is used as a factor 2 times, so $6^2 = 6 \cdot 6 = 36$.

3. False; 1 raised to *any* power is 1. Here, $1^3 = 1 \cdot 1 \cdot 1 = 1$.

5. False; $4 + 3(8 - 2) = 4 + 3 \cdot 6 = 4 + 18 = 22$. The common error leading to 42 is adding 4 to 3 and then multiplying by 6. One must follow the rules for order of operations.

7. $3^2 = 3 \cdot 3 = 9$

9. $7^2 = 7 \cdot 7 = 49$

11. $12^2 = 12 \cdot 12 = 144$

13. $4^3 = 4 \cdot 4 \cdot 4 = 64$

15. $10^3 = 10 \cdot 10 \cdot 10 = 1000$

17. $3^4 = 3 \cdot 3 \cdot 3 \cdot 3 = 81$

19. $4^5 = 4 \cdot 4 \cdot 4 \cdot 4 \cdot 4 = 1024$

21. $\left(\dfrac{1}{6}\right)^2 = \dfrac{1}{6} \cdot \dfrac{1}{6} = \dfrac{1}{36}$

23. $\left(\dfrac{2}{3}\right)^4 = \dfrac{2}{3} \cdot \dfrac{2}{3} \cdot \dfrac{2}{3} \cdot \dfrac{2}{3} = \dfrac{16}{81}$

25. $(0.4)^3 = (0.4)(0.4)(0.4) = 0.064$

27. $64 \div 4 \cdot 2 = (64 \div 4) \cdot 2$
              $= 16 \cdot 2$
              $= 32$

29. $13 + 9 \cdot 5 = 13 + 45$ *Multiply.*
            $= 58$ *Add.*

31. $25.2 - 12.6 \div 4.2 = 25.2 - 3$ *Divide.*
                   $= 22.2$ *Subtract.*

33. $\dfrac{1}{4} \cdot \dfrac{2}{3} + \dfrac{2}{5} \cdot \dfrac{11}{3} = \dfrac{1}{6} + \dfrac{22}{15}$ *Multiply.*

    $= \dfrac{5}{30} + \dfrac{44}{30}$ *LCD = 30*

    $= \dfrac{49}{30}$, or $1\dfrac{19}{30}$ *Add.*

35. $9 \cdot 4 - 8 \cdot 3 = 36 - 24$ *Multiply.*
             $= 12$ *Subtract.*

37. $20 - 4 \cdot 3 + 5 = 20 - 12 + 5$ *Multiply.*
                $= 8 + 5$ *Subtract.*
                $= 13$ *Add.*

39. $10 + 40 \div 5 \cdot 2 = 10 + 8 \cdot 2$ *Divide.*
                $= 10 + 16$ *Multiply.*
                $= 26$ *Add.*

41. $18 - 2(3 + 4) = 18 - 2(7)$ *Add inside parentheses.*

$$= 18 - 14 \quad \textit{Multiply.}$$
$$= 4 \quad\quad\quad \textit{Subtract.}$$

43. $3(4 + 2) + 8 \cdot 3 = 3 \cdot 6 + 8 \cdot 3$ *Add.*

$$= 18 + 24 \quad \textit{Multiply.}$$
$$= 42 \quad\quad\quad \textit{Add.}$$

45. $18 - 4^2 + 3$

$$= 18 - 16 + 3 \quad \textit{Use the exponent.}$$
$$= 2 + 3 \quad\quad\quad \textit{Subtract.}$$
$$= 5 \quad\quad\quad\quad\; \textit{Add.}$$

47. $2 + 3[5 + 4(2)]$

$$= 2 + 3[5 + 8] \quad \textit{Multiply.}$$
$$= 2 + 3[13] \quad\quad \textit{Add.}$$
$$= 2 + 39 \quad\quad\quad \textit{Multiply.}$$
$$= 41 \quad\quad\quad\quad\; \textit{Add.}$$

49. $5[3 + 4(2^2)]$

$$= 5[3 + 4(4)] \quad \textit{Use the exponent.}$$
$$= 5(3 + 16) \quad\quad \textit{Multiply.}$$
$$= 5(19) \quad\quad\quad\; \textit{Add.}$$
$$= 95 \quad\quad\quad\quad\;\; \textit{Multiply.}$$

51. $3^2[(11 + 3) - 4]$

$$= 3^2[14 - 4] \quad \textit{Add inside parentheses.}$$
$$= 3^2[10] \quad\quad\; \textit{Subtract.}$$
$$= 9[10] \quad\quad\;\; \textit{Use the exponent.}$$
$$= 90 \quad\quad\quad\;\; \textit{Multiply.}$$

53. Simplify the numerator and denominator separately; then divide.

$$\frac{6(3^2 - 1) + 8}{8 - 2^2} = \frac{6(9 - 1) + 8}{8 - 4}$$
$$= \frac{6(8) + 8}{4}$$
$$= \frac{48 + 8}{4}$$
$$= \frac{56}{4} = 14$$

55. $\dfrac{4(6 + 2) + 8(8 - 3)}{6(4 - 2) - 2^2} = \dfrac{4(8) + 8(5)}{6(2) - 2^2}$

$$= \frac{4(8) + 8(5)}{6(2) - 4}$$
$$= \frac{32 + 40}{12 - 4}$$
$$= \frac{72}{8} = 9$$

57. $9 \cdot 3 - 11 \le 16$

$$27 - 11 \le 16$$
$$16 \le 16$$

The statement is true since $16 = 16$ is true.

59. $5 \cdot 11 + 2 \cdot 3 \le 60$

$$55 + 6 \le 60$$
$$61 \le 60$$

The statement is false since 61 *is greater than* 60.

61. $0 \ge 12 \cdot 3 - 6 \cdot 6$

$$0 \ge 36 - 36$$
$$0 \ge 0$$

The statement is true since $0 = 0$ is true.

63. $45 \ge 2[2 + 3(2 + 5)]$

$$45 \ge 2[2 + 3(7)]$$
$$45 \ge 2[2 + 21]$$
$$45 \ge 2[23]$$
$$45 \ge 46$$

The statement is false since 45 *is less than* 46.

65. $[3 \cdot 4 + 5(2)] \cdot 3 > 72$

$$[12 + 10] \cdot 3 > 72$$
$$[22] \cdot 3 > 72$$
$$66 > 72$$

The statement is false since 66 *is less than* 72.

67. $\dfrac{3 + 5(4 - 1)}{2 \cdot 4 + 1} \ge 3$

$$\frac{3 + 5(3)}{8 + 1} \ge 3$$
$$\frac{3 + 15}{9} \ge 3$$
$$\frac{18}{9} \ge 3$$
$$2 \ge 3$$

The statement is false since 2 *is less than* 3.

69. $3 \ge \dfrac{2(5 + 1) - 3(1 + 1)}{5(8 - 6) - 4 \cdot 2}$

$$3 \ge \frac{2(6) - 3(2)}{5(2) - 8}$$
$$3 \ge \frac{12 - 6}{10 - 8}$$
$$3 \ge \frac{6}{2}$$
$$3 \ge 3$$

The statement is true since $3 = 3$ is true.

71. $3 \cdot 6 + 4 \cdot 2 = 60$

Listed below are some possibilities. We'll use trial and error until we get the desired result.

$$(3 \cdot 6) + 4 \cdot 2 = 18 + 8 = 26 \neq 60$$
$$(3 \cdot 6 + 4) \cdot 2 = 22 \cdot 2 = 44 \neq 60$$
$$3 \cdot (6 + 4 \cdot 2) = 3 \cdot 14 = 42 \neq 60$$
$$3 \cdot (6 + 4) \cdot 2 = 3 \cdot 10 \cdot 2 = 30 \cdot 2 = 60$$

73. $10 - 7 - 3 = 6$

$$10 - (7 - 3) = 10 - 4 = 6$$

75. $8 + 2^2 = 100$

$$(8 + 2)^2 = 10^2 = 10 \cdot 10 = 100$$

77. "$5 < 17$" means "five is less than seventeen." The statement is true.

79. "$5 \neq 8$" means "five is not equal to eight." The statement is true.

81. "$7 \geq 14$" means "seven is greater than or equal to fourteen." The statement is false.

83. "$15 \leq 15$" means "fifteen is less than or equal to fifteen." The statement is true.

85. "Fifteen is equal to five plus ten" is written

$$15 = 5 + 10.$$

87. "Nine is greater than five minus four" is written

$$9 > 5 - 4.$$

89. "Sixteen is not equal to nineteen" is written

$$16 \neq 19.$$

91. "One-half is less than or equal to two-fourths" is written

$$\frac{1}{2} \leq \frac{2}{4}.$$

93. $5 < 20$ becomes $20 > 5$ when the inequality symbol is reversed.

95. $2.5 \geq 1.3$ becomes $1.3 \leq 2.5$ when the inequality symbol is reversed.

97. **(a)** Substitute "40" for "age" in the expression for women.

$$14.7 - 40 \cdot 0.13$$

(b) $14.7 - 40 \cdot 0.13 = 14.7 - 5.2$ *Multiply.*
$$= 9.5 \qquad\qquad \textit{Subtract.}$$

(c) 85% of $9.5 = 0.85(9.5) = 8.075$

Walking at 5 mph is associated with 8.0 METs, which is the table value closest to 8.075.

99. Answers will vary.

1.3 Variables, Expressions, and Equations

1.3 Now Try Exercises

N1. **(a)** $9k = 9 \cdot k$
$$= 9 \cdot 6 \quad \textit{Let k = 6.}$$
$$= 54 \quad \textit{Multiply.}$$

(b) $4k^2 = 4 \cdot k^2$
$$= 4 \cdot 6^2 \quad \textit{Let k = 6.}$$
$$= 4 \cdot 36 \quad \textit{Square 6.}$$
$$= 144 \quad \textit{Multiply.}$$

N2. Replace x with 4 and y with 7 in each expression.

(a) $3x + 4y = 3(4) + 4(7)$
$$= 12 + 28 \quad \textit{Multiply.}$$
$$= 40 \quad\quad\; \textit{Add.}$$

(b) $\dfrac{6x - 2y}{2y - 9} = \dfrac{6(4) - 2(7)}{2(7) - 9}$

$$= \frac{24 - 14}{14 - 9} \quad \textit{Multiply.}$$

$$= \frac{10}{5} = 2 \quad \textit{Subtract; reduce.}$$

(c) $4x^2 - y^2 = 4 \cdot 4^2 - 7^2$
$$= 4 \cdot 16 - 49 \quad \textit{Use exponents.}$$
$$= 64 - 49 \quad\quad\;\; \textit{Multiply.}$$
$$= 15 \quad\quad\quad\quad\; \textit{Subtract.}$$

N3. **(a)** Using x as the variable to represent the number, "the sum of a number and 10" translates as $x + 10$, or $10 + x$.

(b) "A number divided by 7" translates as $x \div 7$, or $\frac{x}{7}$.

(c) "The difference between 9 and a number" translates as $9 - x$. Thus, "the product of 3 and the difference between 9 and a number" translates as $3(9 - x)$.

N4. $8k + 5 = 61$
$$8 \cdot 7 + 5 \overset{?}{=} 61 \quad \textit{Replace k with 7.}$$
$$56 + 5 \overset{?}{=} 61 \quad \textit{Multiply.}$$
$$61 = 61 \quad \textit{True}$$

The number 7 is a solution of the equation.

N5. Using x as the variable to represent the number, "The sum of a number and nine is equal to the difference between 25 and the number" translates as

$$x + 9 = 25 - x.$$

Now try each number from the set $\{0, 2, 4, 6, 8, 10\}$.

$\boldsymbol{x = 4}$: $4 + 9 \overset{?}{=} 25 - 4$

         $13 = 21$ *False*

$\boldsymbol{x = 6}$: $6 + 9 \overset{?}{=} 25 - 6$

         $15 = 19$ *False*

$\boldsymbol{x = 8}$: $8 + 9 \overset{?}{=} 25 - 8$

         $17 = 17$ *True*

Similarly, $x = 0, 2,$ or 10 result in false statements. Thus, 8 is the only solution.

N6.   **(a)** $2x + 5 = 6$ has an equals symbol, so this is an *equation*.

      (b) $2x + 5 - 6$ has no equals symbol, so this is an *expression*.

1.3 Section Exercises

1.   The expression $8x^2$ means $8 \cdot x \cdot x$. The correct choice is **B**.

3.   The sum of 15 and a number x is represented by the expression $15 + x$. The correct choice is **A**.

5. $2x^3 = 2 \cdot x \cdot x \cdot x$, while $2x \cdot 2x \cdot 2x = (2x)^3$. The last expression is equal to $8x^3$.

7.   The exponent 2 applies only to its base, which is x. (The expression $(5x)^2$ would require multiplying 5 by $x = 4$ first.)

In part (a) of Exercises 9–22, replace x with 4. In part (b), replace x with 6. Then use the order of operations.

9.   **(a)** $x + 7 = 4 + 7$
             $= 11$

      (b) $x + 7 = 6 + 7$
             $= 13$

11.   **(a)** $4x = 4(4) = 16$

      (b) $4x = 4(6) = 24$

13.   **(a)** $4x^2 = 4 \cdot 4^2$
             $= 4 \cdot 16$
             $= 64$

      (b) $4x^2 = 4 \cdot 6^2$
             $= 4 \cdot 36$
             $= 144$

15.   **(a)** $\dfrac{x+1}{3} = \dfrac{4+1}{3}$

              $= \dfrac{5}{3}$

      (b) $\dfrac{x+1}{3} = \dfrac{6+1}{3}$

              $= \dfrac{7}{3}$

17.   **(a)** $\dfrac{3x-5}{2x} = \dfrac{3 \cdot 4 - 5}{2 \cdot 4}$

              $= \dfrac{12 - 5}{8}$

              $= \dfrac{7}{8}$

      (b) $\dfrac{3x-5}{2x} = \dfrac{3 \cdot 6 - 5}{2 \cdot 6}$

              $= \dfrac{18 - 5}{12}$

              $= \dfrac{13}{12}$

19.   **(a)** $3x^2 + x = 3 \cdot 4^2 + 4$
                $= 3 \cdot 16 + 4$
                $= 48 + 4 = 52$

      (b) $3x^2 + x = 3 \cdot 6^2 + 6$
                $= 3 \cdot 36 + 6$
                $= 108 + 6 = 114$

21.   **(a)** $6.459x = 6.459 \cdot 4$
                $= 25.836$

      (b) $6.459x = 6.459 \cdot 6$
                $= 38.754$

In part (a) of Exercises 23–38, replace x with 2 and y with 1. In part (b), replace x with 1 and y with 5.

23.   **(a)** $8x + 3y + 5 = 8(2) + 3(1) + 5$
                    $= 16 + 3 + 5$
                    $= 19 + 5$
                    $= 24$

      (b) $8x + 3y + 5 = 8(1) + 3(5) + 5$
                    $= 8 + 15 + 5$
                    $= 23 + 5$
                    $= 28$

25.   **(a)** $3(x + 2y) = 3(2 + 2 \cdot 1)$
                 $= 3(2 + 2)$
                 $= 3(4)$
                 $= 12$

      (b) $3(x + 2y) = 3(1 + 2 \cdot 5)$
                 $= 3(1 + 10)$
                 $= 3(11)$
                 $= 33$

27. **(a)** $x + \dfrac{4}{y} = 2 + \dfrac{4}{1}$

$= 2 + 4$

$= 6$

(b) $x + \dfrac{4}{y} = 1 + \dfrac{4}{5}$

$= \dfrac{5}{5} + \dfrac{4}{5}$

$= \dfrac{9}{5}$

29. **(a)** $\dfrac{x}{2} + \dfrac{y}{3} = \dfrac{2}{2} + \dfrac{1}{3}$

$= \dfrac{6}{6} + \dfrac{2}{6}$

$= \dfrac{8}{6} = \dfrac{4}{3}$

(b) $\dfrac{x}{2} + \dfrac{y}{3} = \dfrac{1}{2} + \dfrac{5}{3}$

$= \dfrac{3}{6} + \dfrac{10}{6}$

$= \dfrac{13}{6}$

31. **(a)** $\dfrac{2x + 4y - 6}{5y + 2} = \dfrac{2(2) + 4(1) - 6}{5(1) + 2}$

$= \dfrac{4 + 4 - 6}{5 + 2}$

$= \dfrac{8 - 6}{7}$

$= \dfrac{2}{7}$

(b) $\dfrac{2x + 4y - 6}{5y + 2} = \dfrac{2(1) + 4(5) - 6}{5(5) + 2}$

$= \dfrac{2 + 20 - 6}{25 + 2}$

$= \dfrac{22 - 6}{27}$

$= \dfrac{16}{27}$

33. **(a)** $2y^2 + 5x = 2 \cdot 1^2 + 5 \cdot 2$

$= 2 \cdot 1 + 5 \cdot 2$

$= 2 + 10$

$= 12$

(b) $2y^2 + 5x = 2 \cdot 5^2 + 5 \cdot 1$

$= 2 \cdot 25 + 5 \cdot 1$

$= 50 + 5$

$= 55$

35. **(a)** $\dfrac{3x + y^2}{2x + 3y} = \dfrac{3(2) + 1^2}{2(2) + 3(1)}$

$= \dfrac{3(2) + 1}{4 + 3}$

$= \dfrac{6 + 1}{7}$

$= \dfrac{7}{7}$

$= 1$

(b) $\dfrac{3x + y^2}{2x + 3y} = \dfrac{3(1) + 5^2}{2(1) + 3(5)}$

$= \dfrac{3(1) + 25}{2 + 15}$

$= \dfrac{3 + 25}{17}$

$= \dfrac{28}{17}$

37. **(a)** $0.841x^2 + 0.32y^2$

$= 0.841 \cdot 2^2 + 0.32 \cdot 1^2$

$= 0.841 \cdot 4 + 0.32 \cdot 1$

$= 3.364 + 0.32$

$= 3.684$

(b) $0.841x^2 + 0.32y^2$

$= 0.841 \cdot 1^2 + 0.32 \cdot 5^2$

$= 0.841 \cdot 1 + 0.32 \cdot 25$

$= 0.841 + 8$

$= 8.841$

39. "Twelve times a number" translates as $12 \cdot x$ or $12x$.

41. "Added to" indicates addition. "Nine added to a number" translates as $x + 9$.

43. "Four subtracted from a number" translates as $x - 4$.

45. "A number subtracted from seven" translates as $7 - x$.

47. "The difference between a number and 8" translates as $x - 8$.

49. "18 divided by a number" translates as $\frac{18}{x}$.

51. "The product of 6 and four less than a number" translates as $6(x - 4)$.

53. An expression cannot by solved—it indicates a series of operations to perform. An expression is simplified. An equation is solved.

55. $4m + 2 = 6$; 1

$$4(1) + 2 \stackrel{?}{=} 6 \quad \textit{Let m = 1.}$$
$$4 + 2 \stackrel{?}{=} 6$$
$$6 = 6 \quad \textit{True}$$

Because substituting 1 for m results in a true statement, 1 is a solution of the equation.

57. $2y + 3(y - 2) = 14$; 3

$$2 \cdot 3 + 3(3 - 2) \stackrel{?}{=} 14 \quad \textit{Let y = 3.}$$
$$2 \cdot 3 + 3 \cdot 1 \stackrel{?}{=} 14$$
$$6 + 3 \stackrel{?}{=} 14$$
$$9 = 14 \quad \textit{False}$$

Because substituting 3 for y results in a false statement, 3 is not a solution of the equation.

59. $6p + 4p + 9 = 11$; $\frac{1}{5}$

$$6\left(\frac{1}{5}\right) + 4\left(\frac{1}{5}\right) + 9 \stackrel{?}{=} 11 \quad \textit{Let p = }\frac{1}{5}.$$
$$\frac{6}{5} + \frac{4}{5} + 9 \stackrel{?}{=} 11$$
$$\frac{10}{5} + 9 \stackrel{?}{=} 11$$
$$2 + 9 \stackrel{?}{=} 11$$
$$11 = 11 \quad \textit{True}$$

The true result shows that $\frac{1}{5}$ is a solution of the equation.

61. $3r^2 - 2 = 46$; 4

$$3(4)^2 - 2 \stackrel{?}{=} 46 \quad \textit{Let r = 4.}$$
$$3 \cdot 16 - 2 \stackrel{?}{=} 46$$
$$48 - 2 \stackrel{?}{=} 46$$
$$46 = 46 \quad \textit{True}$$

The true result shows that 4 is a solution of the equation.

63. $\frac{3}{8}x + \frac{1}{4} = 1$; 2

$$\frac{3}{8}(2) + \frac{1}{4} \stackrel{?}{=} 1 \quad \textit{Let x = 2.}$$
$$\frac{3}{4} + \frac{1}{4} \stackrel{?}{=} 1$$
$$1 = 1 \quad \textit{True}$$

The true result shows that 2 is a solution of the equation.

65. $0.5(x - 4) = 80$; 20

$$0.5(20 - 4) \stackrel{?}{=} 80 \quad \textit{Let x = 20.}$$
$$0.5(16) \stackrel{?}{=} 80$$
$$8 = 80 \quad \textit{False}$$

The false result shows that 20 is not a solution of the equation.

67. "The sum of a number and 8 is 18" translates as

$$x + 8 = 18.$$

Try each number from the given set, {2, 4, 6, 8, 10}, in turn.

$$
\begin{array}{ll}
x + 8 = 18 & \textit{Given equation} \\
2 + 8 = 18 & \textit{False} \\
4 + 8 = 18 & \textit{False} \\
6 + 8 = 18 & \textit{False} \\
8 + 8 = 18 & \textit{False} \\
10 + 8 = 18 & \textit{True}
\end{array}
$$

The only solution is 10.

69. "Sixteen minus three-fourths of a number is 13" translates as

$$16 - \frac{3}{4}x = 13.$$

Try each number from the given set, {2, 4, 6, 8, 10}, in turn.

$$
\begin{array}{ll}
16 - \frac{3}{4}x = 13 & \textit{Given equation} \\
16 - \frac{3}{4}(2) = 13 & \textit{False} \\
16 - \frac{3}{4}(4) = 13 & \textit{True} \\
16 - \frac{3}{4}(6) = 13 & \textit{False} \\
16 - \frac{3}{4}(8) = 13 & \textit{False} \\
16 - \frac{3}{4}(10) = 13 & \textit{False}
\end{array}
$$

The only solution is 4.

71. "One more than twice a number is 5" translates as

$$2x + 1 = 5.$$

Try each number from the given set. The only resulting true equation is

$$2 \cdot 2 + 1 = 5,$$

So the only solution is 2.

73. "Three times a number is equal to 8 more than twice the number" translates as

$$3x = 2x + 8.$$

Try each number from the given set.

$$
\begin{array}{ll}
3x = 2x + 8 & \textit{Given equation} \\
3(2) = 2(2) + 8 & \textit{False} \\
3(4) = 2(4) + 8 & \textit{False} \\
3(6) = 2(6) + 8 & \textit{False} \\
3(8) = 2(8) + 8 & \textit{True} \\
3(10) = 2(10) + 8 & \textit{False}
\end{array}
$$

The only solution is 8.

75. There is no equals symbol, so $3x + 2(x - 4)$ is an expression.

77. There is an equals symbol, so $7t + 2(t + 1) = 4$ is an equation.

79. There is an equals symbol, so $x + y = 9$ is an equation.

81. $y = 0.212x - 347$
$\quad = 0.212(1943) - 347$
$\quad = 64.916 \approx 64.9$

The life expectancy of an American born in 1943 is about 64.9 years.

83. $y = 0.212x - 347$
$\quad = 0.212(1985) - 347$
$\quad = 73.82 \approx 73.8$

The life expectancy of an American born in 1985 is about 73.8 years.

85. Life expectancy has increased over 13 years during this time.

1.4   Real Numbers and the Number Line

1.4   Now Try Exercises

N1. Since the deepest point is *below* the water's surface, the depth is -136.

N2. $\left\{ -7, -\frac{4}{5}, 0, \sqrt{3}, 2.7, \pi, 13 \right\}$

   (a) The whole numbers are 0, and 13.

   (b) The integers are $-7, 0$, and 13.

   (c) The rational numbers are $-7, -\frac{4}{5}, 0, 2.7$, and 13.

   (d) The irrational numbers are $\sqrt{3}$ and π.

N3. Since -8 lies to the right of -9 on the number line, -8 is greater than -9. Therefore, the statement $-8 \le -9$ is *false*.

N4. **(a)** $|4| = 4$

   (b) $|-4| = -(-4) = 4$

   (c) $-|-4| = -(4) = -4$

N5. The category *new cars* is negative in both years.

1.4   Section Exercises

1. Use the integer 2,866,000 since "increased by 2,866,000" indicates a positive number.

3. Use the integer $-52,000$ since "a decrease of 52,000" indicates a negative number.

5. Use the rational numbers -11.2 and 8.6 since "declined 11.2%" and "rose 8.6%" indicate a

negative number and a positive number, respectively.

7. Use the rational number 82.60 since "closed up 82.60" indicates a positive number.

9. The only integer between 3.6 and 4.6 is 4.

11. There is only one whole number that is not positive and that is less than 1: the number 0.

13. An irrational number that is between $\sqrt{12}$ and $\sqrt{14}$ is $\sqrt{13}$. There are others.

15. True; every natural number is positive.

17. True; every integer is a rational number. For example, 5 can be written as $\frac{5}{1}$.

19. False; if a number is rational, it cannot be irrational, and vice versa.

21. Three examples of positive real numbers that are not integers are $\frac{1}{2}, \frac{5}{8}$, and $1\frac{3}{4}$. Other examples are $0.7, 4\frac{2}{3}$, and 5.1.

23. Three examples of real numbers that are not whole numbers are $-3\frac{1}{2}, -\frac{2}{3}$, and $\frac{3}{7}$. Other examples are $-4.3, -\sqrt{2}$, and $\sqrt{7}$.

25. Three examples of real numbers that are not rational numbers are $\sqrt{5}, \pi$, and $-\sqrt{3}$. All irrational numbers are real numbers that are not rational.

27. $\left\{ -9, -\sqrt{7}, -1\frac{1}{4}, -\frac{3}{5}, 0, 0.\overline{1}, \sqrt{5}, 3, 5.9, 7 \right\}$

   (a) The natural numbers in the given set are 3 and 7, since they are in the natural number set $\{1, 2, 3, \dots\}$.

   (b) The set of whole numbers includes the natural numbers and 0. The whole numbers in the given set are 0, 3, and 7.

   (c) The integers are the set of numbers $\{\dots, -3, -2, -1, 0, 1, 2, 3, \dots\}$. The integers in the given set are $-9, 0, 3$, and 7.

   (d) Rational numbers are the numbers which can be expressed as the quotient of two integers, with denominators not equal to 0.

We can write numbers from the given set in this form as follows:

$$-9 = \frac{-9}{1}, -1\frac{1}{4} = \frac{-5}{4}, -\frac{3}{5} = \frac{-3}{5}, 0 = \frac{0}{1},$$

$$0.\overline{1} = \frac{1}{9}, 3 = \frac{3}{1}, 5.9 = \frac{59}{10}, \text{ and } 7 = \frac{7}{1}.$$

Thus, the rational numbers in the given set are $-9, -1\frac{1}{4}, -\frac{3}{5}, 0, 0.\overline{1}, 3, 5.9$, and 7.

(e) Irrational numbers are real numbers that are not rational. $-\sqrt{7}$ and $\sqrt{5}$ can be represented by points on the number line but cannot be written as a quotient of integers. Thus, the irrational numbers in the given set are $-\sqrt{7}$ and $\sqrt{5}$.

(f) Real numbers are all numbers that can be represented on the number line. All the numbers in the given set are real.

29. Graph 0, 3, -5, and -6.

Place a dot on the number line at the point that corresponds to each number. The order of the numbers from smallest to largest is $-6, -5, 0, 3$.

31. Graph -2, -6, -4, 3, and 4.

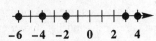

33. Graph $\frac{1}{4}$, $2\frac{1}{2}$, $-3\frac{4}{5}$, -4, and $-1\frac{5}{8}$.

$$-3\tfrac{4}{5} \quad -1\tfrac{5}{8} \quad \tfrac{1}{4} \quad 2\tfrac{1}{2}$$

35. (a) $|-9| = 9$ **(A)**

The distance between -9 and 0 on the number line is 9 units.

(b) $-(-9) = 9$ **(A)**

The opposite of -9 is 9.

(c) $-|-9| = -(9) = -9$ **(B)**

(d) $-|-(-9)| = -|9|$ *Work inside absolute value symbols first*
$$= -(9)$$
$$= -9 \quad \textbf{(B)}$$

37. (a) The opposite of -7 is found by changing the sign of -7. The opposite of -7 is 7.

(b) The absolute value of -7 is the distance between 0 and -7 on the number line.
$$|-7| = 7$$
The absolute value of -7 is 7.

39. (a) The opposite of 8 is -8.

(b) The distance between 0 and 8 on the number line is 8 units, so the absolute value of 8 is 8.

41. (a) The opposite of a number is found by changing the sign of a number, so the opposite of $-\frac{3}{4}$ is $\frac{3}{4}$.

(b) The distance between $-\frac{3}{4}$ and 0 on the number line is $\frac{3}{4}$ unit, so $\left|-\frac{3}{4}\right| = \frac{3}{4}$.

43. Since -6 is a negative number, its absolute value is the additive inverse of -6; that is,
$$|-6| = -(-6) = 6.$$

45. $-|12| = -(12) = -12$

47. $-\left|-\frac{2}{3}\right| = -\left[-\left(-\frac{2}{3}\right)\right] = -\left[\frac{2}{3}\right] = -\frac{2}{3}$

49. $|6 - 3| = |3| = 3$

51. The statement "Absolute value is always positive." is not true. The absolute value of 0 is 0, and 0 is not positive. We could say that *absolute value is never negative,* or *absolute value is always nonnegative.*

53. $-11, -3$
Since -11 is located to the left of -3 on the number line, -11 is the lesser number.

55. $-7, -6$
Since -7 is located to the left of -6 on the number line, -7 is the lesser number.

57. $4, |-5|$
Since $|-5| = 5$, 4 is the lesser of the two numbers.

59. $|-3.5|, |-4.5|$
Since $|-3.5| = 3.5$ and $|-4.5| = 4.5$, $|-3.5|$ or 3.5 is the lesser number.

61. $-|-6|, -|-4|$
Since $-|-6| = -6$ and $-|-4| = -4$, $-|-6|$ is to the left of $-|-4|$ on the number line, so $-|-6|$ or -6 is the lesser number.

63. $|5 - 3|, |6 - 2|$
Since $|5 - 3| = |2| = 2$ and $|6 - 2| = |4| = 4$, $|5 - 3|$ or 2 is the lesser number.

65. $-5 < -2$
Since -5 is to the *left* of -2 on the number line, -5 is *less than* -2, and the statement $-5 < -2$ is true.

67. $-4 \le -(-5)$
Since $-(-5) = 5$ and $-4 < 5$, $-4 \le -(-5)$ is true.

69. $|-6| < |-9|$
Since $|-6| = 6$ and $|-9| = 9$, and $6 < 9$, $|-6| < |-9|$ is true.

71. $-|8| > |-9|$
Since $-|8| = -8$ and $|-9| = -(-9) = 9$, $-|8| < |-9|$, so $-|8| > |-9|$ is false.

73. $-|-5| \ge -|-9|$
Since $-|-5| = -5$, $-|-9| = -9$, and $-5 > -9$, $-|-5| \ge -|-9|$ is true.

75. $|6 - 5| \geq |6 - 2|$
Since $|6 - 5| = |1| = 1$ and $|6 - 2| = |4| = 4$,
$|6 - 5| < |6 - 2|$, so $|6 - 5| \geq |6 - 2|$ is false.

77. The number that represents the greatest percentage increase is 10.6, which corresponds to *fuel and other utilities* from 2004 to 2005.

79. The number with the smallest absolute value in the table is -0.4, so the least change corresponds to *apparel and upkeep* from 2006 to 2007.

1.5 Adding and Subtracting Real Numbers

1.5 Now Try Exercises

N1. **(a)** Start at 0 on a number line. Draw an arrow 3 units to the right to represent the positive number 3. From the right end of this arrow, draw a second arrow 5 units to the right to represent the addition of a positive number. The number below the end of this second arrow is 8, so $3 + 5 = 8$.

(b) Start at 0 on a number line. Draw an arrow 1 unit to the left to represent the negative number -1. From the left end of this arrow, draw a second arrow 3 units to the left. The number below the end of this second arrow is -4, so $-1 + (-3) = -4$.

N2. $-6 + (-11) = -17$

The sum of two negative numbers is negative.

N3. Start at 0 on a number line. Draw an arrow 4 units to the right. From the right end of this arrow, draw a second arrow 8 units to the left. The number below the end of this second arrow is -4, so $4 + (-8) = -4$.

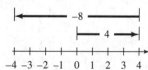

N4. $8 + (-17)$

Since the numbers have different signs, find the difference between their absolute values:

$$17 - 8 = 9.$$

Because -17 has the larger absolute value, the sum is negative:

$$8 + (-17) = -9.$$

N5. **(a)** $\dfrac{2}{3} + \left(-2\dfrac{1}{9}\right) = -1\dfrac{4}{9}$

The number with the greater absolute value is $-2\dfrac{1}{9}$, so the sum will be negative. The answer of $-1\dfrac{4}{9}$ is correct.

(b) $3.7 + (-5.7) = -2$

The number with the greater absolute value is -5.7, so the sum will be negative. The answer of -2 is correct.

N6. **(a)** $-5 - (-11) = -5 + (11)$ *Add the opposite.*
$= 6$

(b) $4 - 15 = 4 + (-15)$ *Add the opposite.*
$= -11$

(c) $-\dfrac{5}{7} - \dfrac{1}{3} = -\dfrac{5}{7} + \left(-\dfrac{1}{3}\right)$ *Add opposite.*

$= -\dfrac{15}{21} + \left(-\dfrac{7}{21}\right)$

$= -\dfrac{22}{21}$, or $-1\dfrac{1}{21}$

N7. **(a)** $8 - [(-3 + 7) - (3 - 9)]$
$= 8 - [(4) - (3 + (-9))]$
$= 8 - [4 - (-6)]$
$= 8 - [4 + 6]$
$= 8 - 10$
$= 8 + (-10)$
$= -2$

(b) $3|6 - 9| - |4 - 12|$
$= 3|6 + (-9)| - |4 + (-12)|$
$= 3|-3| - |-8|$
$= 3(3) - 8$
$= 9 - 8$
$= 1$

N8. "The sum of -3 and 7, increased by 10" is written $(-3 + 7) + 10$.

$$(-3 + 7) + 10 = 4 + 10 = 14$$

N9. **(a)** "The difference between 5 and -8, decreased by 4" is written $[5 - (-8)] - 4$.

$$[5 - (-8)] - 4 = [5 + 8] - 4$$
$$= 13 - 4$$
$$= 9$$

(b) "7 less than -2" is written $-2 - 7$.

$$-2 - 7 = -2 + (-7)$$
$$= -9$$

N10. The difference between a gain of 226 yards and a loss of 7 yards

$$226 - (-7) = 226 + 7$$
$$= 233.$$

The difference is 233 yards.

N11. Subtract the CPI number for 2003 from the CPI number for 2004.

$$119.6 - 119.3 = 0.3$$

A positive result indicates an increase.

1.5 Section Exercises

1. The sum of two negative numbers will always be a *negative* number. In the illustration, we have $-2 + (-3) = -5$.

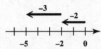

3. When adding a positive number and a negative number, where the negative number has the greater absolute value, the sum will be a *negative* number. In the illustration, the absolute value of -4 is larger than the absolute value of 2, so the sum is a negative number; that is, $-4 + 2 = -2$.

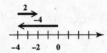

5. By the definition of subtraction, in order to perform the subtraction $-6 - (-8)$, we must add the opposite of $\underline{-8}$ to $\underline{-6}$ to get $\underline{2}$.

7. The expression $x - y$ would have to be *positive* since subtracting a negative number from a positive number is the same as adding a positive number to a positive number, which is a positive number.

9. $|x| = x$, since x is a positive number.

   $y - |x| = y - x$, which is a *negative* number. (See Exercise 8.)

11. $-6 + (-2)$

    The sum of two negative numbers is negative.

    $$-6 + (-2) = -8$$

13. $-5 + (-7)$

    Because the numbers have the same sign, add their absolute values:

    $$5 + 7 = 12.$$

    Because both numbers are negative, their sum is negative:

    $$-5 + (-7) = -12.$$

15. $6 + (-4)$

    To add $6 + (-4)$, find the difference between the absolute values of the numbers.

    $$|6| = 6 \text{ and } |-4| = 4$$
    $$6 - 4 = 2$$

Since $|6| > |-4|$, the sum will be positive:

$$6 + (-4) = 2.$$

17. $4 + (-6)$

    Since the numbers have different signs, find the difference between their absolute values:

    $$6 - 4 = 2.$$

    Because -6 has the larger absolute value, the sum is negative:

    $$4 + (-6) = -2.$$

19. $-3.5 + 12.4$

    Since the numbers have different signs, find the difference between their absolute values:

    $$12.4 - 3.5 = 8.9.$$

    Since 12.4 has the larger absolute value, the answer is positive:

    $$-3.5 + 12.4 = 8.9.$$

21. $4 + [13 + (-5)]$

    Perform the operation inside the brackets first, then add.

    $$4 + [13 + (-5)] = 4 + [8] = 12$$

23. $8 + [-2 + (-1)] = 8 + [-3] = 5$

25. $-2 + [5 + (-1)] = -2 + [4] = 2$

27. $-6 + [6 + (-9)] = -6 + [-3] = -9$

29. $[(-9) + (-3)] + 12 = [-12] + 12 = 0$

31. $-\dfrac{1}{6} + \dfrac{2}{3} = -\dfrac{1}{6} + \dfrac{4}{6} = \dfrac{3}{6} = \dfrac{1}{2}$

33. Since $8 = 2 \cdot 2 \cdot 2$ and $12 = 2 \cdot 2 \cdot 3$, the LCD is $2 \cdot 2 \cdot 2 \cdot 3 = 24$.

    $$\frac{5}{8} + \left(-\frac{17}{12}\right) = \frac{5 \cdot 3}{8 \cdot 3} + \left(-\frac{17 \cdot 2}{12 \cdot 2}\right)$$
    $$= \frac{15}{24} + \left(-\frac{34}{24}\right)$$
    $$= -\frac{19}{24}$$

35. $2\dfrac{1}{2} + \left(-3\dfrac{1}{4}\right) = \dfrac{5}{2} + \left(-\dfrac{13}{4}\right)$
    $$= \frac{10}{4} + \left(-\frac{13}{4}\right)$$
    $$= -\frac{3}{4}$$

37. $-6.1 + [3.2 + (-4.8)] = -6.1 + [-1.6]$
    $$= -7.7$$

39. $[-3 + (-4)] + [5 + (-6)] = [-7] + [-1]$
$$= -8$$

41. $[-4 + (-3)] + [8 + (-1)] = [-7] + [7]$
$$= 0$$

43. $[-4 + (-6)] + [(-3) + (-8)] + [12 + (-11)]$
$$= ([-10] + [-11]) + [1]$$
$$= (-21) + 1$$
$$= -20$$

In Exercises 45–68, use the definition of subtraction to find the differences.

45. $4 - 7 = 4 + (-7) = -3$

47. $5 - 9 = 5 + (-9) = -4$

49. $-7 - 1 = -7 + (-1) = -8$

51. $-8 - 6 = -8 + (-6) = -14$

53. $7 - (-2) = 7 + (2) = 9$

55. $-6 - (-2) = -6 + (2) = -4$

57. $2 - (3 - 5) = 2 - [3 + (-5)]$
$$= 2 - [-2]$$
$$= 2 + (2)$$
$$= 4$$

59. $\dfrac{1}{2} - \left(-\dfrac{1}{4}\right) = \dfrac{1}{2} + \dfrac{1}{4}$
$$= \dfrac{2}{4} + \dfrac{1}{4} = \dfrac{3}{4}$$

61. $-\dfrac{3}{4} - \dfrac{5}{8} = -\dfrac{3}{4} + \left(-\dfrac{5}{8}\right)$
$$= -\dfrac{6}{8} + \left(-\dfrac{5}{8}\right)$$
$$= -\dfrac{11}{8}, \text{ or } -1\dfrac{3}{8}$$

63. $\dfrac{5}{8} - \left(-\dfrac{1}{2} - \dfrac{3}{4}\right)$
$$= \dfrac{5}{8} - \left[-\dfrac{1}{2} + \left(-\dfrac{3}{4}\right)\right]$$
$$= \dfrac{5}{8} - \left[-\dfrac{2}{4} + \left(-\dfrac{3}{4}\right)\right]$$
$$= \dfrac{5}{8} - \left(-\dfrac{5}{4}\right)$$
$$= \dfrac{5}{8} + \dfrac{5}{4}$$
$$= \dfrac{5}{8} + \dfrac{10}{8}$$
$$= \dfrac{15}{8}, \text{ or } 1\dfrac{7}{8}$$

65. $3.4 - (-8.2) = 3.4 + 8.2$
$$= 11.6$$

67. $-6.4 - 3.5 = -6.4 + (-3.5)$
$$= -9.9$$

69. $(4 - 6) + 12 = [4 + (-6)] + 12$
$$= [-2] + 12$$
$$= 10$$

71. $(8 - 1) - 12 = [8 + (-1)] + (-12)$
$$= [7] + (-12)$$
$$= -5$$

73. $6 - (-8 + 3) = 6 - (-5)$
$$= 6 + 5$$
$$= 11$$

75. $2 + (-4 - 8) = 2 + [-4 + (-8)]$
$$= 2 + [-12]$$
$$= -10$$

77. $|-5 - 6| + |9 + 2| = |-5 + (-6)| + |11|$
$$= |-11| + |11|$$
$$= -(-11) + 11$$
$$= 11 + 11$$
$$= 22$$

79. $|-8 - 2| - |-9 - 3|$
$$= |-8 + (-2)| - |-9 + (-3)|$$
$$= |-10| - |-12|$$
$$= -(-10) - [-(-12)]$$
$$= 10 - [12]$$
$$= -2$$

81. $\left(-\dfrac{3}{4} - \dfrac{5}{2}\right) - \left(-\dfrac{1}{8} - 1\right)$
$$= \left(-\dfrac{3}{4} - \dfrac{10}{4}\right) - \left(-\dfrac{1}{8} - \dfrac{8}{8}\right)$$
$$= -\dfrac{13}{4} - \left(-\dfrac{9}{8}\right)$$
$$= -\dfrac{26}{8} + \dfrac{9}{8}$$
$$= -\dfrac{17}{8}, \text{ or } -2\dfrac{1}{8}$$

83. $\left(-\dfrac{1}{2} + 0.25\right) - \left(-\dfrac{3}{4} + 0.75\right)$
$$= \left(-\dfrac{1}{2} + \dfrac{1}{4}\right) - \left(-\dfrac{3}{4} + \dfrac{3}{4}\right)$$
$$= \left(-\dfrac{2}{4} + \dfrac{1}{4}\right) - 0$$
$$= -\dfrac{1}{4}, \text{ or } -0.25$$

85. $-9 + [(3 - 2) - (-4 + 2)]$
$= -9 + [1 - (-2)]$
$= -9 + [1 + 2]$
$= -9 + 3$
$= -6$

87. $-3 + [(-5 - 8) - (-6 + 2)]$
$= -3 + [(-5 + (-8)) - (-4)]$
$= -3 + [-13 + 4]$
$= -3 + [-9]$
$= -12$

89. $-9.1237 + [(-4.8099 - 3.2516) + 11.27903]$
$= -9.1237 + [(-4.8099 + (-3.2516)) + 11.27903]$
$= -9.1237 + [-8.0615 + 11.27903]$
$= -9.1237 + 3.21753$
$= -5.90617$

91. "The sum of -5 and 12 and 6" is written
$-5 + 12 + 6$.
$$-5 + 12 + 6 = [-5 + 12] + 6$$
$$= 7 + 6 = 13$$

93. "14 added to the sum of -19 and -4" is written
$[-19 + (-4)] + 14$.
$$[-19 + (-4)] + 14 = (-23) + 14$$
$$= -9$$

95. "The sum of -4 and -10, increased by 12" is
written $[-4 + (-10)] + 12$.
$$[-4 + (-10)] + 12 = -14 + 12$$
$$= -2$$

97. "$\frac{2}{7}$ more than the sum of $\frac{5}{7}$ and $-\frac{9}{7}$" is written
$\left[\frac{5}{7} + \left(-\frac{9}{7}\right)\right] + \frac{2}{7}$.
$$\left[\frac{5}{7} + \left(-\frac{9}{7}\right)\right] + \frac{2}{7} = -\frac{4}{7} + \frac{2}{7}$$
$$= -\frac{2}{7}$$

99. "The difference between 4 and -8" is written
$4 - (-8)$.
$$4 - (-8) = 4 + 8 = 12$$

101. "8 less than -2" is written $-2 - 8$.
$$-2 - 8 = -2 + (-8) = -10$$

103. "The sum of 9 and -4, decreased by 7" is written
$[9 + (-4)] - 7$.
$$[9 + (-4)] - 7 = 5 + (-7) = -2$$

105. "12 less than the difference between 8 and -5" is
written $[8 - (-5)] - 12$.
$$[8 - (-5)] - 12 = [8 + (5)] - 12$$
$$= 13 - 12$$
$$= 13 + (-12)$$
$$= 1$$

107. $[-5 + (-4)] + (-3) = -9 + (-3)$
$= -12$

The total number of seats that New York,
Pennsylvania, and Ohio are projected to lose is
twelve, which can be represented by the signed
number -12.

109. To find the low temperature, start with 44 and
subtract 100.
$$44 - 100 = 44 + (-100)$$
$$= -56$$
The low temperature was $-56°$F.

111. $33°$F lower than $-36°$F can be represented as
$$-36 - 33 = -36 + (-33)$$
$$= -69.$$
The record low in Utah is $-69°$F.

113. $0 + (-130) + (-54) = -130 + (-54)$
$= -184$

Their new altitude is 184 meters below the
surface, which can be represented by the signed
number -184.

115. (a) $6.9 - (-0.5) = 6.9 + 0.5$
$= 7.4$

The difference is 7.4%.

(b) Americans spent more money than they
earned, which means they had to dip into savings
or increase borrowing.

117. $1879 + 869 - 579 + 1004$
$= 2748 - 579 + 1004$ *Add.*
$= 2169 + 1004$ *Subtract.*
$= 3173$ *Add.*

The average was \$3173.

119. Add the scores of the four turns to get the final
score.
$$-19 + 28 + (-5) + 13 = 9 + (-5) + 13$$
$$= 4 + 13$$
$$= 17$$

Her final score for the four turns was 17.

121. Sum of checks:

$$\$35.84 + \$26.14 + \$3.12 = \$61.98 + \$3.12$$

Sum of deposits: $\quad = \$65.10$

$$\$85.00 + \$120.76 = \$205.76$$

Final balance $=$ Beginning balance $-$ checks $+$ deposits

$$= \$904.89 - \$65.10 + \$205.76$$
$$= \$839.79 + \$205.76$$
$$= \$1045.55$$

Her account balance at the end of August was $1045.55.

123.

$\quad -870.00 \quad$ *amount owed*

$\quad + 185.90 \quad$ *2 return credits*

$\qquad\qquad\quad$ ($35.90 + $150.00)

$\quad \overline{-684.10}$

$\quad -102.50 \quad$ *3 purchases*

$\qquad\qquad\quad$ ($82.50 + $10.00 + $10.00)

$\quad \overline{-786.60}$

$\quad + 500.00 \quad$ *payment*

$\quad \overline{-286.60}$

$\quad -37.23 \quad$ *finance charge*

$\quad \overline{-323.83}$

She still owes $323.83.

125. The outlay for 2005 is $38.7 billion and the outlay for 2006 is $69.1 billion. Thus, the *change in outlay* is

$$69.1 - 38.7 = 69.1 + (-38.7)$$
$$= 30.4$$

billion dollars (an increase).

127. The outlay for 2007 is $39.2 billion and the outlay for 2008 is $42.3 billion. Thus, the *change in outlay* is

$$42.3 - 39.2 = 42.3 + (-39.2)$$
$$= 3.1$$

billion dollars (an increase).

129. $17,400 - (-32,995) = 17,400 + 32,995$
$$= 50,395$$

The difference between the height of Mt. Foraker and the depth of the Philippine Trench is 50,395 feet.

131. $-23,376 - (-24,721) = -23,376 + 24,721$
$$= 1345$$

The Cayman Trench is 1345 feet deeper than the Java Trench.

133. $14,246 - 14,110 = 14,246 + (-14,110)$
$$= 136$$

Mt. Wilson is 136 feet higher than Pikes Peak.

1.6 Multiplying and Dividing Real Numbers

1.6 Now Try Exercises

N1. (a) $-11(9) = -(11 \cdot 9) = -99$

    (b) $3.1(-2.5) = -(3.1 \cdot 2.5) = -7.75$

N2. $-\dfrac{1}{7}\left(-\dfrac{5}{2}\right) = \dfrac{1}{7} \cdot \dfrac{5}{2} = \dfrac{1 \cdot 5}{7 \cdot 2} = \dfrac{5}{14}$

N3. (a) $\dfrac{15}{-3} = 15 \cdot \left(-\dfrac{1}{3}\right) = -5$

    (b) $\dfrac{9.81}{-0.9} = 9.81\left(-\dfrac{1}{0.9}\right) = -10.9$

    (c) $-\dfrac{5}{6} \div \dfrac{17}{9} = -\dfrac{5}{\cancel{6}_{2}} \cdot \dfrac{\cancel{9}^{3}}{17} = -\dfrac{15}{34}$

N4. (a) $\dfrac{-10}{5} = -2$

    (b) $\dfrac{-1.44}{-0.12} = 12$

    (c) $-\dfrac{3}{8} \div \dfrac{7}{10} = -\dfrac{3}{\cancel{8}_{4}} \cdot \dfrac{\cancel{10}^{5}}{7} = -\dfrac{15}{28}$

N5. (a) $-4(6) - (-5)5 = -24 - (-25)$
$$= -24 + 25$$
$$= 1$$

    (b) $\dfrac{12(-4) - 6(-3)}{-4(7 - 16)} = \dfrac{-48 - (-18)}{-4(-9)}$

$$= \dfrac{-48 + 18}{36}$$

$$= \dfrac{-30}{36} = -\dfrac{5}{6}$$

N6. Replace x with -4 and y with -3.

$$\dfrac{3x^2 - 12}{y} = \dfrac{3(-4)^2 - 12}{-3}$$

$$= \dfrac{3(16) - 12}{-3}$$

$$= \dfrac{48 - 12}{-3}$$

$$= \dfrac{36}{-3} = -12$$

N7. (a) "Twice the sum of -10 and 7" is written $2(-10 + 7)$.

$$2(-10 + 7) = 2(-3) = -6$$

    (b) "40% of the difference between 45 and 15" is written $0.40(45 - 15)$.

$$0.40(45 - 15) = 0.40(30) = 12$$

N8. "The quotient of 21 and the sum of 10 and -7" is written $\dfrac{21}{10+(-7)}$.

$$\frac{21}{10+(-7)} = \frac{21}{3} = 7$$

N9. (a) "The sum of a number and -4 is 7" is written

$$x + (-4) = 7.$$

Here, x must be 4 more than 7, so the solution is 11.

$$11 + (-4) = 7, \ldots$$

(b) "The difference between -8 and a number is -11" is written

$$-8 - x = -11.$$

If we start at -8 on a number line, we must move 3 units to the left to get to -11, so the solution is 3.

1.6 Section Exercises

1. The product or the quotient of two numbers with the same sign is _greater than 0_, since the product or quotient of two positive numbers is positive and the product or quotient of two negative numbers is positive.

3. If three negative numbers are multiplied, the product is _less than 0_, since a negative number times a negative number is a positive number, and that positive number times a negative number is a negative number.

5. If a negative number is squared and the result is added to a positive number, the result is _greater than 0_, since a negative number squared is a positive number, and a positive number added to another positive number is a positive number.

7. If three positive numbers, five negative numbers, and zero are multiplied, the product is _equal to 0_. Since one of the numbers is zero, the product is zero (regardless of what the other numbers are).

9. The quotient formed by any nonzero number divided by 0 is _undefined_, and the quotient formed by 0 divided by any nonzero number is _0_. Examples include $\frac{1}{0}$, which is undefined, and $\frac{0}{1}$, which equals 0.

11. $5(-6) = -(5 \cdot 6) = -30$

Note that the product of a positive number and a negative number is negative.

13. $-5(-6) = 5 \cdot 6 = 30$

Note that the product of two negative numbers is positive.

15. $-10(-12) = 10 \cdot 12 = 120$

17. $3(-11) = -(3 \cdot 11) = -33$

19. $-0.5(0) = 0$

21. $-6.8(0.35) = -(6.8 \cdot 0.35) = -2.38$

23.
$$-\frac{3}{8} \cdot \left(-\frac{10}{9}\right) = \frac{3}{8}\left(\frac{10}{9}\right)$$
$$= \frac{3 \cdot 10}{8 \cdot 9}$$
$$= \frac{3 \cdot (2 \cdot 5)}{(4 \cdot 2) \cdot (3 \cdot 3)}$$
$$= \frac{3 \cdot 2 \cdot 5}{4 \cdot 2 \cdot 3 \cdot 3}$$
$$= \frac{5}{4 \cdot 3} = \frac{5}{12}$$

25.
$$\frac{2}{15}\left(-1\frac{1}{4}\right) = \frac{2}{15}\left(-\frac{5}{4}\right)$$
$$= -\frac{2 \cdot 5}{15 \cdot 4}$$
$$= -\frac{2 \cdot 5}{3 \cdot 5 \cdot 2 \cdot 2}$$
$$= -\frac{1}{3 \cdot 2} = -\frac{1}{6}$$

27. $-8\left(-\dfrac{3}{4}\right) = 8\left(\dfrac{3}{4}\right) = \dfrac{24}{4} = 6$

29. Using only positive integer factors, 32 can be written as $1 \cdot 32$, $2 \cdot 16$, or $4 \cdot 8$. Including the negative integer factors, we see that the integer factors of 32 are $-32, -16, -8, -4, -2, -1, 1, 2, 4, 8, 16$, and 32.

31. The integer factors of 40 are $-40, -20, -10, -8, -5, -4, -2, -1, 1, 2, 4, 5, 8, 10, 20$, and 40.

33. The integer factors of 31 are $-31, -1, 1$, and 31.

35. $\dfrac{15}{5} = \dfrac{5 \cdot 3}{5} = \dfrac{3}{1} = 3$

37. $\dfrac{-42}{6} = -\dfrac{2 \cdot 3 \cdot 7}{2 \cdot 3} = -7$

Note that the quotient of two numbers having different signs is negative.

39. $\dfrac{-32}{-4} = \dfrac{4 \cdot 8}{4} = 8$

Note that the quotient of two numbers having the same sign is positive.

41. $\dfrac{96}{-16} = -\dfrac{6 \cdot 16}{16} = -6$

43. Dividing by a fraction (in this case, $-\frac{1}{8}$) is the same as multiplying by the reciprocal of the fraction (in this case, $-\frac{8}{1}$).

$$\left(-\frac{4}{3}\right) \div \left(-\frac{1}{8}\right) = \left(-\frac{4}{3}\right) \cdot \left(-\frac{8}{1}\right)$$
$$= \frac{4 \cdot 8}{3 \cdot 1}$$
$$= \frac{32}{3}, \text{ or } 10\frac{2}{3}$$

45. $\dfrac{-8.8}{2.2} = -\dfrac{4(2.2)}{2.2} = -4$

47. $\dfrac{0}{-5} = 0$, because 0 divided by any nonzero number is 0.

49. $\dfrac{11.5}{0}$ is *undefined* because we cannot divide by 0.

In Exercises 51–68, use the order of operations.

51. $7 - 3 \cdot 6 = 7 - 18$
$$= -11$$

53. $-10 - (-4)(2) = -10 - (-8)$
$$= -10 + 8$$
$$= -2$$

55. $-7(3 - 8) = -7[3 + (-8)]$
$$= -7(-5) = 35$$

57. $7 + 2(4 - 1) = 7 + 2(3)$
$$= 7 + 6$$
$$= 13$$

59. $-4 + 3(2 - 8) = -4 + 3(-6)$
$$= -4 - 18$$
$$= -22$$

61. $(12 - 14)(1 - 4) = (-2)(-3)$
$$= 6$$

63. $(7 - 10)(10 - 4) = (-3)(6)$
$$= -18$$

65. $(-2 - 8)(-6) + 7 = (-10)(-6) + 7$
$$= 60 + 7$$
$$= 67$$

67. $3(-5) + |3 - 10| = -15 + |-7|$
$$= -15 + 7$$
$$= -8$$

69. $\dfrac{-5(-6)}{9 - (-1)} = \dfrac{30}{10}$
$$= \dfrac{3 \cdot 10}{10} = 3$$

71. $\dfrac{-21(3)}{-3 - 6} = \dfrac{-63}{-3 + (-6)}$
$$= \dfrac{-63}{-9} = 7$$

73. $\dfrac{-10(2) + 6(2)}{-3 - (-1)} = \dfrac{-20 + 12}{-3 + 1}$
$$= \dfrac{-8}{-2} = 4$$

75. $\dfrac{3^2 - 4^2}{7(-8 + 9)} = \dfrac{9 - 16}{7(1)} = \dfrac{-7}{7} = -1$

77. $\dfrac{8(-1) - |(-4)(-3)|}{-6 - (-1)} = \dfrac{-8 - |12|}{-6 + 1}$
$$= \dfrac{-8 - 12}{-5}$$
$$= \dfrac{-20}{-5} = 4$$

79. $\dfrac{-13(-4) - (-8)(-2)}{(-10)(2) - 4(-2)}$
$$= \dfrac{52 - 16}{-20 - (-8)}$$
$$= \dfrac{36}{-20 + 8}$$
$$= \dfrac{36}{-12} = -3$$

In Exercises 81–92, replace x with 6, y with -4, and a with 3. Then use the order of operations to evaluate the expression.

81. $5x - 2y + 3a = 5(6) - 2(-4) + 3(3)$
$$= 30 - (-8) + 9$$
$$= 30 + 8 + 9$$
$$= 38 + 9$$
$$= 47$$

83. $(2x + y)(3a) = [2(6) + (-4)][3(3)]$
$$= [12 + (-4)](9)$$
$$= (8)(9)$$
$$= 72$$

85. $\left(\frac{1}{3}x - \frac{4}{5}y\right)\left(-\frac{1}{5}a\right)$

$= \left[\frac{1}{3}(6) - \frac{4}{5}(-4)\right]\left[-\frac{1}{5}(3)\right]$

$= \left[2 - \left(-\frac{16}{5}\right)\right]\left(-\frac{3}{5}\right)$

$= \left(2 + \frac{16}{5}\right)\left(-\frac{3}{5}\right)$

$= \left(\frac{10}{5} + \frac{16}{5}\right)\left(-\frac{3}{5}\right)$

$= \left(\frac{26}{5}\right)\left(-\frac{3}{5}\right)$

$= -\frac{78}{25}$

87. $(-5 + x)(-3 + y)(3 - a)$
$= (-5 + 6)[-3 + (-4)][3 - 3]$
$= (1)(-7)(0)$
$= 0$

89. $-2y^2 + 3a = -2(-4)^2 + 3(3)$
$= -2(16) + 9$
$= -32 + 9$
$= -23$

91. $\frac{2y^2 - x}{a + 10} = \frac{2(-4)^2 - (6)}{3 + 10}$

$= \frac{2(16) - 6}{13}$

$= \frac{32 - 6}{13}$

$= \frac{26}{13}$

$= 2$

93. "The product of -9 and 2, added to 9" is written $9 + (-9)(2)$.

$9 + (-9)(2) = 9 + (-18)$
$= -9$

95. "Twice the product of -1 and 6, subtracted from -4" is written $-4 - 2[(-1)(6)]$.

$-4 - 2[(-1)(6)] = -4 - 2(-6)$
$= -4 - (-12)$
$= -4 + 12 = 8$

97. "Nine subtracted from the product of 1.5 and -3.2 is written $(1.5)(-3.2) - 9$.

$(1.5)(-3.2) - 9 = -4.8 - 9$
$= -4.8 + (-9)$
$= -13.8$

99. "The product of 12 and the difference between 9 and -8" is written $12[9 - (-8)]$.

$12[9 - (-8)] = 12[9 + 8]$
$= 12(17) = 204$

101. "The quotient of -12 and the sum of -5 and -1" is written

$$\frac{-12}{-5 + (-1)},$$

and

$$\frac{-12}{-5 + (-1)} = \frac{-12}{-6} = 2.$$

103. "The sum of 15 and -3, divided by the product of 4 and -3" is written

$$\frac{15 + (-3)}{4(-3)},$$

and

$$\frac{15 + (-3)}{4(-3)} = \frac{12}{-12} = -1.$$

105. "Two-thirds of the difference between 8 and -1" is written

$$\tfrac{2}{3}[8 - (-1)],$$

and

$$\tfrac{2}{3}[8 - (-1)] = \tfrac{2}{3}[8 + (1)] = \tfrac{2}{3}[9] = 6.$$

107. "20% of the product of -5 and 6" is written
$$0.20(-5 \cdot 6),$$
and
$$0.20(-5 \cdot 6) = 0.20(-30) = -6.$$

109. "The sum of $\frac{1}{2}$ and $\frac{5}{8}$, times the difference between $\frac{3}{5}$ and $\frac{1}{3}$" is written

$$\left(\tfrac{1}{2} + \tfrac{5}{8}\right)\left(\tfrac{3}{5} - \tfrac{1}{3}\right),$$

and

$$\left(\tfrac{1}{2} + \tfrac{5}{8}\right)\left(\tfrac{3}{5} - \tfrac{1}{3}\right) = \left(\tfrac{4}{8} + \tfrac{5}{8}\right)\left(\tfrac{9}{15} - \tfrac{5}{15}\right)$$
$$= \tfrac{9}{8}\left(\tfrac{4}{15}\right)$$
$$= \frac{3 \cdot 3 \cdot 4}{2 \cdot 4 \cdot 3 \cdot 5} = \frac{3}{10}.$$

111. "The product of $-\frac{1}{2}$ and $\frac{3}{4}$, divided by $-\frac{2}{3}$" is written $\frac{-\frac{1}{2}\left(\frac{3}{4}\right)}{-\frac{2}{3}}$. Simplifying gives us:

$$\frac{-\frac{1}{2}\left(\frac{3}{4}\right)}{-\frac{2}{3}} = \frac{-\frac{3}{8}}{-\frac{2}{3}}$$
$$= -\frac{3}{8} \cdot \left(-\frac{3}{2}\right)$$
$$= \frac{9}{16}$$

113. "The quotient of a number and 3 is -3" is written

$$\frac{x}{3} = -3.$$

The solution is -9, since

$$\frac{-9}{3} = -3.$$

115. "6 less than a number is 4" is written

$$x - 6 = 4.$$

The solution is 10, since

$$10 - 6 = 4.$$

117. "When 5 is added to a number, the result is -5" is written

$$x + 5 = -5.$$

The solution is -10, since

$$-10 + 5 = -5.$$

119. Add the numbers and divide by 5.

$$\frac{(23 + 18 + 13) + [(-4) + (-8)]}{5}$$

$$= \frac{54 - 12}{5}$$

$$= \frac{42}{5}, \text{ or } 8\frac{2}{5}$$

121. Add the numbers and divide by 4.

$$\frac{(29 + 8) + [(-15) + (-6)]}{4}$$

$$= \frac{37 - 21}{4}$$

$$= \frac{16}{4} = 4$$

123. Add the integers from -10 to 14.

$$(-10) + (-9) + \cdots + 14 = 50$$

[the 3 dots indicate that the pattern continues]

There are 25 integers from -10 to 14 (10 negative, zero, and 14 positive). Thus, the average is

$$\frac{50}{25} = 2.$$

125. (a) 3,473,986 is divisible by 2 because its last digit, 6, is divisible by 2.

(b) 4,336,879 is not divisible by 2 because its last digit, 9, is not divisible by 2.

127. (a) 6,221,464 is divisible by 4 because the number formed by its last two digits, 64, is divisible by 4.

(b) 2,876,335 is not divisible by 4 because the number formed by its last two digits, 35, is not divisible by 4.

129. (a) 1,524,822 is divisible by 2 because its last digit, 2, is divisible by 2. It is also divisible by 3 because the sum of its digits,

$$1 + 5 + 2 + 4 + 8 + 2 + 2 = 24,$$

is divisible by 3.

Because 1,524,822 is divisible by *both* 2 and 3, it is divisible by 6.

(b) 2,873,590 is divisible by 2 because its last digit, 0, is divisible by 2. However, it is not divisible by 3 because the sum of its digits,

$$2 + 8 + 7 + 3 + 5 + 9 + 0 = 34,$$

is not divisible by 3.

Because 2,873,590 is not divisible by *both* 2 and 3, it is not divisible by 6.

131. (a) 4,114,107 is divisible by 9 because the sum of its digits,

$$4 + 1 + 1 + 4 + 1 + 0 + 7 = 18,$$

is divisible by 9.

(b) 2,287,321 is not divisible by 9 because the sum of its digits,

$$2 + 2 + 8 + 7 + 3 + 2 + 1 = 25,$$

is not divisible by 9.

Summary Exercises on Operations with Real Numbers

1. $\begin{aligned} 14 - 3 \cdot 10 &= 14 - 30 \\ &= 14 + (-30) \\ &= -16 \end{aligned}$

3. $\begin{aligned} (3 - 8)(-2) - 10 &= (-5)(-2) - 10 \\ &= 10 - 10 \\ &= 0 \end{aligned}$

5. $\begin{aligned} 7 + 3(2 - 10) &= 7 + 3(-8) \\ &= 7 - 24 \\ &= -17 \end{aligned}$

7. $\begin{aligned} (-4)(7) - (-5)(2) &= (-28) - (-10) \\ &= -28 + (10) \\ &= -18 \end{aligned}$

9. $\begin{aligned} 40 - (-2)[8 - 9] &= 40 - (-2)[-1] \\ &= 40 - (2) \\ &= 38 \end{aligned}$

11. $\dfrac{-3-(-9+1)}{-7-(-6)} = \dfrac{-3-(-8)}{-7+6}$

$= \dfrac{-3+8}{-1}$

$= \dfrac{5}{-1} = -5$

13. $\dfrac{6^2-8}{-2(2)+4(-1)} = \dfrac{36-8}{-4+(-4)}$

$= \dfrac{28}{-8}$

$= -\dfrac{4 \cdot 7}{2 \cdot 4} = -\dfrac{7}{2},$ or $-3\dfrac{1}{2}$

15. $\dfrac{9(-6)-3(8)}{4(-7)+(-2)(-11)} = \dfrac{-54-24}{-28+22}$

$= \dfrac{-78}{-6} = 13$

17. $\dfrac{(2+4)^2}{(5-3)^2} = \dfrac{(6)^2}{(2)^2}$

$= \dfrac{36}{4} = 9$

19. $\dfrac{-9(-6)+(-2)(27)}{3(8-9)} = \dfrac{(54)+(-54)}{3(-1)}$

$= \dfrac{0}{-3} = 0$

21. $\dfrac{6(-10+3)}{15(-2)-3(-9)} = \dfrac{6(-7)}{(-30)-(-27)}$

$= \dfrac{-42}{-30+27}$

$= \dfrac{-42}{-3} = 14$

23. $\dfrac{(-10)^2+10^2}{-10(5)} = \dfrac{100+100}{-50}$

$= \dfrac{200}{-50} = -4$

25. $\dfrac{1}{2} \div \left(-\dfrac{1}{2}\right) = \dfrac{1}{2} \cdot \left(-\dfrac{2}{1}\right)$

$= -\dfrac{2}{2} = -1$

27. $\left[\dfrac{5}{8}-\left(-\dfrac{1}{16}\right)\right]+\dfrac{3}{8} = \left[\dfrac{10}{16}+\dfrac{1}{16}\right]+\dfrac{6}{16}$

$= \left[\dfrac{11}{16}\right]+\dfrac{6}{16}$

$= \dfrac{17}{16},$ or $1\dfrac{1}{16}$

29. $-0.9(-3.7) = 0.9(3.7)$

$= 3.33$

31. $-3^2-2^2 = -(3^2)-(2^2)$

$= -9-4$

$= -13$

33. $40+2[-5-3] = 40+2[-8]$

$= 40-16$

$= 24$

In Exercises 34–42, replace x with -2, y with 3, and a with 4. Then use the order of operations to evaluate the expression.

35. $\cdot (x+6)^3-y^3 = (-2+6)^3-3^3$

$= (4)^3-27$

$= 64-27$

$= 37$

37. $\left(\dfrac{1}{2}x+\dfrac{2}{3}y\right)\left(-\dfrac{1}{4}a\right) = \left(\dfrac{1}{2}(-2)+\dfrac{2}{3}(3)\right)\left(-\dfrac{1}{4}(4)\right)$

$= (-1+2)(-1)$

$= (1)(-1)$

$= -1$

39. $\dfrac{x^2-y^2}{x^2+y^2} = \dfrac{(-2)^2-3^2}{(-2)^2+3^2}$

$= \dfrac{4-9}{4+9}$

$= \dfrac{-5}{13} = -\dfrac{5}{13}$

41. $\left(\dfrac{x}{y}\right)^3 = \left(\dfrac{-2}{3}\right)^3 = \left(-\dfrac{2}{3}\right)\left(-\dfrac{2}{3}\right)\left(-\dfrac{2}{3}\right)$

$= -\dfrac{8}{27}$

1.7 Properties of Real Numbers

1.7 Now Try Exercises

N1. **(a)** $7+(-3) = -3+\underline{7}$

(b) $(-5)4 = 4 \cdot \underline{(-5)}$

N2. **(a)** $-9+(3+7) = \underline{(-9+3)}+7$

(b) $5[(-4) \cdot 9] = \underline{[5 \cdot (-4)]} \cdot 9$

N3. $5+(7+6) = 5+(6+7)$

While the same numbers are grouped inside the two pairs of parentheses, the order of the numbers has been changed. This illustrates a *commutative* property.

N4. (a) $8 + 54 + 7 + 6 + 32$

$\qquad = (8 + 32) + (54 + 6) + 7$

$\qquad = 40 + 60 + 7$

$\qquad = 100 + 7$

$\qquad = 107$

(b) $5(37)(20) = 5(20)(37) = 100(37) = 3700$

N5. (a) $\frac{2}{5} \cdot \underline{1} = \frac{2}{5}$ *Multiplicative identity*

(b) $8 + \underline{0} = 8$ *Additive identity*

N6. (a) $\dfrac{16}{20} = \dfrac{4 \cdot 4}{5 \cdot 4}$ *Factor.*

$\qquad = \dfrac{4}{5} \cdot \dfrac{4}{4}$ *Write as a product.*

$\qquad = \dfrac{4}{5} \cdot 1$ *Divide.*

$\qquad = \dfrac{4}{5}$ *Identity property*

(b) $\dfrac{2}{5} + \dfrac{3}{20} = \dfrac{2}{5} \cdot 1 + \dfrac{3}{20}$ *Identity property*

$\qquad = \dfrac{2}{5} \cdot \dfrac{4}{4} + \dfrac{3}{20}$ *Use 1 = $\frac{4}{4}$ to get a common denominator.*

$\qquad = \dfrac{8}{20} + \dfrac{3}{20}$ *Multiply.*

$\qquad = \dfrac{11}{20}$ *Add.*

N7. (a) $10 + \underline{-10} = 0$ *Inverse property*

(b) $-9 \cdot \underline{\left(-\frac{1}{9}\right)} = 1$ *Inverse property*

N8. $-\dfrac{1}{3}x + 7 + \dfrac{1}{3}x$

$\qquad = \left(-\dfrac{1}{3}x + 7\right) + \dfrac{1}{3}x$ *Order of operations*

$\qquad = \left[7 + \left(-\dfrac{1}{3}x\right)\right] + \dfrac{1}{3}x$ *Commutative property*

$\qquad = 7 + \left[\left(-\dfrac{1}{3}x\right) + \dfrac{1}{3}x\right]$ *Associative property*

$\qquad = 7 + 0$ *Inverse property*

$\qquad = 7$ *Identity property*

N9. (a) $-5(4x + 1) = -5 \cdot 4x + (-5 \cdot 1)$ *Dist. prop.*

$\qquad = -20x - 5$ *Multiply.*

(b) $6(2r + t - 5z) = 6(2r) + 6t + 6(-5z)$

$\qquad = 12r + 6t - 30z$

(c) $5x - 5y = 5(x - y)$

N10. (a) $-(2 - r) = -1(2 - r)$

$\qquad = -2 + r$

(b) $-(2x - 5y - 7) = -1(2x - 5y - 7)$

$\qquad = -2x + 5y + 7$

1.7 Section Exercises

1. (a) B, since 0 is the identity element for addition.

(b) F, since 1 is the identity element for multiplication.

(c) C, since $-a$ is the additive inverse of a.

(d) I, since $\frac{1}{a}$ is the multiplicative inverse, or reciprocal, of any nonzero number a.

(e) B, since 0 is the only number that is equal to its negative; that is, $0 = -0$.

(f) D and F, since -1 has reciprocal $\dfrac{1}{(-1)} = -1$ and 1 has a reciprocal $\dfrac{1}{(1)} = 1$; that is, -1 and 1 are their own multiplicative inverses.

(g) B, since the multiplicative inverse of a number a is $\frac{1}{a}$ and the only number that we *cannot* divide by is 0.

(h) A

(i) G, since we can consider $(5 \cdot 4)$ to be one number, $(5 \cdot 4) \cdot 3$ is the same as $3 \cdot (5 \cdot 4)$ by the commutative property.

(j) H

3. "Washing your face" and "brushing your teeth" *are* commutative.

5. "Preparing a meal" and "eating a meal" *are not* commutative.

7. "Putting on your socks" and "putting on your shoes" *are not* commutative.

9. "(Foreign sales) clerk" is a clerk dealing with foreign sales, whereas "foreign (sales clerk)" is a sales clerk who is foreign.

11. $-15 + 9 = 9 + \underline{(-15)}$

by the *commutative property of addition*.

13. $-8 \cdot 3 = \underline{3} \cdot (-8)$

by the *commutative property of multiplication*.

15. $(3 + 6) + 7 = 3 + (\underline{6} + 7)$

by the *associative property of addition*.

17. $7 \cdot (2 \cdot 5) = (\underline{7} \cdot 2) \cdot 5$

by the *associative property of multiplication*.

19. $25 - (6 - 2) = 25 - (4)$

$\qquad = 21$

$(25 - 6) - 2 = 19 - 2$

$\qquad = 17$

Since $21 \neq 17$, this example shows that subtraction is not associative.

21.

Number	Additive inverse	Multiplicative inverse
5	-5	$\frac{1}{5}$
-10	10	$-\frac{1}{10}$
$-\frac{1}{2}$	$\frac{1}{2}$	-2
$\frac{3}{8}$	$-\frac{3}{8}$	$\frac{8}{3}$
$x\,(x \neq 0)$	$-x$	$\frac{1}{x}$
$-y\,(y \neq 0)$	y	$-\frac{1}{y}$

In general, a number and its additive inverse have <u>opposite</u> signs. A number and its multiplicative inverse have <u>the same</u> signs.

23. $4 + 15 = 15 + 4$

The order of the two numbers has been changed, so this is an example of the commutative property of addition: $a + b = b + a$.

25. $5 \cdot (13 \cdot 7) = (5 \cdot 13) \cdot 7$

The numbers are in the same order but grouped differently, so this is an example of the associative property of multiplication: $(ab)c = a(bc)$.

27. $-6 + (12 + 7) = (-6 + 12) + 7$

The numbers are in the same order but grouped differently, so this is an example of the associative property of addition: $(a + b) + c = a + (b + c)$.

29. $-9 + 9 = 0$

The sum of the two numbers is 0, so they are additive inverses (or opposites) of each other. This is an example of the additive inverse property: $a + (-a) = 0$.

31. $\frac{2}{3}\left(\frac{3}{2}\right) = 1$

The product of the two numbers is 1, so they are multiplicative inverses (or reciprocals) of each other. This is an example of the multiplicative inverse property: $a \cdot \dfrac{1}{a} = 1\ (a \neq 0)$.

33. $1.75 + 0 = 1.75$

The sum of a number and 0 is the original number. This is an example of the identity property of addition: $a + 0 = a$.

35. $(4 + 17) + 3 = 3 + (4 + 17)$

The order of the numbers has been changed, but not the grouping, so this is an example of the commutative property of addition: $a + b = b + a$.

37. $2(x + y) = 2x + 2y$

The number 2 outside the parentheses is "distributed" over the x and y. This is an example of the distributive property.

39. $-\dfrac{5}{9} = -\dfrac{5}{9} \cdot \dfrac{3}{3} = -\dfrac{15}{27}$

$\frac{3}{3}$ is a form of the number 1. We use it to rewrite $-\frac{5}{9}$ as $-\frac{15}{27}$. This is an example of the identity property of multiplication.

41. $4(2x) + 4(3y) = 4(2x + 3y)$

This is an example of the distributive property. The number 4 is "distributed " over $2x$ and $3y$.

43. $97 + 13 + 3 + 37 = (97 + 3) + (13 + 37)$
$$= 100 + 50$$
$$= 150$$

45. $1999 + 2 + 1 + 8 = (1999 + 1) + (2 + 8)$
$$= 2000 + 10$$
$$= 2010$$

47. $159 + 12 + 141 + 88 = (159 + 141) + (12 + 88)$
$$= 300 + 100$$
$$= 400$$

49. $843 + 627 + (-43) + (-27)$
$$= [843 + (-43)] + [627 + (-27)]$$
$$= 800 + 600$$
$$= 1400$$

51. $5(47)(2) = 5(2)(47) = 10(47) = 470$

53. $-4 \cdot 5 \cdot 93 \cdot 5 = -4 \cdot 5 \cdot 5 \cdot 93$
$$= -20 \cdot 5 \cdot 93$$
$$= -100 \cdot 93$$
$$= -9300$$

55. $6t + 8 - 6t + 3$

$= 6t + 8 + (-6t) + 3$	*Definition of subtraction*
$= (6t + 8) + (-6t) + 3$	*Order of operations*
$= (8 + 6t) + (-6t) + 3$	*Commutative property*
$= 8 + [6t + (-6t)] + 3$	*Associative property*
$= 8 + 0 + 3$	*Inverse property*
$= (8 + 0) + 3$	*Order of operations*
$= 8 + 3$	*Identity property*
$= 11$	*Add.*

57. $\dfrac{2}{3}x - 11 + 11 - \dfrac{2}{3}x$

$$= \frac{2}{3}x + (-11) + 11 + \left(-\frac{2}{3}x\right)$$

Definition of subtraction

$$= \left[\frac{2}{3}x + (-11)\right] + 11 + \left(-\frac{2}{3}x\right)$$
Order of operations

$$= \frac{2}{3}x + (-11 + 11) + \left(-\frac{2}{3}x\right)$$
Associative property

$$= \frac{2}{3}x + 0 + \left(-\frac{2}{3}x\right) \qquad \text{Inverse}$$
property

$$= \left(\frac{2}{3}x + 0\right) + \left(-\frac{2}{3}x\right)$$
Order of operations

$$= \frac{2}{3}x + \left(-\frac{2}{3}x\right) \qquad \text{Identity}$$
property

$$= 0 \qquad \text{Inverse}$$
property

59. $\left(\dfrac{9}{7}\right)(-0.38)\left(\dfrac{7}{9}\right)$

$$= \left[\left(\frac{9}{7}\right)(-0.38)\right]\left(\frac{7}{9}\right) \quad \begin{array}{l}\textit{Order of}\\ \textit{operations}\end{array}$$

$$= \left[(-0.38)\left(\frac{9}{7}\right)\right]\left(\frac{7}{9}\right) \quad \begin{array}{l}\textit{Commutative}\\ \textit{property}\end{array}$$

$$= (-0.38)\left[\left(\frac{9}{7}\right)\left(\frac{7}{9}\right)\right] \quad \begin{array}{l}\textit{Associative}\\ \textit{property}\end{array}$$

$$= (-0.38)(1) \qquad \begin{array}{l}\textit{Inverse}\\ \textit{property}\end{array}$$

$$= -0.38 \qquad \begin{array}{l}\textit{Identity}\\ \textit{property}\end{array}$$

61. $t + (-t) + \frac{1}{2}(2)$

$$= t + (-t) + 1 \qquad \textit{Inverse property}$$
$$= [t + (-t)] + 1 \qquad \textit{Order of operations}$$
$$= 0 + 1 \qquad \textit{Inverse property}$$
$$= 1 \qquad \textit{Identity property}$$

63. $-3(4 - 6)$

When distributing a negative number over a quantity, be careful not to "lose" a negative sign. The problem should be worked in the following way.

$$-3(4 - 6) = -3(4) - 3(-6)$$
$$= -12 + 18$$
$$= 6$$

65. $5(9 + 8) = 5 \cdot 9 + 5 \cdot 8$
$$= 45 + 40$$
$$= 85$$

67. $4(t + 3) = 4 \cdot t + 4 \cdot 3$
$$= 4t + 12$$

69. $7(z - 8) = 7[z + (-8)]$
$$= 7z + 7(-8)$$
$$= 7z - 56$$

71. $-8(r + 3) = -8(r) + (-8)(3)$
$$= -8r + (-24)$$
$$= -8r - 24$$

73. $-\frac{1}{4}(8x + 3)$

$$= -\frac{1}{4}(8x) + \left(-\frac{1}{4}\right)(3)$$
$$= \left[\left(-\frac{1}{4}\right) \cdot 8\right]x - \frac{3}{4}$$
$$= -2x - \frac{3}{4}$$

75. $-5(y - 4) = -5(y) + (-5)(-4)$
$$= -5y + 20$$

77. $-\frac{4}{3}(12y + 15z)$

$$= -\frac{4}{3}(12y) + \left(-\frac{4}{3}\right)(15z)$$
$$= \left[\left(-\frac{4}{3}\right) \cdot 12\right]y + \left[\left(-\frac{4}{3}\right) \cdot 15\right]z$$
$$= -16y + (-20)z$$
$$= -16y - 20z$$

79. $8z + 8w = 8(z + w)$

81. $7(2v) + 7(5r) = 7(2v + 5r)$

83. $8(3r + 4s - 5y)$
$$= 8(3r) + 8(4s) + 8(-5y)$$
Distributive property
$$= (8 \cdot 3)r + (8 \cdot 4)s + [8(-5)]y$$
Associative property
$$= 24r + 32s - 40y \quad \textit{Multiply.}$$

85. $-3(8x + 3y + 4z)$
$$= -3(8x) + (-3)(3y) + (-3)(4z)$$
Distributive property
$$= (-3 \cdot 8)x + (-3 \cdot 3)y + (-3 \cdot 4)z$$
Associative property
$$= -24x - 9y - 12z \quad \textit{Multiply.}$$

87. $5x + 15 = 5x + 5 \cdot 3$
$$= 5(x + 3)$$

89. $-(4t + 3m)$
$$= -1(4t + 3m) \qquad \begin{array}{l}\textit{Identity}\\ \textit{property}\end{array}$$
$$= -1(4t) + (-1)(3m) \qquad \begin{array}{l}\textit{Distributive}\\ \textit{property}\end{array}$$
$$= (-1 \cdot 4)t + (-1 \cdot 3)m \qquad \begin{array}{l}\textit{Associative}\\ \textit{property}\end{array}$$
$$= -4t - 3m \qquad \textit{Multiply.}$$

91. $-(-5c - 4d)$

$$= -1(-5c - 4d) \qquad \textit{Identity property}$$

$$= -1(-5c) + (-1)(-4d) \qquad \textit{Distributive property}$$

$$= (-1 \cdot -5)c + (-1 \cdot -4)d \qquad \textit{Associative property}$$

$$= 5c + 4d \qquad \textit{Multiply.}$$

93. $-(-q + 5r - 8s)$

$$= -1(-q + 5r - 8s)$$
$$= -1(-q) + (-1)(5r) + (-1)(-8s)$$
$$= (-1 \cdot -1)q + (-1 \cdot 5)r + (-1 \cdot -8)s$$
$$= q - 5r + 8s$$

1.8   Simplifying Expressions

1.8   Now Try Exercises

N1.   **(a)** $3(2x - 4y) = 3(2x) - 3(4y)$
$$= (3 \cdot 2)x - (3 \cdot 4)y$$
$$= 6x - 12y$$

     (b) $-4 - (-3y + 5) = -4 - 1(-3y + 5)$
$$= -4 - 1(-3y) - 1(5)$$
$$= -4 + 3y + (-5)$$
$$= -4 + (-5) + 3y$$
$$= -9 + 3y, \text{ or } 3y - 9$$

N2.   **(a)** $4x + 6x - 7x = (4 + 6 - 7)x = 3x$

     (b) $z + z = 1z + 1z = (1 + 1)z = 2z$

     (c) $4p^2 - 3p^2 = (4 - 3)p^2 = 1p^2, \text{ or } p^2$

N3.   **(a)** $5k - 6 - (3 - 4k)$
$$= 5k - 6 - 1(3 - 4k)$$
$$= 5k - 6 - 1(3) - 1(-4k)$$
$$= 5k - 6 - 3 + 4k$$
$$= 9k - 9$$

     (b) $\frac{1}{4}x - \frac{2}{3}(x - 9) = \frac{1}{4}x - \frac{2}{3}(x) - \frac{2}{3}(-9)$
$$= \frac{3}{12}x - \frac{8}{12}x + 6$$
$$= -\frac{5}{12}x + 6$$

N4.   "Twice a number, subtracted from the sum of the number and 5" is written $(x + 5) - 2x$.

$$(x + 5) - 2x = x + 5 - 2x$$
$$= -x + 5, \text{ or } 5 - x$$

1.8   Section Exercises

1. $-(6x - 3) = -1(6x - 3)$
$$= -1(6x) - 1(-3)$$
$$= -6x + 3$$
The correct response is **B**.

3.   Examples **A**, **B**, and **D** are pairs of *unlike* terms since either the variables or their powers are different. Example **C** is a pair of *like* terms, since both terms have the same variables (r and y) and the same exponents (both variables are to the first power). Note that we can use the commutative property to rewrite $6yr$ as $6ry$.

5. $4r + 19 - 8 = 4r + 11$

7. $5 + 2(x - 3y) = 5 + 2(x) + 2(-3y)$
$$= 5 + 2x - 6y$$

9. $-2 - (5 - 3p) = -2 - 1(5 - 3p)$
$$= -2 - 1(5) - 1(-3p)$$
$$= -2 - 5 + 3p$$
$$= -7 + 3p$$

11. $6 + (4 - 3x) - 8 = 6 + 4 - 3x - 8$
$$= 10 - 3x - 8$$
$$= 10 - 8 - 3x$$
$$= 2 - 3x$$

13.   The numerical coefficient of the term $-12k$ is -12.

15.   The numerical coefficient of the term $3m^2$ is 3.

17.   Because xw can be written as $1 \cdot xw$, the numerical coefficient of the term xw is 1.

19.   Since $-x = -1x$, the numerical coefficient of the term $-x$ is -1.

21.   Since $\frac{x}{2} = \frac{1}{2}x$, the numerical coefficient of the term $\frac{x}{2}$ is $\frac{1}{2}$.

23.   Since $\frac{2x}{5} = \frac{2}{5}x$, the numerical coefficient of the term $\frac{2x}{5}$ is $\frac{2}{5}$.

25.   The numerical coefficient of the term 10 is 10.

27. $8r$ and $-13r$ are *like* terms since they have the same variable with the same exponent (which is understood to be 1).

29. $5z^4$ and $9z^3$ are *unlike* terms. Although both have the variable z, the exponents are not the same.

31.   All numerical terms (constants) are considered like terms, so 4, 9, and -24 are *like* terms.

33. x and y are *unlike* terms because they do not have the same variable.

35.   The student made a sign error when applying the distributive property.

$$7x - 2(3 - 2x) = 7x - 2(3) - 2(-2x)$$
$$= 7x - 6 + 4x$$
$$= 11x - 6$$

The correct answer is $11x - 6$.

37. $7y + 6y = (7 + 6)y$
$\qquad\quad = 13y$

39. $-6x - 3x = (-6 - 3)x$
$\qquad\qquad = -9x$

41. $12b + b = 12b + 1b$
$\qquad\qquad = (12 + 1)b$
$\qquad\qquad = 13b$

43. $3k + 8 + 4k + 7 = 3k + 4k + 8 + 7$
$\qquad\qquad\qquad\quad = (3 + 4)k + 15$
$\qquad\qquad\qquad\quad = 7k + 15$

45. $-5y + 3 - 1 + 5 + y - 7$
$\quad = (-5y + 1y) + (3 + 5) + (-1 - 7)$
$\quad = (-5 + 1)y + (8) + (-8)$
$\quad = -4y + 8 - 8$
$\quad = -4y$

47. $-2x + 3 + 4x - 17 + 20$
$\quad = (-2x + 4x) + (3 - 17 + 20)$
$\quad = (-2 + 4)x + 6$
$\quad = 2x + 6$

49. $16 - 5m - 4m - 2 + 2m$
$\quad = (16 - 2) + (-5m - 4m + 2m)$
$\quad = 14 + (-5 - 4 + 2)m$
$\quad = 14 - 7m$

51. $-10 + x + 4x - 7 - 4x$
$\quad = (-10 - 7) + (1x + 4x - 4x)$
$\quad = -17 + (1 + 4 - 4)x$
$\quad = -17 + 1x$
$\quad = -17 + x$

53. $1 + 7x + 11x - 1 + 5x$
$\quad = (1 - 1) + (7x + 11x + 5x)$
$\quad = 0 + (7 + 11 + 5)x$
$\quad = 23x$

55. $-\dfrac{4}{3} + 2t + \dfrac{1}{3}t - 8 - \dfrac{8}{3}t$
$\quad = \left(2t + \dfrac{1}{3}t - \dfrac{8}{3}t\right) + \left(-\dfrac{4}{3} - 8\right)$
$\quad = \left(2 + \dfrac{1}{3} - \dfrac{8}{3}\right)t + \left(-\dfrac{4}{3} - 8\right)$
$\quad = \left(\dfrac{6}{3} + \dfrac{1}{3} - \dfrac{8}{3}\right)t + \left(-\dfrac{4}{3} - \dfrac{24}{3}\right)$
$\quad = -\dfrac{1}{3}t - \dfrac{28}{3}$

57. $6y^2 + 11y^2 - 8y^2 = (6 + 11 - 8)y^2$
$\qquad\qquad\qquad\quad = 9y^2$

59. $2p^2 + 3p^2 - 8p^3 - 6p^3$
$\quad = (2p^2 + 3p^2) + (-8p^3 - 6p^3)$
$\quad = (2 + 3)p^2 + (-8 - 6)p^3$
$\quad = 5p^2 - 14p^3 \text{ or } -14p^3 + 5p^2$

61. $2(4x + 6) + 3 = 2(4x) + 2(6) + 3$
$\qquad\qquad\qquad = 8x + 12 + 3$
$\qquad\qquad\qquad = 8x + 15$

63. $100[0.05(x + 3)]$
$\quad = [100(0.05)](x + 3)$ *Associative property*
$\quad = 5(x + 3)$
$\quad = 5(x) + 5(3)$ *Distributive property*
$\quad = 5x + 15$

65. $-6 - 4(y - 7)$
$\quad = -6 - 4(y) + (-4)(-7)$ *Distributive property*
$\quad = -6 - 4y + 28$
$\quad = -4y + 22$

67. $-\dfrac{4}{3}(y - 12) - \dfrac{1}{6}y$
$\quad = -\dfrac{4}{3}y - \dfrac{4}{3}(-12) - \dfrac{1}{6}y$
$\quad = -\dfrac{4}{3}y + 16 - \dfrac{1}{6}y$
$\quad = -\dfrac{4}{3}y - \dfrac{1}{6}y + 16$
$\quad = \left(-\dfrac{8}{6} - \dfrac{1}{6}\right)y + 16$
$\quad = -\dfrac{3}{2}y + 16$ $\left[-\dfrac{9}{6} = -\dfrac{3}{2}\right]$

69. $-5(5y - 9) + 3(3y + 6)$
$\quad = -5(5y) + (-5)(-9) + 3(3y) + 3(6)$
$\qquad\qquad\qquad$ *Distributive property*
$\quad = -25y + 45 + 9y + 18$
$\quad = (-25y + 9y) + (45 + 18)$
$\quad = (-25 + 9)y + 63$
$\quad = -16y + 63$

71. $-3(2r - 3) + 2(5r + 3)$
$\quad = -3(2r) + (-3)(-3) + 2(5r) + 2(3)$
$\qquad\qquad\qquad$ *Distributive property*
$\quad = -6r + 9 + 10r + 6$
$\quad = (-6r + 10r) + (9 + 6)$
$\quad = (-6 + 10)r + 15$
$\quad = 4r + 15$

73. $8(2k - 1) - (4k - 3)$
$\quad = 8(2k - 1) - 1(4k - 3)$
$\qquad\qquad\qquad$ *Replace − with −1.*
$\quad = 8(2k) + 8(-1) + (-1)(4k) + (-1)(-3)$
$\quad = 16k - 8 - 4k + 3$
$\quad = 12k - 5$

75. $-2(-3k + 2) - (5k - 6) - 3k - 5$
$\quad = -2(-3k) + (-2)(2) - 1(5k - 6) - 3k - 5$
$\quad = 6k - 4 + (-1)(5k) + (-1)(-6) - 3k - 5$
$\quad = 6k - 4 - 5k + 6 - 3k - 5$
$\quad = -2k - 3$

77. $-4(-3x + 3) - (6x - 4) - 2x + 1$
$\quad = -4(-3x + 3) - 1(6x - 4) - 2x + 1$
$\quad = 12x - 12 - 6x + 4 - 2x + 1$
$\qquad\qquad\qquad$ *Distributive property*
$\quad = (12x - 6x - 2x) + (-12 + 4 + 1)$
$\qquad\qquad\qquad$ *Group like terms.*
$\quad = 4x - 7 \qquad$ *Combine like terms.*

79. $-7.5(2y + 4) - 2.9(3y - 6)$
$\quad = -7.5(2y) - 7.5(4) - 2.9(3y) - 2.9(-6)$
$\qquad\qquad\qquad$ *Distributive property*
$\quad = -15y - 30 - 8.7y + 17.4$ *Multiply.*
$\quad = -23.7y - 12.6 \qquad$ *Combine like terms.*

81. "Five times a number, added to the sum of the number and three" is written $(x + 3) + 5x$.
$$(x + 3) + 5x = x + 3 + 5x$$
$$= (x + 5x) + 3$$
$$= 6x + 3$$

83. "A number multiplied by -7, subtracted from the sum of 13 and six times the number" is written $(13 + 6x) - (-7x)$.
$$(13 + 6x) - (-7x) = 13 + 6x + 7x$$
$$= 13 + 13x$$

85. "Six times a number added to -4, subtracted from twice the sum of three times the number and 4" is written $2(3x + 4) - (-4 + 6x)$.
$$2(3x + 4) - (-4 + 6x)$$
$$= 2(3x + 4) - 1(-4 + 6x)$$
$$= 6x + 8 + 4 - 6x$$
$$= 6x + (-6x) + 8 + 4$$
$$= 0 + 12 = 12$$

87. For widgets, the fixed cost is $1000 and the variable cost is $5 per widget, so the cost to produce x widgets is

$$1000 + 5x \text{ (dollars)}.$$

88. For gadgets, the fixed cost is $750 and the variable cost is $3 per gadget, so the cost to produce y gadgets is

$$750 + 3y \text{ (dollars)}.$$

89. The total cost to make x widgets and y gadgets is

$$1000 + 5x + 750 + 3y \text{ (dollars)}.$$

90. $1000 + 5x + 750 + 3y$
$\quad = (1000 + 750) + 5x + 3y$
$\quad = 1750 + 5x + 3y,$

so the total cost to make x widgets and y gadgets is

$$1750 + 5x + 3y \text{ (dollars)}.$$

Chapter 1 Review Exercises

1. $\dfrac{8}{5} \div \dfrac{32}{15} = \dfrac{8}{5} \cdot \dfrac{15}{32}$
$\qquad = \dfrac{8 \cdot (3 \cdot 5)}{5 \cdot (8 \cdot 4)}$
$\qquad = \dfrac{8 \cdot 3 \cdot 5}{5 \cdot 8 \cdot 4}$
$\qquad = \dfrac{3}{4}$

2. $2\dfrac{4}{5} \cdot 1\dfrac{1}{4} = \dfrac{14}{5} \cdot \dfrac{5}{4}$
$\qquad = \dfrac{2 \cdot 7 \cdot 5}{5 \cdot 2 \cdot 2}$
$\qquad = \dfrac{7}{2}, \text{ or } 3\dfrac{1}{2}$

3. $\dfrac{5}{8} - \dfrac{1}{6} = \dfrac{5 \cdot 3}{8 \cdot 3} - \dfrac{1 \cdot 4}{6 \cdot 4}$ *LCD = 24*
$\qquad = \dfrac{15}{24} - \dfrac{4}{24}$
$\qquad = \dfrac{11}{24}$

4. $\dfrac{3}{8} + 3\dfrac{1}{2} - \dfrac{3}{16} = \dfrac{3}{8} + \dfrac{7}{2} - \dfrac{3}{16}$
$\qquad = \dfrac{3 \cdot 2}{8 \cdot 2} + \dfrac{7 \cdot 8}{2 \cdot 8} - \dfrac{3}{16}$ *LCD = 16*
$\qquad = \dfrac{6}{16} + \dfrac{56}{16} - \dfrac{3}{16}$
$\qquad = \dfrac{62}{16} - \dfrac{3}{16}$
$\qquad = \dfrac{59}{16}, \text{ or } 3\dfrac{11}{16}$

5. $\dfrac{1}{6} \cdot 7618 = \dfrac{1}{6} \cdot \dfrac{7618}{1}$

$= \dfrac{2 \cdot 3809}{2 \cdot 3}$

$= \dfrac{3809}{3} \approx 1269.7$

About 1270 thousand luxury cars were sold in the United States in 2007.

6. Since the entire pie chart represents $\frac{3}{3}$, this leaves $\frac{3}{3} - \frac{1}{3} = \frac{2}{3}$ of the cars that are *not* small.

$\dfrac{2}{3} \cdot 7618 = \dfrac{2}{3} \cdot \dfrac{7618}{1}$

$= \dfrac{2 \cdot 7618}{3 \cdot 1}$

$= \dfrac{15{,}236}{3} \approx 5078.7$

About 5079 thousand cars sold in the United States in 2007 were *not* small.

7. $5^4 = 5 \cdot 5 \cdot 5 \cdot 5 = 625$

8. $\left(\dfrac{3}{5}\right)^3 = \dfrac{3}{5} \cdot \dfrac{3}{5} \cdot \dfrac{3}{5} = \dfrac{27}{125}$

9. $(0.02)^2 = (0.02)(0.02)$

$= 0.0004$

10. $(0.1)^3 = (0.1)(0.1)(0.1)$

$= 0.001$

11. $8 \cdot 5 - 13 = 40 - 13 = 27$

12. $16 + 12 \div 4 - 2 = 16 + (12 \div 4) - 2$

$= 16 + 3 - 2$

$= 19 - 2$

$= 17$

13. $20 - 2(5 + 3) = 20 - 2(8)$

$= 20 - 16$

$= 4$

14. $7\left[3 + 6(3^2)\right] = 7[3 + 6(9)]$

$= 7(3 + 54)$

$= 7(57)$

$= 399$

15. $\dfrac{9(4^2 - 3)}{4 \cdot 5 - 17} = \dfrac{9(16 - 3)}{20 - 17}$

$= \dfrac{9(13)}{3}$

$= \dfrac{3 \cdot 3 \cdot 13}{3} = 39$

16. $\dfrac{6(5 - 4) + 2(4 - 2)}{3^2 - (4 + 3)} = \dfrac{6(1) + 2(2)}{9 - (4 + 3)}$

$= \dfrac{6 + 4}{9 - 7}$

$= \dfrac{10}{2} = 5$

17. $12 \cdot 3 - 6 \cdot 6 = 36 - 36 = 0$

Since $0 = 0$ is true, so is $0 \le 0$, and therefore, the statement "$12 \cdot 3 - 6 \cdot 6 \le 0$" is true.

18. $3[5(2) - 3] = 3(10 - 3) = 3(7) = 21$

Therefore, the statement "$3[5(2) - 3] > 20$" is true.

19. $4^2 - 8 = 16 - 8 = 8$

Since $9 \le 8$ is false, the statement "$9 \le 4^2 - 8$" is false.

20. "Thirteen is less than seventeen" is written $13 < 17$.

21. "Five plus two is not equal to ten" is written $5 + 2 \ne 10$.

22. "Two-thirds is greater than or equal to four-sixths" is written $\frac{2}{3} \ge \frac{4}{6}$.

In Exercises 23–26, replace x with 6 and y with 3.

23. $2x + 6y = 2(6) + 6(3)$

$= 12 + 18 = 30$

24. $4(3x - y) = 4[3(6) - 3]$

$= 4(18 - 3)$

$= 4(15) = 60$

25. $\dfrac{x}{3} + 4y = \dfrac{6}{3} + 4(3)$

$= 2 + 12 = 14$

26. $\dfrac{x^2 + 3}{3y - x} = \dfrac{6^2 + 3}{3(3) - 6}$

$= \dfrac{36 + 3}{9 - 6}$

$= \dfrac{39}{3} = 13$

27. "Six added to a number" translates as $x + 6$.

28. "A number subtracted from eight" translates as $8 - x$.

29. "Nine subtracted from six times a number" translates as $6x - 9$.

30. "Three-fifths of a number added to 12" translates as $12 + \frac{3}{5}x$.

31. $5x + 3(x + 2) = 22; 2$
$5x + 3(x + 2) = 5(2) + 3(2 + 2)$ *Let x = 2.*
$= 5(2) + 3(4)$
$= 10 + 12 = 22$

Since the left side and the right side are equal, 2 is a solution of the given equation.

32. $\dfrac{t + 5}{3t} = 1; 6$

$\dfrac{t + 5}{3t} = \dfrac{6 + 5}{3(6)}$ *Let t = 6.*

$= \dfrac{11}{18}$

Since the left side, $\frac{11}{18}$, is not equal to the right side, 1, 6 is not a solution of the equation.

33. "Six less than twice a number is 10" is written

$$2x - 6 = 10.$$

Letting x equal 0, 2, 4, 6, and 10 results in a false statement, so those values are not solutions.

Since $2(8) - 6 = 16 - 6 = 10$, the solution is 8.

34. "The product of a number and 4 is 8" is written

$$4x = 8.$$

Since $4(2) = 8$, the solution is 2.

35. $-4, -\frac{1}{2}, 0, 2.5, 5$

Graph these numbers on a number line. They are already arranged in order from smallest to largest.

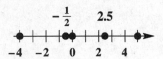

36. $-2, |-3|, -3, |-1|$

Recall that $|-3| = 3$ and $|-1| = 1$. From smallest to largest, the numbers are $-3, -2, |-1|, |-3|$.

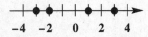

37. Since $\frac{4}{3}$ is the quotient of two integers, it is a *rational number*. Since all rational numbers are also real numbers, $\frac{4}{3}$ is a *real number*.

38. Since the decimal representation of $0.\overline{63}$ repeats, it is a *rational number*. Since all rational numbers are also real numbers, $0.\overline{63}$ is a *real number*.

39. Since 19 is a *natural number*, it is also a *whole number* and an *integer*. We can write it as $\frac{19}{1}$, so it is a *rational number* and hence, a *real number*.

40. Since the decimal representation of $\sqrt{6}$ does not terminate or repeat, it is an *irrational number*. Since all irrational numbers are also real numbers, $\sqrt{6}$ is a *real number*.

41. $-10, 5$
Since any negative number is less than any positive number, -10 is the lesser number.

42. $-8, -9$
Since -9 is to the left of -8 on the number line, -9 is the lesser number.

43. $-\dfrac{2}{3}, -\dfrac{3}{4}$

To compare these fractions, use a common denominator.

$$-\frac{2}{3} = -\frac{8}{12}, \quad -\frac{3}{4} = -\frac{9}{12}$$

Since $-\frac{9}{12}$ is to the left of $-\frac{8}{12}$ on the number line, $-\frac{3}{4}$ is the lesser number.

44. $0, -|23|$
Since $-|23| = -23$ and $-23 < 0$, $-|23|$ is the lesser number.

45. $12 > -13$
This statement is true since 12 is to the right of -13 on the number line.

46. $0 > -5$
This statement is true since 0 is to the right of -5 on the number line.

47. $-9 < -7$
This statement is true since -9 is to the left of -7 on the number line.

48. $-13 \geq -13$
This is a true statement since $-13 = -13$.

49. **(a)** The opposite of the number -9 is its negative; that is, $-(-9) = 9$.

(b) Since $-9 < 0$, the absolute value of the number -9 is $|-9| = -(-9) = 9$.

50. 0 **(a)** $-0 = 0$ **(b)** $|0| = 0$

51. 6 **(a)** $-(6) = -6$ **(b)** $|6| = 6$

52. $-\frac{5}{7}$ **(a)** $-(-\frac{5}{7}) = \frac{5}{7}$

(b) $\left|-\frac{5}{7}\right| = -(-\frac{5}{7}) = \frac{5}{7}$

53. $|-12| = -(-12) = 12$

54. $-|3| = -3$

55. $-|-19| = -[-(-19)] = -19$

56. $-|9 - 2| = -|7| = -7$

57. $-10 + 4 = -6$

58. $14 + (-18) = -4$

59. $-8 + (-9) = -17$

60. $\dfrac{4}{9} + \left(-\dfrac{5}{4}\right) = \dfrac{4 \cdot 4}{9 \cdot 4} + \left(-\dfrac{5 \cdot 9}{4 \cdot 9}\right)$ $LCD = 36$

$= \dfrac{16}{36} + \left(-\dfrac{45}{36}\right)$

$= -\dfrac{29}{36}$

61. $-13.5 + (-8.3) = -21.8$

62. $(-10 + 7) + (-11) = (-3) + (-11)$
$= -14$

63. $[-6 + (-8) + 8] + [9 + (-13)]$
$= \{[-6 + (-8)] + 8\} + (-4)$
$= [(-14) + 8] + (-4)$
$= (-6) + (-4) = -10$

64. $(-4 + 7) + (-11 + 3) + (-15 + 1)$
$= (3) + (-8) + (-14)$
$= [3 + (-8)] + (-14)$
$= (-5) + (-14) = -19$

65. $-7 - 4 = -7 + (-4) = -11$

66. $-12 - (-11) = -12 + (11) = -1$

67. $5 - (-2) = 5 + (2) = 7$

68. $-\dfrac{3}{7} - \dfrac{4}{5} = -\dfrac{3 \cdot 5}{7 \cdot 5} - \dfrac{4 \cdot 7}{5 \cdot 7}$

$= -\dfrac{15}{35} - \dfrac{28}{35}$ $LCD = 35$

$= -\dfrac{15}{35} + \left(-\dfrac{28}{35}\right)$

$= -\dfrac{43}{35}$, or $-1\dfrac{8}{35}$

69. $2.56 - (-7.75) = 2.56 + (7.75)$
$= 10.31$

70. $(-10 - 4) - (-2) = [-10 + (-4)] + 2$
$= (-14) + (2)$
$= -12$

71. $(-3 + 4) - (-1) = (-3 + 4) + 1$
$= 1 + 1$
$= 2$

72. $-(-5 + 6) - 2 = -(1) + (-2)$
$= -1 + (-2)$
$= -3$

73. "19 added to the sum of -31 and 12" is written
$(-31 + 12) + 19 = (-19) + 19$
$= 0.$

74. "13 more than the sum of -4 and -8" is written
$[-4 + (-8)] + 13 = -12 + 13$
$= 1.$

75. "The difference between -4 and -6" is written
$-4 - (-6) = -4 + 6$
$= 2.$

76. "Five less than the sum of 4 and -8" is written
$[4 + (-8)] - 5 = (-4) + (-5)$
$= -9.$

77. $x + (-2) = -4$
Because $(-2) + (-2) = -4$,
the solution is -2.

78. $12 + x = 11$
Because $12 + (-1) = 1$,
the solution is -1.

79. $-23.75 + 50.00 = 26.25$
He now has a positive balance of $26.25.

80. $-26 + 16 = -10$
The high temperature was $-10°F$.

81. $-28 + 13 - 14 = (-28 + 13) - 14$
$= (-28 + 13) + (-14)$
$= -15 + (-14)$
$= -29$
His present financial status is $-$29.

82. $-3 - 7 = -3 + (-7)$
$= -10$
The new temperature is $-10°$.

83. $8 - 12 + 42 = [8 + (-12)] + 42$
$= -4 + 42$
$= 38$
The total net yardage is 38.

84. To get the closing value for the previous day, we can add the amount it was down to the amount at which it closed.
$47.92 + 9496.28 = 9544.20$

85. $(-12)(-3) = 36$

86. $15(-7) = -(15 \cdot 7)$
$= -105$

87. $-\dfrac{4}{3}\left(-\dfrac{3}{8}\right) = \dfrac{4}{3} \cdot \dfrac{3}{8}$

$\qquad\qquad = \dfrac{4 \cdot 3}{3 \cdot 4 \cdot 2}$

$\qquad\qquad = \dfrac{1}{2}$

88. $(-4.8)(-2.1) = 10.08$

89. $5(8 - 12) = 5[8 + (-12)]$

$\qquad\qquad = 5(-4) = -20$

90. $(5 - 7)(8 - 3) = [5 + (-7)][8 + (-3)]$

$\qquad\qquad\qquad = (-2)(5) = -10$

91. $2(-6) - (-4)(-3) = -12 - (12)$

$\qquad\qquad\qquad\quad = -12 + (-12)$

$\qquad\qquad\qquad\quad = -24$

92. $3(-10) - 5 = -30 + (-5) = -35$

93. $\dfrac{-36}{-9} = \dfrac{4 \cdot 9}{9} = 4$

94. $\dfrac{220}{-11} = -\dfrac{20 \cdot 11}{11} = -20$

95. $-\dfrac{1}{2} \div \dfrac{2}{3} = -\dfrac{1}{2} \cdot \dfrac{3}{2} = -\dfrac{3}{4}$

96. $-33.9 \div (-3) = \dfrac{-33.9}{-3} = 11.3$

97. $\dfrac{-5(3) - 1}{8 - 4(-2)} = \dfrac{-15 + (-1)}{8 - (-8)}$

$\qquad\qquad\quad = \dfrac{-16}{8 + 8}$

$\qquad\qquad\quad = \dfrac{-16}{16} = -1$

98. $\dfrac{5(-2) - 3(4)}{-2[3 - (-2)] - 1} = \dfrac{-10 - 12}{-2(3 + 2) - 1}$

$\qquad\qquad\qquad\quad = \dfrac{-10 + (-12)}{-2(5) - 1}$

$\qquad\qquad\qquad\quad = \dfrac{-22}{-10 + (-1)}$

$\qquad\qquad\qquad\quad = \dfrac{-22}{-11} = 2$

99. $\dfrac{10^2 - 5^2}{8^2 + 3^2 - (-2)} = \dfrac{100 - 25}{64 + 9 + 2}$

$\qquad\qquad\qquad\quad = \dfrac{75}{75} = 1$

100. $\dfrac{(0.6)^2 + (0.8)^2}{(-1.2)^2 - (-0.56)} = \dfrac{0.36 + 0.64}{1.44 + 0.56}$

$\qquad\qquad\qquad\qquad = \dfrac{1.00}{2.00} = 0.5$

In Exercises 101–104, replace x with -5, y with 4, and z with -3.

101. $6x - 4z = 6(-5) - 4(-3)$

$\qquad\qquad = -30 - (-12)$

$\qquad\qquad = -30 + 12 = -18$

102. $5x + y - z = 5(-5) + (4) - (-3)$

$\qquad\qquad\qquad = (-25 + 4) + 3$

$\qquad\qquad\qquad = -21 + 3 = -18$

103. $5x^2 = 5(-5)^2$

$\qquad\quad = 5(25)$

$\qquad\quad = 125$

104. $z^2(3x - 8y) = (-3)^2[3(-5) - 8(4)]$

$\qquad\qquad\qquad = 9(-15 - 32)$

$\qquad\qquad\qquad = 9[-15 + (-32)]$

$\qquad\qquad\qquad = 9(-47) = -423$

105. "Nine less than the product of -4 and 5" is written

$$-4(5) - 9 = -20 + (-9)$$
$$= -29.$$

106. "Five-sixths of the sum of 12 and -6" is written

$$\tfrac{5}{6}[12 + (-6)] = \tfrac{5}{6}(6)$$
$$= 5.$$

107. "The quotient of 12 and the sum of 8 and -4" is written

$$\dfrac{12}{8 + (-4)} = \dfrac{12}{4} = 3.$$

108. "The product of -20 and 12, divided by the difference between 15 and -15" is written

$$\dfrac{-20(12)}{15 - (-15)} = \dfrac{-240}{15 + 15}$$
$$= \dfrac{-240}{30} = -8.$$

109. "8 times a number is -24" is written

$$8x = -24.$$

If $x = -3$,

$$8x = 8(-3) = -24.$$

The solution is -3.

110. "The quotient of a number and 3 is -2" is written

$$\frac{x}{3} = -2.$$

If $x = -6$,

$$\frac{x}{3} = \frac{-6}{3} = -2.$$

The solution is -6.

111. Find the average of the eight numbers.

$$\frac{26 + 38 + 40 + 20 + 4 + 14 + 96 + 18}{8}$$

$$= \frac{256}{8} = \frac{8 \cdot 32}{8} = 32$$

112. Find the average of the six numbers.

$$\frac{-12 + 28 + (-36) + 0 + 12 + (-10)}{6}$$

$$= \frac{-18}{6} = -3$$

113. $6 + 0 = 6$

This is an example of an identity property.

114. $5 \cdot 1 = 5$

This is an example of an identity property.

115. $-\frac{2}{3}\left(-\frac{3}{2}\right) = 1$

This is an example of an inverse property.

116. $17 + (-17) = 0$

This is an example of an inverse property.

117. $5 + (-9 + 2) = [5 + (-9)] + 2$

This is an example of an associative property.

118. $w(xy) = (wx)y$

This is an example of an associative property.

119. $3x + 3y = 3(x + y)$

This is an example of the distributive property.

120. $(1 + 2) + 3 = 3 + (1 + 2)$

This is an example of a commutative property.

121. $7y + 14 = 7y + 7 \cdot 2$
$$= 7(y + 2)$$

122. $-12(4 - t) = -12(4) - (-12)(t)$
$$= -48 + 12t$$

123. $3(2s) + 3(5y) = 3(2s + 5y)$

124. $-(-4r + 5s) = -1(-4r + 5s)$
$$= (-1)(-4r) + (-1)(5s)$$
$$= 4r - 5s$$

125. $2m + 9m = (2 + 9)m$ *Distributive property*
$$= 11m$$

126. $15p^2 - 7p^2 + 8p^2$
$$= (15 - 7 + 8)p^2 \quad \textit{Distributive property}$$
$$= 16p^2$$

127. $5p^2 - 4p + 6p + 11p^2$
$$= (5 + 11)p^2 + (-4 + 6)p$$
$$\textit{Distributive property}$$
$$= 16p^2 + 2p$$

128. $-2(3k - 5) + 2(k + 1)$
$$= -6k + 10 + 2k + 2$$
$$\textit{Distributive property}$$
$$= -4k + 12$$

129. $7(2m + 3) - 2(8m - 4)$
$$= 14m + 21 - 16m + 8$$
$$\textit{Distributive property}$$
$$= (14 - 16)m + 29$$
$$= -2m + 29$$

130. $-(2k + 8) - (3k - 7)$
$$= -1(2k + 8) - 1(3k - 7)$$
$$\textit{Replace } - \textit{ with } -1.$$
$$= -2k - 8 - 3k + 7$$
$$\textit{Distributive property}$$
$$= -5k - 1$$

131. "Seven times a number, subtracted from the product of -2 and three times the number" is written

$$-2(3x) - 7x = -6x - 7x = -13x.$$

132. "A number multiplied by 8, added to the sum of 5 and four times the number" is written

$$(5 + 4x) + 8x = 5 + (4x + 8x) = 5 + 12x.$$

133. [1.6] $\dfrac{6(-4) + 2(-12)}{5(-3) + (-3)} = \dfrac{-24 + (-24)}{-15 + (-3)}$
$$= \frac{-48}{-18} = \frac{8 \cdot 6}{3 \cdot 6}$$
$$= \frac{8}{3}, \text{ or } 2\frac{2}{3}$$

134. [1.5] $\dfrac{3}{8} - \dfrac{5}{12} = \dfrac{3 \cdot 3}{8 \cdot 3} - \dfrac{5 \cdot 2}{12 \cdot 2}$
$$= \frac{9}{24} - \frac{10}{24}$$
$$= \frac{9}{24} + \left(-\frac{10}{24}\right)$$
$$= -\frac{1}{24}$$

135. **[1.6]** $\dfrac{8^2 + 6^2}{7^2 + 1^2} = \dfrac{64 + 36}{49 + 1}$

$\qquad\qquad = \dfrac{100}{50} = 2$

136. **[1.6]** $-\dfrac{12}{5} \div \dfrac{9}{7} = -\dfrac{12}{5} \cdot \dfrac{7}{9}$

$\qquad\qquad = -\dfrac{12 \cdot 7}{5 \cdot 9}$

$\qquad\qquad = -\dfrac{3 \cdot 4 \cdot 7}{5 \cdot 3 \cdot 3}$

$\qquad\qquad = -\dfrac{28}{15}, \ \text{or} \ -1\dfrac{13}{15}$

137. **[1.5]** $2\dfrac{5}{6} - 4\dfrac{1}{3} = \dfrac{17}{6} - \dfrac{13}{3}$

$\qquad\qquad = \dfrac{17}{6} - \dfrac{13 \cdot 2}{3 \cdot 2}$

$\qquad\qquad = \dfrac{17}{6} - \dfrac{26}{6}$

$\qquad\qquad = \dfrac{17}{6} + \left(-\dfrac{26}{6}\right)$

$\qquad\qquad = -\dfrac{9}{6} = -\dfrac{3}{2}, \ \text{or} \ -1\dfrac{1}{2}$

138. **[1.6]** $\left(-\dfrac{5}{6}\right)^2 = \left(-\dfrac{5}{6}\right)\left(-\dfrac{5}{6}\right)$

$\qquad\qquad = \dfrac{25}{36}$

139. **[1.5]** $[(-2) + 7 - (-5)] + [-4 - (-10)]$

$\qquad = \{[(-2) + 7] - (-5)\} + (-4 + 10)$

$\qquad = (5 + 5) + 6$

$\qquad = 10 + 6 = 16$

140. **[1.6]** $-16(-3.5) - 7.2(-3)$

$\qquad = 56 - [(7.2)(-3)]$

$\qquad = 56 - (-21.6)$

$\qquad = 56 + 21.6$

$\qquad = 77.6$

141. **[1.5]** $-8 + [(-4 + 17) - (-3 - 3)]$

$\qquad = -8 + \{(13) - [-3 + (-3)]\}$

$\qquad = -8 + [13 - (-6)]$

$\qquad = -8 + (13 + 6)$

$\qquad = -8 + 19 = 11$

142. **[1.8]** $-4(2t + 1) - 8(-3t + 4)$

$\qquad = -4(2t) - 4(1) - 8(-3t) - 8(4)$

$\qquad = -8t - 4 + 24t - 32$

$\qquad = 16t - 36$

143. **[1.8]** $5x^2 - 12y^2 + 3x^2 - 9y^2$

$\qquad = (5x^2 + 3x^2) + (-12y^2 - 9y^2)$

$\qquad = (5 + 3)x^2 + (-12 - 9)y^2$

$\qquad = 8x^2 - 21y^2$

144. **[1.6]** $(-8 - 3) - 5(2 - 9)$

$\qquad = [-8 + (-3)] - 5[2 + (-9)]$

$\qquad = -11 - 5(-7)$

$\qquad = -11 - (-35)$

$\qquad = -11 + 35 = 24$

145. **[1.6]** Dividing 0 *by* a nonzero number gives a quotient of 0. However, dividing a number *by* 0 is undefined.

146. **[1.5]** $118 - 165 = 118 + (-165)$

$\qquad\qquad = -47$

The lowest temperature ever recorded in Iowa was $-47°$F.

147. **[1.5]** The change in enrollment from 1980 to 1985 was $12.39 - 13.23 = -0.84$ million students. Expressed as an integer, this number is $-840,000$.

148. **[1.5]** The change in enrollment from 1985 to 1990 was $11.34 - 12.39 = -1.05$ million students. Expressed as an integer, this number is $-1,050,000$.

149. **[1.5]** The change in enrollment from 1995 to 2000 was $13.52 - 12.50 = 1.02$ million students. Expressed as an integer, this number is $1,020,000$.

150. **[1.5]** The change in enrollment from 2000 to 2005 was $14.91 - 13.52 = 1.39$ million students. Expressed as an integer, this number is $1,390,000$.

Chapter 1 Test

1. $\dfrac{63}{99} = \dfrac{7 \cdot 9}{11 \cdot 9} = \dfrac{7}{11}$

2. The denominators are 8, 12, and 15; or equivalently, 2^3, $2^2 \cdot 3$, and $3 \cdot 5$. So the LCD is $2^3 \cdot 3 \cdot 5 = 120$.

$\dfrac{5}{8} + \dfrac{11}{12} + \dfrac{7}{15}$

$= \dfrac{5 \cdot 15}{8 \cdot 15} + \dfrac{11 \cdot 10}{12 \cdot 10} + \dfrac{7 \cdot 8}{15 \cdot 8}$

$= \dfrac{75}{120} + \dfrac{110}{120} + \dfrac{56}{120}$

$= \dfrac{241}{120}, \ \text{or} \ 2\dfrac{1}{120}$

3. $\dfrac{19}{15} \div \dfrac{6}{5} = \dfrac{19}{15} \cdot \dfrac{5}{6} = \dfrac{19 \cdot 5}{3 \cdot 5 \cdot 6} = \dfrac{19}{18}, \ \text{or} \ 1\dfrac{1}{18}$

4. $4[-20 + 7(-2)] = 4[-20 + (-14)]$

$\qquad\qquad\qquad = 4(-34) = -136$

Since $-136 \le 135$, the statement "$4[-20 + 7(-2)] \le 135$" is true.

5. $-1, -3, |-4|, |-1|$

Recall that $|-4| = 4$ and $|-1| = 1$. From smallest to largest, the numbers are $-3, -1, |-1|, |-4|$.

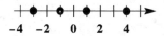

6. The number $-\frac{2}{3}$ can be written as a quotient of two integers with denominator not 0, so it is a *rational number*. Since all rational numbers are real numbers, it is also a *real number*.

7. If -8 and -1 are both graphed on a number line, we see that the point for -8 is to the *left* of the point for -1. This indicates that -8 is *less than* -1.

8. "The quotient of -6 and the sum of 2 and -8" is written $\dfrac{-6}{2 + (-8)}$,

$$\text{and } \dfrac{-6}{2 + (-8)} = \dfrac{-6}{-6} = 1.$$

9. $-2 - (5 - 17) + (-6)$
$= -2 - [5 + (-17)] + (-6)$
$= -2 - (-12) + (-6)$
$= (-2 + 12) + (-6)$
$= 10 + (-6) = 4$

10. $-5\dfrac{1}{2} + 2\dfrac{2}{3} = -\dfrac{11}{2} + \dfrac{8}{3}$

$$= -\dfrac{11 \cdot 3}{2 \cdot 3} + \dfrac{8 \cdot 2}{3 \cdot 2}$$

$$= -\dfrac{33}{6} + \dfrac{16}{6}$$

$$= -\dfrac{17}{6}, \text{ or } -2\dfrac{5}{6}$$

11. $-6 - [-7 + (2 - 3)]$
$= -6 - [-7 + (-1)]$
$= -6 - (-8)$
$= -6 + 8 = 2$

12. $4^2 + (-8) - (2^3 - 6)$
$= 16 + (-8) - (8 - 6)$
$= [16 + (-8)] - 2$
$= 8 - 2 = 6$

13. $(-5)(-12) + 4(-4) + (-8)^2$
$= (-5)(-12) + 4(-4) + 64$
$= [60 + (-16)] + 64$
$= 44 + 64 = 108$

14. $\dfrac{30(-1 - 2)}{-9[3 - (-2)] - 12(-2)}$

$$= \dfrac{30(-3)}{-9(5) - (-24)}$$

$$= \dfrac{-90}{-45 + 24}$$

$$= \dfrac{-90}{-21}$$

$$= \dfrac{30 \cdot 3}{7 \cdot 3} = \dfrac{30}{7}, \text{ or } 4\dfrac{2}{7}$$

15. $-x + 3 = -3$

If $x = 6$,

$$-6 + 3 = -3.$$

Therefore, the solution is 6.

16. $-3x = -12$

If $x = 4$,

$$-3x = -3(4) = -12.$$

Therefore, the solution is 4.

17. $3x - 4y^2$
$= 3(-2) - 4(4^2)$ *Let x = –2, y = 4.*
$= 3(-2) - 4(16)$
$= -6 - 64 = -70$

18. $\dfrac{5x + 7y}{3(x + y)}$

$$= \dfrac{5(-2) + 7(4)}{3(-2 + 4)} \quad \textit{Let x = –2, y = 4.}$$

$$= \dfrac{-10 + 28}{3(2)}$$

$$= \dfrac{18}{6} = 3$$

19. The difference between the highest and lowest elevations is

$6960 - (-40) = 6960 + 40 = 7000$ meters.

20. 4 saves (3 points per save)

$+$ 3 wins (3 points per win)

$+$ 2 losses (-2 points per loss)

$+$ 1 blown save (-2 points per blown save)

$= 4(3) + 3(3) + 2(-2) + 1(-2)$
$= 12 + 9 - 4 - 2$
$= 15$ points

He has a total of 15 points.

21. $2.10 - 3.52 = 2.10 + (-3.52) = -1.42$

As a signed number, the federal budget deficit is $-\$1.42$ trillion.

22. Commutative property

$$(5 + 2) + 8 = 8 + (5 + 2)$$

illustrates a commutative property because the order of the numbers is changed, but not the grouping. The correct response is **B**.

23. Associative property

$$-5 + (3 + 2) = (-5 + 3) + 2$$

illustrates an associative property because the grouping of the numbers is changed, but not the order. The correct response is **D**.

24. Inverse property

$$-\frac{5}{3}\left(-\frac{3}{5}\right) = 1$$

illustrates an inverse property. The correct response is **E**.

25. Identity property

$$3x + 0 = 3x$$

illustrates an identity property. The correct response is **A**.

26. Distributive property

$$-3(x + y) = -3x + (-3y)$$

illustrates the distributive property. The correct response is **C**.

27. $3(x + 1) = 3 \cdot x + 3 \cdot 1$
$ = 3x + 3$

The distributive property is used to rewrite $3(x + 1)$ as $3x + 3$.

28. **(a)** $-6[5 + (-2)] = -6(3) = -18$

(b) $-6[5 + (-2)] = -6(5) + (-6)(-2)$
$ = -30 + 12 = -18$

(c) The distributive property assures us that the answers must be the same, because $a(b + c) = ab + ac$ for all a, b, c.

29. $8x + 4x - 6x + x + 14x$
$= (8 + 4 - 6 + 1 + 14)x$
$= 21x$

30. $5(2x - 1) - (x - 12) + 2(3x - 5)$
$= 5(2x - 1) - 1(x - 12) + 2(3x - 5)$
$= 10x - 5 - x + 12 + 6x - 10$
$= (10 - 1 + 6)x + (-5 + 12 - 10)$
$= 15x - 3$

CHAPTER 2   LINEAR EQUATIONS AND INEQUALITIES IN ONE VARIABLE

2.1 The Addition Property of Equality

2.1 Now Try Exercises

N1. Note: When solving equations we will write "Add 5" as a shorthand notation for "Add 5 to each side" and "Subtract 5" as a notation for "Subtract 5 from each side."

$$x - 13 = 4 \qquad Given$$
$$x - 13 + 13 = 4 + 13 \quad Add\ 13.$$
$$x = 17 \qquad Combine\ like\ terms.$$

We check by substituting 17 for x in the *original* equation.

Check: $x - 13 = 4$ *Original equation*
$$17 - 13 \overset{?}{=} 4 \quad Let\ x = 17.$$
$$4 = 4 \quad True$$

Since a true statement results, $\{17\}$ is the solution set.

N2. $t - 5.7 = -7.2$
$$t - 5.7 + 5.7 = -7.2 + 5.7 \quad Add\ 5.7.$$
$$t = -1.5$$

Check $t = -1.5$: $-7.2 = -7.2$ *True*

This is a shorthand notation for showing that if we substitute -2.2 for m, both sides are equal to -6.3, and hence a true statement results. In practice, this is what you will do, especially if you're using a calculator.

The solution set is $\{-1.5\}$.

N3. $-15 = x + 12$
$$-15 - 12 = x + 12 - 12 \quad Subtract\ 12.$$
$$-27 = x$$

Check $x = -27$: $-15 = -15$ *True*

The solution set is $\{-27\}$.

N4. $\frac{2}{3}x - 4 = \frac{5}{3}x$
$$\frac{2}{3}x - 4 - \frac{2}{3}x = \frac{5}{3}x - \frac{2}{3}x \quad Subtract\ \frac{2}{3}x.$$
$$-4 = \frac{3}{3}x \quad Combine\ terms.$$
$$-4 = x$$

Check $x = -4$: $-\frac{20}{3} = -\frac{20}{3}$ *True*

The solution set is $\{-4\}$.

N5. $6x - 8 = 12 + 5x$ *Given.*
$$6x - 8 - 5x = 12 + 5x - 5x \quad Subtract\ 5x.$$
$$x - 8 = 12 \quad Combine\ terms.$$
$$x - 8 + 8 = 12 + 8 \quad Add\ 8.$$
$$x = 20 \quad Combine\ terms.$$

Check $x = 20$: $120 - 8 = 12 + 100$ *True*

The solution set is $\{20\}$.

N6. $5x - 10 - 12x = 4 - 8x - 9$ *Given*
$$-7x - 10 = -8x - 5 \quad Combine\ terms.$$
$$-7x - 10 + 8x = -8x - 5 + 8x \quad Add\ 8x.$$
$$x - 10 = -5 \quad Combine\ terms.$$
$$x - 10 + 10 = -5 + 10 \quad Add\ 10.$$
$$x = 5 \quad Combine\ terms.$$

Check $x = 5$: $-45 = -45$ *True*

The solution set is $\{5\}$.

N7. $4(3x - 2) - (11x - 4) = 3$ *Given*
$$4(3x - 2) - 1(11x - 4) = 3 \quad -a = -1a$$
$$12x - 8 - 11x + 4 = 3 \quad Distributive\ property$$
$$x - 4 = 3 \quad Combine\ terms.$$
$$x - 4 + 4 = 3 + 4 \quad Add\ 4.$$
$$x = 7$$

Check $x = 7$: $3 = 3$ *True*

The solution set is $\{7\}$.

2.1 Section Exercises

1. **(a)** $5x + 8 - 4x + 7$

This is an expression, not an equation, since there is no equals symbol. It can be simplified by rearranging terms and then combining like terms.

$$5x + 8 - 4x + 7 = 5x - 4x + 8 + 7$$
$$= x + 15$$

(b) $-6y + 12 + 7y - 5$

This is an expression, not an equation, since there is no equals symbol. It can be simplified by rearranging terms and then combining like terms.

$$-6y + 12 + 7y - 5 = -6y + 7y + 12 - 5$$
$$= y + 7$$

(c) $5x + 8 - 4x = 7$

This is an equation because of the equals symbol.

$$5x + 8 - 4x = 7$$
$$x + 8 = 7$$
$$x = -1$$

The solution set is $\{-1\}$.

(d) $-6y + 12 + 7y = -5$

This is an equation because of the equals symbol.

$$-6y + 12 + 7y = -5$$
$$y + 12 = -5$$
$$y = -17$$

The solution set is $\{-17\}$.

3. Equations **A** $x^2 - 5x + 6 = 0$ and **B** $x^3 = x$ are *not* linear equations in one variable because they cannot be written in the form $Ax + B = C$. Note that in a linear equation the exponent on the variable must be 1.

For Exercises 5–52, all solutions should be checked by substituting into the original equation. Checks will be shown here for only a few of the exercises.

5.
$$x - 3 = 9$$
$$x - 3 + 3 = 9 + 3$$
$$x = 12$$

Check this solution by replacing x with 12 in the original equation.

$$x - 3 = 9$$
$$12 - 3 \stackrel{?}{=} 9 \quad \text{Let x = 12.}$$
$$9 = 9 \quad \text{True}$$

Because the final statement is true, $\{12\}$ is the solution set.

7.
$$x - 12 = 19$$
$$x - 12 + 12 = 19 + 12$$
$$x = 31$$

Check $x = 31$:

$$31 - 12 \stackrel{?}{=} 19 \quad \text{Let x = 31.}$$
$$19 = 19 \quad \text{True}$$

Thus, $\{31\}$ is the solution set.

9.
$$x - 6 = -9$$
$$x - 6 + 6 = -9 + 6$$
$$x = -3$$

Checking yields a true statement, so $\{-3\}$ is the solution set.

11.
$$r + 8 = 12$$
$$r + 8 - 8 = 12 - 8$$
$$r = 4$$

Checking yields a true statement, so $\{4\}$ is the solution set.

13.
$$x + 28 = 19$$
$$x + 28 - 28 = 19 - 28$$
$$x = -9$$

Checking yields a true statement, so $\{-9\}$ is the solution set.

15.
$$x + \frac{1}{4} = -\frac{1}{2}$$
$$x + \frac{1}{4} - \frac{1}{4} = -\frac{1}{2} - \frac{1}{4}$$
$$x = -\frac{2}{4} - \frac{1}{4}$$
$$x = -\frac{3}{4}$$

Check $x = -\frac{3}{4}$: $-\frac{1}{2} = -\frac{1}{2}$ *True*

The solution set is $\left\{-\frac{3}{4}\right\}$.

17.
$$7 + r = -3$$
$$r + 7 = -3$$
$$r + 7 - 7 = -3 - 7$$
$$r = -10$$

The solution set is $\{-10\}$.

19.
$$2 = p + 15$$
$$2 - 15 = p + 15 - 15$$
$$-13 = p$$

The solution set is $\{-13\}$.

21.
$$-4 = x - 14$$
$$-4 + 14 = x - 14 + 14$$
$$10 = x$$

The solution set is $\{10\}$.

23.
$$-\frac{1}{3} = x - \frac{3}{5}$$
$$-\frac{1}{3} + \frac{3}{5} = x - \frac{3}{5} + \frac{3}{5}$$
$$-\frac{5}{15} + \frac{9}{15} = x$$
$$\frac{4}{15} = x$$

Check $x = \frac{4}{15}$: $-\frac{5}{15} = \frac{4}{15} - \frac{9}{15}$ *True*

The solution set is $\left\{\frac{4}{15}\right\}$.

25.
$$x - 8.4 = -2.1$$
$$x - 8.4 + 8.4 = -2.1 + 8.4$$
$$x = 6.3$$

The solution set is $\{6.3\}$.

27.
$$t + 12.3 = -4.6$$
$$t + 12.3 - 12.3 = -4.6 - 12.3$$
$$t = -16.9$$

The solution set is $\{-16.9\}$.

29.
$$3x = 2x + 7$$
$$3x - 2x = 2x + 7 - 2x \quad \textit{Subtract } 2x.$$
$$1x = 7 \quad \text{or} \quad x = 7$$

Check $x = 7$: $21 = 21$ *True*

The solution set is $\{7\}$.

31.
$$10x + 4 = 9x$$
$$10x + 4 - 9x = 9x - 9x \quad \textit{Subtract } 9x.$$
$$1x + 4 = 0$$
$$x + 4 - 4 = 0 - 4 \quad\quad \textit{Subtract } 4.$$
$$x = -4$$

Check $x = -4$: $-36 = -36$ *True*

The solution set is $\{-4\}$.

33.
$$3x + 7 = 2x + 4$$
$$3x + 7 - 2x = 2x + 4 - 2x$$
$$x + 7 = 4$$
$$x + 7 - 7 = 4 - 7$$
$$x = -3$$

The solution set is $\{-3\}$.

35.
$$8t + 6 = 7t + 6$$
$$8t + 6 - 7t = 7t + 6 - 7t$$
$$t + 6 = 6$$
$$t + 6 - 6 = 6 - 6$$
$$t = 0$$

The solution set is $\{0\}$.

37.
$$-4x + 7 = -5x + 9$$
$$-4x + 7 + 5x = -5x + 9 + 5x$$
$$x + 7 = 9$$
$$x + 7 - 7 = 9 - 7$$
$$x = 2$$

The solution set is $\{2\}$.

39.
$$\frac{2}{5}w - 6 = \frac{7}{5}w$$
$$\frac{2}{5}w - 6 - \frac{2}{5}w = \frac{7}{5}w - \frac{2}{5}w \quad \textit{Subtract } \frac{2}{5}w.$$
$$-6 = \frac{5}{5}w$$
$$-6 = w$$

The solution set is $\{-6\}$.

41.
$$5.6x + 2 = 4.6x$$
$$5.6x + 2 - 4.6x = 4.6x - 4.6x$$
$$1.0x + 2 = 0$$
$$x + 2 - 2 = 0 - 2$$
$$x = -2$$

The solution set is $\{-2\}$.

43.
$$1.4x - 3 = 0.4x$$
$$1.4x - 3 - 0.4x = 0.4x - 0.4x$$
$$1.0x - 3 = 0$$
$$1.0x - 3 + 3 = 0 + 3$$
$$x = 3$$

The solution set is $\{3\}$.

45.
$$5p = 4p$$
$$5p - 4p = 4p - 4p$$
$$p = 0$$

The solution set is $\{0\}$.

47.
$$1.2y - 4 = 0.2y - 4$$
$$1.2y - 4 - 0.2y = 0.2y - 4 - 0.2y$$
$$1.0y - 4 = -4$$
$$y - 4 + 4 = -4 + 4$$
$$y = 0$$

The solution set is $\{0\}$.

49.
$$\frac{1}{2}x + 5 = -\frac{1}{2}x$$
$$\frac{1}{2}x + \frac{1}{2}x + 5 = -\frac{1}{2}x + \frac{1}{2}x$$
$$x + 5 = 0$$
$$x + 5 - 5 = 0 - 5$$
$$x = -5$$

The solution set is $\{-5\}$.

51.
$$3x + 7 - 2x = 0$$
$$x + 7 = 0$$
$$x + 7 - 7 = 0 - 7$$
$$x = -7$$

The solution set is $\{-7\}$.

53.
$$5t + 3 + 2t - 6t = 4 + 12$$
$$(5 + 2 - 6)t + 3 = 16$$
$$t + 3 - 3 = 16 - 3$$
$$t = 13$$

Check $t = 13$: $16 = 16$ *True*

The solution set is $\{13\}$.

55.
$$6x + 5 + 7x + 3 = 12x + 4$$
$$13x + 8 = 12x + 4$$
$$13x + 8 - 12x = 12x + 4 - 12x$$
$$x + 8 = 4$$
$$x + 8 - 8 = 4 - 8$$
$$x = -4$$

Check $x = -4$: $-44 = -44$ *True*

The solution set is $\{-4\}$.

57.
$$5.2q - 4.6 - 7.1q = -0.9q - 4.6$$
$$-1.9q - 4.6 = -0.9q - 4.6$$
$$-1.9q - 4.6 + 0.9q = -0.9q - 4.6 + 0.9q$$
$$-1.0q - 4.6 = -4.6$$
$$-1.0q - 4.6 + 4.6 = -4.6 + 4.6$$
$$-q = 0$$
$$q = 0$$

Check $q = 0$: $-4.6 = -4.6$ *True*
The solution set is $\{0\}$.

59.
$$\frac{5}{7}x + \frac{1}{3} = \frac{2}{5} - \frac{2}{7}x + \frac{2}{5}$$
$$\frac{5}{7}x + \frac{1}{3} = \frac{4}{5} - \frac{2}{7}x$$
$$\frac{5}{7}x + \frac{2}{7}x + \frac{1}{3} = \frac{4}{5} - \frac{2}{7}x + \frac{2}{7}x \quad Add\ \frac{2}{7}x.$$
$$\frac{7}{7}x + \frac{1}{3} = \frac{4}{5} \qquad \begin{array}{l}Combine\\like\ terms.\end{array}$$
$$1x + \frac{1}{3} - \frac{1}{3} = \frac{4}{5} - \frac{1}{3} \qquad Subtract\ \frac{1}{3}.$$
$$x = \frac{12}{15} - \frac{5}{15} \qquad LCD = 15$$
$$x = \frac{7}{15}$$

Check $x = \frac{7}{15}$: $\frac{2}{3} = \frac{2}{3}$ *True*
The solution set is $\{\frac{7}{15}\}$.

61. $(5y + 6) - (3 + 4y) = 10$
$$5y + 6 - 3 - 4y = 10 \qquad \begin{array}{l}Distributive\\property\end{array}$$
$$y + 3 = 10 \qquad Combine\ terms.$$
$$y + 3 - 3 = 10 - 3 \qquad Subtract\ 3.$$
$$y = 7$$

Check $y = 7$: $10 = 10$ *True*

The solution set is $\{7\}$.

63. $2(p + 5) - (9 + p) = -3$
$$2p + 10 - 9 - p = -3$$
$$p + 1 = -3$$
$$p + 1 - 1 = -3 - 1$$
$$p = -4$$

Check $p = -4$: $-3 = -3$ *True*
The solution set is $\{-4\}$.

65. $-6(2b + 1) + (13b - 7) = 0$
$$-12b - 6 + 13b - 7 = 0$$
$$b - 13 = 0$$
$$b - 13 + 13 = 0 + 13$$
$$b = 13$$

Check $b = 13$: $0 = 0$ *True*

The solution set is $\{13\}$.

67.
$$10(-2x + 1) = -19(x + 1)$$
$$-20x + 10 = -19x - 19$$
$$-20x + 10 + 19x = -19x - 19 + 19x$$
$$-x + 10 = -19$$
$$-x + 10 - 10 = -19 - 10$$
$$-x = -29$$
$$x = 29$$

Check $x = 29$: $-570 = -570$ *True*
The solution set is $\{29\}$.

69. $-2(8p + 2) - 3(2 - 7p) - 2(4 + 2p) = 0$
$$-16p - 4 - 6 + 21p - 8 - 4p = 0$$
$$p - 18 = 0$$
$$p - 18 + 18 = 0 + 18$$
$$p = 18$$

Check $p = 18$: $0 = 0$ *True*

The solution set is $\{18\}$.

71. $4(7x - 1) + 3(2 - 5x) - 4(3x + 5) = -6$
$$28x - 4 + 6 - 15x - 12x - 20 = -6$$
$$x - 18 = -6$$
$$x - 18 + 18 = -6 + 18$$
$$x = 12$$

Check $x = 12$: $-6 = -6$ *True*

The solution set is $\{12\}$.

73.   Answers will vary. One example is $x - 6 = -8$.

75.   "Three times a number is 17 more than twice the number."
$$3x = 2x + 17$$
$$3x - 2x = 2x + 17 - 2x$$
$$x = 17$$

The number is 17 and $\{17\}$ is the solution set.

77.   "If six times a number is subtracted from seven times the number, the result is -9."
$$7x - 6x = -9$$
$$x = -9$$

The number is -9 and $\{-9\}$ is the solution set.

79. $\dfrac{2}{3}\left(\dfrac{3}{2}\right) = \dfrac{2 \cdot 3}{3 \cdot 2} = 1$

81. $-\dfrac{5}{4}\left(-\dfrac{4}{5}x\right) = -\dfrac{5}{4}\left(-\dfrac{4}{5}\right)x$
$$= \dfrac{5 \cdot 4}{4 \cdot 5}x$$
$$= 1x = x$$

83. $9\left(\dfrac{r}{9}\right) = 9\left(\dfrac{1}{9}r\right)$
$$= 9\left(\dfrac{1}{9}\right)r$$
$$= 1r = r$$

2.2 The Multiplication Property of Equality

2.2 Now Try Exercises

N1. $8x = 80$

$$\frac{8x}{8} = \frac{80}{8} \quad \textit{Divide by 8.}$$

$$x = 10$$

Check $x = 10$: $80 = 80$ *True*

The solution set is $\{10\}$.

N2. $10x = -24$

$$\frac{10x}{10} = -\frac{24}{10} \quad \textit{Divide by 10.}$$

$$x = -\frac{24}{10} = -\frac{12}{5} \quad \textit{Write in lowest terms.}$$

Check $x = -\frac{12}{5}$: $-24 = -24$ *True*

The solution set is $\left\{-\frac{12}{5}\right\}$.

N3. $-1.3x = 7.02$

$$\frac{-1.3x}{-1.3} = \frac{7.02}{-1.3} \quad \textit{Divide by } -1.3.$$

$$x = -5.4$$

Check $x = -5.4$: $7.02 = 7.02$ *True*

The solution set is $\{-5.4\}$.

N4.

$$\frac{x}{5} = -7$$

$$\frac{1}{5}x = -7$$

$$5 \cdot \frac{1}{5}x = 5(-7) \quad \textit{Multiply by 5, the reciprocal of } \frac{1}{5}.$$

$$p = -35$$

Check $p = -35$: $-7 = -7$ *True*

The solution set is $\{-35\}$.

N5.

$$\frac{4}{7}z = -16$$

$$\frac{7}{4}\left(\frac{4}{7}z\right) = \frac{7}{4}(-16) \quad \textit{Multiply by } \frac{7}{4}.$$

$$1 \cdot t = \frac{7}{4} \cdot \frac{-16}{1} \quad \begin{array}{l}\textit{Multiplicative}\\\textit{inverse property}\\\textit{Multiplicative}\end{array}$$

$$t = -28 \quad \begin{array}{l}\textit{identity property;}\\\textit{multiply fractions.}\end{array}$$

Check $t = -28$: $-16 = -16$ *True*

The solution set is $\{-28\}$.

N6.

$$-x = 9$$

$$-1 \cdot x = 9 \qquad\qquad -x = -1 \cdot x$$

$$(-1)(-1) \cdot x = (-1)(9) \quad \textit{Multiply by } -1.$$

$$1 \cdot x = -9$$

$$x = -9$$

Check $x = -9$: $9 = 9$ *True*

The solution set is $\{-9\}$.

N7. $9n - 6n = 21$

$$3n = 21 \quad \textit{Combine terms.}$$

$$\frac{3n}{3} = \frac{21}{3} \quad \textit{Divide by 3.}$$

$$n = 7$$

Check $n = 7$: $21 = 21$ *True*

The solution set is $\{7\}$.

2.2 Section Exercises

1. **(a)** multiplication property of equality; to get x alone on the left side of the equation, multiply each side by $\frac{1}{3}$ (or divide each side by 3).

(b) addition property of equality; to get x alone on the left side of the equation, add -3 (or subtract 3) on each side.

(c) multiplication property of equality; to get x alone on the left side of the equation, multiply each side by -1 (or divide each side by -1).

(d) addition property of equality; to get x alone on the right side of the equation, add -6 (or subtract 6) on each side.

3. To find the solution of $-x = 5$, multiply (or divide) each side by -1, or use the rule "If $-x = a$, then $x = -a$."

5. $\frac{4}{5}x = 8$

To get just x on the left side, multiply both sides of the equation by the reciprocal of $\frac{4}{5}$, which is $\frac{5}{4}$.

7. $\frac{x}{10} = 5$

This equation is equivalent to $\frac{1}{10}x = 5$. To get just x on the left side, multiply both sides of the equation by the reciprocal of $\frac{1}{10}$, which is 10.

9. $-\frac{9}{2}x = -4$

To get just x on the left side, multiply both sides of the equation by the reciprocal of $-\frac{9}{2}$, which is $-\frac{2}{9}$.

11. $-x = 0.75$

This equation is equivalent to $-1x = 0.75$. To get just x on the left side, multiply both sides of the equation by the reciprocal of -1, which is -1.

13. $6x = 5$

To get just x on the left side, divide both sides of the equation by the coefficient of x, which is 6.

15. $-4x = 16$

To get just x on the left side, divide both sides of the equation by the coefficient of x, which is -4.

17. $0.12x = 48$

To get just x on the left side, divide both sides of the equation by the coefficient of x, which is 0.12.

19. $-x = 25$

This equation is equivalent to $-1x = 25$. To get just x on the left side, divide both sides of the equation by the coefficient of x, which is -1.

21. $6x = 36$

$\dfrac{6x}{6} = \dfrac{36}{6}$ *Divide by 6.*

$1x = 6$

$x = 6$

Check $x = 6$: $36 = 36$ *True*

The solution set is $\{6\}$.

23. $2m = 15$

$\dfrac{2m}{2} = \dfrac{15}{2}$ *Divide by 2.*

$m = \dfrac{15}{2}$

Check $m = \dfrac{15}{2}$: $15 = 15$ *True*

The solution set is $\left\{\dfrac{15}{2}\right\}$.

25. $4x = -20$

$\dfrac{4x}{4} = \dfrac{-20}{4}$ *Divide by 4.*

$x = -5$

Check $x = -5$: $-20 = -20$ *True*

The solution set is $\{-5\}$.

27. $-7x = 28$

$\dfrac{-7x}{-7} = \dfrac{28}{-7}$ *Divide by -7.*

$x = -4$

Check $x = -4$: $28 = 28$ *True*

The solution set is $\{-4\}$.

29. $10t = -36$

$\dfrac{10t}{10} = \dfrac{-36}{10}$ *Divide by 10.*

$t = -\dfrac{36}{10} = -\dfrac{18}{5}$ *Lowest terms*

Check $t = -\dfrac{18}{5}$: $-36 = -36$ *True*

The solution set is $\left\{-\dfrac{18}{5}\right\}$, or $\{-3.6\}$.

31. $-6x = -72$

$\dfrac{-6x}{-6} = \dfrac{-72}{-6}$ *Divide by -6.*

$x = 12$

Check $x = 12$: $-72 = -72$ *True*

The solution set is $\{12\}$.

33. $4r = 0$

$\dfrac{4r}{4} = \dfrac{0}{4}$ *Divide by 4.*

$r = 0$

Check $r = 0$: $0 = 0$ *True*

The solution set is $\{0\}$.

35. $-x = 12$

$-1 \cdot (-x) = -1 \cdot 12$ *Multiply by -1.*

$x = -12$

Check $x = -12$: $12 = 12$ *True*

The solution set is $\{-12\}$.

37. $-x = -\dfrac{3}{4}$

$-1 \cdot (-x) = -1 \cdot \left(-\dfrac{3}{4}\right)$

$x = \dfrac{3}{4}$

Check $x = \dfrac{3}{4}$: $-\dfrac{3}{4} = -\dfrac{3}{4}$ *True*

The solution set is $\left\{\dfrac{3}{4}\right\}$.

39. $0.2t = 8$

$\dfrac{0.2t}{0.2} = \dfrac{8}{0.2}$

$t = 40$

Check $t = 40$: $8 = 8$ *True*

The solution set is $\{40\}$.

41. $-2.1m = 25.62$

$\dfrac{-2.1m}{-2.1} = \dfrac{25.62}{-2.1}$

$m = -12.2$

Check $m = -12.2$: $25.62 = 25.62$ *True*

The solution set is $\{-12.2\}$.

43. $\dfrac{1}{4}x = -12$

$4 \cdot \dfrac{1}{4}x = 4(-12)$ *Multiply by 4.*

$1x = -48$

$x = -48$

Check $x = -48$: $-12 = -12$ *True*

The solution set is $\{-48\}$.

45. $\dfrac{z}{6} = 12$

$\dfrac{1}{6}z = 12$

$6 \cdot \dfrac{1}{6}z = 6 \cdot 12$

$z = 72$

Check $z = 72$: $12 = 12$ *True*

The solution set is $\{72\}$.

47.
$$\frac{x}{7} = -5$$
$$\frac{1}{7}x = -5$$
$$7\left(\frac{1}{7}x\right) = 7(-5)$$
$$x = -35$$

Check $x = -35$: $-5 = -5$ *True*

The solution set is $\{-35\}$.

49.
$$\frac{2}{7}p = 4$$
$$\frac{7}{2}\left(\frac{2}{7}p\right) = \frac{7}{2}(4)$$ *Multiply by the reciprocal of $\frac{2}{7}$.*
$$p = 14$$

Check $p = 14$: $4 = 4$ *True*

The solution set is $\{14\}$.

51.
$$-\frac{5}{6}t = -15$$
$$-\frac{6}{5}\left(-\frac{5}{6}t\right) = -\frac{6}{5}(-15)$$ *Multiply by the reciprocal of $-\frac{5}{6}$.*
$$t = 18$$

Check $t = 18$: $-15 = -15$ *True*

The solution set is $\{18\}$.

53.
$$-\frac{7}{9}x = \frac{3}{5}$$
$$-\frac{9}{7}\left(-\frac{7}{9}x\right) = -\frac{9}{7} \cdot \frac{3}{5}$$ *Multiply by the reciprocal of $-\frac{7}{9}$.*
$$x = -\frac{27}{35}$$

Check $x = -\frac{27}{35}$: $\frac{3}{5} = \frac{3}{5}$ *True*

The solution set is $\left\{-\frac{27}{35}\right\}$.

55.
$$-0.3x = 9$$
$$\frac{-0.3x}{-0.3} = \frac{9}{-0.3}$$ *Divide by -0.3.*
$$x = -30$$

Check $x = -30$: $9 = 9$ *True*

The solution set is $\{-30\}$.

57.
$$4x + 3x = 21$$
$$7x = 21$$
$$\frac{7x}{7} = \frac{21}{7}$$
$$x = 3$$

Check $x = 3$: $21 = 21$ *True*

The solution set is $\{3\}$.

59.
$$6r - 8r = 10$$
$$-2r = 10$$

$$\frac{-2r}{-2} = \frac{10}{-2}$$
$$r = -5$$

Check $r = -5$: $10 = 10$ *True*

The solution set is $\{-5\}$.

61.
$$\frac{2}{5}x - \frac{3}{10}x = 2$$
$$\frac{4}{10}x - \frac{3}{10}x = 2$$
$$\frac{1}{10}x = 2$$
$$10 \cdot \frac{1}{10}x = 10 \cdot 2$$
$$x = 20$$

Check $x = 20$: $8 - 6 = 2$ *True*

The solution set is $\{20\}$.

63.
$$7m + 6m - 4m = 63$$
$$9m = 63$$
$$\frac{9m}{9} = \frac{63}{9}$$
$$m = 7$$

Check $m = 7$: $63 = 63$ *True*

The solution set is $\{7\}$.

65.
$$-6x + 4x - 7x = 0$$
$$-9x = 0$$
$$\frac{-9x}{-9} = \frac{0}{-9}$$
$$x = 0$$

Check $x = 0$: $0 = 0$ *True*

The solution set is $\{0\}$.

67.
$$8w - 4w + w = -3$$
$$5w = -3$$
$$\frac{5w}{5} = \frac{-3}{5}$$
$$w = -\frac{3}{5}$$

Check $w = -\frac{3}{5}$: $-3 = -3$ *True*

The solution set is $\left\{-\frac{3}{5}\right\}$.

69.
$$\frac{1}{3}x - \frac{1}{4}x + \frac{1}{12}x = 3$$
$$\left(\frac{1}{3} - \frac{1}{4} + \frac{1}{12}\right)x = 3$$ *Distributive property*
$$\left(\frac{4}{12} - \frac{3}{12} + \frac{1}{12}\right)x = 3$$ *LCD = 12*
$$\frac{1}{6}x = 3$$ *Lowest terms*
$$6\left(\frac{1}{6}x\right) = 6(3)$$ *Multiply by 6.*
$$x = 18$$

Check $x = 18$: $6 - 4.5 + 1.5 = 3$ *True*

The solution set is $\{18\}$.

71. Answers will vary. One example is
$$\tfrac{3}{2}x = -6.$$

73. "When a number is multiplied by 4, the result is 6."
$$4x = 6$$
$$\frac{4x}{4} = \frac{6}{4}$$
$$x = \tfrac{3}{2}$$

The number is $\tfrac{3}{2}$ and $\left\{\tfrac{3}{2}\right\}$ is the solution set.

75. "When a number is divided by -5, the result is 2."
$$\frac{x}{-5} = 2$$
$$(-5)\left(-\tfrac{1}{5}x\right) = (-5)(2)$$
$$x = -10$$

The number is -10 and $\{-10\}$ is the solution set.

77. $-(3m + 5)$
$$= -1(3m + 5)$$
$$= -1(3m) - 1(5)$$
$$= -3m - 5$$

79. $4(-5 + 2p) - 3(p - 4)$
$$= 4(-5) + 4(2p) - 3(p) - 3(-4)$$
$$= -20 + 8p - 3p + 12$$
$$= -20 + 12 + 8p - 3p$$
$$= -8 + 5p$$

81. $4x + 5 + 2x = 7x$
$$6x + 5 = 7x \quad \textit{Combine terms.}$$
$$5 = x \quad \textit{Subtract 6x.}$$

Check $x = 5$: $20 + 5 + 10 = 35$ *True*

The solution set is $\{5\}$.

2.3 More on Solving Linear Equations

2.3 Now Try Exercises

N1. *Step 1* (not necessary)

Step 2
$$7 + 2m = -3$$
$$7 + 2m - 7 = -3 - 7 \quad \textit{Subtract 7.}$$
$$2m = -10 \quad \textit{Combine terms.}$$

Step 3
$$\frac{2m}{2} = \frac{-10}{2} \quad \textit{Divide by 2.}$$
$$m = -5$$

Step 4
Check $m = -5$: $7 - 10 = -3$ *True*

The solution set is $\{-5\}$.

N2. *Step 1* (not necessary)

Step 2
$$2q + 3 = 4q - 9$$
$$2q + 3 - 2q = 4q - 9 - 2q \quad \textit{Subtract 2q.}$$
$$3 = 2q - 9 \quad \textit{Combine terms.}$$
$$3 + 9 = 2q - 9 + 9 \quad \textit{Add 9.}$$
$$12 = 2q \quad \textit{Combine terms.}$$

Step 3
$$\frac{12}{2} = \frac{2q}{2} \quad \textit{Divide by 2.}$$
$$6 = q$$

Step 4
Check $q = 6$: $12 + 3 = 24 - 9$ *True*

The solution set is $\{6\}$.

N3. *Step 1*
$$3(z - 6) - 5z = -7z + 7$$
$$3z - 18 - 5z = -7z + 7 \quad \textit{Distributive property}$$
$$-2z - 18 = -7z + 7 \quad \textit{Combine terms.}$$

Step 2
$$-2z - 18 + 18 = -7z + 7 + 18 \quad \textit{Add 18.}$$
$$-2z = -7z + 25$$
$$-2z + 7z = -7z + 25 + 7z \quad \textit{Add 7z.}$$
$$5z = 25$$

Step 3
$$\frac{5z}{5} = \frac{25}{5} \quad \textit{Divide by 5.}$$
$$z = 5$$

Step 4
Check $z = 5$: $-28 = -28$ *True*

The solution set is $\{5\}$.

N4. *Step 1*
$$5x - (x + 9) = x - 4$$
$$5x - x - 9 = x - 4 \quad \textit{Distributive property}$$
$$4x - 9 = x - 4$$

Step 2
$$4x - 9 + 9 = x - 4 + 9 \quad \textit{Add 9.}$$
$$4x = x + 5$$
$$4x - x = x + 5 - x \quad \textit{Subtract x.}$$
$$3x = 5$$

Step 3
$$\frac{3x}{3} = \frac{5}{3} \quad \textit{Divide by 3.}$$
$$x = \tfrac{5}{3}$$

Step 4
Check $x = \tfrac{5}{3}$: $-\tfrac{7}{3} = -\tfrac{7}{3}$ *True*

The solution set is $\left\{\tfrac{5}{3}\right\}$.

N5. *Step 1*

$$24 - 4(7 - 2t) = 4(t - 1)$$
$$24 - 28 + 8t = 4t - 4 \qquad \textit{Dist. prop.}$$
$$-4 + 8t = 4t - 4$$

Step 2

$$-4 + 8t + 4 = 4t - 4 + 4 \qquad \textit{Add 4.}$$
$$8t = 4t$$
$$8t - 4t = 4t - 4t \qquad \textit{Subtract 4t.}$$
$$4t = 0$$

Step 3

$$\frac{4t}{4} = \frac{0}{4} \qquad \textit{Divide by 4.}$$
$$t = 0$$

Step 4

Check $t = 0$: $24 - 4(7) \overset{?}{=} 4(-1)$
$$-4 = -4 \qquad \textit{True}$$

The solution set is $\{0\}$.

N6. *Step 1*

$$\frac{1}{2}x + \frac{5}{8}x = \frac{3}{4}x - 6$$

The LCD of all the fractions in the equation is 8, so multiply each side by 8 to clear the fractions.

$$8\left(\frac{1}{2}x + \frac{5}{8}x\right) = 8\left(\frac{3}{4}x - 6\right)$$
$$8\left(\frac{1}{2}x\right) + 8\left(\frac{5}{8}x\right) = 8\left(\frac{3}{4}x\right) - 8(6)$$

$$\textit{Distributive property}$$
$$4x + 5x = 6x - 48$$
$$9x = 6x - 48$$

Step 2

$$9x - 6x = 6x - 48 - 6x \qquad \textit{Subtract 6x.}$$
$$3x = -48$$

Step 3

$$\frac{3x}{3} = \frac{-48}{3} \qquad \textit{Divide by 3.}$$
$$x = -16$$

Step 4

Check $x = -16$: $-8 - 10 \overset{?}{=} -12 - 6$
$$-18 = -18 \qquad \textit{True}$$

The solution set is $\{-16\}$.

N7. *Step 1*

$$\frac{2}{3}(x + 2) - \frac{1}{2}(3x + 4) = -4$$
$$6\left[\frac{2}{3}(x + 2) - \frac{1}{2}(3x + 4)\right] = 6(-4)$$
$$\textit{Multiply by 6.}$$

$$6\left[\frac{2}{3}(x + 2)\right] - 6\left[\frac{1}{2}(3x + 4)\right] = -24$$
$$\textit{Distributive property}$$
$$4(x + 2) - 3(3x + 4) = -24$$
$$\textit{Multiply.}$$
$$4x + 8 - 9x - 12 = -24$$
$$\textit{Distributive property}$$
$$-5x - 4 = -24$$
$$\textit{Combine terms.}$$

Step 2

$$-5x - 4 + 4 = -24 + 4 \qquad \textit{Add 4.}$$
$$-5x = -20 \qquad \textit{Combine terms.}$$

Step 3

$$\frac{-5x}{-5} = \frac{-20}{-5} \qquad \textit{Divide by -5.}$$
$$x = 4$$

Step 4

Check $x = 4$: $4 - 8 = -4 \quad \textit{True}$

The solution set is $\{4\}$.

N8. *Step 1*

$$0.05(13 - t) - 0.2t = 0.08(30)$$

To clear decimals, multiply both sides by 100.

$$100[0.05(13 - t) - 0.2t] = 100[0.08(30)]$$
$$5(13 - t) - 20t = 8(30)$$
$$65 - 5t - 20t = 240$$
$$65 - 25t = 240$$

Step 2

$$65 - 25t - 65 = 240 - 65$$
$$-25t = 175$$

Step 3

$$\frac{-25t}{-25} = \frac{175}{-25}$$
$$t = -7$$

Step 4

Check $t = -7$: $1 + 1.4 \overset{?}{=} 2.4$
$$2.4 = 2.4 \qquad \textit{True}$$

The solution set is $\{-7\}$.

N9.

$$-3(x - 7) = 2x - 5x + 21$$
$$-3x + 21 = -3x + 21$$
$$-3x + 21 - 21 = -3x + 21 - 21 \qquad \textit{Subtract 21.}$$
$$-3x = -3x$$
$$-3x + 3x = -3x + 3x \qquad \textit{Add 3x.}$$
$$0 = 0 \qquad \textit{True}$$

The variable x has "disappeared," and a *true* statement has resulted. The original equation is an identity. This means that for every real number value of x, the equation is true. Thus, the solution set is $\{$all real numbers$\}$.

N10.
$$-4x + 12 = 3 - 4(x - 3)$$
$$-4x + 12 = 3 - 4x + 12 \qquad \textit{Distributive property}$$
$$-4x + 12 = -4x + 15 \qquad \textit{Combine terms.}$$
$$-4x + 12 + 4x = -4x + 15 + 4x \qquad \textit{Add 4x.}$$
$$12 = 15 \qquad \textit{False}$$

The variable x has "disappeared," and a *false* statement has resulted. This means that for every real number value of x, the equation is false. Thus, the equation has **no solution** and its solution set is the **empty set**, or **null set**, symbolized $\emptyset$.

N11. First, suppose that the sum of two numbers is 18, and one of the numbers is 10. How would you find the other number? You would subtract 10 from 18. Instead of using 10 as one of the numbers, use m. This gives us the expression $18 - m$ for the other number.

2.3 Section Exercises

1. $7x + 8 = 1$

Use the addition property of equality to subtract 8 from each side.

3. $3(2t - 4) = 20 - 2t$

Clear the parentheses by using the distributive property.

5. $\frac{2}{3}x - \frac{1}{6} = \frac{3}{2}x + 1$

Clear fractions by multiplying by the LCD, 6.

7. Equations **A**, **B**, and **C** each have {all real numbers} for their solution set. However, equation **D** gives
$$3x = 2x$$
$$3x - 2x = 2x - 2x$$
$$x = 0.$$

The only solution of this equation is 0, so the correct choice is **D**.

Use the four-step method for solving linear equations as given in the text. The details of these steps will only be shown for a few of the exercises.

9.
$$3x + 2 = 14$$
$$3x + 2 - 2 = 14 - 2 \qquad \textit{Subtract 2.}$$
$$3x = 12 \qquad \textit{Combine terms.}$$
$$\frac{3x}{3} = \frac{12}{3} \qquad \textit{Divide by 3.}$$
$$x = 4$$

Check $x = 4$: $12 + 2 = 14$ *True*

The solution set is $\{4\}$.

11.
$$-5z - 4 = 21$$
$$-5z - 4 + 4 = 21 + 4 \qquad \textit{Add 4.}$$
$$-5z = 25 \qquad \textit{Combine terms.}$$
$$\frac{-5z}{-5} = \frac{25}{-5} \qquad \textit{Divide by } -5.$$
$$z = -5$$

Check $z = -5$: $25 - 4 = 21$ *True*

The solution set is $\{-5\}$.

13.
$$4p - 5 = 2p$$
$$4p - 5 - 4p = 2p - 4p \qquad \textit{Subtract 4p.}$$
$$-5 = -2p \qquad \textit{Combine terms.}$$
$$\frac{-5}{-2} = \frac{-2p}{-2} \qquad \textit{Divide by } -2.$$
$$\frac{5}{2} = p$$

Check $p = \frac{5}{2}$: $10 - 5 = 5$ *True*

The solution set is $\left\{\frac{5}{2}\right\}$.

15.
$$2x + 9 = 4x + 11$$
$$-2x + 9 = 11 \qquad \textit{Subtract 4x.}$$
$$-2x = 2 \qquad \textit{Subtract 9.}$$
$$x = -1 \qquad \textit{Divide by } -2.$$

Check $x = -1$: $7 = 7$ *True*

The solution set is $\{-1\}$.

17.
$$5m + 8 = 7 + 3m$$
For this equation, step 1 is not needed.
Step 2
$$5m + 8 - 8 = 7 + 3m - 8 \qquad \textit{Subtract 8.}$$
$$5m = 3m - 1$$
$$5m - 3m = 3m - 1 - 3m \qquad \textit{Subtract 3m.}$$
$$2m = -1$$
Step 3
$$\frac{2m}{2} = \frac{-1}{2} \qquad \textit{Divide by 2.}$$
$$m = -\frac{1}{2}$$
Step 4

Substitute $-\frac{1}{2}$ for m in the original equation.
$$5m + 8 = 7 + 3m$$
$$5\left(-\tfrac{1}{2}\right) + 8 \overset{?}{=} 7 + 3\left(-\tfrac{1}{2}\right) \qquad \textit{Let } m = -\tfrac{1}{2}.$$
$$-\tfrac{5}{2} + 8 \overset{?}{=} 7 + \left(-\tfrac{3}{2}\right)$$
$$\tfrac{11}{2} = \tfrac{11}{2} \qquad \textit{True}$$

The solution set is $\left\{-\frac{1}{2}\right\}$.

19.
$$-12x - 5 = 10 - 7x$$
$$-12x - 5 + 7x = 10 - 7x + 7x \quad \textit{Add 7x.}$$
$$-5x - 5 = 10$$
$$-5x - 5 + 5 = 10 + 5 \qquad \textit{Add 5.}$$
$$-5x = 15$$
$$\frac{-5x}{-5} = \frac{15}{-5} \qquad \textit{Divide by −5.}$$
$$x = -3$$

Check $x = -3$: $36 - 5 = 10 + 21$ *True*

The solution set is $\{-3\}$.

21.
$$12h - 5 = 11h + 5 - h$$
$$12h - 5 = 10h + 5 \qquad \textit{Combine terms.}$$
$$2h - 5 = 5 \qquad \textit{Subtract 10h.}$$
$$2h = 10 \qquad \textit{Add 5.}$$
$$h = 5 \qquad \textit{Divide by 2.}$$

Check $h = 5$: $55 = 55$ *True*

The solution set is $\{5\}$.

23.
$$7r - 5r + 2 = 5r + 2 - r$$
$$2r + 2 = 4r + 2 \qquad \textit{Combine terms.}$$
$$2 = 2r + 2 \qquad \textit{Subtract 2r.}$$
$$0 = 2r \qquad \textit{Subtract 2.}$$
$$0 = r \qquad \textit{Divide by 2.}$$

Check $r = 0$: $2 = 2$ *True*

The solution set is $\{0\}$.

25.
$$3(4x + 2) + 5x = 30 - x$$
$$12x + 6 + 5x = 30 - x \qquad \textit{Distributive property}$$
$$17x + 6 = 30 - x \qquad \textit{Combine terms.}$$
$$18x + 6 = 30 \qquad \textit{Add 1x.}$$
$$18x = 24 \qquad \textit{Subtract 6.}$$
$$x = \frac{24}{18} = \frac{4}{3} \qquad \textit{Divide by 18.}$$

Check $x = \frac{4}{3}$: $\frac{86}{3} = \frac{86}{3}$ *True*

The solution set is $\left\{\frac{4}{3}\right\}$.

27.
$$-2p + 7 = 3 - (5p + 1)$$
$$-2p + 7 = 3 - 5p - 1 \qquad \textit{Distributive property}$$
$$-2p + 7 = -5p + 2 \qquad \textit{Combine terms.}$$
$$3p + 7 = 2 \qquad \textit{Add 5p.}$$
$$3p = -5 \qquad \textit{Subtract 7.}$$
$$p = -\frac{5}{3}$$

Check $p = -\frac{5}{3}$: $\frac{31}{3} = \frac{31}{3}$ *True*

The solution set is $\left\{-\frac{5}{3}\right\}$.

29.
$$6(3w + 5) = 2(10w + 10)$$
$$18w + 30 = 20w + 20$$
$$18w = 20w - 10 \qquad \textit{Subtract 30.}$$
$$-2w = -10 \qquad \textit{Subtract 20w.}$$
$$w = 5 \qquad \textit{Divide by −2.}$$

Check $w = 5$: $120 = 120$ *True*

The solution set is $\{5\}$.

31.
$$-(4x + 2) - (-3x - 5) = 3$$
$$-1(4x + 2) - 1(-3x - 5) = 3$$
$$-4x - 2 + 3x + 5 = 3$$
$$-x + 3 = 3$$
$$-x = 0$$
$$x = 0$$

Check $x = 0$: $3 = 3$ *True*

The solution set is $\{0\}$.

33.
$$6(4x - 1) = 12(2x + 3)$$
$$24x - 6 = 24x + 36$$
$$-6 = 36 \qquad \textit{Subtract 24x.}$$

The variable has "disappeared," and the resulting equation is false. Therefore, the equation has no solution set, symbolized by $\emptyset$.

35.
$$3(2x - 4) = 6(x - 2)$$
$$6x - 12 = 6x - 12$$
$$-12 = -12 \qquad \textit{Subtract 6x.}$$
$$0 = 0 \qquad \textit{Add 12.}$$

The variable has "disappeared." Since the resulting statement is a *true* one, *any* real number is a solution. We indicate the solution set as {all real numbers}.

37.
$$11x - 5(x + 2) = 6x + 5$$
$$11x - 5x - 10 = 6x + 5$$
$$6x - 10 = 6x + 5$$
$$-10 = 5 \qquad \textit{Subtract 6x.}$$

The variable has "disappeared," and the resulting equation is false. Therefore, the equation has no solution set, symbolized by $\emptyset$.

39.
$$\frac{3}{5}t - \frac{1}{10}t = t - \frac{5}{2}$$

The least common denominator of all the fractions in the equation is 10.

$$10\left(\frac{3}{5}t - \frac{1}{10}t\right) = 10\left(t - \frac{5}{2}\right)$$
$$\textit{Multiply both sides by 10.}$$
$$10\left(\frac{3}{5}t\right) + 10\left(-\frac{1}{10}t\right) = 10t + 10\left(-\frac{5}{2}\right)$$
$$\textit{Distributive property}$$
$$6t - t = 10t - 25$$
$$5t = 10t - 25$$
$$-5t = -25 \qquad \textit{Subtract 10t.}$$
$$\frac{-5t}{-5} = \frac{-25}{-5} \qquad \textit{Divide by −5.}$$
$$t = 5$$

Check $t = 5$: $\frac{5}{2} = \frac{5}{2}$ *True*

The solution set is $\{5\}$.

41. $\frac{3}{4}x - \frac{1}{3}x + 5 = \frac{5}{6}x$

The least common denominator of all the fractions in the equation is 12, so multiply both sides by 12 and solve for x.

$$12\left(\frac{3}{4}x - \frac{1}{3}x + 5\right) = 12\left(\frac{5}{6}x\right)$$
$$9x - 4x + 60 = 10x$$
$$5x + 60 = 10x$$
$$60 = 5x$$
$$\frac{60}{5} = \frac{5x}{5}$$
$$12 = x$$

Check $x = 12$: $9 - 4 + 5 = 10$ *True*

The solution set is $\{12\}$.

43. $\frac{1}{7}(3x + 2) - \frac{1}{5}(x + 4) = 2$

The least common denominator of all the fractions in the equation is 35, so multiply both sides by 35 and solve for x.

$$35\left[\frac{1}{7}(3x + 2) - \frac{1}{5}(x + 4)\right] = 35(2)$$
$$5(3x + 2) - 7(x + 4) = 70$$
$$15x + 10 - 7x - 28 = 70$$
$$8x - 18 = 70$$
$$8x = 88$$
$$\frac{8x}{8} = \frac{88}{8}$$
$$x = 11$$

Check $x = 11$: $5 - 3 = 2$ *True*

The solution set is $\{11\}$.

45. $-\frac{1}{4}(x - 12) + \frac{1}{2}(x + 2) = x + 4$

The LCD of all the fractions is 4.

$$4\left[-\frac{1}{4}(x - 12) + \frac{1}{2}(x + 2)\right] = 4(x + 4)$$
Multiply by 4.
$$4\left(-\frac{1}{4}\right)(x - 12) + 4\left(\frac{1}{2}\right)(x + 2) = 4x + 16$$
Distributive property
$$(-1)(x - 12) + 2(x + 2) = 4x + 16$$
Multiply.
$$-x + 12 + 2x + 4 = 4x + 16$$
Distributive property
$$x + 16 = 4x + 16$$
$$-3x + 16 = 16$$
$$-3x = 0$$
$$\frac{-3x}{-3} = \frac{0}{-3}$$ *Divide by −3.*
$$x = 0$$

Check $x = 0$: $4 = 4$ *True*

The solution set is $\{0\}$.

47. $\frac{2}{3}k - \left(k - \frac{1}{2}\right) = \frac{1}{6}(k - 51)$

The least common denominator of all the fractions in the equation is 6, so multiply both sides by 6 and solve for k.

$$6\left[\frac{2}{3}k - \left(k - \frac{1}{2}\right)\right] = 6\left[\frac{1}{6}(k - 51)\right]$$
$$6\left(\frac{2}{3}k\right) - 6\left(k - \frac{1}{2}\right) = 6\left[\frac{1}{6}(k - 51)\right]$$
Distributive property
$$4k - 6k + 3 = 1(k - 51)$$
$$-2k + 3 = k - 51$$
$$-3k + 3 = -51$$
$$-3k = -54$$
$$k = 18$$

Check $k = 18$: $-\frac{11}{2} = -\frac{11}{2}$ *True*

The solution set is $\{18\}$.

49. $0.2(60) + 0.05x = 0.1(60 + x)$

To eliminate the decimal in 0.2 and 0.1, we need to multiply both sides by 10. But to eliminate the decimal in 0.05, we need to multiply by 100, so we choose 100.

$$100[0.2(60) + 0.05x] = 100[0.1(60 + x)]$$
Multiply by 100.
$$100[0.2(60)] + 100(0.05x) = 100[0.1(60 + x)]$$
Distributive property
$$20(60) + 5x = 10(60 + x)$$
Multiply.
$$1200 + 5x = 600 + 10x$$
$$1200 - 5x = 600$$
$$-5x = -600$$
$$x = \frac{-600}{-5} = 120$$

Check $x = 120$: $18 = 18$ *True*

The solution set is $\{120\}$.

51. $1.00x + 0.05(12 - x) = 0.10(63)$

To clear the equation of decimals, we multiply both sides by 100.

$$100[1.00x + 0.05(12 - x)] = 100[0.10(63)]$$
$$100(1.00x) + 100[0.05(12 - x)] = (100)(0.10)(63)$$
$$100x + 5(12 - x) = 10(63)$$
$$100x + 60 - 5x = 630$$
$$95x + 60 = 630$$
$$95x = 570$$
$$x = \frac{570}{95} = 6$$

Check $x = 6$: $6.3 = 6.3$ *True*

The solution set is $\{6\}$.

53.　$0.6(10,000) + 0.8x = 0.72(10,000 + x)$

$100[0.6(10,000)] + 100(0.8x) =$

$$100[0.72(10,000 + x)]$$

Multiply by both sides by 100.

$60(10,000) + 80x = 72(10,000 + x)$

$600,000 + 80x = 720,000 + 72x$

$600,000 + 8x = 720,000$

$8x = 120,000$

$x = \frac{120,000}{8} = 15,000$

Check $x = 15,000$: $18,000 = 18,000$　*True*

The solution set is $\{15,000\}$.

55.　$10(2x - 1) = 8(2x + 1) + 14$

$20x - 10 = 16x + 8 + 14$

$20x - 10 = 16x + 22$

$4x - 10 = 22$

$4x = 32$

$x = 8$

Check $x = 8$: $150 = 150$　*True*

The solution set is $\{8\}$.

57.　$\frac{1}{2}(x + 2) + \frac{3}{4}(x + 4) = x + 5$

To clear fractions, multiply both sides by the LCD, which is 4.

$4\left[\frac{1}{2}(x + 2) + \frac{3}{4}(x + 4)\right] = 4(x + 5)$

$4\left(\frac{1}{2}\right)(x + 2) + 4\left(\frac{3}{4}\right)(x + 4) = 4x + 20$

$2(x + 2) + 3(x + 4) = 4x + 20$

$2x + 4 + 3x + 12 = 4x + 20$

$5x + 16 = 4x + 20$

$x + 16 = 20$

$x = 4$

Check $x = 4$: $9 = 9$　*True*

The solution set is $\{4\}$.

59.　$0.1(x + 80) + 0.2x = 14$

To eliminate the decimals, multiply both sides by 10.

$10[0.1(x + 80) + 0.2x] = 10(14)$

$1(x + 80) + 2x = 140$

$x + 80 + 2x = 140$

$3x + 80 = 140$

$3x = 60$

$x = 20$

Check $x = 20$: $14 = 14$　*True*

The solution set is $\{20\}$.

61.　$4(x + 8) = 2(2x + 6) + 20$

$4x + 32 = 4x + 12 + 20$

$4x + 32 = 4x + 32$

$4x = 4x$

$0 = 0$

Since $0 = 0$ is a *true* statement, the solution set is {all real numbers}.

63.　$9(v + 1) - 3v = 2(3v + 1) - 8$

$9v + 9 - 3v = 6v + 2 - 8$

$6v + 9 = 6v - 6$

$9 = -6$

Because $9 = -6$ is a *false* statement, the equation has no solution set, symbolized by $\emptyset$.

65.　The sum of q and the other number is 11. To find the other number, you would subtract q from 11, so an expression for the other number is $11 - q$.

67.　The product of x and the other number is 9. To find the other number, you would divide 9 by x, so an expression for the other number is $\frac{9}{x}$.

69.　An expression for the total number of yards is $x + 9$.

71.　If a baseball player gets 65 hits in one season, and h of the hits are in one game, then $65 - h$ of the hits came in the rest of the games.

73.　If Monica is x years old now, then 15 years from now, she will be $x + 15$ years old. Five years ago, she was $x - 5$ years old.

75.　Since the value of each quarter is 25 cents, the value of r quarters is $25r$ cents.

77.　Since each bill is worth 5 dollars, the number of bills is $\frac{t}{5}$.

79.　Since each adult ticket costs x dollars, the cost of 3 adult tickets is $3x$. Since each child's ticket costs y dollars, the cost of 2 children's tickets is $2y$. Therefore, the total cost is $3x + 2y$ (dollars).

81.　"A number added to -6" is written

$$-6 + x.$$

83.　"The difference between -5 and a number" is written

$$-5 - x.$$

85.　"The product of 12 and the difference between a number and 9" is written

$$12(x - 9).$$

Summary Exercises on Solving Linear Equations

1.　$x + 2 = -3$

$x = -5$　*Subtract 2.*

Check $x = -5$: $-3 = -3$　*True*

The solution set is $\{-5\}$.

3. $12.5x = -63.75$

   $x = \frac{-63.75}{12.5}$ *Divide by 12.5.*

   $= -5.1$

Check $x = -5.1$: $-63.75 = -63.75$ *True*

The solution set is $\{-5.1\}$.

5. $\frac{4}{5}x = -20$

   $x = \left(\frac{5}{4}\right)(-20)$ *Multiply by $\frac{5}{4}$.*

   $= -25$

Check $x = -25$: $-20 = -20$ *True*

The solution set is $\{-25\}$.

7. $5x - 9 = 3(x - 3)$

   $5x - 9 = 3x - 9$ *Distributive property*

   $2x - 9 = -9$ *Subtract 3x.*

   $2x = 0$ *Add 9.*

   $x = 0$ *Divide by 2.*

Check $x = 0$: $-9 = -9$ *True*

The solution set is $\{0\}$.

9. $-x = 6$

   $x = -6$ *Multiply by -1.*

Check $x = -6$: $6 = 6$ *True*

The solution set is $\{-6\}$.

11. $4x + 2(3 - 2x) = 6$

   $4x + 6 - 4x = 6$

   $6 = 6$

Since $6 = 6$ is a *true* statement, the solution set is $\{$all real numbers$\}$.

13. $-3(m - 4) + 2(5 + 2m) = 29$

   $-3m + 12 + 10 + 4m = 29$

   $m + 22 = 29$

   $m = 7$

Check $m = 7$: $29 = 29$ *True*

The solution set is $\{7\}$.

15. $0.08x + 0.06(x + 9) = 1.24$

To eliminate the decimals, multiply both sides by 100.

   $100[0.08x + 0.06(x + 9)] = 100(1.24)$

   $8x + 6(x + 9) = 124$

   $8x + 6x + 54 = 124$

   $14x + 54 = 124$

   $14x = 70$

   $x = 5$

Check $x = 5$: $0.4 + 0.84 = 1.24$ *True*

The solution set is $\{5\}$.

17. $7m - (2m - 9) = 39$

   $7m - 2m + 9 = 39$

   $5m + 9 = 39$

   $5m = 30$

   $m = 6$

Check $m = 6$: $39 = 39$ *True*

The solution set is $\{6\}$.

19. $-2t + 5t - 9 = 3(t - 4) - 5$

   $-2t + 5t - 9 = 3t - 12 - 5$

   $3t - 9 = 3t - 17$

   $-9 = -17$

Because $-9 = -17$ is a *false* statement, the equation has no solution set, symbolized by $\emptyset$.

21. $0.2(50) + 0.8r = 0.4(50 + r)$

To eliminate the decimals, multiply both sides by 10.

   $10[0.2(50) + 0.8r] = 10[0.4(50 + r)]$

   $2(50) + 8r = 4(50 + r)$

   $100 + 8r = 200 + 4r$

   $100 + 4r = 200$

   $4r = 100$

   $r = 25$

Check $r = 25$: $10 + 20 = 30$ *True*

The solution set is $\{25\}$.

23. $2(3 + 7x) - (1 + 15x) = 2$

   $6 + 14x - 1 - 15x = 2$

   $-x + 5 = 2$

   $-x = -3$

   $x = 3$

Check $x = 3$: $48 - 46 = 2$ *True*

The solution set is $\{3\}$.

25. $2(4 + 3r) = 3(r + 1) + 11$

   $8 + 6r = 3r + 3 + 11$

   $8 + 6r = 3r + 14$

   $8 + 3r = 14$

   $3r = 6$

   $r = 2$

Check $r = 2$: $20 = 20$ *True*

The solution set is $\{2\}$.

27. $\frac{1}{4}x - 4 = \frac{3}{2}x + \frac{3}{4}x$

To clear fractions, multiply both sides by the LCD, which is 4.

$$4\left(\frac{1}{4}x - 4\right) = 4\left(\frac{3}{2}x + \frac{3}{4}x\right)$$
$$x - 16 = 6x + 3x$$
$$x - 16 = 9x$$
$$-16 = 8x$$
$$x = -2$$

Check $x = -2$: $-4.5 = -3 - 1.5$ *True*

The solution set is $\{-2\}$.

29. $\frac{3}{4}(z - 2) - \frac{1}{3}(5 - 2z) = -2$

To clear fractions, multiply both sides by the LCD, which is 12.

$$12\left[\frac{3}{4}(z-2) - \frac{1}{3}(5-2z)\right] = 12(-2)$$
$$9(z-2) - 4(5-2z) = -24$$
$$9z - 18 - 20 + 8z = -24$$
$$17z - 38 = -24$$
$$17z = 14$$
$$z = \frac{14}{17}$$

Check $z = \frac{14}{17}$: $-\frac{15}{17} - \frac{19}{17} = -2$ *True*

The solution set is $\left\{\frac{14}{17}\right\}$.

2.4 An Introduction to Applications of Linear Equations

2.4 Now Try Exercises

N1. *Step 2*
Let $x =$ the number.

Step 3

If 5 is	added to	a number,	the result is	7 less than 3 times the number.
↓	↓	↓	↓	↓
5	+	x	=	$3x - 7$

Step 4
Solve the equation.

$$5 + x = 3x - 7$$
$$5 + x - 5 = 3x - 7 - 5 \quad \text{Subtract 5.}$$
$$x = 3x - 12$$
$$x - 3x = 3x - 12 - 3x \quad \text{Subtract } 3x.$$
$$-2x = -12$$
$$\frac{-2x}{-2} = \frac{-12}{-2} \quad \text{Divide by } -2.$$
$$x = 6$$

Step 5
The number is 6.

Step 6
5 added to 6 is 11. 3 times 6 is 18, and 7 less than 18 is 11, so 6 is the number.

N2. *Step 2*
Let $x =$ the number of medals Germany won.
Let $x - 7 =$ the number of medals Russia won.

Step 3

The total	is	the number of medals Germany won	plus	the number of medals Russia won.
↓	↓	↓	↓	↓
51	=	x	+	$(x - 7)$

Step 4
Solve this equation.

$$51 = 2x - 7$$
$$51 + 7 = 2x - 7 + 7$$
$$58 = 2x$$
$$\frac{58}{2} = \frac{2x}{2}$$
$$29 = x$$

Step 5
Germany won 29 medals and Russia won $29 - 7 = 22$ medals.

Step 6
22 is 7 fewer than 29 and the sum of 22 and 29 is 51.

N3. *Step 2*
Let $x =$ the number of orders for chocolate scones.
Then $\frac{2}{3}x =$ the number of orders for bagels.

Step 3

The total	is	orders for choc. scones	plus	orders for bagels.
↓	↓	↓	↓	↓
525	=	x	+	$\frac{2}{3}x$

Step 4
Solve this equation.

$$525 = 1x + \frac{2}{3}x \quad x = 1x$$
$$525 = \frac{3}{3}x + \frac{2}{3}x \quad LCD = 3$$
$$525 = \frac{5}{3}x \quad \text{Combine like terms.}$$
$$\frac{3}{5}(525) = \frac{3}{5}\left(\frac{5}{3}x\right) \quad \text{Multiply by } \frac{3}{5}.$$
$$315 = x$$

Step 5
The number of orders for chocolate scones was 315, so the number of orders for bagels was $\frac{2}{3}(315) = 210$.

Step 6
Two-thirds of 315 is 210 and the sum of 315 and 210 is 525.

N4. *Step 2*

Let x = the number of residents.

Then $4x$ = the number of guests.

(If each resident brought four guests, there would be four times as many guests as residents.)

Step 3

Number of residents	plus	number of guests	is	the total in attendance
↓	↓	↓	↓	↓
x	+	$4x$	=	175

Step 4

Solve this equation.

$$x + 4x = 175$$
$$5x = 175$$
$$\frac{5x}{5} = \frac{175}{5}$$
$$x = 35$$

Step 5

There were 35 residents and $4 \cdot 35 = 140$ guests.

Step 6

140 is four times as much as 35 and the sum of 35 and 140 is 175.

N5. *Step 2*

Let x = the time spent practicing free throws.

Then $2x$ = the time spent lifting weights

and $x + 2$ = the time spent watching game films.

Step 3

Time spent practicing free throws	plus	time spent lifting weights
↓	↓	↓
x	+	$2x$

plus	time spent watching game films	is	total time.
↓	↓	↓	↓
+	$x + 2$	=	6

Step 4

Solve this equation.

$$x + 2x + (x + 2) = 6$$
$$4x + 2 = 6$$
$$4x = 4$$
$$\frac{4x}{4} = \frac{4}{4}$$
$$x = 1$$

Step 5

The time spent practicing free throws is 1 hour, the time spent lifting weights is $2(1) = 2$ hours,

and the time spent watching game films is $1 + 2 = 3$ hours.

Step 6

Since 2 hours is twice as much time as 1 hour, and 3 hours is 2 more hours than 1 hour, and the sum of the times is $1 + 2 + 3 = 6$ hours (the total time spent), the answers are correct.

N6. *Step 2*

Let x = the lesser page number.

Then $x + 1$ = the greater page number.

Step 3

Because the sum of the page numbers is 593, an equation is

$$x + (x + 1) = 593$$

Step 4 $2x + 1 = 593$ *Combine like terms.*
$$2x = 592 \quad \textit{Subtract 1.}$$
$$x = 296 \quad \textit{Divide by 2.}$$

Step 5

The lesser page number is 296, and the greater page number is $296 + 1 = 297$.

Step 6

297 is one more than 296 and the sum of 296 and 297 is 593.

N7. Let x = the lesser odd integer.

Then $x + 2$ = the greater odd integer.

From the given information, we have

$$2 \cdot x + 3(x + 2) = 191.$$

Solve this equation.

$$2x + 3x + 6 = 191$$
$$5x + 6 = 191$$
$$5x = 185$$
$$x = 37$$

The lesser odd integer is 37 and the greater consecutive odd integer is $37 + 2 = 39$. Two times 37 is 74, three times 39 is 117, and 74 plus 117 is 191.

N8. Let x = the degree measure of the angle.

Then $90 - x$ = the degree measure of its complement.

The complement	is	twice the angle.
↓	↓	↓
$90 - x$	=	$2x$

Solve this equation.

$$90 - x = 2x$$
$$90 = 3x$$
$$30 = x$$

The measure of the angle is 30°.

N9. *Step 2*
Let $x =$ the degree measure of the angle.
Then $90 - x =$ the degree measure of its complement, and $180 - x =$ the degree measure of its supplement.

Step 3

supplement	equals	$46°$ less than 3 times its complement
↓	↓	↓
$180 - x$	$=$	$3(90 - x) - 46$

Step 4

$$180 - x = 3(90 - x) - 46$$

$180 - x = 270 - 3x - 46$	*Dist. property*
$180 - x = 224 - 3x$	*Combine terms.*
$180 + 2x = 224$	*Add 3x.*
$2x = 44$	*Subtract 180.*
$x = 22$	*Divide by 2.*

Step 5
The measure of the angle is $22°$.

Step 6
The complement of $22°$ is $90° - 22° = 68°$ and $46°$ less than 3 times $68°$ is
$3(68°) - 46° = 204° - 46° = 158°$.
The supplement is $180° - 22° = 158°$.

2.4 Section Exercises

1. Choice **D**, $6\frac{1}{2}$, is *not* a reasonable answer in an applied problem that requires finding the number of cars on a dealer's lot, since you cannot have $\frac{1}{2}$ of a car. The number of cars must be a whole number.

3. Choice **A**, -10, is *not* a reasonable answer since distance cannot be negative.

The applied problems in this section should be solved by using the six-step method shown in the text. These steps will only be listed in a few of the solutions, but all of the solutions are based on this method.

5. *Step 2*
Let $x =$ the number.

Step 3

The product of 8,	and a number increased by 6,	is	104.
↓	↓	↓	↓
8	$\cdot \quad (x + 6)$	$=$	104

Step 4
Solve this equation.

$$8(x + 6) = 104$$
$$8x + 48 = 104$$
$$8x = 56$$
$$x = 7$$

Step 5
The number is 7.

Step 6
7 increased by 6 is 13. The product of 8 and 13 is 104, so 7 is the number.

7. *Step 2*
Let $x =$ the unknown number. Then $5x + 2$ represents "2 is added to five times a number," and $4x + 5$ represents "5 more than four times the number."

Step 3 $\quad 5x + 2 = 4x + 5$

Step 4 $\quad \begin{aligned} 5x + 2 &= 4x + 5 \\ x + 2 &= 5 \\ x &= 3 \end{aligned}$

Step 5
The number is 3.

Step 6
Check that 3 is the correct answer by substituting this result into the words of the original problem. 2 added to five times a number is $2 + 5(3) = 17$ and 5 more than four times the number is $5 + 4(3) = 17$. The values are equal, so the number 3 is the correct answer.

9. *Step 2*
Let $x =$ the unknown number. Then $x - 2$ is two subtracted from the number, $3(x - 2)$ is triple the difference, and $x + 6$ is six more than the number.

Step 3 $\quad 3(x - 2) = x + 6$

Step 4 $\quad \begin{aligned} 3x - 6 &= x + 6 \\ 2x - 6 &= 6 \\ 2x &= 12 \\ x &= 6 \end{aligned}$

Step 5
The number is 6.

Step 6
Check that 6 is the correct answer by substituting this result into the words of the original problem. Two subtracted from the number is $6 - 2 = 4$. Triple this difference is $3(4) = 12$, which is equal to 6 more than the number, since $6 + 6 = 12$.

11. *Step 2*
Let $x =$ the unknown number. Then $3x$ is three times the number, $x + 7$ is 7 more than the number, $2x$ is twice the number, and $-11 - 2x$ is the difference between -11 and twice the number.

Step 3 $\quad 3x + (x + 7) = -11 - 2x$

Step 4 $\quad \begin{aligned} 4x + 7 &= -11 - 2x \\ 6x + 7 &= -11 \\ 6x &= -18 \\ x &= -3 \end{aligned}$

Step 5
The number is -3.

Step 6
Check that -3 is the correct answer by substituting this result into the words of the original problem. The sum of three times a number and 7 more than the number is $3(-3) + (-3 + 7) = -5$ and the difference between -11 and twice the number is $-11 - 2(-3) = -5$. The values are equal, so the number -3 is the correct answer.

13. Let $x =$ the number of drive-in movie screens in Ohio.

Then $x + 2 =$ the number of drive-in movie screens in Pennsylvania.

Since the total number of screens was 68, we can write the equation
$$x + (x + 2) = 68.$$
Solve this equation.
$$2x + 2 = 68$$
$$2x = 66$$
$$x = 33$$

Since $x = 33$, $x + 2 = 35$.

There were 33 drive-in movie screens in Ohio and 35 in Pennsylvania. Since 35 is 2 more than 33 and $33 + 35 = 68$, this answer checks.

15. Let $x =$ the number of Republicans.

Then $x + 18 =$ the number of Democrats.

The number of Republicans	plus	the number of Democrats
↓	↓	↓
x	$+$	$(x + 18)$

equals	the total number of Republican and Democrat members in the Senate.
↓	↓
$=$	98

Solve the equation.
$$x + (x + 18) = 98$$
$$2x + 18 = 98$$
$$2x = 80$$
$$x = 40$$

There were 40 Republicans and $40 + 18 = 58$ Democrats.

17. Let $x =$ revenue from ticket sales for Bon Jovi.
Then $x - 6.1 =$ revenue from ticket sales for Bruce Springsteen.

Since the total revenue from ticket sales was \$415.3 (all numbers in millions), we can write the equation
$$x + (x - 6.1) = 415.3.$$
Solve this equation.
$$2x - 6.1 = 415.3$$
$$2x = 421.4$$
$$x = 210.7$$

Since $x = 210.7$, $x - 6.1 = 204.6$.

Bon Jovi took in \$210.7 million and Bruce Springsteen took in \$204.6 million. Since 204.6 is 6.1 less than 210.7 and $210.7 + 204.6 = 415.3$, this answer checks.

19. Let $x =$ the number of games the Celtics lost.
Then $3x + 2 =$ the number of games the Celtics won.

Since the total number of games played was 82, we can write the equation
$$x + (3x + 2) = 82.$$
Solve this equation.
$$4x + 2 = 82$$
$$4x = 80$$
$$x = 20$$

Since $x = 20$, $3x + 2 = 62$.

The Celtics won 62 games and lost 20 games. Since $62 + 20 = 82$, this answer checks.

21. Let $x =$ the number of mg of vitamin C in a one-cup serving of pineapple juice.
Then $4x - 3 =$ the number of mg of vitamin C in a one-cup serving of orange juice.

Since the total amount of vitamin C in a serving of the two juices is 122 mg, we can write
$$x + (4x - 3) = 122.$$
Solve this equation.
$$5x - 3 = 122$$
$$5x = 125$$
$$x = 25$$

Since $x = 25$, $4x - 3 = 97$.

A one-cup serving of pineapple juice has 25 mg of vitamin C and a one-cup serving of orange juice has 97 mg of vitamin C. Since 97 is 3 less than four times 25 and $25 + 97 = 122$, this answer checks.

23. Let $x =$ the number of CDs sold.
Then $\frac{8}{5}x =$ the number of DVDs sold.

The total number of CDs and DVDs sold was 273, so
$$x + \frac{8}{5}x = 273.$$

Solve this equation.

$$1x + \tfrac{8}{5}x = 273$$
$$\tfrac{13}{5}x = 273$$
$$\tfrac{5}{13}\left(\tfrac{13}{5}x\right) = \tfrac{5}{13}(273)$$
$$x = \frac{5}{\cancel{13}_{\,1}} \cdot \frac{\cancel{273}^{\,21}}{1} = 105$$

Since $x = 105$, $\tfrac{8}{5}x = \tfrac{8}{5}(105) = 168$.

There were 168 DVDs sold.

25. Let $x =$ the number of kg of onions.
Then $6.6x =$ the number of kg of grilled steak.
The total weight of these two ingredients was 617.6 kg, so
$$x + 6.6x = 617.6.$$
Solve this equation.
$$1x + 6.6x = 617.6$$
$$7.6x = 617.6$$
$$x = \tfrac{617.6}{7.6} \approx 81.3$$

Since $x = \tfrac{617.6}{7.6}$, $6.6x = 6.6\left(\tfrac{617.6}{7.6}\right) \approx 536.3$.

To the nearest tenth of a kilogram, 81.3 kg of onions and 536.3 kg of grilled steak were used to make the taco.

27. Let $x =$ the value of the 1945 nickel.
Then $2x =$ the value of the 1950 nickel.
The total value of the two coins is 24.00, so
$$x + 2x = 24.$$
Solve this equation.
$$3x = 24$$
$$x = 8 \qquad \textit{Divide by 8.}$$

Since $x = 8$, $2x = 2(8) = 16$.

The value of the 1945 Philadelphia nickel is 8.00 and the value of the 1950 Denver nickel is 16.00.

29. Let $x =$ the number of ounces of rye flour.
Then $4x =$ the number of ounces of whole wheat flour.

The total number of ounces would be 32, so
$$x + 4x = 32.$$

Solve this equation.
$$5x = 32$$
$$x = \tfrac{32}{5} = 6.4$$

Since $x = 6.4$, $4x = 4(6.4) = 25.6$. To make a loaf of bread weighing 32 oz, use 6.4 oz of rye flour and 25.6 oz of whole wheat flour.

31. Let $x =$ the number of tickets booked on United Airlines.
Then $x + 7 =$ the number of tickets booked on American Airlines, and
$2x + 4 =$ the number of tickets booked on Southwest Airlines.

The total number of tickets booked was 55, so
$$x + (x + 7) + (2x + 4) = 55.$$
Solve this equation.
$$4x + 11 = 55$$
$$4x = 44$$
$$x = \tfrac{44}{4} = 11$$

Since $x = 11$, $x + 7 = 11 + 7 = 18$, and $2x + 4 = 2(11) + 4 = 26$. He booked 18 tickets on American, 11 tickets on United, and 26 tickets on Southwest.

33. Let $x =$ the length of the shortest piece.
Then $x + 5 =$ the length of the middle piece, and $x + 9 =$ the length of the longest piece.

The total length is 59 inches, so
$$x + (x + 5) + (x + 9) = 59.$$
Solve this equation.
$$3x + 14 = 59$$
$$3x = 45$$
$$x = 15$$

Since $x = 15$, $x + 5 = 20$, and $x + 9 = 24$.

The shortest piece should be 15 inches, the middle piece should be 20 inches, and the longest piece should be 24 inches. The answer checks since
$$15 + 20 + 24 = 59.$$

35. Let $x =$ the distance of Mercury from the sun.
Then $x + 31.2 =$ the distance of Venus from the sun, and
$x + 57 =$ the distance of Earth from the sun.

Since the total of the distances from these three planets is 196.2 (all distances in millions of miles), we can write the equation
$$x + (x + 31.2) + (x + 57) = 196.2.$$

Solve this equation.
$$3x + 88.2 = 196.2$$
$$3x = 108$$
$$x = 36$$

Mercury is 36 million miles from the sun, Venus is $36 + 31.2 = 67.2$ million miles from the sun, and Earth is $36 + 57 = 93$ million miles from the sun. The answer checks since
$$36 + 67.2 + 93 = 196.2.$$

37. Let x = the measure of angles A and B.
Then $x + 60$ = the measure of angle C.

The sum of the measures of the angles of any triangle is $180°$, so

$$x + x + (x + 60) = 180.$$

Solve this equation.

$$3x + 60 = 180$$
$$3x = 120$$
$$x = 40$$

Angles A and B have measures of 40 degrees, and angle C has a measure of $40 + 60 = 100$ degrees. The answer checks since

$$40 + 40 + 100 = 180.$$

39. Let x = the number on the first locker.
Then $x + 1$ = the number on the next locker.

Since the numbers have a sum of 137, we can write the equation

$$x + (x + 1) = 137.$$

Solve the equation.

$$2x + 1 = 137$$
$$2x = 136$$
$$x = \frac{136}{2} = 68$$

Since $x = 68$, $x + 1 = 69$.

The lockers have numbers 68 and 69. Since $68 + 69 = 137$, this answer checks.

41. Because the two pages are back-to-back, they must have page numbers that are consecutive integers.

Let x = the lesser page number.
Then $x + 1$ = the greater page number.

$$x + (x + 1) = 203$$
$$2x + 1 = 203$$
$$2x = 202$$
$$x = 101$$

Since $x = 101$, $x + 1 = 102$.

The page numbers are 101 and 102. This answer checks since the sum is 203.

43. Let x = the lesser even integer.
Then $x + 2$ = the greater even integer.

The lesser added to three times the greater gives a sum of 46 can be written as

$$x + 3(x + 2) = 46.$$
$$x + 3x + 6 = 46$$
$$4x + 6 = 46$$
$$4x = 40$$
$$x = 10$$

Since $x = 10$, $x + 2 = 12$.

The integers are 10 and 12. This answer checks since $10 + 3(12) = 46$.

45. Let x = the lesser integer.
Then $x + 1$ = the greater integer.

$$x + 3(x + 1) = 43$$
$$x + 3x + 3 = 43$$
$$4x + 3 = 43$$
$$4x = 40$$
$$x = 10$$

Since $x = 10$, $x + 1 = 11$.

The integers are 10 and 11. This answer checks since $10 + 3(11) = 43$.

47. Let x = the first even integer.
Then $x + 2$ = the second even integer, and $x + 4$ = the third even integer.

$$x + (x + 2) + (x + 4) = 60$$
$$3x + 6 = 60$$
$$3x = 54$$
$$x = 18$$

Since $x = 18$, $x + 2 = 20$, and $x + 4 = 22$.

The first even integer is 18. This answer checks since $18 + 20 + 22 = 60$.

49. Let x = the first odd integer.
Then $x + 2$ = the second odd integer, and $x + 4$ = the third odd integer.

$$2[(x + 4) - 6] = [x + 2(x + 2)] - 23$$
$$2(x - 2) = x + 2x + 4 - 23$$
$$2x - 4 = 3x - 19$$
$$-4 = x - 19$$
$$15 = x$$

Since $x = 15$, $x + 2 = 17$, and $x + 4 = 19$.

The integers are 15, 17, and 19.

51. Let x = the measure of the angle.
Then $90 - x$ = the measure of its complement.

The "complement is four times its measure" can be written as

$$90 - x = 4x.$$

Solve this equation.

$$90 = 5x$$
$$x = \frac{90}{5} = 18$$

The measure of the angle is $18°$. The complement is $90° - 18° = 72°$, which is four times $18°$.

53. Let $x = $ the measure of the angle.
Then $180 - x = $ the measure of its supplement.

The "supplement is eight times its measure" can be written as

$$180 - x = 8x.$$

Solve this equation.

$$180 = 9x$$
$$x = \frac{180}{9} = 20$$

The measure of the angle is 20°. The supplement is $180° - 20° = 160°$, which is eight times 20°.

55. Let $x = $ the measure of the angle. Then $90 - x = $ the measure of its complement, and $180 - x = $ the measure of its supplement.

Its supplement	measures	39°
↓	↓	↓
$180 - x$	$=$	39

	more than	twice its complement.
	↓	↓
	$+$	$2(90 - x)$

Solve the equation.

$$180 - x = 39 + 2(90 - x)$$
$$180 - x = 39 + 180 - 2x$$
$$180 - x = 219 - 2x$$
$$x + 180 = 219$$
$$x = 39$$

The measure of the angle is 39°. The complement is $90° - 39° = 51°$. Now 39° more than twice its complement is $39° + 2(51°) = 141°$, which is the supplement of 39° since $180° - 39° = 141°$.

57. Let $x = $ the measure of the angle. Then $180 - x = $ the measure of its supplement, and $90 - x = $ the measure of its complement.

	The difference	

between the measure of its supplement and		three times the measure of its complement
↓	↓	↓
$(180 - x)$	$-$	$3(90 - x)$

	is	10°.
	↓	↓
	$=$	10

Solve the equation.

$$(180 - x) - 3(90 - x) = 10$$
$$180 - x - 270 + 3x = 10$$
$$2x - 90 = 10$$
$$2x = 100$$
$$x = 50$$

The measure of the angle is 50°. The supplement is $180° - 50° = 130°$ and the complement is $90° - 50° = 40°$. The answer checks since $130° - 3(40°) = 10°$.

59. $L = 6$ and $W = 4$, so $LW = 6 \cdot 4 = 24$.

61. $L = 8$ and $W = 2$, so
$$2L + 2W = 2(8) + 2(2) = 16 + 4 = 20.$$

2.5 Formulas and Additional Applications from Geometry

2.5 Now Try Exercises

N1. $P = 2a + 2b$
   $78 = 2(12) + 2b$ *Let P = 78 and a = 12.*
   $78 = 24 + 2b$
   $54 = 2b$ *Subtract 24.*
   $27 = b$ *Divide by 2.*

N2. The fence will enclose the perimeter of the rectangular garden, so use the formula for the perimeter of a rectangle. Find the width of the garden by substituting $P = 160$ and $L = 2W - 10$ into the formula and solving for W.

$$P = 2L + 2W$$
$$160 = 2(2W - 10) + 2W$$
$$160 = 4W - 20 + 2W$$
$$180 = 6W$$
$$30 = W$$

Since $W = 30$, $L = 2(30) - 10 = 50$. The dimensions of the garden are 50 ft by 30 ft.

N3. Let $s = $ the length of the medium side, in feet; $s + 1 = $ the length of the longest side, and, $s - 7 = $ the length of the shortest side.

The perimeter is 30 feet, so

$$s + (s + 1) + (s - 7) = 30.$$
$$3s - 6 = 30$$
$$3s = 36$$
$$s = 12$$

Since $s = 12$, $s + 1 = 13$, and $s - 7 = 5$. The lengths of the sides are 5, 12, and 13 feet. The perimeter is $5 + 12 + 13 = 30$, as required.

N4. Use the formula for the area of a triangle.

$$A = \tfrac{1}{2}bh$$
$$77 = \tfrac{1}{2}(14)h \quad \textit{Let A = 77, h = 14.}$$
$$77 = 7h$$
$$11 = h$$

The length of the height is 11 centimeters.

N5. Since the marked angles are vertical angles, they have equal measures.

$$6x + 2 = 8x - 8$$
$$2 = 2x - 8$$
$$10 = 2x$$
$$5 = x$$

If $x = 5$, $6x + 2 = 6(5) + 2 = 32$
and $8x - 8 = 8(5) - 8 = 32$.

The measure of the angles is $32°$.

N6. Solve $W = Fd$ for F.

$$\frac{W}{d} = \frac{Fd}{d} \qquad \textit{Divide by d.}$$
$$\frac{W}{d} = F, \ \text{ or } \ F = \frac{W}{d}$$

N7. Solve $Ax + By = C$ for A.

$$Ax = C - By \quad \textit{Subtract By.}$$
$$\frac{Ax}{x} = \frac{C - By}{x} \quad \textit{Divide by x.}$$
$$A = \frac{C - By}{x}$$

N8. Solve $x = u + zs$ for z.

$$x - u = u + zs - u \qquad \textit{Subtract u.}$$
$$x - u = zs$$
$$\frac{x - u}{s} = \frac{zs}{s} \qquad \textit{Divide by s.}$$
$$\frac{x - u}{s} = z, \ \text{ or } \ z = \frac{x - u}{s}$$

N9. Solve $S = \tfrac{1}{2}(a + b + c)$ for a.

$$2S = a + b + c \quad \textit{Multiply by 2.}$$
$$2S - b - c = a \qquad \textit{Subtract b and c.}$$

2.5 Section Exercises

1. **(a)** The perimeter of a plane geometric figure is the distance around the figure. It can be found by adding up the lengths of all the sides. Perimeter is a one-dimensional (linear) measurement, so it is given in linear units (inches, centimeters, feet, etc.).

(b) The area of a plane geometric figure is the measure of the surface covered or enclosed by the figure. Area is a two-dimensional measurement, so it is given in square units (square centimeters, square feet, etc.).

3. **(a)** The measure of a straight angle is <u>180°</u>.

(b) Vertical angles have <u>the same</u> measures.

5. Carpeting for a bedroom covers the surface of the bedroom floor, so *area* would be used.

7. To measure fencing for a yard, use *perimeter* since you would need to measure the lengths of the sides of the yard.

9. Tile for a bathroom covers the surface of the bathroom floor, so *area* would be used.

11. To determine the cost of replacing a linoleum floor with a wood floor, use *area* since you need to know the measure of the surface covered by the wood.

In Exercises 13–32, substitute the given values into the formula and then solve for the remaining variable.

13. $P = 2L + 2W$; $L = 8$, $W = 5$

$$P = 2L + 2W$$
$$= 2(8) + 2(5)$$
$$= 16 + 10$$
$$P = 26$$

15. $A = \tfrac{1}{2}bh$; $b = 8$, $h = 16$

$$A = \tfrac{1}{2}bh$$
$$= \tfrac{1}{2}(8)(16)$$
$$A = 64$$

17. $P = a + b + c$; $P = 12$, $a = 3$, $c = 5$

$$P = a + b + c$$
$$12 = 3 + b + 5$$
$$12 = b + 8$$
$$4 = b$$

19. $d = rt$; $d = 252$, $r = 45$

$$d = rt$$
$$252 = 45t$$
$$\frac{252}{45} = \frac{45t}{45}$$
$$5.6 = t$$

21. $I = prt$; $p = 7500$, $r = 0.035$, $t = 6$

$$I = prt$$
$$= (7500)(0.035)(6)$$
$$I = 1575$$

23. $A = \tfrac{1}{2}h(b + B)$; $A = 91$, $h = 7$, $b = 12$

$$A = \tfrac{1}{2}h(b + B)$$
$$91 = \tfrac{1}{2}(7)(12 + B)$$
$$182 = (7)(12 + B)$$
$$12 + B = \tfrac{1}{7}(182)$$
$$B = 26 - 12 = 14$$

25. $C = 2\pi r$; $C = 16.328$, $\pi = 3.14$

$$C = 2\pi r$$
$$16.328 = 2(3.14)r$$
$$16.328 = 6.28r$$
$$2.6 = r$$

27. $C = 2\pi r$; $C = 20\pi$

$$C = 2\pi r$$
$$20\pi = 2\pi r$$
$$10 = r \qquad \text{Divide by } 2\pi.$$

29. $A = \pi r^2$; $r = 4$, $\pi = 3.14$

$$A = \pi r^2$$
$$= 3.14(4)^2$$
$$= 3.14(16)$$
$$A = 50.24$$

31. $S = 2\pi rh$; $S = 120\pi$, $h = 10$

$$S = 2\pi rh$$
$$120\pi = 2\pi r(10)$$
$$120\pi = 20\pi r$$
$$6 = r \qquad \text{Divide by } 20\pi.$$

In Exercises 33–38, substitute the given values into the formula and then evaluate V.

33. $V = LWH$; $L = 10, W = 5, H = 3$

$$V = LWH$$
$$= (10)(5)(3)$$
$$V = 150$$

35. $V = \frac{1}{3}Bh$; $B = 12$, $h = 13$

$$V = \frac{1}{3}Bh$$
$$= \frac{1}{3}(12)(13)$$
$$V = 52$$

37. $V = \frac{4}{3}\pi r^3$; $r = 12$, $\pi = 3.14$

$$V = \frac{4}{3}\pi r^3$$
$$= \frac{4}{3}(3.14)(12)^3$$
$$= \frac{4}{3}(3.14)(1728)$$
$$V = 7234.56$$

39.
$$P = 2L + 2W$$
$$54 = 2(W + 9) + 2W \quad \text{Let } L = W + 9.$$
$$54 = 2W + 18 + 2W$$
$$54 = 4W + 18$$
$$36 = 4W$$
$$9 = W$$

The width is 9 inches and the length is $9 + 9 = 18$ inches.

41. $P = 2l + 2w$
$$36 = 2(3w + 2) + 2w \quad \text{Let } l = 3w + 2.$$
$$36 = 6w + 4 + 2w$$

$$36 = 6w + 4 + 2w \quad \textit{equation repeated}$$
$$36 = 8w + 4$$
$$32 = 8w$$
$$4 = w$$

The width is 4 meters and the length is $3(4) + 2 = 14$ meters.

43. Let $s =$ the length of the shortest side, in inches;

$s + 2 =$ the length of the medium side, and,

$s + 3 =$ the length of the longest side.

The perimeter is 20 inches, so

$$s + (s + 2) + (s + 3) = 20.$$
$$3s + 5 = 20$$
$$3s = 15$$
$$s = 5$$

Since $s = 5$, $s + 2 = 7$, and $s + 3 = 8$. The lengths of the sides are 5, 7, and 8 inches. The perimeter is $5 + 7 + 8 = 20$, as required.

45. Let $s =$ the length of the two sides that have equal length, in meters; and $2s - 4 =$ the length of the third side.

The perimeter is 24 meters, so

$$s + s + (2s - 4) = 24.$$
$$4s - 4 = 24$$
$$4s = 28$$
$$s = 7$$

Since $s = 7$, $2s - 4 = 10$. The lengths of the sides are 7, 7, and 10 meters. The perimeter is $7 + 7 + 10 = 24$, as required.

47. The diameter of the circle is 443 feet, so its radius is $\frac{443}{2} = 221.5$ ft. Use the area of a circle formula to find the enclosed area.

$$A = \pi r^2$$
$$= \pi(221.5)^2$$
$$\approx 154{,}133.6 \text{ ft}^2,$$

or about $154{,}000$ ft^2. (If 3.14 is used for π, the value is $154{,}055.465$.)

49. The page is a rectangle with length 1.5 m and width 1.2 m, so use the formulas for the perimeter and area of a rectangle.

$$P = 2L + 2W$$
$$= 2(1.5) + 2(1.2)$$
$$= 3 + 2.4$$
$$P = 5.4 \text{ meters}$$

$$A = LW$$
$$= (1.5)(1.2)$$
$$A = 1.8 \text{ square meters}$$

51. Use the formula for the area of a triangle with $A = 70$ and $b = 14$.

$$A = \tfrac{1}{2}bh$$
$$70 = \tfrac{1}{2}(14)h$$
$$70 = 7h$$
$$10 = h$$

The height of the sign is 10 feet.

53. To find the area of the drum face, use the formula for the area of a circle, $A = \pi r^2$. Since the diameter of the circle is 15.74 feet, the radius is $(\tfrac{1}{2})(15.74) = 7.87$ feet.

$$A = \pi r^2$$
$$\approx (3.14)(7.87)^2$$
$$= (3.14)(61.9369)$$
$$A \approx 194.48$$

The area of the drum face is about 194.48 square feet.

Use the circumference of a circle formula.

$$C \approx 2(3.14)(7.87)$$
$$= 49.4236$$

The circumference is about 49.42 feet.

55. Use the formula for the area of a trapezoid with $B = 115.80$, $b = 171.00$, and $h = 165.97$.

$$A = \tfrac{1}{2}(B + b)h$$
$$= \tfrac{1}{2}(115.80 + 171.00)(165.97)$$
$$= \tfrac{1}{2}(286.80)(165.97)$$
$$= 23{,}800.098$$

To the nearest hundredth of a square foot, the combined area of the two lots is 23,800.10 square feet.

57. The girth is $4 \cdot 18 = 72$ inches. Since the length plus the girth is 108, we have

$$L + G = 108$$
$$L + 72 = 108$$
$$L = 36 \text{ in.}$$

The volume of the box is

$$V = LWH$$
$$= (36)(18)(18)$$
$$= 11{,}664 \text{ in.}^3$$

59. The two angles are supplementary, so the sum of their measures is 180°.

$$(x + 1) + (4x - 56) = 180$$
$$5x - 55 = 180$$
$$5x = 235$$
$$x = 47$$

Since $x = 47$, $x + 1 = 47 + 1 = 48$, and $4x - 56 = 4(47) - 56 = 132$.

The measures of the angles are 48° and 132°.

61. In the figure, the two angles are complementary, so their sum is 90°.

$$(8x - 1) + 5x = 90$$
$$13x - 1 = 90$$
$$13x = 91$$
$$x = 7$$

Since $x = 7$, $8x - 1 = 8(7) - 1 = 55$, and $5x = 5(7) = 35$.

The two angle measures are 55° and 35°.

63. The two angles are vertical angles, which have equal measures. Set their measures equal to each other and solve for x.

$$5x - 129 = 2x - 21$$
$$3x - 129 = -21$$
$$3x = 108$$
$$x = 36$$

Since $x = 36$, $5x - 129 = 5(36) - 129 = 51$, and $2x - 21 = 2(36) - 21 = 51$.

The measure of each angle is 51°.

65. The angles are vertical angles, so their measures are equal.

$$12x - 3 = 10x + 15$$
$$2x - 3 = 15$$
$$2x = 18$$
$$x = 9$$

Since $x = 9$, $12x - 3 = 12(9) - 3 = 105$, and $10x + 15 = 10(9) + 15 = 105$.

The measure of each angle is 105°.

67. $d = rt$ for t

$$\frac{d}{r} = \frac{rt}{r} \quad \textit{Divide by } r.$$
$$\frac{d}{r} = t \quad \text{or} \quad t = \frac{d}{r}$$

69. $A = bh$ for b

$$\frac{A}{h} = \frac{bh}{h} \quad \textit{Divide by } h.$$
$$\frac{A}{h} = b \quad \text{or} \quad b = \frac{A}{h}$$

71. $C = \pi d$ for d

$$\frac{C}{\pi} = \frac{\pi d}{\pi} \quad \textit{Divide by } \pi.$$
$$\frac{C}{\pi} = d \quad \text{or} \quad d = \frac{C}{\pi}$$

73. $V = LWH$ for H

$$\frac{V}{LW} = \frac{LWH}{LW} \quad \text{Divide by } LW.$$

$$\frac{V}{LW} = H \quad \text{or} \quad H = \frac{V}{LW}$$

75. $I = prt$ for r

$$\frac{I}{pt} = \frac{prt}{pt} \quad \text{Divide by } pt.$$

$$\frac{I}{pt} = r \quad \text{or} \quad r = \frac{I}{pt}$$

77. $A = \frac{1}{2}bh$ for h

$$2A = 2\left(\frac{1}{2}bh\right) \quad \text{Multiply by 2.}$$

$$2A = bh$$

$$\frac{2A}{b} = \frac{bh}{b} \quad \text{Divide by } b.$$

$$\frac{2A}{b} = h \quad \text{or} \quad h = \frac{2A}{b}$$

79. $V = \frac{1}{3}\pi r^2 h$ for h

$$3V = 3\left(\frac{1}{3}\right)\pi r^2 h \quad \text{Multiply by 3.}$$

$$3V = \pi r^2 h$$

$$\frac{3V}{\pi r^2} = \frac{\pi r^2 h}{\pi r^2} \quad \text{Divide by } \pi r^2.$$

$$\frac{3V}{\pi r^2} = h \quad \text{or} \quad h = \frac{3V}{\pi r^2}$$

81. $P = a + b + c$ for b

$$P - a - c = a + b + c - a - c$$

$$\text{Subtract } a \text{ and } c.$$

$$P - a - c = b \quad \text{or} \quad b = P - a - c$$

83. $P = 2L + 2W$ for W

$$P - 2L = 2L + 2W - 2L \quad \text{Subtract } 2L.$$

$$P - 2L = 2W$$

$$\frac{P - 2L}{2} = \frac{2W}{2} \quad \text{Divide by 2.}$$

$$\frac{P - 2L}{2} = W \quad \text{or} \quad W = \frac{P - 2L}{2}$$

85. $y = mx + b$ for m

$$y - b = mx + b - b \quad \text{Subtract } b.$$

$$y - b = mx$$

$$\frac{y - b}{x} = \frac{mx}{x} \quad \text{Divide by } x.$$

$$\frac{y - b}{x} = m \quad \text{or} \quad m = \frac{y - b}{x}$$

87. $Ax + By = C$ for y

$$By = C - Ax \quad \text{Subtract } Ax.$$

$$\frac{By}{B} = \frac{C - Ax}{B} \quad \text{Divide by } B.$$

$$y = \frac{C - Ax}{B}$$

89. $M = C(1 + r)$ for r

$$M = C + Cr \quad \text{Distributive Property}$$

$$M - C = Cr \quad \text{Subtract } C.$$

$$\frac{M - C}{C} = \frac{Cr}{C} \quad \text{Divide by } C.$$

$$\frac{M - C}{C} = r \quad \text{or} \quad r = \frac{M - C}{C}$$

Alternative solution:

$$M = C(1 + r)$$

$$\frac{M}{C} = 1 + r \quad \text{Divide by } C.$$

$$\frac{M}{C} - 1 = r \quad \text{Subtract 1.}$$

91. $P = 2(a + b)$ for a

$$P = 2a + 2b \quad \text{Distributive property}$$

$$P - 2b = 2a \quad \text{Subtract } 2b.$$

$$\frac{P - 2b}{2} = \frac{2a}{2} \quad \text{Divide by 2.}$$

$$\frac{P - 2b}{2} = a$$

93.

$$0.06x = 300$$

$$x = \frac{300}{0.06} = 5000$$

The solution set is $\{5000\}$.

95.

$$\frac{3}{4}x = 21$$

$$\frac{4}{3}\left(\frac{3}{4}x\right) = \frac{4}{3}(21)$$

$$x = 28$$

The solution set is $\{28\}$.

97.

$$-3x = \frac{1}{4}$$

$$\frac{-3x}{-3} = \frac{\frac{1}{4}}{-3}$$

$$x = \frac{1}{4}\left(-\frac{1}{3}\right) = -\frac{1}{12}$$

The solution set is $\left\{-\frac{1}{12}\right\}$.

2.6 Ratio, Proportion, and Percent

2.6 Now Try Exercises

N1. (a) The ratio of 7 inches to 4 inches is

$$\frac{7 \text{ inches}}{4 \text{ inches}} = \frac{7}{4}.$$

(b) 45 seconds $= \frac{45}{60} = \frac{3}{4}$ minute

The ratio of 45 seconds to 2 minutes is then

$$\frac{45 \text{ seconds}}{2 \text{ minutes}} = \frac{3}{4} \div 2 = \frac{3}{4} \cdot \frac{1}{2} = \frac{3}{8}.$$

N2. The results in the following table are rounded to the nearest thousandth.

Size	Unit Cost (dollars per oz)
150 oz	$\frac{\$19.97}{150} = \$0.133$
100 oz	$\frac{\$13.97}{100} = \$0.140$
75 oz	$\frac{\$8.94}{75} = \$0.119$ (*)

Because the 75 oz size produces the lowest unit cost, it is the best buy. The unit cost, to the nearest thousandth, is $0.119 per oz.

N3. **(a)** $\dfrac{1}{3} = \dfrac{33}{100}$

Compare the cross products.

$$1 \cdot 100 = 100$$
$$3 \cdot 33 = 99$$

The cross products are *different*, so the proportion is *false*.

(b) $\dfrac{4}{13} = \dfrac{16}{52}$

Check to see whether the cross products are equal.

$$4 \cdot 52 = 208$$
$$13 \cdot 16 = 208$$

The cross products are *equal*, so the proportion is *true*.

N4. $\dfrac{9}{7} = \dfrac{x}{56}$

$$
\begin{aligned}
7x &= 9 \cdot 56 & &\textit{Cross products} \\
x &= \frac{9 \cdot 56}{7} & &\textit{Divide by 7.} \\
&= \frac{9 \cdot 7 \cdot 8}{7} & &\textit{Factor.} \\
&= 72 & &\textit{Cancel.}
\end{aligned}
$$

The solution set is $\{72\}$.

Note: We could have multiplied $9 \cdot 56$ to get 504 and then divided 504 by 7 to get 72. This may be the best approach if you are doing these calculations on a calculator. The factor and cancel method is preferable if you're not using a calculator.

N5. $\dfrac{k-3}{6} = \dfrac{3k+2}{4}$

$$
\begin{aligned}
4(k-3) &= 6(3k+2) & &\textit{Cross products} \\
4k - 12 &= 18k + 12 & &\textit{Distributive prop.} \\
-14k - 12 &= 12 & &\textit{Subtract 18k.} \\
-14k &= 24 & &\textit{Add 12.} \\
a &= -\frac{24}{14} = -\frac{12}{7} & &\textit{Divide by } -14.
\end{aligned}
$$

The solution set is $\left\{-\frac{12}{7}\right\}$.

N6. Let $x =$ the cost for 27 gallons.

$$
\begin{aligned}
\frac{\$49.80}{20} &= \frac{x}{27} \\
20x &= 27(49.80) & &\textit{Cross products} \\
20x &= 1344.60 & &\textit{Multiply.} \\
x &= 67.23 & &\textit{Divide by 20.}
\end{aligned}
$$

It would cost $67.23.

N7. **(a)** $16\% = 16 \cdot 1\% = 16 \cdot 0.01 = 0.16$

(b) $1.5 = 150 \cdot 0.01 = 150 \cdot 1\% = 150\%$

N8. **(a)** What is 20% of 70?

$$20\% \cdot 70 = 0.20 \cdot 70 = 14$$

(b) 40% of what number is 130?

As in Example 8(b), let n denote the number.

$$
\begin{aligned}
0.40 \cdot n &= 130 \\
n &= \frac{130}{0.40} & &\textit{Divide by 0.40.} \\
&= 325 & &\textit{Simplify.}
\end{aligned}
$$

40% of 325 is 130.

(c) 121 is what percent of 484?

As in Example 8(c), let p denote the percent.

$$
\begin{aligned}
121 &= p \cdot 484 \\
p &= \frac{121}{484} = 0.25 = 25\%
\end{aligned}
$$

121 is 25% of 484.

N9. We can think of this problem as

$$48 \text{ is what percent of } 120?$$

Let p denote the percent.

$$
\begin{aligned}
48 &= p \cdot 120 \\
p &= \frac{48}{120} = 0.40 = 40\%
\end{aligned}
$$

The coat cost 40% of the regular price, so the savings is $100\% - 40\% = 60\%$ of the regular price.

2.6   Section Exercises

1.   **(a)** 75 to 100 is $\dfrac{75}{100} = \dfrac{3}{4}$ or 3 to 4.

The answer is **C**.

(b) 5 to 4 or $\dfrac{5}{4} = \dfrac{5 \cdot 3}{4 \cdot 3} = \dfrac{15}{12}$ or 15 to 12.

The answer is **D**.

(c) $\dfrac{1}{2} = \dfrac{1 \cdot 50}{2 \cdot 50} = \dfrac{50}{100}$ or 50 to 100

The answer is **B**.

(d) 4 to 5 or $\dfrac{4}{5} = \dfrac{4 \cdot 20}{5 \cdot 20} = \dfrac{80}{100}$ or 80 to 100.

The answer is **A**.

3. The ratio of 40 miles to 30 miles is

$$\frac{40 \text{ miles}}{30 \text{ miles}} = \frac{40}{30} = \frac{4}{3}.$$

5. The ratio of 120 people to 90 people is

$$\frac{120 \text{ people}}{90 \text{ people}} = \frac{4 \cdot 30}{3 \cdot 30} = \frac{4}{3}.$$

7. To find the ratio of 20 yards to 8 feet, first convert 20 yards to feet.

$$20 \text{ yards} = 20 \text{ yards} \cdot \frac{3 \text{ feet}}{1 \text{ yard}} = 60 \text{ feet}$$

The ratio of 20 yards to 8 feet is then

$$\frac{60 \text{ feet}}{8 \text{ feet}} = \frac{60}{8} = \frac{15 \cdot 4}{2 \cdot 4} = \frac{15}{2}.$$

9. Convert 2 hours to minutes.

$$2 \text{ hours} = 2 \text{ hours} \cdot \frac{60 \text{ minutes}}{1 \text{ hour}}$$
$$= 120 \text{ minutes}$$

The ratio of 24 minutes to 2 hours is then

$$\frac{24 \text{ minutes}}{120 \text{ minutes}} = \frac{24}{120} = \frac{1 \cdot 24}{5 \cdot 24} = \frac{1}{5}.$$

11. 2 yards $= 2 \cdot 3 = 6$ feet
6 feet $= 6 \cdot 12 = 72$ inches
The ratio of 60 inches to 2 yards is then

$$\frac{60 \text{ inches}}{72 \text{ inches}} = \frac{5 \cdot 12}{6 \cdot 12} = \frac{5}{6}.$$

In Exercises 13–22, to find the best buy, divide the price by the number of units to get the unit cost. Each result was found by using a calculator and rounding the answer to three decimal places. The *best buy* (based on price per unit) is the smallest unit cost.

13.

Size	Unit Cost (dollars per lb)
4 lb	$\frac{\$1.78}{4} = \$0.445$
10 lb	$\frac{\$4.29}{10} = \$0.429$ (∗)

The 10 lb size is the best buy.

15.

Size	Unit Cost (dollars per oz)
16 oz	$\frac{\$2.44}{16} = \$0.153$
32 oz	$\frac{\$2.98}{32} = \$0.093$ (∗)
48 oz	$\frac{\$4.95}{48} = \$0.103$

The 32 oz size is the best buy.

17.

Size	Unit Cost (dollars per oz)
16 oz	$\frac{\$1.66}{16} = \$0.104$
32 oz	$\frac{\$2.59}{32} = \$0.081$
64 oz	$\frac{\$4.29}{64} = \$0.067$
128 oz	$\frac{\$6.49}{128} = \$0.051$ (∗)

The 128-oz size is the best buy.

19.

Size	Unit Cost (dollars per oz)
14 oz	$\frac{\$1.39}{14} = \$0.099$
24 oz	$\frac{\$1.55}{24} = \$0.065$
36 oz	$\frac{\$1.78}{36} = \$0.049$ (∗)
64 oz	$\frac{\$3.99}{64} = \$0.062$

The 36 oz size is the best buy.

21.

Size	Unit Cost (dollars per oz)
87 oz	$\frac{\$7.88}{87} = \$0.091$
131 oz	$\frac{\$10.98}{131} = \$0.084$
263 oz	$\frac{\$19.96}{263} = \$0.076$ (∗)

The 263 oz size is the best buy.

23. $\dfrac{5}{35} = \dfrac{8}{56}$

Check to see whether the cross products are equal.

$$5 \cdot 56 = 280$$
$$35 \cdot 8 = 280$$

The cross products are *equal*, so the proportion is *true*.

25. $\dfrac{120}{82} = \dfrac{7}{10}$

Compare the cross products.

$$120 \cdot 10 = 1200$$
$$82 \cdot 7 = 574$$

The cross products are *different*, so the proportion is *false*.

27. $\frac{\frac{1}{2}}{5} = \frac{1}{10}$

Compare the cross products.

$$\frac{1}{2} \cdot 10 = 5$$
$$5 \cdot 1 = 5$$

The cross products are *equal*, so the proportion is *true*.

29. $\frac{k}{4} = \frac{175}{20}$

$20k = 4(175)$ *Cross products are equal.*

$20k = 700$

$\frac{20k}{20} = \frac{700}{20}$ *Divide by 20.*

$k = 35$

The solution set is $\{35\}$.

31. $\frac{49}{56} = \frac{z}{8}$

$56z = 49(8)$ *Cross products are equal.*

$56z = 392$

$\frac{56z}{56} = \frac{392}{56}$ *Divide by 56.*

$z = 7$

The solution set is $\{7\}$.

33. $\frac{x}{24} = \frac{15}{16}$

$16x = 24(15)$ *Cross products are equal.*

$16x = 360$

$\frac{16x}{16} = \frac{360}{16}$ *Divide by 16.*

$x = \frac{45 \cdot 8}{2 \cdot 8} = \frac{45}{2}$

The solution set is $\left\{\frac{45}{2}\right\}$.

35. $\frac{z}{2} = \frac{z+1}{3}$

$3z = 2(z+1)$ *Cross products are equal.*

$3z = 2z + 2$ *Distributive property*

$z = 2$ *Subtract 2z.*

The solution set is $\{2\}$.

37. $\frac{3y - 2}{5} = \frac{6y - 5}{11}$

$11(3y - 2) = 5(6y - 5)$ *Cross products are equal.*

$33y - 22 = 30y - 25$ *Dist. prop.*

$3y - 22 = -25$ *Subtract 30y.*

$3y = -3$ *Add 22.*

$y = -1$ *Divide by 3.*

The solution set is $\{-1\}$.

39. $\frac{5k + 1}{6} = \frac{3k - 2}{3}$

$3(5k + 1) = 6(3k - 2)$ *Cross products*

$15k + 3 = 18k - 12$ *Dist. prop.*

$-3k + 3 = -12$ *Subtract 18k.*

$-3k = -15$ *Subtract 3.*

$k = 5$ *Divide by −3.*

The solution set is $\{5\}$.

41. $\frac{2p + 7}{3} = \frac{p - 1}{4}$

$4(2p + 7) = 3(p - 1)$ *Cross products*

$8p + 28 = 3p - 3$ *Dist. prop.*

$5p + 28 = -3$ *Subtract 3p.*

$5p = -31$ *Subtract 28.*

$p = -\frac{31}{5}$ *Divide by 5.*

The solution set is $\left\{-\frac{31}{5}\right\}$.

43. Let $x =$ the cost of 24 candy bars.
Set up a proportion.

$$\frac{x}{24} = \frac{\$20.00}{16}$$
$$16x = 24(20)$$
$$16x = 480$$
$$x = 30$$

The cost of 24 candy bars is $30.00.

45. Let $x =$ the cost of 5 quarts of oil.
Set up a proportion.

$$\frac{x}{5} = \frac{\$14.00}{8}$$
$$8x = 5(14)$$
$$8x = 70$$
$$x = 8.75$$

The cost of five quarts of oil is $8.75.

47. Let $x =$ the cost of five pairs of jeans.

$$\frac{9 \text{ pairs}}{\$121.50} = \frac{5 \text{ pairs}}{x}$$
$$9x = 5(121.50)$$
$$9x = 607.5$$
$$\frac{9x}{9} = \frac{607.5}{9}$$
$$x = 67.5$$

The cost of five pairs is $67.50.

49. Let $x =$ the cost for filling a 15-gallon tank.
Set up a proportion.

$$\frac{x \text{ dollars}}{\$19.56} = \frac{15 \text{ gallons}}{6 \text{ gallons}}$$
$$6x = 15(19.56)$$
$$6x = 293.4$$
$$x = 48.90$$

It would cost $48.90 to completely fill a 15-gallon tank.

51. Let $x =$ the distance between Memphis and Philadelphia on the map (in feet).

Set up a proportion with one ratio involving map distances and the other involving actual distances.

$$\frac{x \text{ feet}}{2.4 \text{ feet}} = \frac{1000 \text{ miles}}{600 \text{ miles}}$$
$$\frac{x}{2.4} = \frac{1000}{600}$$
$$600x = (2.4)(1000)$$
$$600x = 2400$$
$$x = 4$$

The distance on the map between Memphis and Philadelphia would be 4 feet.

53. Let $x =$ the number of inches between St. Louis and Des Moines on the map. Set up a proportion.

$$\frac{8.5 \text{ inches}}{x \text{ inches}} = \frac{1040 \text{ miles}}{333 \text{ miles}}$$
$$1040x = 8.5(333)$$
$$1040x = 2830.5$$
$$x \approx 2.72$$

St. Louis and Des Moines are about 2.7 inches apart on the map.

55. Let $x =$ the number of inches between Moscow and Berlin on the globe. Set up a proportion.

$$\frac{12.4 \text{ inches}}{x \text{ inches}} = \frac{10,080 \text{ km}}{1610 \text{ km}}$$
$$10,080x = 12.4(1610)$$
$$10,080x = 19,964$$
$$x \approx 1.98$$

Moscow and Berlin are about 2.0 inches apart on the globe.

57. Let $x =$ the number of cups of cleaner. Set up a proportion with one ratio involving the number of cups of cleaner and the other involving the number of gallons of water.

$$\frac{x \text{ cups}}{\frac{1}{4} \text{ cup}} = \frac{10\frac{1}{2} \text{ gallons}}{1 \text{ gallon}}$$
$$x \cdot 1 = \frac{1}{4}\left(10\frac{1}{2}\right)$$
$$x = \frac{1}{4}\left(\frac{21}{2}\right) = \frac{21}{8}$$

The amount of cleaner needed is $2\frac{5}{8}$ cups.

59. Let $x =$ the number of U.S. dollars Ashley exchanged. Set up a proportion.

$$\frac{\$1.4294}{x \text{ dollars}} = \frac{1 \text{ euro}}{300 \text{ euros}}$$
$$x \cdot 1 = 1.4294(300)$$
$$x = 428.82$$

She exchanged $428.82.

61. Let $x =$ the number of fish in North Bay.

Set up a proportion with one ratio involving the sample and the other involving the total number of fish.

$$\frac{7 \text{ fish}}{700 \text{ fish}} = \frac{500 \text{ fish}}{x \text{ fish}}$$
$$7x = (700)(500)$$
$$7x = 350,000$$
$$x = 50,000$$

We estimate that there are 50,000 fish in North Bay.

63. $\frac{x}{12} = \frac{3}{9}$
$$9x = 12 \cdot 3 = 36$$
$$x = 4$$

Other possibilities for the proportion are:

$$\frac{12}{x} = \frac{9}{3}, \quad \frac{x}{12} = \frac{5}{15}, \quad \frac{12}{x} = \frac{15}{5}$$

65. Ratios of corresponding sides are equal, so $\frac{2}{3} = \frac{x}{3}$, and we must have $x = 2$.

67. $\frac{x}{3} = \frac{2}{6}$ $\frac{y}{\frac{4}{3}} = \frac{6}{2}$
$$6x = 3 \cdot 2 = 6 \qquad 2y = 6\left(\frac{4}{3}\right) = 8$$
$$x = 1 \qquad\qquad y = 4$$

69. (a)

(b) These two triangles are similar, so their sides are proportional.

$$\frac{x}{12} = \frac{18}{4}$$
$$4x = 18(12)$$
$$4x = 216$$
$$x = 54$$

The chair is 54 feet tall.

71. Let $x =$ the 1997 price of electricity.

$$\frac{1995 \text{ price}}{1995 \text{ index}} = \frac{1997 \text{ price}}{1997 \text{ index}}$$

$$\frac{225}{152.4} = \frac{x}{160.5}$$

$$152.4x = 225(160.5)$$

$$x = \frac{225(160.5)}{152.4} \approx 236.96$$

The 1997 price would be about \$237.

73. Let $x =$ the 2003 price of electricity.

$$\frac{1995 \text{ price}}{1995 \text{ index}} = \frac{2003 \text{ price}}{2003 \text{ index}}$$

$$\frac{225}{152.4} = \frac{x}{184.0}$$

$$152.4x = 225(184.0)$$

$$x = \frac{225(184.0)}{152.4} \approx 271.65$$

The 2003 price would be about \$272.

75. $53\% = 53 \cdot 1\% = 53(0.01) = 0.53$

An alternative method is

$53 = 53 \cdot 1\% = 53\left(\frac{1}{100}\right) = \frac{53}{100} = 0.53.$

77. $96\% = 96 \cdot 1\% = 96(0.01) = 0.96$

79. $9\% = 9 \cdot 1\% = 9(0.01) = 0.09$

81. $129\% = 129 \cdot 1\% = 129(0.01) = 1.29$

83. $0.80 = 80(0.01) = 80 \cdot 1\% = 80\%$

85. $0.02 = 2(0.01) = 2 \cdot 1\% = 2\%$

87. $0.125 = 12.5(0.01) = 12.5 \cdot 1\% = 12.5\%$

89. $2.2 = 220(0.01) = 220 \cdot 1\% = 220\%$

91. What is 14% of 780?

$$14\% \cdot 780 = 0.14 \cdot 780 = 109.2$$

93. 42% of what number is 294?

As in Example 8(b), let n denote the number.

$$0.42 \cdot n = 294$$

$$n = \frac{294}{0.42} \quad \textit{Divide by 0.42.}$$

$$= 700 \quad \textit{Simplify.}$$

42% of 700 is 294.

95. 120% of what number is 510?

As in Example 8(b), let n denote the number.

$$1.20 \cdot n = 510$$

$$n = \frac{510}{1.20} \quad \textit{Divide by 1.20.}$$

$$= 425 \quad \textit{Simplify.}$$

120% of 425 is 510.

97. 4 is what percent of 50?

As in Example 8(c), let p denote the percent.

$$4 = p \cdot 50$$

$$p = \frac{4}{50} = 0.08 = 8\%$$

4 is 8% of 50.

99. What percent of 30 is 36?

As in Example 8(c), let p denote the percent.

$$36 = p \cdot 30$$

$$p = \frac{36}{30} = 1.2 = 120\%$$

36 is 120% of 30.

101. The discount is 15% of \$795.

$$15\% \cdot 795 = 0.15 \cdot 795 = 119.25$$

The discount is \$119.25.

The sale price is $795 - \$119.25 = \$675.75.

103. 48 is what percent of 60?

As in Example 8(c), let p denote the percent.

$$48 = p \cdot 60$$

$$p = \frac{48}{60} = 0.8 = 80\%$$

He earned 80% of the total points.

105. 65% of what number is 1950?

As in Example 8(b), let n denote the number.

$$0.65 \cdot n = 1950$$

$$n = \frac{1950}{0.65} \quad \textit{Divide by 0.65.}$$

$$= 3000 \quad \textit{Simplify.}$$

She needs \$3000 for the car.

107.
$$0.15x + 0.30(3) = 0.20(3 + x)$$
$$100[0.15x + 0.30(3)] = 100[0.20(3 + x)]$$
$$15x + 30(3) = 20(3 + x)$$
$$15x + 90 = 60 + 20x$$
$$90 = 60 + 5x$$
$$30 = 5x$$
$$6 = x$$

The solution set is $\{6\}$.

109.
$$0.92x + 0.98(12 - x) = 0.96(12)$$
$$100[0.92x + 0.98(12 - x)] = 100[0.96(12)]$$
$$92x + 98(12 - x) = 96(12)$$
$$92x + 1176 - 98x = 1152$$
$$1176 - 6x = 1152$$
$$-6x = -24$$
$$x = 4$$

The solution set is $\{4\}$.

2.7 Further Applications of Linear Equations

2.7 Now Try Exercises

N1. **(a)** The amount of pure alcohol in 70 L of a 20% alcohol solution is

$$
\underset{\substack{\uparrow \\ \text{Amount} \\ \text{of} \\ \text{solution}}}{70 \text{ L}} \times \underset{\substack{\uparrow \\ \text{Rate} \\ \text{of} \\ \text{concentration}}}{0.20} = \underset{\substack{\uparrow \\ \text{Amount} \\ \text{of pure} \\ \text{alcohol}}}{14 \text{ L.}}
$$

(b) If $3200 is invested for one year at 2% simple interest, the amount of interest earned is

$$
\underset{\substack{\uparrow \\ \text{Principal}}}{\$3200} \times \underset{\substack{\uparrow \\ \text{Interest} \\ \text{Rate}}}{0.02} = \underset{\substack{\uparrow \\ \text{Interest} \\ \text{earned}}}{\$64.}
$$

N2. Let $x =$ the number of ounces of seasoning that is 70% salt.

Then $x + 30 =$ the number of ounces of seasoning that is 50% salt.

Number of ounces of seasoning	x		30		$x + 30$
Percent of salt in seasoning	0.70	$+$	0.10	$=$	0.50

$$
\underset{\substack{\downarrow \\ 0.70x}}{\substack{\text{Salt in} \\ \text{70\% seasoning}}} \underset{\downarrow}{\overset{\text{plus}}{+}} \underset{\substack{\downarrow \\ 0.10(30)}}{\substack{\text{salt in} \\ \text{10\% seasoning}}}
$$

$$
\underset{\substack{\downarrow \\ = \quad 0.50(x+30)}}{\overset{\text{is}}{}\substack{\text{salt in} \\ \text{50\% seasoning.}}}
$$

Solve the equation.

$$0.70x + 0.10(30) = 0.50x + 15$$

Multiply by 10 to clear decimals.

$$
\begin{aligned}
7x + 1(30) &= 5x + 150 \\
7x + 30 &= 5x + 150 \\
2x + 30 &= 150 \\
2x &= 120 \\
x &= 60
\end{aligned}
$$

60 ounces of seasoning that is 70% salt are needed.

Check $x = 60$:

LS and RS refer to the left side and right side of the original equation.

LS: $0.70(60) + 0.10(30) = 42 + 3 = 45$
RS: $\quad 0.50(60 + 30) = 0.50(90) = 45$

N3. Let $x =$ the amount invested at 3%.
Then $x + 5000 =$ the amount invested at 4%.

Amount invested (in dollars)	Rate of interest	Interest for one year
x	0.03	$0.03x$
$x + 5000$	0.04	$0.04(x + 5000)$

Since the total annual interest is $410, the equation is

$$0.03x + 0.04(x + 5000) = 410.$$

Multiply both sides by 100 to eliminate the decimals.

$$
\begin{aligned}
3x + 4(x + 5000) &= 100(410) \\
3x + 4x + 20{,}000 &= 41{,}000 \\
7x + 20{,}000 &= 41{,}000 \\
7x &= 21{,}000 \\
x &= 3000
\end{aligned}
$$

The financial advisor should invest $3000 at 3% and $8000 at 4%.

N4. Let $x =$ the number of dimes.
Then $x + 10 =$ the number of quarters.

$$
\underset{\substack{\downarrow \\ 0.10x}}{\substack{\text{The value} \\ \text{of dimes}}} \overset{\text{plus}}{\underset{\downarrow}{+}} \underset{\substack{\downarrow \\ 0.25(x+10)}}{\substack{\text{the value} \\ \text{of quarters}}} \overset{\text{is }\$5.65.}{\underset{\substack{\downarrow \downarrow \\ = 5.65}}{}}
$$

$$\textit{Multiply by 100.}$$

$$
\begin{aligned}
10x + 25(x + 10) &= 565 \\
10x + 25x + 250 &= 565 \\
35x + 250 &= 565 \\
35x &= 315 \\
x &= 9
\end{aligned}
$$

Since $x = 9$, $x + 10 = 19$.
Clayton has 19 quarters and 9 dimes.

N5. $r = \dfrac{d}{t} = \dfrac{400 \text{ miles}}{6 \text{ hours}} \approx 66.6667$

The rate was about 66.67 miles per hour.

N6. Let $t =$ the time it takes for the bicyclists to be 5 miles apart. Use the formula $d = rt$.

$$
\begin{aligned}
d_{\text{faster}} - d_{\text{slower}} &= d_{\text{total}} \\
20t - 18t &= 5 \\
2t &= 5 \\
t &= \tfrac{5}{2} = 2.5
\end{aligned}
$$

It will take 2.5 hours for the bicyclists to be 5 miles apart.

N7. Let x = the rate of the slower car.
Then $x + 6$ = the rate of the faster car.

Use the formula $d = rt$ and the fact that each car travels for $\frac{1}{4}$ hour.

$$d_{\text{slower}} + d_{\text{faster}} = d_{\text{total}}$$
$$(x + 6)\left(\tfrac{1}{4}\right) + x\left(\tfrac{1}{4}\right) + = 35$$
$$\tfrac{1}{4}x + \tfrac{3}{2} + \tfrac{1}{4}x = \tfrac{70}{2}$$
$$\tfrac{2}{4}x + \tfrac{3}{2} = \tfrac{70}{2}$$
$$\tfrac{1}{2}x = \tfrac{67}{2}$$
$$\tfrac{2}{1}\left(\tfrac{1}{2}x\right) = \tfrac{2}{1}\left(\tfrac{67}{2}\right)$$
$$x = 67$$

Since $x = 67$, $x + 6 = 73$. The slower car had a rate of 67 mph and the faster car had a rate of 73 mph.

Check: The slower car traveled $67\left(\tfrac{1}{4}\right) = \tfrac{67}{4}$ miles and the faster car traveled $73\left(\tfrac{1}{4}\right) = \tfrac{73}{4}$ miles. The total miles traveled is $\tfrac{67}{4} + \tfrac{73}{4} = \tfrac{140}{4} = 35$, as required.

2.7 Section Exercises

1. The amount of pure alcohol in 150 liters of a 30% alcohol solution is

$$150 \quad \times \quad 0.30 \quad = 45 \text{ liters.}$$

Amount of solution	Rate of concentration	Amount of pure alcohol
↑	↑	↑

3. If \$25,000 is invested at 3% simple interest for one year, the amount of interest earned is

$$\$25{,}000 \quad \times \quad 0.03 \quad \times \quad 1 \quad = \$750.$$

Principal	Interest Rate	Time	Interest earned
↑	↑	↑	↑

5. The monetary value of 35 half-dollars is

$$35 \quad \times \quad \$0.50 \quad = \$17.50.$$

Number of coins	Denomination	Monetary value
↑	↑	↑

7. 44.4% is about 45%. The population of 1,917,000 is about 1.9 million, and 45% of 1.9 million is 855,000. So choice **A** is the correct answer.

9. **(a)** 14% of 3.8 million $= 0.14(3{,}800{,}000)$
$= 532{,}000$ white cars.

(b) 21% of 3.8 million $= 0.21(3{,}800{,}000)$
$= 798{,}000$ silver cars.

(c) 13% of 3.8 million $= 0.13(3{,}800{,}000)$
$= 494{,}000$ red cars.

11. The concentration of the new solution could not be more than the strength of the stronger of the original solutions, so the correct answer is **D**, since 32% is stronger than both 20% and 30%.

13. *Step 2*
Let x = the number of liters of 25% acid solution to be used.

Step 3
Use the box diagram in the textbook to write the equation.

Pure acid in 25% solution	Pure acid in 40% solution	Pure acid in 30% solution
↓	↓	↓
$0.25x$ +	$0.40(80)$ =	$0.30(x + 80)$

Step 4
Multiply by 100 to clear decimals.

$$25x + 40(80) = 30(x + 80)$$
$$25x + 3200 = 30x + 2400$$
$$25x + 800 = 30x$$
$$800 = 5x$$
$$160 = x$$

Step 5
160 liters of 25% acid solution must be added.

Step 6
25% of 160 liters plus 40% of 80 liters is 40 liters plus 32 liters, or 72 liters, of pure acid; which is equal to 30% of $(160 + 80)$ liters.
$[0.30(240) = 72]$

15. Let x = the number of liters of 5% drug solution.

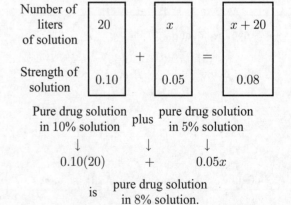

$$0.10(20) \quad + \quad 0.05x$$

$$\text{is} \quad \text{pure drug solution in 8\% solution.}$$

$$= \quad 0.08(x + 20)$$

Solve the equation.

$$0.10(20) + 0.05x = 0.08(x + 20)$$
$$10(20) + 5x = 8(x + 20)$$
$$200 + 5x = 8x + 160$$
$$200 = 3x + 160$$
$$40 = 3x$$
$$x = \frac{40}{3} = 13\frac{1}{3}$$

The pharmacist needs $13\frac{1}{3}$ liters of 5% drug solution.

Check $x = 13\frac{1}{3}$:

LS and RS refer to the left side and right side of the original equation.

LS: $0.10(20) + 0.05\left(13\frac{1}{3}\right) = 2\frac{2}{3}$
RS: $0.08\left(13\frac{1}{3} + 20\right) = 2\frac{2}{3}$

17. Let $x =$ the number of liters of the 20% alcohol solution.
Complete the table.

Strength	Liters of solution	Liters of pure alcohol
12%	12	$0.12(12) = 1.44$
20%	x	$0.20x$
14%	$x + 12$	$0.14(x + 12)$

From the last column, we can formulate an equation that compares the number of liters of pure alcohol.

Alcohol in 12% solution + alcohol in 20% solution = alcohol in 14% solution.

$$1.44 + 0.20x = 0.14(x + 12)$$
$$1.44 + 0.20x = 0.14x + 1.68$$
$$0.06x = 0.24$$
$$x = 4$$

4 L of the 20% alcohol solution are needed.

19. Let $x =$ the amount of water to be added.
Then $20 + x =$ the amount of 2% solution.

There is no minoxidil in water.

Number of milliliters of solution	x		20		$20 + x$
Concentration of solution	0	+	0.04	=	0.02

Pure minoxidil in x milliliters of water plus pure minoxidil in 4% solution is pure minoxidil in 2% solution

$$0(x) \quad + \quad 0.04(20) \quad = \quad 0.02(20 + x)$$
Solve the equation.

$$0x + 0.04(20) = 0.02(20 + x)$$
$$4(20) = 2(20 + x)$$
$$80 = 40 + 2x$$
$$40 = 2x$$
$$20 = x$$

20 milliliters of water should be used.

Check $x = 20$:
LS: $0(20) + 0.04(20) = 0.8$
RS: $0.02(20 + 20) = 0.8$

This answer should make common sense; that is, equal amounts of 0% and 4% solutions should produce a 2% solution.

21. Let $x =$ the number of liters of 60% acid solution.
Then $20 - x =$ the number of liters of 75% acid solution.

Number of liters of solution	x		$20 - x$		20
Concentration of acid	0.60	+	0.75	=	0.72

Pure acid in 60% solution plus pure acid in 75% solution

$$0.60x \quad + \quad 0.75(20 - x)$$

is pure acid in 72% solution.

$$= \quad 0.72(20)$$

Solve the equation.

$$0.60x + 0.75(20 - x) = 0.72(20)$$
$$60x + 75(20 - x) = 72(20)$$
$$60x + 1500 - 75x = 1440$$
$$1500 - 15x = 1440$$
$$-15x = -60$$
$$x = 4$$

4 liters of 60% acid solution must used.

Check $x = 4$:
LS: $0.60(4) + 0.75(20 - 4) = 14.4$
RS: $0.72(20) = 14.4$

23. Let $x =$ the amount invested at 5% (in dollars).
Then $x - 1200 =$ the amount invested at 4% (in dollars).

Amount invested (in dollars)	Rate of interest	Interest for one year
x	0.05	$0.05x$
$x - 1200$	0.04	$0.04(x - 1200)$

Since the total annual interest was $141, the equation is

$$0.05x + 0.04(x - 1200) = 141.$$
$$5x + 4(x - 1200) = 100(141)$$
$$5x + 4x - 4800 = 14{,}100$$
$$9x - 4800 = 14{,}100$$
$$9x = 18{,}900$$
$$x = 2100$$

Since $x = 2100$, $x - 1200 = 900$.
Arlene invested $2100 at 5% and $900 at 4%.

25. Let $x =$ the amount invested at 6%.
Then $3x + 6000 =$ the amount invested at 5%.

$$0.06x + 0.05(3x + 6000) = 825$$
$$6x + 5(3x + 6000) = 100(825)$$
$$6x + 15x + 30{,}000 = 82{,}500$$
$$21x + 30{,}000 = 82{,}500$$
$$21x = 52{,}500$$
$$x = 2500$$

Since $x = 2500$, $3x + 6000 = 13{,}500$.
The artist invested $2500 at 6% and $13,500 at 5%.

27. Let $x =$ the number of nickels.
Then $x + 2 =$ the number of dimes.

The value of nickels	plus	the value of dimes		is	$1.70
$\downarrow$	$\downarrow$	$\downarrow$		$\downarrow$	$\downarrow$
$0.05x$	$+$	$0.10(x + 2)$		$=$	1.70

$$5x + 10(x + 2) = 100(1.70)$$
$$5x + 10x + 20 = 170$$
$$15x + 20 = 170$$
$$15x = 150$$
$$x = 10$$

The collector has 10 nickels.

29. Let $x =$ the number of 44-cent stamps.
Then $45 - x =$ the number of 17-cent stamps.

The value of the 44-cent stamps is $0.44x$ and the value of the 17-cent stamps is $0.17(45 - x)$. The total value is $14.40, so

$$0.44x + 0.17(45 - x) = 14.40.$$
$$44x + 17(45 - x) = 1440$$
$$44x + 765 - 17x = 1440$$
$$27x + 765 = 1440$$
$$27x = 675$$
$$x = 25$$

Since $x = 25$, $45 - x = 20$. She bought 25 44-cent stamps valued at $11.00 and 20 17-cent stamps valued at $3.40, for a total value of $14.40.

31. Let $x =$ the number of pounds of Colombian Decaf beans.
Then $2x =$ the number of pounds of Arabian Mocha beans.

Number of Pounds	Cost per Pound	Total Value (in $)
x	$8.00	$8x$
$2x$	$8.50	$8.5(2x)$

The total value is $87.50, so

$$8x + 8.5(2x) = 87.50.$$
$$8x + 17x = 87.50$$
$$25x = 87.50$$
$$x = 3.5$$

Since $x = 3.5$, $2x = 7$. She can buy 3.5 pounds of Colombian Decaf and 7 pounds of Arabian Mocha.

33. To estimate the average rate of the trip, round 405 to 400 and 8.2 to 8.

Use $r = \frac{d}{t}$ with $d = 400$ and $t = 8$.

$$r = \frac{d}{t} = \frac{400}{8} = 50$$

The best estimate is **A**, 50 miles per hour.

35. Use the formula $d = rt$ with $r = 53$ and $t = 10$.

$$d = rt$$
$$= (53)(10)$$
$$= 530$$

The distance between Memphis and Chicago is 530 miles.

37. Use $d = rt$ with $d = 500$ and $r = 143.567$.

$$d = rt$$
$$500 = 143.567t$$
$$t = \frac{500}{143.567} \approx 3.483$$

His time was about 3.483 hours.

39. $r = \dfrac{d}{t} = \dfrac{100 \text{ meters}}{12.54 \text{ seconds}} \approx 7.974$

Her rate was about 7.97 meters per second.

41. $r = \dfrac{d}{t} = \dfrac{400 \text{ meters}}{47.25 \text{ seconds}} \approx 8.466$

His rate was about 8.47 meters per second.

43. Let t = the number of hours until John and Pat meet.

The distance John travels and the distance Pat travels total 440 miles.

John's distance and Pat's distance equal total distance.
$$\downarrow \quad \downarrow \quad \downarrow \quad \downarrow \quad \downarrow$$
$$60t \quad + \quad 28t \quad = \quad 440$$
$$88t = 440$$
$$t = 5$$

It will take 5 hours for them to meet.

45. Let t = the number of hours until the trains are 315 kilometers apart.

Distance of northbound train plus distance of southbound train is total distance
$$\downarrow \qquad \downarrow \qquad \downarrow \quad \downarrow \qquad \downarrow$$
$$85t \qquad + \qquad 95t \quad = \qquad 315$$
$$180t = 315$$
$$t = \frac{315}{180} = \frac{7}{4}$$

It will take $1\frac{3}{4}$ hours for the trains to be 315 kilometers apart.

47. Let t = the number of hours Marco and Celeste traveled.

Make a chart using the formula $d = rt$.

	r	t	d
Marco	10	t	$10t$
Celeste	12	t	$12t$

Marco's distance minus Celeste's distance is 15.
$$\downarrow \quad \downarrow \quad \downarrow \quad \downarrow \ \downarrow$$
$$12t \quad - \quad 10t \quad = 15$$
$$12t - 10t = 15$$
$$2t = 15$$
$$t = \frac{15}{2} \text{ or } 7\frac{1}{2}$$

They will be 15 miles apart in $7\frac{1}{2}$ hours.

49. Let x = the rate of the westbound plane.
Then $x - 150$ = the rate of the eastbound plane.

Using the formula $d = rt$ and the chart in the text, we see that

$$d_{\text{west}} + d_{\text{east}} = d_{\text{total}}$$
$$x(3) + (x - 150)(3) = 2250$$
$$3x + 3x - 450 = 2250$$
$$6x = 2700$$
$$x = 450$$

Since $x = 450$, $x - 150 = 300$.

The rate of the westbound plane is 450 mph and the rate of the eastbound plane is 300 mph.

51. Let x = the rate of the slower car.
Then $x + 20$ = the rate of the faster car.

Use the formula $d = rt$ and the fact that each car travels for 4 hours.

$$d_{\text{faster}} + d_{\text{slower}} = d_{\text{total}}$$
$$(x + 20)(4) + (x)(4) = 400$$
$$4x + 80 + 4x = 400$$
$$8x = 320$$
$$x = 40$$

The rate of the slower car is 40 mph and the rate of the faster car is 60 mph.

53. Let x = Bob's current age.
Then $3x$ = Kevin's current age.

Three years ago, Bob's age was $x - 3$ and Kevin's age was $3x - 3$, and this sum was 22.

$$(x - 3) + (3x - 3) = 22$$
$$4x - 6 = 22$$
$$4x = 28$$
$$x = 7$$

Bob is 7 years old and Kevin is $3(7) = 21$ years old.

55. Let w = the width of the table.
Then $3w$ = the length of the table.

If we subtract 3 feet from the length $(3w - 3)$ and add 3 feet to the width $(w + 3)$, then the length and the width would be equal.

$$3w - 3 = w + 3$$
$$3w = w + 6$$
$$2w = 6$$
$$w = 3$$

The width is 3 feet and the length is $3(3) = 9$ feet.

57. Let x = her gross pay (pay before deductions).

gross pay − deductions = take-home pay
$$x - 0.10(x) = 585$$
$$0.90x = 585$$
$$x = \frac{585}{0.90} = 650$$

She was paid $650 before deductions.

59. Since $6 = 6$, the statement $6 > 6$ is *false*.

61. Since -4 is to the left of -3 on the real number line, the statement $-4 \le -3$ is *true*.

63. Since 0 is to the right of $-\frac{1}{2}$ on the real number line, the statement $0 > -\frac{1}{2}$ is *true*.

2.8   Solving Linear Inequalities

2.8 Now Try Exercises

N1. (a) The statement $x < -1$ says that x can represent any number less than -1. The interval is written as $(-\infty, -1)$. Graph this inequality by placing a parenthesis at -1 on a number line and drawing an arrow to the left.

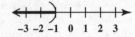

(b) The statement $-2 \le x$ is the same as $x \ge -2$. The interval is written as $[-2, \infty)$. Graph this inequality by placing a bracket at -2 on a number line and drawing an arrow to the right.

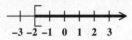

N2.
$$5 + 5x \ge 4x + 3$$
$$5 + 5x - 5 \ge 4x + 3 - 5 \quad \textit{Subtract 5.}$$
$$5x \ge 4x - 2$$
$$5x - 4x \ge 4x - 2 - 4x \quad \textit{Subtract 4x.}$$
$$x \ge -2$$

Graph the solution set $[-2, \infty)$.

To graph this inequality, place a bracket at -2 on a number line and draw an arrow to the right.

N3. $-5k \ge 15$
$$\frac{-5k}{-5} \le \frac{15}{-5} \quad \begin{array}{l}\textit{Divide by } -5; \\ \textit{reverse the symbol.}\end{array}$$
$$k \le -3$$

Graph the solution set $(-\infty, -3]$.

N4. $6 - 2t + 5t \le 8t - 4$
$$3t + 6 \le 8t - 4$$
$$3t + 6 - 8t \le 8t - 4 - 8t \quad \textit{Subtract 8t.}$$
$$-5t + 6 \le -4$$
$$-5t + 6 - 6 \le -4 - 6 \quad \textit{Subtract 6.}$$
$$-5t \le -10$$

$$\frac{-5t}{-5} \ge \frac{-10}{-5} \quad \begin{array}{l}\textit{Divide by } -5; \\ \textit{reverse the} \\ \textit{symbol.}\end{array}$$
$$t \ge 2$$

Graph the solution set $[2, \infty)$.

N5. $2x - 3(x - 6) < 4(x + 7)$
$$2x - 3x + 18 < 4x + 28 \quad \begin{array}{l}\textit{Distributive} \\ \textit{property}\end{array}$$
$$-x + 18 < 4x + 28$$
$$-x + 18 + x < 4x + 28 + x \quad \textit{Add x.}$$
$$18 < 5x + 28$$
$$18 - 28 < 5x + 28 - 28 \quad \textit{Subtract 28.}$$
$$-10 < 5x$$
$$\frac{-10}{5} < \frac{5x}{5} \quad \textit{Divide by 5.}$$
$$-2 < x$$
$$x > -2$$

Graph the solution set $(-2, \infty)$.

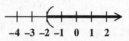

N6. Let $x =$ Kristine's score on the third test.

The average   is at least   90.

$$\frac{98 + 85 + x}{3} \qquad \ge \qquad 90$$

Solve the inequality.

$$3\left(\frac{183 + x}{3}\right) \ge 3(90) \quad \begin{array}{l}\textit{Add in the} \\ \textit{numerator;} \\ \textit{multiply by 3.}\end{array}$$
$$183 + x \ge 270$$
$$183 + x - 183 \ge 270 - 183 \quad \textit{Subtract 183.}$$
$$x \ge 87 \qquad\qquad \textit{Combine terms.}$$

She must score 87 or more on the third test to have an average of *at least* 90.

N7. The statement $0 \le x < 2$ says that x can represent any number between 0 and 2, including 0 and excluding 2. To graph the inequality, place a bracket at 0 and a parenthesis at 2 and draw a line segment between them. The interval is written as $[0, 2)$.

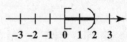

N8. $-4 \le \quad \frac{3}{2}x - 1 \quad \le 0$

$-4 + 1 \le \frac{3}{2}x - 1 + 1 \quad \le 0 + 1$ *Add 1 to each part.*

$-3 \le \quad \frac{3}{2}x \quad \le 1$

$\frac{2}{3}(-3) \le \quad \frac{2}{3}(\frac{3}{2}x) \quad \le \frac{2}{3}(1)$ *Mult. each part by $\frac{2}{3}$.*

$-2 \le \quad x \quad \le \frac{2}{3}$

Graph the solution set $[-2, \frac{2}{3}]$.

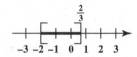

2.8 Section Exercises

1. When graphing an inequality, use a parenthesis if the inequality symbol is $>$ or $<$. Use a square bracket if the inequality symbol is $\ge$ or $\le$. Examples:

A parenthesis would be used for the inequalities $x < 2$ and $x > 3$. A square bracket would be used for the inequalities $x \le 2$ and $x \ge 3$.

3. In interval notation, the set $\{x \mid x > 0\}$ is written $(0, \infty)$.

5. The set of numbers graphed corresponds to the inequality $x > -4$.

7. The set of numbers graphed corresponds to the inequality $x \le 4$.

9. The statement $z \le 4$ says that z can represent any number less than or equal to 4. The interval is written as $(-\infty, 4]$. To graph the inequality, place a square bracket at 4 (to show that 4 is part of the graph) and draw an arrow extending to the left.

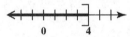

11. The statement $x < -3$ says that x can represent any number less than -3. The interval is written as $(-\infty, -3)$. To graph the inequality, place a parenthesis at -3 (to show that -3 is *not* part of the graph) and draw an arrow extending to the left.

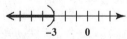

13. The statement $t > 4$ says that t can represent any number greater than 4. The interval is written $(4, \infty)$. To graph the inequality, place a parenthesis at 4 (to show that 4 is *not* part of the graph) and draw an arrow extending to the right.

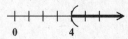

15. The statement $0 \ge x$ (or $x \le 0$) says that x can represent any number less than or equal to 0. The interval is written as $(-\infty, 0]$. To graph the inequality, place a bracket at 0 (to show that 0 is part of the graph) and draw an arrow extending to the left.

17. The statement $-\frac{1}{2} \le x$ (or $x \ge -\frac{1}{2}$) says that x can represent any number greater than or equal to $-\frac{1}{2}$. The interval is written $[-\frac{1}{2}, \infty)$. To graph the inequality, place a bracket at $-\frac{1}{2}$ (to show that $-\frac{1}{2}$ is part of the graph) and draw an arrow extending to the right.

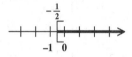

19. $z - 8 \ge -7$

$z - 8 + 8 \ge -7 + 8$ *Add 8.*

$z \ge 1$

Graph the solution set $[1, \infty)$.

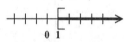

21. $2k + 3 \ge k + 8$

$2k + 3 - k \ge k + 8 - k$ *Subtract k.*

$k + 3 \ge 8$

$k + 3 - 3 \ge 8 - 3$ *Subtract 3.*

$k \ge 5$

Graph the solution set $[5, \infty)$.

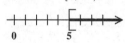

23. $3n + 5 < 2n - 6$

$3n - 2n + 5 < 2n - 2n - 6$ *Subtract 2n.*

$n + 5 < -6$

$n + 5 - 5 < -6 - 5$ *Subtract 5.*

$n < -11$

Graph the solution set $(-\infty, -11)$.

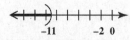

25. The inequality symbol must be reversed when one is multiplying or dividing by a negative number.

27. $3x < 18$

$\dfrac{3x}{3} < \dfrac{18}{3}$ *Divide by 3.*

$x < 6$

Graph the solution set $(-\infty, 6)$.

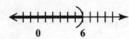

29. $2y \geq -20$

$\dfrac{2y}{2} \geq \dfrac{-20}{2}$ *Divide by 2.*

$y \geq -10$

Graph the solution set $[-10, \infty)$.

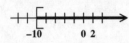

31. $-8t > 24$

$\dfrac{-8t}{-8} < \dfrac{24}{-8}$ *Divide by -8; reverse the symbol from $>$ to $<$.*

$t < -3$

Graph the solution set $(-\infty, -3)$.

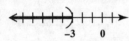

33. $-x \geq 0$

$-1x \geq 0$

$\dfrac{-1x}{-1} \leq \dfrac{0}{-1}$ *Divide by -1; reverse the symbol from $\geq$ to $\leq$.*

$x \leq 0$

Graph the solution set $(-\infty, 0]$.

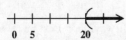

35. $-\frac{3}{4}r < -15$

$\left(-\frac{4}{3}\right)\left(-\frac{3}{4}r\right) > \left(-\frac{4}{3}\right)(-15)$

Multiply by $-\frac{4}{3}$ (the reciprocal of $-\frac{3}{4}$); reverse the symbol from $<$ to $>$.

$r > 20$

Graph the solution set $(20, \infty)$.

37. $-0.02x \leq 0.06$

$\dfrac{-0.02x}{-0.02} \geq \dfrac{0.06}{-0.02}$ *Divide by -0.02; reverse the symbol from $\leq$ to $\geq$.*

$x \geq -3$

Graph the solution set $[-3, \infty)$.

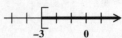

39. $8x + 9 \leq -15$

$8x \leq -24$ *Subtract 9.*

$\dfrac{8x}{8} \leq \dfrac{-24}{8}$ *Divide by 8.*

$x \leq -3$

Graph the solution set $(-\infty, -3]$.

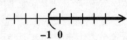

41. $-4x - 3 < 1$

$-4x < 4$ *Add 3.*

$\dfrac{-4x}{-4} > \dfrac{4}{-4}$ *Divide by -4; reverse the symbol from $<$ to $>$.*

$x > -1$

Graph the solution set $(-1, \infty)$.

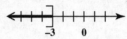

43. $5r + 1 \geq 3r - 9$

$2r + 1 \geq -9$ *Subtract 3r.*

$2r \geq -10$ *Subtract 1.*

$r \geq -5$ *Divide by 2.*

Graph the solution set $[-5, \infty)$.

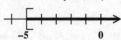

45. $6x + 3 + x < 2 + 4x + 4$

$7x + 3 < 4x + 6$ *Combine like terms.*

$3x + 3 < 6$ *Subtract 4x.*

$3x < 3$ *Subtract 3.*

$x < 1$ *Divide by 3.*

Graph the solution set $(-\infty, 1)$.

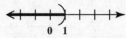

47. $-x + 4 + 7x \le -2 + 3x + 6$

$6x + 4 \le 4 + 3x$

$3x + 4 \le 4$

$3x \le 0$

$x \le 0$

Graph the solution set $(-\infty, 0]$.

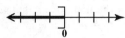

49. $5(t - 1) > 3(t - 2)$

$5t - 5 > 3t - 6$

$2t - 5 > -6$

$2t > -1$

$t > -\frac{1}{2}$

Graph the solution set $\left(-\frac{1}{2}, \infty\right)$.

51. $5(x + 3) - 6x \le 3(2x + 1) - 4x$

$5x + 15 - 6x \le 6x + 3 - 4x$

$-x + 15 \le 2x + 3$

$-3x + 15 \le 3$

$-3x \le -12$

$\dfrac{-3x}{-3} \ge \dfrac{-12}{-3}$ *Divide by –3;*
reverse the symbol.

$x \ge 4$

Graph the solution set $[4, \infty)$.

53. $\frac{2}{3}(p + 3) > \frac{5}{6}(p - 4)$

$6\left(\frac{2}{3}\right)(p + 3) > 6\left(\frac{5}{6}\right)(p - 4)$

           Multiply by 6, the LCD.

$4(p + 3) > 5(p - 4)$

$4p + 12 > 5p - 20$

$-p + 12 > -20$

$-p > -32$

$\dfrac{-p}{-1} < \dfrac{-32}{-1}$ *Divide by –1;*
reverse the symbol.

$p < 32$

Graph the solution set $(-\infty, 32)$.

55. $4x - (6x + 1) \le 8x + 2(x - 3)$

$4x - 6x - 1 \le 8x + 2x - 6$

$-2x - 1 \le 10x - 6$

$-12x - 1 \le -6$

$-12x \le -5$

$\dfrac{-12x}{-12} \ge \dfrac{-5}{-12}$ *Divide by –12;*
reverse the symbol.

$x \ge \frac{5}{12}$

Graph the solution set $\left[\frac{5}{12}, \infty\right)$.

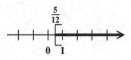

57. $5(2k + 3) - 2(k - 8) > 3(2k + 4) + k - 2$

$10k + 15 - 2k + 16 > 6k + 12 + k - 2$

$8k + 31 > 7k + 10$

$k + 31 > 10$

$k > -21$

Graph the solution set $(-21, \infty)$.

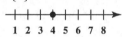

59. $3x + 2 = 14$

$3x = 12$

$x = 4$

Solution set: $\{4\}$

60. $3x + 2 > 14$

$3x > 12$

$x > 4$

Solution set: $(4, \infty)$

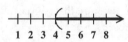

61. $3x + 2 < 14$

$3x < 12$

$x < 4$

Solution set: $(-\infty, 4)$

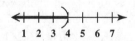

62. If you were to graph all the solutions from Exercises 59–61 on the same number line, the graph would be the complete number line, that is, all real numbers.

63. The statement "You must be at least 18 yr old to vote" translates as $x \ge 18$.

65. The statement "Chicago received more than 5 in. of snow" translates as $x > 5$.

67. The statement "Tracy could spend at most $20 on a gift" translates as $x \le 20$.

69. Let x = the score on the third test.

$$\begin{array}{ccc} \text{The average} & \text{is at} & 80. \\ \text{of the} & \text{least} & \\ \text{three tests} & & \\ \downarrow & \downarrow & \downarrow \\ \dfrac{76+81+x}{3} & \geq & 80 \end{array}$$

$$\frac{157+x}{3} \geq 80$$

$$3\left(\frac{157+x}{3}\right) \geq 3(80)$$

$$157+x \geq 240$$

$$x \geq 83$$

In order to average at least 80, Christy's score on her third test must be 83 or greater.

71. Let n = the number.
"When 2 is added to the difference between six times a number and 5, the result is greater than 13 added to five times the number" translates to

$$(6n-5)+2 > 5n+13.$$

Solve the inequality.

$$\begin{array}{rl} 6n-5+2 > 5n+13 & \\ 6n-3 > 5n+13 & \\ n-3 > 13 & \textit{Subtract 5n.} \\ n > 16 & \textit{Add 3.} \end{array}$$

All numbers greater than 16 satisfy the given condition.

73. The Fahrenheit temperature must correspond to a Celsius temperature that is greater than or equal to -25 degrees.

$$C = \frac{5}{9}(F-32) \geq -25$$

$$\frac{9}{5}\left[\frac{5}{9}(F-32)\right] \geq \frac{9}{5}(-25)$$

$$F-32 \geq -45$$

$$F \geq -13$$

The temperature in Minneapolis on a certain winter day is never less than $-13°$ Fahrenheit.

75. $P = 2L + 2W;\ P \geq 400$
From the figure, we have $L = 4x+3$ and $W = x+37$. Thus, we have the inequality

$$2(4x+3) + 2(x+37) \geq 400.$$

Solve this inequality.

$$\begin{array}{r} 8x+6+2x+74 \geq 400 \\ 10x+80 \geq 400 \\ 10x \geq 320 \\ x \geq 32 \end{array}$$

The rectangle will have a perimeter of at least 400 if the value of x is 32 or greater.

77.
$$2+0.30x \leq 5.60$$
$$10(2+0.30x) \leq 10(5.60)$$
$$20+3x \leq 56$$
$$3x \leq 36$$
$$x \leq 12$$

Alan can use the phone for a maximum of 12 minutes after the first three minutes. This means that the maximum *total* time he can use the phone is 15 minutes.

79. "Revenue from the sales of the DVDs is $5 per DVD less sales costs of $100" translates to

$$R = 5x - 100,$$

where x represents the number of DVDs to be produced.

81. $P = R - C$
$$= (5x-100) - (125+4x)$$
$$= 5x - 100 - 125 - 4x$$
$$= x - 225$$

We can use this expression for P to solve the inequality.

$$P > 0$$
$$x - 225 > 0$$
$$x > 225$$

83. The graph corresponds to the inequality $-1 < x < 2$, excluding both -1 and 2.

85. The graph corresponds to the inequality $-1 < x \leq 2$, excluding -1 but including 2.

87. The statement $8 \leq x \leq 10$ says that x can represent any number between 8 and 10, including 8 and 10. To graph the inequality, place brackets at 8 and 10 (to show that 8 and 10 are part of the graph) and draw a line segment between the brackets. The interval is written as $[8, 10]$.

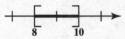

89. The statement $0 < y \leq 10$ says that y can represent any number between 0 and 10, excluding 0 and including 10. To graph the inequality, place a parenthesis at 0 and a bracket at 10 and draw a line segment between them. The interval is written as $(0, 10]$.

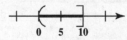

91. The statement $4 > x > -3$ can be written as $-3 < x < 4$. Graph the solution set $(-3, 4)$.

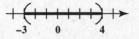

93. $-5 \leq 2x - 3 \leq 9$

$-5 + 3 \leq 2x - 3 + 3 \leq 9 + 3$ *Add 3 to each part.*

$-2 \leq 2x \leq 12$

$\dfrac{-2}{2} \leq \dfrac{2x}{2} \leq \dfrac{12}{2}$ *Divide each part by 2.*

$-1 \leq x \leq 6$

Graph the solution set $[-1, 6]$.

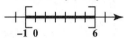

95. $5 < 1 - 6m < 12$

$5 - 1 < 1 - 6m - 1 < 12 - 1$ *Subtract 1 from each part.*

$4 < -6m < 11$

$\dfrac{4}{-6} > \dfrac{-6m}{-6} > \dfrac{11}{-6}$ *Divide each part by -6; reverse both symbols.*

$-\dfrac{2}{3} > m > -\dfrac{11}{6}$

or $-\dfrac{11}{6} < m < -\dfrac{2}{3}$ *Equivalent inequality*

Graph the solution set $\left(-\dfrac{11}{6}, -\dfrac{2}{3}\right)$.

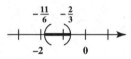

97. $10 < 7p + 3 < 24$

$7 < 7p < 21$ *Subtract 3.*

$1 < p < 3$ *Divide by 7.*

Graph the solution set $(1, 3)$.

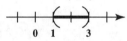

99. $-12 \leq \frac{1}{2}z + 1 \leq 4$

$2(-12) \leq 2\left(\frac{1}{2}z + 1\right) \leq 2(4)$ *Multiply each part by 2.*

$-24 \leq z + 2 \leq 8$

$-26 \leq z \leq 6$ *Subtract 2.*

Note: We could have started this solution by subtracting 1 from each part.

Graph the solution set $[-26, 6]$.

101. $1 \leq 3 + \frac{2}{3}p \leq 7$

$3(1) \leq 3\left(3 + \frac{2}{3}p\right) \leq 3(7)$ *Multiply by 3.*

$3 \leq 9 + 2p \leq 21$

$-6 \leq 2p \leq 12$ *Subtract 9.*

$-3 \leq p \leq 6$ *Divide by 2.*

Graph the solution set $[-3, 6]$.

103. $-7 \leq \frac{5}{4}r - 1 \leq -1$

$-6 \leq \frac{5}{4}r \leq 0$ *Add 1.*

$\frac{4}{5}(-6) \leq \frac{4}{5}\left(\frac{5}{4}r\right) \leq \frac{4}{5}(0)$ *Multiply by $\frac{4}{5}$.*

$-\frac{24}{5} \leq r \leq 0$

Graph the solution set $\left[-\frac{24}{5}, 0\right]$.

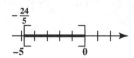

In part (a) of Exercises 105–110, replace x with -2. In part (b), replace x with 4. Then use the order of operations.

105. (a) $y = 5x + 3$
$y = 5(-2) + 3$
$y = -10 + 3$
$y = -7$

(b) $y = 5x + 3$
$y = 5(4) + 3$
$y = 20 + 3$
$y = 23$

107. (a) $6x - 2 = y$
$6(-2) - 2 = y$
$-12 - 2 = y$
$-14 = y$

(b) $6x - 2 = y$
$6(4) - 2 = y$
$24 - 2 = y$
$22 = y$

109. (a) $2x - 5y = 10$
$2(-2) - 5y = 10$
$-4 - 5y = 10$
$-5y = 14$
$y = -\frac{14}{5}$

(b) $2x - 5y = 10$
$2(4) - 5y = 10$
$8 - 5y = 10$
$-5y = 2$
$y = -\frac{2}{5}$

Chapter 2 Review Exercises

1. $x - 5 = 1$
$x = 6$ *Add 5.*

The solution set is $\{6\}$.

2. $x + 8 = -4$
$x = -12$ *Subtract 8.*

The solution set is $\{-12\}$.

3. $3t + 1 = 2t + 8$
$t + 1 = 8$ *Subtract $2t$.*
$t = 7$ *Subtract 1.*

The solution set is $\{7\}$.

4. $5z = 4z + \frac{2}{3}$
$z = \frac{2}{3}$ *Subtract $4z$.*

The solution set is $\left\{\frac{2}{3}\right\}$.

5. $(4r - 2) - (3r + 1) = 8$
$(4r - 2) - 1(3r + 1) = 8$ *Replace $-$ with -1.*
$4r - 2 - 3r - 1 = 8$ *Dist. prop.*
$r - 3 = 8$
$r = 11$ *Add 3.*

The solution set is $\{11\}$.

6. $3(2x - 5) = 2 + 5x$

$\quad\ 6x - 15 = 2 + 5x$ *Dist. prop.*

$\quad\quad x - 15 = 2$ *Subtract 5x.*

$\quad\quad\quad\ x = 17$ *Add 15.*

The solution set is $\{17\}$.

7. $7x = 35$

$\quad x = 5$ *Divide by 7.*

The solution set is $\{5\}$.

8. $12r = -48$

$\quad r = -4$ *Divide by 12.*

The solution set is $\{-4\}$.

9. $2p - 7p + 8p = 15$

$\quad\quad\quad\quad 3p = 15$

$\quad\quad\quad\quad\ p = 5$ *Divide by 3.*

The solution set is $\{5\}$.

10. $\dfrac{x}{12} = -1$

$\quad x = -12$ *Multiply by 12.*

The solution set is $\{-12\}$.

11. $\dfrac{5}{8}q = 8$

$\dfrac{8}{5}\left(\dfrac{5}{8}q\right) = \dfrac{8}{5}(8)$ *Multiply by $\frac{8}{5}$.*

$\quad\quad\ q = \dfrac{64}{5}$

The solution set is $\left\{\dfrac{64}{5}\right\}$.

12. $12m + 11 = 59$

$\quad 12m = 48$ *Subtract 11.*

$\quad\quad m = 4$ *Divide by 12.*

The solution set is $\{4\}$.

13. $3(2x + 6) - 5(x + 8) = x - 22$

$\quad 6x + 18 - 5x - 40 = x - 22$

$\quad\quad\quad\quad\quad x - 22 = x - 22$

This is a true statement, so the solution set is $\{$all real numbers$\}$.

14. $5x + 9 - (2x - 3) = 2x - 7$

$\quad 5x + 9 - 2x + 3 = 2x - 7$

$\quad\quad\quad\ 3x + 12 = 2x - 7$

$\quad\quad\quad\quad x + 12 = -7$

$\quad\quad\quad\quad\quad\ x = -19$

The solution set is $\{-19\}$.

15. $\dfrac{1}{2}r - \dfrac{r}{3} = \dfrac{r}{6}$

$6\left(\dfrac{1}{2}r\right) - 6\left(\dfrac{r}{3}\right) = 6\left(\dfrac{r}{6}\right)$ *Multiply by 6.*

$\quad\quad 3r - 2r = r$

$\quad\quad\quad\quad\ r = r$

This is a true statent, so the solution set is $\{$all real numbers$\}$.

16. $0.1(x + 80) + 0.2x = 14$

$10[0.1(x + 80) + 0.2x] = 10(14)$ *Multiply by 10.*

$\quad\ (x + 80) + 2x = 140$ *Dist. prop.*

$\quad\quad\quad 3x + 80 = 140$

$\quad\quad\quad\quad\quad 3x = 60$

$\quad\quad\quad\quad\quad\ x = 20$

The solution set is $\{20\}$.

17. $3x - (-2x + 6) = 4(x - 4) + x$

$\quad 3x + 2x - 6 = 4x - 16 + x$

$\quad\quad\ 5x - 6 = 5x - 16$

$\quad\quad\quad\quad -6 = -16$

This statement is false, so there is no solution set, symbolized by $\emptyset$.

18. $\dfrac{1}{2}(x + 3) - \dfrac{2}{3}(x - 2) = 3$

Multiply both sides by 6 to eliminate fractions.

$6\left[\dfrac{1}{2}(x + 3) - \dfrac{2}{3}(x - 2)\right] = 6(3)$

$6\left[\dfrac{1}{2}(x + 3)\right] - 6\left[\dfrac{2}{3}(x - 2)\right] = 18$

$\quad\quad 3(x + 3) - 4(x - 2) = 18$

$\quad\quad 3x + 9 - 4x + 8 = 18$

$\quad\quad\quad\quad\quad -x + 17 = 18$

$\quad\quad\quad\quad\quad\quad\ -x = 1$

$\quad\quad\quad\quad\quad\quad\ x = -1$

The solution set is $\{-1\}$.

19. Let x represent the number.

$\quad 5x + 7 = 3x$

$\quad\quad\quad\ 7 = -2x$ *Subtract 5x.*

$\quad -\dfrac{7}{2} = x$ *Divide by -2.*

The number is $-\dfrac{7}{2}$.

20. *Step 2*

Let $x =$ the number of Republicans.

Then $x + 22 =$ the number of Democrats.

Step 3 $x + (x + 22) = 118$

Step 4 $2x + 22 = 118$

$\quad\quad\quad\quad\quad\ 2x = 96$

$\quad\quad\quad\quad\quad\ x = 48$

Step 5

Since $x = 48$, $x + 22 = 70$.

There were 70 Democrats and 48 Republicans.

Step 6

There are 22 more Democrats than Republicans and the total is 118.

21. *Step 2*

Let $x =$ the land area of Rhode Island.

Then $x + 5213 =$ land area of Hawaii.

Step 3
The areas total 7637 square miles, so

$$x + (x + 5213) = 7637.$$

Step 4 $2x + 5213 = 7637$
$$2x = 2424$$
$$x = 1212$$

Step 5
Since $x = 1212$, $x + 5213 = 6425$. The land area of Rhode Island is 1212 square miles and that of Hawaii is 6425 square miles.

Step 6
The land area of Hawaii is 5213 square miles greater than the land area of Rhode Island and the total is 7637 square miles.

22.   *Step 2*
Let x = the height of Twin Falls.
Then $\frac{5}{2}x$ = the height of Seven Falls.

Step 3
The sum of the heights is 420 feet, so

$$x + \frac{5}{2}x = 420.$$

Step 4 $2(x + \frac{5}{2}x) = 2(420)$
$$2x + 5x = 840$$
$$7x = 840$$
$$x = 120$$

Step 5
Since $x = 120$, $\frac{5}{2}x = \frac{5}{2}(120) = 300$. The height of Twin Falls is 120 feet and that of Seven Falls is 300 feet.

Step 6
The height of Seven Falls is $\frac{5}{2}$ the height of Twin Falls and the sum is 420.

23.   *Step 2*
Let x = the measure of the angle.
Then $90 - x$ = the measure of its complement and $180 - x$ = the measure of its supplement.

Step 3 $180 - x = 10(90 - x)$

Step 4 $180 - x = 900 - 10x$
$$9x + 180 = 900$$
$$9x = 720$$
$$x = 80$$

Step 5
The measure of the angle is 80°.
Its complement measures $90° - 80° = 10°$, and its supplement measures $180° - 80° = 100°$.

Step 6
The measure of the supplement is 10 times the measure of the complement.

24.   *Step 2*
Let x = lesser odd integer.
Then $x + 2$ = greater odd integer.

Step 3 $x + 2(x + 2) = (x + 2) + 24$

Step 4 $x + 2x + 4 = x + 26$
$$3x + 4 = x + 26$$
$$2x + 4 = 26$$
$$2x = 22$$
$$x = 11$$

Step 5
Since $x = 11$, $x + 2 = 13$. The consecutive odd numbers are 11 and 13.

Step 6
The lesser plus twice the greater is $11 + 2(13) = 37$, which is 24 more than the greater.

In Exercises 25–28, substitute the given values into the given formula and then solve for the remaining variable.

25. $A = \frac{1}{2}bh; A = 44, b = 8$

$$A = \frac{1}{2}bh$$
$$44 = \frac{1}{2}(8)h$$
$$44 = 4h$$
$$11 = h$$

26. $A = \frac{1}{2}h(b + B); h = 8, b = 3, B = 4$

$$A = \frac{1}{2}h(b + B)$$
$$A = \frac{1}{2}(8)(3 + 4)$$
$$= \frac{1}{2}(8)(7)$$
$$= (4)(7)$$
$$A = 28$$

27. $C = 2\pi r; C = 29.83, \pi = 3.14$

$$C = 2\pi r$$
$$29.83 = 2(3.14)r$$
$$29.83 = 6.28r$$
$$\frac{29.83}{6.28} = \frac{6.28r}{6.28}$$
$$4.75 = r$$

28. $V = \frac{4}{3}\pi r^3; r = 6, \pi = 3.14$

$$V = \frac{4}{3}\pi r^3$$
$$= \frac{4}{3}(3.14)(6)^3$$
$$= \frac{4}{3}(3.14)(216)$$
$$= \frac{4}{3}(678.24)$$
$$V = 904.32$$

29. $A = bh$ for h

$$\frac{A}{b} = \frac{bh}{b} \quad \text{Divide by } b.$$
$$\frac{A}{b} = h, \text{ or } h = \frac{A}{b}$$

30. $A = \frac{1}{2}h(b + B)$ for h

$$2A = 2\left[\frac{1}{2}h(b + B)\right] \quad \textit{Multiply by 2.}$$
$$2A = h(b + B)$$
$$\frac{2A}{(b + B)} = \frac{h(b + B)}{(b + B)} \quad \textit{Divide by } b + B.$$
$$\frac{2A}{b + B} = h, \text{ or } h = \frac{2A}{b + B}$$

31. Because the two angles are supplementary,

$$(8x - 1) + (3x - 6) = 180.$$
$$11x - 7 = 180$$
$$11x = 187$$
$$x = 17$$

Since $x = 17$, $8x - 1 = 135$, and $3x - 6 = 45$. The measures of the two angles are $135°$ and $45°$.

32. The angles are vertical angles, so their measures are equal.

$$3x + 10 = 4x - 20$$
$$10 = x - 20$$
$$30 = x$$

Since $x = 30$, $3x + 10 = 100$, and $4x - 20 = 100$. Each angle has a measure of $100°$.

33. Let $W =$ the width of the rectangle. Then $W + 12 =$ the length of the rectangle.

The perimeter of the rectangle is 16 times the width can be written as

$$2L + 2W = 16W$$

since the perimeter is $2L + 2W$.

Because $L = W + 12$, we have

$$2(W + 12) + 2W = 16W.$$
$$2W + 24 + 2W = 16W$$
$$4W + 24 = 16W$$
$$-12W + 24 = 0$$
$$-12W = -24$$
$$W = 2$$

The width is 2 cm and the length is $2 + 12 = 14$ cm.

34. First, use the formula for the circumference of a circle to find the value of r.

$$C = 2\pi r$$
$$62.5 \approx 2(3.14)(r) \quad \begin{array}{l} \textit{Let } C = 62.5, \\ \pi = 3.14. \end{array}$$
$$62.5 = 6.28r$$
$$\frac{62.5}{6.28} = \frac{6.28r}{6.28}$$
$$9.95 \approx r$$

The radius of the turntable is approximately 9.95 feet. The diameter is twice the radius, so the diameter is approximately 19.9 feet.

Now use the formula for the area of a circle.

$$A = \pi r^2$$
$$\approx (3.14)(9.95)^2 \quad \textit{Let } \pi = 3.14, \ r = 9.95.$$
$$= (3.14)(99.0025)$$
$$A \approx 311$$

The area of the turntable is approximately 311 square feet.

35. The sum of the three marked angles in the triangle is $180°$.

$$45° + (x + 12.2)° + (3x + 2.8)° = 180°$$
$$4x + 60 = 180$$
$$4x = 120$$
$$x = 30$$

Since $x = 30$, $(x + 12.2)° = 42.2°$, and $(3x + 2.8)° = 92.8°$.

36. The ratio of 60 centimeters to 40 centimeters is

$$\frac{60 \text{ cm}}{40 \text{ cm}} = \frac{3 \cdot 20}{2 \cdot 20} = \frac{3}{2}.$$

37. To find the ratio of 5 days to 2 weeks, first convert 2 weeks to days.

$$2 \text{ weeks} = 2 \cdot 7 = 14 \text{ days}$$

Thus, the ratio of 5 days to 2 weeks is $\frac{5}{14}$.

38. To find the ratio of 90 inches to 10 feet, first convert 10 feet to inches.

$$10 \text{ feet} = 10 \cdot 12 = 120 \text{ inches}$$

Thus, the ratio of 90 inches to 10 feet is

$$\frac{90}{120} = \frac{3 \cdot 30}{4 \cdot 30} = \frac{3}{4}.$$

39. $\dfrac{p}{21} = \dfrac{5}{30}$

$$30p = 105 \quad \textit{Cross products are equal.}$$
$$\frac{30p}{30} = \frac{105}{30} \quad \textit{Divide by 30.}$$
$$p = \frac{105}{30} = \frac{7 \cdot 15}{2 \cdot 15} = \frac{7}{2}$$

The solution set is $\left\{\frac{7}{2}\right\}$.

40. $\dfrac{5 + x}{3} = \dfrac{2 - x}{6}$

$$6(5 + x) = 3(2 - x) \quad \begin{array}{l} \textit{Cross products} \\ \textit{are equal.} \end{array}$$
$$30 + 6x = 6 - 3x \quad \begin{array}{l} \textit{Distributive} \\ \textit{property} \end{array}$$
$$30 + 9x = 6 \quad \textit{Add 3x.}$$
$$9x = -24 \quad \textit{Subtract 30.}$$
$$x = \frac{-24}{9} = -\frac{8}{3}$$

The solution set is $\left\{-\frac{8}{3}\right\}$.

41. Let $x =$ the tax on a $36.00 item.
Set up a proportion with one ratio involving sales tax and the other involving the costs of the items.

$$\frac{x \text{ dollars}}{\$2.04} = \frac{\$36}{\$24}$$
$$24x = (2.04)(36) = 73.44$$
$$x = \frac{73.44}{24} = 3.06$$

The sales tax on a $36.00 item is $3.06.

42. Let $x =$ the actual distance between the second pair of cities (in kilometers).

Set up a proportion with one ratio involving map distances and the other involving actual distances.

$$\frac{x \text{ kilometers}}{150 \text{ kilometers}} = \frac{80 \text{ centimeters}}{32 \text{ centimeters}}$$
$$32x = (150)(80) = 12,000$$
$$x = \frac{12,000}{32} = 375$$

The cities are 375 kilometers apart.

43. Let $x =$ the number of bronze medals earned by Japan.

$$\frac{x \text{ bronze medals}}{25 \text{ medals}} = \frac{2 \text{ bronze medals}}{5 \text{ medals}}$$
$$5x = 2(25) = 50$$
$$x = 10$$

At the 2008 Olympics, 10 bronze medals were earned by Japan.

In Exercise 44, to find the best buy, divide the price by the number of units to get the unit cost. Each result was found by using a calculator and rounding the answer to three decimal places. The *best buy* (based on price per unit) is the smallest unit cost.

44.

Size	Unit Cost (dollars per oz)
15 oz	$\frac{\$2.69}{15} = \$0.179$
20 oz	$\frac{\$3.29}{20} = \$0.165$
25.5 oz	$\frac{\$3.49}{25.5} = \$0.137$ $(*)$

The 25.5 oz size is the best buy.

45. What is 8% of 75?

$$8\% \cdot 75 = 0.08 \cdot 75 = 6$$

46. What percent of 12 is 21?

Let p denote the percent.

$$21 = p \cdot 12$$
$$p = \tfrac{21}{12} = 1.75 = 175\%$$

Thus, 175% of 12 is 21.

47. 36% of what number is 900?

Let n denote the number.

$$0.36 \cdot n = 900$$
$$n = \frac{900}{0.36} \quad \textit{Divide by 0.36.}$$
$$= 2500 \quad \textit{Simplify.}$$

Thus, 36% of 2500 is 900.

48. Let $x =$ the number of liters of the 60% solution to be used.
Then $x + 15 =$ the number of liters of the 20% solution.

Liters of solution	15		x		$x + 15$
Strength of solution	0.10	$+$	0.60	$=$	0.20

Drug amount in 10% solution **plus** drug amount in 60% solution **is** drug amount in 20% solution.

$$0.10(15) \quad + \quad 0.60(x) \quad = \quad 0.20(x + 15)$$

Multiply by 10 to clear decimals.

$$1(15) + 6x = 2(x + 15)$$
$$15 + 6x = 2x + 30$$
$$15 + 4x = 30$$
$$4x = 15$$
$$x = \tfrac{15}{4} = 3.75$$

3.75 liters of 60% solution are needed.

49. Let $x =$ the amount invested at 5%.
Then $10,000 - x =$ the amount invested at 6%.

Interest at 5% **plus** interest at 6% **equals $550.**

$$0.05x \quad + \quad 0.06(10,000 - x) \quad = \quad 550$$

$$5x + 6(10,000 - x) = 100(550)$$
$$5x + 60,000 - 6x = 55,000$$
$$-x = -5000$$
$$x = 5000$$

Robert invested $5000 at 5% and $10,000 - 5000 = \$5000$ at 6%.

50. Use the formula $d = rt$ or $r = \frac{d}{t}$.

$$r = \frac{d}{t} = \frac{3150}{384} \approx 8.203$$

Rounded to the nearest tenth, the *Yorkshire's* average rate was 8.2 mph.

51. Use the formula $d = rt$ or $t = \frac{d}{r}$.

$$t = \frac{d}{r} = \frac{819}{63} = 13$$

Janet drove for 13 hours.

52. Let $t =$ the number of hours until the planes are 1925 miles apart.

Use $d = rt$.

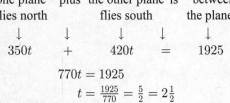

The distance one plane flies north	plus	the distance the other plane flies south	is	the distance between the planes.
$350t$	$+$	$420t$	$=$	1925

$$770t = 1925$$
$$t = \frac{1925}{770} = \frac{5}{2} = 2\frac{1}{2}$$

The planes will be 1925 miles apart in $2\frac{1}{2}$ hours.

53. The statement $x \geq -4$ can be written as $[-4, \infty)$.

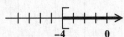

54. The statement $x < 7$ can be written as $(-\infty, 7)$.

55. The statement $-5 \leq x < 6$ can be written as $[-5, 6)$.

56. By examining the choices, we see that $-4x \leq 36$ is the only inequality that has a negative coefficient of x. Thus, **B** is the only inequality that requires a reversal of the inequality symbol when it is solved.

57. $x + 6 \geq 3$

$\quad x \geq -3$ *Subtract 6.*

Graph the solution set $[-3, \infty)$.

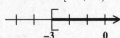

58. $5x < 4x + 2$

$\quad x < 2$ *Subtract 4x.*

Graph the solution set $(-\infty, 2)$.

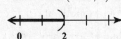

59. $-6x \leq -18$

$\dfrac{-6x}{-6} \geq \dfrac{-18}{-6}$ *Divide by –6;*
$\qquad\qquad$ *reverse the symbol.*

$\quad x \geq 3$

Graph the solution set $[3, \infty)$.

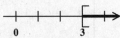

60. $8(x - 5) - (2 + 7x) \geq 4$

$\quad 8x - 40 - 2 - 7x \geq 4$

$\qquad\qquad x - 42 \geq 4$

$\qquad\qquad\quad x \geq 46$

Graph the solution set $[46, \infty)$.

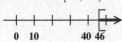

61. $4x - 3x > 10 - 4x + 7x$

$\qquad x > 10 + 3x$

$\quad -2x > 10$

$\dfrac{-2x}{-2} < \dfrac{10}{-2}$ *Divide by –2;*
$\qquad\qquad$ *reverse the symbol.*

$\quad x < -5$

Graph the solution set $(-\infty, -5)$.

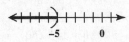

62. $3(2x + 5) + 4(8 + 3x) < 5(3x + 7)$

$\quad 6x + 15 + 32 + 12x < 15x + 35$

$\qquad\qquad 18x + 47 < 15x + 35$

$\qquad\qquad\quad 3x + 47 < 35$

$\qquad\qquad\qquad 3x < -12$

$\qquad\qquad\qquad\quad x < -4$

Graph the solution set $(-\infty, -4)$.

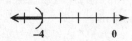

63. $-3 \leq 2x + 1 \leq 4$

$-4 \leq \quad 2x \quad \leq 3$ *Subtract 1.*

$-2 \leq \quad x \quad \leq \dfrac{3}{2}$ *Divide by 2.*

Graph the solution set $[-2, \frac{3}{2}]$.

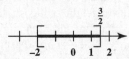

64. $9 < 3x + 5 \leq 20$

$4 < \quad 3x \quad \leq 15$ *Subtract 5.*

$\dfrac{4}{3} < \quad x \quad \leq 5$ *Divide by 3.*

Graph the solution set $\left(\frac{4}{3}, 5\right]$.

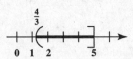

65. Let $x =$ the score on the third test.

The average of the three tests	is at least	90.
$\downarrow$	$\downarrow$	$\downarrow$
$\dfrac{94 + 88 + x}{3}$	$\geq$	90

$$\frac{182 + x}{3} \geq 90$$
$$3\left(\frac{182 + x}{3}\right) \geq 3(90)$$
$$182 + x \geq 270$$
$$x \geq 88$$

In order to average at least 90, Awilda's score on her third test must be 88 or more.

66. Let $n =$ the number.
"If nine times a number is added to 6, the result is at most 3" can be written as

$$9n + 6 \leq 3.$$

Solve the inequality.

$$9n \leq -3 \quad \textit{Subtract 6.}$$
$$n \leq \frac{-3}{9} \quad \textit{Divide by 9.}$$

All numbers less or equal to $-\frac{1}{3}$ satisfy the given condition.

67. **[2.6]** $\quad \dfrac{x}{7} = \dfrac{x - 5}{2}$
$$2x = 7(x - 5) \quad \textit{Cross products are equal.}$$
$$2x = 7x - 35$$
$$-5x = -35$$
$$x = 7$$

The solution set is $\{7\}$.

68. **[2.5]** $I = prt$ for r

$$\frac{I}{pt} = \frac{prt}{pt} \quad \textit{Divide by pt.}$$

$$\frac{I}{pt} = r \quad \text{or} \quad r = \frac{I}{pt}$$

69. **[2.8]** $\quad -2x > -4$
$$\frac{-2x}{-2} < \frac{-4}{-2} \quad \begin{array}{l} \textit{Divide by -2;} \\ \textit{reverse the symbol.} \end{array}$$
$$x < 2$$

The solution set is $(-\infty, 2)$.

70. **[2.3]** $\quad 2k - 5 = 4k + 13$
$$-2k - 5 = 13 \qquad \textit{Subtract 4k.}$$
$$-2k = 18 \qquad \textit{Add 5.}$$
$$k = -9 \qquad \textit{Divide by -2.}$$

The solution set is $\{-9\}$.

71. **[2.2]** $0.05x + 0.02x = 4.9$
To clear decimals, multiply both sides by 100.

$$100(0.05x + 0.02x) = 100(4.9)$$
$$5x + 2x = 490$$
$$7x = 490$$
$$x = 70$$

The solution set is $\{70\}$.

72. **[2.3]** $\quad 2 - 3(x - 5) = 4 + x$
$$2 - 3x + 15 = 4 + x$$
$$17 - 3x = 4 + x$$
$$17 - 4x = 4$$
$$-4x = -13$$
$$x = \frac{-13}{-4} = \frac{13}{4}$$

The solution set is $\left\{\frac{13}{4}\right\}$.

73. **[2.3]** $\quad 9x - (7x + 2) = 3x + (2 - x)$
$$9x - 7x - 2 = 3x + 2 - x$$
$$2x - 2 = 2x + 2$$
$$-2 = 2$$

Because $-2 = 2$ is a false statement, the given equation has no solution, symbolized by $\emptyset$.

74. **[2.3]** $\frac{1}{3}s + \frac{1}{2}s + 7 = \frac{5}{6}s + 5 + 2$
$$\frac{1}{3}s + \frac{1}{2}s = \frac{5}{6}s \qquad \textit{Subtract 7.}$$
The least common denominator is 6.
$$6\left(\frac{1}{3}s + \frac{1}{2}s\right) = 6\left(\frac{5}{6}s\right)$$
$$2s + 3s = 5s$$
$$5s = 5s$$

Because $5s = 5s$ is a true statement, the solution set is {all real numbers}.

75. **[2.6]** The problem may be stated as follows:

"What is 8% of \$3800?"

$$8\% \cdot 3800 = 0.08 \cdot 3800 = 304$$

The family plans to spend \$304 on entertainment.

76. **[2.6]** Let $x =$ the number of calories a 175-pound athlete can consume.
Set up a proportion with one ratio involving calories and the other involving pounds.

$$\frac{x \text{ calories}}{50 \text{ calories}} = \frac{175 \text{ pounds}}{2.2 \text{ pounds}}$$
$$2.2x = 50(175)$$
$$x = \frac{8750}{2.2} \approx 3977.3$$

To the nearest hundred calories, a 175-pound athlete in a vigorous training program can consume 4000 calories per day.

77. **[2.4]** Let $x =$ the length of the Brooklyn Bridge. Then $x + 2604 =$ the length of the Golden Gate Bridge.

$$x + (x + 2604) = 5796$$
$$2x + 2604 = 5796$$
$$2x = 3192$$
$$x = 1596$$

Since $x = 1596$, $x + 2604 = 4200$.

The length of the Brooklyn Bridge is 1596 feet and that of the Golden Gate Bridge is 4200 feet.

78. [2.6]

Size	Unit Cost (dollars per oz)
50 oz	$\frac{\$4.69}{50} = \$0.094$
100 oz	$\frac{\$5.98}{100} = \$0.060$ (*)
200 oz	$\frac{\$13.68}{200} = \$0.068$

The 100-oz size is the best buy.

79. [2.6] Let $x =$ the number of quarts of oil needed for 192 quarts of gasoline.
Set up a proportion with one ratio involving oil and the other involving gasoline.

$$\frac{x \text{ quarts}}{1 \text{ quart}} = \frac{192 \text{ quarts}}{24 \text{ quarts}}$$

$$\begin{aligned} x \cdot 24 &= 1 \cdot 192 && \textit{Cross products} \\ x &= 8 && \textit{Divide by 24.} \end{aligned}$$

The amount of oil needed is 8 quarts.

80. [2.7] Let $x =$ the rate of the slower train.
Then $x + 30 =$ the rate of the faster train.

	r	t	d
Slower train	x	3	$3x$
Faster train	$x + 30$	3	$3(x+30)$

The sum of the distances traveled by the two trains is 390 miles, so

$$\begin{aligned} 3x + 3(x + 30) &= 390. \\ 3x + 3x + 90 &= 390 \\ 6x + 90 &= 390 \\ 6x &= 300 \\ x &= 50 \end{aligned}$$

Since $x = 50$, $x + 30 = 80$.

The rate of the slower train is 50 miles per hour and the rate of the faster train is 80 miles per hour.

81. [2.5] Let $x =$ the length of the first side.
Then $2x =$ the length of the second side.

Use the formula for the perimeter of a triangle, $P = a + b + c$, with perimeter 96 and third side 30.

$$\begin{aligned} x + 2x + 30 &= 96 \\ 3x + 30 &= 96 \\ 3x &= 66 \\ x &= 22 \end{aligned}$$

The sides have lengths 22 meters, 44 meters, and 30 meters. The length of the longest side is 44 meters.

82. [2.8] Let $s =$ the length of a side of the square.
The formula for the perimeter of a square is $P = 4s$.

The perimeter cannot be greater than 200.

$$\begin{array}{ccc} \downarrow & \downarrow & \downarrow \\ 4s & \le & 200 \end{array}$$

$$\begin{aligned} 4s &\le 200 \\ s &\le 50 \end{aligned}$$

The length of a side is 50 meters or less.

Chapter 2 Test

1.
$$\begin{aligned} 5x + 9 &= 7x + 21 \\ -2x + 9 &= 21 && \textit{Subtract 7x.} \\ -2x &= 12 && \textit{Subtract 9.} \\ x &= -6 && \textit{Divide by -2.} \end{aligned}$$

The solution set is $\{-6\}$.

2.
$$\begin{aligned} -\tfrac{4}{7}x &= -12 \\ \left(-\tfrac{7}{4}\right)\left(-\tfrac{4}{7}x\right) &= \left(-\tfrac{7}{4}\right)(-12) \\ x &= 21 \end{aligned}$$

The solution set is $\{21\}$.

3.
$$\begin{aligned} 7 - (x - 4) &= -3x + 2(x + 1) \\ 7 - x + 4 &= -3x + 2x + 2 \\ -x + 11 &= -x + 2 \\ 11 &= 2 \end{aligned}$$

Because the last statement is false, the equation has no solution set, symbolized by $\emptyset$.

4. $0.6(x + 20) + 0.8(x - 10) = 46$
To clear decimals, multiply both sides by 10.

$$\begin{aligned} 10[0.6(x + 20) + 0.8(x - 10)] &= 10(46) \\ 6(x + 20) + 8(x - 10) &= 460 \\ 6x + 120 + 8x - 80 &= 460 \\ 14x + 40 &= 460 \\ 14x &= 420 \\ x &= 30 \end{aligned}$$

The solution set is $\{30\}$.

5.
$$\begin{aligned} -8(2x + 4) &= -4(4x + 8) \\ -16x - 32 &= -16x - 32 \end{aligned}$$

Because the last statement is true, the solution set is {all real numbers}.

6. Let $x =$ the number of games the Angels lost.
Then $2x - 24 =$ the number of games the Angels won.

The total number of games played was 162.

$$\begin{aligned} x + (2x - 24) &= 162 \\ 3x - 24 &= 162 \\ 3x &= 186 \\ x &= 62 \end{aligned}$$

Since $x = 62$, $2x - 24 = 100$.
The Angels won 100 games and lost 62 games.

7. Let x = the area of Kauai (in square miles).
Then $x + 177$ = the area of Maui
(in square miles), and
$(x + 177) + 3293 = x + 3470$ = the area of Hawaii.

$$x + (x + 177) + (x + 3470) = 5300$$
$$3x + 3647 = 5300$$
$$3x = 1653$$
$$x = 551$$

Since $x = 551$, $x + 177 = 728$, and $x + 3470 = 4021$.

The area of Hawaii is 4021 square miles, the area of Maui is 728 square miles, and the area of Kauai is 551 square miles.

8. Let x = the measure of the angle.
Then $90 - x$ = the measure of its complement, and $180 - x$ = the measure of its supplement.

$$180 - x = 3(90 - x) + 10$$
$$180 - x = 270 - 3x + 10$$
$$180 - x = 280 - 3x$$
$$180 + 2x = 280$$
$$2x = 100$$
$$x = 50$$

The measure of the angle is $50°$. The measure of its supplement, $130°$, is $10°$ more than three times its complement, $40°$.

9. **(a)** Solve $P = 2L + 2W$ for W.

$$P - 2L = 2W$$
$$\frac{P - 2L}{2} = W \text{ or } W = \frac{P - 2L}{2}$$

(b) Substitute 116 for P and 40 for L in the formula obtained in part (a).

$$W = \frac{P - 2L}{2}$$
$$= \frac{116 - 2(40)}{2}$$
$$= \frac{116 - 80}{2} = \frac{36}{2} = 18$$

10. The angles are vertical angles, so their measures are equal.

$$3x + 15 = 4x - 5$$
$$15 = x - 5$$
$$20 = x$$

Since $x = 20$, $3x + 15 = 75$ and $4x - 5 = 75$.

Both angles have measure $75°$.

11. $\dfrac{z}{8} = \dfrac{12}{16}$

$$16z = 8(12) \quad \textit{Cross products are equal.}$$
$$16z = 96$$
$$\frac{16z}{16} = \frac{96}{16} \quad \textit{Divide by 16.}$$
$$z = 6$$

The solution set is $\{6\}$.

12. $\dfrac{x + 5}{3} = \dfrac{x - 3}{4}$

$$4(x + 5) = 3(x - 3)$$
$$4x + 20 = 3x - 9$$
$$x + 20 = -9$$
$$x = -29$$

The solution set is $\{-29\}$.

13. The results in the following table are rounded to the nearest thousandth.

Size	Unit Cost (dollars per oz)
8 oz	$\frac{\$2.79}{8} = \$0.349$
16 oz	$\frac{\$4.99}{16} = \$0.312$
32 oz	$\frac{\$7.99}{32} = \$0.250$ $(*)$

The best buy is the 32-oz size.

14. Let x = the actual distance between Seattle and Cincinnati.

$$\frac{x \text{ miles}}{1050 \text{ miles}} = \frac{92 \text{ inches}}{42 \text{ inches}}$$
$$42x = 92(1050) = 96{,}600$$
$$x = \frac{96{,}600}{42} = 2300$$

The actual distance between Seattle and Cincinnati is 2300 miles.

15. Let x = the amount invested at 3%.
Then $x + 6000$ = the amount invested at 4.5%.

Amount invested (in dollars)	Rate of interest	Interest for one year
x	0.03	$0.03x$
$x + 6000$	0.045	$0.045(x + 6000)$

$$0.03x + 0.045(x + 6000) = 870$$
$$1000(0.03x) + 1000[0.045(x + 6000)] = 1000(870)$$
$$30x + 45(x + 6000) = 870{,}000$$
$$30x + 45x + 270{,}000 = 870{,}000$$
$$75x + 270{,}000 = 870{,}000$$
$$75x = 600{,}000$$
$$x = 8000$$

Since $x = 8000$, $x + 6000 = 14{,}000$.

Carlos invested \$8000 at 3% and \$14,000 at 4.5%.

16. Use the formula $d = rt$ and let t be the number of hours they traveled.

	r	t	d
First car	50	t	$50t$
Second car	65	t	$65t$

First car's distance	and	second car's distance
↓	↓	↓
$50t$	$+$	$65t$

is	total distance.
↓	↓
$=$	460

$$50t + 65t = 460$$
$$115t = 460$$
$$t = 4$$

The two cars will be 460 miles apart in 4 hours.

17. $-4x + 2(x - 3) \geq 4x - (3 + 5x) - 7$

$-4x + 2x - 6 \geq 4x - 3 - 5x - 7$

$-2x - 6 \geq -x - 10$

$-x - 6 \geq -10$

$-x \geq -4$

$\dfrac{-1x}{-1} \leq \dfrac{-4}{-1}$ *Divide by –1;*

reverse the symbol

$x \leq 4$

Graph the solution set $(-\infty, 4]$.

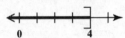

18. $-10 < 3x - 4 \leq 14$

$-6 < 3x \leq 18$ *Add 4.*

$-2 < x \leq 6$ *Divide by 3.*

Graph the solution set $(-2, 6]$.

19. Let $x =$ the score on the third test.

The average of the three tests	is at least	80.
↓	↓	↓
$\dfrac{76 + 81 + x}{3}$	$\geq$	80

$$\frac{157 + x}{3} \geq 80$$

$$3\left(\frac{157 + x}{3}\right) \geq 3(80)$$

$$157 + x \geq 240$$

$$x \geq 83$$

In order to average at least 80, Susan's score on her third test must be 83 or more.

20. When an inequality is multiplied or divided by a negative number, the direction of the inequality symbol must be reversed.

Cumulative Review Exercises (Chapters 1–2)

1. $\dfrac{5}{6} + \dfrac{1}{4} - \dfrac{7}{15} = \dfrac{50}{60} + \dfrac{15}{60} - \dfrac{28}{60}$

$= \dfrac{65 - 28}{60}$

$= \dfrac{37}{60}$

2. $\dfrac{9}{8} \cdot \dfrac{16}{3} \div \dfrac{5}{8} = \dfrac{9}{8} \cdot \dfrac{16}{3} \cdot \dfrac{8}{5}$

$= \dfrac{3 \cdot 3 \cdot 16 \cdot 8}{8 \cdot 3 \cdot 5}$

$= \dfrac{48}{5}$

3. "The difference between half a number and 18" is written

$$\tfrac{1}{2}x - 18.$$

4. "The quotient of 6 and 12 more than a number is 2" is written

$$\frac{6}{x + 12} = 2.$$

5. $\dfrac{8(7) - 5(6 + 2)}{3 \cdot 5 + 1} \geq 1$

$\dfrac{8(7) - 5(8)}{3 \cdot 5 + 1} \geq 1$

$\dfrac{56 - 40}{15 + 1} \geq 1$

$\dfrac{16}{16} \geq 1$

$1 \geq 1$

The statement is *true*.

6. $\dfrac{-4(9)(-2)}{-3^2} = \dfrac{-36(-2)}{-1 \cdot 3^2}$

$= \dfrac{72}{-9}$

$= -8$

7. $(-7 - 1)(-4) + (-4) = (-8)(-4) + (-4)$

$= 32 + (-4)$

$= 28$

8. $\dfrac{3x^2 - y^3}{-4z} = \dfrac{3(-2)^2 - (-4)^3}{-4(3)}$ *Let $x = -2$,*

$y = -4, z = 3$.

$= \dfrac{3(4) - (-64)}{-12}$

$= \dfrac{12 + 64}{-12}$

$= \dfrac{76}{-12}$

$= -\dfrac{19}{3}$

9. $7(p + q) = 7p + 7q$

The multiplication of 7 is distributed over the sum, which illustrates the distributive property.

10. $3 + (5 + 2) = 3 + (2 + 5)$

The order of the numbers added in the parentheses is changed, which illustrates the commutative property.

11. $2r - 6 = 8r$
$\quad -6 = 6r$
$\quad -1 = r$

Check $r = -1$: $-8 = -8$ *True*

The solution set is $\{-1\}$.

12. $4 - 5(s + 2) = 3(s + 1) - 1$
$4 - 5s - 10 = 3s + 3 - 1$
$\quad -5s - 6 = 3s + 2$
$\quad -8s - 6 = 2$
$\quad -8s = 8$
$\quad s = -1$

Check $s = -1$: $-1 = -1$ *True*

The solution set is $\{-1\}$.

13. $\frac{2}{3}x + \frac{3}{4}x = -17$
$12\left(\frac{2}{3}x + \frac{3}{4}x\right) = 12(-17)$ *LCD = 12*
$8x + 9x = -204$
$17x = -204$
$x = -12$

Check $x = -12$: $-17 = -17$ *True*

The solution set is $\{-12\}$.

14. $\dfrac{2x + 3}{5} = \dfrac{x - 4}{2}$
$(2x + 3)(2) = (5)(x - 4)$
$4x + 6 = 5x - 20$
$6 = x - 20$
$26 = x$

Check $x = 26$: $11 = 11$ *True*

The solution set is $\{26\}$.

15. Solve $3x + 4y = 24$ for y.
$4y = 24 - 3x$
$y = \dfrac{24 - 3x}{4}$

16. $6(r - 1) + 2(3r - 5) \le -4$
$6r - 6 + 6r - 10 \le -4$
$12r - 16 \le -4$
$12r \le 12$
$r \le 1$

Graph the solution set $(-\infty, 1]$.

17. $-18 \le -9z < 9$
$2 \ge z > -1$ *Divide by –9; reverse the symbols.*
or $-1 < z \le 2$

Graph the solution set $(-1, 2]$.

18. Let $x =$ the length of the middle-sized piece. Then $3x =$ the length of the longest piece, and $x - 5 =$ the length of the shortest piece.

$x + 3x + (x - 5) = 40$
$5x - 5 = 40$
$5x = 45$
$x = 9$

The length of the middle-sized piece is 9 centimeters, of the longest piece is 27 centimeters, and of the shortest piece is 4 centimeters.

19. Let $r =$ the radius and use 3.14 for π. Using the formula for circumference, $C = 2\pi r$, and $C = 78$, we have

$2\pi r = 78.$
$r = \dfrac{78}{2\pi} \approx 12.4204$

To the nearest hundredth, the radius is 12.42 cm.

20. Let $x =$ rate of slower car. Then $x + 20 =$ rate of faster car.

Use the formula $d = rt$.

$d_{\text{slower}} + d_{\text{faster}} = d_{\text{total}}$
$(x)(4) + (x + 20)(4) = 400$
$4x + 4x + 80 = 400$
$8x + 80 = 400$
$8x = 320$
$x = 40$

The rates are 40 mph and 60 mph.

CHAPTER 3 LINEAR EQUATIONS AND INEQUALITIES IN TWO VARIABLES; FUNCTIONS

3.1 Linear Equations in Two Variables; The Rectangular Coordinate System

3.1 Now Try Exercises

N1. **(a)** Move up from 2006 on the horizontal scale to the point plotted for 2006. Looking across at the vertical scale, this point is close to the line on the vertical scale for $2.60. So, it cost about $2.60 for a gallon of gasoline in 2006.

(b) As in part (a), we see that the point for 2008 is about one-fourth of the way between the lines on the vertical scale for $3.20 and $3.40. Halfway between the lines for $3.20 and $3.40 would be $3.30. So, it cost about $3.25 for a gallon of gasoline in 2008. The increase is $3.25 − $2.60 = $0.65.

N2. **(a)** $(3, 4)$

$$3x - 7y = 19$$
$$3(3) - 7(4) \stackrel{?}{=} 19 \quad \textit{Let x = 3, y = 4.}$$
$$9 - 28 \stackrel{?}{=} 19$$
$$-19 = 19 \quad \textit{False}$$

No, $(3, 4)$ is not a solution.

(b) $(-3, -4)$

$$3x - 7y = 19$$
$$3(-3) - 7(-4) \stackrel{?}{=} 19 \quad \textit{Let x = − 3, y = − 4.}$$
$$-9 + 28 \stackrel{?}{=} 19$$
$$19 = 19 \quad \textit{True}$$

Yes, $(-3, -4)$ is a solution.

N3. **(a)** In the ordered pair $(4, \_\_)$, $x = 4$.

$$y = 3x - 12$$
$$y = 3(4) - 12 \quad \textit{Let x = 4.}$$
$$y = 12 - 12$$
$$y = 0$$

The ordered pair is $(4, 0)$.

(b) In the ordered pair $(\_\_, 3)$, $y = 3$. Find the corresponding value of x by replacing y with 3 in the given equation.

$$y = 3x - 12$$
$$3 = 3x - 12 \quad \textit{Let y = 3.}$$
$$15 = 3x \quad \textit{Add 12.}$$
$$5 = x \quad \textit{Divide by 3.}$$

The ordered pair is $(5, 3)$.

N4. To complete the first ordered pair, let $x = 0$.

$$5x - 4y = 20$$
$$5(0) - 4y = 20 \quad \textit{Let x = 0.}$$
$$-4y = 20$$
$$y = -5$$

The first ordered pair is $(0, -5)$.

To complete the second ordered pair, let $y = 0$.

$$5x - 4y = 20$$
$$5x - 4(0) = 20 \quad \textit{Let y = 0.}$$
$$5x = 20$$
$$x = 4$$

The second ordered pair is $(4, 0)$.

To complete the third ordered pair, let $x = 2$.

$$5x - 4y = 20$$
$$5(2) - 4y = 20 \quad \textit{Let x = 2.}$$
$$10 - 4y = 20$$
$$-4y = 10$$
$$y = -\frac{5}{2}$$

The third ordered pair is $(2, -\frac{5}{2})$.

The completed table of ordered pairs follows.

x	y
0	-5
4	0
2	$-\frac{5}{2}$

N5. To plot the ordered pairs, start at the origin in each case.

(a) $(-3, 1)$

Go 3 units to the left along the x-axis; then go up 1 unit, parallel to the y-axis.

(b) $(2, -4)$

Go 2 units to the right along the x-axis; then go down 4 units.

(c) $(0, -1)$

Go down 1 unit. This point is on the y-axis since the x-coordinate is 0.

(d) $(\frac{5}{2}, 3)$

Go 2.5 units to the right; then go up 3 units.

(e) $(-4, -3)$

Go 4 units to the left; then go down 3 units.

(f) $(-4, 0)$

Go 4 units to the left. This point is on the x-axis since the y-coordinate is 0.

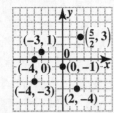

N6. To find y when $x = 2004$, substitute 2004 for x into the equation and use a calculator.

$$y = 3.049x - 5979.0$$
$$y = 3.049(2004) - 5979.0$$
$$y \approx 131$$

There were about 131,000 twin births in the United States in 2004.

3.1 Section Exercises

1. The line segments between 2003 and 2004, 2004 and 2005, and 2005 and 2006 fall, so these are the pairs of consecutive years in which the unemployment rate decreased.

3. For 2003, the unemployment rate was about 6%. For 2004, the unemployment rate was about 5.5%. The decline is $6\% - 5.5\% = 0.5\%$.

5. The symbol (x, y) *does* represent an ordered pair, while the symbols $[x, y]$ and $\{x, y\}$ *do not* represent ordered pairs. (Note that only parentheses are used to write ordered pairs.)

7. All points having coordinates in the form

(negative, positive)

are in quadrant II, so the point whose graph has coordinates $(-4, 2)$ is in quadrant II.

9. All ordered pairs that are solutions of the equation $y = 3$ have y-coordinates equal to 3, so the ordered pair $(4, 3)$ is a solution of the equation $y = 3$.

11. $x + y = 8$; $(0, 8)$

To determine whether $(0, 8)$ is a solution of the given equation, substitute 0 for x and 8 for y.

$$x + y = 8$$
$$0 + 8 \stackrel{?}{=} 8 \quad \text{Let } x = 0, y = 8.$$
$$8 = 8 \quad \text{True}$$

The result is true, so $(0, 8)$ is a solution of the given equation $x + y = 8$.

13. $2x + y = 5$; $(3, -1)$

Substitute 3 for x and -1 for y.

$$2x + y = 5$$
$$2(3) + (-1) \stackrel{?}{=} 5 \quad \text{Let } x = 3, y = -1.$$
$$6 - 1 \stackrel{?}{=} 5$$
$$5 = 5 \quad \text{True}$$

The result is true, so $(3, -1)$ is a solution of $2x + y = 5$.

15. $5x - 3y = 15$; $(5, 2)$

Substitute 5 for x and 2 for y.

$$5x - 3y = 15$$
$$5(5) - 3(2) \stackrel{?}{=} 15 \quad \text{Let } x = 5, y = 2.$$
$$25 - 6 \stackrel{?}{=} 15$$
$$19 = 15 \quad \text{False}$$

The result is false, so $(5, 2)$ is not a solution of $5x - 3y = 15$.

17. $x = -4y$; $(-8, 2)$

Substitute -8 for x and 2 for y.

$$x = -4y$$
$$-8 \stackrel{?}{=} -4(2) \quad \text{Let } x = -8, y = 2.$$
$$-8 = -8 \quad \text{True}$$

The result is true, so $(-8, 2)$ is a solution of $x = -4y$.

19. $y = 2$; $(4, 2)$

Since x does not appear in the equation, we just substitute 2 for y.

$$y = 2$$
$$2 = 2 \quad \text{Let } y = 2; \text{ true}$$

The result is true, so $(4, 2)$ is a solution of $y = 2$.

21. $x - 6 = 0$; $(4, 2)$

Since y does not appear in the equation, we just substitute 4 for x.

$$x - 6 = 0$$
$$4 - 6 \stackrel{?}{=} 0 \quad \text{Let } x = 4.$$
$$-2 = 0 \quad \text{False}$$

The result is false, so $(4, 2)$ is not a solution of $x - 6 = 0$.

23. $y = 2x + 7$; $(5, \underline{\quad})$

In this ordered pair, $x = 5$. Find the corresponding value of y by replacing x with 5 in the given equation.

$$y = 2x + 7$$
$$y = 2(5) + 7 \quad \text{Let } x = 5.$$
$$y = 10 + 7$$
$$y = 17$$

The ordered pair is $(5, 17)$.

25. $y = 2x + 7$; (__ , -3)

In this ordered pair, $y = -3$. Find the corresponding value of x by replacing y with -3 in the given equation.

$$y = 2x + 7$$
$$-3 = 2x + 7 \quad \textit{Let y = -3.}$$
$$-10 = 2x$$
$$-5 = x$$

The ordered pair is $(-5, -3)$.

27. $y = -4x - 4$; (__ , 0)

$$y = -4x - 4$$
$$0 = -4x - 4 \quad \textit{Let y = 0.}$$
$$4 = -4x$$
$$-1 = x$$

The ordered pair is $(-1, 0)$.

29. $y = -4x - 4$; (__ , 24)

$$y = -4x - 4$$
$$24 = -4x - 4 \quad \textit{Let y = 24.}$$
$$28 = -4x$$
$$-7 = x$$

The ordered pair is $(-7, 24)$.

31. $4x + 3y = 24$

If $x = 0$,

$$4(0) + 3y = 24$$
$$0 + 3y = 24$$
$$y = 8.$$

If $y = 0$,

$$4x + 3(0) = 24$$
$$4x + 0 = 24$$
$$x = 6.$$

If $y = 4$,

$$4x + 3(4) = 24$$
$$4x + 12 = 24$$
$$4x = 12$$
$$x = 3.$$

The completed table of values is shown below.

x	y	ordered pair
0	8	$(0, 8)$
6	0	$(6, 0)$
3	4	$(3, 4)$

33. $4x - 9y = -36$

If $y = 0$,

$$4x - 9(0) = -36$$
$$4x - 0 = -36$$
$$4x = -36$$
$$x = -9.$$

If $x = 0$,

$$4(0) - 9y = -36$$
$$0 - 9y = -36$$
$$-9y = -36$$
$$y = 4.$$

If $y = 8$,

$$4x - 9(8) = -36$$
$$4x - 72 = -36$$
$$4x = 36$$
$$x = 9.$$

The completed table of values is shown below.

x	y	ordered pair
-9	0	$(-9, 0)$
0	4	$(0, 4)$
9	8	$(9, 8)$

35. $x = 12$

No matter which value of y is chosen, the value of x will always be 12. Each ordered pair can be completed by placing 12 in the first position.

x	y	ordered pair
12	3	$(12, 3)$
12	8	$(12, 8)$
12	0	$(12, 0)$

37. $y = -10$

No matter which value of x is chosen, the value of y will always be -10. Each ordered pair can be completed by placing -10 in the second position.

x	y	ordered pair
4	-10	$(4, -10)$
0	-10	$(0, -10)$
-4	-10	$(-4, -10)$

39. The given equation, $y + 2 = 0$, may be written $y = -2$. For any value of x, the value of y will always be -2.

x	y	ordered pair
9	-2	$(9, -2)$
2	-2	$(2, -2)$
0	-2	$(0, -2)$

41. The given equation, $x - 4 = 0$, may be written $x = 4$. For any value of y, the value of x will always be 4.

x	y	ordered pair
4	4	$(4, 4)$
4	0	$(4, 0)$
4	-4	$(4, -4)$

43. No, the ordered pair $(3, 4)$ represents the point 3 units to the right of the origin and 4 units up from the x-axis. The ordered pair $(4, 3)$ represents the point 4 units to the right of the origin and 3 units up from the x-axis.

45. Point A was plotted by starting at the origin, going 2 units to the right along the x-axis, and then going up 4 units on a line parallel to the y-axis. The ordered pair for this point is $(2, 4)$, which is in quadrant I.

Point B was plotted by starting at the origin, going 3 units to the left along the x-axis, and then going up 2 units on a line parallel to the y-axis. The ordered pair for this point is $(-3, 2)$, which is in quadrant II.

Point C was plotted by starting at the origin, going 5 units to the left along the x-axis, and then going up 4 units on a line parallel to the y-axis. The ordered pair for this point is $(-5, 4)$, which is in quadrant II.

Point D was plotted by starting at the origin, going 5 units to the left along the x-axis, and then going down 2 units on a line parallel to the y-axis. The ordered pair for this point is $(-5, -2)$, which is in quadrant III.

Point E was plotted by starting at the origin, going 3 units to the right along the x-axis. The ordered pair for this point is $(3, 0)$, which is not in any quadrant.

Point F is located on the y-axis. It was plotted by starting at the origin and going down 2 units along the y-axis. The ordered pair for this point is $(0, -2)$, which is not in any quadrant.

For Exercises 47–58, the ordered pairs are plotted on the graph following the solution for Exercise 57.

47. To plot $(6, 2)$, start at the origin, go 6 units to the right, and then go up 2 units.

49. To plot $(-4, 2)$, start at the origin, go 4 units to the left, and then go up 2 units.

51. To plot $\left(-\frac{4}{5}, -1\right)$, start at the origin, go $\frac{4}{5}$ unit to the left, and then go down 1 unit.

53. To plot $(3, -1.75)$, start at the origin, go 3 units to the right, and then go down 1.75 (or $1\frac{3}{4}$) units.

55. To plot $(0, 4)$, start at the origin and go up 4 units. The point lies on the y-axis.

57. To plot $(4, 0)$, start at the origin and go right 4 units. The point lies on the x-axis.

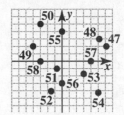

59. The point with coordinates (x, y) is in quadrant III if x is _negative_ and y is _negative_.

61. The point with coordinates (x, y) is in quadrant IV if x is _positive_ and y is _negative_.

63. If $xy < 0$, then either $x < 0$ and $y > 0$ or $x > 0$ and $y < 0$. If $x < 0$ and $y > 0$, then the point lies in quadrant II. If $x > 0$ and $y < 0$, then the point lies in quadrant IV.

65. $x - 2y = 6$

x	y
0	
	0
2	
	-1

Substitute the given values to complete the ordered pairs.

$$x - 2y = 6$$
$$0 - 2y = 6 \quad \textit{Let x = 0.}$$
$$-2y = 6$$
$$y = -3$$

$$x - 2y = 6$$
$$x - 2(0) = 6 \quad \textit{Let y = 0.}$$
$$x - 0 = 6$$
$$x = 6$$

$$x - 2y = 6$$
$$2 - 2y = 6 \quad \textit{Let x = 2.}$$
$$-2y = 4$$
$$y = -2$$

$$x - 2y = 6$$
$$x - 2(-1) = 6 \quad \textit{Let y = -1.}$$
$$x + 2 = 6$$
$$x = 4$$

The completed table of values follows.

x	y
0	-3
6	0
2	-2
4	-1

Plot the points $(0, -3)$, $(6, 0)$, $(2, -2)$, and $(4, -1)$ on a coordinate system.

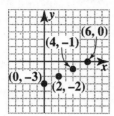

67. $3x - 4y = 12$

x	y
0	
	0
-4	
	-4

Substitute the given values to complete the ordered pairs.

$$3x - 4y = 12$$
$$3(0) - 4y = 12 \qquad \textit{Let x = 0.}$$
$$0 - 4y = 12$$
$$-4y = 12$$
$$y = -3$$

$$3x - 4y = 12$$
$$3x - 4(0) = 12 \qquad \textit{Let y = 0.}$$
$$3x - 0 = 12$$
$$3x = 12$$
$$x = 4$$

$$3x - 4y = 12$$
$$3(-4) - 4y = 12 \qquad \textit{Let x = -4.}$$
$$-12 - 4y = 12$$
$$-4y = 24$$
$$y = -6$$

$$3x - 4y = 12$$
$$3x - 4(-4) = 12 \qquad \textit{Let y = -4.}$$
$$3x + 16 = 12$$
$$3x = -4$$
$$x = -\frac{4}{3}$$

The completed table is as follows.

x	y
0	-3
4	0
-4	-6
$-\frac{4}{3}$	-4

Plot the points $(0, -3)$, $(4, 0)$, $(-4, -6)$, and $\left(-\frac{4}{3}, -4\right)$ on a coordinate system.

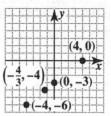

69. The given equation, $y + 4 = 0$, can be written as $y = -4$. So regardless of the value of x, the value of y is -4.

x	y
0	-4
5	-4
-2	-4
-3	-4

Plot the points $(0, -4)$, $(5, -4)$, $(-2, -4)$, and $(-3, -4)$ on a coordinate system.

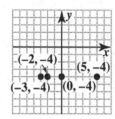

71. The points in each graph appear to lie on a straight line.

73. **(a)** When $x = 5$, $y = 45$. The ordered pair is $(5, 45)$.

     (b) When $y = 50$, $x = 6$. The ordered pair is $(6, 50)$.

75. **(a)** We can write the results from the table as ordered pairs (x, y).

     $(2002, 31.6)$, $(2003, 30.1)$, $(2004, 29.0)$, $(2005, 27.5)$, $(2006, 26.6)$, $(2007, 26.9)$

     (b) $(2007, 26.9)$ means that 26.9 percent of 2-year college students in 2007 received a degree within 3 years.

     (c)

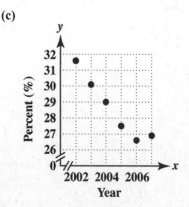

(d) With the exception of the point for 2007, the points lie approximately on a straight line. Rates at which 2-year college students complete a degree within 3 years were generally decreasing.

77. (a) Substitute $x = 20, 40, 60, 80$ in the equation $y = -0.65x + 143$.

$$y = -0.65(20) + 143 = -13 + 143 = 130$$
$$y = -0.65(40) + 143 = -26 + 143 = 117$$
$$y = -0.65(60) + 143 = -39 + 143 = 104$$
$$y = -0.65(80) + 143 = -52 + 143 = 91$$

The completed table follows.

Age	Heartbeats (per minute)
20	130
40	117
60	104
80	91

(b) In the ordered pairs (x, y), x represents age and y represents heartbeats. $(20, 130)$, $(40, 117)$, $(60, 104)$, $(80, 91)$

(c)

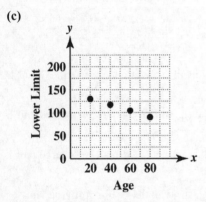

Yes, the points lie in a linear pattern.

79. Refer to the tables and locate age 20 and age 40 in each table. For age 20, the target heart rate is between 130 and 170. For age 40, the target heart rate is between 117 and 153.

81. $3x + 6 = 0$
$$3x = -6$$
$$x = \frac{-6}{3} = -2$$

Check $x = -2$: $-6 + 6 = 0$

The solution set is $\{-2\}$.

83. $9 - x = -4$
$$-x = -13$$
$$x = 13$$

Check $x = 13$: $9 - 13 = -4$
The solution set is $\{13\}$.

3.2   Graphing Linear Equations in Two Variables

3.2   Now Try Exercises

N1. $2x - 4y = 8$

Find the x-intercept by letting $y = 0$.

$$2x - 4y = 8$$
$$2x - 4(0) = 8 \quad \textit{Let y = 0.}$$
$$2x - 0 = 8$$
$$2x = 8$$
$$x = 4$$

The x-intercept is $(4, 0)$.

Find the y-intercept by letting $x = 0$.

$$2x - 4y = 8$$
$$2(0) - 4y = 8 \quad \textit{Let x = 0.}$$
$$0 - 4y = 8$$
$$-4y = 8$$
$$y = -2$$

The y-intercept is $(0, -2)$.

Get a third point as a check. For example, choosing $x = -2$ gives $y = -3$. Plot $(4, 0)$, $(0, -2)$, and $(-2, -3)$ and draw a line through them.

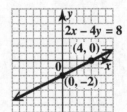

N2. $y = \frac{1}{3}x + 1$

Find the x-intercept by letting $y = 0$.

$$y = \frac{1}{3}x + 1$$
$$0 = \frac{1}{3}x + 1 \quad \textit{Let y = 0.}$$
$$-1 = \frac{1}{3}x$$
$$3(-1) = 3\left(\frac{1}{3}x\right) \quad \textit{Multiply by 3.}$$
$$-3 = x$$

The x-intercept is $(-3, 0)$.

Find the y-intercept by letting $x = 0$.

$$y = \frac{1}{3}x + 1$$
$$y = \frac{1}{3}(0) + 1 \quad \textit{Let x = 0.}$$
$$y = 0 + 1$$
$$y = 1$$

The y-intercept is $(0, 1)$.

Get a third point as a check. For example, choosing $x = 3$ gives $y = 2$. Plot $(-3, 0)$, $(0, 1)$, and $(3, 2)$ and draw a line through them.

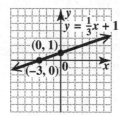

N3. $x + 2y = 2$

Find the x-intercept by letting $y = 0$.

$$x + 2y = 2$$
$$x + 2(0) = 2 \quad \textit{Let y = 0.}$$
$$x + 0 = 2$$
$$x = 2$$

The x-intercept is $(2, 0)$.

Find the y-intercept by letting $x = 0$.

$$x + 2y = 2$$
$$0 + 2y = 2 \quad \textit{Let x = 0.}$$
$$2y = 2$$
$$y = 1$$

The y-intercept is $(0, 1)$.

Get a third point as a check. For example, choosing $x = 4$ gives $y = -1$. Plot $(2, 0)$, $(0, 1)$, and $(4, -1)$ and draw a line through them.

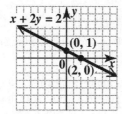

N4. $2x + y = 0$

The x-intercept and the y-intercept are the *same* point, $(0, 0)$. Select two other values for x or y to find two other points on the graph, as in the following table.

x	y
0	0
1	-2
-1	2

Graph the equation by plotting these points and drawing a line through them.

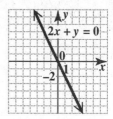

N5. $y = 2$

No matter what value we choose for x, y is always 2, as shown in the table of ordered pairs.

x	y
0	2
-5	2
3	2

The graph is a horizontal line through $(0, 2)$.

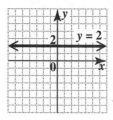

N6. $x + 4 = 0$ or $x = -4$

x is always -4 regardless of the value of y, as shown in the table of ordered pairs.

x	y
-4	0
-4	-5
-4	2

The graph is a vertical line through $(-4, 0)$.

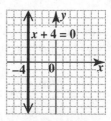

N7. **(a)** $x = 0$ represents 2000, so $x = 6$ represents 2006. On the graph, find 6 on the horizontal axis, move up to the graphed line, then across to the vertical axis. It appears that credit card debt in 2006 was about 875 billion dollars.

(b) To use the equation, substitute 6 for x.

$$y = 32.0x + 684$$
$$y = 32.0(6) + 684 \quad \textit{Let x = 6.}$$
$$y = 876 \text{ billion dollars}$$

3.2 Section Exercises

1. $x + y = 5$

$(0, \_\_), (\_\_, 0), (2, \_\_)$

If $x = 0$,   If $y = 0$,
$0 + y = 5$ $x + 0 = 5$
$\quad y = 5.$ $x = 5.$

If $x = 2$,
$2 + y = 5$
$\quad y = 3.$

The ordered pairs are $(0, 5)$, $(5, 0)$, and $(2, 3)$. Plot the corresponding points and draw a line through them.

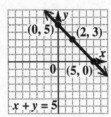

3. $y = \frac{2}{3}x + 1$

$(0, \_\_), (3, \_\_), (-3, \_\_)$

If $x = 0$,   If $x = 3$,
$y = \frac{2}{3}(0) + 1$ $y = \frac{2}{3}(3) + 1$
$y = 0 + 1$ $y = 2 + 1$
$y = 1.$ $y = 3.$

If $x = -3$,
$y = \frac{2}{3}(-3) + 1$
$y = -2 + 1$
$y = -1.$

The ordered pairs are $(0, 1)$, $(3, 3)$, and $(-3, -1)$. Plot the corresponding points and draw a line through them.

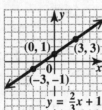

5. $3x = -y - 6$

$(0, \_\_), (\_\_, 0), \left(-\frac{1}{3}, \_\_\right)$

If $x = 0$,   If $y = 0$,
$3(0) = -y - 6$ $3x = -0 - 6$
$0 = -y - 6$ $3x = -6$
$y = -6.$ $x = -2.$

If $x = -\frac{1}{3}$,
$3\left(-\frac{1}{3}\right) = -y - 6$
$-1 = -y - 6$
$y - 1 = -6$
$\quad y = -5.$

The ordered pairs are $(0, -6)$, $(-2, 0)$, and $\left(-\frac{1}{3}, -5\right)$. Plot the corresponding points and draw a line through them.

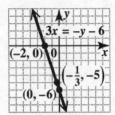

7. **(a)** To determine which equation has y-intercept $(0, -4)$, set x equal to 0 and see which equation is equivalent to $y = -4$. Choice **A** is correct since

$$3x + y = -4$$
$$3(0) + y = -4$$
$$y = -4$$

(b) If the graph of the equation goes through the origin, then substituting 0 for x and 0 for y will result in a true statement. Choice **C** is correct since

$$y = 4x$$
$$0 = 4(0)$$
$$0 = 0$$

is a true statement.

(c) Choice **D** is correct since the graph of $y = 4$ is a horizontal line.

(d) To determine which equation has x-intercept $(4, 0)$, set y equal to 0 and see which equation is equivalent to $x = 4$. Choice **B** is correct since

$$x - 4 = 0$$
$$x = 4$$

9. The dot at 4 on the x-axis is the x-intercept, that is, $(4, 0)$. The dot at -4 on the y-axis is the y-intercept, that is, $(0, -4)$.

11. The dot at -2 on the x-axis is the x-intercept, that is, $(-2, 0)$. The dot at -3 on the y-axis is the y-intercept, that is, $(0, -3)$.

13. To find the x-intercept, let $y = 0$.

$$x - y = 8$$
$$x - 0 = 8$$
$$x = 8$$

The x-intercept is $(8, 0)$.

To find the y-intercept, let $x = 0$.

$$x - y = 8$$
$$0 - y = 8$$
$$-y = 8$$
$$y = -8$$

The y-intercept is $(0, -8)$.

15. To find the x-intercept, let $y = 0$.

$$5x - 2y = 20$$
$$5x - 2(0) = 20$$
$$5x = 20$$
$$x = 4$$

The x-intercept is $(4, 0)$.

To find the y-intercept, let $x = 0$.

$$5x - 2y = 20$$
$$5(0) - 2y = 20$$
$$-2y = 20$$
$$y = -10$$

The y-intercept is $(0, -10)$.

17. To find the x-intercept, let $y = 0$.

$$x + 6y = 0$$
$$x + 6(0) = 0$$
$$x + 0 = 0$$
$$x = 0$$

The x-intercept is $(0, 0)$. Since we have found the point with x equal to 0, this is also the y-intercept.

19. To find the x-intercept, let $y = 0$.

$$y = -2x + 4$$
$$0 = -2x + 4$$
$$2x = 4$$
$$x = 2$$

The x-intercept is $(2, 0)$.

To find the y-intercept, let $x = 0$.

$$y = -2x + 4$$
$$y = -2(0) + 4$$
$$y = 0 + 4$$
$$y = 4$$

The y-intercept is $(0, 4)$.

21. To find the x-intercept, let $y = 0$.

$$y = \tfrac{1}{3}x - 2$$
$$0 = \tfrac{1}{3}x - 2$$
$$2 = \tfrac{1}{3}x$$
$$6 = x$$

The x-intercept is $(6, 0)$.

To find the y-intercept, let $x = 0$.

$$y = \tfrac{1}{3}x - 2$$
$$y = \tfrac{1}{3}(0) - 2$$
$$y = 0 - 2$$
$$y = -2$$

The y-intercept is $(0, -2)$.

23. To find the x-intercept, let $y = 0$.

$$2x - 3y = 0$$
$$2x - 3(0) = 0$$
$$2x - 0 = 0$$
$$2x = 0$$
$$x = 0$$

The x-intercept is $(0, 0)$. Since we have found the point with x equal to 0, this is also the y-intercept.

25. $x - 4 = 0$ is equivalent to $x = 4$. This is an equation of a vertical line. Its x-intercept is $(4, 0)$ and there is no y-intercept.

27. $y = 2.5$ is an equation of a horizontal line. Its y-intercept is $(0, 2.5)$ and there is no x-intercept.

29. **(a)** The graph of $x = -2$ is a vertical line with x-intercept $(-2, 0)$, which is choice **D**.

(b) The graph of $y = -2$ is a horizontal line with y-intercept $(0, -2)$, which is choice **C**.

(c) The graph of $x = 2$ is a vertical line with x-intercept $(2, 0)$, which is choice **B**.

(d) The graph of $y = 2$ is a horizontal line with y-intercept $(0, 2)$, which is choice **A**.

31. Begin by finding the intercepts.

$$x = y + 2 \qquad\qquad x = y + 2$$
$$x = 0 + 2 \ \textit{Let y = 0.} \qquad 0 = y + 2 \ \textit{Let x = 0.}$$
$$x = 2 \qquad\qquad\qquad -2 = y$$

The x-intercept is $(2, 0)$ and the y-intercept is $(0, -2)$. To find a third point, choose $y = 1$.

$$x = y + 2$$
$$x = 1 + 2 \ \textit{Let y = 1.}$$
$$x = 3$$

This gives the ordered pair $(3, 1)$. Plot $(2, 0)$, $(0, -2)$, and $(3, 1)$ and draw a line through them.

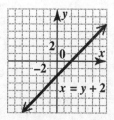

33. Find the intercepts.

$$x - y = 4 \qquad\qquad x - y = 4$$
$$x - 0 = 4 \ \textit{Let y = 0.} \qquad 0 - y = 4 \ \textit{Let x = 0.}$$
$$x = 4 \qquad\qquad\qquad y = -4$$

The x-intercept is $(4, 0)$ and the y-intercept is $(0, -4)$. To find a third point, choose $y = 1$.

$$x - y = 4$$
$$x - 1 = 4 \ \textit{Let y = 1.}$$
$$x = 5$$

This gives the ordered pair $(5, 1)$. Plot $(4, 0)$, $(0, -4)$, and $(5, 1)$ and draw a line through them.

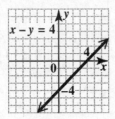

35. Find the intercepts.

$$2x + y = 6 \qquad\qquad 2x + y = 6$$
$$2x + 0 = 6 \ \textit{Let y = 0.} \quad 2(0) + y = 6 \ \textit{Let x = 0.}$$
$$2x = 6 \qquad\qquad\qquad 0 + y = 6$$
$$x = 3 \qquad\qquad\qquad\quad y = 6$$

The x-intercept is $(3, 0)$ and the y-intercept is $(0, 6)$. To find a third point, choose $x = 1$.

$$2x + y = 6$$
$$2(1) + y = 6 \ \textit{Let x = 1.}$$
$$2 + y = 6$$
$$y = 4$$

This gives the ordered pair $(1, 4)$. Plot $(3, 0)$, $(0, 6)$, and $(1, 4)$ and draw a line through them.

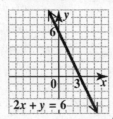

37. Find the intercepts.

$$y = 2x - 5 \qquad\qquad y = 2x - 5$$
$$0 = 2x - 5 \ \textit{Let y = 0.} \quad y = 2(0) - 5 \ \textit{Let x = 0.}$$
$$5 = 2x \qquad\qquad\qquad y = 0 - 5$$
$$\tfrac{5}{2} = x \qquad\qquad\qquad y = -5$$

The x-intercept is $\left(\tfrac{5}{2}, 0\right)$ and the y-intercept is $(0, -5)$. To find a third point, choose $x = 1$.

$$y = 2x - 5$$
$$y = 2(1) - 5 \ \textit{Let x = 1.}$$
$$y = 2 - 5$$
$$y = -3$$

This gives the ordered pair $(1, -3)$. Plot $\left(\tfrac{5}{2}, 0\right)$, $(0, -5)$, and $(1, -3)$ and draw a line through them.

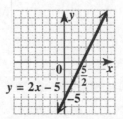

39. Find the intercepts.

$$3x + 7y = 14$$
$$3x + 7(0) = 14 \ \textit{Let y = 0.}$$
$$3x + 0 = 14$$
$$3x = 14$$
$$x = \tfrac{14}{3}$$

$$3x + 7y = 14$$
$$3(0) + 7y = 14 \ \textit{Let x = 0.}$$
$$0 + 7y = 14$$
$$7y = 14$$
$$y = 2$$

The x-intercept is $\left(\tfrac{14}{3}, 0\right)$ and the y-intercept is $(0, 2)$. To find a third point, choose $x = 2$.

$$3x + 7y = 14$$
$$3(2) + 7y = 14 \ \textit{Let x = 2.}$$
$$6 + 7y = 14$$
$$7y = 8$$
$$y = \tfrac{8}{7}$$

This gives the ordered pair $\left(2, \tfrac{8}{7}\right)$. Plot $\left(\tfrac{14}{3}, 0\right)$, $(0, 2)$, and $\left(2, \tfrac{8}{7}\right)$. Writing $\tfrac{14}{3}$ as the mixed number $4\tfrac{2}{3}$ and $\tfrac{8}{7}$ as $1\tfrac{1}{7}$ will be helpful for plotting. Draw a line through these three points.

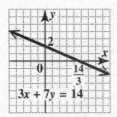

41. $y = -\frac{3}{4}x + 3$

x-intercept $(y = 0)$

$0 = -\frac{3}{4}x + 3$

$\frac{3}{4}x = 3$

$x = \frac{4}{3}(3) = 4$ **$(4, 0)$**

y-intercept $(x = 0)$

$y = -\frac{3}{4}(0) + 3$

$y = 0 + 3$

$y = 3$ **$(0, 3)$**

third point $(x = 2)$

$y = -\frac{3}{4}(2) + 3$

$y = -\frac{3}{2} + 3$

$y = \frac{3}{2}$ **$\left(2, \frac{3}{2}\right)$**

Plot $(4, 0)$, $(0, 3)$, and $\left(2, \frac{3}{2}\right)$ and draw a line through these points.

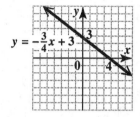

43. $y - 2x = 0$

If $y = 0$, $x = 0$. Both intercepts are the origin, $(0, 0)$. Find two additional points.

$y - 2x = 0$

$y - 2(1) = 0$ *Let x = 1.*

$y - 2 = 0$

$y = 2$

$y - 2x = 0$

$y - 2(-3) = 0$ *Let x = -3.*

$y + 6 = 0$

$y = -6$

Plot $(0, 0)$, $(1, 2)$, and $(-3, -6)$ and draw a line through them.

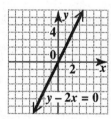

45. $y = -6x$

Find three points on the line.

If $x = 0$, $y = -6(0) = 0$.

If $x = 1$, $y = -6(1) = -6$.

If $x = -1$, $y = -6(-1) = 6$.

Plot $(0, 0)$, $(1, -6)$, and $(-1, 6)$ and draw a line through these points.

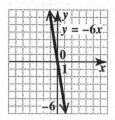

47. $y = -1$

For any value of x, the value of y is -1. Three ordered pairs are $(-4, -1)$, $(0, -1)$, and $(3, -1)$. Plot these points and draw a line through them. The graph is a horizontal line.

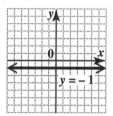

49. $x + 2 = 0$

$x = -2$

For any value of y, the value of x is -2. Three ordered pairs are $(-2, 3)$, $(-2, 0)$, and $(-2, -4)$. Plot these points and draw a line through them. The graph is a vertical line.

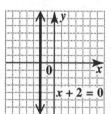

51. $-3y = 15$

$y = \frac{15}{-3} = -5$

For any value of x, the value of y is -5. Three ordered pairs are $(-2, -5)$, $(0, -5)$, and $(1, -5)$. Plot these points and draw a line through them. The graph is a horizontal line.

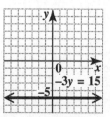

53. $x + 2 = 8$

$x = 6$

For any value of y, the value of x is 6. Three ordered pairs are $(6, -2)$, $(6, 0)$, and $(6, 1)$. Plot these points and draw a line through them. The graph is a vertical line.

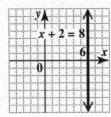

55. $3x = y - 9$

If $x = 0$,	If $y = 0$,
$3(0) = y - 9$	$3x = 0 - 9$
$0 = y - 9$	$3x = -9$
$9 = y.$	$x = -3.$

The graph of this equation is a line with x-intercept $(-3, 0)$ and y-intercept $(0, 9)$.

57. $x - 10 = 1$
$\qquad x = 11$ *Add 10.*

The graph of this equation is a vertical line with x-intercept $(11, 0)$.

59. $3y = -6$
$\qquad y = -2$ *Divide by 3.*

The graph of this equation is a horizontal line with y-intercept $(0, -2)$.

61. $2x = 4y$ can be written as $2x - 4y = 0$, so the graph of this equation passes through the origin, $(0, 0)$, which is the x-intercept and the y-intercept. If we let $x = 2$, then $4 = 4y$, so $y = 1$; and if we let $x = 4$, then $8 = 4y$, so $y = 2$. Thus, the graph also passes through the points $(2, 1)$ and $(4, 2)$.

63. The points $(3, 5)$, $(3, 0)$, and $(3, -3)$ all have x-coordinate 3, so $x = 3$ is the equation of the line.

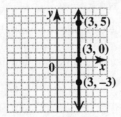

65. The points $(-3, -3)$, $(0, -3)$, and $(4, -3)$ all have y-coordinate -3, so $y = -3$ is the equation of the line.

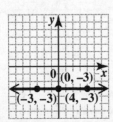

67. **(a)** $y = 5.5x - 220$

$y = 5.5(62) - 220 = 121$ *Let x = 62.*
$y = 5.5(66) - 220 = 143$ *Let x = 66.*
$y = 5.5(72) - 220 = 176$ *Let x = 72.*

The approximate weights of men whose heights are 62 in., 66 in., and 72 in. are 121 lb, 143 lb, and 176 lb, respectively.

(b) The three ordered pairs from part (a) are $(62, 121)$, $(66, 143)$, and $(72, 176)$.

(c) Plot the points $(62, 121)$, $(66, 143)$, and $(72, 176)$ and connect them with a smooth line.

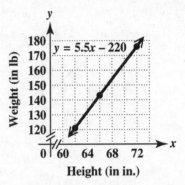

Height (in in.)

(d) Locate 155 on the vertical scale, then move across to the line, then down to the horizontal scale. From the graph, the height of a man who weighs 155 lb is about 68 in. Now substitute 155 for y in the equation.

$$y = 5.5x - 220$$
$$155 = 5.5x - 220$$
$$375 = 5.5x$$
$$x = \frac{375}{5.5} \approx 68.18$$

From the equation, the height is 68 in. to the nearest inch.

69. **(a)** Let $x = 50$.　　　Let $x = 100$.

$y = 0.75x + 25$	$y = 0.75x + 25$
$y = 0.75(50) + 25$	$y = 0.75(100) + 25$
$y = 62.50$	$y = 100$

Thus, 50 posters cost \$62.50 and 100 posters cost \$100.

(b) Let $y = 175$.　　$175 = 0.75x + 25$
$\qquad\qquad\qquad 150 = 0.75x$
$$x = \frac{150}{0.75} = 200$$

Thus, 200 posters cost \$175.

(c) The three ordered pairs from parts (a) and (b) are $(50, 62.50)$, $(100, 100)$, and $(200, 175)$.

(d) Plot the points in part (c) and connect them with a smooth line.

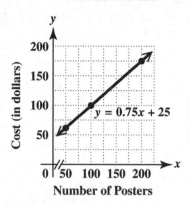

71. **(a)** At $x = 0$, $y = 30,000$.
The initial value of the SUV is $30,000.

(b) At $x = 3$, $y = 15,000$.
$30,000 - 15,000 = 15,000$
The depreciation after the first 3 years is $15,000.

(c) The line connecting consecutive years from 0 to 1, 1 to 2, 2 to 3, and so on, drops 5000 for each segment. Therefore, the annual depreciation in each of the first 5 years is $5000.

(d) $(5, 5000)$ means after 5 years the SUV has a value of $5000.

73. **(a)** $y = 3.018x + 72.52$

For 2000, let $x = 0$.
$y = 3.018(0) + 72.52 = 72.52$

For 2004, let $x = 4$.
$y = 3.018(4) + 72.52 = 84.592$

For 2006, let $x = 6$.
$y = 3.018(6) + 72.52 = 90.628$

The approximate sales for 2000, 2004, and 2006 are $73 billion, $85 billion, and $91 billion, respectively.

(b) Locate 0, 4, and 6 on the horizontal scale, then find the corresponding value on the vertical scale.

The approximate sales for 2000, 2004, and 2006 are $74 billion, $86 billion, and $91 billion, respectively.

(c) The corresponding values are quite close.

75. $\dfrac{4 - 2}{8 - 5} = \dfrac{2}{3}$

77. $\dfrac{-2 - (-4)}{3 - (-1)} = \dfrac{-2 + 4}{3 + 1} = \dfrac{2}{4} = \dfrac{1}{2}$

3.3 The Slope of a Line

3.3 Now Try Exercises

N1. The given points are $(-2, -2)$ and $(-1, 1)$.

$$\text{slope} = \dfrac{\text{change in } y \text{ (rise)}}{\text{change in } x \text{ (run)}}$$
$$= \dfrac{1 - (-2)}{-1 - (-2)} = \dfrac{1 + 2}{-1 + 2} = \dfrac{3}{1} = 3$$

N2. Use the slope formula with $(4, -5)$ and $(-2, -4)$.

$$\text{slope } m = \dfrac{\text{change in } y}{\text{change in } x} = \dfrac{y_2 - y_1}{x_2 - x_1}$$
$$= \dfrac{-4 - (-5)}{-2 - 4} = \dfrac{1}{-6} = -\dfrac{1}{6}$$

N3. Use the slope formula with $(1, -3)$ and $(4, -3)$.

$$\text{slope } m = \dfrac{\text{change in } y}{\text{change in } x} = \dfrac{-3 - (-3)}{4 - 1} = \dfrac{0}{3} = 0$$

N4. Use the slope formula with $(-2, 1)$ and $(-2, -4)$.

$$\text{slope } m = \dfrac{\text{change in } y}{\text{change in } x} = \dfrac{-4 - 1}{-2 - (-2)} = \dfrac{-5}{0},$$

which is *undefined*.

N5. Solve the equation for y.

$$3x + 5y = -1$$
$$5y = -3x - 1 \quad \text{Subtract } 3x.$$
$$y = -\tfrac{3}{5}x - \tfrac{1}{5} \quad \text{Divide by 5.}$$

The slope is given by the coefficient of x, so the slope is $-\tfrac{3}{5}$.

N6. Find the slope of each line by first solving each equation for y.

$$2x - 3y = 1 \qquad\qquad 4x + 6y = 5$$
$$-3y = -2x + 1 \qquad\quad 6y = -4x + 5$$
$$y = \tfrac{2}{3}x - \tfrac{1}{3} \qquad\quad y = -\tfrac{2}{3}x + \tfrac{5}{6}$$
Slope is $\tfrac{2}{3}$. $\qquad\qquad$ Slope is $-\tfrac{2}{3}$.

The slopes are not equal, so the lines are not parallel. The product of the slopes is not -1, so the lines are not perpendicular. Thus, the lines are *neither* parallel nor perpendicular.

3.3 Section Exercises

1. The indicated points have coordinates $(-1, -4)$ and $(1, 4)$.

$$\text{slope} = \dfrac{\text{change in } y \text{ (rise)}}{\text{change in } x \text{ (run)}}$$
$$= \dfrac{4 - (-4)}{1 - (-1)} = \dfrac{8}{2} = 4$$

3. The indicated points have coordinates $(-3, 2)$ and $(5, -2)$.

$$\text{slope} = \dfrac{\text{change in } y \text{ (rise)}}{\text{change in } x \text{ (run)}}$$
$$= \dfrac{-2 - 2}{5 - (-3)} = \dfrac{-4}{8} = -\dfrac{1}{2}$$

5. The indicated points have coordinates $(-2, -4)$ and $(4, -4)$.

$$\text{slope} = \frac{\text{change in } y \text{ (rise)}}{\text{change in } x \text{ (run)}}$$

$$= \frac{-4 - (-4)}{4 - (-2)} = \frac{-4 + 4}{4 + 2} = \frac{0}{6} = 0$$

7. Rise is the vertical change between two different points on a line.

Run is the horizontal change between two different points on a line.

9. **(a)** The indicated points have coordinates $(-1, 2)$ and $(2, 0)$.

$$\text{slope} = \frac{\text{change in } y \text{ (rise)}}{\text{change in } x \text{ (run)}}$$

$$= \frac{0 - 2}{2 - (-1)} = \frac{-2}{3} = -\frac{2}{3} \quad (\text{choice } \mathbf{C})$$

(b) The indicated points have coordinates $(-1, 2)$ and $(-4, 0)$.

$$\text{slope} = \frac{\text{change in } y \text{ (rise)}}{\text{change in } x \text{ (run)}} = \frac{0 - 2}{-4 - (-1)}$$

$$= \frac{-2}{-4 + 1} = \frac{-2}{-3} = \frac{2}{3} \quad (\text{choice } \mathbf{A})$$

(c) The indicated points have coordinates $(-1, 2)$ and $(1, -1)$.

$$\text{slope} = \frac{\text{change in } y \text{ (rise)}}{\text{change in } x \text{ (run)}}$$

$$= \frac{-1 - 2}{1 - (-1)} = \frac{-3}{2} = -\frac{3}{2} \quad (\text{choice } \mathbf{D})$$

(d) The indicated points have coordinates $(-1, 2)$ and $(-3, -1)$.

$$\text{slope} = \frac{\text{change in } y \text{ (rise)}}{\text{change in } x \text{ (run)}}$$

$$= \frac{-1 - 2}{-3 - (-1)} = \frac{-3}{-2} = \frac{3}{2} \quad (\text{choice } \mathbf{B})$$

11. Negative slope

Sketches will vary. The line must fall from left to right. One such line is shown in the following graph.

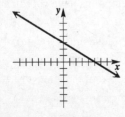

13. Undefined slope

Sketches will vary. The line must be vertical. One such line is shown in the following graph.

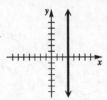

15. **(a)** Because the line *falls* from left to right, its slope is *negative*.

(b) Because the line intersects the y-axis *at* the origin, the y-value of its y-intercept is *zero*.

17. **(a)** Because the line *rises* from left to right, its slope is *positive*.

(b) Because the line intersects the y-axis *below* the origin, the y-value of its y-intercept is *negative*.

19. **(a)** The line is *horizontal*, so its slope is *zero*.

(b) The line intersects the y-axis *below* the origin, so the y-value of its y-intercept is *negative*.

21. The slope (or grade) of the hill is the ratio of the rise to the run, or the ratio of the vertical change to the horizontal change. Since the rise is 32 and the run is 108, the slope is

$$\frac{32}{108} = \frac{8 \cdot 4}{27 \cdot 4} = \frac{8}{27}.$$

23. $$\text{slope} = \frac{\text{vertical change (rise)}}{\text{horizontal change (run)}}$$

$$= \frac{-8}{12} \quad (\text{"drops" indicates negative})$$

$$= -\frac{2}{3}$$

25. Because he found the difference $3 - 5 = -2$ in the numerator, he should have subtracted in the same order in the denominator to get $-1 - 2 = -3$. The correct slope is $\frac{-2}{-3} = \frac{2}{3}$. Note that the student's slope and the correct slope are opposites of one another.

27. Use the slope formula with $(1, -2) = (x_1, y_1)$ and $(-3, -7) = (x_2, y_2)$.

$$\text{slope } m = \frac{\text{change in } y}{\text{change in } x} = \frac{y_2 - y_1}{x_2 - x_1}$$

$$= \frac{-7 - (-2)}{-3 - 1} = \frac{-5}{-4} = \frac{5}{4}$$

29. Use the slope formula with $(0, 3) = (x_1, y_1)$ and $(-2, 0) = (x_2, y_2)$.

$$\text{slope } m = \frac{\text{change in } y}{\text{change in } x} = \frac{y_2 - y_1}{x_2 - x_1}$$

$$= \frac{0 - 3}{-2 - 0} = \frac{-3}{-2} = \frac{3}{2}$$

31. Use the slope formula with $(4, 3) = (x_1, y_1)$ and $(-6, 3) = (x_2, y_2)$.

$$\text{slope } m = \frac{\text{change in } y}{\text{change in } x} = \frac{y_2 - y_1}{x_2 - x_1}$$

$$= \frac{3 - 3}{-6 - 4} = \frac{0}{-10} = 0$$

33. Use the slope formula with $(-2, 4) = (x_1, y_1)$ and $(-3, 7) = (x_2, y_2)$.

$$\text{slope } m = \frac{\text{change in } y}{\text{change in } x} = \frac{y_2 - y_1}{x_2 - x_1}$$

$$= \frac{7 - 4}{-3 - (-2)} = \frac{3}{-1} = -3$$

35. Use the slope formula with $(-12, 3) = (x_1, y_1)$ and $(-12, -7) = (x_2, y_2)$.

$$\text{slope } m = \frac{\text{change in } y}{\text{change in } x} = \frac{y_2 - y_1}{x_2 - x_1}$$

$$= \frac{-7 - 3}{-12 - (-12)} = \frac{-10}{0},$$

which is *undefined*.

37. Use the slope formula with $(4.8, 2.5) = (x_1, y_1)$ and $(3.6, 2.2) = (x_2, y_2)$.

$$\text{slope } m = \frac{\text{change in } y}{\text{change in } x} = \frac{y_2 - y_1}{x_2 - x_1}$$

$$= \frac{2.2 - 2.5}{3.6 - 4.8} = \frac{-0.3}{-1.2} = \frac{1}{4}$$

39. Use the slope formula with $\left(-\frac{7}{5}, \frac{3}{10}\right) = (x_1, y_1)$ and $\left(\frac{1}{5}, -\frac{1}{2}\right) = (x_2, y_2)$.

$$\text{slope } m = \frac{\text{change in } y}{\text{change in } x} = \frac{y_2 - y_1}{x_2 - x_1}$$

$$= \frac{-\frac{1}{2} - \frac{3}{10}}{\frac{1}{5} - \left(-\frac{7}{5}\right)} = \frac{-\frac{5}{10} - \frac{3}{10}}{\frac{1}{5} + \frac{7}{5}} = \frac{-\frac{8}{10}}{\frac{8}{5}}$$

$$= \left(-\frac{8}{10}\right)\left(\frac{5}{8}\right) = -\frac{5}{10} = -\frac{1}{2}$$

41. $y = 5x + 12$

Since the equation is already solved for y, the slope is given by the coefficient of x, which is 5. Thus, the slope of the line is 5.

43. Solve the equation for y.

$$4y = x + 1$$
$$y = \tfrac{1}{4}x + \tfrac{1}{4} \quad \textit{Divide by 4.}$$

The slope of the line is given by the coefficient of x, so the slope is $\frac{1}{4}$.

45. Solve the equation for y.

$$3x - 2y = 3$$
$$-2y = -3x + 3 \quad \textit{Subtract 3x.}$$
$$y = \tfrac{3}{2}x - \tfrac{3}{2} \quad \textit{Divide by } -2.$$

The slope of the line is given by the coefficient of x, so the slope is $\frac{3}{2}$.

47. Solve the equation for y.

$$-3x + 2y = 5$$
$$2y = 3x + 5 \quad \textit{Add 3x.}$$
$$y = \tfrac{3}{2}x + \tfrac{5}{2} \quad \textit{Divide by 2.}$$

The slope of the line is given by the coefficient of x, so the slope is $\frac{3}{2}$.

49. $y = -5$

This is an equation of a horizontal line. Its slope is 0. (This equation may be rewritten in the form $y = 0x - 5$, where the coefficient of x gives the slope.)

51. $x = 6$

This is an equation of a vertical line. Its slope is *undefined.*

53. Solve the equation for y.

$$3x + y = 7$$
$$y = -3x + 7 \quad \textit{Subtract 3x.}$$

The slope of the given line is -3, so the slope of a line whose graph is parallel to the graph of the given line is also -3.

The slope of a line whose graph is perpendicular to the graph of the given line is the negative reciprocal of -3, that is, $\frac{1}{3}$.

55. If two lines are both vertical or both horizontal, they are *parallel.* Choice **A** is correct.

57. Find the slope of each line by solving the equations for y.

$$2x + 5y = 4$$
$$5y = -2x + 4 \quad \textit{Subtract 2x.}$$
$$y = -\tfrac{2}{5}x + \tfrac{4}{5} \quad \textit{Divide by 5.}$$

The slope of the first line is $-\frac{2}{5}$.

$$4x + 10y = 1$$
$$10y = -4x + 1 \quad \textit{Subtract 4x.}$$
$$y = -\tfrac{4}{10}x + \tfrac{1}{10} \quad \textit{Divide by 10.}$$
$$y = -\tfrac{2}{5}x + \tfrac{1}{10} \quad \textit{Lowest terms}$$

The slope of the second line is $-\frac{2}{5}$.

The slopes are equal, so the lines are *parallel.*

59. Find the slope of each line by solving the equations for y.

$$8x - 9y = 6$$
$$-9y = -8x + 6 \quad \textit{Subtract 8x.}$$
$$y = \tfrac{8}{9}x - \tfrac{2}{3} \quad \textit{Divide by -9.}$$

The slope of the first line is $\tfrac{8}{9}$.

$$8x + 6y = -5$$
$$6y = -8x - 5 \quad \textit{Subtract 8x.}$$
$$y \doteq -\tfrac{4}{3}x - \tfrac{5}{6} \quad \textit{Divide by 6.}$$

The slope of the second line is $-\tfrac{4}{3}$.

The slopes are not equal, so the lines are not parallel. The slopes are not negative reciprocals (the negative reciprocal of $\tfrac{8}{9}$ is $-\tfrac{9}{8}$), so the lines are not perpendicular. Thus, the lines are *neither* parallel nor perpendicular.

61. Find the slope of each line by solving the equations for y.

$$3x - 2y = 6$$
$$-2y = -3x + 6 \quad \textit{Subtract 3x.}$$
$$y = \tfrac{3}{2}x - 3 \quad \textit{Divide by -2.}$$

The slope of the first line is $\tfrac{3}{2}$.

$$2x + 3y = 3$$
$$3y = -2x + 3 \quad \textit{Subtract 2x.}$$
$$y = -\tfrac{2}{3}x + 1 \quad \textit{Divide by 3.}$$

The slope of the second line is $-\tfrac{2}{3}$.

The product of the slopes is

$$\tfrac{3}{2}\left(-\tfrac{2}{3}\right) = -1,$$

so the lines are *perpendicular*.

63. Find the slope of each line by solving the equations for y.

$$5x - y = 1$$
$$-y = -5x + 1 \quad \textit{Subtract 5x.}$$
$$y = 5x - 1 \quad \textit{Divide by -1.}$$

The slope of the first line is 5.

$$x - 5y = -10$$
$$-5y = -x - 10 \quad \textit{Subtract x.}$$
$$y = \tfrac{1}{5}x + 2 \quad \textit{Divide by -5.}$$

The slope of the second line is $\tfrac{1}{5}$.

The slopes are not equal, so the lines are not parallel. The slopes are not negative reciprocals (the negative reciprocal of 5 is $-\tfrac{1}{5}$), so the lines are not perpendicular. Thus, the lines are *neither* parallel nor perpendicular.

65. We use the points with coordinates $(1990, 11{,}338)$ and $(2005, 14{,}818)$.

$$m = \frac{14{,}818 \text{ thousand } - 11{,}338 \text{ thousand}}{2005 - 1990}$$
$$= \frac{3480 \text{ thousand}}{15}$$
$$= 232 \text{ thousand} \quad \text{or} \quad 232{,}000.$$

66. The slope of the line in Figure A is *positive*. This means that during the period represented, enrollment *increased* in grades 9–12.

67. The increase is approximately 232,000 students per year.

68. We use the points with coordinates $(1990, 20)$ and $(2007, 3.8)$.

$$m = \frac{3.8 - 20}{2007 - 1990}$$
$$= \frac{-16.2}{17}$$
$$\approx -0.95 \text{ students per computer}$$

69. The slope of the line in Figure B is *negative*. This means that the number of students per computer *decreased* during the period shown.

70. The decrease is 0.95 students per computer per year.

71. **(a)** The ordered pairs that represent music purchases are $(2004, 817)$ and $(2008, 1513)$.

(b) $m = \dfrac{1513 - 817}{2008 - 2004} = \dfrac{696}{4} = 174$

(c) Music purchases increased by 696 million units in 4 years, or 174 million units per year.

73. Two of the ordered pairs have coordinates $(-12, -0.8)$ and $(0, 4)$. Therefore, the slope is

$$m = \frac{4 - (-0.8)}{0 - (-12)} = \frac{4.8}{12} = 0.4.$$

75. From the table, when $X = 0$, $Y_1 = 4$. Therefore, the y-intercept is $(0, 4)$.

77.
$$2x + 5y = 15$$
$$5y = -2x + 15$$
$$y = -\tfrac{2}{5}x + 3$$

79.
$$10x = 30 + 3y$$
$$10x - 30 = 3y$$
$$\tfrac{10}{3}x - 10 = y$$

81.
$$y - (-8) = 2(x - 4)$$
$$y + 8 = 2x - 8$$
$$y = 2x - 16$$

3.4   Writing and Graphing Equations of Lines

3.4 Now Try Exercises

N1. **(a)** $y = -\frac{3}{5}x - 9$

The slope is the coefficient of x, that is, $-\frac{3}{5}$. The y-intercept has x-coordinate 0 and y-coordinate -9, that is, $(0, -9)$.

(b) $y = -\frac{x}{3} + \frac{7}{3}$ or $y = -\frac{1}{3}x + \frac{7}{3}$

The slope is $-\frac{1}{3}$ and the y-intercept is $\left(0, \frac{7}{3}\right)$.

N2. The slope-intercept form of the equation of a line with slope -4 and y-intercept $(0, 2)$ is

$$y = mx + b$$
$$y = -4x + 2.$$

N3. *Step 1*
Begin by solving for y.

$$
\begin{array}{ll}
3x + 2y = 8 & \textit{Given equation} \\
2y = -3x + 8 & \textit{Subtract 3x.} \\
y = -\frac{3}{2}x + 4 & \textit{Divide by 2.}
\end{array}
$$

Step 2
The y-intercept is $(0, 4)$. Graph this point.

Step 3
The slope is $-\dfrac{3}{2} = \dfrac{\text{change in } y \text{ (rise)}}{\text{change in } x \text{ (run)}}$.

Starting at the y-intercept, we count 3 units down and 2 units right to obtain another point on the graph, $(2, 1)$.

Step 4
Draw the line through the points $(0, 4)$ and $(2, 1)$ to obtain the graph of $3x + 2y = 8$.

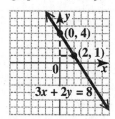

N4. Graph the line through $(-3, -4)$ with slope $\frac{5}{2}$.

To graph the line, remember the slope is

$$m = \frac{\text{change in } y \text{ (rise)}}{\text{change in } x \text{ (run)}} = \frac{5}{2}.$$

Locate $(-3, -4)$. Count 5 units up and 2 units to the right. Draw a line through this point, $(-1, 1)$, and $(-3, -4)$.

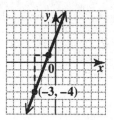

N5.
$$
\begin{array}{ll}
y = mx + b & \textit{slope-intercept form} \\
1 = 3(-2) + b & \textit{Let } x = -2,\, y = 1, \\
& \textit{and } m = 3. \\
1 = -6 + b & \textit{Multiply.} \\
7 = b & \textit{Add 6.}
\end{array}
$$

The y-intercept is $(0, 7)$ and the slope-intercept form of the equation of the line is

$$y = 3x + 7.$$

N6. Write an equation of the line through $(3, -1)$ with slope $-\frac{2}{5}$.

Use the point-slope form of the equation of a line with $x_1 = 3$, $y_1 = -1$, and $m = -\frac{2}{5}$.

$$
\begin{aligned}
y - y_1 &= m(x - x_1) \\
y - (-1) &= -\tfrac{2}{5}(x - 3) \\
y + 1 &= -\tfrac{2}{5}x + \tfrac{6}{5} \\
y &= -\tfrac{2}{5}x + \tfrac{6}{5} - \tfrac{5}{5} \\
y &= -\tfrac{2}{5}x + \tfrac{1}{5}
\end{aligned}
$$

N7. Write an equation of the line through the points $(4, 1)$ and $(6, -2)$.

(a) First, find the slope of the line.

$$m = \frac{-2 - 1}{6 - 4} = \frac{-3}{2} = -\frac{3}{2}$$

Now use the point $(4, 1)$ for (x_1, y_1) and $m = -\frac{3}{2}$ in the point-slope form.

$$
\begin{aligned}
y - y_1 &= m(x - x_1) \\
y - 1 &= -\tfrac{3}{2}(x - 4) \\
y - 1 &= -\tfrac{3}{2}x + 6 \\
y &= -\tfrac{3}{2}x + 7
\end{aligned}
$$

(b)
$$
\begin{array}{ll}
y = -\tfrac{3}{2}x + 7 & \\
2y = -3x + 14 & \textit{Multiply by 2.} \\
3x + 2y = 14 & \textit{Add x.}
\end{array}
$$

N8. Let $(x_1, y_1) = (3, 4645)$ and $(x_2, y_2) = (5, 5491)$.

$$
\begin{aligned}
m &= \frac{y_2 - y_1}{x_2 - x_1} \\
&= \frac{5491 - 4645}{5 - 3} \\
&= \frac{846}{2} \\
&= 423
\end{aligned}
$$

Now use this slope and the point $(3, 4645)$ in the

point-slope form to find an equation of the line.

$$y - y_1 = m(x - x_1)$$
$$y - 4645 = 423(x - 3)$$
$$y - 4645 = 423x - 1269$$
$$y = 423x + 3376$$

For 2007, let $x = 7$.

$$y = 423(7) + 3376$$
$$y = 6337$$

The equation gives $y = 6337$ when $x = 7$, which approximates the cost given in the table, 6185, reasonable well.

3.4 Section Exercises

1. The point-slope form of the equation of a line with slope -2 passing through the point $(4, 1)$ is

$$y - 1 = -2(x - 4).$$

Choice **E** is correct.

3. A line that passes through the points $(0, 0)$ [its y-intercept] and $(4, 1)$ has slope

$$m = \frac{\text{rise}}{\text{run}} = \frac{1 - 0}{4 - 0} = \frac{1}{4}.$$

Its slope-intercept form is

$$y = \tfrac{1}{4}x + 0 \quad \text{or} \quad y = \tfrac{1}{4}x.$$

Choice **B** is correct.

5. $y = x + 3 = 1x + 3$

The graph of this equation is a line with slope 1 and y-intercept $(0, 3)$. The only graph which has a *positive* slope and intersects the y-axis *above* the origin is **C**.

7. $y = x - 3 = 1x - 3$

The graph of this equation is a line with slope 1 and y-intercept $(0, -3)$. The only graph which has a *positive* slope and intersects the y-axis *below* the origin is **A**.

9. $y = \frac{5}{2}x - 4$

The slope is $\frac{5}{2}$ and the y-intercept is $(0, -4)$.

11. $y = -x + 9 \quad \text{or} \quad y = -1x + 9$

The slope is -1 and the y-intercept is $(0, 9)$.

13. $y = \frac{x}{5} - \frac{3}{10} \quad \text{or} \quad y = \frac{1}{5}x - \frac{3}{10}$

The slope is $\frac{1}{5}$ and the y-intercept is $\left(0, -\frac{3}{10}\right)$.

15. The rise is 3 and the run is 1, so the slope is given by

$$m = \frac{\text{rise}}{\text{run}} = \frac{3}{1} = 3.$$

The y-intercept is $(0, -3)$, so $b = -3$. The equation of the line, written in slope-intercept form, is

$$y = 3x - 3.$$

17. Since the line falls from left to right, the "rise" is negative. For this line, the rise is -3 and the run is 3, so the slope is

$$m = \frac{\text{rise}}{\text{run}} = \frac{-3}{3} = -1.$$

The y-intercept is $(0, 3)$, so $b = 3$. The slope-intercept form of the equation of the line is

$$y = -1x + 3$$
$$y = -x + 3.$$

19. Since the line falls from left to right, the "rise" is negative. For this line, the rise is -2 and the run is 4, so the slope is

$$m = \frac{\text{rise}}{\text{run}} = \frac{-2}{4} = -\frac{1}{2}.$$

The y-intercept is $(0, 2)$, so $b = 2$. The slope-intercept form of the equation of the line is

$$y = -\tfrac{1}{2}x + 2.$$

21. $m = 4, (0, -3)$

Since the y-intercept is $(0, -3)$, we have $b = -3$. Use the slope-intercept form.

$$y = mx + b$$
$$y = 4x + (-3)$$
$$y = 4x - 3$$

23. $m = -1, (0, -7)$

Since the y-intercept is $(0, -7)$, we have $b = -7$. Use the slope-intercept form.

$$y = mx + b$$
$$y = -1x - 7$$
$$y = -x - 7$$

25. $m = 0, (0, 3)$

Since the y-intercept is $(0, 3)$, we have $b = 3$. Use the slope-intercept form.

$$y = mx + b$$
$$y = 0x + 3$$
$$y = 3$$

27. Undefined slope, $(0, -2)$

Since the slope is undefined, the line is vertical and has equation $x = k$. Because the line goes through a point $(x, y) = (0, -2)$, we must have $x = 0$.

29. $y = 3x + 2$

The slope is

$$3 = \frac{3}{1} = \frac{\text{change in } y}{\text{change in } x},$$

and the y-intercept is $(0, 2)$. Graph that point and count up 3 units and right 1 unit to get to the point $(1, 5)$. Draw a line through the points.

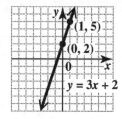

31. $y = -\frac{1}{3}x + 4$

The slope is

$$-\frac{1}{3} = \frac{-1}{3} = \frac{\text{change in } y}{\text{change in } x},$$

and the y-intercept is $(0, 4)$. Graph that point and count down 1 unit and right 3 units to get to the point $(3, 3)$. Draw a line through the points.

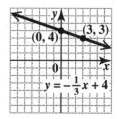

33. $2x + y = -5$, so $y = -2x - 5$

The slope is

$$-2 = \frac{-2}{1} = \frac{\text{change in } y}{\text{change in } x},$$

and the y-intercept is $(0, -5)$. Graph that point and count down 2 units and right 1 unit to get to the point $(1, -7)$. Draw a line through the points.

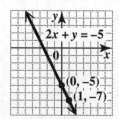

35. Solve the equation for y.

$$\begin{aligned} 4x - 5y &= 20 && \textit{Given equation} \\ -5y &= -4x + 20 && \textit{Subtract } 4x. \\ y &= \tfrac{4}{5}x - 4 && \textit{Divide by } -5. \end{aligned}$$

The slope is the coefficient of x, $\frac{4}{5}$.

The slope is $\dfrac{4}{5} = \dfrac{\text{change in } y}{\text{change in } x}$, and the y-intercept is $(0, -4)$. Graph that point and count up 4 units and right 5 units to get to the point $(5, 0)$. Draw a line through the points.

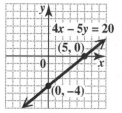

37. $(0, 1)$, $m = 4$

First, locate the point $(0, 1)$, which is the y-intercept of the line to be graphed. Write the slope as

$$m = \frac{\text{rise}}{\text{run}} = \frac{4}{1}.$$

Locate another point by counting 4 units up and then 1 unit to the right. Draw a line through this new point, $(1, 5)$, and the given point, $(0, 1)$.

39. $(1, -5)$, $m = -\frac{2}{5}$

First, locate the point $(1, -5)$. Write the slope as

$$m = \frac{\text{rise}}{\text{run}} = \frac{-2}{5}.$$

Locate another point by counting 2 units down (because of the negative sign) and then 5 units to the right. Draw a line through this new point, $(6, -7)$, and the given point, $(1, -5)$.

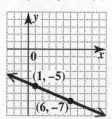

41. $(-1, 4), m = \frac{2}{5}$

First, locate the point $(-1, 4)$. The slope is

$$m = \frac{\text{rise}}{\text{run}} = \frac{2}{5}.$$

Locate another point by counting 2 units up and then 5 units to the right. Draw a line through this new point, $(4, 6)$, and the given point, $(-1, 4)$.

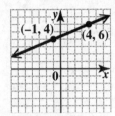

43. $(0, 0), m = -2$

First, locate the point $(0, 0)$. The slope is

$$m = \frac{\text{rise}}{\text{run}} = \frac{-2}{1}.$$

Locate another point by counting 2 units down and then 1 unit to the right. Draw a line through this new point, $(1, -2)$, and the given point, $(0, 0)$.

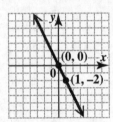

45. $(-2, 3), m = 0$

First, locate the point $(-2, 3)$. Since the slope is 0, the line will be horizontal. Draw a horizontal line through the point $(-2, 3)$.

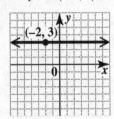

47. $(2, 4)$, undefined slope

First, locate the point $(2, 4)$. Since the slope is undefined, the line will be vertical. Draw the vertical line through the point $(2, 4)$.

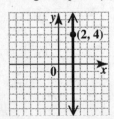

49. The common name given to a vertical line whose x-intercept is the origin is the y-axis.

51. $(4, 1), m = 2$

The given point is $(4, 1)$, so $x_1 = 4$ and $y_1 = 1$. Also, $m = 2$. Substitute these values into the point-slope form. Then solve for y to obtain the slope-intercept form.

$$\begin{aligned} y - y_1 &= m(x - x_1) \\ y - 1 &= 2(x - 4) \\ y - 1 &= 2x - 8 \quad \text{Dist. property} \\ y &= 2x - 7 \quad \text{Add 1.} \end{aligned}$$

53. $(-1, 3), m = -4$

The given point is $(-1, 3)$, so $x_1 = -1$ and $y_1 = 3$. Also, $m = -4$. Substitute these values into the point-slope form. Then solve for y to obtain the slope-intercept form.

$$\begin{aligned} y - y_1 &= m(x - x_1) \\ y - 3 &= -4[x - (-1)] \\ y - 3 &= -4(x + 1) \\ y - 3 &= -4x - 4 \quad \text{Dist. property} \\ y &= -4x - 1 \quad \text{Add 3.} \end{aligned}$$

55. $(9, 3), m = 1$

Use the values $x_1 = 9$, $y_1 = 3$, and $m = 1$ in the point-slope form.

$$\begin{aligned} y - y_1 &= m(x - x_1) \\ y - 3 &= 1(x - 9) \\ y - 3 &= 1x - 9 \quad \text{Dist. property} \\ y &= x - 6 \quad \text{Add 3.} \end{aligned}$$

57. $(-4, 1), m = \frac{3}{4}$

Use the values $x_1 = -4$, $y_1 = 1$, and $m = \frac{3}{4}$ in the point-slope form.

$$\begin{aligned} y - y_1 &= m(x - x_1) \\ y - 1 &= \tfrac{3}{4}[x - (-4)] \\ y - 1 &= \tfrac{3}{4}(x + 4) \\ y - 1 &= \tfrac{3}{4}x + 3 \\ y &= \tfrac{3}{4}x + 4 \end{aligned}$$

59. $(-2, 5), m = \frac{2}{3}$

Use the values $x_1 = -2$, $y_1 = 5$, and $m = \frac{2}{3}$ in the point-slope form.

$$\begin{aligned} y - y_1 &= m(x - x_1) \\ y - 5 &= \tfrac{2}{3}[x - (-2)] \\ y - 5 &= \tfrac{2}{3}(x + 2) \\ y - 5 &= \tfrac{2}{3}x + \tfrac{4}{3} \\ y &= \tfrac{2}{3}x + \tfrac{19}{3} \quad \text{Add } 5 = \tfrac{15}{3}. \end{aligned}$$

61. $(6, -3)$, $m = -\frac{4}{5}$

Use the values $x_1 = 6$, $y_1 = -3$, and $m = -\frac{4}{5}$ in the point-slope form.

$$y - y_1 = m(x - x_1)$$
$$y - (-3) = -\frac{4}{5}(x - 6)$$
$$y + 3 = -\frac{4}{5}x + \frac{24}{5}$$
$$y = -\frac{4}{5}x + \frac{9}{5} \quad \textit{Subtract 3} = \frac{15}{5}.$$

63. **A.** 　$y = \frac{2}{3}x - 2$　　　　*Given*

$3(y) = 3(\frac{2}{3}x - 2)$　　*Multiply by 3.*

$3y = 3(\frac{2}{3}x) - 3(2)$　*Distributive property*

$3y = 2x - 6$　　　　*Multiply.*

$6 = 2x - 3y$　　　　*Standard form*

B. 　　$-2x + 3y = -6$　　　*Given*

$-1(-2x + 3y) = -1(-6)$　*Mult. by −1.*

$-1(-2x) - 1(3y) = -1(-6)$　*Dist. property*

$2x - 3y = 6$　　　　*Multiply.*

C. 　　$y = -\frac{3}{2}x + 3$　　　*Given*

$2(y) = 2(-\frac{3}{2}x + 3)$　　*Multiply by 2.*

$2y = 2(-\frac{3}{2}x) + 2(3)$　*Dist. property*

$2y = -3x + 6$　　　*Multiply.*

$3x + 2y = 6$　　　*Standard form*

D. 　　$y - 2 = \frac{2}{3}(x - 6)$　　*Given*

$3(y - 2) = 3[\frac{2}{3}(x - 6)]$　*Multiply by 3.*

$3y - 6 = 2(x - 6)$　　*Distributive prop.*
　　　　　　　　　　Associative prop.

$3y - 6 = 2x - 12$　　*Distributive prop.*

$6 = 2x - 3y$　　　*Standard form*

So **A**, **B**, and **D** are equivalent to $2x - 3y = 6$.

65. **(a)** $(4, 10)$ and $(6, 12)$

First, find the slope of the line.

$$m = \frac{12 - 10}{6 - 4} = \frac{2}{2} = 1$$

Now use the point $(4, 10)$ for (x_1, y_1) and $m = 1$ in the point-slope form.

$$y - y_1 = m(x - x_1)$$
$$y - 10 = 1(x - 4)$$
$$y - 10 = x - 4$$
$$y = x + 6$$

The same result would be found by using $(6, 12)$ for (x_1, y_1).

(b) 　　$y = x + 6$

$-x + y = 6$　　*Subtract x.*

$x - y = -6$　　*Multiply by −1.*

67. **(a)** $(-4, 0)$ and $(0, 2)$

$$m = \frac{2 - 0}{0 - (-4)} = \frac{2}{4} = \frac{1}{2}$$

Use the point-slope form with $(x_1, y_1) = (-4, 0)$ and $m = \frac{1}{2}$.

$$y - y_1 = m(x - x_1)$$
$$y - 0 = \frac{1}{2}[x - (-4)]$$
$$y = \frac{1}{2}(x + 4)$$
$$y = \frac{1}{2}x + 2$$

(b) 　　$y = \frac{1}{2}x + 2$

$2y = x + 4$　　*Multiply by 2.*

$-x + 2y = 4$　　*Subtract x.*

$x - 2y = -4$　　*Multiply by −1.*

69. **(a)** $(-2, -1)$ and $(3, -4)$

$$m = \frac{-4 - (-1)}{3 - (-2)} = \frac{-3}{5} = -\frac{3}{5}$$

Use the point-slope form with $(x_1, y_1) = (3, -4)$ and $m = -\frac{3}{5}$.

$$y - y_1 = m(x - x_1)$$
$$y - (-4) = -\frac{3}{5}(x - 3)$$
$$y + 4 = -\frac{3}{5}(x - 3)$$
$$y + 4 = -\frac{3}{5}x + \frac{9}{5}$$
$$y = -\frac{3}{5}x - \frac{11}{5}$$

(b) 　　$y = -\frac{3}{5}x - \frac{11}{5}$

$5y = -3x - 11$　*Multiply by 5.*

$3x + 5y = -11$　*Add 3x.*

71. **(a)** $\left(-\frac{2}{3}, \frac{8}{3}\right)$ and $\left(\frac{1}{3}, \frac{7}{3}\right)$

$$m = \frac{\frac{7}{3} - \frac{8}{3}}{\frac{1}{3} - \left(-\frac{2}{3}\right)} = \frac{-\frac{1}{3}}{\frac{3}{3}} = \frac{-\frac{1}{3}}{1} = -\frac{1}{3}$$

Use the point-slope form with $(x_1, y_1) = \left(\frac{1}{3}, \frac{7}{3}\right)$ and $m = -\frac{1}{3}$.

$$y - y_1 = m(x - x_1)$$
$$y - \frac{7}{3} = -\frac{1}{3}\left(x - \frac{1}{3}\right)$$
$$y - \frac{7}{3} = -\frac{1}{3}x + \frac{1}{9}$$
$$y = -\frac{1}{3}x + \frac{22}{9}$$

(b) 　　$y = -\frac{1}{3}x + \frac{22}{9}$

$9y = -3x + 22$　*Multiply by 9.*

$3x + 9y = 22$　　*Add 3x.*

73. Solve the equation for y.

$$x - 2y = 7$$
$$-2y = -x + 7$$
$$y = \tfrac{1}{2}x - \tfrac{7}{2}$$

The slope is $\tfrac{1}{2}$. A line perpendicular to this line has slope -2 $\left(\text{the negative reciprocal of } \tfrac{1}{2}\right)$. Now use the slope-intercept form with $m = -2$ and y-intercept $(0, -3)$.

$$y = mx + b$$
$$y = -2x - 3$$

75. Solve the equation for y.

$$4x - y = -2$$
$$4x + 2 = y$$

The slope is 4. A line parallel to this line has the same slope. Now use the point-slope form with $m = 4$ and $(x_1, y_1) = (2, 3)$.

$$y - y_1 = m(x - x_1)$$
$$y - 3 = 4(x - 2)$$
$$y - 3 = 4x - 8$$
$$y = 4x - 5$$

77. Solve the equation for y.

$$3x = 4y + 5$$
$$-4y = -3x + 5$$
$$y = \tfrac{3}{4}x - \tfrac{5}{4}$$

The slope is $\tfrac{3}{4}$. A line parallel to this line has the same slope. Now use the point-slope form with $m = \tfrac{3}{4}$ and $(x_1, y_1) = (2, -3)$.

$$y - y_1 = m(x - x_1)$$
$$y - (-3) = \tfrac{3}{4}(x - 2)$$
$$y + 3 = \tfrac{3}{4}x - \tfrac{3}{2}$$
$$y = \tfrac{3}{4}x - \tfrac{3}{2} - \tfrac{6}{2}$$
$$y = \tfrac{3}{4}x - \tfrac{9}{2}$$

79. **(a)** The fixed cost is $400.

(b) The variable cost is $0.25.

(c) Substitute $m = 0.25$ and $b = 400$ into $y = mx + b$ to get the cost equation

$$y = 0.25x + 400.$$

(d) Let $x = 100$ in the cost equation.

$$y = 0.25(100) + 400$$
$$y = 25 + 400$$
$$y = 425$$

The cost to produce 100 snow cones will be $425.

(e) Let $y = 775$ in the cost equation.

$$775 = 0.25x + 400$$
$$375 = 0.25x \qquad \textit{Subtract 400.}$$
$$x = \tfrac{375}{0.25} = 1500 \quad \textit{Divide by 0.25.}$$

If the total cost is $775, 1500 snow cones will be produced.

81. **(a)** x represents the year and y represents the cost in the ordered pairs (x, y). The ordered pairs are $(1, 2079)$, $(2, 2182)$, $(3, 2272)$, $(4, 2361)$, and $(5, 2402)$.

(b)

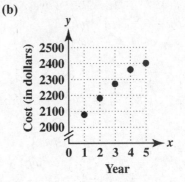

Yes, the points lie approximately in a straight line.

(c) Find the slope using

$(x_1, y_1) = (1, 2079)$ and $(x_2, y_2) = (4, 2361)$.

$$m = \frac{y_2 - y_1}{x_2 - x_1} = \frac{2361 - 2079}{4 - 1} = \frac{282}{3} = 94$$

Now use the point-slope form with $m = 94$ and $(x_1, y_1) = (1, 2079)$.

$$y - y_1 = m(x - x_1)$$
$$y - 2079 = 94(x - 1)$$
$$y - 2079 = 94x - 94$$
$$y = 94x + 1985$$

(d) Since year 1 represents 2004, year 0 represents 2003.

For 2009, $x = 2009 - 2003 = 6$.

$$y = 94x + 1985$$
$$y = 94(6) + 1985$$
$$y = 2549$$

In 2009, the estimate of the average annual cost at 2-year colleges is $2549.

83. Find the slope using

$(x_1, y_1) = (2002, 52)$ and $(x_2, y_2) = (2007, 127)$.

$$m = \frac{y_2 - y_1}{x_2 - x_1} = \frac{127 - 52}{2007 - 2002} = \frac{75}{5} = 15$$

Now use the point-slope form with $m = 15$ and $(x_2, y_2) = (2007, 127)$.

$$y - y_1 = m(x - x_1)$$
$$y - 127 = 15(x - 2007)$$
$$y - 127 = 15x - 30{,}105$$
$$y = 15x - 29{,}978$$

85. From the calculator screens, we see that the points with coordinates $(-1, 9)$ and $(4, -6)$ are on the line. First, find the slope.

$$m = \frac{-6 - 9}{4 - (-1)} = \frac{-15}{5} = -3$$

Use the point-slope form with $(x_1, y_1) = (-1, 9)$ and $m = -3$.

$$y - 9 = -3[x - (-1)]$$
$$y - 9 = -3(x + 1)$$
$$y - 9 = -3x - 3$$
$$y = -3x + 6$$

87. $3x + 8 > -1$

$\quad 3x > -9 \quad$ *Subtract 8.*

$\quad x > -3 \quad$ *Divide by 3.*

Graph the solution set $(-3, \infty)$.

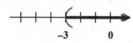

89. $5 - 3x \le -10$

$\quad -3x \le -15 \quad$ *Subtract 5.*

$\quad x \ge 5 \quad$ *Divide by -3; reverse the symbol from $\le$ to $\ge$.*

Graph the solution set $[5, \infty)$.

Summary Exercises on Linear Equations and Graphs

1. Solve the equation for y.

$\quad x - 2y = -4 \qquad$ *Given equation*

$\quad -2y = -x - 4 \quad$ *Subtract x.*

$\quad y = \frac{1}{2}x + 2 \quad$ *Divide by -2.*

The slope is the coefficient of x, $\frac{1}{2}$.

The slope is $\frac{1}{2} = \dfrac{\text{change in } y}{\text{change in } x}$, and the y-intercept is $(0, 2)$. Graph that point and count up 1 unit and right 2 units to get to the point $(2, 3)$. Draw a line through the points.

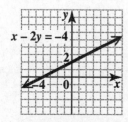

3. $m = 1$, y-intercept $(0, -2)$

The slope is $1 = \dfrac{1}{1} = \dfrac{\text{change in } y}{\text{change in } x}$, and the y-intercept is $(0, -2)$. Graph that point and count up 1 unit and right 1 unit to get to the point $(1, -1)$. Draw a line through the points.

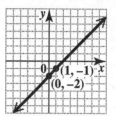

5. $m = -\frac{2}{3}$, passes through $(3, -4)$

The slope is $-\dfrac{2}{3} = \dfrac{2}{-3} = \dfrac{\text{change in } y}{\text{change in } x}$. Graph the point $(3, -4)$ and count up 2 units and left 3 units to get to the point $(0, -2)$.

Draw a line through the points.

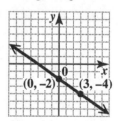

7. Solve the equation for y.

$\quad x - 4y = 0 \qquad$ *Given equation*

$\quad -4y = -x \quad$ *Subtract x.*

$\quad y = \frac{1}{4}x \quad$ *Divide by -4.*

The slope is the coefficient of x, $\frac{1}{4}$.

The slope is $\dfrac{1}{4} = \dfrac{\text{change in } y}{\text{change in } x}$, and the y-intercept is $(0, 0)$. Graph that point and count up 1 unit and right 4 units to get to the point $(4, 1)$. Draw a line through the points.

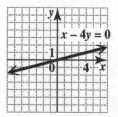

9. $8x = 6y + 24$

x-intercept $(y = 0)$ **y-intercept $(x = 0)$**
$8x = 6(0) + 24$ $8(0) = 6y + 24$
$8x = 24$ $-24 = 6y$
$x = 3$ $\underline{(3, 0)}$ $-4 = y$ $\underline{(0, -4)}$

third point $(x = 2)$
$8(2) = 6y + 24$
$16 = 6y + 24$
$-8 = 6y$
$-\frac{4}{3} = y$ $\underline{\left(2, -\frac{4}{3}\right)}$

Plot $(3, 0)$, $(0, -4)$, and $\left(2, -\frac{4}{3}\right)$ and draw a line through these points.

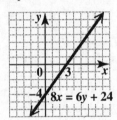

11.   Solve the equation for y.

$5x + 2y = 10$ *Given equation*
$2y = -5x + 10$ *Subtract 5x.*
$y = -\frac{5}{2}x + 5$ *Divide by 2.*

The slope is the coefficient of x, $-\frac{5}{2}$.

The slope is $-\frac{5}{2} = \frac{-5}{2} = \frac{\text{change in } y}{\text{change in } x}$, and the y-intercept is $(0, 5)$. Graph that point and count down 5 units and right 2 units to get to the point $(2, 0)$. Draw a line through the points.

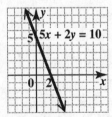

13. $m = 0$, passes through $\left(0, \frac{3}{2}\right)$

The slope is 0, so the graph is a horizontal line. Graph the y-intercept, $\left(0, \frac{3}{2}\right)$, and draw a horizontal line through it.

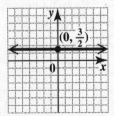

15. $y = -x + 6$

The slope is $-1 = \frac{-1}{1} = \frac{\text{change in } y}{\text{change in } x}$, and the y-intercept is $(0, 6)$. Graph that point and count down 1 unit and right 1 unit to get to the point $(1, 5)$. Draw a line through the points.

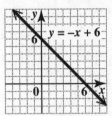

17. $x + 4 = 0$
$x = -4$ *Subtract 4.*

This is an equation of the vertical line with x-intercept $(-4, 0)$. The slope of the line is undefined.

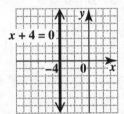

19.   **(a)** "Slope -0.5" indicates that $m = -\frac{1}{2}$ and "$b = -2$" indicates that the y-intercept is $(0, -2)$. Given the slope and y-intercept, we know that one form of the equation of the line is $y = -\frac{1}{2}x - 2$, choice **B**.

(b) The slope of the line passing through the x-intercept of $(4, 0)$ and the y-intercept of $(0, 2)$ is

$$m = \frac{2 - 0}{0 - 4} = \frac{2}{-4} = -\frac{1}{2}.$$

The slope-intercept form is $y = -\frac{1}{2}x + 2$. None of the choices has this equation, so change the form of $y = -\frac{1}{2}x + 2$ by multiplying each side by 2 to clear fractions.

$$2(y) = 2\left(-\frac{1}{2}x + 2\right)$$
$$2y = -x + 4$$

Add x to both sides to get $x + 2y = 4$, choice **D**.

(c) The slope of the line passing through $(4, -2)$ and $(0, 0)$ is

$$m = \frac{0 - (-2)}{0 - 4} = \frac{2}{-4} = -\frac{1}{2}.$$

The slope-intercept form is $y = -\frac{1}{2}x$, choice **A**.

(d) $m = \frac{1}{2}$; passes through $(-2, -2)$

$y = mx + b$	*Slope-intercept form*
$-2 = \frac{1}{2}(-2) + b$	*Let $x = -2$, $y = -2$, $m = \frac{1}{2}$.*
$-2 = -1 + b$	*Multiply.*
$-1 = b$	*Add 1.*

The line has equation $y = \frac{1}{2}x - 1$. Multiply each side by 2 to get $2y = x - 2$, which is the same as $2 = x - 2y$, choice **C**.

21. $m = -3, b = -6$

$$y = mx + b$$
$$y = -3x - 6$$

23. Through $(1, -7)$ and $(-2, 5)$

$$m = \frac{5 - (-7)}{-2 - 1} = \frac{12}{-3} = -4$$

$$y - y_1 = m(x - x_1)$$
$$y - (-7) = -4(x - 1)$$
$$y + 7 = -4x + 4$$
$$y = -4x - 3$$

25. Through $(0, 0)$, undefined slope

A line with undefined slope has equation $x = x\text{-intercept value}$; in this case, $x = 0$.

27. Through $(0, 0)$ and $(3, 2)$

$$m = \frac{2 - 0}{3 - 0} = \frac{2}{3}$$

$$y = mx + b$$
$$y = \frac{2}{3}x + 0$$
$$y = \frac{2}{3}x$$

29. Through $(5, 0)$ and $(0, -5)$

$$m = \frac{-5 - 0}{0 - 5} = \frac{-5}{-5} = 1$$

$$y = mx + b$$
$$y = 1x - 5$$
$$y = x - 5$$

31. $m = \frac{5}{3}$, through $(-3, 0)$

$$y - y_1 = m(x - x_1)$$
$$y - 0 = \frac{5}{3}[x - (-3)]$$
$$y = \frac{5}{3}(x + 3)$$
$$y = \frac{5}{3}x + 5$$

3.5 Graphing Linear Inequalities in Two Variables

3.5 Now Try Exercises

N1. Graph $x + 3y \leq 6$.

Start by graphing the equation

$$x + 3y = 6.$$

The intercepts are $(6, 0)$ and $(0, 2)$. Draw a solid line through these points to show that the points on the line are solutions to the inequality

$$x + 3y \leq 6.$$

Choose a test point not on the line, such as $(0, 0)$.

$$x + 3y \leq 6$$
$$(0) + 3(0) \overset{?}{\leq} 6 \quad \textit{Let x = 0, y = 0.}$$
$$0 + 0 \overset{?}{\leq} 6$$
$$0 \leq 6 \quad \textit{True}$$

Since the last statement is true, shade the region that includes the test point $(0, 0)$, that is, the region below the line. The shaded region, along with the boundary, is the desired graph.

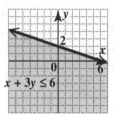

N2. Graph $2x - 4y > 8$.

Start by graphing the equation

$$2x - 4y = 8.$$

The intercepts are $(4, 0)$ and $(0, -2)$. Draw a dashed line through these points to show that the points on the line are not solutions to the inequality

$$2x - 4y > 8.$$

Choose a test point not on the line, such as $(0, 0)$.

$$2x - 4y > 8$$
$$2(0) - 4(0) \overset{?}{>} 8 \quad \textit{Let x = 0, y = 0.}$$
$$0 > 8 \quad \textit{False}$$

Since the last statement is false, shade the region that does *not* include the test point, $(0, 0)$, that is, the region below the line. The dashed line shows that the boundary is not part of the graph.

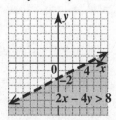

N3. $x > 2$

First graph $x = 2$, a vertical line through the point $(2, 0)$. Use a dashed line because of the $>$ symbol. Choose $(0, 0)$ as a test point.

$$x > 2$$
$$0 > 2 \quad Let\ x = 0. \quad False$$

Since the last statement is false, shade the region that does *not* include the test point $(0, 0)$, that is, the region to the right of line. The dashed line shows that the boundary is not part of the graph.

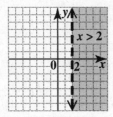

N4. $y \leq -2x$

Graph $y = -2x$ as a solid line through $(0, 0)$ and $(1, -2)$. We cannot use $(0, 0)$ as a test point because $(0, 0)$ is on the line $y = -2x$. Instead, we choose a test point off the line, $(1, 0)$.

$$y \leq -2x$$
$$0 \overset{?}{\leq} -2(1) \quad Let\ x = 1,\ y = 0.$$
$$0 \leq -2 \quad False$$

Since the last statement is false, shade the region that does *not* include the test point $(1, 0)$, that is, the region below the line. The shaded region, along with the boundary, is the desired graph.

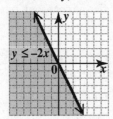

3.5 Section Exercises

1. The key phrase is "more than" (occurs twice), so use the symbol $>$ twice.

3. The key phrase is "at most," so use the symbol $\leq$.

5. The key phrase is "less than," so use the symbol $<$.

7. For the point $(4, 0)$, substitute 4 for x and 0 for y in the inequality.

$$3x - 4y < 12$$
$$3(4) - 4(0) \overset{?}{<} 12$$
$$12 - 0 \overset{?}{<} 12$$
$$12 < 12 \quad False$$

The *false* result shows that $(4, 0)$ *is not* a solution of the given inequality.

9. $3x - 2y \geq 0$

(i) For the point $(4, 1)$, substitute 4 for x and 1 for y in the given inequality.

$$3x - 2y \geq 0$$
$$3(4) - 2(1) \overset{?}{\geq} 0$$
$$12 - 2 \overset{?}{\geq} 0$$
$$10 \geq 0 \quad True$$

The *true* result shows that $(4, 1)$ *is* a solution of the given inequality.

(ii) For the point $(0, 0)$, substitute 0 for x and 0 for y in the given inequality.

$$3x - 2y \geq 0$$
$$3(0) - 2(0) \overset{?}{\geq} 0$$
$$0 \geq 0 \quad True$$

The *true* result shows that $(0, 0)$ *is* a solution of the given inequality.

Since (i) and (ii) are true, the given statement is *true*.

11. $x + 2y \geq 7$

Use $(0, 0)$ as a test point.

$$x + 2y \geq 7$$
$$0 + 2(0) \overset{?}{\geq} 7 \quad Let\ x = 0,\ y = 0.$$
$$0 \geq 7 \quad False$$

Because the last statement is *false*, we shade the region that does *not* include the test point $(0, 0)$. This is the region above the line.

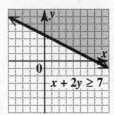

13. $-3x + 4y > 12$

Use $(0, 0)$ as a test point.

$$-3x + 4y > 12$$
$$-3(0) + 4(0) \overset{?}{>} 12 \quad Let\ x = 0,\ y = 0.$$
$$0 > 12 \quad False$$

Because the last statement is *false*, we shade the region that does *not* include the test point $(0, 0)$. This is the region above the line.

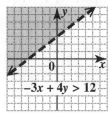

15. $y < -1$

Use $(0, 0)$ as a test point.

$$y < -1$$
$$0 < -1 \quad \textit{Let y = 0; False}$$

Because $0 < -1$ is *false*, shade the region *not* containing $(0, 0)$. This is the region below the line.

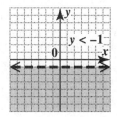

17. Use a *dashed* line if the symbol is $<$ or $>$. Use a *solid* line if the symbol is $\leq$ or $\geq$.

19. $x + y \leq 5$

Step 1
Graph the boundary of the region, the line with equation $x + y = 5$.

If $y = 0$, $x = 5$, so the x-intercept is $(5, 0)$.
If $x = 0$, $y = 5$, so the y-intercept is $(0, 5)$.

Draw the line through these intercepts. Make the line solid because of the $\leq$ sign.

Step 2
Choose the point $(0, 0)$ as a test point.

$$x + y \leq 5$$
$$0 + 0 \overset{?}{\leq} 5 \quad \textit{Let x = 0, y = 0.}$$
$$0 \leq 5 \quad \textit{True}$$

Because $0 \leq 5$ is true, shade the region containing the origin. The shaded region, along with the boundary, is the desired graph.

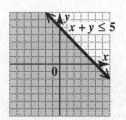

21. $2x + 3y > -6$

The boundary is the line with equation $2x + 3y = -6$. Draw this line through its intercepts, $(-3, 0)$ and $(0, -2)$. The line should be dashed because of the $>$ sign. Choose $(0, 0)$ as a test point. Since $2(0) + 3(0) > -6$ is true, shade the region containing the origin. The dashed line shows that the boundary is not part of the graph.

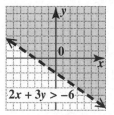

23. $y \geq 2x + 1$

The boundary is the line with equation $y = 2x + 1$. This line has slope 2 and y-intercept $(0, 1)$. It may be graphed by starting at $(0, 1)$ and going 2 units up and then 1 unit to the right to reach the point $(1, 3)$. Draw a solid line through $(0, 1)$ and $(1, 3)$.

Using $(0, 0)$ as a test point will result in the inequality $0 \geq 1$, which is false. Shade the region *not* containing the origin, that is, the region above the line. The solid line shows that the boundary is part of the graph.

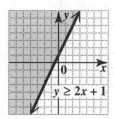

25. $x < -2$

The boundary is the line with equation $x = -2$. This is a vertical line through $(-2, 0)$. Make this line dashed because of the $<$ sign.

Using $(0, 0)$ as a test point will result in the inequality $0 < -2$, which is false. Shade the region *not* containing the origin. This is the region to the left of the boundary. The dashed line shows that the boundary is not part of the graph.

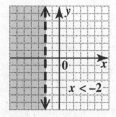

27. $y \leq 5$

The boundary is the line with equation $y = 5$. This is the horizontal line through $(0, 5)$. Make this line solid because of the $\leq$ sign.

Using $(0, 0)$ as a test point will result in the inequality $0 \leq 5$, which is true. Shade the region containing the origin, that is, the region below the line. The solid line shows that the boundary is part of the graph.

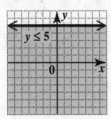

29. $y \geq 4x$

The boundary has the equation $y = 4x$. This line goes through the points $(0, 0)$ and $(1, 4)$. Make the line solid because of the $\geq$ sign. Because the boundary passes through the origin, we cannot use $(0, 0)$ as a test point.

Using $(2, 0)$ as a test point will result in the inequality $0 \geq 8$, which is false. Shade the region *not* containing $(2, 0)$. The solid line shows that the boundary is part of the graph.

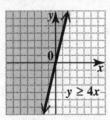

31. Every point in quadrant IV has a positive x-value and a negative y-value. Substituting into $y > x$ would imply that a negative number is greater than a positive number, which is always false. Thus, the graph of $y > x$ cannot lie in quadrant IV.

33. $x + y \geq 500$

(a) Graph the inequality.

Step 1
Graph the line $x + y = 500$.

If $x = 0$, then $y = 500$, so the y-intercept is $(0, 500)$.
If $y = 0$, then $x = 500$, so the x-intercept is $(500, 0)$.

Graph the line with these intercepts.

The line is solid because of the $\geq$ sign.

Step 2
Use $(0, 0)$ as a test point.

$$x + y \geq 500 \quad \textit{Original inequality}$$
$$0 + 0 \overset{?}{\geq} 500 \quad \textit{Let x = 0, y = 0.}$$
$$0 \geq 500 \quad \textit{False}$$

Since $0 \geq 500$ is false, shade the side of the boundary not containing $(0, 0)$. Because of the restrictions $x \geq 0$ and $y \geq 0$ in this applied problem, only the portion of the graph that lies in quadrant I is included.

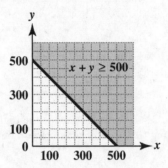

(b) Any point in the shaded region or on the boundary satisfies the inequality. Some ordered pairs are $(500, 0)$, $(200, 400)$, and $(400, 200)$. There are many other ordered pairs that will also satisfy the inequality.

For exercises 35–40, use the expression
$$3x^2 + 8x + 5.$$

35. $3(0)^2 + 8(0) + 5$
$$= 0 + 0 + 5$$
$$= 5$$

37. $3(4)^2 + 8(4) + 5$
$$= 3(16) + 8(4) + 5$$
$$= 48 + 32 + 5$$
$$= 80 + 5$$
$$= 85$$

39. $3(1)^2 + 8(1) + 5$
$$= 3 + 8 + 5$$
$$= 11 + 5$$
$$= 16$$

3.6 Introduction to Functions

3.6 Now Try Exercises

N1. $\{(-2, 3), (0, 7), (2, 8), (2, 10)\}$

The domain is the set of all first components in the ordered pairs, $\{-2, 0, 2\}$.

The range is the set of all second components in the ordered pairs, $\{3, 7, 8, 10\}$.

N2. (a) $\{(-1, 2), (0, 1), (1, 0), (4, 3), (4, 5)\}$

The first component 4 appears in two ordered pairs, and corresponds to more than one second component. Therefore, this relation *is not a function*.

(b) $\{(-1,-3),(0,2),(3,1),(8,1)\}$

Notice that each first component appears once and only once. Because of this, the relation *is a function*.

N3. (a) $y = x - 5$

This linear equation is in the form $y = mx + b$. Since the graph of this equation is a line that is not vertical, the equation defines a function.

(b) Use the vertical line test. A vertical line could intersect the graph twice, so this graph *is not* the graph of a function.

N4. $y = x^2 - 2$

Any number may be used for x, so the domain is the set of all real numbers, written $(-\infty, \infty)$.

The second power of a real number cannot be negative, and since $y = x^2 - 2$, the values of y cannot be less than -2. The range is the set of all real numbers greater than or equal to -2, written $[-2, \infty)$.

N5. $f(x) = x^3 - 7$
$f(-2) = (-2)^3 - 7$ *Let x = −2.*
$\quad = -8 - 7$
$\quad = -15$

N6. (a) Since 2006 corresponds with 14.9,

$$f(2006) = 14.9.$$

The population of Asian-Americans was 14.9 million in 2006.

(b) $f(x)$ is equal to 9.7 for the x-value 1996.

3.6 Section Exercises

1. If $x = 1$, then $x + 2 = 3$. Since $f(x) = x + 2$, $f(x)$ is also equal to 3. The ordered pair (x, y) is $(1, 3)$.

3. If $x = 3$, then $x + 2 = 5$. Since $f(x) = x + 2$, $f(x)$ is also equal to 5. The ordered pair (x, y) is $(3, 5)$.

5. If the domain of the function f in Exercises 1–4 is $\{0, 1, 2, 3\}$, then the graph of f consists of the four points $(0, 2)$, $(1, 3)$, $(2, 4)$, and $(3, 5)$.

7. $\{(-4,3),(-2,1),(0,5),(-2,-8)\}$

The domain is the set of all first components in the ordered pairs, $\{-4, -2, 0\}$.

The range is the set of all second components in the ordered pairs, $\{3, 1, 5, -8\}$.

This relation *is not a function* since one value of x, namely -2, corresponds to two values of y, namely 1 and -8.

9. The domain is $\{A, B, C, D, E\}$.
The range is $\{2, 3, 6, 4\}$.

The relation *is a function* since each of the first components A, B, C, D, and E corresponds to exactly one second component.

11. The graph consists of the following set of six ordered pairs:

$$\{(-4,1),(-2,0),(-2,2),(0,-2),(2,1),(3,3)\}$$

The domain is the set of all first components in the ordered pairs, $\{-4, -2, 0, 2, 3\}$.

The range is the set of all second components in the ordered pairs, $\{1, 0, 2, -2, 3\}$.

This relation *is not a function* since one value of x, namely -2, corresponds to two values of y, namely 0 and 2.

13. Any vertical line will intersect the graph in only one point. The graph passes the vertical line test, so this is the graph of a function.

15. A vertical line can cross the graph twice, so this is not the graph of a function.

17. $y = 5x + 3$

Every value of x will give one and only one value of y, so the equation defines a function.

19. $y = x^2$

Every value of x will give one and only one value of y, so the equation defines a function.

21. $x = y^2$

Every positive value of x will give two values of y. For example, if $x = 16$, $16 = y^2$ and $y = +4$ or -4. Therefore, the equation does not define a function.

23. $y = 3x - 2$

Any number may be used for x, so the domain is the set of all real numbers, written $(-\infty, \infty)$.

In $y = 3x - 2$, any number may be used for y, so the range is also $(-\infty, \infty)$.

25. $y = x^2 + 2$

Any number may be used for x, so the domain is the set of all real numbers, written $(-\infty, \infty)$.

The second power of a real number cannot be negative, and since $y = x^2 + 2$, the values of y cannot be less than 2. The range is the set of all real numbers greater than or equal to 2, written $[2, \infty)$.

27. $f(x) = \sqrt{x}$

Any real number greater than or equal to 0 has a real square root, so the domain is $[0, \infty)$. The symbol $\sqrt{x}$ represents the nonnegative square root of x, and takes on all such values, so the range is also $[0, \infty)$.

29. If $f(2) = 4$, one point on the line has coordinates $(2, 4)$.

30. If $f(-1) = -4$, then another point on the line has coordinates $(-1, -4)$.

31. Using the points $(2, 4)$ and $(-1, -4)$, we obtain

$$m = \frac{-4 - 4}{-1 - 2} = \frac{-8}{-3} = \frac{8}{3}.$$

32. Start with the point-slope form of a line using $m = \frac{8}{3}$ and $(x_1, y_1) = (2, 4)$.

$$y - y_1 = m(x - x_1)$$
$$y - 4 = \frac{8}{3}(x - 2)$$

Now solve for y to write this equation in slope-intercept form.

$$y - 4 = \frac{8}{3}x - \frac{16}{3}$$
$$y = \frac{8}{3}x - \frac{16}{3} + \frac{12}{3}$$
$$y = \frac{8}{3}x - \frac{4}{3}$$

Therefore,

$$f(x) = \frac{8}{3}x - \frac{4}{3}.$$

33. $f(x) = 4x + 3$

(a) $f(2) = 4(2) + 3 = 8 + 3 = 11$

(b) $f(0) = 4(0) + 3 = 0 + 3 = 3$

(c) $f(-3) = 4(-3) + 3 = -12 + 3 = -9$

35. $f(x) = x^2 - x + 2$

(a) $f(2) = (2)^2 - (2) + 2 = 4 - 2 + 2 = 4$

(b) $f(0) = (0)^2 - (0) + 2 = 0 - 0 + 2 = 2$

(c) $f(-3) = (-3)^2 - (-3) + 2$
$= 9 + 3 + 2 = 14$

37. $f(x) = |x|$

(a) $f(2) = |2| = 2$

(b) $f(0) = |0| = 0$

(c) $f(-3) = |-3| = -(-3) = 3$

39. Write the information in the graph as a set of ordered pairs of the form (year, population). The set is $\{(1970, 9.6), (1980, 14.1), (1990, 19.8), (2000, 28.4), (2007, 37.3)\}$. Since each year corresponds to exactly one number, the set defines a function.

41. $g(1980) = 14.1$; $g(2000) = 28.4$

43. $g(2007) = 37.3$, so $x = 2007$.

45. (a) The domain value is 4.

(b) The range value is $\sqrt{4}$, which is 2.

47. When $X = 3$, $Y_1 = 4$ in the calculator-generated table.

49. When $Y_1 = 2$, $X = 1$.

51. (a) We choose two points from the table, $(0, 1)$ and $(2, 3)$.

$$m = \frac{3 - 1}{2 - 0} = \frac{2}{2} = 1$$

(b) The y-intercept of a line is $(0, b)$. Since the ordered pair $(0, 1)$ occurs in the table, the y-intercept is $(0, 1)$.

53. $2x + 3y = 12$

If $x = 0$, $y = 4$. If $y = 0$, $x = 6$. Graph a line through the intercepts, $(0, 4)$ and $(6, 0)$.

$4x - 2y = 8$

If $x = 0$, $y = -4$. If $y = 0$, $x = 2$. Graph a line through the intercepts, $(0, -4)$ and $(2, 0)$.

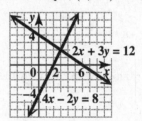

55. $-5x + 2y = 10$

If $x = 0$, $y = 5$. If $y = 0$, $x = -2$. Graph a line through the intercepts, $(0, 5)$ and $(-2, 0)$.

$2y = -4 + 5x$

If $x = 0$, $y = -2$. If $y = 0$, $x = \frac{4}{5}$. Graph a line through the intercepts, $(0, -2)$ and $(\frac{4}{5}, 0)$.

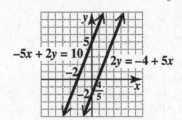

Chapter 3 Review Exercises

1. (a) The graph rises from 2002 to 2003, 2003 to 2004, 2004 to 2005, and 2006 to 2007, so the number of real trees purchased increased from 2002 to 2005 and 2006 to 2007.

(b) The graph falls from 2005 to 2006, so the number of real trees purchased decreased from 2005 to 2006.

(c) Locate 2005 on the horizontal scale and follow the line up to the line graph. Then move across to read the value on the vertical scale. The number of real trees purchased in 2005 is about 33 million. Similarly, we get about 29 million for 2006.

(d) The change from 2005 to 2006 was

29 million − 33 million = −4 million.

This represents a decrease of about 4 million.

2. $y = 3x + 2; \ (-1, \_\_) \ (0, \_\_) \ (\_\_, 5)$

$$y = 3x + 2$$
$$y = 3(-1) + 2 \quad \textit{Let x = −1.}$$
$$y = -3 + 2$$
$$y = -1$$

$$y = 3x + 2$$
$$y = 3(0) + 2 \quad \textit{Let x = 0.}$$
$$y = 0 + 2$$
$$y = 2$$

$$y = 3x + 2$$
$$5 = 3x + 2 \quad \textit{Let y = 5.}$$
$$3 = 3x$$
$$1 = x$$

The ordered pairs are $(-1, -1)$, $(0, 2)$, and $(1, 5)$.

3. $4x + 3y = 6; \ (0, \_\_) \ (\_\_, 0) \ (-2, \_\_)$

$$4x + 3y = 6$$
$$4(0) + 3y = 6 \quad \textit{Let x = 0.}$$
$$3y = 6$$
$$y = 2$$

$$4x + 3y = 6$$
$$4x + 3(0) = 6 \quad \textit{Let y = 0.}$$
$$4x = 6$$
$$x = \tfrac{6}{4} = \tfrac{3}{2}$$

$$4x + 3y = 6$$
$$4(-2) + 3y = 6 \quad \textit{Let x = −2.}$$
$$-8 + 3y = 6$$
$$3y = 14$$
$$y = \tfrac{14}{3}$$

The ordered pairs are $(0, 2)$, $(\tfrac{3}{2}, 0)$, and $(-2, \tfrac{14}{3})$.

4. $x = 3y; \ (0, \_\_) \ (8, \_\_) \ (\_\_, -3)$

$$x = 3y$$
$$0 = 3y \quad \textit{Let x = 0.}$$
$$0 = y$$

$$x = 3y$$
$$8 = 3y \quad \textit{Let x = 8.}$$
$$\tfrac{8}{3} = y$$

$$x = 3y$$
$$x = 3(-3) \quad \textit{Let y = −3.}$$
$$x = -9$$

The ordered pairs are $(0, 0)$, $(8, \tfrac{8}{3})$, and $(-9, -3)$.

5. $x - 7 = 0; \ (\_\_, -3) \ (\_\_, 0) \ (\_\_, 5)$
The given equation may be written $x = 7$. For any value of y, the value of x will always be 7. The ordered pairs are $(7, -3)$, $(7, 0)$, and $(7, 5)$.

6. $x + y = 7; \ (2, 5)$
Substitute 2 for x and 5 for y in the given equation.

$$x + y = 7$$
$$2 + 5 \overset{?}{=} 7$$
$$7 = 7 \quad \textit{True}$$

Yes, $(2, 5)$ is a solution of the given equation.

7. $2x + y = 5; \ (-1, 3)$
Substitute -1 for x and 3 for y in the given equation.

$$2x + y = 5$$
$$2(-1) + 3 \overset{?}{=} 5$$
$$-2 + 3 \overset{?}{=} 5$$
$$1 = 5 \quad \textit{False}$$

No, $(-1, 3)$ is not a solution of the given equation.

8. $3x - y = 4; \ (\tfrac{1}{3}, -3)$
Substitute $\tfrac{1}{3}$ for x and -3 for y in the given equation.

$$3x - y = 4$$
$$3(\tfrac{1}{3}) - (-3) \overset{?}{=} 4$$
$$1 + 3 \overset{?}{=} 4$$
$$4 = 4 \quad \textit{True}$$

Yes, $(\tfrac{1}{3}, -3)$ is a solution of the given equation.

9. To plot $(2, 3)$, start at the origin, go 2 units to the right, and then go up 3 units. The point lies in quadrant I. (Graph follows Exercise 12.)

10. To plot $(-4, 2)$, start at the origin, go 4 units to the left, and then go up 2 units. The point lies in quadrant II.

11. To plot $(3, 0)$, start at the origin, go 3 units to the right. The point lies on the x-axis (not in any quadrant).

12. To plot $(0, -6)$, start at the origin, go down 6 units. The point lies on the y-axis (not in any quadrant).

Graph for Exercises 9–12

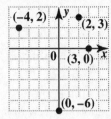

13. The product of two numbers is positive whenever the two numbers have the same sign. If $xy > 0$, either $x > 0$ and $y > 0$, so that (x, y) lies in quadrant I, or $x < 0$ and $y < 0$, so that (x, y) lies in quadrant III.

14. To find the x-intercept, let $y = 0$.

$$y = 2x + 5$$
$$0 = 2x + 5 \quad \text{Let } y = 0.$$
$$-2x = 5$$
$$x = -\frac{5}{2}$$

The x-intercept is $\left(-\frac{5}{2}, 0\right)$.

To find the y-intercept, let $x = 0$.

$$y = 2x + 5$$
$$y = 2(0) + 5 \quad \text{Let } x = 0.$$
$$y = 5$$

The y-intercept is $(0, 5)$.

To find a third point, choose $x = -1$.

$$y = 2x + 5$$
$$y = 2(-1) + 5 \quad \text{Let } x = -1.$$
$$y = 3$$

This gives the ordered pair $(-1, 3)$. Plot $\left(-\frac{5}{2}, 0\right)$, $(0, 5)$, and $(-1, 3)$ and draw a line through them.

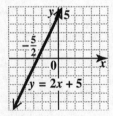

15. To find the x-intercept, let $y = 0$.

$$3x + 2y = 8$$
$$3x + 2(0) = 8 \quad \text{Let } y = 0.$$
$$3x = 8$$
$$x = \frac{8}{3}$$

The x-intercept is $\left(\frac{8}{3}, 0\right)$.

To find the y-intercept, let $x = 0$.

$$3x + 2y = 8$$
$$3(0) + 2y = 8 \quad \text{Let } x = 0.$$
$$2y = 8$$
$$y = 4$$

The y-intercept is $(0, 4)$.

To find a third point, choose $x = 1$.

$$3x + 2y = 8$$
$$3(1) + 2y = 8 \quad \text{Let } x = 1.$$
$$2y = 5$$
$$y = \frac{5}{2}$$

This gives the ordered pair $\left(1, \frac{5}{2}\right)$. Plot $\left(\frac{8}{3}, 0\right)$, $(0, 4)$, and $\left(1, \frac{5}{2}\right)$ and draw a line through them.

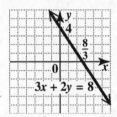

16. $x + 2y = -4$

Find the intercepts.

If $y = 0$, $x = -4$, so the x-intercept is $(-4, 0)$.

If $x = 0$, $y = -2$, so the y-intercept is $(0, -2)$.

To find a third point, choose $x = 2$.

$$x + 2y = -4$$
$$2 + 2y = -4 \quad \text{Let } x = 2.$$
$$2y = -6$$
$$y = -3$$

This gives the ordered pair $(2, -3)$. Plot $(-4, 0)$, $(0, -2)$, and $(2, -3)$ and draw a line through them.

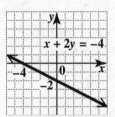

17. Let $(2, 3) = (x_1, y_1)$ and $(-4, 6) = (x_2, y_2)$.

$$\text{slope } m = \frac{\text{change in } y}{\text{change in } x} = \frac{y_2 - y_1}{x_2 - x_1}$$

$$= \frac{6 - 3}{-4 - 2} = \frac{3}{-6} = -\frac{1}{2}$$

18. Let $(2, 5) = (x_1, y_1)$ and $(2, 8) = (x_2, y_2)$.

$$\text{slope } m = \frac{\text{change in } y}{\text{change in } x} = \frac{y_2 - y_1}{x_2 - x_1}$$
$$= \frac{8 - 5}{2 - 2} = \frac{3}{0},$$

which is *undefined*.

19. $y = 3x - 4$

The equation is already solved for y, so the slope of the line is given by the coefficient of x. Thus, the slope is 3.

20. $y = 5$ is an equation of a horizontal line. Its slope is 0.

21. The indicated points have coordinates $(0, -2)$ and $(2, 1)$. Use the definition of slope with $(0, -2) = (x_1, y_1)$ and $(2, 1) = (x_2, y_2)$.

$$m = \frac{\text{change in } y}{\text{change in } x} = \frac{y_2 - y_1}{x_2 - x_1}$$
$$= \frac{1 - (-2)}{2 - 0} = \frac{3}{2}$$

Note: We could also simply count grid marks to get $m = \dfrac{\text{rise}}{\text{run}} = \dfrac{+3}{+2} = \dfrac{3}{2}$.

22. The indicated points have coordinates $(0, 1)$ and $(3, 0)$. Use the definition of slope with $(0, 1) = (x_1, y_1)$ and $(3, 0) = (x_2, y_2)$.

$$m = \frac{\text{change in } y}{\text{change in } x} = \frac{y_2 - y_1}{x_2 - x_1}$$
$$= \frac{0 - 1}{3 - 0} = \frac{-1}{3} = -\frac{1}{3}$$

Note: We could also simply count grid marks to get $m = \dfrac{\text{rise}}{\text{run}} = \dfrac{-1}{+3} = -\dfrac{1}{3}$.

23. From the table, we choose the two points $(0, 1)$ and $(2, 4)$. Therefore,

$$\text{slope } m = \frac{\text{change in } y}{\text{change in } x} = \frac{y_2 - y_1}{x_2 - x_1}$$
$$= \frac{4 - 1}{2 - 0} = \frac{3}{2}.$$

24. **(a)** Because parallel lines have equal slopes and the slope of the graph of $y = 2x + 3$ is 2, the slope of a line parallel to it will also be 2.

(b) Because perpendicular lines have slopes which are negative reciprocals of each other and the slope of the graph of $y = -3x + 3$ is -3, the slope of a line perpendicular to it will be

$$-\frac{1}{-3} = \frac{1}{3}.$$

25. Find the slope of each line by solving the equations for y.

$$3x + 2y = 6$$
$$2y = -3x + 6 \quad \text{Subtract 3x.}$$
$$y = -\tfrac{3}{2}x + 3 \quad \text{Divide by 2.}$$

The slope of the first line is $-\frac{3}{2}$.

$$6x + 4y = 8$$
$$4y = -6x + 8 \quad \text{Subtract 6x.}$$
$$y = -\tfrac{6}{4}x + 2 \quad \text{Divide by 4.}$$
$$y = -\tfrac{3}{2}x + 2 \quad \text{Lowest terms}$$

The slope of the second line is $-\frac{3}{2}$. The slopes are equal so the lines are *parallel*.

26. Find the slope of each line by solving the equations for y.

$$x - 3y = 1$$
$$-3y = -x + 1 \quad \text{Subtract x.}$$
$$y = \tfrac{1}{3}x - \tfrac{1}{3} \quad \text{Divide by –3.}$$

The slope of the first line is $\frac{1}{3}$.

$$3x + y = 4$$
$$y = -3x + 4 \quad \text{Subtract 3x.}$$

The slope of the second line is -3.

The product of the slopes is

$$\tfrac{1}{3}(-3) = -1,$$

so the lines are *perpendicular*.

27. Find the slope of each line by solving the equations for y.

$$x - 2y = 8$$
$$-2y = -x + 8 \quad \text{Subtract x.}$$
$$y = \tfrac{1}{2}x - 4 \quad \text{Divide by –2.}$$

The slope of the first line is $\frac{1}{2}$.

$$x + 2y = 8$$
$$2y = -x + 8 \quad \text{Subtract x.}$$
$$y = -\tfrac{1}{2}x + 4 \quad \text{Divide by 2.}$$

The slope of the second line is $-\frac{1}{2}$.

The slopes are not equal and their product is

$$\left(\tfrac{1}{2}\right)\left(-\tfrac{1}{2}\right) = -\tfrac{1}{4} \neq -1,$$

so the lines are *neither* parallel nor perpendicular.

28. $m = -1, b = \frac{2}{3}$

Use the slope-intercept form, $y = mx + b$.

$$y = -1 \cdot x + \tfrac{2}{3} \quad \text{or} \quad y = -x + \tfrac{2}{3}$$

29. Through $(2, 3)$ and $(-4, 6)$

$$m = \frac{6 - 3}{-4 - 2} = \frac{3}{-6} = -\frac{1}{2}$$

Use $(2, 3)$ and $m = -\frac{1}{2}$ in the point-slope form.

$$y - y_1 = m(x - x_1)$$
$$y - 3 = -\frac{1}{2}(x - 2)$$
$$y - 3 = -\frac{1}{2}x + 1$$
$$y = -\frac{1}{2}x + 4$$

30. Through $(4, -3)$, $m = 1$

Use the point-slope form.

$$y - y_1 = m(x - x_1)$$
$$y - (-3) = 1(x - 4)$$
$$y + 3 = x - 4$$
$$y = x - 7$$

31. Through $(-1, 4)$, $m = \frac{2}{3}$

Use the point-slope form.

$$y - y_1 = m(x - x_1)$$
$$y - 4 = \frac{2}{3}[x - (-1)]$$
$$y - 4 = \frac{2}{3}(x + 1)$$
$$y - 4 = \frac{2}{3}x + \frac{2}{3}$$
$$y = \frac{2}{3}x + \frac{14}{3}$$

32. Through $(1, -1)$, $m = -\frac{3}{4}$

Use the point-slope form.

$$y - (-1) = -\frac{3}{4}(x - 1)$$
$$y + 1 = -\frac{3}{4}x + \frac{3}{4}$$
$$y = -\frac{3}{4}x - \frac{1}{4}$$

33. $m = -\frac{1}{4}$, $b = \frac{3}{2}$

Use the slope-intercept form, $y = mx + b$.

$$y = -\frac{1}{4}x + \frac{3}{2}$$

34. Slope 0, through $(-4, 1)$

Horizontal lines have 0 slope and equations of the form $y = k$. In this case, k must equal 1 since the line goes through $(-4, 1)$, so the equation is $y = 1$.

35. Through $(\frac{1}{3}, -\frac{5}{4})$ with undefined slope

Vertical lines have undefined slope and equations of the form $x = k$. In this case, k must equal $\frac{1}{3}$ since the line goes through $(\frac{1}{3}, -\frac{5}{4})$, so the equation is $x = \frac{1}{3}$.

It is not possible to express this equation as $y = mx + b$.

36. $3x + 5y > 9$

To graph the boundary, which is the line $3x + 5y = 9$, find its intercepts.

$3x + 5y = 9$	$3x + 5y = 9$
$3x + 5(0) = 9$	$3(0) + 5y = 9$
Let y = 0.	*Let x = 0.*
$3x = 9$	$5y = 9$
$x = 3$	$y = \frac{9}{5}$

The x-intercept is $(3, 0)$ and the y-intercept is $(0, \frac{9}{5})$. (A third point may be used as a check.) Draw a dashed line through these points. In order to determine which side of the line should be shaded, use $(0, 0)$ as a test point. Substituting 0 for x and 0 for y will result in the inequality $0 > 9$, which is false. Shade the region *not* containing the origin. This is the region above the line. The dashed line shows that the boundary is not part of the graph.

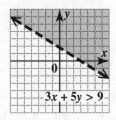

37. $2x - 3y > -6$

Use intercepts to graph the boundary, $2x - 3y = -6$.

If $y = 0$, $x = -3$, so the x-intercept is $(-3, 0)$.

If $x = 0$, $y = 2$, so the y-intercept is $(0, 2)$.

Draw a dashed line through $(-3, 0)$ and $(0, 2)$.

Using $(0, 0)$ as a test point will result in the inequality $0 > -6$, which is true. Shade the region containing the origin. This is the region below the line. The dashed line shows that the boundary is not part of the graph.

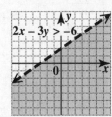

38. $x - 2y \geq 0$

The equation of the boundary is $x - 2y = 0$. This line goes through the origin, so both intercepts are $(0, 0)$. Two other points on this line are $(2, 1)$ and $(-2, -1)$. Draw a solid line through $(0, 0)$, $(2, 1)$, and $(-2, -1)$.

Because $(0,0)$ lies on the boundary, we must choose another point as the test point. Using $(0,3)$ results in the inequality $-6 \geq 0$, which is false. Shade the region *not* containing the test point $(0,3)$.

This is the region below the line.

The solid line shows that the boundary is part of the graph.

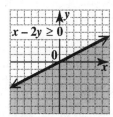

39. $\{(-2,4), (0,8), (2,5), (2,3)\}$

Since $x = 2$ appears in two ordered pairs, one value of x yields more than one value of y. Hence, this relation is not a function.

The domain is the set of first components of the ordered pairs, $\{-2, 0, 2\}$.

The range is the set of second components of the ordered pairs, $\{4, 8, 5, 3\}$.

40. $\{(8,3), (7,4), (6,5), (5,6), (4,7)\}$

Since each first component of the ordered pairs corresponds to exactly one second component, the relation is a function.

The domain is $\{8, 7, 6, 5, 4\}$.

The range is $\{3, 4, 5, 6, 7\}$.

41. Since a vertical line may cross the graph twice, this is not the graph of a function.

42. Any vertical line will cross this graph at exactly one point, so it is the graph of a function.

43. $2x + 3y = 12$

Solve the equation for y.
$$3y = -2x + 12$$
$$y = -\tfrac{2}{3}x + 4$$

Since one value of x will lead to only one value of y, $2x + 3y = 12$ is a function.

44. $y = x^2$

Each value of x will lead to only one value of y, so $y = x^2$ is a function.

45. $f(x) = 3x + 2$

(a) $f(2) = 3(2) + 2 = 6 + 2 = 8$

(b) $f(-1) = 3(-1) + 2 = -3 + 2 = -1$

46. $f(x) = 2x^2 - 1$

(a) $f(2) = 2(2)^2 - 1$
$$= 2(4) - 1 = 8 - 1 = 7$$

(b) $f(-1) = 2(-1)^2 - 1$
$$= 2(1) - 1 = 2 - 1 = 1$$

47. $f(x) = |x + 3|$

(a) $f(2) = |2 + 3| = |5| = 5$

(b) $f(-1) = |-1 + 3| = |2| = 2$

48. **[3.3]** Vertical lines have undefined slopes. The answer is **A**.

49. **[3.2]** Two graphs pass through $(0, -3)$. **C** and **D** are the answers.

50. **[3.2]** Three graphs pass through the point $(-3, 0)$. **A**, **B**, and **D** are the answers.

51. **[3.3]** Lines that fall from left to right have negative slope. The answer is **D**.

52. **[3.2]** $y = -3$ is a horizontal line passing through $(0, -3)$. **C** is the answer.

53. **[3.3]** **B** is the only graph that has a positive slope, so it is the only one we need to investigate. **B** passes through the points $(0, 3)$ and $(-3, 0)$. Find the slope.
$$m = \frac{0 - 3}{-3 - 0} = \frac{-3}{-3} = 1$$

B is the answer.

54. **[3.3]** $y = -2x - 5$

The equation is in the slope-intercept form, so the slope is -2 and the y-intercept is $(0, -5)$. To find the x-intercept, let $y = 0$.
$$0 = -2x - 5 \quad \textit{Let y = 0.}$$
$$2x = -5$$
$$x = -\tfrac{5}{2}$$

The x-intercept is $(-\tfrac{5}{2}, 0)$.

Graph the line using the intercepts.

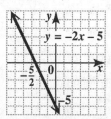

55. **[3.3]** $x + 3y = 0$

Solve the equation for y.

$$x + 3y = 0$$
$$3y = -x \quad \textit{Subtract x.}$$
$$y = -\frac{1}{3}x \quad \textit{Divide by 3.}$$

From this slope-intercept form, we see that the slope is $-\frac{1}{3}$ and the y-intercept is $(0, 0)$, which is also the x-intercept. To find another point, let $y = 1$.

$$x + 3(1) = 0 \quad \textit{Let y = 1.}$$
$$x = -3$$

So the point $(-3, 1)$ is on the graph. Graph the line through $(0, 0)$ and $(-3, 1)$.

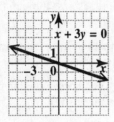

56. **[3.3]** $y - 5 = 0$ or $y = 5$

This is a horizontal line passing through the point $(0, 5)$, which is the y-intercept.

There is no x-intercept.
Horizontal lines have slopes of 0.

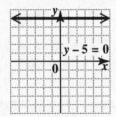

57. **[3.4]** $m = -\frac{1}{4}, b = -\frac{5}{4}$

Substitute the values $m = -\frac{1}{4}$ and $b = -\frac{5}{4}$ into the slope-intercept form.

$$y = mx + b$$
$$y = -\frac{1}{4}x - \frac{5}{4}$$

58. **[3.4]** Through $(8, 6), m = -3$

Use the point-slope form with $(x_1, y_1) = (8, 6)$ and $m = -3$.

$$y - y_1 = m(x - x_1)$$
$$y - 6 = -3(x - 8)$$
$$y - 6 = -3x + 24$$
$$y = -3x + 30$$

59. **[3.4]** Through $(3, -5)$ and $(-4, -1)$

First, find the slope of the line.

$$m = \frac{-1 - (-5)}{-4 - 3} = \frac{4}{-7} = -\frac{4}{7}$$

Now use either point and the slope in the point-slope form. If we use $(3, -5)$, we get the following.

$$y - y_1 = m(x - x_1)$$
$$y - (-5) = -\frac{4}{7}(x - 3)$$
$$y + 5 = -\frac{4}{7}x + \frac{12}{7}$$
$$y = -\frac{4}{7}x - \frac{23}{7} \quad \textit{Subtract 5} = \frac{35}{7}.$$

60. **[3.4]** Slope 0, through $(5, -5)$

Since the slope is 0, we have a horizontal line. Knowing that one y-value is -5, tells us that all y-values are -5, so the equation is $y = -5$.

61. **[3.5]** $x - 2y \leq 6$

This is a linear inequality, so its graph will be a shaded region. Graph the boundary, $x - 2y = 6$, as a solid line through the intercepts $(0, -3)$ and $(6, 0)$.

Using $(0, 0)$ as a test point results in the true statement $0 \leq 6$, so shade the region containing the origin. This is the region above the line. The solid line shows that the boundary is part of the graph.

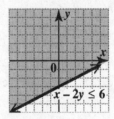

62. **[3.5]** $y < -4x$

This is a linear inequality, so its graph will be a shaded region. Graph the boundary, $y = -4x$, as a dashed line through $(0, 0)$, $(1, -4)$, and $(-1, 4)$.

Choose a test point that does not lie on the line. Using $(2, 0)$ results in the statement $0 < -8$, which is false, so shade the region *not* containing $(2, 0)$. This is the region below the line. The dashed line shows that the boundary is not part of the graph.

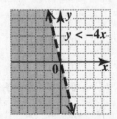

63. **[3.3]** Since the graph rises from left to right, the slope is positive.

64. **[3.1]** Let x represent the year and y represent the percent in the ordered pairs (x, y). The ordered pairs are $(2002, 41.2)$ and $(2007, 43.7)$.

65. **[3.4]** Find the slope using $(x_1, y_1) = (2002, 41.2)$ and $(x_2, y_2) = (2007, 43.7)$.

$$
\begin{aligned}
m &= \frac{y_2 - y_1}{x_2 - x_1} \\
&= \frac{43.7 - 41.2}{2007 - 2002} \\
&= \frac{2.5}{5} \\
&= 0.5
\end{aligned}
$$

Now use the point-slope form with $m = 0.5$ and $(x_1, y_1) = (2002, 41.2)$.

$$
\begin{aligned}
y - y_1 &= m(x - x_1) \\
y - 41.2 &= 0.5(x - 2002) \\
y - 41.2 &= 0.5x - 1001 \\
y &= 0.5x - 959.8
\end{aligned}
$$

66. **[3.1]** Substitute 2003, 2004, 2005, and 2006 for x in $y = 0.5x - 959.8$ to complete the table.

Year (x)	Percent (y)
2003	41.7
2004	42.2
2005	42.7
2006	43.2

Chapter 3 Test

1. $3x + 5y = -30$; $(0, \_\_)$, $(\_\_, 0)$, $(\_\_, -3)$

$$
\begin{aligned}
3x + 5y &= -30 \\
3(0) + 5y &= -30 \quad \textit{Let x = 0.} \\
5y &= -30 \\
y &= -6
\end{aligned}
$$

$$
\begin{aligned}
3x + 5y &= -30 \\
3x + 5(0) &= -30 \quad \textit{Let y = 0.} \\
3x &= -30 \\
x &= -10
\end{aligned}
$$

$$
\begin{aligned}
3x + 5y &= -30 \\
3x + 5(-3) &= -30 \quad \textit{Let y = -3.} \\
3x - 15 &= -30 \\
3x &= -15 \\
x &= -5
\end{aligned}
$$

The ordered pairs are $(0, -6)$, $(-10, 0)$, and $(-5, -3)$.

2.
$$
\begin{aligned}
4x - 7y &= 9 \\
4(4) - 7(-1) &\overset{?}{=} 9 \quad \textit{Let x = 4, y = -1.} \\
16 + 7 &\overset{?}{=} 9 \\
23 &\overset{?}{=} 9 \quad \textit{No}
\end{aligned}
$$

So $(4, -1)$ is not a solution of $4x - 7y = 9$.

3. To find the x-intercept, let $y = 0$ and solve for x. To find the y-intercept, let $x = 0$ and solve for y.

4. $3x + y = 6$

If $y = 0$, $x = 2$, so the x-intercept is $(2, 0)$.
If $x = 0$, $y = 6$, so the y-intercept is $(0, 6)$.

A third point, such as $(1, 3)$, can be used as a check. Draw a line through $(0, 6)$, $(1, 3)$, and $(2, 0)$.

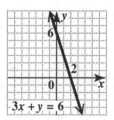

5. $y - 2x = 0$
Solving for y gives us the slope-intercept form of the line, $y = 2x$. We see that the y-intercept is $(0, 0)$ and so the x-intercept is also $(0, 0)$. The slope is 2 and can be written as

$$
m = \frac{\text{rise}}{\text{run}} = \frac{2}{1}.
$$

Starting at the origin and moving to the right 1 unit and then up 2 units gives us the point $(1, 2)$. Draw a line through $(0, 0)$ and $(1, 2)$.

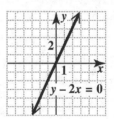

6. $x + 3 = 0$ can also be written as $x = -3$. Its graph is a vertical line with x-intercept $(-3, 0)$. There is no y-intercept.

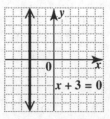

7. $y = 1$ is the graph of a horizontal line with y-intercept $(0, 1)$. There is no x-intercept.

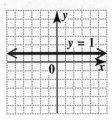

8. $x - y = 4$

If $y = 0$, $x = 4$, so the x-intercept is $(4, 0)$.
If $x = 0$, $y = -4$, so the y-intercept is $(0, -4)$.

A third point, such as $(2, -2)$, can be used as a check. Draw a line through $(0, -4)$, $(2, -2)$, and $(4, 0)$.

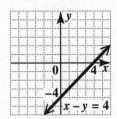

9. Through $(-4, 6)$ and $(-1, -2)$

Use the definition of slope with $(x_1, y_1) = (-4, 6)$ and $(x_2, y_2) = (-1, -2)$.

$$\text{slope } m = \frac{\text{change in } y}{\text{change in } x} = \frac{y_2 - y_1}{x_2 - x_1}$$
$$= \frac{-2 - 6}{-1 - (-4)} = \frac{-8}{3} = -\frac{8}{3}$$

10. $2x + y = 10$

To find the slope, solve the given equation for y.

$$2x + y = 10$$
$$y = -2x + 10$$

The equation is now written in $y = mx + b$ form, so the slope is given by the coefficient of x, which is -2.

11. $x + 12 = 0$ can also be written as $x = -12$. Its graph is a vertical line with x-intercept $(-12, 0)$. The slope is undefined.

12. $y - 4 = 6$ can also be written as $y = 10$. Its graph is a horizontal line with y-intercept $(0, 10)$. Its slope is 0, as is the slope of any line parallel to it.

13. The indicated points are $(0, -4)$ and $(2, 1)$. Use the definition of slope with $(x_1, y_1) = (0, -4)$ and $(x_2, y_2) = (2, 1)$.

$$\text{slope } m = \frac{\text{change in } y}{\text{change in } x} = \frac{y_2 - y_1}{x_2 - x_1}$$
$$= \frac{1 - (-4)}{2 - 0} = \frac{5}{2}$$

14. Through $(-1, 4)$; $m = 2$

Let $x_1 = -1$, $y_1 = 4$, and $m = 2$ in the point-slope form.

$$y - y_1 = m(x - x_1)$$
$$y - 4 = 2[x - (-1)]$$
$$y - 4 = 2(x + 1)$$
$$y - 4 = 2x + 2$$
$$y = 2x + 6$$

15. The indicated points are $(0, -4)$ and $(2, 1)$. The slope of the line through those points is

$$m = \frac{1 - (-4)}{2 - 0} = \frac{5}{2}.$$

The y-intercept is $(0, -4)$, so the slope-intercept form is

$$y = \tfrac{5}{2}x - 4.$$

16. Through $(2, -6)$ and $(1, 3)$

The slope of the line through these points is

$$m = \frac{3 - (-6)}{1 - 2} = \frac{9}{-1} = -9.$$

Use the point-slope form of a line with $(1, 3) = (x_1, y_1)$ and $m = -9$.

$$y - y_1 = m(x - x_1)$$
$$y - 3 = -9(x - 1)$$
$$y - 3 = -9x + 9$$
$$y = -9x + 12$$

17. $x + y \le 3$

Graph the boundary, $x + y = 3$, as a solid line through the intercepts $(3, 0)$ and $(0, 3)$.

Using $(0, 0)$ as a test point results in the true statement $0 \le 3$, so shade the region containing the origin. This is the region below the line. The solid line shows that the boundary is part of the graph.

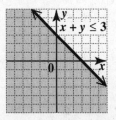

18. $3x - y > 0$

The boundary, $3x - y = 0$, goes through the origin, so both intercepts are $(0,0)$. Two other points on this line are $(1,3)$ and $(-1,-3)$. Draw the boundary as a dashed line.

Choose a test point which is not on the boundary. Using $(3,0)$ results in the true statement $9 > 0$, so shade the region containing $(3,0)$. This is the region below the line. The dashed line shows that the boundary is not part of the graph.

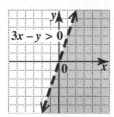

19. The slope is negative since worldwide snowmobile sales are decreasing, indicated by the line which falls from left to right.

20. Two ordered pairs are $(0, 209)$ and $(7, 160)$. Use these points to find the slope.

$$m = \frac{y_2 - y_1}{x_2 - x_1} = \frac{160 - 209}{7 - 0} = \frac{-49}{7} = -7$$

The slope is -7. Using the ordered pair $(0, 209)$, which is the y-intercept, and the slope of -7, we get the slope-intercept form

$$y = -7x + 209.$$

21. For 2005, $x = 2005 - 2000 = 5$.

$$y = -7x + 209$$
$$y = -7(5) + 209$$
$$y = 174$$

This represents worldwide snowmobile sales of 174 thousand in 2005. The equation gives a good approximation of the actual sales, which are 173.7 thousand.

22. The ordered pair $(7, 160)$ means that in the year 2007, worldwide snowmobile sales were 160 thousand.

23. **(a)** $\{(2,3),(2,4),(2,5)\}$
Since $x = 2$ appears in two ordered pairs, one value of x yields more than one value of y. Hence, this relation is not a function. **(b)** $\{(0,2),(1,2),(2,2)\}$
Since each first component of the ordered pairs corresponds to exactly one second component, the relation is a function. The domain is $\{0, 1, 2\}$ and the range is $\{2\}$.

24. The vertical line test shows that this graph is not the graph of a function; a vertical line could cross the graph twice.

25. $f(x) = 3x + 7$
$f(-2) = 3(-2) + 7 = -6 + 7 = 1$

Cumulative Review Exercises (Chapters 1–3)

1. $10\frac{5}{8} - 3\frac{1}{10} = \frac{85}{8} - \frac{31}{10}$

$$= \frac{425}{40} - \frac{124}{40}$$

$$= \frac{301}{40}, \text{ or } 7\frac{21}{40}$$

2. $\dfrac{3}{4} \div \dfrac{1}{8} = \dfrac{3}{4} \cdot \dfrac{8}{1} = \dfrac{3 \cdot 2 \cdot 4}{4 \cdot 1} = 3 \cdot 2 = 6$

3. $5 - (-4) + (-2) = 9 + (-2) = 7$

4. $\dfrac{(-3)^2 - (-4)(2^4)}{5(2) - (-2)^3}$

$$= \frac{9 - (-4)(16)}{10 - (-8)} \quad \textit{Do exponents first.}$$

$$= \frac{9 - (-64)}{10 - (-8)} \quad \textit{Multiply.}$$

$$= \frac{9 + 64}{10 + 8} = \frac{73}{18}, \text{ or } 4\frac{1}{18}$$

5. $\dfrac{4(3 - 9)}{2 - 6} \overset{?}{\geq} 6$

$$\frac{4(-6)}{-4} \overset{?}{\geq} 6$$

$$\frac{-24}{-4} \overset{?}{\geq} 6$$

$$6 \geq 6$$

The statement is *true* since $6 = 6$.

6. $xz^3 - 5y^2 = (-2)(-1)^3 - 5(-3)^2$
$$\textit{Let } x = -2, \, y = -3, \, z = -1.$$
$$= (-2)(-1) + (-5)(9)$$
$$= 2 + (-45)$$
$$= -43$$

7. $3(-2 + x) = 3 \cdot (-2) + 3(x)$
$$= -6 + 3x$$
illustrates the *distributive property*.

8. $-4p - 6 + 3p + 8 = (-4p + 3p) + (-6 + 8)$
$$= -p + 2$$

9. $V = \frac{1}{3}\pi r^2 h$ for h
$$3V = \pi r^2 h \qquad \textit{Multiply by 3.}$$
$$\frac{3V}{\pi r^2} = h \qquad \textit{Divide by } \pi r^2.$$

10. $6 - 3(1 + x) = 2(x + 5) - 2$

$6 - 3 - 3x = 2x + 10 - 2$

$3 - 3x = 2x + 8$

$-5x = 5$

$x = -1$

The solution set is $\{-1\}$.

11. $-(m - 3) = 5 - 2m$

$\quad -m + 3 = 5 - 2m \quad$ *Distributive property*

$\quad m + 3 = 5 \qquad$ *Add 2m.*

$\quad m = 2 \qquad\quad$ *Subtract 3.*

The solution set is $\{2\}$.

12. $\dfrac{x - 2}{3} = \dfrac{2x + 1}{5}$

$(x - 2)(5) = (3)(2x + 1) \quad$ *Cross products*

$5x - 10 = 6x + 3$

$-10 = x + 3$

$-13 = x$

The solution set is $\{-13\}$.

13. $-2.5x < 6.5$

$\dfrac{-2.5x}{-2.5} > \dfrac{6.5}{-2.5}$

Divide by –2.5; reverse the symbol.

$x > -2.6$

Thus, the solution set is $(-2.6, \infty)$.

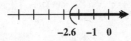

14. $4(x + 3) - 5x < 12$

$\quad 4x + 12 - 5x < 12 \quad$ *Distributive property*

$\qquad\qquad\qquad\qquad$ *Combine*

$\quad -x + 12 < 12 \qquad$ *like terms.*

$\qquad -x < 0 \qquad$ *Subtract 12.*

$\qquad x > 0 \qquad$ *Divide by –1;*

$\qquad\qquad\qquad$ *reverse the symbol.*

Thus, the solution set is $(0, \infty)$.

15. $\frac{2}{3}x - \frac{1}{6}x \le -2$

$6\left(\frac{2}{3}x - \frac{1}{6}x\right) \le 6(-2) \quad$ *Multiply by 6.*

$6\left(\frac{2}{3}x\right) - 6\left(\frac{1}{6}x\right) \le 6(-2) \quad$ *Dist. property*

$4x - x \le -12$

$3x \le -12$

$x \le -4$

Thus, the solution set is $(-\infty, -4]$.

16. Let x = average annual earnings for a person with a high school diploma.

Then $x + 20{,}124$ = average annual earnings for a person with a bachelor's degree.

$x + (x + 20{,}124) = 81{,}588$

$2x + 20{,}124 = 81{,}588$

$2x = 61{,}464$

$x = 30{,}732$

A person with a high school diploma can expect to earn \$30,732 per year while a person with a bachelor's degree can expect to earn

$\$30{,}732 + \$20{,}124 = \$50{,}856$ per year.

17. **(a)** Multiply 14% (or 0.14) by the total of \$50,000.

$0.14(50{,}000) = 7000$

\$7000 is expected to go toward home purchase.

(b) Multiply 20% (or 0.20) by the total of \$50,000.

$0.20(50{,}000) = 10{,}000$

\$10,000 is expected to go toward retirement.

18. Since the sector for paying off debt or funding children's education is about three times larger than the sector for retirement, 3(\$10,000) or about \$30,000 is expected to go toward paying off debt or funding children's education.

19. To find the x-intercept, let $y = 0$.

$-3x + 4y = 12$

$-3x + 4(0) = 12$

$-3x = 12$

$x = -4$

The x-intercept is $(-4, 0)$.

To find the y-intercept, let $x = 0$.

$-3x + 4y = 12$

$-3(0) + 4y = 12$

$4y = 12$

$y = 3$

The y-intercept is $(0, 3)$.

20. To find the slope of the line, solve the equation for y.

$-3x + 4y = 12$

$4y = 3x + 12$

$y = \frac{3}{4}x + 3$

The slope is the coefficient of x, $\frac{3}{4}$.

21. To find a third point, let $x = 4$.

$$-3x + 4y = 12$$
$$-3(4) + 4y = 12$$
$$4y = 12 + 12 = 24$$
$$y = 6$$

Plot the points $(-4, 0)$, $(0, 3)$, and $(4, 6)$ and draw a line through them.

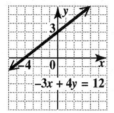

22. $x + 5y = -6$
$$5y = -x - 6$$
$$y = -\frac{1}{5}x - \frac{6}{5}$$

The slope of the first line is $-\frac{1}{5}$.

The slope of the second line, $y = 5x - 8$, is 5. Since $-\frac{1}{5}$ is the negative reciprocal of 5, the lines are *perpendicular*.

23. Through $(2, -5)$ with slope 3.

Use the point-slope form of a line.

$$y - y_1 = m(x - x_1)$$
$$y - (-5) = 3(x - 2)$$
$$y + 5 = 3x - 6$$
$$y = 3x - 11$$

24. Through $(0, 4)$ and $(2, 4)$

$$\text{slope } m = \frac{\text{change in } y}{\text{change in } x} = \frac{4 - 4}{2 - 0} = \frac{0}{2} = 0$$

Since the slope is 0, the line is horizontal. Horizontal lines have equations of the form $y = k$. An equation of the line is $y = 4$.

CHAPTER 4   SYSTEMS OF LINEAR EQUATIONS AND INEQUALITIES

4.1   Solving Systems of Linear Equations by Graphing

4.1   Now Try Exercises

N1. **(a)** $(5, 2)$ $2x + 5y = 20$
                       $x - y = 7$

To decide whether $(5, 2)$ is a solution, substitute 5 for x and 2 for y in each equation.

$$2x + 5y = 20$$
$$2(5) + 5(2) \overset{?}{=} 20$$
$$10 + 10 \overset{?}{=} 20$$
$$20 = 20 \quad \textit{True}$$

$$x - y = 7$$
$$5 - 2 \overset{?}{=} 7$$
$$3 = 7 \quad \textit{False}$$

Since $(5, 2)$ does not satisfy the second equation, it is not a solution of the system. **No**

(b) $(5, 2)$ $3x - y = 13$
                       $2x + y = 12$

To decide whether $(5, 2)$ is a solution, substitute 5 for x and 2 for y in each equation.

$$3x - y = 13$$
$$3(5) - 2 \overset{?}{=} 13$$
$$15 - 2 \overset{?}{=} 13 \quad \textit{Multiply.}$$
$$13 = 13 \quad \textit{True}$$

$$2x + y = 12$$
$$2(5) + 2 \overset{?}{=} 12$$
$$10 + 2 \overset{?}{=} 12 \quad \textit{Multiply.}$$
$$12 = 12 \quad \textit{True}$$

Since $(5, 2)$ satisfies both equations, it is a solution of the system. **Yes**

N2. $x - 2y = 4$
        $2x + y = 3$

Graph each equation by plotting several points for each line. The intercepts are good choices.

Graph $x - 2y = 4$.   Graph $2x + y = 3$.

x	y	x	y
0	-2	0	3
2	-1	1	1
4	0	$\frac{3}{2}$	0

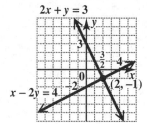

As suggested by the figure, the point at which the graphs of the two lines intersect is $(2, -1)$. Check by substituting 2 for x and -1 for y in both equations of the system. The solution set is $\{(2, -1)\}$.

N3. **(a)** $5x - 3y = 2$
                 $10x - 6y = 4$

Notice that if we multiply each side of the first equation by 2, we obtain the second equation. Thus, the graphs of these two equations are the same line.

Graph $5x - 3y = 2$.

x	y
0	$-\frac{2}{3}$
$\frac{2}{5}$	0
1	1

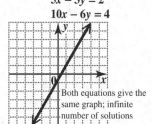

Both equations give the same graph; infinite number of solutions

In this case, every point on the line is a solution of the system, and the solution set contains an infinite number of ordered pairs, each of which satisfies both equations of the system. We write the solution set as $\{(x, y) \mid 5x - 3y = 2\}$.

(b) $4x + y = 7$
$12x + 3y = 10$

Graph $4x + y = 7$.

x	y
$\frac{7}{4}$	0
0	7
1	3

Graph $12x + 3y = 10$.

x	y
$\frac{5}{6}$	0
0	$\frac{10}{3}$
1	$-\frac{2}{3}$

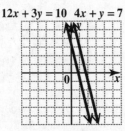

$12x + 3y = 10$ $4x + y = 7$

The lines each have slope -4 and hence, are parallel. Since they do not intersect, there is *no solution*. We write the solution set as $\emptyset$.

N4. **(a)** Solve each equation for y.

$$5x - 8y = 4 \quad \Big| \quad x - \tfrac{8}{5}y = \tfrac{4}{5}$$
$$-8y = -5x + 4 \quad \Big| \quad -\tfrac{8}{5}y = -x + \tfrac{4}{5}$$
$$y = \tfrac{5}{8}x - \tfrac{1}{2} \quad \Big| \quad y = \tfrac{5}{8}x - \tfrac{1}{2}$$

The slope-intercept forms are the same, so the equations represent the same line. The system has an infinite number of solutions.

(b) Solve each equation for y.

$$2x + y = 7 \quad \Big| \quad 3y = -6x - 12$$
$$y = -2x + 7 \quad \Big| \quad y = -2x - 4$$

The equations have the same slope, -2, but different y-intercepts, 7 and -4. The equations represent parallel lines. The system has no solution.

(c) Solve each equation for y.

$$y - 3x = 7 \quad \Big| \quad 3y - x = 0$$
$$y = 3x + 7 \quad \Big| \quad 3y = x$$
$$\quad \Big| \quad y = \tfrac{1}{3}x$$

The slopes, 3 and $\frac{1}{3}$, are different, so the graphs of these equations are neither parallel nor the same line. The system has exactly one solution.

4.1   Section Exercises

1. From the graph, the ordered pair that is a solution of the system is in the second quadrant. Choice **A**, $(-4, -4)$, is the only ordered pair given that is not in quadrant II, so it is the only valid choice.

3. $(2, -3)$ $x + y = -1$
$2x + 5y = 19$

To decide whether $(2, -3)$ is a solution of the system, substitute 2 for x and -3 for y in each equation.

$$x + y = -1$$
$$2 + (-3) \overset{?}{=} -1$$
$$-1 = -1 \quad True$$

$$2x + 5y = 19$$
$$2(2) + 5(-3) \overset{?}{=} 19$$
$$4 + (-15) \overset{?}{=} 19$$
$$-11 = 19 \quad False$$

The ordered pair $(2, -3)$ satisfies the first equation but not the second. Because it does not satisfy *both* equations, it is not a solution of the system.

5. $(-1, -3)$ $3x + 5y = -18$
$4x + 2y = -10$

Substitute -1 for x and -3 for y in each equation.

$$3x + 5y = -18$$
$$3(-1) + 5(-3) \overset{?}{=} -18$$
$$-3 - 15 \overset{?}{=} -18$$
$$-18 = -18 \quad True$$

$$4x + 2y = -10$$
$$4(-1) + 2(-3) \overset{?}{=} -10$$
$$-4 - 6 \overset{?}{=} -10$$
$$-10 = -10 \quad True$$

Since $(-1, -3)$ satisfies both equations, it is a solution of the system.

7. $(7, -2)$ $4x = 26 - y$
$3x = 29 + 4y$

Substitute 7 for x and -2 for y in each equation.

$$4x = 26 - y$$
$$4(7) \overset{?}{=} 26 - (-2)$$
$$28 \overset{?}{=} 26 + 2$$
$$28 = 28 \quad True$$

$$3x = 29 + 4y$$
$$3(7) \overset{?}{=} 29 + 4(-2)$$
$$21 \overset{?}{=} 29 - 8$$
$$21 = 21 \quad True$$

Since $(7, -2)$ satisfies both equations, it is a solution of the system.

9. $(6, -8)$ $-2y = x + 10$
      $3y = 2x + 30$

Substitute 6 for x and -8 for y in each equation.

$$-2y = x + 10$$
$$-2(-8) \overset{?}{=} 6 + 10$$
$$16 = 16 \qquad \textit{True}$$

$$3y = 2x + 30$$
$$3(-8) \overset{?}{=} 2(6) + 30$$
$$-24 \overset{?}{=} 12 + 30$$
$$-24 = 42 \qquad \textit{False}$$

The ordered pair $(6, -8)$ satisfies the first equation but not the second. Because it does not satisfy *both* equations, it is not a solution of the system.

11. $(0, 0)$ $4x + 2y = 0$
      $x + y = 0$

Substitute 0 for x and 0 for y in each equation.

$$4x + 2y = 0$$
$$4(0) + 2(0) \overset{?}{=} 0$$
$$0 = 0 \quad \textit{True}$$

$$x + y = 0$$
$$0 + 0 \overset{?}{=} 0$$
$$0 = 0 \quad \textit{True}$$

Since $(0, 0)$ satisfies both equations, it is a solution of the system.

13. **(a)** $(3, 4)$ is in quadrant I—choice **B** is correct.

(b) $(-2, 3)$ is in quadrant II—choice **C** is correct since choice **D** is the only other choice with intersection point in quadrant II, but it's x-coordinate of the intersection point is -3, not -2.

(c) $(-3, 2)$ is in quadrant II—choice **D** is correct.

(d) $(5, -2)$ is in quadrant IV—choice **A** is correct.

15. $x - y = 2$
      $x + y = 6$

To graph the equations, find the intercepts.

$\boldsymbol{x - y = 2}$:   Let $y = 0$; then $x = 2$.
          Let $x = 0$; then $y = -2$.

Plot the intercepts, $(2, 0)$ and $(0, -2)$, and draw the line through them.

$\boldsymbol{x + y = 6}$:   Let $y = 0$; then $x = 6$.
          Let $x = 0$; then $y = 6$.

Plot the intercepts, $(6, 0)$ and $(0, 6)$, and draw the line through them.

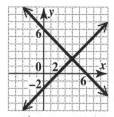

It appears that the lines intersect at the point $(4, 2)$. Check this by substituting 4 for x and 2 for y in both equations. Since $(4, 2)$ satisfies both equations, the solution set of this system is $\{(4, 2)\}$.

17. $x + y = 4$
      $y - x = 4$

To graph the equations, find the intercepts.

$\boldsymbol{x + y = 4}$:   Let $y = 0$; then $x = 4$.
          Let $x = 0$; then $y = 4$.

Plot the intercepts, $(4, 0)$ and $(0, 4)$, and draw the line through them.

$\boldsymbol{y - x = 4}$:   Let $y = 0$; then $x = -4$.
          Let $x = 0$; then $y = 4$.

Plot the intercepts, $(-4, 0)$ and $(0, 4)$, and draw the line through them.

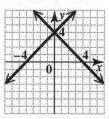

The lines intersect at their common y-intercept, $(0, 4)$, so $\{(0, 4)\}$ is the solution set of the system.

19. $x - 2y = 6$
$x + 2y = 2$

To graph the equations, find the intercepts.

$\boldsymbol{x - 2y = 6}$: Let $y = 0$; then $x = 6$.
Let $x = 0$; then $y = -3$.

Plot the intercepts, $(6, 0)$ and $(0, -3)$, and draw the line through them.

$\boldsymbol{x + 2y = 2}$: Let $y = 0$; then $x = 2$.
Let $x = 0$; then $y = 1$.

Plot the intercepts, $(2, 0)$ and $(0, 1)$, and draw the line through them.

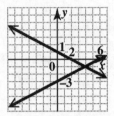

It appears that the lines intersect at the point $(4, -1)$. Since $(4, -1)$ satisfies both equations, the solution set of this system is $\{(4, -1)\}$.

21. $3x - 2y = -3$
$-3x - y = -6$

To graph the equations, find the intercepts.

$\boldsymbol{3x - 2y = -3}$: Let $y = 0$; then $x = -1$.
Let $x = 0$; then $y = \frac{3}{2}$.

Plot the intercepts, $(-1, 0)$ and $\left(0, \frac{3}{2}\right)$, and draw the line through them.

$\boldsymbol{-3x - y = -6}$: Let $y = 0$; then $x = 2$.
Let $x = 0$; then $y = 6$.

Plot the intercepts, $(2, 0)$ and $(0, 6)$, and draw the line through them.

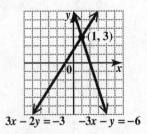

It appears that the lines intersect at the point $(1, 3)$. Since $(1, 3)$ satisfies both equations, the solution set of this system is $\{(1, 3)\}$.

23. $2x - 3y = -6$
$y = -3x + 2$

To graph the first line, find the intercepts.

$\boldsymbol{2x - 3y = -6}$: Let $y = 0$; then $x = -3$.
Let $x = 0$; then $y = 2$.

Plot the intercepts, $(-3, 0)$ and $(0, 2)$, and draw the line through them.

To graph the second line, start by plotting the y-intercept, $(0, 2)$. From this point, go 3 units down and 1 unit to the right (because the slope is -3) to reach the point $(1, -1)$. Draw the line through $(0, 2)$ and $(1, -1)$.

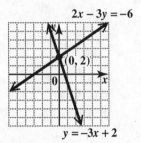

The lines intersect at their common y-intercept, $(0, 2)$, so $\{(0, 2)\}$ is the solution set of the system.

25. $2x - y = 6$
$4x - 2y = 8$

Graph the line $2x - y = 6$ through its intercepts, $(3, 0)$ and $(0, -6)$.

Graph the line $4x - 2y = 8$ through its intercepts, $(2, 0)$ and $(0, -4)$.

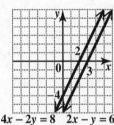

Solving each equation for y gives us $y = 2x - 6$ and $y = 2x - 4$, so the lines each have slope 2, and hence, are parallel. Since they do not intersect, there is no solution. This is an inconsistent system and the solution set is $\emptyset$.

27. $3x + y = 5$
$6x + 2y = 10$

$\boldsymbol{3x + y = 5}$: Let $y = 0$; then $x = \frac{5}{3}$.
Let $x = 0$; then $y = 5$.

Plot these intercepts, $\left(\frac{5}{3}, 0\right)$ and $(0, 5)$, and draw the line through them.

$\boldsymbol{6x + 2y = 10}$: Let $y = 0$; then $x = \frac{5}{3}$.
Let $x = 0$; then $y = 5$.

Plot these intercepts, $\left(\frac{5}{3}, 0\right)$ and $(0, 5)$, and draw the line through them.

Since both equations have the same intercepts, they are equations of the same line.

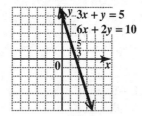

There is an infinite number of solutions. The equations are dependent equations and the solution set contains an infinite number of ordered pairs. The solution set is $\{(x, y) \mid 3x + y = 5\}$.

29. $3x - 4y = 24$
$y = -\frac{3}{2}x + 3$

Graph the line $3x - 4y = 24$ through its intercepts, $(8, 0)$ and $(0, -6)$.

To graph the line $y = -\frac{3}{2}x + 3$, plot the y-intercept $(0, 3)$ and then go 3 units down and 2 units to the right (because the slope is $-\frac{3}{2}$) to reach the point $(2, 0)$. Draw the line through $(0, 3)$ and $(2, 0)$.

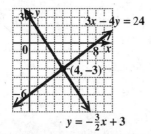

It appears that the lines intersect at the point $(4, -3)$. Since $(4, -3)$ satisfies both equations, the solution set of this system is $\{(4, -3)\}$.

31. $2x = y - 4$
$4x + 4 = 2y$

Graph the line $2x = y - 4$ through its intercepts, $(-2, 0)$ and $(0, 4)$.

Graph the line $4x + 4 = 2y$ through its intercepts, $(-1, 0)$ and $(0, 2)$.

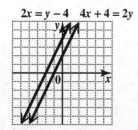

Solving each equation for y gives us $y = 2x + 4$ and $y = 2x + 2$, so the lines each have slope 2, and hence, are parallel. Since they do not

intersect, there is no solution. This is an inconsistent system and the solution set is $\emptyset$.

33. Graph both lines on the same axes. For $2x + 3y = 6$, use the intercepts $(3, 0)$ and $(0, 2)$. For $x - 3y = 5$, use the intercepts $(5, 0)$ and $(0, -\frac{5}{3})$. The two lines do not intersect at integer coordinates. It is difficult to read the exact coordinates of the solution. Thus, the solution cannot be checked.

35. $y - x = -5$
$x + y = 1$

Write the equations in slope-intercept form.

$y - x = -5$	$x + y = 1$
$y = x - 5$	$y = -x + 1$
$m = 1$	$m = -1$

The lines have different slopes.

(a) The system is consistent because it has a solution. The equations are independent because they have different graphs. Therefore, the answer is "neither."

(b) The graph is a pair of intersecting lines.

(c) The system has one solution.

37. $x + 2y = 0$
$4y = -2x$

Write the equations in slope-intercept form.

$x + 2y = 0$	$4y = -2x$
$2y = -x$	$y = -\frac{1}{2}x$
$y = -\frac{1}{2}x$	

For both lines, $m = -\frac{1}{2}$ and $b = 0$.

(a) Since the equations have the same slope and y-intercept, they are dependent.

(b) The graph is one line.

(c) The system has an infinite number of solutions.

39. $x - 3y = 5$
$2x + y = 8$

Write the equations in slope-intercept form.

$x - 3y = 5$	$2x + y = 8$
$-3y = -x + 5$	$y = -2x + 8$
$y = \frac{1}{3}x - \frac{5}{3}$	$m = -2$
$m = \frac{1}{3}$	

The lines have different slopes.

(a) The system is consistent because it has a solution. The equations are independent because they have different graphs. Therefore, the answer is "neither."

(b) The graph is a pair of intersecting lines.

(c) The system has one solution.

41. $5x + 4y = 7$

$10x + 8y = 4$

Write the equations in slope-intercept form.

$$
\begin{array}{l|l}
5x + 4y = 7 & 10x + 8y = 4 \\
\quad 4y = -5x + 7 & \quad 8y = -10x + 4 \\
\quad y = -\frac{5}{4}x + \frac{7}{4} & \quad y = -\frac{10}{8}x + \frac{4}{8} \\
\quad m = -\frac{5}{4}, b = \frac{7}{4} & \quad y = -\frac{5}{4}x + \frac{1}{2} \\
& \quad m = -\frac{5}{4}, b = \frac{1}{2}
\end{array}
$$

The lines have the same slope but different y-intercepts.

(a) The system is inconsistent because it has no solution.

(b) The graph is a pair of parallel lines.

(c) The system has no solution.

43. **(a)** The graph representing evening dailies is above the graph representing morning dailies for the years 1980 to 2000, so there were more evening dailies than morning dailies during the years 1980–2000.

(b) The graphs intersect when the year is about 2001. The intersection point is about $\frac{3}{4}$ of the way between 600 and 800, so there were about 750 newspapers of each type in 2001.

45. **(a)** The graph representing ATM use is above the graphs representing credit card *and* debit card use for the years 1997–2002.

(b) The graph representing debit card use rises above the graph for credit card use in about 2001.

(c) The graph representing debit card use rises above the graph for ATM use in about 2002.

(d) The y-value is approximately 30, so the ordered pair is $(1998, 30)$.

47. Supply equals demand at the point where the two lines intersect, or when $p = 30$.

49. For $x + y = 4$, the x-intercept is $(4, 0)$ and the y-intercept is $(0, 4)$. The only screen with a graph having those intercepts is choice **B**. The point of intersection, $(3, 1)$ is listed at the bottom of the screen. Check $(3, 1)$ in the system of equations. Since it satisfies the equations, $\{(3, 1)\}$ is the solution set.

51. The graph of $x - y = 0$ goes through the origin. The only screen with a graph that goes through the origin is choice **A**.

53. $3x + y = 4$

$\qquad y = -3x + 4$

55. $9x - 2y = 4$

$\quad -2y = -9x + 4$

$\qquad y = \frac{9}{2}x - 2$

57. $-2(x - 2) + 5x = 10$

$\quad -2x + 4 + 5x = 10$

$\qquad\quad 3x + 4 = 10$

$\qquad\qquad 3x = 6$

$\qquad\qquad\ x = 2$

Check $x = 2$: $0 + 10 = 10$

The solution set is $\{2\}$.

59. $4x - 2(1 - 3x) = 6$

$\quad 4x - 2 + 6x = 6$

$\qquad 10x - 2 = 6$

$\qquad\quad 10x = 8$

$\qquad\qquad x = \frac{8}{10} = \frac{4}{5}$

Check $x = \frac{4}{5}$: $\frac{16}{5} - 2(-\frac{7}{5}) = 6$

$\qquad\qquad\qquad \frac{16}{5} + \frac{14}{5} = 6$

The solution set is $\left\{\frac{4}{5}\right\}$.

4.2 Solving Systems of Linear Equations by Substitution

4.2 Now Try Exercises

N1. $2x - 4y = 28$

$\quad\ y = -3x$

The second equation is already solved for y. Substitute $-3x$ for x in the first equation.

$$
\begin{array}{ll}
2x - 4y = 28 & \\
2x - 4(-3x) = 28 & \textit{Let } y = -3x. \\
2x + 12x = 28 & \textit{Multiply.} \\
14x = 28 & \textit{Combine like terms.} \\
x = 2 & \textit{Divide by 2.}
\end{array}
$$

Now substitute 2 for x in the second equation.

$$y = -3x = -3(2) = -6$$

Check $x = 2, y = -6$:

$$
\begin{array}{c}
2x - 4y = 28 \\
2(2) - 4(-6) \stackrel{?}{=} 28 \\
4 + 24 \stackrel{?}{=} 28 \\
28 = 28 \quad \textit{True}
\end{array}
$$

Check $x = 2, y = -6$:

$$
\begin{array}{c}
y = -3x \\
-6 \stackrel{?}{=} -3(2) \\
-6 = -6 \quad \textit{True}
\end{array}
$$

Since $(2, -6)$ satisfies both equations, the solution set is $\{(2, -6)\}$.

N2. $4x + 9y = 1$

$\quad\quad x = y - 3$

Substitute $y - 3$ for x in the first equation, and solve for y.

$$4x + 9y = 1$$
$$4(y - 3) + 9y = 1 \quad\text{Let } x = y - 3.$$
$$4y - 12 + 9y = 1$$
$$13y - 12 = 1$$
$$13y = 13$$
$$y = 1$$

To find x, use $x = y - 3$ and $y = 1$.

$$x = y - 3 = 1 - 3 = -2$$

The solution set is $\{(-2, 1)\}$.

N3. $\quad 2y = x - 2 \quad (1)$

$\quad 4x - 5y = -4 \quad (2)$

Solve (1) for x.

$$2y = x - 2$$
$$2y + 2 = x \quad\quad\quad (3)$$

Substitute $2y + 2$ for x in equation (2) and solve for y.

$$4x - 5y = -4 \quad\text{Let } x = 2y + 2.$$
$$4(2y + 2) - 5y = -4$$
$$8y + 8 - 5y = -4$$
$$3y + 8 = -4$$
$$3y = -12$$
$$y = -4$$

To find x, let $y = -4$ in equation (3).

$$x = 2y + 2 = 2(-4) + 2$$
$$= -8 + 2 = -6$$

Check that $(-6, -4)$ is the solution.

$$2y = x - 2 \quad (1)$$
$$2(-4) \stackrel{?}{=} -6 - 2$$
$$-8 = -8 \quad\quad\text{True}$$

$$4x - 5y = -4 \quad (2)$$
$$4(-6) - 5(-4) \stackrel{?}{=} -4$$
$$-4 = -4 \quad\text{True}$$

The solution set of the system is $\{(-6, -4)\}$.

N4. $8x - 2y = 1 \quad (1)$

$\quad\, y = 4x - 8 \quad (2)$

Substitute $4x - 8$ for y in equation (1).

$$8x - 2y = 1$$
$$8x - 2(4x - 8) = 1 \quad\text{Let } y = 4x - 8.$$
$$8x - 8x + 16 = 8$$
$$16 = 8 \quad\text{False}$$

This false result indicates that the system has no solution, so the solution set is $\emptyset$.

N5. $\quad 5x - \;\; y = \;\;\; 6 \quad (1)$

$\quad -10x + 2y = -12 \quad (2)$

Solve equation (1) for y.

$$5x - y = 6$$
$$5x - 6 = y$$

Substitute $5x - 6$ for y in equation (2) and solve the resulting equation.

$$-10x + 2y = -12 \quad (2)$$
$$-10x + 2(5x - 6) = -12 \quad\text{Let } y = 5x - 6.$$
$$-10x + 10x - 12 = -12 \quad\text{Distributive property}$$
$$0 = 0 \quad\quad\text{Add 12; combine like terms.}$$

This true result means that every solution of one equation is also a solution of the other, so the system has an infinite number of solutions. The solution set is $\{(x, y) \mid 5x - y = 6\}$.

N6. $\quad x + \frac{1}{2}y = \frac{1}{2} \quad (1)$

$\quad \frac{1}{6}x - \frac{1}{3}y = \frac{4}{3} \quad (2)$

First, clear all fractions.

Equation (1):

$$2\left(x + \tfrac{1}{2}y\right) = 2\left(\tfrac{1}{2}\right) \quad\text{Multiply by the LCD, 2.}$$
$$2(x) + 2\left(\tfrac{1}{2}y\right) = 1 \quad\text{Distributive property}$$
$$2x + y = 1 \quad (3)$$

Equation (2):

$$6\left(\tfrac{1}{6}x - \tfrac{1}{3}y\right) = 6\left(\tfrac{4}{3}\right) \quad\text{Multiply by the LCD, 6.}$$
$$6\left(\tfrac{1}{6}x\right) - 6\left(\tfrac{1}{3}y\right) = 8 \quad\text{Distributive property}$$
$$x - 2y = 8 \quad (4)$$

The system has been simplified to

$$2x + y = 1 \quad (3)$$
$$x - 2y = 8 \quad (4)$$

Solve this system by the substitution method.

$$x = 2y + 8 \quad (5) \quad\text{Solve (4) for x.}$$

$$2(2y + 8) + y = 1 \quad \textit{Substitute for x in (3).}$$
$$4y + 16 + y = 1$$
$$5y + 16 = 1$$
$$5y = -15$$
$$y = -3$$

To find x, let $y = -3$ in equation (5).

$$x = 2(-3) + 8 = -6 + 8 = 2$$

The solution set is $\{(2, -3)\}$.

N7. $0.2x + 0.3y = 0.5$ (1)

$0.3x - 0.1y = 1.3$ (2)

First, clear all decimals.

Equation (1):

$$10(0.2x + 0.3y) = 10(0.5) \quad \textit{Multiply by 10.}$$
$$10(0.2x) + 10(0.3y) = 10(0.5) \quad \textit{Dist. property}$$
$$2x + 3y = 5 \quad (3)$$

Equation (2):

$$10(0.3x - 0.1y) = 10(1.3) \quad \textit{Multiply by 10.}$$
$$10(0.3x) - 10(0.1y) = 10(1.3) \quad \textit{Dist. property}$$
$$3x - y = 13 \quad (4)$$

The system has been simplified to

$$2x + 3y = 5 \quad (3)$$
$$3x - y = 13 \quad (4)$$

Solve this system by the substitution method.

$$y = 3x - 13 \quad (5) \quad \textit{Solve (4) for y.}$$

$$2x + 3(3x - 13) = 5 \quad \textit{Substitute for y in (3).}$$
$$2x + 9x - 39 = 5$$
$$11x - 39 = 5$$
$$11x = 44$$
$$x = 4$$

To find y, let $x = 4$ in equation (5).

$$y = 3(4) - 13 = 12 - 13 = -1$$

The solution set is $\{(4, -1)\}$.

4.2   Section Exercises

1.   The student must find the value of y and write the solution as an ordered pair. Substituting 3 for x in the first equation, $5x - y = 15$, gives us $15 - y = 15$, so $y = 0$. The correct solution set is $\{(3, 0)\}$.

In this section, all solutions should be checked by substituting in *both* equations of the original system. Checks will not be shown here.

3. $x + y = 12$ (1)

$y = 3x$ (2)

Equation (2) is already solved for y. Substitute $3x$ for y in equation (1) and solve the resulting equation for x.

$$x + y = 12$$
$$x + 3x = 12 \quad \textit{Let y = 3x.}$$
$$4x = 12$$
$$x = 3$$

To find the y-value of the solution, substitute 3 for x in equation (2).

$$y = 3x$$
$$y = 3(3) \quad \textit{Let x = 3.}$$
$$= 9$$

The solution set is $\{(3, 9)\}$.

To check this solution, substitute 3 for x and 9 for y in both equations of the given system.

5. $3x + 2y = 27$ (1)

$x = y + 4$ (2)

Equation (2) is already solved for x. Substitute $y + 4$ for x in equation (1).

$$3x + 2y = 27$$
$$3(y + 4) + 2y = 27$$
$$3y + 12 + 2y = 27$$
$$5y = 15$$
$$y = 3$$

To find x, substitute 3 for y in equation (2).

$$x = y + 4$$
$$x = 3 + 4 = 7$$

The solution set is $\{(7, 3)\}$.

7. $3x + 4 = -y$ (1)

$2x + y = 0$ (2)

Solve equation (1) for y.

$$3x + 4 = -y$$
$$y = -3x - 4 \quad (3)$$

Substitute $-3x - 4$ for y in equation (2) and solve for x.

$$2x + y = 0$$
$$2x + (-3x - 4) = 0$$
$$-x - 4 = 0$$
$$-x = 4$$
$$x = -4$$

To find y, substitute -4 for x in equation (3).

$$y = -3x - 4$$
$$= -3(-4) - 4 = 8$$

The solution set is $\{(-4, 8)\}$.

9. $7x + 4y = 13$ $\quad$ (1)
$\quad\quad x + y = 1$ $\quad\quad$ (2)

Solve equation (2) for y.

$$x + y = 1$$
$$y = 1 - x \quad (3)$$

Substitute $1 - x$ for y in equation (1).

$$7x + 4y = 13$$
$$7x + 4(1 - x) = 13$$
$$7x + 4 - 4x = 13$$
$$3x + 4 = 13$$
$$3x = 9$$
$$x = 3$$

To find y, substitute 3 for x in equation (3).

$$y = 1 - x$$
$$y = 1 - 3 = -2$$

The solution set is $\{(3, -2)\}$.

11. $3x + 5y = 25$ $\quad$ (1)
$\quad\; x - 2y = -10$ $\quad$ (2)

Solve equation (2) for x since its coefficient is 1.

$$x - 2y = -10$$
$$x = 2y - 10 \quad (3)$$

Substitute $2y - 10$ for x in equation (1) and solve for y.

$$3x + 5y = 25$$
$$3(2y - 10) + 5y = 25$$
$$6y - 30 + 5y = 25$$
$$11y = 55$$
$$y = 5$$

To find x, substitute 5 for y in equation (3).

$$x = 2y - 10$$
$$x = 2(5) - 10 = 0$$

The solution set is $\{(0, 5)\}$.

13. $3x - y = 5$ $\quad$ (1)
$\quad\;\; y = 3x - 5$ $\quad$ (2)

Equation (2) is already solved for y, so we substitute $3x - 5$ for y in equation (1).

$$3x - (3x - 5) = 5$$
$$3x - 3x + 5 = 5$$
$$5 = 5 \quad \textit{True}$$

This true result means that every solution of one equation is also a solution of the other, so the system has an infinite number of solutions. The solution set is $\{(x, y) \mid 3x - y = 5\}$.

15. $2x + y = 0$ $\quad$ (1)
$\quad\; 4x - 2y = 2$ $\quad$ (2)

Solve equation (1) for y.

$$2x + y = 0$$
$$y = -2x \quad (3)$$

Substitute $-2x$ for y in equation (2) and solve for x.

$$4x - 2(-2x) = 2$$
$$4x + 4x = 2$$
$$8x = 2$$
$$x = \tfrac{1}{4}$$

To find y, let $x = \tfrac{1}{4}$ in equation (3).

$$y = -2x$$
$$y = -2\left(\tfrac{1}{4}\right)$$
$$= -\tfrac{1}{2}$$

The solution set is $\left\{\left(\tfrac{1}{4}, -\tfrac{1}{2}\right)\right\}$.

17. $2x + 8y = 3$ $\quad$ (1)
$\quad\quad\; x = 8 - 4y$ $\quad$ (2)

Equation (2) is already solved for x, so substitute $8 - 4y$ for x in equation (1).

$$2(8 - 4y) + 8y = 3$$
$$16 - 8y + 8y = 3$$
$$16 = 3 \quad \textit{False}$$

This false result means that the system is inconsistent and its solution set is $\emptyset$.

19. $\quad 2y = 4x + 24$ $\quad$ (1)
$\quad\; 2x - y = -12$ $\quad\quad$ (2)

Solve equation (2) for y.

$$2x - y = -12$$
$$2x = y - 12$$
$$2x + 12 = y$$

Substitute $2x + 12$ for y in (1) and solve for x.

$$2(2x + 12) = 4x + 24$$
$$4x + 24 = 4x + 24 \quad \textit{True}$$

This true result means that every solution of one equation is also a solution of the other, so the system has an infinite number of solutions. The solution set is $\{(x, y) \mid 2x - y = -12\}$.

21. $\frac{1}{2}x + \frac{1}{3}y = 3$ (1)

$\quad\ \ y = 3x$ (2)

First, clear all fractions in equation (1).

$6\left(\frac{1}{2}x + \frac{1}{3}y\right) = 6(3)$ *Multiply by the LCD, 6.*

$6\left(\frac{1}{2}x\right) + 6\left(\frac{1}{3}y\right) = 18$ *Distributive property*

$\qquad\qquad 3x + 2y = 18$ (3)

From equation (2), substitute $3x$ for y in equation (3).

$$3x + 2(3x) = 18$$
$$3x + 6x = 18$$
$$9x = 18$$
$$x = 2$$

To find y, let $x = 2$ in equation (2).

$$y = 3(2) = 6$$

The solution set is $\{(2, 6)\}$.

23. $\frac{1}{2}x + \frac{1}{3}y = -\frac{1}{3}$ (1)

$\quad\ \ \frac{1}{2}x + 2y = -7$ (2)

First, clear all fractions.

Equation (1):

$6\left(\frac{1}{2}x + \frac{1}{3}y\right) = 6\left(-\frac{1}{3}\right)$ *Multiply by the LCD, 6.*

$6\left(\frac{1}{2}x\right) + 6\left(\frac{1}{3}y\right) = -2$ *Distributive property*

$\qquad\qquad 3x + 2y = -2$ (3)

Equation (2):

$2\left(\frac{1}{2}x + 2y\right) = 2(-7)$ *Multiply by 2.*

$\qquad\quad x + 4y = -14$ (4)

The system has been simplified to

$$3x + 2y = -2 \qquad (3)$$
$$x + 4y = -14 \qquad (4)$$

Solve this system by the substitution method.

$x = -4y - 14$ (5)   *Solve (4) for x.*

$3(-4y - 14) + 2y = -2$ *Substitute for x in (3).*

$$-12y - 42 + 2y = -2$$
$$-10y - 42 = -2$$
$$-10y = 40$$
$$y = -4$$

To find x, let $y = -4$ in equation (5).

$$x = -4(-4) - 14 = 16 - 14 = 2$$

The solution set is $\{(2, -4)\}$.

25. $\frac{x}{5} + 2y = \frac{8}{5}$ (1)

$\quad\ \ \frac{3x}{5} + \frac{y}{2} = -\frac{7}{10}$ (2)

First, clear all fractions.

Equation (1):

$5\left(\frac{x}{5} + 2y\right) = 5\left(\frac{8}{5}\right)$ *Multiply by 5.*

$\qquad\quad x + 10y = 8$ (3)

Equation (2):

$10\left(\frac{3x}{5} + \frac{y}{2}\right) = 10\left(-\frac{7}{10}\right)$ *Multiply by 10.*

$\qquad\quad 6x + 5y = -7$ (4)

The system has been simplified to

$$x + 10y = 8 \qquad (3)$$
$$6x + 5y = -7. \qquad (4)$$

Solve this system by the substitution method. Solve equation (3) for x.

$$x = 8 - 10y \quad (5)$$

Substitute $8 - 10y$ for x in equation (4).

$$6(8 - 10y) + 5y = -7$$
$$48 - 60y + 5y = -7$$
$$-55y = -55$$
$$y = 1$$

To find x, let $y = 1$ in equation (5).

$$x = 8 - 10(1) = -2$$

The solution set is $\{(-2, 1)\}$.

27. $\frac{1}{6}x + \frac{1}{3}y = 8$ (1)

$\quad\ \ \frac{1}{4}x + \frac{1}{2}y = 12$ (2)

Multiply equation (1) by 6.

$6\left(\frac{1}{6}x + \frac{1}{3}y\right) = 6(8)$

$\qquad\quad x + 2y = 48$ (3)

Multiply equation (2) by 4.

$4\left(\frac{1}{4}x + \frac{1}{2}y\right) = 4(12)$

$\qquad\quad x + 2y = 48$ (4)

Equations (3) and (4) are identical.

This means that every solution of one equation is also a solution of the other, so the system has an infinite number of solutions. The solution set is $\{(x, y) \mid x + 2y = 48\}$.

29. $0.2x - 1.3y = -3.2$ (1)
$-0.1x + 2.7y = 9.8$ (2)

Multiply equation (1) and equation (2) by 10 to eliminate the decimals. Then solve the resulting equations by the substitution method.

$2x - 13y = -32$ (3)
$-x + 27y = 98$ (4)

Solve equation (4) for x.

$27y - 98 = x$ (5)

Substitute $27y - 98$ for x in equation (3).

$2x - 13y = -32$
$2(27y - 98) - 13y = -32$ *Let $x = 27y - 98$.*
$54y - 196 - 13y = -32$
$41y = 164$
$y = 4$

To find x, let $y = 4$ in equation (5).

$x = 27(4) - 98 = 10$

The solution set is $\{(10, 4)\}$. Check in both of the original equations.

31. $0.3x - 0.1y = 2.1$ (1)
$0.6x + 0.3y = -0.3$ (2)

Multiply equation (1) and equation (2) by 10 to eliminate the decimals. Then solve the resulting equations by the substitution method.

$3x - y = 21$ (3)
$6x + 3y = -3$ (4)

Solve equation (3) for y.

$y = 3x - 21$ (5)

Substitute $3x - 21$ for y in equation (4).

$6x + 3(3x - 21) = -3$
$6x + 9x - 63 = -3$ *Let $y = 3x - 21$.*
$15x - 63 = -3$
$15x = 60$
$x = 4$

To find y, let $x = 4$ in equation (5).

$y = 3(4) - 21 = -9$

The solution set is $\{(4, -9)\}$. Check in both of the original equations.

33.   To find the total cost, multiply the number of bicycles (x) by the cost per bicycle ($400), and add the fixed cost ($5000). Thus, $y_1 = 400x + 5000$ gives the total cost (in dollars).

34.   Since each bicycle sells for $600, the total revenue for selling x bicycles is $600x$ (in dollars). Thus, $y_2 = 600x$ gives the total revenue.

35. $y_1 = 400x + 5000$ (1)
$y_2 = 600x$ (2)

To solve this system by the substitution method, substitute $600x$ for y_1 in equation (1).

$600x = 400x + 5000$
$200x = 5000$
$x = 25$

If $x = 25$, $y_2 = 600(25) = 15,000$.

The solution set is $\{(25, 15,000)\}$.

36.   The value of x from Exercise 35 is the number of bikes it takes to break even. When 25 bikes are sold, the break-even point is reached. At that point, you have spent 15,000 dollars and taken in 15,000 dollars.

37. $y = 6 - x$ (1)
$y = 2x$ (2)

Substitute $2x$ for y in equation (1).

$2x = 6 - x$
$3x = 6$
$x = 2$

Substituting 2 for x in either of the original equations gives $y = 4$.

The solution set is $\{(2, 4)\}$.

Input the equations $Y_1 = 6 - X$ and $Y_2 = 2X$. Next, graph the equations using a standard window. On the TI-82/3/4, just press ZOOM 6.

Now we'll let the calculator find the coordinates of the point of intersection of the graphs. Press 2nd CALC 5 ENTER ENTER to indicate the graphs for which we're trying to find the point of intersection. Now press the right cursor key, ▷, nine times to get close to the point of intersection. Lastly, press ENTER to produce the following graph.

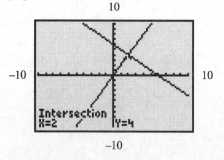

The display at the bottom of the last figure indicates that the solution set is $\{(2, 4)\}$.

39. $y = -\frac{4}{3}x + \frac{19}{3}$ (1)
$y = \frac{15}{2}x - \frac{5}{2}$ (2)

Substitute the expression from equation (2) into equation (1).

$$\frac{15}{2}x - \frac{5}{2} = -\frac{4}{3}x + \frac{19}{3}$$

Multiply by 6 to clear fractions.

$$6\left(\frac{15}{2}x - \frac{5}{2}\right) = 6\left(-\frac{4}{3}x + \frac{19}{3}\right)$$
$$45x - 15 = -8x + 38$$
$$53x = 53$$
$$x = 1$$

To find y, let $x = 1$ in equation (1).

$$y = -\frac{4}{3}(1) + \frac{19}{3}$$
$$= -\frac{4}{3} + \frac{19}{3} = \frac{15}{3} = 5$$

The solution set is $\{(1, 5)\}$.

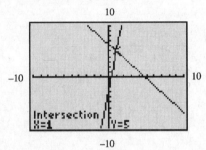

41. $4x + 5y = 5$ (1)
$2x + 3y = 1$ (2)

Solve equation (2) for y.

$$2x + 3y = 1$$
$$3y = 1 - 2x$$
$$y = \frac{1 - 2x}{3}$$ (3)

Substitute for y in equation (1).

$$4x + 5\left(\frac{1 - 2x}{3}\right) = 5$$
$$4x + \frac{5(1 - 2x)}{3} = 5$$

Multiply by 3 to clear fractions.

$$3\left[4x + \frac{5(1 - 2x)}{3}\right] = 3(5)$$
$$12x + 5(1 - 2x) = 15$$
$$12x + 5 - 10x = 15$$
$$2x = 10$$
$$x = 5$$

To find y, let $x = 5$ in equation (3).

$$y = \frac{1 - 2(5)}{3} = \frac{-9}{3} = -3$$

The solution set is $\{(5, -3)\}$.

To graph the original system on a graphing calculator, each equation must be solved for y.

Equation (1):

$$4x + 5y = 5$$
$$5y = 5 - 4x$$
$$y = \frac{5 - 4x}{5}$$

Equation (2) was solved for y — see (3).

Thus, the equations to input are

$$Y_1 = \frac{5 - 4X}{5} = (5 - 4X)/5$$

and

$$Y_2 = \frac{1 - 2X}{3} = (1 - 2X)/3.$$

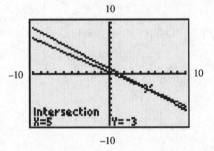

43. $(14x - 3y) + (2x + 3y)$
$= 14x - 3y + 2x + 3y$
$= (14 + 2)x + (-3 + 3)y$
$= 16x + 0y = 16x$

45. $(-x + 7y) + (3y + x)$
$= -x + 7y + 3y + x$
$= (-1 + 1)x + (7 + 3)y$
$= 0x + 10y = 10y$

47. To get a sum of 0, we must add the additive inverse of $-4x$, that is, $4x$.

49. The desired term times $4y$ must equal $-8y$, which is the additive inverse of $8y$. By trial and error, -2 times $4y$ equals $-8y$, so $4y$ must be multiplied by -2.

4.3 Solving Systems of Linear Equations by Elimination

4.3 Now Try Exercises

N1. $x - y = 4$
$3x + y = 8$

To eliminate y, add equations (1) and (2).

$$
\begin{array}{rcrl}
x & - & y & = & 4 & (1) \\
3x & + & y & = & 8 & (2) \\
\hline
4x & & & = & 12 & \textit{Add (1) and (2).} \\
& & x & = & 3
\end{array}
$$

This result gives the x-value of the solution. To find the y-value of the solution, substitute 2 for x in either equation. We will use equation (2).

$$3x + y = 8 \quad (2)$$
$$3(3) + y = 8 \quad \textit{Let x = 3.}$$
$$9 + y = 8$$
$$y = -1$$

Check by substituting 3 for x and -1 for y in both equations of the original system.

Check Equation (1)

$$x - y = 4$$
$$3 - (-1) \overset{?}{=} 4 \quad \textit{Let x = 3, y = -1.}$$
$$3 + 1 \overset{?}{=} 4$$
$$4 = 4 \quad \textit{True}$$

Check Equation (2)

$$3x + y = 8$$
$$3(3) + (-1) \overset{?}{=} 8 \quad \textit{Let x = 3, y = -1.}$$
$$9 - 1 \overset{?}{=} 8$$
$$8 = 8 \quad \textit{True}$$

Since both results are true, the solution set of the system is $\{(3, -1)\}$.

N2. $\quad 2x - 6 = -3y$
$\quad\quad 5x - 3y = -27$

First, rewrite the first equation in the standard form, $Ax + By = C$.

$$
\begin{array}{rcrcrl}
2x & + & 3y & = & 6 & (1) \\
5x & - & 3y & = & -27 & (2) \\
\hline
7x & & & = & -21 & \textit{Add (1) and (2).} \\
& & x & = & -3 &
\end{array}
$$

Substitute -3 for x in equation (1).

$$2x + 3y = 6$$
$$2(-3) + 3y = 6$$
$$-6 + 3y = 6$$
$$3y = 12$$
$$y = 4$$

The solution set is $\{(-3, 4)\}$.

N3. $\quad 3x - 5y = 25 \quad (1)$
$\quad\quad 2x + 8y = -6 \quad (2)$

If we simply add the equations, we will not eliminate either variable. To eliminate y, multiply equation (1) by 8, equation (2) by 5, and add the resulting equations.

$$
\begin{array}{rcrcrl}
24x & - & 40y & = & 200 & (3) \quad 8 \times \text{Eq. (1)} \\
10x & + & 40y & = & -30 & (4) \quad 5 \times \text{Eq. (2)} \\
\hline
34x & & & = & 170 & \textit{Add (3) and (4).} \\
& & x & = & 5 &
\end{array}
$$

Substitute 5 for x in equation (1).

$$3x - 5y = 25$$
$$3(5) - 5y = 25$$
$$15 - 5y = 25$$
$$-5y = 10$$
$$y = -2$$

Check $x = 5, y = -2$:

Eq. (1): $\quad 15 - (-10) = 25 \quad$ *True*
Eq. (2): $\quad 10 + (-16) = -6 \quad$ *True*

The solution set is $\{(5, -2)\}$.

N4. $\quad 4x + 9y = 3 \quad\quad$ (A)
$\quad\quad\quad 5y = 6 - 3x \quad\quad$ (B)

Rewrite in standard form.

$$4x + 9y = 3 \quad (1)$$
$$3x + 5y = 6 \quad (2)$$

Solve for y by using elimination. The least common multiple of 4 and 3 is 12.

$$
\begin{array}{rcrcrl}
-12x & - & 27y & = & -9 & (3) \quad -3 \times \text{Eq. (1)} \\
12x & + & 20y & = & 24 & (4) \quad 4 \times \text{Eq. (2)} \\
\hline
& & -7y & = & 15 & \textit{Add (3) and (4).} \\
& & y & = & -\frac{15}{7} &
\end{array}
$$

Solve for x by using elimination. The least common multiple of 9 and 5 is 45.

$$
\begin{array}{rcrcrl}
20x & + & 45y & = & 15 & (5) \quad 5 \times \text{Eq. (1)} \\
-27x & - & 45y & = & -54 & (6) \quad -9 \times \text{Eq. (2)} \\
\hline
-7x & & & = & -39 & \textit{Add (5) and (6).} \\
& & x & = & \frac{39}{7} &
\end{array}
$$

Check $x = \frac{39}{7}, y = -\frac{15}{7}$:

Eq. (A): $\quad \frac{156}{7} - \frac{135}{7} = \frac{21}{7} \quad$ *True*
Eq. (B): $\quad -\frac{75}{7} = \frac{42}{7} - \frac{117}{7} \quad$ *True*

The solution set is $\left\{\left(\frac{39}{7}, -\frac{15}{7}\right)\right\}$.

N5. (a) $\quad x - y = 2 \quad\quad (1)$
$\quad\quad\quad 5x - 5y = 10 \quad\quad (2)$

$$
\begin{array}{rcrcrl}
-5x & + & 5y & = & -10 & (3) \quad -5 \times \text{Eq. (1)} \\
5x & - & 5y & = & 10 & (2) \\
\hline
& & 0 & = & 0 & \textit{Add (3) and (2).}
\end{array}
$$

The true statement, $0 = 0$, indicates that there are an infinite number of solutions and the solution set is $\{(x, y) \mid x - y = 2\}$.

(b)

$$
\begin{array}{rcrcrl}
4x & + & 3y & = & 0 & (1) \\
-4x & - & 3y & = & -1 & (2) \\
\hline
& & 0 & = & -1 & \textit{Add (1) and (2).}
\end{array}
$$

This false statement, $0 = -1$, indicates that the given system has solution set $\emptyset$.

4.3 Section Exercises

1. The statement is *false*; if the elimination method leads to the equation $0 = -1$, the solution set of the system is $\emptyset$.

In Exercises 3–42, check your answers by substituting into *both* of the original equations. The check will be shown only for Exercise 3.

3. To eliminate y, add equations (1) and (2).

$$
\begin{array}{rcrcrl}
x & - & y & = & -2 & (1) \\
x & + & y & = & 10 & (2) \\
\hline
2x & & & = & 8 & \textit{Add } (1) \textit{ and } (2). \\
& & x & = & 4 &
\end{array}
$$

This result gives the x-value of the solution. To find the y-value of the solution, substitute 4 for x in either equation. We will use equation (2).

$$
\begin{aligned}
x + y &= 10 \\
4 + y &= 10 \quad \textit{Let x = 4.} \\
y &= 6
\end{aligned}
$$

Check by substituting 4 for x and 6 for y in both equations of the original system.

Check Equation (1)

$$
\begin{aligned}
x - y &= -2 \\
4 - 6 &\overset{?}{=} -2 \quad \textit{Let x = 4, y = 6.} \\
-2 &= -2 \quad \textit{True}
\end{aligned}
$$

Check Equation (2)

$$
\begin{aligned}
x + y &= 10 \\
4 + 6 &\overset{?}{=} 10 \quad \textit{Let x = 4, y = 6.} \\
10 &= 10 \quad \textit{True}
\end{aligned}
$$

The solution set is $\{(4, 6)\}$.

5.
$$
\begin{array}{rcrcrl}
2x & + & y & = & -5 & (1) \\
x & - & y & = & 2 & (2) \\
\hline
3x & & & = & -3 & \textit{Add } (1) \textit{ and } (2). \\
& & x & = & -1 &
\end{array}
$$

Substitute -1 for x in equation (1) to find the y-value of the solution.

$$
\begin{aligned}
2x + y &= -5 \\
2(-1) + y &= -5 \quad \textit{Let x = -1.} \\
-2 + y &= -5 \\
y &= -3
\end{aligned}
$$

The solution set is $\{(-1, -3)\}$.

7. First, rewrite both equations in the standard form, $Ax + By = C$. We can write the first equation, $2y = -3x$, as $3x + 2y = 0$ by adding $3x$ to each side.

$$
\begin{array}{rcrcrl}
3x & + & 2y & = & 0 & (1) \\
-3x & - & y & = & 3 & (2) \\
\hline
& & y & = & 3 & \textit{Add } (1) \textit{ and } (2).
\end{array}
$$

Substitute 3 for y in equation (1).

$$
\begin{aligned}
3x + 2y &= 0 \\
3x + 2(3) &= 0 \\
3x + 6 &= 0 \\
3x &= -6 \\
x &= -2
\end{aligned}
$$

The solution set is $\{(-2, 3)\}$.

9.
$$
\begin{aligned}
6x - y &= -1 \\
5y &= 17 + 6x
\end{aligned}
$$

Rewrite in standard form.

$$
\begin{array}{rcrcrl}
6x & - & y & = & -1 & (1) \\
-6x & + & 5y & = & 17 & (2) \\
\hline
& & 4y & = & 16 & \textit{Add } (1) \textit{ and } (2). \\
& & y & = & 4 & \textit{Solve for y.}
\end{array}
$$

Substitute 4 for y in equation (1).

$$
\begin{aligned}
6x - y &= -1 \\
6x - 4 &= -1 \\
6x &= 3 \\
x &= \tfrac{3}{6} = \tfrac{1}{2}
\end{aligned}
$$

The solution set is $\left\{ \left(\tfrac{1}{2}, 4 \right) \right\}$.

11.
$$
\begin{aligned}
2x - y &= 12 \quad (1) \\
3x + 2y &= -3 \quad (2)
\end{aligned}
$$

If we simply add the equations, we will not eliminate either variable. To eliminate y, multiply equation (1) by 2 and add the result to equation (2).

$$
\begin{array}{rcrcrl}
4x & - & 2y & = & 24 & (3) \\
3x & + & 2y & = & -3 & (2) \\
\hline
7x & & & = & 21 & \textit{Add } (3) \textit{ and } (2). \\
& & x & = & 3 &
\end{array}
$$

Substitute 3 for x in equation (1).

$$
\begin{aligned}
2x - y &= 12 \\
2(3) - y &= 12 \\
-y &= 6 \\
y &= -6
\end{aligned}
$$

The solution set is $\{(3, -6)\}$.

13.
$$
\begin{aligned}
x + 4y &= 16 \quad (1) \\
3x + 5y &= 20 \quad (2)
\end{aligned}
$$

To eliminate x, multiply equation (1) by -3 and add the result to equation (2).

$$
\begin{array}{rcll}
-3x & - & 12y & = & -48 & (3) \\
3x & + & 5y & = & 20 & (2) \\
\hline
& & -7y & = & -28 & \textit{Add (3) and (2).} \\
& & y & = & 4
\end{array}
$$

Substitute 4 for y in equation (1).

$$
\begin{aligned}
x + 4y &= 16 \\
x + 4(4) &= 16 \\
x + 16 &= 16 \\
x &= 0
\end{aligned}
$$

The solution set is $\{(0, 4)\}$.

15. $\quad 2x - 8y = 0 \qquad (1)$
$\qquad 4x + 5y = 0 \qquad (2)$

To eliminate x, multiply equation (1) by -2 and add the result to equation (2).

$$
\begin{array}{rcll}
-4x & + & 16y & = & 0 & (3) \\
4x & + & 5y & = & 0 & (2) \\
\hline
& & 21y & = & 0 & \textit{Add (3) and (2).} \\
& & y & = & 0
\end{array}
$$

Substitute 0 for y in equation (1).

$$
\begin{aligned}
2x - 8y &= 0 \\
2x - 8(0) &= 0 \\
2x &= 0 \\
x &= 0
\end{aligned}
$$

The solution set is $\{(0, 0)\}$.

17. $\quad 3x + 3y = 33 \qquad (1)$
$\qquad 5x - 2y = 27 \qquad (2)$

To eliminate y, multiply equation (1) by 2 and equation (2) by 3.

$$
\begin{array}{rcll}
6x & + & 6y & = & 66 & (3) \\
15x & - & 6y & = & 81 & (4) \\
\hline
21x & & & = & 147 & \textit{Add (3) and (4).} \\
& & x & = & 7
\end{array}
$$

Substitute 7 for x in equation (2).

$$
\begin{aligned}
5x - 2y &= 27 \\
5(7) - 2y &= 27 \\
35 - 2y &= 27 \\
-2y &= -8 \\
y &= 4
\end{aligned}
$$

The solution set is $\{(7, 4)\}$.

Note that we could have reduced equation (1) by dividing by 3 in the beginning.

19. $\quad 5x + 4y = 12 \qquad (1)$
$\qquad 3x + 5y = 15 \qquad (2)$

To eliminate x, we could multiply equation (1) by $-\frac{3}{5}$, but that would introduce fractions and make the solution more complicated. Instead, we'll work with the least common multiple of the

coefficients of x, which is 15, and choose suitable multipliers of these coefficients so that the new coefficients are opposites.

In this case, we could pick -3 times equation (1) and 5 times equation (2) *or* 3 times equation (1) and -5 times equation (2). If we wanted to eliminate y, we could multiply equation (1) by -5 and equation (2) by 4 *or* equation (1) by 5 and equation (2) by -4.

$$
\begin{array}{rclll}
-15x & - & 12y & = & -36 & (3)\;\; -3 \times \text{Eq.(1)} \\
15x & + & 25y & = & 75 & (4)\;\;\;\; 5 \times \text{Eq.(2)} \\
\hline
& & 13y & = & 39 & \textit{Add (3) and (4).} \\
& & y & = & 3
\end{array}
$$

Substitute 3 for y in equation (1).

$$
\begin{aligned}
5x + 4y &= 12 \\
5x + 4(3) &= 12 \\
5x + 12 &= 12 \\
5x &= 0 \\
x &= 0
\end{aligned}
$$

The solution set is $\{(0, 3)\}$.

21. $\quad 5x - 4y = 15 \qquad (1)$
$\qquad -3x + 6y = -9 \qquad (2)$

$$
\begin{array}{rclll}
15x & - & 12y & = & 45 & (3)\;\; 3 \times \text{Eq.(1)} \\
-15x & + & 30y & = & -45 & (4)\;\; 5 \times \text{Eq.(2)} \\
\hline
& & 18y & = & 0 & \textit{Add (3) and (4).} \\
& & y & = & 0
\end{array}
$$

Substitute 0 for y in equation (1).

$$
\begin{aligned}
5x - 4y &= 15 \\
5x - 4(0) &= 15 \\
5x &= 15 \\
x &= 3
\end{aligned}
$$

The solution set is $\{(3, 0)\}$.

23. $\quad -x + 3y = 4 \qquad (1)$
$\qquad -2x + 6y = 8 \qquad (2)$

$$
\begin{array}{rclll}
2x & - & 6y & = & -8 & (3)\;\; -2 \times \text{Eq.(1)} \\
-2x & + & 6y & = & 8 & (2) \\
\hline
& & 0 & = & 0 & \textit{Add (3) and (2).}
\end{array}
$$

Since $0 = 0$ is a *true* statement, the equations are equivalent. This result indicates that every solution of one equation is also a solution of the other; there are an *infinite number of solutions*. The solution set is $\{(x, y) \mid x - 3y = -4\}$.

25. $\quad 5x - 2y = 3 \qquad (1)$
$\qquad 10x - 4y = 5 \qquad (2)$

$$
\begin{array}{rclll}
-10x & + & 4y & = & -6 & (3)\;\; -2 \times \text{Eq.(1)} \\
10x & - & 4y & = & 5 & (2) \\
\hline
& & 0 & = & -1 & \textit{Add (3) and (2).}
\end{array}
$$

Since $0 = -1$ is a *false* statement, there are no solutions of the system and the solution set is $\emptyset$.

27.
$$6x - 2y = -22 \quad (1)$$
$$-3x + 4y = 17 \quad (2)$$

$$\begin{array}{rl} 6x - 2y = -22 & (1) \\ \underline{-6x + 8y = 34} & (3) \quad 2 \times \text{Eq.}(2) \\ 6y = 12 & \quad Add\ (1)\ and\ (3). \\ y = 2 & \end{array}$$

Substitute 2 for y in equation (1).

$$6x - 2y = -22$$
$$6x - 2(2) = -22$$
$$6x - 4 = -22$$
$$6x = -18$$
$$x = -3$$

The solution set is $\{(-3, 2)\}$.

29.
$$3x = 3 + 2y \quad (1)$$
$$-\tfrac{4}{3}x + y = \tfrac{1}{3} \quad (2)$$

Rewrite equation (1) in standard form and multiply equation (2) by 3 to clear fractions.

$$3x - 2y = 3 \quad (3)$$
$$-4x + 3y = 1 \quad (4)$$

$$\begin{array}{rl} 12x - 8y = 12 & (5) \quad 4 \times \text{Eq.}(3) \\ \underline{-12x + 9y = 3} & (6) \quad 3 \times \text{Eq.}(4) \\ y = 15 & \quad Add\ (5)\ and\ (6). \end{array}$$

$$\begin{array}{rl} 9x - 6y = 9 & (7) \quad 3 \times \text{Eq.}(3) \\ \underline{-8x + 6y = 2} & (8) \quad 2 \times \text{Eq.}(4) \\ x = 11 & \quad Add\ (7)\ and\ (8). \end{array}$$

The solution set is $\{(11, 15)\}$.

31.
$$\tfrac{1}{5}x + y = \tfrac{6}{5} \quad (1)$$
$$\tfrac{1}{10}x + \tfrac{1}{3}y = \tfrac{5}{6} \quad (2)$$

First, clear all fractions.

Equation (1):

$$5\left(\tfrac{1}{5}x + y\right) = 5\left(\tfrac{6}{5}\right) \quad \textit{Multiply by LCD, 5.}$$
$$x + 5y = 6 \quad (3)$$

Equation (2):

$$30\left(\tfrac{1}{10}x + \tfrac{1}{3}y\right) = 30\left(\tfrac{5}{6}\right) \quad \textit{Multiply by LCD, 30.}$$
$$3x + 10y = 25 \quad (4)$$

The system has been simplified to

$$x + 5y = 6 \quad (3)$$
$$3x + 10y = 25 \quad (4)$$

To eliminate y, add -2 times equation (3) to (4).

$$\begin{array}{rl} -2x - 10y = -12 & (5) \quad -2 \times \text{Eq.}(3) \\ \underline{3x + 10y = 25} & (4) \\ x = 13 & \quad Add\ (5)\ and\ (4). \end{array}$$

Substitute 13 for x in equation (3).

$$x + 5y = 6 \quad (3)$$
$$13 + 5y = 6$$
$$5y = -7$$
$$y = -\tfrac{7}{5}$$

The solution set is $\left\{\left(13, -\tfrac{7}{5}\right)\right\}$.

33.
$$2.4x + 1.7y = 7.6 \quad (1)$$
$$1.2x - 0.5y = 9.2 \quad (2)$$

Multiply equation (1) and equation (2) by 10 to eliminate the decimals. Then solve the resulting equations by the elimination method.

$$24x + 17y = 76 \quad (3)$$
$$12x - 5y = 92 \quad (4)$$

Multiply equation (2) by -2.

$$\begin{array}{rl} 24x + 17y = 76 & (3) \\ \underline{-24x + 10y = -184} & (5) \\ 27y = -108 & \quad Add. \\ y = -4 & \end{array}$$

Substitute -4 for y in equation (4).

$$12x - 5y = 92$$
$$12x - 5(-4) = 92 \quad \textit{Let } y = -4.$$
$$12x + 20 = 92$$
$$12x = 72$$
$$x = 6 \quad \textit{Divide by 12.}$$

The solution set is $\{(6, -4)\}$.

35.
$$x + 3y = 6$$
$$-2x + 12 = 6y$$

Rewrite in standard form.

$$x + 3y = 6 \quad (1)$$
$$-2x - 6y = -12 \quad (2)$$

To eliminate x, multiply equation (1) by 2 and add the result to equation (2).

$$\begin{array}{rl} 2x + 6y = 12 & (3) \\ \underline{-2x - 6y = -12} & (2) \\ 0 = 0 & \quad Add\ (3)\ and\ (2). \end{array}$$

The true statement, $0 = 0$, indicates that there are an infinite number of solutions and the solution set is $\{(x, y) \mid x + 3y = 6\}$.

37.
$$4x - 3y = 1$$
$$8x = 3 + 6y$$

Rewrite in standard form.

$$4x - 3y = 1 \quad (1)$$
$$8x - 6y = 3 \quad (2)$$

To eliminate x, multiply equation (1) by -2 and add the result to equation (2).

$$
\begin{array}{rcll}
-8x + 6y &=& -2 & (3) \\
8x - 6y &=& 3 & (2) \\
\hline
0 &=& 1 & \text{Add } (3) \text{ and } (2).
\end{array}
$$

The false statement, $0 = 1$, indicates that the given system has solution set $\emptyset$.

39.
$$4x = 3y - 2$$
$$5x + 3 = 2y$$

Rewrite in standard form.

$$
\begin{array}{rcll}
-4x + 3y &=& 2 & (1) \\
5x - 2y &=& -3 & (2)
\end{array}
$$

$$
\begin{array}{rcll}
-8x + 6y &=& 4 & (3) \quad 2 \times \text{Eq.}(1) \\
15x - 6y &=& -9 & (4) \quad 3 \times \text{Eq.}(2) \\
\hline
7x &=& -5 & \text{Add } (3) \text{ and } (4). \\
x &=& -\frac{5}{7} &
\end{array}
$$

Rather than substitute $-\frac{5}{7}$ for x in (1) or (2), we will eliminate x by multiplying equation (1) by 5, equation (2) by 4, and adding the results.

$$
\begin{array}{rcll}
-20x + 15y &=& 10 & (5) \quad 5 \times \text{Eq.}(1) \\
20x - 8y &=& -12 & (6) \quad 4 \times \text{Eq.}(2) \\
\hline
7y &=& -2 & \text{Add } (5) \text{ and } (6). \\
y &=& -\frac{2}{7} &
\end{array}
$$

Solving the system in this fashion reduces the chance of making an arithmetic error.

The solution set is $\left\{ \left(-\frac{5}{7}, -\frac{2}{7} \right) \right\}$.

When you get a solution that has non-integer components, it is sometimes more difficult to check the problem than it was to solve it. A graphing calculator can be very helpful in this case. Just store the values for x and y in their respective memory locations, and then type the expressions as shown in the following screen. The results 2 and -3 (the right sides of the *equations* (1) and (2)) indicate that we have found the correct solution.

```
-5/7→X: -2/7→Y
          -.2857142857
-4X+3Y
                    2
5X-2Y
                   -3
```

41.
$$
\begin{array}{rcll}
24x + 12y &=& -7 & (1) \\
16x - 18y &=& 17 & (2)
\end{array}
$$

$$
\begin{array}{rcll}
48x + 24y &=& -14 & (3) \quad 2 \times \text{Eq.}(1) \\
-48x + 54y &=& -51 & (4) \quad -3 \times \text{Eq.}(2) \\
\hline
78y &=& -65 & \text{Add } (3) \text{ and } (4). \\
y &=& \frac{-65}{78} = -\frac{5}{6} &
\end{array}
$$

$$
\begin{array}{rcll}
72x + 36y &=& -21 & (5) \quad 3 \times \text{Eq.}(1) \\
32x - 36y &=& 34 & (6) \quad 2 \times \text{Eq.}(2) \\
\hline
104x &=& 13 & \text{Add } (5) \text{ and } (6). \\
x &=& \frac{13}{104} = \frac{1}{8} &
\end{array}
$$

The solution set is $\left\{ \left(\frac{1}{8}, -\frac{5}{6} \right) \right\}$.

43.
$$y = ax + b$$
$$5.39 = a(2000) + b \quad \text{Let } x = 2000, \, y = 5.39.$$
$$5.39 = 2000a + b$$

44. As in Exercise 43,
$$7.18 = 2008a + b.$$

45.
$$
\begin{array}{rcll}
2000a + b &=& 5.39 & (1) \\
2008a + b &=& 7.18 & (2)
\end{array}
$$

Multiply equation (1) by -1 and add the result to equation (2),

$$
\begin{array}{rcll}
-2000a - b &=& -5.39 & \\
2008a + b &=& 7.18 & \\
\hline
8a &=& 1.79 & \\
a &=& 0.22375 &
\end{array}
$$

Substitute 0.22375 for a in equation (1).

$$2000(0.22375) + b = 5.39$$
$$447.5 + b = 5.39$$
$$b = -442.11$$

The solution set is $\{(0.22375, -442.11)\}$.

46. **(a)** An equation of the segment PQ is
$$y = 0.22375x - 442.11$$
for $2000 \leq x \leq 2008$.

(b) $y = 0.22375x - 442.11$
$$y = 0.22375(2007) - 442.11 \quad \text{Let } x = 2007.$$
$$= 449.06625 - 442.11$$
$$= 6.95625 \approx 6.96$$

This is a bit more ($\$0.08$) than the actual figure.

47. Let $x =$ the number of goals.
Then $x + 1069 =$ the number of assists.
The total is 2857, so

$$x + (x + 1069) = 2857.$$
$$2x + 1069 = 2857$$
$$2x = 1788$$
$$x = 894$$

Since $x = 894$, $x + 1069 = 1963$. He had 894 goals and 1963 assists for a total of 2857 points.

49. Let $x =$ the number of twenties.
Then $x + 6 =$ the number of tens.
There are 32 bills altogether, so

$$x + (x + 6) = 32.$$
$$2x + 6 = 32$$
$$2x = 26$$
$$x = 13$$

There are 13 twenties.

Summary Exercises on Solving Systems of Linear Equations

1. **(a)** $3x + 5y = 69$
$\quad\quad\quad\ y = 4x$

Use substitution since the second equation is solved for y.

(b) $3x + y = -7$
$\quad\ x - y = -5$

Use elimination since the coefficients of the y-terms are opposites.

(c) $3x - 2y = 0$
$\quad\ 9x + 8y = 7$

Use elimination since the equations are in standard form with no coefficients of 1 or -1. Solving by substitution would involve fractions.

3. $4x - 3y = -8 \quad (1)$
$\quad\ x + 3y = 13 \quad (2)$

(a) Solve the system by the elimination method.

$$
\begin{array}{rrrrl}
4x & - & 3y & = & -8 \quad (1) \\
x & + & 3y & = & 13 \quad (2) \\
\hline
5x & & & = & 5 \quad \textit{Add (1) and (2).} \\
& & x & = & 1
\end{array}
$$

To find y, let $x = 1$ in equation (2).

$$x + 3y = 13$$
$$1 + 3y = 13$$
$$3y = 12$$
$$y = 4$$

The solution set is $\{(1, 4)\}$.

(b) To solve this system by the substitution method, begin by solving equation (2) for x.

$$x + 3y = 13$$
$$x = -3y + 13$$

Substitute $-3y + 13$ for x in equation (1).

$$4(-3y + 13) - 3y = -8$$
$$-12y + 52 - 3y = -8$$
$$-15y = -60$$
$$y = 4$$

To find x, let $y = 4$ in equation (2).

$$x + 3y = 13$$
$$x + 3(4) = 13$$
$$x + 12 = 13$$
$$x = 1$$

The solution set is $\{(1, 4)\}$.

(c) For this particular system, the elimination method is preferable because both equations are already written in the form $Ax + By = C$, and the equations can be added without multiplying either by a constant. Comparing solutions by the two methods, we see that the elimination method requires fewer steps than the substitution method for this system.

5. $3x + 5y = 69 \quad (1)$
$\quad\quad\quad\ y = 4x \quad\ (2)$

Equation (2) is already solved for y, so we'll use the substitution method. Substitute $4x$ for y in equation (1) and solve the resulting equation for x.

$$3x + 5y = 69$$
$$3x + 5(4x) = 69 \quad \textit{Let y = 4x.}$$
$$3x + 20x = 69$$
$$23x = 69$$
$$x = \tfrac{69}{23} = 3$$

To find the y-value of the solution, substitute 3 for x in equation (2).

$$y = 4x$$
$$y = 4(3) \quad \textit{Let x = 3.}$$
$$= 12$$

The solution set is $\{(3, 12)\}$.

7. $3x - 2y = 0 \quad (1)$
$\quad\ 9x + 8y = 7 \quad (2)$

$$
\begin{array}{rrrrll}
12x & - & 8y & = & 0 & \quad (3) \quad 4 \times \text{Eq. (1)} \\
9x & + & 8y & = & 7 & \quad (2) \\
\hline
21x & & & = & 7 & \quad \textit{Add (3) and (2).} \\
& & x & = & \tfrac{7}{21} = \tfrac{1}{3} &
\end{array}
$$

$$
\begin{array}{rrrrll}
-9x & + & 6y & = & 0 & \quad (4) \quad -3 \times \text{Eq.(1)} \\
9x & + & 8y & = & 7 & \quad (2) \\
\hline
& & 14y & = & 7 & \quad \textit{Add (4) and (2).} \\
& & y & = & \tfrac{7}{14} = \tfrac{1}{2} &
\end{array}
$$

The solution set is $\left\{ \left(\tfrac{1}{3}, \tfrac{1}{2} \right) \right\}$.

9. $6x + 7y = 4$ (1)
$5x + 8y = -1$ (2)

$\begin{array}{rcl} -30x - 35y &=& -20 \\ 30x + 48y &=& -6 \end{array}$ (3) $-5 \times$ Eq.(1)
(4) $6 \times$ Eq.(2)
$\begin{array}{rcl} 13y &=& -26 \end{array}$ *Add* (3) *and* (4).
$y = -2$

Substitute -2 for y in equation (1).

$$6x + 7y = 4$$
$$6x + 7(-2) = 4 \quad \textit{Let } y = -2.$$
$$6x - 14 = 4$$
$$6x = 18$$
$$x = 3$$

The solution set is $\{(3, -2)\}$.

11. $4x - 6y = 10$ (1)
$-10x + 15y = -25$ (2)

Divide equation (1) by 2 and equation (2) by 5.

$\begin{array}{rcl} 2x - 3y &=& 5 \\ -2x + 3y &=& -5 \end{array}$ (3)
(4)
$\begin{array}{rcl} 0 &=& 0 \end{array}$ *Add* (3) *and* (4).

Since $0 = 0$ is a *true* statement, there are an *infinite number of solutions*. The solution set is $\{(x, y) \mid 2x - 3y = 5\}$.

13. $5x = 7 + 2y$
$5y = 5 - 3x$

Rewrite in standard form.

$5x - 2y = 7$ (1)
$3x + 5y = 5$ (2)

$\begin{array}{rcl} 25x - 10y &=& 35 \\ 6x + 10y &=& 10 \end{array}$ (3) $5 \times$ Eq.(1)
(4) $2 \times$ Eq.(2)
$\begin{array}{rcl} 31x &=& 45 \end{array}$ *Add* (3) *and* (4).
$x = \frac{45}{31}$

$\begin{array}{rcl} -15x + 6y &=& -21 \\ 15x + 25y &=& 25 \end{array}$ (5) $-3 \times$ Eq.(1)
(6) $5 \times$ Eq.(2)
$\begin{array}{rcl} 31y &=& 4 \end{array}$ *Add* (5) *and* (6).
$y = \frac{4}{31}$

The solution set is $\left\{ \left(\frac{45}{31}, \frac{4}{31} \right) \right\}$.

15. $2x - 3y = 7$ (1)
$-4x + 6y = 14$ (2)

$\begin{array}{rcl} 4x - 6y &=& 14 \\ -4x + 6y &=& 14 \end{array}$ (3) $2 \times$ Eq.(1)
(2)
$\begin{array}{rcl} 0 &=& 28 \end{array}$ *Add* (3) *and* (2).

Since $0 = 28$ is *false*, the solution set is $\emptyset$.

17. $2x + 5y = 4$ (1)
$x + y = -1$ (2)

Solve equation (2) for x.

$$x = -y - 1 \quad (3)$$

Substitute $-y - 1$ for x in equation (1).

$$2(-y - 1) + 5y = 4 \quad \textit{Let } x = -y - 1.$$
$$-2y - 2 + 5y = 4$$
$$3y - 2 = 4$$
$$3y = 6$$
$$y = 2$$

Substitute 2 for y in equation (3).

$$x = -(2) - 1 \quad \textit{Let } y = 2.$$
$$= -3$$

The solution set is $\{(-3, 2)\}$.

19. $7x - 4y = 0$
$3x = 2y$

Rewrite in standard form.

$7x - 4y = 0$ (1)
$3x - 2y = 0$ (2)

$\begin{array}{rcl} 7x - 4y &=& 0 \\ -6x + 4y &=& 0 \end{array}$ (1)
(3) $-2 \times$ Eq.(2)
$\begin{array}{rcl} x &=& 0 \end{array}$ *Add* (1) *and* (3).

Substitute 0 for x in equation (1).

$$7x - 4y = 0$$
$$7(0) - 4y = 0$$
$$-4y = 0$$
$$y = 0$$

The solution set is $\{(0, 0)\}$.

21. $\frac{1}{6}x + \frac{1}{6}y = 2$ (1)
$-\frac{1}{2}x - \frac{1}{3}y = -8$ (2)

Multiply each side of equation (1) by 6 to clear fractions.

$$6\left(\tfrac{1}{6}x + \tfrac{1}{6}y\right) = 6(2)$$
$$6\left(\tfrac{1}{6}x\right) + 6\left(\tfrac{1}{6}y\right) = 6(2)$$
$$x + y = 12 \quad (3)$$

Multiply each side of equation (2) by the LCD, 6, to clear fractions.

$$6\left(-\tfrac{1}{2}x - \tfrac{1}{3}y\right) = 6(-8)$$
$$6\left(-\tfrac{1}{2}x\right) + 6\left(-\tfrac{1}{3}y\right) = 6(-8)$$
$$-3x - 2y = -48 \qquad (4)$$

The given system of equations has been simplified as follows.

$$x + y = 12 \qquad (3)$$
$$-3x - 2y = -48 \qquad (4)$$

Multiply equation (3) by 3 and add the result to equation (4).

$$\begin{array}{rcrcr} 3x & + & 3y & = & 36 \\ -3x & - & 2y & = & -48 \\ \hline & & y & = & -12 \end{array}$$

To find x, let $y = -12$ in equation (3).

$$x + y = 12$$
$$x + (-12) = 12$$
$$x - 12 = 12$$
$$x = 24$$

The solution set is $\{(24, -12)\}$.

23. $\dfrac{x}{2} - \dfrac{y}{3} = 9 \qquad (1)$

$\dfrac{x}{5} - \dfrac{y}{4} = 5 \qquad (2)$

To clear fractions, multiply each side of equation (1) by the LCD, 6.

$$6\left(\dfrac{x}{2} - \dfrac{y}{3}\right) = 6(9)$$
$$3x - 2y = 54 \qquad (3)$$

To clear fractions, multiply each side of equation (2) by the LCD, 20.

$$20\left(\dfrac{x}{5} - \dfrac{y}{4}\right) = 20(5)$$
$$4x - 5y = 100 \qquad (4)$$

We now have the simplified system

$$3x - 2y = 54 \qquad (3)$$
$$4x - 5y = 100. \qquad (4)$$

To solve this system by the elimination method, multiply equation (3) by 5 and equation (4) by -2; then add the results.

$$\begin{array}{rcrcr} 15x & - & 10y & = & 270 \\ -8x & + & 10y & = & -200 \\ \hline 7x & & & = & 70 \\ & & x & = & 10 \end{array}$$

To find y, let $x = 10$ in equation (3).

$$3x - 2y = 54$$
$$3(10) - 2y = 54$$
$$30 - 2y = 54$$
$$-2y = 24$$
$$y = -12$$

The solution set is $\{(10, -12)\}$.

25. $0.2x - 0.3y = 0.1 \quad (1)$
$0.3x - 0.2y = 0.9 \quad (2)$

Multiply equation (1) and equation (2) by 10 to eliminate the decimals. Then solve the resulting equations by the elimination method.

$$2x - 3y = 1 \quad (3)$$
$$3x - 2y = 9 \quad (4)$$

Multiply equation (3) by -3 and equation (4) by 2.

$$\begin{array}{rcrcrl} -6x & + & 9y & = & -3 & (5) \\ 6x & - & 4y & = & 18 & (6) \\ \hline & & 5y & = & 15 & \textit{Add}. \\ & & y & = & 3 \end{array}$$

Substitute 3 for y in equation (4).

$$3x - 2y = 9$$
$$3x - 2(3) = 9 \qquad \textit{Let } y = 3.$$
$$3x - 6 = 9$$
$$3x = 15$$
$$x = 5 \qquad \textit{Divide by 3.}$$

The solution set is $\{(5, 3)\}$.

4.4 Applications of Linear Systems

4.4 Now Try Exercises

N1. Let $x =$ the amount per month that was spent on electricity, and $y =$ the amount per month that was spent on rent. The total was $1150, so

$$x + y = 1150. \quad (1)$$

They spent $650 more on rent than on electricity, so

$$y = x + 650. \quad (2)$$

Substitute $x + 650$ for y in equation (1).

$$x + y = 1150$$
$$x + (x + 650) = 1150$$
$$2x + 650 = 1150$$
$$2x = 500$$
$$x = 250$$

$250 was spent on electricity per month and $250 + $650 = $900 was spent on rent per month.

N2. *Step 2*
Let $x =$ the number of adult tickets sold;
$y =$ the number of children's tickets sold.

Kind of Ticket	Number Sold	Cost of Each (in dollars)	Total Value (in dollars)
Adult	x	19	$19x$
Child	y	16	$16y$
Total	27	XXXXXX	$462

Step 3
The total number of tickets sold was 27, so

$$x + y = 27. \qquad (1)$$

Since the total value was $462, the final column leads to

$$19x + 16y = 462. \qquad (2)$$

Step 4
Multiply both sides of equation (1) by -16 and add this result to equation (2).

$$
\begin{array}{rcrcr}
-16x & - & 16y & = & -432 \\
19x & + & 16y & = & 462 \\
\hline
3x & & & = & 30 \\
& & x & = & 10
\end{array}
$$

From (1), $y = 27 - 10 = 17$.

Step 5
There were 10 adults and 17 children at the game.

Step 6
Check: The total number of tickets sold was $10 + 17 = 27$. Since 10 adults paid $19 each and 17 children paid $16 each, the value of tickets sold should be $10(19) + 17(16) = 462$, or $462. The result agrees with the given information.

N3. Let $x =$ the number of liters of 10% saline solution needed and
$y =$ the number of liters of 35% saline solution needed.

Summarize the information in a table.

Liters	Percent	Liters of Pure Alcohol
x	0.10	$0.10x$
y	0.35	$0.35y$
80	0.30	$0.30(80)$

$x + y = 80$ *From first column*

$0.10x + 0.35y = 0.30(80)$ *From third column*

To eliminate the x-terms, multiply the first equation by -0.10. Then add the result to the second equation.

$$
\begin{array}{rcrcrl}
-0.10x & - & 0.10y & = & -8 & \\
0.10x & + & 0.35y & = & 24 & \\
\hline
& & 0.25y & = & 16 & \textit{Add.} \\
& & y & = & 64 &
\end{array}
$$

To find x, substitute 64 for y in the first equation of the original system.

$$
\begin{aligned}
x + y &= 80 \\
x + 64 &= 80 \quad \textit{Let } y = 64. \\
x &= 16
\end{aligned}
$$

The solution is $x = 16$, $y = 64$. Mix 16 L of 10% solution with 64 L of 35% solution.

N4. Let $x =$ the rate of the faster truck, and
$y =$ the rate of the slower truck.

Summarize the information in a table.

	r	t	d
Faster Truck	x	3	$3x$
Slower Truck	y	3	$3y$

Write a system of equations.

$$
\begin{array}{ll}
3x + 3y = 405 & \textit{Total distance} \\
x = y + 5 & \textit{Faster truck is} \\
& \textit{5 mph faster.}
\end{array}
$$

Substitute $y + 5$ for x in the first equation, and solve for y.

$$
\begin{aligned}
3x + 3y &= 405 \\
3(y + 5) + 3y &= 405 \quad \textit{Let } x = y + 5. \\
3y + 15 + 3y &= 405 \\
6y + 15 &= 405 \\
6y &= 390 \\
y &= 65
\end{aligned}
$$

To find x, use $x = y + 5$ and $y = 65$.

$$x = y + 5 = 65 + 5 = 70$$

The faster truck's rate is 70 miles per hour and the slower truck's rate is 65 miles per hour.

4.4 Section Exercises

1. To represent the monetary value of x 5-dollar bills, multiply 5 times x. The answer is **D**, $5x$ dollars.

3. The amount of interest earned on d dollars at an interest rate of 3% (0.03) is choice **B**, $0.03d$ dollars.

5. If a cheetah's rate is 70 mph and it runs at that rate for x hours, then the distance covered is choice **D**, $70x$ miles.

7. Since the plane is traveling *with* the wind, add the rate of the plane, 650 miles per hour, to the rate of the wind, r miles per hour. The answer is **C**, $(650 + r)$ mph.

9. *Step 2*
Let $x =$ the first number and
let $y = $ the second number.

Step 3
First equation: $x + y = 98$
Second equation: $x - y = 48$

Step 4
Add the two equations.

$$\begin{array}{rcrcr} x & + & y & = & 98 \\ x & - & y & = & 48 \\ \hline 2x & & & = & 146 \\ & & x & = & 73 \end{array}$$

Substitute 73 for x in either equation to find $y = 25$.

Step 5
The two numbers are 73 and 25.

Step 6
The sum of 73 and 25 is 98. The difference between 73 and 25 is 48. The solution satisfies the conditions of the problem.

11. *Step 2*
Let $x =$ the number of performances of *Cats*; $y =$ the number of performances of *Phantom of the Opera*.

Step 3
The total number of performances was 16,088, so one equation is

$$x + y = 16{,}088. \quad (1)$$

Phantom of the Opera had 1118 more performances than *Cats*, so another equation is

$$y = x + 1118. \quad (2)$$

Step 4
Substitute $x + 1118$ for y in equation (1).

$$\begin{aligned} x + y &= 16{,}088 \\ x + (x + 1118) &= 16{,}088 \\ 2x + 1118 &= 16{,}088 \\ 2x &= 14{,}970 \\ x &= 7485 \end{aligned}$$

Substitute 7485 for x in (2) to find $y = 7485 + 1118 = 8603$.

Step 5
Cats had 7485 performances and *Phantom of the Opera* had 8603 performances.

Step 6
The sum of 7485 and 8603 is 16,088 and 8603 is 1118 more than 7485.

13. *Step 2*
Let $x =$ the amount earned by *Avatar*; $y =$ the amount earned by *Transformers 2: Revenge of the Fallen*.

Step 3
Transformers 2: Revenge of the Fallen grossed $26.9 million less than *Avatar*, so

$$y = x - 26.9. \quad (1)$$

The total earned by these two films was $831.1 million, so

$$x + y = 831.1 \quad (2)$$

Step 4
Substitute $x - 26.9$ for y in equation (2).

$$\begin{aligned} x + (x - 26.9) &= 831.1 \\ 2x - 26.9 &= 831.1 \\ 2x &= 858 \\ x &= 429 \end{aligned}$$

To find y, let $x = 429$ in equation (1).

$$\begin{aligned} y &= x - 26.9 \\ y &= 429 - 26.9 = 402.1 \end{aligned}$$

Step 5
Avatar earned $429 million and *Transformers 2: Revenge of the Fallen* earned $402.1 million.

Step 6
The sum of 429 and 402.1 is 831.1 and 402.1 is 26.9 less than 429.

15. **(a)** $C = 85x + 900;\ R = 105x;$
No more than 38 units can be sold.

To find the break-even quantity, let $C = R$.

$$\begin{aligned} 85x + 900 &= 105x \\ 900 &= 20x \\ x &= \frac{900}{20} = 45 \end{aligned}$$

The break-even quantity is 45 units.

(b) Since no more than 38 units can be sold, do not produce the product (since $38 < 45$). The product will lead to a loss.

17. Multiply the Number of Coins by the Value per Coin to get the Total Value in the table.

Number of Coins	Value per Coin	Total Value
x	$0.25	$0.25x$
y	$0.10	$0.10y$
39	XXXXXX	$7.50

Step 3
From the first and third columns of the table, we obtain the system

$$\begin{aligned} x + y &= 39 \quad (1) \\ 0.25x + 0.10y &= 7.50. \quad (2) \end{aligned}$$

Step 4
To solve this system by the elimination method, multiply equation (1) by -10, equation (2) by 100, and add the resulting equations.

$$\begin{array}{rcrcr} -10x & - & 10y & = & -390 \\ 25x & + & 10y & = & 750 \\ \hline 15x & & & = & 360 \\ & & x & = & 24 \end{array}$$

To find y, let $x = 24$ for y in equation (1).

$$x + y = 39$$
$$24 + y = 39$$
$$y = 15$$

Step 5

Jonathan had 24 quarters and 15 dimes.

Step 6

24 quarters and 15 dimes give us 39 coins worth $7.50.

19. *Step 2*

Let x = the number of *The Blind Side* DVDs;

y = the number of Beyoncé CDs

Type of Gift	Number Bought	Cost of each (in dollars)	Total Value
DVD	x	14.95	$14.95x$
CD	y	16.88	$16.88y$
Totals	7	XXXXXX	$114.30

Step 3

From the second and fourth columns of the table, we obtain the system

$$x + y = 7 \qquad (1)$$
$$14.95x + 16.88y = 114.30. \quad (2)$$

Step 4

Multiply equation (1) by -14.95 and add the result to equation (2).

$$\begin{array}{rrrr} -14.95x & - & 14.95y & = & -104.65 \\ 14.95x & + & 16.88y & = & 114.30 \\ \hline & & 1.93y & = & 9.65 \\ & & y & = & 5 \end{array}$$

From (1) with $y = 5$, $x = 2$.

Step 5

She bought 2 DVDs of *The Blind Side* and 5 Beyoncé CDs.

Step 6

Two $14.95 DVDs and five $16.88 CDs give us 7 gifts worth $114.30.

21. *Step 2*

Let x = the amount invested at 5%;

y = the amount invested at 4%.

Amount Invested	Interest Rate (as a decimal)	Interest Income (yearly)
x	0.05	$0.05x$
y	0.04	$0.04y$
XXXXX	XXXXXX	$350

Step 3

Karen has invested twice as much at 5% as at 4%, so

$$x = 2y. \quad (1)$$

Her total interest income is $350, so

$$0.05x + 0.04y = 350. \quad (2)$$

Step 4

Solve the system by substitution. Substitute $2y$ for x in equation (2).

$$0.05(2y) + 0.04y = 350$$
$$0.14y = 350$$
$$y = \frac{350}{0.14} = 2500$$

Substitute 2500 for y in equation (1).

$$x = 2y$$
$$x = 2(2500) = 5000$$

Step 5

Karen has $5000 invested at 5% and $2500 invested at 4%.

Step 6

$5000 is twice as much as $2500. 5% of $5000 is $250 and 4% of $2500 is $100, which is a total of $350 in interest.

23. *Step 2*

Let x = the average ticket cost for The Police;

y = the average ticket cost for Madonna.

Step 3

Six tickets for The Police plus five tickets for Madonna cost $1297, so

$$6x + 5y = 1297. \quad (1)$$

Three tickets for The Police plus four tickets for Madonna cost $854, so

$$3x + 4y = 854. \quad (2)$$

Step 4

Multiply (2) by -2 and add the results.

$$\begin{array}{rrrr} 6x & + & 5y & = & 1297 \\ -6x & - & 8y & = & -1708 \\ \hline & & -3y & = & -411 \\ & & y & = & 137 \end{array}$$

Substitute $y = 137$ in (1).

$$6x + 5(137) = 1297$$
$$6x + 685 = 1297$$
$$6x = 612$$
$$x = 102$$

Step 5

The average ticket cost for The Police is $102 and the average ticket cost for Madonna is $137.

Step 6

Six tickets (on average) for The Police cost $612 and five tickets for Madonna cost $685; a sum of $1297. Three tickets for The Police cost $306 and four tickets for Madonna cost $548; a sum of $854.

25. *Step 2*

Let $x =$ the amount of 40% dye solution;

$y =$ the amount of 70% dye solution.

Liters of Solution	Percent (as a decimal)	Liters of Pure Dye
x	0.40	$0.40x$
y	0.70	$0.70y$
120	0.50	$0.50(120) = 60$

Step 3

The total number of liters in the final mixture is 120, so

$$x + y = 120. \quad (1)$$

The amount of pure dye in the 40% solution added to the amount of pure dye in the 70% solution is equal to the amount of pure dye in the 50% mixture, so

$$0.40x + 0.70y = 60. \quad (2)$$

Multiply equation (2) by 10 to clear decimals.

$$4x + 7y = 600 \quad (3)$$

We now have the system

$$x + y = 120 \quad (1)$$
$$4x + 7y = 600. \quad (3)$$

Step 4

Solve this system by the elimination method.

$$
\begin{array}{rcll}
-4x & - & 4y & = & -480 & \quad \textit{Multiply (1) by } -4.\\
4x & + & 7y & = & 600 \\
\hline
& & 3y & = & 120 \\
& & y & = & 40
\end{array}
$$

$$x + 40 = 120 \qquad \textit{Let } y = 40 \textit{ in (1).}$$
$$x = 120 - 40 = 80$$

Step 5

80 liters of 40% solution should be mixed with 40 liters of 70% solution.

Step 6

Since $80 + 40 = 120$ and $0.40(80) + 0.70(40) = 60$, this mixture will give the 120 liters of 50 percent solution, as required in the original problem.

27. *Step 2*

Let $x =$ the number of pounds of coffee worth $6 per pound;

$y =$ the number of pounds of coffee worth $3 per pound.

Complete the table given in the textbook.

Pounds	Dollars per Pound	Cost
x	6	$6x$
y	3	$3y$
90	4	$90(4) = \$360$

Step 3

The mixture contains 90 pounds, so

$$x + y = 90. \quad (1)$$

The cost of the mixture is $360, so

$$6x + 3y = 360. \quad (2)$$

Equation (2) may be simplified by dividing each side by 3.

$$2x + y = 120 \quad (3)$$

We now have the system

$$x + y = 90 \qquad (1)$$
$$2x + y = 120. \qquad (3)$$

Step 4

To solve this system by the elimination method, multiply equation (1) by -1 and add the result to equation (3).

$$
\begin{array}{rcll}
-x & - & y & = & -90 \\
2x & + & y & = & 120 \\
\hline
x & & & = & 30
\end{array}
$$

From (1), $y = 60$.

Step 5

Deoraj will need to mix 30 pounds of coffee at $6 per pound with 60 pounds at $3 per pound.

Step 6

Since $30 + 60 = 90$ and $\$6(30) + \$3(60) = \$360$, this mixture will give the 90 pounds worth $4 per pound, as required in the original problem.

29. *Step 2*

Let $x =$ the number of pounds of nuts selling for $6 per pound;

$y =$ the number of pounds of raisins selling for $3 per pound.

Complete a table.

Pounds	Dollars per Pound	Cost
x	6	$6x$
y	3	$3y$
60	5	$60(5) = \$300$

Step 3

The mixture contains 60 pounds, so

$$x + y = 60. \quad (1)$$

The cost of the mixture is $300, so

$$6x + 3y = 300. \quad (2)$$

Equation (2) may be simplified by dividing each side by 3.

$$2x + y = 100 \quad (3)$$

We now have the system

$$\begin{aligned} x + y &= 60 \quad (1) \\ 2x + y &= 100. \quad (3) \end{aligned}$$

Step 4
To solve this system by the elimination method, multiply equation (1) by -1 and add the result to equation (3).

$$\begin{array}{rcr} -x \ - \ y &=& -60 \\ 2x \ + \ y &=& 100 \\ \hline x \qquad\;\; &=& 40 \end{array}$$

From (1), $y = 20$.

Step 5
Theresa will need to mix 40 pounds of nuts at \$6 per pound with 20 pounds of raisins at \$3 per pound.

Step 6
Since $40 + 20 = 60$ and $\$6(40) + \$3(20) = \$300$, this mixture will give the 60 pounds worth \$5 per pound, as required in the original problem.

31.   *Step 2*
Let $x =$ the average rate of the slower train;
$y =$ the average rate of the faster train.

	r	t	d
Slower train	x	4.5	$4.5x$
Faster train	y	4.5	$4.5y$

Step 3
The total distance is 495 miles, so

$$4.5x + 4.5y = 495 \quad (1)$$

or, dividing equation (1) by 4.5,

$$x + y = 110. \quad (2)$$

The faster train traveled 10 mph faster than the slower train, so

$$y = x + 10. \quad (3)$$

Step 4
From (3), substitute $x + 10$ for y in (2).

$$\begin{aligned} x + (x + 10) &= 110 \\ 2x + 10 &= 110 \\ 2x &= 100 \\ x &= 50 \end{aligned}$$

From (3) with $x = 50$, $y = 50 + 10 = 60$.

Step 5
The average rate of the slower train was 50 mph

and the average rate of the faster train was 60 mph.

Step 6
60 mph is 10 mph faster than 50 mph. The slower train traveled $4.5(50) = 225$ miles and the faster train traveled $4.5(60) = 270$ miles. The sum is $225 + 270 = 495$, as required.

33.   *Step 2*
Let $x =$ the average rate of the slower car;
$y =$ the average rate of the faster car.

	r	t	d
Slower car	x	6	$6x$
Faster car	y	6	$6y$

Step 3
The total distance is 600 miles, so

$$6x + 6y = 600 \quad (1)$$

or, dividing equation (1) by 6,

$$x + y = 100. \quad (2)$$

The faster car travels 30 miles per hour faster than the slower car, so

$$y = x + 30. \quad (3)$$

Step 4
From (3), substitute $x + 30$ for y in (2).

$$\begin{aligned} x + (x + 30) &= 100 \\ 2x &= 70 \\ x &= 35 \end{aligned}$$

From (3), $y = 65$.

Step 5
The faster car travels at 65 miles per hour and the slower car travels at 35 miles per hour.

Step 6
65 miles per hour is 30 miles per hour faster than 35 miles per hour. The slower car travels $35(6) = 210$ miles and the faster car travels $65(6) = 390$ miles. The total distance traveled is $210 + 390 = 600$, as required.

35.   *Step 2*
Let $x =$ the average rate of the bicycle;
$y =$ the average rate of the car.

	r	t	d
Bicycle	x	7.5	$7.5x$
Car	y	7.5	$7.5y$

Step 3
The total distance is 471 miles, so

$$7.5x + 7.5y = 471 \quad (1)$$

or, dividing equation (1) by 7.5,

$$x + y = 62.8. \quad (2)$$

The car traveled 35.8 mph faster than the bicycle, so

$$y = x + 35.8. \quad (3)$$

Step 4
From (3), substitute $x + 35.8$ for y in (2).

$$x + (x + 35.8) = 62.8$$
$$2x + 35.8 = 62.8$$
$$2x = 27.0$$
$$x = 13.5$$

From (3) with $x = 13.5$, $y = 13.5 + 35.8 = 49.3$.

Step 5
The average rate of the bicycle was 13.5 mph and the average rate of the car was 49.3 mph.

Step 6
49.3 mph is 35.8 mph faster than 13.5 mph. The bicycle traveled $7.5(13.5) = 101.25$ miles and the car traveled $7.5(49.3) = 369.75$ miles. The sum is $101.25 + 369.75 = 471$, as required.

37. *Step 2*
Let $x =$ the rate of the boat in still water;
$y =$ the rate of the current.

	r	t	d
Downstream	$x + y$	3	$3(x + y)$
Upstream	$x - y$	3	$3(x - y)$

Step 3
Use the formula $d = rt$ and the completed table to write the system of equations.

$$3(x + y) = 36 \quad (1) \quad \textit{Distance downstream}$$
$$3(x - y) = 24 \quad (2) \quad \textit{Distance upstream}$$

Step 4
Divide equations (1) and (2) by 3.

$$x + y = 12 \quad (3)$$
$$x - y = 8 \quad (4)$$

Now add equations (3) and (4).

$$
\begin{array}{rcrcll}
x & + & y & = & 12 & (3) \\
x & - & y & = & 8 & (4) \\
\hline
2x & & & = & 20 & \textit{Add.} \\
& & x & = & 10 &
\end{array}
$$

From (3), $y = 2$.

Step 5
The rate of the current is 2 miles per hour; the rate of the boat in still water is 10 miles per hour.

Step 6
Traveling downstream for 3 hours at $10 + 2 = 12$ miles per hour gives us a 36-mile trip. Traveling upstream for 3 hours at $10 - 2 = 8$ miles per hour gives us a 24-mile trip, as required.

39. *Step 2*
Let $x =$ the rate of the plane in still air;
$y =$ the rate of the wind.

Step 3
The rate of the plane with the wind is $x + y$, so

$$x + y = 500. \quad (1)$$

The rate of the plane into the wind is $x - y$, so

$$x - y = 440. \quad (2)$$

Step 4
To solve the system by the elimination method, add equations (1) and (2).

$$
\begin{array}{rcrcl}
x & + & y & = & 500 \\
x & - & y & = & 440 \\
\hline
2x & & & = & 940 \\
& & x & = & 470
\end{array}
$$

From (1), $y = 30$.

Step 5
The rate of the wind is 30 miles per hour; the rate of the plane in still air is 470 miles per hour.

Step 6
The plane travels $470 + 30 = 500$ miles per hour with the wind and $470 - 30 = 440$ miles per hour into the wind, as required.

41. *Step 2*
Let $x =$ Yady's rate;
$y =$ Dane's rate.

Step 3
Use the formula $d = rt$ to complete two tables.

Riding in same direction

	r	t	d
Yady	x	6	$6x$
Dane	y	6	$6y$

Yady rode 30 miles farther than Dane, so

$$6x = 6y + 30$$
$$\text{or} \quad x - y = 5. \quad (1)$$

Riding toward each other

	r	t	d
Yady	x	1	$1x$
Dane	y	1	$1y$

Yady and Dane rode a total of 30 miles, so

$$x + y = 30 \quad (2)$$

We have the system

$$x - y = 5 \qquad (1)$$
$$x + y = 30. \qquad (2)$$

Step 4

To solve the system by the elimination method, add equations (1) and (2).

$$
\begin{array}{rcl}
x & - & y & = & 5 \\
x & + & y & = & 30 \\
\hline
2x & & & = & 35 \\
& & x & = & 17.5
\end{array}
$$

From (2) with $x = 17.5$, $y = 12.5$.

Step 5

Yady's rate is 17.5 miles per hour and Dane's rate is 12.5 miles per hour.

Step 6

Riding in the same direction, Yady rides $6(17.5) = 105$ miles and Dane rides $6(12.5) = 75$ miles. The difference is 30 miles, as required.

Riding toward each other, Yady rides $1(17.5) = 17.5$ miles and Dane rides $1(12.5) = 12.5$ miles. The sum is 30 miles, as required.

43. $x + y \leq 4$

The boundary is the line with equation $x + y = 4$. Graph this as a solid line through its intercepts, $(4, 0)$ and $(0, 4)$.

Using $(0, 0)$ as a test point will result in the inequality $0 \leq 4$, which is true. Shade the region containing the test point $(0, 0)$. The solid line indicates that the boundary is part of the graph.

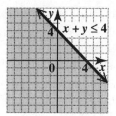

45. $3x + 2y < 0$

The boundary is the line with equation $3x + 2y = 0$. Graph this as a dashed line through $(0, 0)$ and $(2, -3)$. We cannot use $(0, 0)$ as a test point because it lies on the boundary.

Using $(4, 0)$ as a test point will result in the inequality $12 < 0$, which is false. Shade the region not containing the test point $(4, 0)$. The dashed line indicates that the boundary is not part of the graph.

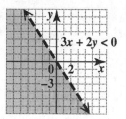

4.5 Solving Systems of Linear Inequalities

4.5 Now Try Exercises

N1. $4x - 2y \leq 8$
$x + 3y \geq 3$

The graph of $4x - 2y = 8$ has intercepts $(2, 0)$ and $(0, -4)$. The graph of $x + 3y = 3$ has intercepts $(3, 0)$ and $(0, 1)$. Both are graphed as solid lines because of the $\leq$ and $\geq$ signs. Use $(0, 0)$ as a test point in each case.

$$
\begin{array}{l}
4x - 2y \leq 8 \\
4(0) - 2(0) \leq 8 \quad \textit{Let x = 0, y = 0.} \\
\qquad\qquad 0 \leq 8 \quad \textit{True}
\end{array}
$$

Shade the side of the graph for $4x - 2y = 8$ that contains $(0, 0)$.

$$
\begin{array}{l}
x + 3y \geq 3 \\
0 + 3(0) \geq 3 \quad \textit{Let x = 0, y = 0.} \\
\qquad\quad 0 \geq 3 \quad \textit{False}
\end{array}
$$

Shade the side of the graph for $x + 3y = 3$ that does not contain $(0, 0)$.

The solution set of this system is the intersection (overlap) of the two shaded regions, and includes the portions of the boundary lines that bound this region.

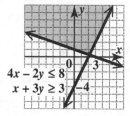

N2. $2x + 5y > 10$
$x - 2y < 0$

Graph $2x + 5y = 10$ as a dashed line through its intercepts, $(5, 0)$ and $(0, 2)$. Use $(0, 0)$ as a test point.

$$
\begin{array}{l}
2x + 5y > 10 \\
2(0) + 5(0) > 10 \quad \textit{Let x = 0, y = 0.} \\
\qquad\qquad 0 > 10 \quad \textit{False}
\end{array}
$$

The solution is the region that does not include $(0, 0)$.

Then graph $x - 2y = 0$ as a dashed line through the points $(0, 0)$ and $(2, 1)$. Use $(0, 3)$ as a test point.

$$x - 2y < 0$$
$$0 - 2(3) < 0 \quad \textit{Let } x = 0, y = 3.$$
$$-6 < 0 \quad \textit{True}$$

The solution is the region that includes $(0, 3)$.

The solution set of the system is the intersection of the two shaded regions. Because the inequality signs are $>$ and $<$, the solution set does not include the boundary lines.

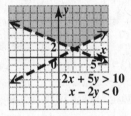

N3. $x - y < 2$
$\quad\quad x \geq -2$
$\quad\quad y \leq 4$

Graph $x - y = 2$ as a dashed line through its intercepts, $(2, 0)$ and $(0, -2)$. Use $(0, 0)$ as a test point to get $0 < 2$, a true statement, so shade above the line.

Recall that $x = -2$ is a vertical line through the point $(-2, 0)$, and $y = 4$ is a horizontal line through the point $(0, 4)$. Graph these as solid lines. Shade to the right of $x = -2$ and below $y = 4$.

The solution set of this system is the intersection (overlap) of the three shaded regions, and includes the portions of the horizontal and vertical boundary lines that bound this region.

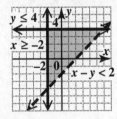

4.5 Section Exercises

1. $x \geq 5$ is the region to the right of the vertical line $x = 5$ and includes the line. $y \leq -3$ is the region below the horizontal line $y = -3$ and includes the line. The correct choice is **C**.

3. $x > 5$ is the region to the right of the vertical line $x = 5$. $y < -3$ is the region below the horizontal line $y = -3$. The correct choice is **B**.

5. $x + y \leq 6$
$\quad\quad x - y \geq 1$

Graph the boundary $x + y = 6$ as a solid line through its intercepts, $(6, 0)$ and $(0, 6)$. Using $(0, 0)$ as a test point will result in the true statement $0 \leq 6$, so shade the region containing the origin.

Graph the boundary $x - y = 1$ as a solid line through its intercepts, $(1, 0)$ and $(0, -1)$. Using $(0, 0)$ as a test point will result in the false statement $0 \geq 1$, so shade the region *not* containing the origin.

The solution set of this system is the intersection (overlap) of the two shaded regions, and includes the portions of the boundary lines that bound this region.

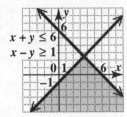

7. $4x + 5y \geq 20$
$\quad\quad x - 2y \leq 5$

Graph the boundary $4x + 5y = 20$ as a solid line through its intercepts, $(5, 0)$ and $(0, 4)$. Using $(0, 0)$ as a test point will result in the false statement $0 \geq 20$, so shade the region *not* containing the origin.

Graph the boundary $x - 2y = 5$ as a solid line through $(5, 0)$ and $(1, -2)$. Using $(0, 0)$ as a test point will result in the true statement $0 \leq 5$, so shade the region containing the origin.

The solution set of this system is the intersection of the two shaded regions, and includes the portions of the boundary lines that bound the region.

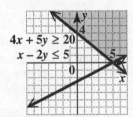

9. $2x + 3y < 6$
$\quad\quad x - y < 5$

Graph $2x + 3y = 6$ as a dashed line through $(3, 0)$ and $(0, 2)$. Using $(0, 0)$ as a test point will result in the true statement $0 < 6$, so shade the region containing the origin.

Now graph $x - y = 5$ as a dashed line through $(5, 0)$ and $(0, -5)$. Using $(0, 0)$ as a test point will result in the true statement $0 < 5$, so shade the region containing the origin.

The solution set of the system is the intersection of the two shaded regions. Because the inequality signs are both $<$, the solution set does not include the boundary lines.

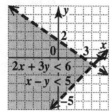

11. $y \le 2x - 5$
$x < 3y + 2$

Graph $y = 2x - 5$ as a solid line through $(0, -5)$ and $(3, 1)$. Using $(0, 0)$ as a test point will result in the false statement $0 \le -5$, so shade the region *not* containing the origin.

Now graph $x = 3y + 2$ as a dashed line through $(2, 0)$ and $(-1, -1)$. Using $(0, 0)$ as a test point will result in the true statement $0 < 2$, so shade the region containing the origin.

The solution set of the system is the intersection of the two shaded regions. It includes the portion of the line $y = 2x - 5$ that bounds the region, but not the portion of the line $x = 3y + 2$.

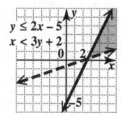

13. $4x + 3y < 6$
$x - 2y > 4$

Graph $4x + 3y = 6$ as a dashed line through $\left(\frac{3}{2}, 0\right)$ and $(0, 2)$. Using $(0, 0)$ as a test point will result in the true statement $0 < 6$, so shade the region containing the origin.

Now graph $x - 2y = 4$ as a dashed line through $(4, 0)$ and $(0, -2)$. Using $(0, 0)$ as a test point will result in the false statement $0 > 4$, so shade the region *not* containing the origin.

The solution set of the system is the intersection of the two shaded regions. It does not include the boundary lines.

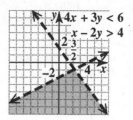

15. $x \le 2y + 3$
$x + y < 0$

Graph $x = 2y + 3$ as a solid line through $(3, 0)$ and $(7, 2)$. Using $(0, 0)$ as a test point will result in the true statement $0 \le 3$, so shade the region containing the origin.

Now graph $x + y = 0$ as a dashed line through $(0, 0)$ and $(1, -1)$. Using $(1, 0)$ as a test point will result in the false statement $1 < 0$, so shade the region *not* containing $(1, 0)$.

The solution set of the system is the intersection of the two shaded regions. It includes the portion of the line $x = 2y + 3$ that bounds the region, but not the portion of the line $x + y = 0$.

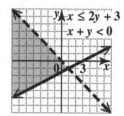

17. $-3x + y \ge 1$
$6x - 2y \ge -10$

Graph $-3x + y = 1$ as a solid line through $\left(-\frac{1}{3}, 0\right)$ and $(0, 1)$. Using $(0, 0)$ as a test point will result in the false statement $0 \ge 1$, so shade the region *not* containing the origin. This is the region above the line.

Now graph $6x - 2y = -10$ as a solid line through $\left(-\frac{5}{3}, 0\right)$ and $(0, 5)$. Using $(0, 0)$ as a test point will result in the true statement $0 \ge -10$, so shade the region containing the origin. This is the region below the line.

The solution set of the system is the intersection of the two shaded regions. This is the region between the two parallel lines (both lines have slope 3). These boundary lines are included in the solution set.

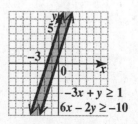

19. $x - 3y \leq 6$
$\quad\quad x \geq -4$

Graph $x - 3y = 6$ as a solid line through $(6, 0)$ and $(0, -2)$. Using $(0, 0)$ as a test point will result in the true statement $0 \leq 6$, so shade the region containing the origin.

Now graph $x = -4$ as a solid vertical line through $(-4, 0)$ and $(-4, 3)$. Using $(0, 0)$ as a test point will result in the true statement $0 \geq -4$, so shade the region containing the origin.

The solution set of the system is the intersection of the two shaded regions, and includes the portions of the two lines that bound the region.

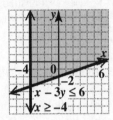

21. $4x + 5y < 8$
$\quad\quad y > -2$
$\quad\quad x > -4$

Graph $4x + 5y = 8$, $y = -2$, and $x = -4$ as dashed lines. All three inequalities are true for $(0, 0)$. Shade the region bounded by the three lines, which contains the test point $(0, 0)$.

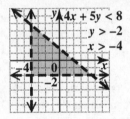

23. $3x - 2y \geq 6$
$\quad\quad x + y \leq 4$
$\quad\quad x \geq 0$
$\quad\quad y \geq -4$

Graph $3x - 2y = 6$, $x + y = 4$, $x = 0$, and $y = -4$ as solid lines. All four inequalities are true for $(2, -2)$. Shade the region bounded by the four lines, which contains the test point $(2, -2)$.

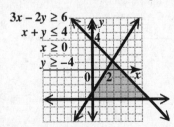

25. $y \geq x$
$\quad\quad y \leq 2x - 3$

The graph of the solution set will be the region above the graph of $y = x$ and below the graph of $y = 2x - 3$. This is the calculator-generated graph **D**.

27. $y \geq -x$
$\quad\quad y \leq 2x - 3$

The graph of the solution set will be the region above the graph of $y = -x$ and below the graph of $y = 2x - 3$. This is the calculator-generated graph **A**.

29. $2 \cdot 2 \cdot 2 \cdot 2 \cdot 2 \cdot 2 = 64$

31. $5 \cdot 5 \cdot 5 \cdot 5 = 625$

33. $\frac{2}{3} \cdot \frac{2}{3} \cdot \frac{2}{3} = \frac{8}{27}$

Chapter 4 Review Exercises

1. $(3, 4)$ $\quad\quad 4x - 2y = 4$
$\quad\quad\quad\quad\quad\quad 5x + y = 19$

To decide whether $(3, 4)$ is a solution of the system, substitute 3 for x and 4 for y in each equation.

$$4x - 2y = 4$$
$$4(3) - 2(4) \overset{?}{=} 4$$
$$12 - 8 \overset{?}{=} 4$$
$$4 = 4 \quad True$$

$$5x + y = 19$$
$$5(3) + 4 \overset{?}{=} 19$$
$$15 + 4 \overset{?}{=} 19$$
$$19 = 19 \quad True$$

Since $(3, 4)$ satisfies both equations, it is a solution of the system.

2. $(-5, 2)$ $\quad\quad x - 4y = -13 \quad (1)$
$\quad\quad\quad\quad\quad\quad 2x + 3y = 4 \quad\quad (2)$

Substitute -5 for x and 2 for y in equation (2).

$$2x + 3y = 4$$
$$2(-5) + 3(2) \overset{?}{=} 4$$
$$-10 + 6 \overset{?}{=} 4$$
$$-4 = 4 \quad False$$

Since $(-5, 2)$ is not a solution of the second equation, it cannot be a solution of the system.

3. $x + y = 4$
$2x - y = 5$

To graph the equations, find the intercepts.

$x + y = 4$: Let $y = 0$;   then $x = 4$.
Let $x = 0$;   then $y = 4$.

Plot the intercepts, $(4, 0)$ and $(0, 4)$, and draw the line through them.

$2x - y = 5$: Let $y = 0$;   then $x = \frac{5}{2}$.
Let $x = 0$;   then $y = -5$.

Plot the intercepts, $\left(\frac{5}{2}, 0\right)$ and $(0, -5)$, and draw the line through them.

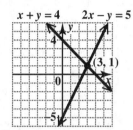

It appears that the lines intersect at the point $(3, 1)$. Check this by substituting 3 for x and 1 for y in both equations. Since $(3, 1)$ satisfies both equations, the solution set of this system is $\{(3, 1)\}$.

4. $x - 2y = 4$
$2x + y = -2$

To graph the equations, find the intercepts.

$x - 2y = 4$: Let $y = 0$;   then $x = 4$.
Let $x = 0$;   then $y = -2$.

Plot the intercepts, $(4, 0)$ and $(0, -2)$, and draw the line through them.

$2x + y = -2$: Let $y = 0$;   then $x = -1$.
Let $x = 0$;   then $y = -2$.

Plot the intercepts, $(-1, 0)$ and $(0, -2)$, and draw the line through them.

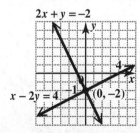

The lines intersect at their common y-intercept, $(0, -2)$, so $\{(0, -2)\}$ is the solution set of the system.

5. $2x + 4 = 2y$
$y - x = -3$

Graph the line $2x + 4 = 2y$ through its intercepts, $(-2, 0)$ and $(0, 2)$. Graph the line $y - x = -3$ through its intercepts, $(3, 0)$ and $(0, -3)$.

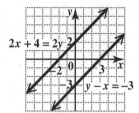

Solving each equation for y gives us $y = x + 2$ and $y = x - 3$, so the lines are parallel (both lines have slope 1). Since they do not intersect, there is no solution. This is an inconsistent system and the solution set is $\emptyset$.

6. $x - 2 = 2y$
$2x - 4y = 4$

Graph the line $x - 2 = 2y$ through its intercepts, $(2, 0)$ and $(0, -1)$.

Graph the line $2x - 4y = 4$ through its intercepts, $(2, 0)$ and $(0, -1)$.

Since both equations have the same intercepts, they are equations of the same line.

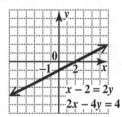

There is an infinite number of solutions. The equations are dependent equations and the solution set contains an infinite number of ordered pairs. The solution set is $\{(x, y) \mid x - 2 = 2y\}$.

7. It would be easiest to solve for x in the second equation because its coefficient is -1. No fractions would be involved.

8. The true statement $0 = 0$ is an indication that the system has an infinite number of solutions. Write the solution set using set-builder notation and the equation of the system that is in standard form with integer coefficients having greatest common factor 1.

9. $3x + y = 7$ $\quad$ (1)
$\quad\quad x = 2y$ $\quad\quad$ (2)

Substitute $2y$ for x in equation (1) and solve the resulting equation for y.

$$3x + y = 7$$
$$3(2y) + y = 7$$
$$6y + y = 7$$
$$7y = 7$$
$$y = 1$$

To find x, let $y = 1$ in equation (2).

$$x = 2y$$
$$x = 2(1) = 2$$

The solution set is $\{(2, 1)\}$.

10. $2x - 5y = -19$ $\quad$ (1)
$\quad\quad y = x + 2$ $\quad\quad$ (2)

Substitute $x + 2$ for y in equation (1).

$$2x - 5y = -19$$
$$2x - 5(x + 2) = -19$$
$$2x - 5x - 10 = -19$$
$$-3x - 10 = -19$$
$$-3x = -9$$
$$x = 3$$

To find y, let $x = 3$ in equation (2).

$$y = x + 2$$
$$y = 3 + 2 = 5$$

The solution set is $\{(3, 5)\}$.

11. $4x + 5y = 44$ $\quad$ (1)
$\quad\quad x + 2 = 2y$ $\quad$ (2)

Solve equation (2) for x.

$$x + 2 = 2y$$
$$x = 2y - 2 \quad (3)$$

Substitute $2y - 2$ for x in equation (1) and solve the resulting equation for y.

$$4x + 5y = 44$$
$$4(2y - 2) + 5y = 44$$
$$8y - 8 + 5y = 44$$
$$13y - 8 = 44$$
$$13y = 52$$
$$y = 4$$

To find x, let $y = 4$ in equation (3).

$$x = 2y - 2$$
$$x = 2(4) - 2 = 6$$

The solution set is $\{(6, 4)\}$.

12. $5x + 15y = 30$ $\quad$ (1)
$\quad\quad x + 3y = 6$ $\quad\quad$ (2)

Solve equation (2) for x.

$$x + 3y = 6$$
$$x = 6 - 3y \quad (3)$$

Substitute $6 - 3y$ for x in equation (1).

$$5x + 15y = 30$$
$$5(6 - 3y) + 15y = 30$$
$$30 - 15y + 15y = 30$$
$$30 = 30 \quad \textit{True}$$

This true result means that every solution of one equation is also a solution of the other, so the system has an infinite number of solutions. The solution set is $\{(x, y) \mid x + 3y = 6\}$.

13. If we simply add the given equations without first multiplying one or both equations by a constant, choice **C** is the only system in which a variable will be eliminated. If we add the equations in **C** we get $3x = 17$. (The variable y was eliminated.)

14. $2x + 12y = 7$ $\quad$ (1)
$\quad\quad 3x + 4y = 1$ $\quad$ (2)

(a) If we multiply equation (1) by -3, the first term will become $-6x$. To eliminate x, we need to change the first term on the left side of equation (2) from $3x$ to $6x$. In order to do this, we must multiply equation (2) by 2.

(b) If we multiply equation (1) by -3, the second term will become $-36y$. To eliminate y, we need to change the second term on the left side of equation (2) from $4y$ to $36y$. In order to do this, we must multiply equation (2) by 9.

15. $2x \quad - \quad y \quad = \quad 13$ $\quad$ (1)
$\quad\quad\underline{x \quad + \quad y \quad = \quad 8}$ $\quad\quad$ (2)
$\quad\quad 3x \quad\quad\quad\quad = \quad 21$ $\quad$ *Add* (1) *and* (2).
$\quad\quad\quad\quad\quad x \quad = \quad 7$

From (2), $y = 1$.

The solution set is $\{(7, 1)\}$.

16. $-4x + 3y = 25$ $\quad$ (1)
$\quad\quad 6x - 5y = -39$ $\quad$ (2)

Multiply equation (1) by 3 and equation (2) by 2; then add the results.

$\quad -12x \quad + \quad 9y \quad = \quad 75$
$\quad\underline{12x \quad - \quad 10y \quad = \quad -78}$
$\quad\quad\quad\quad\quad\quad -y \quad = \quad -3$
$\quad\quad\quad\quad\quad\quady \quad = \quad 3$

To find x, let $y = 3$ in equation (1).

$$-4x + 3y = 25$$
$$-4x + 3(3) = 25$$
$$-4x + 9 = 25$$
$$-4x = 16$$
$$x = -4$$

The solution set is $\{(-4, 3)\}$.

17. $3x - 4y = 9$ (1)
      $6x - 8y = 18$ (2)

Multiply equation (1) by -2 and add the result to equation (2).

$$
\begin{array}{rcrcr}
-6x & + & 8y & = & -18 \\
6x & - & 8y & = & 18 \\
\hline
& & 0 & = & 0 \quad \textit{True}
\end{array}
$$

This result indicates that all solutions of equation (1) are also solutions of equation (2). The given system has an infinite number of solutions. The solution set is $\{(x, y) \mid 3x - 4y = 9\}$.

18. $2x + y = 3$ (1)
      $-4x - 2y = 6$ (2)

Multiply equation (1) by 2 and add the result to equation (2).

$$
\begin{array}{rcrcr}
4x & + & 2y & = & 6 \\
-4x & - & 2y & = & 6 \\
\hline
& & 0 & = & 12 \quad \textit{False}
\end{array}
$$

This result indicates that the given system has solution set $\emptyset$.

19. $2x + 3y = -5$ (1)
      $3x + 4y = -8$ (2)

Multiply equation (1) by -3 and equation (2) by 2; then add the results.

$$
\begin{array}{rcrcr}
-6x & - & 9y & = & 15 \\
6x & + & 8y & = & -16 \\
\hline
& & -y & = & -1 \\
& & y & = & 1
\end{array}
$$

To find x, let $y = 1$ in equation (1).

$$
\begin{aligned}
2x + 3y &= -5 \\
2x + 3(1) &= -5 \\
2x + 3 &= -5 \\
2x &= -8 \\
x &= -4
\end{aligned}
$$

The solution set is $\{(-4, 1)\}$.

20. $6x - 9y = 0$ (1)
      $2x - 3y = 0$ (2)

Multiply equation (2) by -3 and add the result to equation (1).

$$
\begin{array}{rcrcr}
6x & - & 9y & = & 0 \\
-6x & + & 9y & = & 0 \\
\hline
& & 0 & = & 0 \quad \textit{True}
\end{array}
$$

This result indicates that the system has an infinite number of solutions. The solution set is $\{(x, y) \mid 2x - 3y = 0\}$.

21. $x - 2y = 5$ (1)
      $y = x - 7$ (2)

From (2), substitute $x - 7$ for y in equation (1).

$$
\begin{aligned}
x - 2y &= 5 \\
x - 2(x - 7) &= 5 \\
x - 2x + 14 &= 5 \\
-x &= -9 \\
x &= 9
\end{aligned}
$$

Let $x = 9$ in equation (2) to find y.

$$
y = 9 - 7 = 2
$$

The solution set is $\{(9, 2)\}$.

22. $\dfrac{x}{2} + \dfrac{y}{3} = 7$ (1)

      $\dfrac{x}{4} + \dfrac{2y}{3} = 8$ (2)

Multiply equation (1) by 6 to clear fractions.

$$
\begin{aligned}
6\left(\frac{x}{2} + \frac{y}{3}\right) &= 6(7) \\
3x + 2y &= 42 \quad (3)
\end{aligned}
$$

Multiply equation (2) by 12 to clear fractions.

$$
\begin{aligned}
12\left(\frac{x}{4} + \frac{2y}{3}\right) &= 12(8) \\
3x + 8y &= 96 \quad (4)
\end{aligned}
$$

To solve this system by the elimination method, multiply equation (3) by -1 and add the result to equation (4).

$$
\begin{array}{rcrcr}
-3x & - & 2y & = & -42 \\
3x & + & 8y & = & 96 \\
\hline
& & 6y & = & 54 \\
& & y & = & 9
\end{array}
$$

To find x, let $y = 9$ in equation (3).

$$
\begin{aligned}
3x + 2y &= 42 \\
3x + 2(9) &= 42 \\
3x + 18 &= 42 \\
3x &= 24 \\
x &= 8
\end{aligned}
$$

The solution set is $\{(8, 9)\}$.

23. $\frac{3}{4}x - \frac{1}{3}y = \frac{7}{6}$ (1)
      $\frac{1}{2}x + \frac{2}{3}y = \frac{5}{3}$ (2)

Multiply equation (1) by 12 to clear fractions.

$$
\begin{aligned}
12(\tfrac{3}{4}x) - 12(\tfrac{1}{3}y) &= 12(\tfrac{7}{6}) \\
9x - 4y &= 14 \quad (3)
\end{aligned}
$$

Multiply equation (2) by 6 to clear fractions.

$$
\begin{aligned}
6(\tfrac{1}{2}x) + 6(\tfrac{2}{3}y) &= 6(\tfrac{5}{3}) \\
3x + 4y &= 10 \quad (4)
\end{aligned}
$$

Add equations (3) and (4) to eliminate y.

$$
\begin{array}{rcrcr}
9x & - & 4y & = & 14 \\
3x & + & 4y & = & 10 \\
\hline
12x & & & = & 24 \\
& & x & = & 2
\end{array}
$$

To find y, let $x = 2$ in equation (4).

$$
\begin{aligned}
3(2) + 4y &= 10 \\
6 + 4y &= 10 \\
4y &= 4 \\
y &= 1
\end{aligned}
$$

The solution set is $\{(2, 1)\}$.

24. $0.4x - 0.5y = -2.2$ (1)
$0.3x + 0.2y = -0.5$ (2)

Multiply equation (1) and equation (2) by 10 to eliminate the decimals. Then solve the resulting equations by the substitution method.

$$
\begin{aligned}
4x - 5y &= -22 \quad (3) \\
3x + 2y &= -5 \quad (4)
\end{aligned}
$$

To solve this system by the elimination method, multiply equation (3) by 2, multiply equation (4) by 5, and add the resulting equations.

$$
\begin{array}{rcrcr}
8x & - & 10y & = & -44 \\
15x & + & 10y & = & -25 \\
\hline
23x & & & = & -69 \\
& & x & = & -3
\end{array}
$$

To find y, let $x = -3$ in equation (4).

$$
\begin{aligned}
3x + 2y &= -5 \\
3(-3) + 2y &= -5 \\
-9 + 2y &= -5 \\
2y &= 4 \\
y &= 2
\end{aligned}
$$

The solution set is $\{(-3, 2)\}$.

25. *Step 2*
Let $x =$ the number of Domino's restaurants;
   $y =$ the number of Pizza Hut restaurants.

Step 3
Pizza Hut had 6118 more locations than Domino's, so

$$
y = x + 6118. \quad (1)
$$

Together, the two chains had 23,400 locations, so

$$
x + y = 23,400. \quad (2)
$$

Step 4
Substitute $x + 6118$ for y in (2).

$$
\begin{aligned}
x + (x + 6118) &= 23,400 \\
2x + 6118 &= 23,400 \\
2x &= 17,282 \\
x &= 8641
\end{aligned}
$$

From (1), $y = 8641 + 6118 = 14,759$.

Step 5
In July 2009, Pizza Hut had 14,759 locations and Domino's had 8641 locations.

Step 6
14,759 is 6118 more than 8641 and the sum of 14,759 and 8641 is 23,400.

26. *Step 2*
Let $x =$ the circulation figure for
   Reader's Digest;
   $y =$ the circulation figure for *People.*

Step 3
The total circulation was 12.9 million, so

$$
x + y = 12.9. \quad (1)
$$

People's circulation was 5.7 million less than that of *Reader's Digest*, so

$$
y = x - 5.7. \quad (2)
$$

Step 4
Substitute $x - 5.7$ for y in (1).

$$
\begin{aligned}
x + (x - 5.7) &= 12.9 \\
2x - 5.7 &= 12.9 \\
2x &= 18.6 \\
x &= 9.3
\end{aligned}
$$

From (2), $y = 9.3 - 5.7 = 3.6$.

Step 5
The average circulation for *Reader's Digest* was 9.3 million and for *People* it was 3.6 million.

Step 6
3.6 is 5.7 less than 9.3 and the sum of 3.6 and 9.3 is 12.9.

27. *Step 2*
Let $x =$ the length of the rectangle;
   $y =$ the width of the rectangle.

Step 3
The perimeter is 90 meters, so

$$
2x + 2y = 90. \quad (1)
$$

The length is $1\frac{1}{2}$ (or $\frac{3}{2}$) times the width, so

$$
x = \frac{3}{2}y. \quad (2)
$$

Step 4
Substitute $\frac{3}{2}y$ for x in equation (1).

$$2x + 2y = 90$$
$$2(\tfrac{3}{2}y) + 2y = 90$$
$$3y + 2y = 90$$
$$5y = 90$$
$$y = 18$$

From (2), $x = \tfrac{3}{2}(18) = 27$.

Step 5
The length is 27 meters and the width is 18 meters.

Step 6
27 is $1\tfrac{1}{2}$ times 18 and the perimeter is
$2(27) + 2(18) = 90$ meters.

28. *Step 2*
Let $x =$ the number of \$20 bills;
$y =$ the number of \$10 bills.

Step 3
The total number of bills is 20, so

$$x + y = 20. \quad (1)$$

The total value of the money is \$330, so

$$20x + 10y = 330. \quad (2)$$

We may simplify equation (2) by dividing each side by 10.

$$2x + y = 33 \quad (3)$$

Step 4
To solve this equation by the elimination method, multiply equation (1) by -1 and add the result to equation (3).

$$
\begin{array}{rcrcr}
-x & - & y & = & -20 \\
2x & + & y & = & 33 \\
\hline
x & & & = & 13
\end{array}
$$

From (1), $y = 7$.

Step 5
She has 13 twenties and 7 tens.

Step 6
There are $13 + 7 = 20$ bills worth
$13(\$20) + 7(\$10) = \$330$.

29. *Step 2*
Let $x =$ the number of pounds of
\$1.30 per pound candy;
$y =$ the number of pounds of
\$0.90 per pound candy.

Number of pounds	Cost per pound (in dollars)	Total value (in dollars)
x	1.30	$1.30x$
y	0.90	$0.90y$
100	1.00	$100(1) = \$100$

Step 3
From the first and third columns of the table, we obtain the system

$$x + y = 100 \quad (1)$$
$$1.30x + 0.90y = 100. \quad (2)$$

Step 4
Multiply equation (2) by 10 (to clear decimals) and equation (1) by -9.

$$
\begin{array}{rcrcr}
-9x & - & 9y & = & -900 \\
13x & + & 9y & = & 1000 \\
\hline
4x & & & = & 100 \\
& & x & = & 25
\end{array}
$$

From (1), $y = 75$.

Step 5
25 pounds of candy at \$1.30 per pound and 75 pounds of candy at \$0.90 per pound should be used.

Step 6
The value of the mixture is
$25(1.30) + 75(0.90) = \$100$, giving us 100 pounds of candy that can sell for \$1 per pound.

30. *Step 2*
Let $x =$ the number of liters of 40%
antifreeze solution;
$y =$ the number of liters of 70%
antifreeze solution.

Liters of Solution	Percent (as a decimal)	Amount of Pure Antifreeze
x	0.40	$0.40x$
y	0.70	$0.70y$
90	0.50	$0.50(90) = 45$

Step 3
From the first and third columns of the table, we obtain the system

$$x + y = 90 \quad (1)$$
$$0.40x + 0.70y = 45. \quad (2)$$

Step 4
Multiply equation (2) by 10 (to clear decimals) and equation (1) by -4.

$$
\begin{array}{rcrcr}
-4x & - & 4y & = & -360 \\
4x & + & 7y & = & 450 \\
\hline
& & 3y & = & 90 \\
& & y & = & 30
\end{array}
$$

From (1), $x = 60$.

Step 5
In order to make 90 liters of 50% antifreeze solution, 60 liters of 40% solution and 30 liters of 70% solution will be needed.

Step 6
60 liters added to 30 liters will give us the desired
90 liters. 40% of 60 liters gives us 24 liters of
pure antifreeze and 70% of 30 liters gives us 21
liters of pure antifreeze. This is a total of 45 liters
of pure antifreeze, as required.

31. *Step 2*
Let x = the amount invested at 3%;
y = the amount invested at 4%.

Amount of Principal	Rate	Interest
x	0.03	$0.03x$
y	0.04	$0.04y$
$18,000	XXXX	$650

Step 3
From the chart, we obtain the equations

$$x + y = 18{,}000 \qquad (1)$$
$$0.03x + 0.04y = 650. \qquad (2)$$

Step 4
To clear decimals, multiply each side of equation
(2) by 100.

$$100(0.03x + 0.04y) = 100(650)$$
$$3x + 4y = 65{,}000 \qquad (3)$$

Multiply equation (1) by -3 and add the result to
equation (3).

$$\begin{array}{rcr} -3x \ - \ 3y &=& -54{,}000 \\ 3x \ + \ 4y &=& 65{,}000 \\ \hline y &=& 11{,}000 \end{array}$$

From (1), $x = 7000$.

Step 5
She invested $7000 at 3% and $11,000 at 4%.

Step 6
The sum of $7000 and $11,000 is $18,000.
3% of $7000 is $210 and 4% of $11,000 is $440.
This gives us $210 + $440 = $650 in interest, as
required.

32. *Step 2*
Let x = the rate of the plane in still air;
y = the rate of the wind.

	d	r	t
With wind	540	$x + y$	2
Against wind	690	$x - y$	3

Step 3
Use the formula $d = rt$.

$$540 = (x + y)(2)$$
$$270 = x + y \qquad (1) \quad \textit{Divide by 2.}$$

$$690 = (x - y)(3)$$
$$230 = x - y \qquad (2) \quad \textit{Divide by 3.}$$

Step 4
Solve the system by the elimination method.

$$\begin{array}{rcll} 270 &=& x + y & (1) \\ 230 &=& x - y & (2) \\ \hline 500 &=& 2x & \textit{Add } (1) \textit{ and } (2). \\ 250 &=& x \end{array}$$

From (1), $y = 20$.

Step 5
The rate of the wind is 20 miles per hour; the rate
of the plane in still air is 250 miles per hour.

Step 6
Flying with the wind at $250 + 20 = 270$ miles per
hour for 2 hours results in a trip of 540 miles.
Flying against the wind at $250 - 20 = 230$ miles
per hour for 3 hours results in a trip of 690 miles.

33. $x + y \geq 2$
$x - y \leq 4$

Graph $x + y = 2$ as a solid line through its
intercepts, $(2, 0)$ and $(0, 2)$. Using $(0, 0)$ as a test
point will result in the false statement $0 \geq 2$, so
shade the region *not* containing the origin.

Graph $x - y = 4$ as a solid line through its
intercepts, $(4, 0)$ and $(0, -4)$. Using $(0, 0)$ as a
test point will result in the true statement $0 \leq 4$, so
shade the region containing the origin.

The solution set of this system is the intersection
of the two shaded regions, and includes the
portions of the two lines that bound this region.

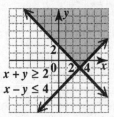

$x + y \geq 2$
$x - y \leq 4$

34. $y \geq 2x$
$2x + 3y \leq 6$

Graph $y = 2x$ as a solid line through $(0, 0)$ and
$(1, 2)$. This line goes through the origin, so a
different test point must be used. Choosing
$(-4, 0)$ as a test point will result in the true
statement $0 \geq -8$, so shade the region containing
$(-4, 0)$.

Graph $2x + 3y = 6$ as a solid line through its
intercepts, $(3, 0)$ and $(0, 2)$. Choosing $(0, 0)$ as a
test point will result in the true statement $0 \leq 6$, so
shade the region containing the origin.

The solution set of this system is the intersection
of the two shaded regions, and includes the
portions of the two lines that bound this region.

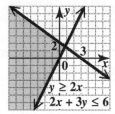

35. $x + y < 3$
$\qquad$ $2x > y$

Graph $x + y = 3$ as a dashed line through $(3, 0)$ and $(0, 3)$. Using $(0, 0)$ as a test point will result in the true statement $0 < 3$, so shade the region containing the origin.

Graph $2x = y$ as a dashed line through $(0, 0)$ and $(1, 2)$. Choosing $(0, -3)$ as a test point will result in the true statement $0 \geq -3$, so shade the region containing $(0, -3)$. The solution set of this system is the intersection of the two shaded regions. It does not contain the boundary lines.

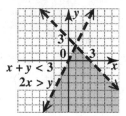

36. $3x - y \leq 3$
$\qquad$ $x \geq -1$
$\qquad$ $y \leq 2$

Graph the line $3x - y = 3$ through its intercepts $(1, 0)$ and $(0, -3)$, the vertical line $x = -1$, and the horizontal line $y = 2$. All of these lines should be solid because of the $\leq$ and $\geq$ signs. All three inequalities are true for $(0, 0)$. Shade the triangular region bounded by the three lines, which contains the test point $(0, 0)$. The solution includes the portions of the three lines that bound the region and form the sides of the triangle.

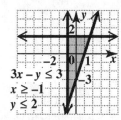

37.   **[4.4]** **(a)** The graph of the fixed rate mortgage is above the graph of the variable rate mortgage for years 0 to 6, so that is when the monthly payment for the fixed rate mortgage is more than the monthly payment for the variable rate mortgage.

(b) The graphs intersect at year 6, so that's when the payments would be the same. The monthly payment at that time appears to be about $650.

38.   **[4.2]** System **B** is easier to solve by the substitution method than system **A** because the bottom equation in system **B** is already solved for y.

Solving system **A** would require our solving one of the equations for one of the variables before substituting, and the expression to be substituted would involve fractions.

39.   **[4.5]** The shaded region is to the left of the vertical line $x = 3$, above the horizontal line $y = 1$, and includes both lines. This is the graph of the system

$$x \leq 3$$
$$y \geq 1,$$

so the answer is **B**.

40.   **[4.3]** $\dfrac{2x}{3} + \dfrac{y}{4} = \dfrac{14}{3}$ (1)

$\qquad\qquad$ $\dfrac{x}{2} + \dfrac{y}{12} = \dfrac{8}{3}$ (2)

To clear fractions, multiply both sides of each equation by 12.

$$12\left(\frac{2x}{3} + \frac{y}{4}\right) = 12\left(\frac{14}{3}\right)$$
$$8x + 3y = 56 \qquad (3)$$

$$12\left(\frac{x}{2} + \frac{y}{12}\right) = 12\left(\frac{8}{3}\right)$$
$$6x + y = 32 \qquad (4)$$

To solve this system by the elimination method, multiply both sides of equation (4) by -3 and add the result to equation (3).

$$
\begin{array}{rrrcr}
8x & + & 3y & = & 56 \\
-18x & - & 3y & = & -96 \\
\hline
-10x & & & = & -40 \\
& & x & = & 4
\end{array}
$$

To find y, let $x = 4$ in equation (4).

$$6x + y = 32$$
$$6(4) + y = 32$$
$$24 + y = 32$$
$$y = 8$$

The solution set is $\{(4, 8)\}$.

41.   **[4.3]** $\qquad x = y + 6$ (1)

$\qquad\qquad$ $2y - 2x = -12$ (2)

Rewrite equations in the $Ax + By = C$ form. Equation (1) becomes

$$x - y = 6, \quad (3)$$

and equation (2) becomes

$$-x + y = -6, \quad (4)$$

after dividing by 2.

To solve this system by the elimination method, add equations (3) and (4).

$$
\begin{array}{rcrcr}
x & - & y & = & 6 \\
-x & + & y & = & -6 \\
\hline
 & & 0 & = & 0 \quad \textit{True}
\end{array}
$$

This result indicates that the system has an infinite number of solutions. The solution set is $\{(x, y) \mid x - y = 6\}$.

42. **[4.3]** $3x + 4y = 6$ (1)
         $4x - 5y = 8$ (2)

To solve this system by the elimination method, multiply equation (1) by -4 and equation (2) by 3; then add the results.

$$
\begin{array}{rcrcr}
-12x & - & 16y & = & -24 \\
12x & - & 15y & = & 24 \\
\hline
 & & -31y & = & 0 \\
 & & y & = & 0
\end{array}
$$

To find x, substitute 0 for y in equation (1).

$$
\begin{aligned}
3x + 4(0) &= 6 \\
3x &= 6 \\
x &= 2
\end{aligned}
$$

The solution set is $\{(2, 0)\}$.

43. **[4.3]** $0.4x - 0.9y = 0.7$ (1)
         $0.3x + 0.2y = 1.4$ (2)

Multiply equation (1) and equation (2) by 10 to eliminate the decimals.

$$
\begin{aligned}
4x - 9y &= 7 \quad (3) \\
3x + 2y &= 14 \quad (4)
\end{aligned}
$$

To solve this system by the elimination method, multiply equation (3) by 2, multiply equation (4) by 9, and add the resulting equations.

$$
\begin{array}{rcrcr}
8x & - & 18y & = & 14 \\
27x & + & 18y & = & 126 \\
\hline
35x & & & = & 140 \\
 & & x & = & 4
\end{array}
$$

To find y, let $x = 4$ in equation (4).

$$
\begin{aligned}
3x + 2y &= 14 \\
3(4) + 2y &= 14 \\
12 + 2y &= 14 \\
2y &= 2 \\
y &= 1
\end{aligned}
$$

The solution set is $\{(4, 1)\}$.

44. **[4.5]** $x + y < 5$
         $x - y \geq 2$

Graph $x + y = 5$ as a dashed line through its intercepts, $(5, 0)$ and $(0, 5)$. Using $(0, 0)$ as a test point will result in the true statement $0 < 5$, so shade the region containing the origin.

Graph $x - y = 2$ as a solid line through its intercepts, $(2, 0)$ and $(0, -2)$. Using $(0, 0)$ as a test point will result in the false statement $0 \geq 2$, so shade the region *not* containing the origin.

The solution set of this system is the intersection of the two shaded regions. It includes the portion of the line $x - y = 2$ that bounds this region, but not the line $x + y = 5$.

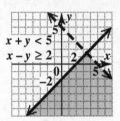

45. **[4.5]** $y \leq 2x$
         $x + 2y > 4$

Graph $y = 2x$ as a solid line through $(0, 0)$ and $(1, 2)$. Shade the region below the line.

Graph $x + 2y = 4$ as a dashed line through its intercepts, $(4, 0)$ and $(0, 2)$. Using $(0, 0)$ as a test point will result in the false statement $0 > 4$, so shade the region *not* containing the origin.

The solution set of this system is the intersection of the shaded regions. It includes the portion of the line $y = 2x$ that bounds this region, but not the line $x + 2y = 4$.

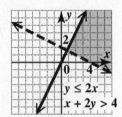

46. **[4.4]** Let $x =$ the length of each of the two equal sides;
         $y =$ the length of the longer third side.

The perimeter is 29 inches, so

$$
\begin{aligned}
x + x + y &= 29 \\
\text{or} \quad 2x + y &= 29. \quad (1)
\end{aligned}
$$

The third side is 5 inches longer than each of the two equal sides, so

$$
y = x + 5. \quad (2)
$$

Substitute $x + 5$ for y in (1).

$$2x + y = 29$$
$$2x + (x + 5) = 29$$
$$3x + 5 = 29$$
$$3x = 24$$
$$x = 8$$

From (2), $y = 8 + 5 = 13$.

The lengths of the sides of the triangle are 8 inches, 8 inches, and 13 inches.

47. **[4.4]** Let $x =$ the number of people who visited the Statue of Liberty;

$y =$ the number of people who visited the National World War II Memorial.

The total was 7.5 million, so

$$x + y = 7.5. \quad (1)$$

The Statue of Liberty had 0.7 million fewer visitors than the National World War II Memorial, so

$$x = y - 0.7. \quad (2)$$

Substitute $y - 0.7$ for x in (1).

$$(y - 0.7) + y = 7.5$$
$$2y - 0.7 = 7.5$$
$$2y = 8.2$$
$$y = 4.1$$

From (2), $x = 4.1 - 0.7 = 3.4$.

In 2007, 4.1 million people visited the National World War II Memorial and 3.4 million people visited the Statue of Liberty.

48. **[4.4]** Let $x =$ the rate of the slower car;

$y =$ the rate of the faster car.

One car travels 30 miles per hour faster than the other, so

$$y = x + 30. \quad (1)$$

In $2\frac{1}{2}$ (or $\frac{5}{2}$) hours, the slower car travels $\left(\frac{5}{2}\right)(x)$ miles and the faster car travels $\left(\frac{5}{2}\right)(y)$ miles. The cars will be 265 miles apart, so

$$\frac{5}{2}x + \frac{5}{2}y = 265. \quad (2)$$

Clear fractions from equation (2).

$$2\left(\frac{5}{2}x + \frac{5}{2}y\right) = 2(265) \qquad \textit{Multiply by 2.}$$
$$5x + 5y = 530$$
$$x + y = 106 \qquad \textit{Divide by 5.}$$

We now have the system

$$y = x + 30 \qquad (1)$$
$$x + y = 106. \qquad (3)$$

Substitute $x + 30$ for y in equation (3).

$$x + y = 106$$
$$x + (x + 30) = 106$$
$$2x + 30 = 106$$
$$2x = 76$$
$$x = 38$$

From (1), $y = 38 + 30 = 68$.

The slower car went 38 miles per hour, and the faster car went 68 miles per hour.

Chapter 4 Test

1. $2x + y = -3 \qquad (1)$
    $x - y = -9 \qquad (2)$

(a) $(1, -5)$

Substitute 1 for x and -5 for y in (1) and (2).

$$(1) \qquad 2(1) + (-5) \overset{?}{=} -3$$
$$-3 = -3 \quad \textit{True}$$

$$(2) \qquad (1) - (-5) \overset{?}{=} -9$$
$$6 = -9 \quad \textit{False}$$

Since $(1, -5)$ does not satisfy both equations, it *is not* a solution of the system.

(b) $(1, 10)$

$$(1) \qquad 2(1) + (10) \overset{?}{=} -3$$
$$12 = -3 \quad \textit{False}$$

The false result indicates that $(1, 10)$ *is not* a solution of the system.

(c) $(-4, 5)$

$$(1) \qquad 2(-4) + (5) \overset{?}{=} -3$$
$$-3 = -3 \quad \textit{True}$$

$$(2) \qquad (-4) - (5) \overset{?}{=} -9$$
$$-9 = -9 \quad \textit{True}$$

Since $(-4, 5)$ satisfies both equations, it *is* a solution of the system.

2. $x + 2y = 6$
    $-2x + y = -7$

Graph the line $x + 2y = 6$ through its intercepts, $(6, 0)$ and $(0, 3)$. Because the x-intercept of the line $-2x + y = -7$, which is $\left(\frac{7}{2}, 0\right)$, has a fractional coordinate, we can graph the line more accurately by using its slope and y-intercept. Rewrite the equation as

$$y = 2x - 7.$$

Start by plotting the y-intercept, $(0, -7)$. From this point, go 2 units up and 1 unit to the right to reach the point $(1, -5)$. Draw the line through $(0, -7)$ and $(1, -5)$.

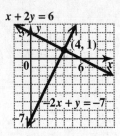

$x + 2y = 6$

$(4, 1)$

$-2x + y = -7$

It appears that the lines intersect at the point $(4, 1)$. Since $(4, 1)$ satisfies both equations, the solution set of this system is $\{(4, 1)\}$.

3. $2x + y = -4$ (1)

   $\quad\; x = y + 7$ (2)

Substitute $y + 7$ for x in equation (1), and solve for y.

$$2x + y = -4$$
$$2(y + 7) + y = -4$$
$$2y + 14 + y = -4$$
$$3y = -18$$
$$y = -6$$

From (2), $x = -6 + 7 = 1$.

The solution set is $\{(1, -6)\}$.

4. $4x + 3y = -35$ (1)

   $\quad x + y = 0$ (2)

Solve equation (2) for y.

$$y = -x \quad (3)$$

Substitute $-x$ for y in equation (1) and solve for x.

$$4x + 3y = -35$$
$$4x + 3(-x) = -35$$
$$4x - 3x = -35$$
$$x = -35$$

From (3), $y = -(-35) = 35$.

The solution set is $\{(-35, 35)\}$.

5. $\begin{array}{rcrcl} 2x & - & y & = & 4 \quad (1) \\ 3x & + & y & = & 21 \quad (2) \\ \hline 5x & & & = & 25 \quad \textit{Add (1) and (2).} \\ & & x & = & 5 \end{array}$

To find y, let $x = 5$ in equation (2).

$$3x + y = 21$$
$$3(5) + y = 21$$
$$15 + y = 21$$
$$y = 6$$

The solution set is $\{(5, 6)\}$.

6. $4x + 2y = 2$ (1)

   $5x + 4y = 7$ (2)

Multiply equation (1) by -2 and add the result to equation (2).

$$\begin{array}{rcrcl} -8x & - & 4y & = & -4 \\ 5x & + & 4y & = & 7 \\ \hline -3x & & & = & 3 \\ & & x & = & -1 \end{array}$$

To find y, let $x = -1$ in equation (1).

$$4x + 2y = 2$$
$$4(-1) + 2y = 2$$
$$-4 + 2y = 2$$
$$2y = 6$$
$$y = 3$$

The solution set is $\{(-1, 3)\}$.

7. $3x + 4y = 9$ (1)

   $2x + 5y = 13$ (2)

$$\begin{array}{rcrcll} 6x & + & 8y & = & 18 & (3) \; 2 \times \text{Eq.(1)} \\ -6x & - & 15y & = & -39 & (4) \, -3 \times \text{Eq.(2)} \\ \hline & & -7y & = & -21 & \textit{Add (3) and (4).} \\ & & y & = & 3 \end{array}$$

Substitute 3 for y in (1).

$$3x + 4y = 9$$
$$3x + 4(3) = 9$$
$$3x + 12 = 9$$
$$3x = -3$$
$$x = -1$$

The solution set is $\{(-1, 3)\}$.

8. $4x + 5y = 2$ (1)

   $-8x - 10y = 6$ (2)

Multiply equation (1) by 2 and add the result to equation (2).

$$\begin{array}{rcrcll} 8x & + & 10y & = & 4 \\ -8x & - & 10y & = & 6 \\ \hline & & 0 & = & 10 & \textit{False} \end{array}$$

This result indicates that the system has solution set $\emptyset$.

9. $6x - 5y = 0$ (1)

   $-2x + 3y = 0$ (2)

Multiply equation (2) by 3 and add the result to equation (1).

$$\begin{array}{rcrcl} 6x & - & 5y & = & 0 \\ -6x & + & 9y & = & 0 \\ \hline & & 4y & = & 0 \\ & & y & = & 0 \end{array}$$

To find x, let $y = 0$ in equation (1).

$$6x - 5y = 0$$
$$6x - 5(0) = 0$$
$$6x = 0$$
$$x = 0$$

The solution set is $\{(0,0)\}$.

10.
$$4y = -3x + 5 \qquad (1)$$
$$6x = -8y + 10 \qquad (2)$$

Rewrite (1) and (2) in standard form.

$$3x + 4y = 5 \qquad (3)$$
$$6x + 8y = 10 \qquad (4)$$

Multiply equation (3) by -2 and add the result to (4).

$$
\begin{array}{rcrcr}
-6x & - & 8y & = & -10 \\
6x & + & 8y & = & 10 \\
\hline
& & 0 & = & 0 \quad \textit{True}
\end{array}
$$

The true result indicates that this system has an infinite number of solutions.

The solution set is $\{(x,y) \mid 3x + 4y = 5\}$.

11.
$$\tfrac{6}{5}x - \tfrac{1}{3}y = -20 \qquad (1)$$
$$-\tfrac{2}{3}x + \tfrac{1}{6}y = 11 \qquad (2)$$

To clear fractions, multiply by the LCD for each equation. Multiply equation (1) by 15.

$$15\left(\tfrac{6}{5}x - \tfrac{1}{3}y\right) = 15(-20)$$
$$18x - 5y = -300 \qquad (3)$$

Multiply equation (2) by 6.

$$6\left(-\tfrac{2}{3}x + \tfrac{1}{6}y\right) = 6(11)$$
$$-4x + y = 66 \qquad (4)$$

To solve this system by the elimination method, multiply equation (4) by 5 and add the result to equation (3).

$$
\begin{array}{rcrcr}
18x & - & 5y & = & -300 \\
-20x & + & 5y & = & 330 \\
\hline
-2x & & & = & 30 \\
& & x & = & -15
\end{array}
$$

To find y, let $x = -15$ in equation (4).

$$-4x + y = 66$$
$$-4(-15) + y = 66$$
$$60 + y = 66$$
$$y = 6$$

The solution set is $\{(-15, 6)\}$.

12. *Step 2*
Let $x =$ the distance between Memphis and Atlanta (in miles);
$y =$ the distance between Minneapolis and Houston (in miles).

Step 3
Since the distance between Memphis and Atlanta is 782 miles less than the distance between Minneapolis and Houston,

$$x = y - 782. \quad (1)$$

Together the two distances total 1570 miles, so

$$x + y = 1570. \quad (2)$$

Step 4
Substitute $y - 782$ for x in equation (2).

$$(y - 782) + y = 1570$$
$$2y - 782 = 1570$$
$$2y = 2352$$
$$y = 1176$$

From (1), $x = 1176 - 782 = 394$.

Step 5
The distance between Memphis and Atlanta is 394 miles, while the distance between Minneapolis and Houston is 1176 miles.

Step 6
394 is 782 less than 1176 and the sum of 394 and 1176 is 1570, as required.

13. *Step 2*
Let $x =$ the number of visitors to the Magic Kingdom (in millions);
$y =$ the number of visitors to Disneyland (in millions).

Step 3
Disneyland had 2.4 million fewer visitors than the Magic Kingdom, so

$$y = x - 2.4. \quad (1)$$

Together they had 31.8 million visitors, so

$$x + y = 31.8. \quad (2)$$

Step 4
From (1), substitute $x - 2.4$ for y in (2).

$$x + (x - 2.4) = 31.8$$
$$2x - 2.4 = 31.8$$
$$2x = 34.2$$
$$x = 17.1$$

From (1), $y = 17.1 - 2.4 = 14.7$.

Step 5
In 2008, the Magic Kingdom had 17.1 million visitors and Disneyland had 14.7 million visitors.

Step 6
14.7 is 2.4 fewer than 17.1 and the sum of 14.7 and 17.1 is 31.8, as required.

14. *Step 2*

Let $x =$ the number of liters of 25%
    alcohol solution;

$y =$ the number of liters of 40%
    alcohol solution.

Liters of Solution	Percent (as a decimal)	Liters of Pure Alcohol
x	0.25	$0.25x$
y	0.40	$0.40y$
50	0.30	$0.30(50) = 15$

Step 3

From the first and third columns of the table, we obtain the equations

$$x + y = 50 \qquad (1)$$
$$0.25x + 0.40y = 15. \qquad (2)$$

To clear decimals, multiply both sides of equation (2) by 100.

$$25x + 40y = 1500 \qquad (3)$$

Step 4

To solve this system by the elimination method, multiply equation (1) by -25 and add the result to equation (3).

$$
\begin{array}{rcr}
-25x - 25y &=& -1250 \\
25x + 40y &=& 1500 \\
\hline
15y &=& 250 \\
\end{array}
$$
$$y = \frac{250}{15} = \frac{50}{3} = 16\frac{2}{3}$$

From (1), $x = 33\frac{1}{3}$.

Step 5

To get 50 liters of a 30% alcohol solution, $33\frac{1}{3}$ liters of a 25% solution and $16\frac{2}{3}$ liters of a 40% solution should be used.

Step 6

Since $33\frac{1}{3} + 16\frac{2}{3} = 50$ and
$0.25(33\frac{1}{3}) + 0.40(16\frac{2}{3})$
$= \frac{1}{4}(\frac{100}{3}) + \frac{2}{5}(\frac{50}{3})$
$= \frac{25}{3} + \frac{20}{3} = \frac{45}{3} = 15$, this mixture will give the 50 liters of 30% solution, as required.

15. *Step 2*

Let $x =$ the rate of the faster car;

$y =$ the rate of the slower car.

Use $d = rt$ in constructing the table.

	r	t	d
Faster car	x	3	$3x$
Slower car	y	3	$3y$

Step 3

The faster car travels $1\frac{1}{3}$ times as fast as the other car, so

$$x = 1\frac{1}{3}y$$
$$\text{or} \quad x = \frac{4}{3}y. \qquad (1)$$

After 3 hours, they are 45 miles apart, that is, the difference between their distances is 45.

$$3x - 3y = 45 \quad (2)$$

Step 4

To solve the system by substitution, substitute $\frac{4}{3}y$ for x in equation (2).

$$3(\tfrac{4}{3}y) - 3y = 45$$
$$4y - 3y = 45$$
$$y = 45$$

From (1), $x = \frac{4}{3}(45) = 60.$

Step 5

The rate of the faster car is 60 miles per hour, and the rate of the slower car is 45 miles per hour.

Step 6

60 is one and one-third times 45. In three hours, the faster car travels $3(60) = 180$ miles and the slower car travels $3(45) = 135$ miles. The cars are $180 - 135 = 45$ miles apart, as required.

16. $2x + 7y \le 14$
    $x - y \ge 1$

Graph $2x + 7y = 14$ as a solid line through its intercepts, $(7, 0)$ and $(0, 2)$. Choosing $(0, 0)$ as a test point will result in the true statement $0 \le 14$, so shade the side of the line containing the origin.

Graph $x - y = 1$ as a solid line through its intercepts, $(1, 0)$ and $(0, -1)$. Choosing $(0, 0)$ as a test point will result in the false statement $0 \ge 1$, so shade the side of the line *not* containing the origin.

The solution set of the given system is the intersection of the two shaded regions, and includes the portions of the two lines that bound this region.

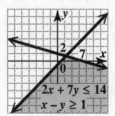

17. $2x - y > 6$
$4y + 12 \geq -3x$

Graph $2x - y = 6$ as a dashed line through its intercepts, $(3, 0)$ and $(0, -6)$. Choosing $(0, 0)$ as a test point will result in the false statement $0 > 6$, so shade the side of the line *not* containing the origin.

Graph $4y + 12 = -3x$ as a solid line through its intercepts, $(-4, 0)$ and $(0, -3)$. Choosing $(0, 0)$ as a test point will result in the true statement $12 \geq 0$, so shade the side of the line containing the origin.

The solution set of the given system is the intersection of the two shaded regions. It includes the portion of the line $4y + 12 = -3x$ that bounds this region, but not the line $2x - y = 6$.

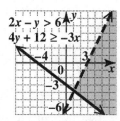

18. It is impossible for the sum of any two numbers to be both greater than 4 and less than 3. Therefore, system **B** has no solution.

Cumulative Review Exercises (Chapters 1–4)

1. The integer factors of 40 are $-1, 1, -2, 2, -4, 4,$
$-5, 5, -8, 8, -10, 10, -20, 20, -40,$ and 40.

2. $-2 + 6[3 - (4 - 9)] = -2 + 6[3 - (-5)]$
$\qquad\qquad\qquad\quad = -2 + 6(8)$
$\qquad\qquad\qquad\quad = -2 + 48$
$\qquad\qquad\qquad\quad = 46$

3. $\dfrac{3x^2 + 2y^2}{10y + 3} = \dfrac{3 \cdot 1^2 + 2 \cdot 5^2}{10(5) + 3}$ *Let x = 1, y = 5.*

$\qquad\qquad = \dfrac{3 \cdot 1 + 2 \cdot 25}{50 + 3}$

$\qquad\qquad = \dfrac{3 + 50}{50 + 3} = \dfrac{53}{53} = 1$

4. $r(s - k) = rs - rk$

This is an example of the distributive property.

5. $2 - 3(6x + 2) = 4(x + 1) + 18$
$2 - 18x - 6 = 4x + 4 + 18$
$-18x - 4 = 4x + 22$
$-22x = 26$
$x = \dfrac{26}{-22} = -\dfrac{13}{11}$

The solution set is $\left\{ -\dfrac{13}{11} \right\}$.

6. $\dfrac{3}{2}\left(\dfrac{1}{3}x + 4\right) = 6\left(\dfrac{1}{4} + x\right)$

Multiply each side by 2.

$2\left[\dfrac{3}{2}\left(\dfrac{1}{3}x + 4\right)\right] = 2\left[6\left(\dfrac{1}{4} + x\right)\right]$
$3\left(\dfrac{1}{3}x + 4\right) = 12\left(\dfrac{1}{4} + x\right)$

Use the distributive property to remove parentheses.

$x + 12 = 3 + 12x$
$-11x = -9$
$x = \dfrac{9}{11}$

The solution set is $\left\{ \dfrac{9}{11} \right\}$.

7. $P = \dfrac{kT}{V}$ for T

$PV = kT$ \qquad\qquad *Multiply by V.*

$\dfrac{PV}{k} = \dfrac{kT}{k}$ \qquad\qquad *Divide by k.*

$\dfrac{PV}{k} = T$ or $T = \dfrac{PV}{k}$

8. $-\dfrac{5}{6}x < 15$

Multiply each side by the reciprocal of $-\dfrac{5}{6}$, which is $-\dfrac{6}{5}$, and reverse the direction of the inequality symbol.

$-\dfrac{6}{5}\left(-\dfrac{5}{6}x\right) > -\dfrac{6}{5}(15)$
$x > -18$

The solution set is $(-18, \infty)$.

9. $-8 < 2x + 3$
$-11 < 2x$
$-\dfrac{11}{2} < x$ or $x > -\dfrac{11}{2}$

The solution set is $\left(-\dfrac{11}{2}, \infty\right)$.

10. 80.4% of 2500 is $0.804(2500) = 2010$
72.5% of 2500 is $0.725(2500) = 1812.5 \approx 1813$

$\dfrac{1570}{2500} = 0.628$ or 62.8%

$\dfrac{1430}{2500} = 0.572$ or 57.2%

Product or Company	Percent	Actual Number
Charmin	80.4%	2010
Wheaties	72.5%	1813
Budweiser	62.8%	1570
State Farm	57.2%	1430

11. Let x = number of "against" votes;
y = number of "in favor" votes.

There were 99 total votes and there were 37 more "in favor" votes than "against" votes, so we have:

$$x + y = 99 \qquad (1)$$
$$y = x + 37 \qquad (2)$$

Substitute $x + 37$ for y in equation (1).

$$x + (x + 37) = 99$$
$$2x + 37 = 99$$
$$2x = 62$$
$$x = 31$$

From (2), $y = 31 + 37 = 68$.
There were 68 "in favor" votes and 31 "against" votes.

12. Let x = the measure of the equal angles.
Then $2x - 4$ = the measure of the third angle.

The sum of the measures of the angles in a triangle is $180°$.

$$x + x + (2x - 4) = 180$$
$$4x - 4 = 180$$
$$4x = 184$$
$$x = 46$$

The third angle has measure

$$2x - 4 = 2(46) - 4 = 92 - 4 = 88.$$

The measures of the angles are $46°$, $46°$, and $88°$.

13. $x - y = 4$

To graph this line, find the intercepts.

If $y = 0$, $x = 4$, so the x-intercept is $(4, 0)$.
If $x = 0$, $y = -4$, so the y-intercept is $(0, -4)$.

Graph the line through these intercepts. A third point, such as $(5, 1)$, may be used as a check.

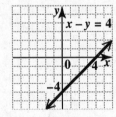

14. $3x + y = 6$

If $y = 0$, $x = 2$, so the x-intercept is $(2, 0)$.
If $x = 0$, $y = 6$, so the y-intercept is $(0, 6)$.

Graph the line through these intercepts. A third point, such as $(1, 3)$, may be used as a check.

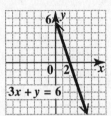

15. The slope m of the line passing through the points $(-5, 6)$ and $(1, -2)$ is

$$m = \frac{y_2 - y_1}{x_2 - x_1} = \frac{-2 - 6}{1 - (-5)} = \frac{-8}{6} = -\frac{4}{3}.$$

16. The slope of the line $y = 4x - 3$ is 4. The slope of the line whose graph is perpendicular to that of $y = 4x - 3$ is the negative reciprocal of 4, namely $-\frac{1}{4}$.

17. Through $(-4, 1)$ with slope $\frac{1}{2}$
Use the point-slope form of a line.

$$y - y_1 = m(x - x_1)$$
$$y - 1 = \tfrac{1}{2}[x - (-4)]$$
$$y - 1 = \tfrac{1}{2}(x + 4)$$
$$y - 1 = \tfrac{1}{2}x + 2$$
$$y = \tfrac{1}{2}x + 3 \qquad \textit{Slope-intercept form}$$

18. Through the points $(1, 3)$ and $(-2, -3)$
Find the slope.

$$m = \frac{y_2 - y_1}{x_2 - x_1} = \frac{-3 - 3}{-2 - 1} = \frac{-6}{-3} = 2$$

Use the point-slope form of a line.

$$y - y_1 = m(x - x_1)$$
$$y - 3 = 2(x - 1)$$
$$y - 3 = 2x - 2$$
$$y = 2x + 1 \qquad \textit{Slope-intercept form}$$

19. **(a)** On the vertical line through $(9, -2)$, the x-coordinate of every point is 9. Therefore, an equation of this line is $x = 9$.

(b) On the horizontal line through $(4, -1)$, the y-coordinate of every point is -1. Therefore, an equation of this line is $y = -1$.

20. $2x - y = -8$ (1)
$x + 2y = 11$ (2)

To solve this system by the elimination method, multiply equation (1) by 2 and add the result to equation (2) to eliminate y.

$$
\begin{array}{rrrrr}
4x & - & 2y & = & -16 \\
x & + & 2y & = & 11 \\
\hline
5x & & & = & -5 \\
& & x & = & -1
\end{array}
$$

To find y, let $x = -1$ in equation (2).

$$
\begin{aligned}
-1 + 2y &= 11 \\
2y &= 12 \\
y &= 6
\end{aligned}
$$

The solution set is $\{(-1, 6)\}$.

21. $4x + 5y = -8$ (1)
$3x + 4y = -7$ (2)

Multiply equation (1) by -3 and equation (2) by 4. Add the resulting equations to eliminate x.

$$
\begin{array}{rrrrr}
-12x & - & 15y & = & 24 \\
12x & + & 16y & = & -28 \\
\hline
& & y & = & -4
\end{array}
$$

To find x, let $y = -4$ in equation (1).

$$
\begin{aligned}
4x + 5y &= -8 \\
4x + 5(-4) &= -8 \\
4x - 20 &= -8 \\
4x &= 12 \\
x &= 3
\end{aligned}
$$

The solution set is $\{(3, -4)\}$.

22. $3x + 4y = 2$ (1)
$6x + 8y = 1$ (2)

Multiply equation (1) by -2 and add the result to equation (2).

$$
\begin{array}{rrrrrl}
-6x & - & 8y & = & -4 & \\
6x & + & 8y & = & 1 & \\
\hline
& & 0 & = & -3 & \textit{False}
\end{array}
$$

Since $0 = -3$ is a false statement, the solution set is $\emptyset$.

23. *Step 2*

Let $x =$ the number of adult tickets sold;
$y =$ the number of child tickets sold.

Kind of Ticket	Number Sold	Cost of Each (in dollars)	Total Value (in dollars)
Adult	x	6	$6x$
Child	y	2	$2y$
Total	454	XXXXXX	$2528

Step 3

The total number of tickets sold was 454, so

$$x + y = 454. \quad (1)$$

Since the total value was \$2528, the final column leads to

$$6x + 2y = 2528. \quad (2)$$

Step 4

Multiply both sides of equation (1) by -2 and add this result to equation (2).

$$
\begin{array}{rrrrr}
-2x & - & 2y & = & -908 \\
6x & + & 2y & = & 2528 \\
\hline
4x & & & = & 1620 \\
& & x & = & 405
\end{array}
$$

From (1), $y = 454 - 405 = 49$.

Step 5

There were 405 adult tickets and 49 child tickets sold.

Step 6

The total number of tickets sold was $405 + 49 = 454$. Since 405 adults paid \$6 each and 49 children paid \$2 each, the value of tickets sold should be $405(6) + 49(2) = 2528$, or \$2528. The result agrees with the given information.

24. Let $x =$ the number of liters of 20% alcohol solution;

$y =$ the number of liters of 50% alcohol solution.

Liters of Solution	Percent (as a decimal)	Liters of Pure Alcohol
x	0.20	$0.20x$
y	0.50	$0.50y$
12	0.40	$0.40(12) = 4.8$

Since 12 L of the mixture are needed,

$$x + y = 12. \quad (1)$$

Since the amount of pure alcohol in the 20% solution plus the amount of pure alcohol in the 50% solution must equal the amount of pure alcohol in the mixture,

$$0.20x + 0.50y = 4.8.$$

Multiply by 10 to clear the decimals.

$$2x + 5y = 48 \quad (2)$$

Multiply equation (1) by -2 and add the result to equation (2).

$$
\begin{array}{rcr}
-2x - 2y &=& -24 \\
2x + 5y &=& 48 \\
\hline
3y &=& 24 \\
y &=& 8
\end{array}
$$

From (1), $x = 12 - 8 = 4$.

4 L of 20% alcohol solution and 8 L of 50% alcohol solution will be needed.

25. $x + 2y \leq 12$
$2x - y \leq 8$

Graph the boundary with equation $x + 2y = 12$ as a solid line through its intercepts, $(12, 0)$ and $(0, 6)$. Using $(0, 0)$ as a test point results in a true statement, $0 \leq 12$. Shade the region containing the origin.

Graph the boundary with equation $2x - y = 8$ as a solid line through its intercepts, $(4, 0)$ and $(0, -8)$. Using $(0, 0)$ as a test point results in a true statement, $0 \leq 8$. Shade the region containing the origin.

The solution is the intersection of the two shaded regions and includes the portions of the lines that bound this region.

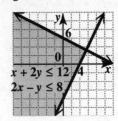

CHAPTER 5   EXPONENTS AND POLYNOMIALS

5.1   The Product Rule and Power Rules for Exponents

5.1   Now Try Exercises

N1. $4 \cdot 4 \cdot 4 = 4^3$ 4 occurs as a factor 3 times.
$$= 64$$

N2.   **(a)** $(-3)^4 = (-3)(-3)(-3)(-3)$
$$= 81$$

Base: -3; exponent: 4

(b) $-3^4 = -1 \cdot 3^4$
$$= -1 \cdot (3 \cdot 3 \cdot 3 \cdot 3)$$
$$= -81$$

Base: 3; exponent: 4

N3.   **(a)** $(-5)^2(-5)^4 = (-5)^{2+4}$ *Product rule*
$$= (-5)^6$$

(b) $y^2 \cdot y \cdot y^5 = y^2 \cdot y^1 \cdot y^5 = y^{2+1+5} = y^8$

(c) $(2x^3)(4x^7) = (2 \cdot 4) \cdot (x^3 \cdot x^7)$
$$= 8(x^{3+7})$$
$$= 8x^{10}$$

(d) $2^4 \cdot 5^3$: the product rule does not apply because the bases are different.
$$2^4 \cdot 5^3 = 16 \cdot 125 = 2000$$

(e) $3^2 + 3^3$: the product rule does not apply because it is a sum, not a product.
$$3^2 + 3^3 = 9 + 27 = 36$$

N4.   **(a)** $(4^7)^5 = 4^{7 \cdot 5}$ *Power rule (a)*
$$= 4^{35}$$

(b) $(y^4)^7 = y^{4 \cdot 7}$ *Power rule (a)*
$$= y^{28}$$

N5.   **(a)** $(-5ab)^3$
$$= (-1 \cdot 5 \cdot a \cdot b)^3 \qquad -x = -1 \cdot x$$
$$= (-1)^3 \cdot 5^3 \cdot a^3 \cdot b^3 \quad \textit{Power rule (b)}$$
$$= -5^3 a^3 b^3 \quad \text{or} \quad -125a^3 b^3$$

(b) $3(4t^3 p^5)^2 = 3 \cdot 4^2 (t^3)^2 (p^5)^2$ *Power rule (b)*
$$= 48t^6 p^{10} \qquad\qquad \textit{Power rule (a)}$$

N6.   **(a)** $\left(\dfrac{p}{q}\right)^5 = \dfrac{p^5}{q^5}$ *Power rule (c), $q \neq 0$*

(b) $\left(\dfrac{1}{4}\right)^3 = \dfrac{1^3}{4^3}$ *Power rule (c)*
$$= \dfrac{1}{64}$$

N7.   **(a)** $\left(\dfrac{3}{5}\right)^3 \cdot 3^2 = \dfrac{3^3}{5^3} \cdot \dfrac{3^2}{1}$ *Power rule (c)*
$$= \dfrac{3^{3+2}}{5^3} \qquad\qquad \textit{Power rule (b)}$$
$$= \dfrac{3^5}{5^3} \quad \text{or} \quad \dfrac{243}{125}$$

(b) $(8k)^5(8k)^4 = (8k)^{5+4}$ *Product rule*
$$= (8k)^9$$
$$= 8^9 k^9 \qquad\qquad \textit{Power rule (b)}$$

(c) $(x^4 y)^5 (-2x^2 y^5)^3$
$$= (x^4)^5 y^5 (-1)^3 2^3 (x^2)^3 (y^5)^3 \; \textit{Power rule (b)}$$
$$= x^{4 \cdot 5} y^5 (-1)^3 2^3 x^{2 \cdot 3} y^{5 \cdot 3} \quad \textit{Power rule (a)}$$
$$= -2^3 x^{20} y^5 x^6 y^{15}$$
$$= -2^3 x^{20+6} y^{5+15} \qquad\qquad \textit{Product rule}$$
$$= -2^3 x^{26} y^{20} \quad \text{or} \quad -8x^{26} y^{20}$$

N8.   Use the formula for the area of a triangle, $A = \frac{1}{2}bh$, with $b = 10x^8$ and $h = 3x^7$.
$$A = \tfrac{1}{2}(10x^8)(3x^7) \qquad \textit{Area formula}$$
$$= \tfrac{1}{2} \cdot 10 \cdot 3 \cdot x^{8+7} \quad \textit{Product rule}$$
$$= 15x^{15}$$

5.1   Section Exercises

1. $3^3 = 3 \cdot 3 \cdot 3 = 27$, so the statement $3^3 = 9$ is *false*.

3. $(x^2)^3 = x^{2(3)} = x^6$, so the statement $(x^2)^3 = x^5$ is *false*.

5. $\underbrace{w \cdot w \cdot w \cdot w \cdot w \cdot w}_{6w\text{'s}} = w^6$

7. $\left(\frac{1}{2}\right)\left(\frac{1}{2}\right)\left(\frac{1}{2}\right)\left(\frac{1}{2}\right)\left(\frac{1}{2}\right)\left(\frac{1}{2}\right) = \left(\frac{1}{2}\right)^6$

9. $(-4)(-4)(-4)(-4) = (-4)^4$

11. $(-7y)(-7y)(-7y)(-7y) = (-7y)^4$

13.   In $(-3)^4$, -3 is the base.
$$(-3)^4 = (-3)(-3)(-3)(-3) = 81$$
In -3^4, 3 is the base.
$$-3^4 = -(3 \cdot 3 \cdot 3 \cdot 3) = -81$$

15.   In the exponential expression 3^5, the base is 3 and the exponent is 5.
$$3^5 = 3 \cdot 3 \cdot 3 \cdot 3 \cdot 3 = 243$$

17.   In the expression $(-3)^5$, the base is -3 and the exponent is 5.
$$(-3)^5 = (-3)(-3)(-3)(-3)(-3) = -243$$

19.   In the expression $(-6x)^4$, the base is $-6x$ and the exponent is 4.

21. In the expression $-6x^4$, -6 is not part of the base. The base is x and the exponent is 4.

23. The product rule does not apply to $5^2 + 5^3$ because the expression is a sum, not a product. The product rule would apply if we had $5^2 \cdot 5^3$.
$$5^2 + 5^3 = 25 + 125 = 150$$

25. $5^2 \cdot 5^6 = 5^{2+6} = 5^8$

27. $4^2 \cdot 4^7 \cdot 4^3 = 4^{2+7+3} = 4^{12}$

29. $(-7)^3(-7)^6 = (-7)^{3+6} = (-7)^9$

31. $t^3 \cdot t^8 \cdot t^{13} = t^{3+8+13} = t^{24}$

33. $\left(-8r^4\right)\left(7r^3\right) = -8 \cdot 7 \cdot r^4 \cdot r^3$
$$= -56r^{4+3}$$
$$= -56r^7$$

35. $\left(-6p^5\right)\left(-7p^5\right) = (-6)(-7)p^5 \cdot p^5$
$$= 42p^{5+5}$$
$$= 42p^{10}$$

37. $\left(5x^2\right)\left(-2x^3\right)\left(3x^4\right)$
$$= (5)(-2)(3)x^2 \cdot x^3 \cdot x^4$$
$$= (-10)(3)x^{2+3+4}$$
$$= -30x^9$$

39. $3^8 + 3^9$ is a sum, so the product rule does not apply.

41. $5^8 \cdot 3^9$ is a product with different bases, so the product rule does not apply.

43. $(4^3)^2 = 4^{3 \cdot 2}$ *Power rule (a)*
$$= 4^6$$

45. $(t^4)^5 = t^{4 \cdot 5} = t^{20}$ *Power rule (a)*

47. $(7r)^3 = 7^3 r^3$ *Power rule (b)*

49. $(5xy)^5 = 5^5 x^5 y^5$ *Power rule (b)*

51. $(-5^2)^6 = (-1 \cdot 5^2)^6$
$$= (-1)^6 \cdot (5^2)^6 \quad \textit{Power rule (b)}$$
$$= 1 \cdot 5^{2 \cdot 6} \quad \textit{Power rule (a)}$$
$$= 1 \cdot 5^{12} = 5^{12}$$

53. $(-8^3)^5 = (-1 \cdot 8^3)^5$
$$= (-1)^5 \cdot (8^3)^5 \quad \textit{Power rule (b)}$$
$$= -1 \cdot 8^{3 \cdot 5} \quad \textit{Power rule (a)}$$
$$= -8^{15}$$

55. $8(qr)^3 = 8q^3 r^3$ *Power rule (b)*

57. $\left(\dfrac{9}{5}\right)^8 = \dfrac{9^8}{5^8}$ *Power rule (c)*

59. $\left(\dfrac{1}{2}\right)^3 = \dfrac{1^3}{2^3} = \dfrac{1}{2^3}$ *Power rule (c)*

61. $\left(\dfrac{a}{b}\right)^3 = \dfrac{a^3}{b^3}$ *Power rule (c), $(b \neq 0)$*

63. $\left(\dfrac{x}{2}\right)^3 = \dfrac{x^3}{2^3}$ *Power rule (c)*

65. $\left(\dfrac{5}{2}\right)^3 \cdot \left(\dfrac{5}{2}\right)^2 = \left(\dfrac{5}{2}\right)^{3+2}$ *Product rule*
$$= \left(\dfrac{5}{2}\right)^5$$
$$= \dfrac{5^5}{2^5} \quad \textit{Power rule (c)}$$

67. $\left(\dfrac{9}{8}\right)^3 \cdot 9^2 = \dfrac{9^3}{8^3} \cdot \dfrac{9^2}{1}$ *Power rule (c)*
$$= \dfrac{9^3 \cdot 9^2}{8^3 \cdot 1} \quad \textit{Multiply fractions}$$
$$= \dfrac{9^{3+2}}{8^3} \quad \textit{Product rule}$$
$$= \dfrac{9^5}{8^3}$$

69. $(2x)^9(2x)^3 = (2x)^{9+3}$ *Product rule*
$$= (2x)^{12}$$
$$= 2^{12} x^{12} \quad \textit{Product rule (b)}$$

71. $(-6p)^4(-6p)$
$$= (-6p)^4(-6p)^1$$
$$= (-6p)^5 \quad \textit{Product rule}$$
$$= (-1)^5 6^5 p^5 \quad \textit{Power rule (b)}$$
$$= -6^5 p^5$$

73. $(6x^2 y^3)^5 = 6^5(x^2)^5(y^3)^5$ *Power rule (b)*
$$= 6^5 x^{2 \cdot 5} y^{3 \cdot 5} \quad \textit{Power rule (a)}$$
$$= 6^5 x^{10} y^{15}$$

75. $(x^2)^3(x^3)^5 = x^6 \cdot x^{15}$ *Power rule (a)*
$$= x^{21} \quad \textit{Product rule}$$

77. $(2w^2 x^3 y)^2(x^4 y)^5$
$$= \left[2^2(w^2)^2(x^3)^2 y^2\right]\left[(x^4)^5 y^5\right] \quad \begin{array}{l}\textit{Power} \\ \textit{rule (b)}\end{array}$$
$$= \left(2^2 w^4 x^6 y^2\right)\left(x^{20} y^5\right) \quad \begin{array}{l}\textit{Power} \\ \textit{rule (a)}\end{array}$$
$$= 2^2 w^4 \left(x^6 x^{20}\right)\left(y^2 y^5\right)$$
Commutative and associative properties
$$= 4w^4 x^{26} y^7$$

79. $(-r^4 s)^2(-r^2 s^3)^5$
$$= \left[(-1)r^4 s\right]^2\left[(-1)r^2 s^3\right]^5$$
$$= \left[(-1)^2(r^4)^2 s^2\right]\left[(-1)^5(r^2)^5(s^3)^5\right]$$
$$\quad\quad\quad\quad\quad\quad\quad\quad \textit{Power rule (b)}$$

$$= \left[(-1)^2 r^8 s^2\right]\left[(-1)^5 r^{10} s^{15}\right] \quad \text{\textit{Power rule (a)}}$$

$$= (-1)^7 r^{18} s^{17} \quad \text{\textit{Product rule}}$$

$$= -r^{18} s^{17}$$

81. $\left(\dfrac{5a^2 b^5}{c^6}\right)^3 \quad (c \neq 0)$

$$= \frac{(5a^2 b^5)^3}{(c^6)^3} \qquad \text{\textit{Power rule (c)}}$$

$$= \frac{5^3 (a^2)^3 (b^5)^3}{(c^6)^3} \qquad \text{\textit{Power rule (b)}}$$

$$= \frac{125 a^6 b^{15}}{c^{18}} \qquad \text{\textit{Power rule (a)}}$$

83. To simplify $(10^2)^3$ as 1000^6 is not correct. Using power rule (a) to simplify $(10^2)^3$, we obtain

$$(10^2)^3 = 10^{2 \cdot 3}$$
$$= 10^6$$
$$= 10 \cdot 10 \cdot 10 \cdot 10 \cdot 10 \cdot 10$$
$$= 1{,}000{,}000.$$

85. Use the formula for the area of a rectangle, $A = LW$, with $L = 4x^3$ and $W = 3x^2$.

$$A = (4x^3)(3x^2)$$
$$= 4 \cdot 3 \cdot x^3 \cdot x^2$$
$$= 12 x^5$$

87. Use the formula for the area of a parallelogram, $A = bh$, with $b = 2p^5$ and $h = 3p^2$.

$$A = (2p^5)(3p^2)$$
$$= 2 \cdot 3 \cdot p^5 \cdot p^2$$
$$= 6p^7$$

89. Use the formula for the volume of a cube, $V = e^3$, with $e = 5x^2$.

$$V = (5x^2)^3$$
$$= 5^3 (x^2)^3$$
$$= 125 x^6$$

91. If a is a number greater than 1, then:

(1) $-(-a)^3 = -1(-a^3) = a^3$ is positive
(2) $-a^3 = -1(a^3)$ is negative
(3) $(-a)^4 = a^4$ is positive
(4) $-a^4 = -1(a^4)$ is negative

Since $a^4 > a^3$, $-a^4 < -a^3$, so in order from least to greatest, we have

$$-a^4, \ -a^3, \ -(-a)^3, \ (-a)^4.$$

Another way to determine the order is to choose a number greater than 1 and substitute it for a in each expression, and then arrange the terms from least to greatest.

93. Use the formula $A = P(1 + r)^n$ with $P = \$250$, $r = 0.04$, and $n = 5$.

$$A = 250(1 + 0.04)^5$$
$$= 250(1.04)^5$$
$$\approx 304.16$$

The amount of money in the account will be $304.16.

95. Use the formula $A = P(1 + r)^n$ with $P = \$1500$, $r = 0.035$, and $n = 6$.

$$A = 1500(1 + 0.035)^6$$
$$= 1500(1.035)^6$$
$$\approx 1843.88$$

The amount of money in the account will be $1843.88.

97. The reciprocal of 9 is $\frac{1}{9}$.

99. The reciprocal of $-\frac{1}{8}$ is $\frac{1}{-\frac{1}{8}} = 1\left(-\frac{8}{1}\right) = -8$.

101. $8 - (-4) = 8 + 4 = 12$

103. "Subtract -6 from -3" translates and simplifies as:

$$-3 - (-6) = -3 + 6$$
$$= 3$$

5.2 Integer Exponents and the Quotient Rule

5.2 Now Try Exercises

N1. (a) $6^0 = 1$ *Definition of zero exponent*

(b) $-12^0 = -(12^0) = -1$

(c) $(-12x)^0 = 1 \ (x \neq 0)$

(d) $14^0 - 12^0 = 1 - 1 = 0$

N2. (a) $2^{-3} = \dfrac{1}{2^3}$ *Definition of negative exponent*

$$= \frac{1}{8}$$

(b) $\left(\frac{1}{7}\right)^{-2} = \left(\frac{7}{1}\right)^2$ *$\frac{1}{7}$ and 7 are reciprocals.*

$$= 49$$

(c) $\left(\frac{3}{2}\right)^{-4} = \left(\frac{2}{3}\right)^4 = \frac{16}{81}$

(d) $3^{-2} + 4^{-2} = \dfrac{1}{3^2} + \dfrac{1}{4^2}$

$$= \frac{1}{9} + \frac{1}{16}$$

$$= \frac{16}{144} + \frac{9}{144} = \frac{25}{144}$$

(e) $p^{-4} = \dfrac{1}{p^4}, \ p \neq 0$

N3. (a) $\dfrac{5^{-3}}{6^{-2}} = \dfrac{6^2}{5^3}$, or $\dfrac{36}{125}$

(b) $m^2 n^{-4} = \dfrac{m^2}{1} \cdot \dfrac{1}{n^4} = \dfrac{m^2}{n^4}$

(c) $\dfrac{x^2 y^{-3}}{5z^{-4}} = \dfrac{x^2}{5} \cdot \dfrac{y^{-3}}{z^{-4}} = \dfrac{x^2}{5} \cdot \dfrac{z^4}{y^3} = \dfrac{x^2 z^4}{5y^3}$

N4. **(a)** $\dfrac{6^3}{6^4} = 6^{3-4} = 6^{-1} = \dfrac{1}{6^1} = \dfrac{1}{6}$

(b) $\dfrac{t^4}{t^{-5}} = t^{4-(-5)} = t^{4+5} = t^9$

(c) $\dfrac{(p+q)^{-3}}{(p+q)^{-7}} = (p+q)^{-3-(-7)}$

$$= (p+q)^{-3+7}$$

$$= (p+q)^4$$

(d) $\dfrac{5^2 xy^{-3}}{3^{-1} x^{-2} y^2} = \dfrac{5^2}{3^{-1}} \cdot \dfrac{x}{x^{-2}} \cdot \dfrac{y^{-3}}{y^2}$

$$= 5^2 3^1 \cdot x^{1-(-2)} \cdot y^{-3-2}$$

$$= 25 \cdot 3 \cdot x^3 \cdot y^{-5}$$

$$= \dfrac{75}{1} \cdot \dfrac{x^3}{1} \cdot \dfrac{1}{y^5}$$

$$= \dfrac{75x^3}{y^5}$$

N5. **(a)** $\dfrac{3^{15}}{(3^3)^4} = \dfrac{3^{15}}{3^{12}} = 3^{15-12} = 3^3$, or 27

(b) $(4t)^5 (4t)^{-3} = (4t)^{5+(-3)}$

$$= (4t)^2$$

$$= 4^2 t^2, \text{ or } 16t^2$$

(c) $\left(\dfrac{7y^4}{10}\right)^{-3} = \left(\dfrac{10}{7y^4}\right)^3$

$$= \dfrac{10^3}{(7y^4)^3}$$

$$= \dfrac{10^3}{7^3 (y^4)^3}$$

$$= \dfrac{10^3}{7^3 y^{12}}, \text{ or } \dfrac{1000}{343 y^{12}}$$

(d) $\dfrac{(a^2 b^{-2} c)^{-3}}{(2ab^3 c^{-4})^5} = \dfrac{(a^2)^{-3}(b^{-2})^{-3} c^{-3}}{2^5 a^5 (b^3)^5 (c^{-4})^5}$

$$= \dfrac{a^{-6} b^6 c^{-3}}{2^5 a^5 b^{15} c^{-20}}$$

$$= \dfrac{b^6 c^{20}}{2^5 a^5 a^6 b^{15} c^3}$$

$$= \dfrac{c^{20-3}}{2^5 a^{5+6} b^{15-6}}$$

$$= \dfrac{c^{17}}{2^5 a^{11} b^9}, \text{ or } \dfrac{c^{17}}{32 a^{11} b^9}$$

5.2   Section Exercises

1. By definition, $a^0 = 1 \, (a \neq 0)$, so $9^0 = 1$.

3. $(-2)^0 = 1$ *Definition of zero exponent*

5. $-8^0 = -(8^0) = -(1) = -1$

7. $-(-6)^0 = -1 \cdot (-6)^0 = -1 \cdot 1 = -1$

9. $(-4)^0 - 4^0 = 1 - 1 = 0$

11. $\dfrac{0^{10}}{12^0} = \dfrac{0}{1} = 0$

13. $8^0 - 12^0 = 1 - 1 = 0$

15. $\dfrac{0^2}{2^0 + 0^2} = \dfrac{0}{1+0} = \dfrac{0}{1} = 0$

17. **(a)** $x^0 = 1$ (Choice **B**)

(b) $-x^0 = -1 \cdot x^0 = -1 \cdot 1 = -1$ (Choice **C**)

(c) $7x^0 = 7 \cdot x^0 = 7 \cdot 1 = 7$ (Choice **D**)

(d) $(7x)^0 = 1$ (Choice **B**)

(e) $-7x^0 = -7 \cdot x^0 = -7 \cdot 1 = -7$ (Choice **E**)

(f) $(-7x)^0 = 1$ (Choice **B**)

19. $6^0 + 8^0 = 1 + 1 = 2$

21. $4^{-3} = \dfrac{1}{4^3}$ *Definition of negative exponent*

$$= \dfrac{1}{64}$$

23. When we evaluate a fraction raised to a negative exponent, we can use a shortcut. Note that

$$\left(\dfrac{a}{b}\right)^{-n} = \dfrac{1}{\left(\dfrac{a}{b}\right)^n} = \dfrac{1}{\dfrac{a^n}{b^n}} = \dfrac{b^n}{a^n} = \left(\dfrac{b}{a}\right)^n.$$

In words, a fraction raised to the negative of a number is equal to its reciprocal raised to the number. We will use the simple phrase "$\frac{a}{b}$ and $\frac{b}{a}$ are reciprocals" to indicate our use of this evaluation shortcut.

$$\left(\dfrac{1}{2}\right)^{-4} = 2^4 = 16 \quad \begin{array}{l}\frac{1}{2} \text{ and 2 are} \\ \text{reciprocals.}\end{array}$$

25. $\left(\dfrac{6}{7}\right)^{-2} = \left(\dfrac{7}{6}\right)^2$ $\quad \begin{array}{l}\frac{6}{7} \text{ and } \frac{7}{6} \text{ are} \\ \text{reciprocals.}\end{array}$

$$= \dfrac{7^2}{6^2} \qquad \textit{Power rule (c)}$$

$$= \dfrac{49}{36}$$

27. $(-3)^{-4} = \dfrac{1}{(-3)^4}$

$$= \dfrac{1}{81}$$

29. $5^{-1} + 3^{-1} = \dfrac{1}{5} + \dfrac{1}{3}$

$$= \dfrac{3}{15} + \dfrac{5}{15} = \dfrac{8}{15}$$

31. $3^{-2} - 2^{-1} = \dfrac{1}{3^2} - \dfrac{1}{2^1}$

$\qquad\qquad = \dfrac{1}{9} - \dfrac{1}{2}$

$\qquad\qquad = \dfrac{2}{18} - \dfrac{9}{18} = -\dfrac{7}{18}$

33. $\left(\dfrac{1}{2}\right)^{-1} + \left(\dfrac{2}{3}\right)^{-1} = \left(\dfrac{2}{1}\right)^1 + \left(\dfrac{3}{2}\right)^1$

$\qquad\qquad\qquad = \dfrac{4}{2} + \dfrac{3}{2}$

$\qquad\qquad\qquad = \dfrac{7}{2}$

35. $\dfrac{5^8}{5^5} = 5^{8-5} = 5^3$, or 125   *Quotient rule*

37. $\dfrac{3^{-2}}{5^{-3}} = \dfrac{5^3}{3^2}$,   *Changing from negative*
to positive exponents

$\qquad$ or $\dfrac{125}{9}$

39. $\dfrac{5}{5^{-1}} = \dfrac{5^1}{5^{-1}} = 5^1 \cdot 5^1$ *Changing from negative*
to positive exponents

$\qquad\qquad = 5^{1+1} = 5^2$, or 25

41. $\dfrac{x^{12}}{x^{-3}} = x^{12} \cdot x^3$ *Changing from negative*
to positive exponents

$\qquad = x^{12+3} = x^{15}$

43. $\dfrac{1}{6^{-3}} = 6^3$, or 216   *Changing from negative*
to positive exponents

45. $\dfrac{2}{r^{-4}} = 2r^4$ *Changing from negative*
to positive exponents

47. $\dfrac{4^{-3}}{5^{-2}} = \dfrac{5^2}{4^3}$, or $\dfrac{25}{64}$ *Changing from negative*
to positive exponents

49. $p^5 q^{-8} = \dfrac{p^5}{q^8}$ *Changing from negative*
to positive exponents

51. $\dfrac{r^5}{r^{-4}} = r^5 \cdot r^4 = r^{5+4} = r^9$

Or we can use the quotient rule:

$\dfrac{r^5}{r^{-4}} = r^{5-(-4)} = r^{5+4} = r^9$

53. $\dfrac{x^{-3}y}{4z^{-2}} = \dfrac{yz^2}{4x^3}$

55. Treat the expression in parentheses as a single variable; that is, treat $(a + b)$ as you would treat x.

$\dfrac{(a+b)^{-3}}{(a+b)^{-4}} = (a+b)^{-3-(-4)}$

$\qquad\qquad\quad = (a+b)^{-3+4}$

$\qquad\qquad\quad = (a+b)^1 = a+b$

Another Method:

$\dfrac{(a+b)^{-3}}{(a+b)^{-4}} = \dfrac{(a+b)^4}{(a+b)^3}$

$\qquad\qquad\quad = (a+b)^{4-3}$

$\qquad\qquad\quad = (a+b)^1 = a+b$

57. $\dfrac{(x+2y)^{-3}}{(x+2y)^{-5}} = (x+2y)^{-3-(-5)}$

$\qquad\qquad\quad = (x+2y)^{-3+5}$

$\qquad\qquad\quad = (x+2y)^2$

59. In simplest form, $\dfrac{25}{25} = 1$.

60. $\dfrac{25}{25} = \dfrac{5^2}{5^2}$

61. $\dfrac{5^2}{5^2} = 5^{2-2} = 5^0$

62. $1 = 5^0$; this supports the definition of 0 as an exponent.

63. $\dfrac{(7^4)^3}{7^9} = \dfrac{7^{4 \cdot 3}}{7^9}$ *Power rule (a)*

$\qquad = \dfrac{7^{12}}{7^9}$

$\qquad = 7^{12-9}$ *Quotient rule*

$\qquad = 7^3$, or 343

65. $x^{-3} \cdot x^5 \cdot x^{-4}$

$\qquad = x^{-3+5+(-4)}$ *Product rule*

$\qquad = x^{-2}$

$\qquad = \dfrac{1}{x^2}$ *Definition of*
negative exponent

67. $\dfrac{(3x)^{-2}}{(4x)^{-3}} = \dfrac{(4x)^3}{(3x)^2}$ *Changing from*
negative to
positive exponents

$\qquad = \dfrac{4^3 x^3}{3^2 x^2}$ *Power rule (b)*

$\qquad = \dfrac{4^3 x^{3-2}}{3^2}$ *Quotient rule*

$\qquad = \dfrac{4^3 x}{3^2}$

$\qquad = \dfrac{64x}{9}$

69. $\left(\dfrac{x^{-1}y}{z^2}\right)^{-2} = \dfrac{(x^{-1}y)^{-2}}{(z^2)^{-2}}$ *Power rule (c)*

$\qquad = \dfrac{(x^{-1})^{-2}y^{-2}}{(z^2)^{-2}}$ *Power rule (b)*

$\qquad = \dfrac{x^2 y^{-2}}{z^{-4}}$ *Power rule (a)*

$\qquad = \dfrac{x^2 z^4}{y^2}$ *Definition of*
negative exponent

71. $(6x)^4(6x)^{-3} = (6x)^{4+(-3)}$ *Product rule*
$$= (6x)^1 = 6x$$

73. $\dfrac{(m^7n)^{-2}}{m^{-4}n^3} = \dfrac{(m^7)^{-2}n^{-2}}{m^{-4}n^3}$
$$= \dfrac{m^{7(-2)}n^{-2}}{m^{-4}n^3}$$
$$= \dfrac{m^{-14}n^{-2}}{m^{-4}n^3}$$
$$= m^{-14-(-4)}n^{-2-3}$$
$$= m^{-10}n^{-5}$$
$$= \dfrac{1}{m^{10}n^5}$$

75. $\dfrac{(x^{-1}y^2z)^{-2}}{(x^{-3}y^3z)^{-1}} = \dfrac{(x^{-1})^{-2}(y^2)^{-2}z^{-2}}{(x^{-3})^{-1}(y^3)^{-1}z^{-1}}$
$$= \dfrac{x^2y^{-4}z^{-2}}{x^3y^{-3}z^{-1}}$$
$$= \dfrac{x^2y^3z^1}{x^3y^4z^2}$$
$$= \dfrac{1}{xyz}$$

77. $\left(\dfrac{xy^{-2}}{x^2y}\right)^{-3} = \dfrac{x^{-3}(y^{-2})^{-3}}{(x^2)^{-3}y^{-3}}$
$$= \dfrac{x^{-3}y^6}{x^{-6}y^{-3}}$$
$$= \dfrac{x^6y^6y^3}{x^3}$$
$$= x^3y^9$$

79. $\dfrac{(4a^2b^3)^{-2}(2ab^{-1})^3}{(a^3b)^{-4}}$
$$= \dfrac{(a^3b)^4(2ab^{-1})^3}{(4a^2b^3)^2}$$
$$= \dfrac{(a^3)^4b^42^3a^3(b^{-1})^3}{4^2(a^2)^2(b^3)^2}$$
$$= \dfrac{a^{12}b^48a^3b^{-3}}{16a^4b^6}$$
$$= \dfrac{8a^{15}b^1}{16a^4b^6}$$
$$= \dfrac{a^{11}}{2b^5}$$

81. $\dfrac{(2y^{-1}z^2)^2(3y^{-2}z^{-3})^3}{(y^3z^2)^{-1}}$
$$= \dfrac{2^2y^{-2}z^43^3y^{-6}z^{-9}}{y^{-3}z^{-2}}$$
$$= \dfrac{4 \cdot 27y^{-8}z^{-5}}{y^{-3}z^{-2}}$$
$$= 108y^{-5}z^{-3}$$
$$= \dfrac{108}{y^5z^3}$$

83. $\dfrac{(9^{-1}z^{-2}x)^{-1}(4z^2x^4)^{-2}}{(5z^{-2}x^{-3})^2}$
$$= \dfrac{9^1z^2x^{-1}4^{-2}z^{-4}x^{-8}}{5^2z^{-4}x^{-6}}$$
$$= \dfrac{9z^{-2}x^{-9}}{25 \cdot 4^2z^{-4}x^{-6}}$$
$$= \dfrac{9z^2x^{-3}}{400}$$
$$= \dfrac{9z^2}{400x^3}$$

85. The student attempted to use the quotient rule with unequal bases. The correct way to simplify this expression is
$$\dfrac{16^3}{2^2} = \dfrac{(2^4)^3}{2^2} = \dfrac{2^{12}}{2^2} = 2^{10} = 1024.$$

87. $10 = 10^1$; move the decimal point to the *right* 1 place when *multiplying*.
$$10(6428) = 64{,}280$$

89. $1000(1.53) = 1530$ $(1000 = 10^3)$

91. $10 = 10^1$; move the decimal point to the *left* 1 place when *dividing*.
$$38 \div 10 = 3.8$$

93. $277 \div 1000 = 0.277$ $(1000 = 10^3)$

Summary Exercises on the Rules for Exponents

1. $(10x^2y^4)^2(10xy^2)^3$
$$= 10^2(x^2)^2(y^4)^2 \cdot 10^3x^3(y^2)^3$$
$$= 10^2x^4y^810^3x^3y^6$$
$$= 10^5x^7y^{14}$$

3. $\left(\dfrac{9wx^3}{y^4}\right)^3 = \dfrac{(9wx^3)^3}{(y^4)^3}$

$= \dfrac{9^3 w^3 (x^3)^3}{y^{12}}$

$= \dfrac{729 w^3 x^9}{y^{12}}$

5. $\dfrac{c^{11}(c^2)^4}{(c^3)^3(c^2)^{-6}} = \dfrac{c^{11}c^8}{c^9 c^{-12}}$

$= \dfrac{c^{19}}{c^{-3}}$

$= c^{19-(-3)} = c^{22}$

7. $5^{-1} + 6^{-1} = \dfrac{1}{5^1} + \dfrac{1}{6^1}$

$= \dfrac{6}{30} + \dfrac{5}{30}$

$= \dfrac{11}{30}$

9. $\dfrac{(2xy^{-1})^3}{2^3 x^{-3}y^2} = \dfrac{2^3 x^3 y^{-3}}{2^3 x^{-3}y^2}$

$= x^6 y^{-5}$

$= \dfrac{x^6}{y^5}$

11. $(z^4)^{-3}(z^{-2})^{-5}$

$= z^{-12} z^{10} = z^{-2} = \dfrac{1}{z^2}$

13. $\dfrac{(3^{-1}x^{-3}y)^{-1}(2x^2 y^{-3})^2}{(5x^{-2}y^2)^{-2}}$

$= \dfrac{(5x^{-2}y^2)^2 (2x^2 y^{-3})^2}{(3^{-1}x^{-3}y)^1}$

$= \dfrac{5^2 x^{-4} y^4 2^2 x^4 y^{-6}}{3^{-1} x^{-3} y}$

$= \dfrac{25 x^0 y^{-2} \cdot 4}{3^{-1} x^{-3} y}$

$= 100 x^3 y^{-3} \cdot 3$

$= \dfrac{300 x^3}{y^3}$

15. $\left(\dfrac{-9x^{-2}}{9x^2}\right)^{-2} = \left(\dfrac{9x^2}{-9x^{-2}}\right)^2$

$= \left(\dfrac{x^4}{-1}\right)^2 = \dfrac{(x^4)^2}{(-1)^2}$

$= \dfrac{x^8}{1} = x^8$

17. $\dfrac{(a^{-2}b^3)^{-4}}{(a^{-3}b^2)^{-2}(ab)^{-4}}$

$= \dfrac{a^8 b^{-12}}{a^6 b^{-4} a^{-4} b^{-4}}$

$= \dfrac{a^8 b^{-12}}{a^2 b^{-8}}$

$= a^6 b^{-4}$

$= \dfrac{a^6}{b^4}$

19. $5^{-2} + 6^{-2} = \dfrac{1}{5^2} + \dfrac{1}{6^2}$

$= \dfrac{1}{25} + \dfrac{1}{36}$

$= \dfrac{36}{25 \cdot 36} + \dfrac{25}{25 \cdot 36}$

$= \dfrac{36 + 25}{900} = \dfrac{61}{900}$

21. $\left(\dfrac{7a^2 b^3}{2}\right)^3 = \dfrac{(7a^2 b^3)^3}{2^3}$

$= \dfrac{7^3 a^6 b^9}{8} = \dfrac{343 a^6 b^9}{8}$

23. $-(-13)^0 = -1$

25. $\dfrac{(2xy^{-3})^{-2}}{(3x^{-2}y^4)^{-3}}$

$= \dfrac{(3x^{-2}y^4)^3}{(2xy^{-3})^2}$

$= \dfrac{3^3 x^{-6} y^{12}}{2^2 x^2 y^{-6}}$

$= \dfrac{27 x^{-8} y^{18}}{4}$

$= \dfrac{27 y^{18}}{4x^8}$

27. $(6x^{-5}z^3)^{-3}$

$= 6^{-3} x^{15} z^{-9}$

$= \dfrac{x^{15}}{6^3 z^9}$

$= \dfrac{x^{15}}{216 z^9}$

29. $\dfrac{(xy)^{-3}(xy)^5}{(xy)^{-4}} = (xy)^{-3+5-(-4)}$

$= (xy)^6 = x^6 y^6$

31. $\dfrac{(7^{-1}x^{-3})^{-2}(x^4)^{-6}}{7^{-1}x^{-3}}$

$= \dfrac{7^2x^6x^{-24}}{7^{-1}x^{-3}}$

$= 7^{2-(-1)}x^{6-24-(-3)}$

$= 7^3x^{-15} = \dfrac{343}{x^{15}}$

33. $(5p^{-2}q)^{-3}(5pq^3)^4$

$= 5^{-3}p^6q^{-3}5^4p^4q^{12}$

$= 5p^{10}q^9$

35. $\left[\dfrac{4r^{-6}s^{-2}t}{2r^8s^{-4}t^2}\right]^{-1}$

$= \left[\dfrac{2s^2}{r^{14}t}\right]^{-1}$

$= \dfrac{r^{14}t}{2s^2}$

37. $\dfrac{(8pq^{-2})^4}{(8p^{-2}q^{-3})^3}$

$= \dfrac{8^4p^4q^{-8}}{8^3p^{-6}q^{-9}}$

$= 8^{4-3}p^{4-(-6)}q^{-8-(-9)}$

$= 8p^{10}q$

39. $-(-8^0)^0 = -1$

5.3 An Application of Exponents: Scientific Notation

5.3 Now Try Exercises

N1. **(a)** $12{,}600{,}000 = 1.26 \times 10^7$

The decimal point has been moved 7 places to put it after the first nonzero digit. Since 1.26 is less than the original number, it must be multiplied by a number greater than 1 to get 12,600,000. Thus, the exponent on 10 must be positive.

(b) $0.00027 = 2.7 \times 10^{-4}$

The decimal point has been moved 4 places to put it after the first nonzero digit. Since 2.7 is greater than the original number, it must be multiplied by a number less than 1 to get 0.00027. Thus, the exponent on 10 must be negative.

(c) $-0.0000341 = -3.41 \times 10^{-5}$

Work with the absolute value of the number, then apply the negative sign to the answer.

N2. **(a)** $5.71 \times 10^4 = 57{,}100$

Since the exponent is positive, move the decimal point 4 places to the *right*.

(b) $2.72 \times 10^{-5} = 0.0000272$

Since the exponent is negative, move the decimal point 5 places to the *left*.

N3. **(a)** $(6 \times 10^7)(7 \times 10^{-4})$

$= (6 \times 7)(10^7 \times 10^{-4})$

$= 42 \times 10^3$

$= (4.2 \times 10^1) \times 10^3$

$= 4.2 \times (10^1 \times 10^3)$

$= 4.2 \times 10^4$, or 42,000

(b) $\dfrac{18 \times 10^{-3}}{6 \times 10^4} = \dfrac{18}{6} \times \dfrac{10^{-3}}{10^4}$

$= 3 \times 10^{-3-4}$

$= 3 \times 10^{-7}$, or 0.0000003

N4. $(8{,}000{,}000)(0.00000003937)$

$= (8 \times 10^6) \times (3.937 \times 10^{-8})$

$= (8 \times 3.937) \times (10^6 \times 10^{-8})$

$= 31.496 \times 10^{6+(-8)}$

$= 31.496 \times 10^{-2}$

$= 3.1496 \times 10^{-1}$, or 0.31496

Thus, 8,000,000 nanometers would measure 3.1496×10^{-1} in., or 0.31496 in.

N5. Divide the land area by the population.

$\dfrac{1.6 \times 10^5}{4 \times 10^7} = \dfrac{1.6}{4} \times \dfrac{10^5}{10^7}$

$= 0.4 \times 10^{5-7}$

$= 0.4 \times 10^{-2}$

$= 4 \times 10^{-3}$

$= 0.004$

In 2008, the number of square miles per person living in California was about 4×10^{-3}, or 0.004.

5.3 Section Exercises

1. **(a)** Move the decimal point to the left 4 places due to the exponent, -4.

$$4.6 \times 10^{-4} = 0.00046$$

Choice **C** is correct.

(b) $4.6 \times 10^4 = 46{,}000$

Choice **A** is correct.

(c) Move the decimal point to the right 5 places due to the exponent, 5.

$$4.6 \times 10^5 = 460{,}000$$

Choice **B** is correct.

(d) $4.6 \times 10^{-5} = 0.000046$

Choice **D** is correct.

3. 4.56×10^4 is written in scientific notation because 4.56 is between 1 and 10, and 10^4 is a power of 10.

5. 5,600,000 is not in scientific notation. It can be written in scientific notation as 5.6×10^6.

7. 0.8×10^2 is not in scientific notation because $|0.8| = 0.8$ is not greater than or equal to 1 and less than 10. It can be written in scientific notation as 8×10^1.

9. 0.004 is not in scientific notation because $|0.004| = 0.004$ is not between 1 and 10. It can be written in scientific notation as 4×10^{-3}.

11. A number is written in scientific notation if it is written as a product of two numbers, the first of which has absolute value less than 10 and greater than or equal to 1 and the second of which is a power of 10. Some examples are 2.3×10^{-4} and 6.02×10^{23}.

13. 5,876,000,000

Move the decimal point to the right of the first nonzero digit and count the number of places the decimal point was moved.

$$5,876,000,000 \quad 9 \; places$$

Because moving the decimal point to the *left* made the number *smaller*, we must multiply by a *positive* power of 10 so that the product 5.876×10^n will equal the larger number. Thus, $n = 9$, and

$$5,876,000,000 = 5.876 \times 10^9.$$

15. 82,350

Move the decimal point left 4 places so it is to the right of the first nonzero digit.

$$8.2350 \quad 4 \; places$$

Since the number got smaller, multiply by a positive power of 10.

$$82,350 = 8.2350 \times 10^4 = 8.235 \times 10^4$$

(Note that the final zero need not be written.)

17. 0.000 007

Move the decimal point to the right of the first nonzero digit.

$$0.000007. \quad 6 \; places$$

Since moving the decimal point to the *right* made the number *larger*, we must multiply by a *negative* power of 10 so that the product 7×10^n will equal the smaller number. Thus, $n = -6$, and

$$0.000\,007 = 7 \times 10^{-6}.$$

19. 0.002 03

To move the decimal point to the right of the first nonzero digit, we move it 3 places. Since 2.03 is larger than 0.00203, the exponent on 10 must be negative.

$$0.002\,03 = 2.03 \times 10^{-3}$$

21. −13,000,000

Move the decimal point to the right of the first nonzero digit and count the number of places the decimal point was moved.

$$-13,000,000 \quad 7 \; places$$

Because moving the decimal point to the *left* made the number *smaller*, we must multiply by a *positive* power of 10 so that the product -1.3×10^n will equal the larger number. Thus, $n = 7$, and

$$-13,000,000 = -1.3 \times 10^7.$$

23. −0.006

To move the decimal point to the right of the first nonzero digit, we move it 3 places. Since 6 is larger than 0.006, the exponent on 10 must be negative. Remember to include the negative sign.

$$-0.006 = -6 \times 10^{-3}$$

25. 7.5×10^5

Since the exponent is positive, make 7.5 larger by moving the decimal point 5 places to the right.

$$7.5 \times 10^5 = 750,000$$

27. 5.677×10^{12}

Since the exponent is positive, make 5.677 larger by moving the decimal point 12 places to the right. We need to add 9 zeros.

$$5.677 \times 10^{12} = 5,677,000,000,000$$

29. 1×10^{12}

Since the exponent is positive, make 1 larger by moving the decimal point 12 places to the right. We need to add 12 zeros.

$$1 \times 10^{12} = 1,000,000,000,000$$

31. 6.21×10^0

Because the exponent is 0, the decimal point should not be moved.

$$6.21 \times 10^0 = 6.21$$

We know this result is correct because $10^0 = 1$.

33. 7.8×10^{-4}

Since the exponent is negative, make 7.8 smaller by moving the decimal point 4 places to the left.

$$7.8 \times 10^{-4} = 0.000\,78$$

35. 5.134×10^{-9}

Since the exponent is negative, make 5.134 smaller by moving the decimal point 9 places to the left.

$$5.134 \times 10^{-9} = 0.000\,000\,005\,134$$

37. -4×10^{-3}

Move the decimal point 3 places to the left.

$$-4 \times 10^{-3} = -0.004$$

39. -8.1×10^5

Move the decimal point 5 places to the right.

$$-8.1 \times 10^5 = -810,000$$

41. **(a)** $(2 \times 10^8)(3 \times 10^3)$

$$= (2 \times 3)(10^8 \times 10^3) \quad \begin{array}{l} \textit{Commutative} \\ \textit{and associative} \\ \textit{properties} \\ \textit{Product rule} \\ \textit{for exponents} \end{array}$$

$$= 6 \times 10^{11}$$

(b) $6 \times 10^{11} = 600,000,000,000$

43. **(a)** $(5 \times 10^4)(3 \times 10^2)$

$$= (5 \times 3)(10^4 \times 10^2)$$
$$= 15 \times 10^6$$
$$= 1.5 \times 10^7$$

(b) $15 \times 10^6 = 15,000,000$

45. **(a)** $(3 \times 10^{-4})(-2 \times 10^8)$

$$= (3 \times (-2))(10^{-4} \times 10^8)$$
$$= -6 \times 10^4$$

(b) $-6 \times 10^4 = -60,000$

47. **(a)** $(6 \times 10^3)(4 \times 10^{-2})$

$$= (6 \times 4)(10^3 \times 10^{-2})$$
$$= 24 \times 10^1$$
$$= 2.4 \times 10^2$$

(b) $2.4 \times 10^2 = 240$

49. **(a)** $(9 \times 10^4)(7 \times 10^{-7})$

$$= (9 \times 7)(10^4 \times 10^{-7})$$
$$= 63 \times 10^{-3}$$
$$= 6.3 \times 10^{-2}$$

(b) $6.3 \times 10^{-2} = 0.063$

51. **(a)** $\dfrac{9 \times 10^{-5}}{3 \times 10^{-1}} = \dfrac{9}{3} \times \dfrac{10^{-5}}{10^{-1}}$

$$= 3 \times 10^{-5-(-1)}$$
$$= 3 \times 10^{-4}$$

(b) $3 \times 10^{-4} = 0.0003$

53. **(a)** $\dfrac{8 \times 10^3}{-2 \times 10^2} = \dfrac{8}{-2} \times \dfrac{10^3}{10^2}$

$$= -4 \times 10^1 = -4 \times 10$$

(b) $-4 \times 10^1 = -40$

55. **(a)** $\dfrac{2.6 \times 10^{-3}}{2 \times 10^2} = \dfrac{2.6}{2} \times \dfrac{10^{-3}}{10^2}$

$$= 1.3 \times 10^{-5}$$

(b) $1.3 \times 10^{-5} = 0.000\,013$

57. **(a)** $\dfrac{4 \times 10^5}{8 \times 10^2} = \dfrac{4}{8} \times \dfrac{10^5}{10^2}$

$$= 0.5 \times 10^{5-2}$$
$$= 0.5 \times 10^3$$
$$= 5 \times 10^2$$

(b) $5 \times 10^2 = 500$

59. **(a)** $\dfrac{-4.5 \times 10^4}{1.5 \times 10^{-2}} = \dfrac{-4.5}{1.5} \times \dfrac{10^4}{10^{-2}}$

$$= -3 \times 10^{4-(-2)}$$
$$= -3 \times 10^6$$

(b) $-3 \times 10^6 = -3,000,000$

61. **(a)** $\dfrac{-8 \times 10^{-4}}{-4 \times 10^3} = \dfrac{-8}{-4} \times \dfrac{10^{-4}}{10^3}$

$$= 2 \times 10^{-4-3}$$
$$= 2 \times 10^{-7}$$

(b) $2 \times 10^{-7} = 0.000\,000\,2$

63. To work in scientific mode on the TI-83/84 Plus, press MODE and then change from "Normal" to "Sci."

$$0.000\,000\,47 = 4.7 \times 10^{-7}$$

Prediction: $4.7\,\text{E}\,{-}7$

65. $(8\,\text{E}5)/(4\,\text{E}\,{-}2) = \dfrac{8 \times 10^5}{4 \times 10^{-2}}$

$$= \tfrac{8}{4} \times 10^{5-(-2)}$$
$$= 2 \times 10^7$$

Prediction: $2\,\text{E}7$

67. $(2\,\text{E}6)*(2\,\text{E}\,{-}3)/(4\,\text{E}2)$

$$= \dfrac{(2 \times 10^6)(2 \times 10^{-3})}{4 \times 10^2}$$

$$= \dfrac{(2 \times 2) \times (10^6 \times 10^{-3})}{4 \times 10^2}$$

$$= \dfrac{4 \times 10^3}{4 \times 10^2}$$

$$= \tfrac{4}{4} \times 10^{3-2}$$

$$= 1 \times 10^1$$

Prediction: 1 E1

69. $$\frac{650{,}000{,}000(0.000\,003\,2)}{0.000\,02}$$

$$= \frac{(6.5 \times 10^8)(3.2 \times 10^{-6})}{2 \times 10^{-5}}$$

$$= \frac{20.8 \times 10^2}{2 \times 10^{-5}}$$

$$= \frac{20.8}{2} \times \frac{10^2}{10^{-5}}$$

$$= 10.4 \times 10^7$$

$$= 1.04 \times 10^8$$

71. $$\frac{0.000\,000\,72(0.000\,23)}{0.000\,000\,018}$$

$$= \frac{\left(7.2 \times 10^{-7}\right)\left(2.3 \times 10^{-4}\right)}{1.8 \times 10^{-8}}$$

$$= \frac{16.56 \times 10^{-11}}{1.8 \times 10^{-8}}$$

$$= 9.2 \times 10^{-3}$$

73. $$\frac{0.000\,001\,6(240{,}000{,}000)}{0.000\,02(0.0032)}$$

$$= \frac{(1.6 \times 10^{-6})(2.4 \times 10^8)}{(2 \times 10^{-5})(3.2 \times 10^{-3})}$$

$$= \frac{3.84 \times 10^2}{6.4 \times 10^{-8}}$$

$$= 0.6 \times 10^{10}$$

$$= 6 \times 10^9$$

75. $$10 \text{ billion} = 10{,}000{,}000{,}000$$
$$= 1 \times 10^{10}$$

77. $$2 \times 10^9 = 2{,}000{,}000{,}000$$

79. $$3{,}305{,}000{,}000 = 3.305 \times 10^9$$

81. $$\$28{,}100{,}000{,}000 = \$2.81 \times 10^{10}$$

83. $$1200(2.3 \times 10^{-4}) = (1.2 \times 10^3)(2.3 \times 10^{-4})$$

$$= (1.2 \times 2.3) \times (10^3 \times 10^{-4})$$

$$= 2.76 \times 10^{3+(-4)}$$

$$= 2.76 \times 10^{-1}$$

$$= 0.276$$

There is about 2.76×10^{-1}, or 0.276, lb of copper in 1200 such people.

85. Use the formula $d = rt$.

distance = rate $\times$ time

$$6.68 \times 10^7 = (1.86 \times 10^5) \times \text{time}$$

$$\frac{6.68 \times 10^7}{1.86 \times 10^5} = \text{time}$$

$$\text{time} = \frac{6.68}{1.86} \times \frac{10^7}{10^5}$$

$$\approx 3.59 \times 10^2$$

It takes about 359 seconds for light to travel from the sun to Venus.

87. To get the average ticket price, divide the total gross by the attendance.

$$\frac{\$9.38 \times 10^8}{1.23 \times 10^7} = \frac{\$9.38}{1.23} \times \frac{10^8}{10^7}$$

$$\approx \$7.626 \times 10^1$$

The average ticket price was about $76.26.

89. To find the debt per person, divide the debt by the population.

$$\frac{\$1.2 \times 10^{13}}{300 \text{ million}} = \frac{\$1.2 \times 10^{13}}{300{,}000{,}000}$$

$$= \frac{\$1.2 \times 10^{13}}{3 \times 10^8}$$

$$= \$0.4 \times 10^5$$

$$= \$4 \times 10^4$$

The debt will be about $40,000 for every person.

91. To find the number of miles, multiply the number of light-years by the number of miles per light-year.

$$(25{,}000)(6{,}000{,}000{,}000{,}000)$$

$$= (2.5 \times 10^4)(6 \times 10^{12})$$

$$= (2.5 \times 6)(10^4 \times 10^{12})$$

$$= 15 \times 10^{16}$$

$$= 1.5 \times 10^{17}$$

The nebula is about 1.5×10^{17} miles from Earth.

93. $$\$4013(304{,}000{,}000)$$

$$= (\$4.013 \times 10^3)(3.04 \times 10^8)$$

$$= (\$4.013 \times 3.04)(10^3 \times 10^8)$$

$$\approx \$12.2 \times 10^{11}$$

$$= \$1.22 \times 10^{12}$$

In 2008, the U.S. government collected about $1,220,000,000,000 in taxes.

95. To find the amount each person would have to contribute, divide the amount by the population.

$$\frac{\$1{,}000{,}000{,}000{,}000}{307.5 \text{ million}} = \frac{\$1 \times 10^{12}}{307{,}500{,}000}$$

$$= \frac{\$1 \times 10^{12}}{3.075 \times 10^8}$$

$$= \frac{\$1}{3.075} \times \frac{10^{12}}{10^8}$$

$$\approx \$0.3252 \times 10^4$$

$$= \$3.252 \times 10^3$$

Each person would have to contribute about $3252.

97. $-3(2x + 4) + 4(2x - 6)$
$= -6x - 12 + 8x - 24$
$= 2x - 36$

99. $2x^2 - 3x + 10$
$= 2(3)^2 - 3(3) + 10$ *Let x = 3.*
$= 2(9) - 9 + 10$
$= 19$

101. $4x^3 - 5x^2 + 2x - 5$
$= 4(3)^3 - 5(3)^2 + 2(3) - 5$ *Let x = 3.*
$= 4(27) - 5(9) + 6 - 5$
$= 108 - 45 + 6 - 5$
$= 63 + 6 - 5$
$= 69 - 5$
$= 64$

5.4 Adding and Subtracting Polynomials; Graphing Simple Polynomials

5.4 Now Try Exercises

N1. The coefficient of t, or $1t$, is 1 and the coefficient of $-10t^2$, is -10.

N2. $3x^2 - x^2 + 2x$
$= (3x^2 - 1x^2) + 2x$ *Associative property*
$= (3 - 1)x^2 + 2x$ *Distributive property*
$= 2x^2 + 2x$

N3. $x^2 + 4x - 2x - 8 = x^2 + (4 - 2)x - 8$ simplifies to $x^2 + 2x - 8$. This polynomial has three terms, and the largest exponent on the variable x is 2. So the polynomial has degree 2 and is a trinomial.

N4. $4t^3 - t^2 - t$ *Replace t with −3.*
$= 4(-3)^3 - (-3)^2 - (-3)$
$= 4(-27) - (9) + 3$
$= -108 - 9 + 3$
$= -117 + 3$
$= -114$

N5. Add, column by column.

$4y^3 - 2y^2 + y - 1$ *First polynomial*
$\underline{y^3 + 0y^2 - y - 7}$ *Second polynomial*
$5y^3 - 2y^2 \quad\; - 8$ *Sum*

N6. Combine like terms.

$(10x^4 - 3x^2 - x) + (x^4 - 3x^2 + 5x)$
$= (10x^4 + x^4) + (-3x^2 - 3x^2) + (-x + 5x)$
$= 11x^4 - 6x^2 + 4x$

N7. Change subtraction to addition.

$(4t^4 - t^2 + 7) - (5t^4 - 3t^2 + 1)$
$= (4t^4 - t^2 + 7) + (-5t^4 + 3t^2 - 1)$

Combine like terms.

$= (4t^4 - 5t^4) + (-t^2 + 3t^2) + (7 - 1)$
$= -t^4 + 2t^2 + 6$

Check by adding.

$5t^4 - 3t^2 + 1$ *Second polynomial*
$\underline{-t^4 + 2t^2 + 6}$ *Answer*
$4t^4 - \; t^2 + 7$ *First polynomial*

N8. Arrange like terms in columns. Insert zeros for missing terms.

$12x^2 - 9x + 4$
$\underline{-10x^2 - 3x + 7}$

Change all signs in the second row, and then add.

$12x^2 - 9x + 4$
$\underline{10x^2 + 3x - 7}$
$22x^2 - 6x - 3$

N9. $(4x^2 - 2xy + y^2) - (6x^2 - 7xy + 2y^2)$
$= 4x^2 - 2xy + y^2 - 6x^2 + 7xy - 2y^2$
$= (4x^2 - 6x^2) + (-2xy + 7xy) + (y^2 - 2y^2)$
$= -2x^2 + 5xy - y^2$

N10. Graph $y = -x^2 - 1$.

Make a table of ordered pairs whose x-values are on either side of the vertex's x-value of $x = 0$.

x	$y = -x^2 - 1$
0	$-(0)^2 - 1 = 0 - 1 = -1$
± 1	$-(\pm 1)^2 - 1 = -1 - 1 = -2$
± 2	$-(\pm 2)^2 - 1 = -4 - 1 = -5$

Plot these seven ordered pairs and connect them with a smooth curve.

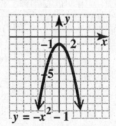

5.4 Section Exercises

1. In the term $4x^6$, the coefficient is $\underline{4}$ and the exponent is $\underline{6}$.

3. The degree of the term $-3x^9$ is $\underline{9}$, the exponent.

5. When $x^2 + 10$ is evaluated for $x = 3$, the result is
$3^2 + 10 = 9 + 10 = \underline{19}$.

7. $-3xy - 2xy + 5xy = (-3 - 2 + 5)xy$
$= (0)xy$
$= \underline{0}$

9. The polynomial $6x^4$ has one term. The coefficient of this term is 6.

11. The polynomial t^4 has one term. Since $t^4 = 1 \cdot t^4$, the coefficient of this term is 1.

13. The polynomial $-19r^2 - r$ has two terms. The coefficient of r^2 is -19 and the coefficient of r is -1.

15. The polynomial $x + 8x^2 + 5x^3$ has three terms. The coefficient of x is 1, the coefficient of x^2 is 8, and the coefficient of x^3 is 5.

In Exercises 17–28, use the distributive property to add like terms.

17. $\begin{aligned} -3m^5 + 5m^5 &= (-3 + 5)m^5 \\ &= 2m^5 \end{aligned}$

19. $\begin{aligned} 2r^5 + (-3r^5) &= [2 + (-3)]r^5 \\ &= -1r^5 = -r^5 \end{aligned}$

21. The polynomial $0.2m^5 - 0.5m^2$ cannot be simplified. The two terms are unlike because the exponents on the variables are different, so they cannot be combined.

23. $\begin{aligned} -3x^5 + 3x^5 - 5x^5 &= (-3 + 3 - 5)x^5 \\ &= -5x^5 \end{aligned}$

25. $\begin{aligned} -4p^7 + 8p^7 + 5p^9 &= (-4 + 8)p^7 + 5p^9 \\ &= 4p^7 + 5p^9 \end{aligned}$

In descending powers of the variable, this polynomial is written $5p^9 + 4p^7$.

27. $\begin{aligned} -4xy^2 + 3xy^2 &- 2xy^2 + xy^2 \\ &= (-4 + 3 - 2 + 1)xy^2 \\ &= -2xy^2 \end{aligned}$

29. $6x^4 - 9x$

This polynomial has no like terms, so it is already simplified. It is already written in descending powers of the variable x. The highest degree of any nonzero term is 4, so the degree of the polynomial is 4. There are two terms, so this is a *binomial.*

31. $\begin{aligned} 5m^4 &- 3m^2 + 6m^4 - 7m^3 \\ &= (5m^4 + 6m^4) + (-7m^3) + (-3m^2) \\ &= 11m^4 - 7m^3 - 3m^2 \end{aligned}$

The resulting polynomial is a *trinomial* of degree 4.

33. $\begin{aligned} \tfrac{5}{3}x^4 - \tfrac{2}{3}x^4 &= \left(\tfrac{5}{3} - \tfrac{2}{3}\right)x^4 \\ &= \tfrac{3}{3}x^4 = x^4 \end{aligned}$

The resulting polynomial is a *monomial* of degree 4.

35. $\begin{aligned} 0.8x^4 &- 0.3x^4 - 0.5x^4 + 7 \\ &= (0.8 - 0.3 - 0.5)x^4 + 7 \\ &= 0x^4 + 7 = 7 \end{aligned}$

Since 7 can be written as $7x^0$, the degree of the polynomial is 0. The simplified polynomial has one term, so it is a *monomial.*

37. **(a)** $\begin{aligned} 2x^2 &- 3x - 5 \\ &= 2(2)^2 - 3(2) - 5 \quad \textit{Let x = 2.} \\ &= 2(4) - 6 - 5 \\ &= 8 - 6 - 5 \\ &= 2 - 5 \\ &= -3 \end{aligned}$

(b) $\begin{aligned} 2x^2 &- 3x - 5 \\ &= 2(-1)^2 - 3(-1) - 5 \quad \textit{Let x = -1.} \\ &= 2(1) + 3 - 5 \\ &= 2 + 3 - 5 \\ &= 5 - 5 \\ &= 0 \end{aligned}$

39. **(a)** $\begin{aligned} -3x^2 &+ 14x - 2 \\ &= -3(2)^2 + 14(2) - 2 \quad \textit{Let x = 2.} \\ &= -3(4) + 28 - 2 \\ &= -12 + 28 - 2 \\ &= 16 - 2 \\ &= 14 \end{aligned}$

(b) $\begin{aligned} -3x^2 &+ 14x - 2 \\ &= -3(-1)^2 + 14(-1) - 2 \quad \textit{Let x = -1.} \\ &= -3(1) - 14 - 2 \\ &= -3 - 14 - 2 \\ &= -17 - 2 \\ &= -19 \end{aligned}$

41. **(a)** $\begin{aligned} 2x^5 &- 4x^4 + 5x^3 - x^2 \\ &= 2(2)^5 - 4(2)^4 + 5(2)^3 - (2)^2 \quad \textit{Let x = 2.} \\ &= 2(32) - 4(16) + 5(8) - 4 \\ &= 64 - 64 + 40 - 4 \\ &= 36 \end{aligned}$

(b) $\begin{aligned} 2x^5 &- 4x^4 + 5x^3 - x^2 \\ &= 2(-1)^5 - 4(-1)^4 + 5(-1)^3 - (-1)^2 \\ &\qquad\qquad\qquad\qquad \textit{Let x = -1.} \\ &= 2(-1) - 4(1) + 5(-1) - 1 \\ &= -2 - 4 - 5 - 1 \\ &= -12 \end{aligned}$

43. Add, column by column.

$$\begin{array}{r} 2x^2 - 4x \\ 3x^2 + 2x \\ \hline 5x^2 - 2x \end{array}$$

45. Add, column by column.

$$3m^2 + 5m + 6$$
$$2m^2 - 2m - 4$$
$$\overline{5m^2 + 3m + 2}$$

47. Add, column by column.

$$\tfrac{2}{3}x^2 + \tfrac{1}{5}x + \tfrac{1}{6}$$
$$\tfrac{1}{2}x^2 - \tfrac{1}{3}x + \tfrac{2}{3}$$

Rewrite the fractions so that the fractions in each column have a common denominator; then add column by column.

$$\tfrac{4}{6}x^2 + \tfrac{3}{15}x + \tfrac{1}{6}$$
$$\tfrac{3}{6}x^2 - \tfrac{5}{15}x + \tfrac{4}{6}$$
$$\overline{\tfrac{7}{6}x^2 - \tfrac{2}{15}x + \tfrac{5}{6}}$$

49. $(9m^3 - 5m^2 + 4m - 8) + (-3m^3 + 6m^2 - 6)$
$= (9m^3 - 3m^3) + (-5m^2 + 6m^2) + 4m$
$\quad + (-8 - 6)$
$= 6m^3 + m^2 + 4m - 14$

51. Subtract, column by column.

$$5y^3 - 3y^2$$
$$2y^3 + 8y^2$$

Change all signs in the second row, and then add.

$$5y^3 - 3y^2$$
$$-2y^3 - 8y^2$$
$$\overline{3y^3 - 11y^2}$$

53. Subtract, column by column.

$$12x^4 - x^2 + x$$
$$8x^4 + 3x^2 - 3x$$

Change all signs in the second row, and then add.

$$12x^4 - x^2 + x$$
$$-8x^4 - 3x^2 + 3x$$
$$\overline{4x^4 - 4x^2 + 4x}$$

55. Subtract, column by column.

$$12m^3 - 8m^2 + 6m + 7$$
$$-3m^3 + 5m^2 - 2m - 4$$

Change all signs in the second row, and then add.

$$12m^3 - 8m^2 + 6m + 7$$
$$3m^3 - 5m^2 + 2m + 4$$
$$\overline{15m^3 - 13m^2 + 8m + 11}$$

57. Vertical addition and subtraction of polynomials is preferable because like terms are arranged in columns.

59. $(8m^2 - 7m) - (3m^2 + 7m - 6)$
$= (8m^2 - 7m) + (-3m^2 - 7m + 6)$
$= (8 - 3)m^2 + (-7 - 7)m + 6$
$= 5m^2 - 14m + 6$

61. $(16x^3 - x^2 + 3x) + (-12x^3 + 3x^2 + 2x)$
$= 16x^3 - x^2 + 3x - 12x^3 + 3x^2 + 2x$
$= (16 - 12)x^3 + (-1 + 3)x^2 + (3 + 2)x$
$= 4x^3 + 2x^2 + 5x$

63. $(7y^4 + 3y^2 + 2y) - (18y^4 - 5y^2 + y)$
$= (7y^4 + 3y^2 + 2y) + (-18y^4 + 5y^2 - y)$
$= (7 - 18)y^4 + (3 + 5)y^2 + (2 - 1)y$
$= -11y^4 + 8y^2 + y$

65. $(9a^4 - 3a^2 + 2) + (4a^4 - 4a^2 + 2)$
$\quad + (-12a^4 + 6a^2 - 3)$
$= (9a^4 + 4a^4 - 12a^4)$
$\quad + (-3a^2 - 4a^2 + 6a^2) + (2 + 2 - 3)$
$= a^4 - a^2 + 1$

67. $[(8m^2 + 4m - 7) - (2m^2 - 5m + 2)]$
$\quad - (m^2 + m + 1)$
$= (8m^2 + 4m - 7) + (-2m^2 + 5m - 2)$
$\quad + (-m^2 - m - 1)$
$= (8 - 2 - 1)m^2 + (4 + 5 - 1)m$
$\quad + (-7 - 2 - 1)$
$= 5m^2 + 8m - 10$

69. $[(3x^2 - 2x + 7) - (4x^2 + 2x - 3)]$
$\quad - [(9x^2 + 4x - 6) + (-4x^2 + 4x + 4)]$
$= [(3 - 4)x^2 + (-2 - 2)x + (7 + 3)]$
$\quad - [(9 - 4)x^2 + (4 + 4)x + (-6 + 4)]$
$= (-x^2 - 4x + 10) - (5x^2 + 8x - 2)$
$= -x^2 - 4x + 10 - 5x^2 - 8x + 2$
$= -6x^2 - 12x + 12$

71. The coefficients of the x^2 terms are -4, $-(-2)$, and -8. The sum of these numbers is

$$-4 + 2 - 8 = -10.$$

73. $(6b + 3c) + (-2b - 8c)$
$= (6b - 2b) + (3c - 8c)$
$= 4b - 5c$

75. $(4x + 2xy - 3) - (-2x + 3xy + 4)$
$= (4x + 2xy - 3) + (2x - 3xy - 4)$
$= (4x + 2x) + (2xy - 3xy) + (-3 - 4)$
$= 6x - xy - 7$

77. $(5x^2y - 2xy + 9xy^2)$
$\qquad - (8x^2y + 13xy + 12xy^2)$
$\quad = (5x^2y - 2xy + 9xy^2)$
$\qquad + (-8x^2y - 13xy - 12xy^2)$
$\quad = (5x^2y - 8x^2y) + (-2xy - 13xy)$
$\qquad + (9xy^2 - 12xy^2)$
$\quad = -3x^2y - 15xy - 3xy^2$

79. Use the formula for the perimeter of a rectangle, $P = 2L + 2W$, with length $L = 4x^2 + 3x + 1$ and width $W = x + 2$.

$P = 2L + 2W$
$\quad = 2(4x^2 + 3x + 1) + 2(x + 2)$
$\quad = 8x^2 + 6x + 2 + 2x + 4$
$\quad = 8x^2 + 8x + 6$

A polynomial that represents the perimeter of the rectangle is $8x^2 + 8x + 6$.

81. Use the formula for the perimeter of a square, $P = 4s$, with $s = \frac{1}{2}x^2 + 2x$.

$P = 4s$
$\quad = 4(\frac{1}{2}x^2 + 2x)$
$\quad = 4(\frac{1}{2}x^2) + 4(2x)$
$\quad = 2x^2 + 8x$

A polynomial that represents the perimeter of the square is $2x^2 + 8x$.

83. Use the formula for the perimeter of a triangle, $P = a + b + c$, with $a = 3t^2 + 2t + 7$, $b = 5t^2 + 2$, and $c = 6t + 4$.

$P = (3t^2 + 2t + 7) + (5t^2 + 2) + (6t + 4)$
$\quad = (3t^2 + 5t^2) + (2t + 6t) + (7 + 2 + 4)$
$\quad = 8t^2 + 8t + 13$

A polynomial that represents the perimeter of the triangle is $8t^2 + 8t + 13$.

85. **(a)** Use the formula for the perimeter of a triangle, $P = a + b + c$, with $a = 2y - 3t$, $b = 5y + 3t$, and $c = 16y + 5t$.

$P = (2y - 3t) + (5y + 3t) + (16y + 5t)$
$\quad = (2y + 5y + 16y) + (-3t + 3t + 5t)$
$\quad = 23y + 5t$

The perimeter of the triangle is $23y + 5t$.

(b) Use the fact that the sum of the angles of any triangle is $180°$.

$(7x - 3)° + (5x + 2)° + (2x - 1)° = 180°$
$(7x + 5x + 2x) + (-3 + 2 - 1) = 180$
$\qquad\qquad\qquad 14x - 2 = 180$
$\qquad\qquad\qquad\quad 14x = 182$
$\qquad\qquad\qquad\quad\ x = \frac{182}{14} = 13$

If $x = 13$,

$7x - 3 = 7(13) - 3 = 88,$
$5x + 2 = 5(13) + 2 = 67,$
and $\ 2x - 1 = 2(13) - 1 = 25.$

The measures of the angles are $25°$, $67°$, and $88°$.

87. First find the two sums and then find their difference.

$[(5x^2 + 2x - 3) + (x^2 - 8x + 2)]$
$\quad - [(7x^2 - 3x + 6) + (-x^2 + 4x - 6)$
$= (6x^2 - 6x - 1) - (6x^2 + x)$
$= (6x^2 - 6x - 1) + (-6x^2 - x)$
$= (6x^2 - 6x^2) + (-6x - x) + (-1)$
$= -7x - 1$

89. $y = x^2 - 4$

x	$y = x^2 - 4$
-2	$(-2)^2 - 4 = 4 - 4 = 0$
-1	$(-1)^2 - 4 = 1 - 4 = -3$
0	$(0)^2 - 4 = 0 - 4 = -4$
1	$(1)^2 - 4 = 1 - 4 = -3$
2	$(2)^2 - 4 = 4 - 4 = 0$

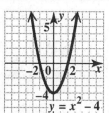

91. $y = 2x^2 - 1$

x	$y = 2x^2 - 1$
-2	$2(-2)^2 - 1 = 2 \cdot 4 - 1 = 7$
-1	$2(-1)^2 - 1 = 2 \cdot 1 - 1 = 1$
0	$2(0)^2 - 1 = 2 \cdot 0 - 1 = -1$
1	$2(1)^2 - 1 = 2 \cdot 1 - 1 = 1$
2	$2(2)^2 - 1 = 2 \cdot 4 - 1 = 7$

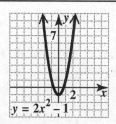

93. $y = -x^2 + 4$

x	$y = -x^2 + 4$
-2	$-(-2)^2 + 4 = -4 + 4 = 0$
-1	$-(-1)^2 + 4 = -1 + 4 = 3$
0	$-(0)^2 + 4 = -0 + 4 = 4$
1	$-(1)^2 + 4 = -1 + 4 = 3$
2	$-(2)^2 + 4 = -4 + 4 = 0$

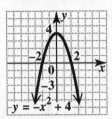

95. $y = (x + 3)^2$

x	$y = (x + 3)^2$
-5	$(-5 + 3)^2 = (-2)^2 = 4$
-4	$(-4 + 3)^2 = (-1)^2 = 1$
-3	$(-3 + 3)^2 = (0)^2 = 0$
-2	$(-2 + 3)^2 = (1)^2 = 1$
-1	$(-1 + 3)^2 = (2)^2 = 4$

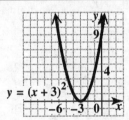

97. If $x = 9$, then $y = 7x = 7(9) = 63$. If a dog is 9 in dog years, then it is 63 in human years.

98. $-0.0545x^2 + 5.047x + 11.78$

$= -0.0545(3)^2 + 5.047(3) + 11.78$ *Let x = 3.*

$= -0.4905 + 15.141 + 11.78$

$= 26.4305$

If a dog is 3 in dog years, then it is about 26 in human years.

99. $-16x^2 + 60x + 80$

$= -16(2.5)^2 + 60(2.5) + 80$ *Let x = 2.5.*

$= -16(6.25) + 150 + 80$

$= -100 + 150 + 80$

$= 130$

If 2.5 seconds have elapsed, the height of the object is 130 feet.

100. If $x = 6$,

$2x + 15 = 2(6) + 15 = 12 + 15 = 27$.

If the saw is rented for 6 days, the cost is $27.

101. $5(x + 4) = 5x + 5(4) = 5x + 20$

103. $4(2a + 6b) = 4(2a) + 4(6b)$
$= 8a + 24b$

105. $(2a)(-5ab) = (2)(-5)(a \cdot a)b$
$= -10a^2b$

107. $(-m^2)(m^5) = -m^{2+5} = -m^7$

5.5 Multiplying Polynomials

5.5 Now Try Exercises

N1. $-3x^5(2x^3 - 5x^2 + 10)$

$= -3x^5(2x^3) + (-3x^5)(-5x^2) + (-3x^5)(10)$

   Distributive property

$= -6x^8 + 15x^7 + (-30x^5)$

   Multiply monomials.

$= -6x^8 + 15x^7 - 30x^5$

N2. $(x^2 - 4)(2x^2 - 5x + 3)$

$= x^2(2x^2) + x^2(-5x) + x^2(3)$

$\quad - 4(2x^2) - 4(-5x) - 4(3)$

$= 2x^4 - 5x^3 + 3x^2 - 8x^2 + 20x - 12$

$= 2x^4 - 5x^3 - 5x^2 + 20x - 12$

N3.
$$
\begin{array}{r}
5t^2 - 7t + 4 \\
2t - 6 \\
\hline
-30t^2 + 42t - 24 \leftarrow -6(5t^2 - 7t + 4) \\
10t^3 - 14t^2 + 8t \qquad\quad \leftarrow 2t(5t^2 - 7t + 4) \\
\hline
10t^3 - 44t^2 + 50t - 24 \leftarrow \text{Add like terms.}
\end{array}
$$

N4.
$$
\begin{array}{r}
9x^3 - 12x^2 + 3 \\
\tfrac{1}{3}x^2 - \tfrac{2}{3} \\
\hline
-6x^3 + 8x^2 - 2 \quad (\textit{times } -\tfrac{2}{3}) \\
3x^5 - 4x^4 \qquad\quad + x^2 \qquad (\textit{times } \tfrac{1}{3}x^2) \\
\hline
3x^5 - 4x^4 - 6x^3 + 9x^2 - 2 \quad \textit{Add.}
\end{array}
$$

The product is $3x^5 - 4x^4 - 6x^3 + 9x^2 - 2$.

N5. $(t - 6)(t + 5)$

F: $t(t) = t^2$

O: $t(5) = 5t$

I: $-6(t) = -6t$

L: $-6(5) = -30$

$(t - 6)(t + 5) = t^2 + 5t - 6t - 30$

$= t^2 - t - 30$

N6. $(7y - 3)(2x + 5)$

F: $7y(2x) = 14yx$

O: $7y(5) = 35y$

I: $-3(2x) = -6x$

L: $-3(5) = -15$

The product $(7y - 3)(2x + 5)$ is equal to $14yx + 35y - 6x - 15$.

N7. **(a)** $(3p - 5q)(4p - q)$

F: $\quad 3p(4p) = 12p^2$

O: $\quad 3p(-q) = -3pq$

I: $\quad -5q(4p) = -20pq$

L: $\quad -5q(-q) = 5q^2$

$(3p - 5q)(4p - q)$

$\quad = 12p^2 - 3pq - 20pq + 5q^2$

$\quad = 12p^2 - 23pq + 5q^2$

(b) $5x^2(3x + 1)(x - 5)$

First multiply the binomials and then multiply that result by $5x^2$.

F: $\quad 3x(x) = 3x^2$

O: $\quad 3x(-5) = -15x$

I: $\quad 1(x) = x$

L: $\quad 1(-5) = -5$

$5x^2(3x + 1)(x - 5) = 5x^2(3x^2 - 15x + x - 5)$

$\quad\quad\quad\quad\quad\quad\quad\quad = 5x^2(3x^2 - 14x - 5)$

$\quad\quad\quad\quad\quad\quad\quad\quad = 15x^4 - 70x^3 - 25x^2$

5.5 Section Exercises

1. **(a)** $5x^3(6x^7)$

$\quad = 5 \cdot 6x^{3+7}$

$\quad = 30x^{10}$

Choice **B** is correct.

(b) $-5x^7(6x^3)$

$\quad = -5 \cdot 6x^{7+3}$

$\quad = -30x^{10}$

Choice **D** is correct.

(c) $(5x^7)^3 = (5)^3(x^7)^3$

$\quad\quad\quad\quad = 125x^{7 \cdot 3}$

$\quad\quad\quad\quad = 125x^{21}$

Choice **A** is correct.

(d) $(-6x^3)^3 = (-6)^3(x^3)^3$

$\quad\quad\quad\quad\quad = -216x^{3 \cdot 3}$

$\quad\quad\quad\quad\quad = -216x^9$

Choice **C** is correct.

3. $5y^4(3y^7) = 5(3)y^{4+7}$

$\quad\quad\quad\quad = 15y^{11}$

5. $-15a^4(-2a^5) = -15(-2)a^{4+5}$

$\quad\quad\quad\quad\quad\quad = 30a^9$

7. $5p(3q^2) = 5(3)pq^2$

$\quad\quad\quad = 15pq^2$

9. $-6m^3(3n^2) = -6(3)m^3n^2$

$\quad\quad\quad\quad\quad = -18m^3n^2$

11. $y^5 \cdot 9y \cdot y^4 = 9y^{5+1+4}$

$\quad\quad\quad\quad\quad = 9y^{10}$

13. $(4x^3)(2x^2)(-x^5) = (-4 \cdot 2)x^{3+2+5}$

$\quad\quad\quad\quad\quad\quad\quad\quad = -8x^{10}$

15. $2m(3m + 2) = 2m(3m) + 2m(2)$

$\quad\quad\quad\quad\quad = 6m^2 + 4m$

17. $3p(-2p^3 + 4p^2) = 3p(-2p^3) + 3p(4p^2)$

$\quad\quad\quad\quad\quad\quad\quad = -6p^4 + 12p^3$

19. $-8z(2z + 3z^2 + 3z^3)$

$\quad = -8z(2z) + (-8z)(3z^2) + (-8z)(3z^3)$

$\quad = -16z^2 - 24z^3 - 24z^4$

21. $2y^3(3 + 2y + 5y^4)$

$\quad = 2y^3(3) + 2y^3(2y) + 2y^3(5y^4)$

$\quad = 6y^3 + 4y^4 + 10y^7$

23. $-4r^3(-7r^2 + 8r - 9)$

$\quad = -4r^3(-7r^2) + (-4r^3)(8r) + (-4r^3)(-9)$

$\quad = 28r^5 - 32r^4 + 36r^3$

25. $3a^2(2a^2 - 4ab + 5b^2)$

$\quad = 3a^2(2a^2) + 3a^2(-4ab) + 3a^2(5b^2)$

$\quad = 6a^4 - 12a^3b + 15a^2b^2$

27. $7m^3n^2(3m^2 + 2mn - n^3)$

$\quad = 7m^3n^2(3m^2) + 7m^3n^2(2mn) + 7m^3n^2(-n^3)$

$\quad = 21m^5n^2 + 14m^4n^3 - 7m^3n^5$

In Exercises 29–40, we can multiply the polynomials horizontally or vertically. The following solutions illustrate these two methods.

29. $(6x + 1)(2x^2 + 4x + 1)$

$\quad = 6x(2x^2) + 6x(4x) + 6x(1)$

$\quad\quad + 1(2x^2) + 1(4x) + 1(1)$

$\quad = 12x^3 + 24x^2 + 6x + 2x^2 + 4x + 1$

$\quad = 12x^3 + 26x^2 + 10x + 1$

31. $(9y - 2)(8y^2 - 6y + 1)$

Multiply vertically.

$$
\begin{array}{rrrr}
 & 8y^2 & - \; 6y & + \; 1 \\
 & & 9y & - \; 2 \\
\hline
 & - \; 16y^2 & + \; 12y & - \; 2 \\
72y^3 & - \; 54y^2 & + \; 9y & \\
\hline
72y^3 & - \; 70y^2 & + \; 21y & - \; 2 \\
\end{array}
$$

33. $(4m + 3)(5m^3 - 4m^2 + m - 5)$

Multiply vertically.

$$
\begin{array}{r}
5m^3 - 4m^2 + m - 5 \\
4m + 3 \\
\hline
15m^3 - 12m^2 + 3m - 15 \\
20m^4 - 16m^3 + 4m^2 - 20m \\
\hline
20m^4 - m^3 - 8m^2 - 17m - 15
\end{array}
$$

35. $(2x - 1)(3x^5 - 2x^3 + x^2 - 2x + 3)$

Multiply vertically.

$$
\begin{array}{r}
3x^5 \qquad -2x^3 + x^2 - 2x + 3 \\
2x - 1 \\
\hline
-3x^5 \qquad +2x^3 - x^2 + 2x - 3 \\
6x^6 \qquad -4x^4 + 2x^3 - 4x^2 + 6x \\
\hline
6x^6 - 3x^5 - 4x^4 + 4x^3 - 5x^2 + 8x - 3
\end{array}
$$

37. $(5x^2 + 2x + 1)(x^2 - 3x + 5)$

Multiply vertically.

$$
\begin{array}{r}
5x^2 + 2x + 1 \\
x^2 - 3x + 5 \\
\hline
25x^2 + 10x + 5 \\
-15x^3 - 6x^2 - 3x \\
5x^4 + 2x^3 + x^2 \\
\hline
5x^4 - 13x^3 + 20x^2 + 7x + 5
\end{array}
$$

39. $(6x^4 - 4x^2 + 8x)(\frac{1}{2}x + 3)$

$= 6x^4(\frac{1}{2}x) + -4x^2(\frac{1}{2}x) + 8x(\frac{1}{2}x)$

$\quad + 6x^4(3) + -4x^2(3) + 8x(3)$

$= 3x^5 - 2x^3 + 4x^2 + 18x^4 - 12x^2 + 24x$

$= 3x^5 + 18x^4 - 2x^3 - 8x^2 + 24x$

41. $(m + 7)(m + 5)$

$$
\begin{array}{cccc}
\mathbf{F} & \mathbf{O} & \mathbf{I} & \mathbf{L}
\end{array}
$$

$= m(m) + m(5) + 7(m) + 7(5)$

$= m^2 + 5m + 7m + 35$

$= m^2 + 12m + 35$

43. $(n - 1)(n + 4)$

$$
\begin{array}{cccc}
\mathbf{F} & \mathbf{O} & \mathbf{I} & \mathbf{L}
\end{array}
$$

$= n(n) + n(4) + (-1)(n) + (-1)(4)$

$= n^2 + 4n + (-1n) + (-4)$

$= n^2 + 3n - 4$

45. $(x + 5)(x - 5)$

$$
\begin{array}{cccc}
\mathbf{F} & \mathbf{O} & \mathbf{I} & \mathbf{L}
\end{array}
$$

$= x(x) + x(-5) + 5(x) + 5(-5)$

$= x^2 - 5x + 5x - 25$

$= x^2 - 25$

47. $(2x + 3)(6x - 4)$

$$
\begin{array}{cccc}
\mathbf{F} & \mathbf{O} & \mathbf{I} & \mathbf{L}
\end{array}
$$

$= 2x(6x) + 2x(-4) + 3(6x) + 3(-4)$

$= 12x^2 - 8x + 18x - 12$

$= 12x^2 + 10x - 12$

49. $(9 + t)(9 - t)$

$$
\begin{array}{cccc}
\mathbf{F} & \mathbf{O} & \mathbf{I} & \mathbf{L}
\end{array}
$$

$= 9(9) + 9(-t) + t(9) + t(-t)$

$= 81 - 9t + 9t - t^2$

$= 81 - t^2$

51. $(3x - 2)(3x - 2)$

$$
\begin{array}{cccc}
\mathbf{F} & \mathbf{O} & \mathbf{I} & \mathbf{L}
\end{array}
$$

$= 3x(3x) + 3x(-2) + (-2)(3x) + (-2)(-2)$

$= 9x^2 - 6x - 6x + 4$

$= 9x^2 - 12x + 4$

53. $(5a + 1)(2a + 7)$

$$
\begin{array}{cccc}
\mathbf{F} & \mathbf{O} & \mathbf{I} & \mathbf{L}
\end{array}
$$

$= 5a(2a) + 5a(7) + 1(2a) + 1(7)$

$= 10a^2 + 35a + 2a + 7$

$= 10a^2 + 37a + 7$

55. $(6 - 5m)(2 + 3m)$

$$
\begin{array}{cccc}
\mathbf{F} & \mathbf{O} & \mathbf{I} & \mathbf{L}
\end{array}
$$

$= 6(2) + 6(3m) + (-5m)(2) + (-5m)(3m)$

$= 12 + 18m - 10m - 15m^2$

$= 12 + 8m - 15m^2$

57. $(5 - 3x)(4 + x)$

$$
\begin{array}{cccc}
\mathbf{F} & \mathbf{O} & \mathbf{I} & \mathbf{L}
\end{array}
$$

$= 5(4) + 5(x) + (-3x)(4) + (-3x)(x)$

$= 20 + 5x - 12x - 3x^2$

$= 20 - 7x - 3x^2$

59. $(3t - 4s)(t + 3s)$

$$
\begin{array}{cccc}
\mathbf{F} & \mathbf{O} & \mathbf{I} & \mathbf{L}
\end{array}
$$

$= 3t(t) + 3t(3s) + (-4s)(t) + (-4s)(3s)$

$= 3t^2 + 9st - 4st - 12s^2$

$= 3t^2 + 5st - 12s^2$

61. $(4x + 3)(2y - 1)$

$$
\begin{array}{cccc}
\mathbf{F} & \mathbf{O} & \mathbf{I} & \mathbf{L}
\end{array}
$$

$= 4x(2y) + 4x(-1) + 3(2y) + 3(-1)$

$= 8xy - 4x + 6y - 3$

63. $(3x + 2y)(5x - 3y)$

$$
\begin{array}{cccc}
\mathbf{F} & \mathbf{O} & \mathbf{I} & \mathbf{L}
\end{array}
$$

$= 3x(5x) + 3x(-3y) + 2y(5x) + 2y(-3y)$

$= 15x^2 - 9xy + 10xy - 6y^2$

$= 15x^2 + xy - 6y^2$

65. $3y^3(2y+3)(y-5)$

$(2y+3)(y-5)$

$\quad$ **F** $\quad$ **O** $\quad$ **I** $\quad$ **L**

$= 2y(y) + 2y(-5) + 3(y) + 3(-5)$

$= 2y^2 - 10y + 3y - 15$

$= 2y^2 - 7y - 15$

Now multiply this result by $3y^3$.

$3y^3(2y^2 - 7y - 15)$

$= 3y^3(2y^2) + 3y^3(-7y) + 3y^3(-15)$

$= 6y^5 - 21y^4 - 45y^3$

67. $-8r^3(5r^2+2)(5r^2-2)$

$(5r^2+2)(5r^2-2)$

$\quad$ **F** $\quad$ **O** $\quad$ **I** $\quad$ **L**

$= 5r^2(5r^2) + 5r^2(-2) + 2(5r^2) + 2(-2)$

$= 25r^4 - 10r^2 + 10r^2 - 4$

$= 25r^4 - 4$

Now multiply this result by $-8r^3$.

$-8r^3(25r^4 - 4)$

$= -8r^3(25r^4) + (-8r^3)(-4)$

$= -200r^7 + 32r^3$

69. **(a)** Use the formula for the area of a rectangle, $A = LW$, with $L = 3y + 7$ and $W = y + 1$.

$A = (3y+7)(y+1)$

$= (3y+7)(y) + (3y+7)(1)$

$= 3y^2 + 7y + 3y + 7$

$= 3y^2 + 10y + 7$

(b) Use the formula for the perimeter of a rectangle, $P = 2L + 2W$, with $L = 3y + 7$ and $W = y + 1$.

$P = 2(3y+7) + 2(y+1)$

$= 2(3y) + 2(7) + 2(y) + 2(1)$

$= 6y + 14 + 2y + 2$

$= 8y + 16$

71. $\left(3p + \frac{5}{4}q\right)\left(2p - \frac{5}{3}q\right)$ $\quad$ *Use FOIL.*

$= 3p(2p) + 3p\left(-\frac{5}{3}q\right) + \frac{5}{4}q(2p) + \frac{5}{4}q\left(-\frac{5}{3}q\right)$

$= 6p^2 - 5pq + \frac{5}{2}pq - \frac{25}{12}q^2$

$= 6p^2 + \left(-\frac{10}{2} + \frac{5}{2}\right)pq - \frac{25}{12}q^2$

$= 6p^2 - \frac{5}{2}pq - \frac{25}{12}q^2$

73. $(x+7)^2 = (x+7)(x+7)$ $\quad$ *Use FOIL.*

$= x^2 + 7x + 7x + 49$

$= x^2 + 14x + 49$

75. $(a-4)(a+4)$ $\quad$ *Use FOIL.*

$= a^2 + 4a - 4a - 16$

$= a^2 - 16$

77. $(2p-5)^2 = (2p-5)(2p-5)$ $\quad$ *Use FOIL.*

$= 4p^2 - 10p - 10p + 25$

$= 4p^2 - 20p + 25$

79. $(5k+3q)^2 = (5k+3q)(5k+3q)$ *Use FOIL.*

$= 25k^2 + 15kq + 15kq + 9q^2$

$= 25k^2 + 30kq + 9q^2$

81. Recall that a^3 means $(a)(a)(a)$, so $(m-5)^3 = (m-5)(m-5)(m-5)$. We'll start by finding

$(m-5)(m-5) = m^2 - 5m - 5m + 25$

$= m^2 - 10m + 25.$

Now multiply that result by $m - 5$.

$$
\begin{array}{rrrr}
m^2 & -\ 10m & +\ 25 & \\
 & & m & -\ 5 \\
\hline
-\ & 5m^2 & +\ 50m & -\ 125 \\
m^3 & -\ 10m^2 & +\ 25m & \\
\hline
m^3 & -\ 15m^2 & +\ 75m & -\ 125
\end{array}
$$

83. $(2a+1)^3 = (2a+1)(2a+1)(2a+1)$

$(2a+1)(2a+1) = 4a^2 + 2a + 2a + 1$

$= 4a^2 + 4a + 1$

Now multiply vertically.

$$
\begin{array}{rrrr}
4a^2 & +\ 4a & +\ 1 & \\
 & 2a & +\ 1 & \\
\hline
4a^2 & +\ 4a & +\ 1 & \\
8a^3 + & 8a^2 & +\ 2a & \\
\hline
8a^3 + & 12a^2 & +\ 6a & +\ 1
\end{array}
$$

85. $-3a(3a+1)(a-4)$

$= -3a(3a^2 - 12a + a - 4)$ $\quad$ *FOIL*

$= -3a(3a^2 - 11a - 4)$

$= -9a^3 + 33a^2 + 12a$

87. $7(4m-3)(2m+1)$

$= 7(8m^2 + 4m - 6m - 3)$ $\quad$ *FOIL*

$= 7(8m^2 - 2m - 3)$

$= 56m^2 - 14m - 21$

89. $(3r - 2s)^4 = (3r - 2s)^2(3r - 2s)^2$

First we find $(3r - 2s)^2$.

$$(3r - 2s)^2 = (3r - 2s)(3r - 2s)$$
$$= 9r^2 - 6rs - 6rs + 4s^2$$
$$= 9r^2 - 12rs + 4s^2$$

Now multiply this result by itself.

$$
\begin{array}{r}
9r^2 \quad - 12rs \quad + 4s^2 \\
9r^2 \quad - 12rs \quad + 4s^2 \\
\hline
36r^2s^2 \quad - 48rs^3 \quad + 16s^4 \\
- 108r^3s \quad + 144r^2s^2 \quad - 48rs^3 \\
81r^4 \quad - 108r^3s \quad + 36r^2s^2 \\
\hline
81r^4 \quad - 216r^3s \quad + 216r^2s^2 \quad - 96rs^3 \quad + 16s^4
\end{array}
$$

91. $3p^3(2p^2 + 5p)(p^3 + 2p + 1)$

$$= [3p^3(2p^2) + 3p^3(5p)](p^3 + 2p + 1)$$
$$\textit{Distributive property}$$
$$= (6p^5 + 15p^4)(p^3 + 2p + 1)$$

Now multiply vertically.

$$
\begin{array}{r}
p^3 \quad + 2p \quad +1 \\
6p^5 \quad + 15p^4 \\
\hline
15p^7 \quad + 30p^5 \quad + 15p^4 \\
6p^8 \quad + 12p^6 \quad + 6p^5 \\
\hline
6p^8 \quad + 15p^7 \quad + 12p^6 \quad + 36p^5 \quad + 15p^4
\end{array}
$$

93. $-2x^5(3x^2 + 2x - 5)(4x + 2)$

$$= \left[-2x^5(3x^2) + (-2x^5)(2x)\right.$$
$$\left. + (-2x^5)(-5)\right](4x + 2)$$
$$\textit{Distributive property}$$
$$= (-6x^7 - 4x^6 + 10x^5)(4x + 2)$$

Now multiply vertically.

$$
\begin{array}{r}
-6x^7 \quad - 4x^6 \quad + 10x^5 \\
4x \quad + 2 \\
\hline
- 12x^7 \quad - 8x^6 \quad + 20x^5 \\
-24x^8 \quad - 16x^7 \quad + 40x^6 \\
\hline
-24x^8 \quad - 28x^7 \quad + 32x^6 \quad + 20x^5
\end{array}
$$

95. The area $\mathcal{A}$ of the shaded region is the difference between the area of the larger square, which has sides of length $x + 7$, and the area of the smaller square, which has sides of length x.

$$\mathcal{A} = (x + 7)^2 - (x)^2$$
$$= (x + 7)(x + 7) - x^2$$
$$= (x^2 + 7x + 7x + 49) - x^2$$
$$= x^2 + 14x + 49 - x^2$$
$$= 14x + 49$$

97. The area $\mathcal{A}$ of the shaded region is the difference between the area of the circle, which has radius x, and the area of the square, which has sides of length 3.

$$\mathcal{A} = \pi r^2 - s^2$$
$$= \pi(x)^2 - (3)^2$$
$$= \pi x^2 - 9$$

99. $(3m)^2 = 3^2 m^2 = 9m^2$

101. $(-2r)^2 = (-2)^2 r^2 = 4r^2$

103. $(4x^2)^2 = 4^2(x^2)^2 = 16x^4$

5.6 Special Products

5.6 Now Try Exercises

N1. $(x + 5)^2 = x^2 + 2(x)(5) + 5^2$
$$= x^2 + 10x + 25$$

N2. **(a)** $(3x - 1)^2 = (3x)^2 - 2(3x)(1) + 1^2$
$$= 9x^2 - 6x + 1$$

(b) $(4p - 5q)^2 = (4p)^2 - 2(4p)(5q) + (5q)^2$
$$= 16p^2 - 40pq + 25q^2$$

(c) $(6t - \frac{1}{3})^2 = (6t)^2 - 2(6t)(\frac{1}{3}) + (\frac{1}{3})^2$
$$= 36t^2 - 4t + \frac{1}{9}$$

(d) $m(2m + 3)^2 = m(4m^2 + 12m + 9)$
$$= 4m^3 + 12m^2 + 9m$$

N3. $(t + 10)(t - 10) = t^2 - 10^2$
$$= t^2 - 100$$

N4. **(a)** $(4x - 6)(4x + 6) = (4x)^2 - 6^2$
$$= 16x^2 - 36$$

(b) $(5r - \frac{4}{5})(5r + \frac{4}{5}) = (5r)^2 - (\frac{4}{5})^2$
$$= 25r^2 - \frac{16}{25}$$

(c) $y(3y + 1)(3y - 1)$

First use the rule for the product of the sum and difference of two terms.

$$(3y + 1)(3y - 1) = (3y)^2 - 1^2$$
$$= 9y^2 - 1$$

Now multiply this result by y.

$$y(9y^2 - 1) = y(9y^2) + y(-1)$$
$$= 9y^3 - y$$

N5. $(2m - 1)^3$

Since $(2m - 1)^3 = (2m - 1)^2(2m - 1)$, the first step is to find the product $(2m - 1)^2$.

$$(2m - 1)^2 = (2m)^2 - 2(2m)(1) + 1^2$$
$$= 4m^2 - 4m + 1$$

Now multiply this result by $2m - 1$.

$$(2m - 1)^3$$
$$= (2m - 1)(4m^2 - 4m + 1)$$
$$= 8m^3 - 8m^2 + 2m - 4m^2 + 4m - 1$$
$$= 8m^3 - 12m^2 + 6m - 1$$

5.6 Section Exercises

1. $(4x + 3)^2$

(a) The square of the first term is
$$(4x)^2 = (4x)(4x) = 16x^2.$$

(b) Twice the product of the two terms is
$$2(4x)(3) = 24x.$$

(c) The square of the last term is
$$3^2 = 9.$$

(d) The final product is the trinomial
$$16x^2 + 24x + 9.$$

In Exercises 3–20, use one of the following formulas for the square of a binomial:
$$(x + y)^2 = x^2 + 2xy + y^2$$
$$(x - y)^2 = x^2 - 2xy + y^2$$

3. $(m + 2)^2 = m^2 + 2(m)(2) + 2^2$
$$= m^2 + 4m + 4$$

5. $(r - 3)^2 = r^2 - 2(r)(3) + 3^2$
$$= r^2 - 6r + 9$$

7. $(x + 2y)^2 = x^2 + 2(x)(2y) + (2y)^2$
$$= x^2 + 4xy + 4y^2$$

9. $(5p + 2q)^2 = (5p)^2 + 2(5p)(2q) + (2q)^2$
$$= 25p^2 + 20pq + 4q^2$$

11. $(4a + 5b)^2 = (4a)^2 + 2(4a)(5b) + (5b)^2$
$$= 16a^2 + 40ab + 25b^2$$

13. $(6m - \frac{4}{5}n)^2 = (6m)^2 - 2(6m)(\frac{4}{5}n) + (\frac{4}{5}n)^2$
$$= 36m^2 - \frac{48}{5}mn + \frac{16}{25}n^2$$

15. $t(3t - 1)^2$
$$(3t - 1)^2 = (3t)^2 - 2(3t)(1) + 1^2$$
$$= 9t^2 - 6t + 1$$

Now multiply by t.
$$t(9t^2 - 6t + 1) = 9t^3 - 6t^2 + t$$

17. $3t(4t + 1)^2$
$$(4t + 1)^2 = (4t)^2 + 2(4t)(1) + 1^2$$
$$= 16t^2 + 8t + 1$$

Now multiply by $3t$.
$$3t(16t^2 + 8t + 1) = 48t^3 + 24t^2 + 3t$$

19. $-(4r - 2)^2$
$$(4r - 2)^2 = (4r)^2 - 2(4r)(2) + 2^2$$
$$= 16r^2 - 16r + 4$$

Now multiply by -1.
$$-1(16r^2 - 16r + 4) = -16r^2 + 16r - 4$$

21. **(a)** $7x(7x) = 49x^2$

(b) $7x(-3y) + 3y(7x)$
$$= -21xy + 21xy = 0$$

(c) $3y(-3y) = -9y^2$

(d) $49x^2 - 9y^2$

The sum found in part (b) is omitted because it is 0. Adding 0, the identity element for addition, would not change the answer.

In Exercises 23–40, use the formula for the product of the sum and difference of two terms.
$$(x + y)(x - y) = x^2 - y^2$$

23. $(k + 5)(k - 5) = k^2 - 5^2$
$$= k^2 - 25$$

25. $(4 - 3t)(4 + 3t) = 4^2 - (3t)^2$
$$= 16 - 9t^2$$

27. $(5x + 2)(5x - 2) = (5x)^2 - 2^2$
$$= 25x^2 - 4$$

29. $(5y + 3x)(5y - 3x) = (5y)^2 - (3x)^2$
$$= 25y^2 - 9x^2$$

31. $(10x + 3y)(10x - 3y) = (10x)^2 - (3y)^2$
$$= 100x^2 - 9y^2$$

33. $(2x^2 - 5)(2x^2 + 5) = (2x^2)^2 - 5^2$
$$= 4x^4 - 25$$

35. $(\frac{3}{4} - x)(\frac{3}{4} + x) = (\frac{3}{4})^2 - x^2$
$$= \frac{9}{16} - x^2$$

37. $(9y + \frac{2}{3})(9y - \frac{2}{3}) = (9y)^2 - (\frac{2}{3})^2$
$$= 81y^2 - \frac{4}{9}$$

39. $q(5q - 1)(5q + 1)$
$$(5q - 1)(5q + 1) = (5q)^2 - 1^2$$
$$= 25q^2 - 1$$

Now multiply by q.
$$q(25q^2 - 1) = 25q^3 - q$$

41. No. In general, $(a + b)^2$ equals $a^2 + 2ab + b^2$, which is not equivalent to $a^2 + b^2$.

43. $(x + 1)^3$
$$= (x + 1)^2(x + 1) \qquad a^3 = a^2 \cdot a$$
$$= (x^2 + 2x + 1)(x + 1) \qquad \text{Square the binomial.}$$
$$= x^3 + 2x^2 + x + x^2 + 2x + 1 \qquad \text{Multiply polynomials.}$$
$$= x^3 + 3x^2 + 3x + 1 \qquad \text{Combine like terms.}$$

45. $(t-3)^3$

$\quad = (t-3)^2(t-3)$ $\qquad a^3 = a^2 \cdot a$

$\quad = (t^2 - 6t + 9)(t-3)$ $\quad$ *Square the binomial.*

$\quad = t^3 - 6t^2 + 9t - 3t^2 + 18t - 27$ $\quad$ *Multiply polynomials.*

$\quad = t^3 - 9t^2 + 27t - 27$ $\quad$ *Combine like terms.*

47. $(r+5)^3$

$\quad = (r+5)^2(r+5)$ $\qquad a^3 = a^2 \cdot a$

$\quad = (r^2 + 10r + 25)(r+5)$ $\quad$ *Square the binomial.*

$\quad = r^3 + 10r^2 + 25r + 5r^2 + 50r + 125$

$\qquad\qquad$ *Multiply polynomials.*

$\quad = r^3 + 15r^2 + 75r + 125$ $\quad$ *Combine like terms.*

49. $(2a+1)^3$

$\quad = (2a+1)^2(2a+1)$ $\qquad a^3 = a^2 \cdot a$

$\quad = (4a^2 + 4a + 1)(2a+1)$ $\quad$ *Square the binomial.*

$\quad = 8a^3 + 8a^2 + 2a + 4a^2 + 4a + 1$ $\quad$ *Multiply polynomials.*

$\quad = 8a^3 + 12a^2 + 6a + 1$ $\quad$ *Combine like terms.*

51. $(4x-1)^4$

$\quad = (4x-1)^2(4x-1)^2$ $\qquad a^4 = a^2 \cdot a^2$

$\quad = (16x^2 - 8x + 1)(16x^2 - 8x + 1)$

$\qquad\qquad$ *Square each binomial.*

$\quad = 256x^4 - 128x^3 + 16x^2 - 128x^3 + 64x^2$

$\quad\quad - 8x + 16x^2 - 8x + 1$

$\qquad\qquad$ *Multiply polynomials.*

$\quad = 256x^4 - 256x^3 + 96x^2 - 16x + 1$

$\qquad\qquad$ *Combine like terms.*

53. $(3r-2t)^4$

$\quad = (3r-2t)^2(3r-2t)^2$ $\qquad a^4 = a^2 \cdot a^2$

$\quad = (9r^2 - 12rt + 4t^2)(9r^2 - 12rt + 4t^2)$

$\qquad\qquad$ *Square each binomial.*

$\quad = 81r^4 - 108r^3t + 36r^2t^2 - 108r^3t$

$\quad\quad + 144r^2t^2 - 48rt^3 + 36r^2t^2 - 48rt^3 + 16t^4$

$\qquad\qquad$ *Multiply polynomials.*

$\quad = 81r^4 - 216r^3t + 216r^2t^2 - 96rt^3 + 16t^4$

$\qquad\qquad$ *Combine like terms.*

55. $2x(x+1)^3$

$\quad = 2x(x+1)(x+1)^2$

$\quad = 2x(x+1)(x^2 + 2x + 1)$

$\quad = 2x(x^3 + 2x^2 + x + x^2 + 2x + 1)$

$\quad = 2x(x^3 + 3x^2 + 3x + 1)$

$\quad = 2x^4 + 6x^3 + 6x^2 + 2x$

57. $-4t(t+3)^3$

$\quad = -4t(t+3)(t+3)^2$

$\quad = -4t(t+3)(t^2 + 6t + 9)$

$\quad = -4t(t^3 + 6t^2 + 9t + 3t^2 + 18t + 27)$

$\quad = -4t(t^3 + 9t^2 + 27t + 27)$

$\quad = -4t^4 - 36t^3 - 108t^2 - 108t$

59. $(x+y)^2(x-y)^2$

$\quad = [(x+y)(x-y)]^2$ $\qquad a^2b^2 = (ab)^2$

$\quad = (x^2 - y^2)^2$

$\quad = (x^2)^2 - 2x^2y^2 + (y^2)^2$

$\quad = x^4 - 2x^2y^2 + y^4$

Another method is to expand both binomials first and then multiply those trinomials.

61. The large square has sides of length $a + b$, so its area is $(a+b)^2$.

62. The red square has sides of length a, so its area is a^2.

63. Each blue rectangle has length a and width b, so each has an area of ab. Thus, the sum of the areas of the blue rectangles is

$$ab + ab = 2ab.$$

64. The yellow square has sides of length b, so its area is b^2.

65. Sum $= a^2 + 2ab + b^2$

66. The area of the largest square equals the sum of the areas of the two smaller squares and the two rectangles. Therefore, $(a+b)^2$ must equal $a^2 + 2ab + b^2$.

67. $35^2 = (35)(35)$

$$\begin{array}{r} 35 \\ 35 \\ \hline 175 \\ 105 \\ \hline 1225 \end{array}$$

68. $(a+b)^2 = a^2 + 2ab + b^2$

$\quad (30+5)^2 = 30^2 + 2(30)(5) + 5^2$

69. $30^2 + 2(30)(5) + 5^2$

$\quad = 900 + 60(5) + 25$

$\quad = 900 + 300 + 25$

$\quad = 1225$

70. The answers are equal.

71. $101 \times 99 = (100+1)(100-1)$

$\quad = 100^2 - 1^2$

$\quad = 10{,}000 - 1$

$\quad = 9999$

73. $201 \times 199 = (200 + 1)(200 - 1)$

$= 200^2 - 1^2$

$= 40,000 - 1$

$= 39,999$

75. $20\frac{1}{2} \times 19\frac{1}{2} = (20 + \frac{1}{2})(20 - \frac{1}{2})$

$= 20^2 - (\frac{1}{2})^2$

$= 400 - \frac{1}{4}$

$= 399\frac{3}{4}$

77. Use the formula for area the of a triangle, $A = \frac{1}{2}bh$, with $b = m + 2n$ and $h = m - 2n$.

$A = \frac{1}{2}(m + 2n)(m - 2n)$

$= \frac{1}{2}[m^2 - (2n)^2]$

$= \frac{1}{2}(m^2 - 4n^2)$

$= \frac{1}{2}m^2 - 2n^2$

79. Use the formula for the area of a parallelogram, $A = bh$, with $b = 3a + 2$ and $h = 3a - 2$.

$A = (3a + 2)(3a - 2)$

$= (3a)^2 - 2^2$

$= 9a^2 - 4$

81. Use the formula for the area of a circle, $A = \pi r^2$, with $r = x + 2$.

$A = \pi(x + 2)^2$

$= \pi(x^2 + 4x + 4)$

$= \pi x^2 + 4\pi x + 4\pi$

83. Use the formula for the volume of a cube, $V = e^3$, with $e = x + 2$.

$V = (x + 2)^3$

$= (x + 2)^2(x + 2)$

$= (x^2 + 4x + 4)(x + 2)$

$= x^3 + 4x^2 + 4x + 2x^2 + 8x + 8$

$= x^3 + 6x^2 + 12x + 8$

85. $\frac{1}{2p}(4p^2 + 2p + 8)$

$= \frac{1}{2p}(4p^2) + \frac{1}{2p}(2p) + \frac{1}{2p}(8)$

$= 2p + 1 + \frac{4}{p}$

87. $\frac{1}{3m}(m^3 + 9m^2 - 6m)$

$= \frac{1}{3m}(m^3) + \frac{1}{3m}(9m^2) + \frac{1}{3m}(-6m)$

$= \frac{m^2}{3} + 3m - 2$

89. $-3k(8k^2 - 12k + 2)$

$= -3k(8k^2) + (-3k)(-12k) + (-3k)(2)$

$= -24k^3 + 36k^2 - 6k$

91. $(-2k + 1)(8k^2 + 9k + 3)$

$= -2k(8k^2) + (-2k)(9k) + (-2k)(3)$

$\quad + 1(8k^2) + 1(9k) + 1(3)$

$= -16k^3 - 18k^2 - 6k + 8k^2 + 9k + 3$

$= -16k^3 - 10k^2 + 3k + 3$

93. Subtract.

$$
\begin{array}{rrrr}
5t^2 & + 2t & - 6 \\
5t^2 & - 3t & - 9 \\
\hline
\end{array}
$$

Change all signs in the second row, and then add.

$$
\begin{array}{rrrr}
5t^2 & + 2t & - 6 \\
-5t^2 & + 3t & + 9 \\
\hline
& 5t & + 3
\end{array}
$$

5.7 Dividing Polynomials

5.7 Now Try Exercises

N1. $\dfrac{16a^6 - 12a^4}{4a^2}$

$= \dfrac{16a^6}{4a^2} - \dfrac{12a^4}{4a^2}$

$= 4a^4 - 3a^2$

N2. $\dfrac{36x^5 + 24x^4 - 12x^3}{6x^4}$

$= \dfrac{36x^5}{6x^4} + \dfrac{24x^4}{6x^4} - \dfrac{12x^3}{6x^4}$

$= 6x + 4 - \dfrac{2}{x}$

N3. Divide $7y^4 - 40y^5 + 100y^2$ by $-5y^2$.

$\dfrac{7y^4 - 40y^5 + 100y^2}{-5y^2}$

$= \dfrac{7y^4}{-5y^2} - \dfrac{40y^5}{-5y^2} + \dfrac{100y^2}{-5y^2}$

$= -\dfrac{7y^2}{5} + 8y^3 - 20 \quad \text{or} \quad 8y^3 - \dfrac{7y^2}{5} - 20$

N4. Divide $35m^5n^4 - 49m^2n^3 + 12mn$ by $7m^2n$.

$\dfrac{35m^5n^4 - 49m^2n^3 + 12mn}{7m^2n}$

$= \dfrac{35m^5n^4}{7m^2n} - \dfrac{49m^2n^3}{7m^2n} + \dfrac{12mn}{7m^2n}$

$= 5m^3n^3 - 7n^2 + \dfrac{12}{7m}$

N5. Divide. $\dfrac{4x^2 + x - 18}{x - 2}$

Step 1 (see the division following Step 4)
$4x^2$ divided by x is $4x$;
$4x(x - 2) = 4x^2 - 8x$.

Step 2
Subtract $4x^2 - 8x$ from $4x^2 + x$.
Bring down -18.

Step 3

$9x$ divided by x is 9;

$9(x - 2) = 9x - 18$.

Step 4

Subtract $9x - 18$ from $9x - 18$.

The remainder is 0.

The answer is the quotient, $4x + 9$.

$$
\begin{array}{r}
4x + 9 \\
x - 2 \overline{\smash{)}4x^2 + x - 18} \\
\underline{4x^2 - 8x} \\
9x - 18 \\
\underline{9x - 18} \\
0
\end{array}
$$

Check: Multiply the divisor, $x - 2$, by the quotient, $4x + 9$. The product must be the original dividend, $4x^2 + x - 18$.

$$(x - 2)(4x + 9)$$
$$= 4x^2 - 8x + 9x - 18$$
$$= 4x^2 + x - 18$$

N6. Divide $6k^3 - 20k - k^2 + 1$ by $2k - 3$.

Write the numerator in descending powers and divide.

$$
\begin{array}{r}
3k^2 + 4k - 4 \\
2k - 3 \overline{\smash{)}6k^3 - k^2 - 20k + 1} \\
\underline{6k^3 - 9k^2} \\
8k^2 - 20k \\
\underline{8k^2 - 12k} \\
-8k + 1 \\
\underline{-8k + 12} \\
-11
\end{array}
$$

$\dfrac{6k^3}{2k} = 3k^2$

$\dfrac{8k^2}{2k} = 4k$

$\dfrac{-8k}{2k} = -4$

The remainder is -11.

The answer is $3k^2 + 4k - 4 + \dfrac{-11}{2k - 3}$.

N7. Divide $m^3 - 1000$ by $m - 10$.

Use 0 as the coefficient of the missing m^2- and m-terms.

$$
\begin{array}{r}
m^2 + 10m + 100 \\
m - 10 \overline{\smash{)}m^3 + 0m^2 + 0m - 1000} \\
\underline{m^3 - 10m^2} \\
10m^2 + 0m \\
\underline{10m^2 - 100m} \\
100m - 1000 \\
\underline{100m - 1000} \\
0
\end{array}
$$

$(m^3 - 1000)$ divided by $(m - 10)$ is

$$m^2 + 10m + 100.$$

N8. Divide $y^4 - 5y^3 + 6y^2 + y - 4$ by $y^2 + 2$.

$$
\begin{array}{r}
y^2 - 5y + 4 \\
y^2 + 0y + 2 \overline{\smash{)}y^4 - 5y^3 + 6y^2 + y - 4} \\
\underline{y^4 + 0y^3 + 2y^2} \\
-5y^3 + 4y^2 + y \\
\underline{-5y^3 - 0y^2 - 10y} \\
4y^2 + 11y - 4 \\
\underline{4y^2 + 0y + 8} \\
11y - 12
\end{array}
$$

$(y^4 - 5y^3 + 6y^2 + y - 4)$ divided by $(y^2 + 2)$ is

$y^2 - 5y + 4 + \dfrac{11y - 12}{y^2 + 3}$.

N9. Divide $10x^3 + 21x^2 + 5x - 8$ by $2x + 4$.

$$
\begin{array}{r}
5x^2 + \tfrac{1}{2}x + \tfrac{3}{2} \\
2x + 4 \overline{\smash{)}10x^3 + 21x^2 + 5x - 8} \\
\underline{10x^3 + 20x^2} \\
x^2 + 5x \\
\underline{x^2 + 2x} \\
3x - 8 \\
\underline{3x + 6} \\
-14
\end{array}
$$

$\dfrac{x^2}{2x} = \tfrac{1}{2}x$

$\dfrac{3x}{2x} = \tfrac{3}{2}$

$(10x^3 + 21x^2 + 5x - 8)$ divided by $(2x + 4)$ is

$$5x^2 + \frac{1}{2}x + \frac{3}{2} + \frac{-14}{2x + 4}.$$

5.7   Section Exercises

1. In the statement $\dfrac{10x^2 + 8}{2} = 5x^2 + 4$, $\underline{10x^2 + 8}$ is the dividend, $\underline{2}$ is the divisor, and $\underline{5x^2 + 4}$ is the quotient.

3. To check the division shown in Exercise 1, multiply $\underline{5x^2 + 4}$ by $\underline{2}$ (or $\underline{2}$ by $\underline{5x^2 + 4}$) and show that the product is $\underline{10x^2 + 8}$.

5. In this section, we are dividing a polynomial by a monomial. The problem

$$\frac{16m^3 - 12m^2}{4m}$$

is an example of such a division. However, in the problem

$$\frac{4m}{16m^3 - 12m^2},$$

we are dividing a monomial by a binomial. Therefore, the methods of this section do not apply.

7. $\dfrac{60x^4 - 20x^2 + 10x}{2x}$

$= \dfrac{60x^4}{2x} - \dfrac{20x^2}{2x} + \dfrac{10x}{2x}$

$= \dfrac{60}{2}x^{4-1} - \dfrac{20}{2}x^{2-1} + \dfrac{10}{2}$

$= 30x^3 - 10x + 5$

9. $\dfrac{20m^5 - 10m^4 + 5m^2}{5m^2}$

$= \dfrac{20m^5}{5m^2} - \dfrac{10m^4}{5m^2} + \dfrac{5m^2}{5m^2}$

$= \dfrac{20}{5}m^{5-2} - \dfrac{10}{5}m^{4-2} + \dfrac{5}{5}$

$= 4m^3 - 2m^2 + 1$

11. $\dfrac{8t^5 - 4t^3 + 4t^2}{2t}$

$= \dfrac{8t^5}{2t} - \dfrac{4t^3}{2t} + \dfrac{4t^2}{2t}$

$= 4t^4 - 2t^2 + 2t$

13. $\dfrac{4a^5 - 4a^2 + 8}{4a}$

$= \dfrac{4a^5}{4a} - \dfrac{4a^2}{4a} + \dfrac{8}{4a}$

$= a^4 - a + \dfrac{2}{a}$

15. $\dfrac{18p^5 + 12p^3 - 6p^2}{-6p^3}$

$= \dfrac{18p^5}{-6p^3} + \dfrac{12p^3}{-6p^3} - \dfrac{6p^2}{-6p^3}$

$= -3p^2 - 2 + \dfrac{1}{p}$

17. $\dfrac{-7r^7 + 6r^5 - r^4}{-r^5}$

$= \dfrac{-7r^7}{-r^5} + \dfrac{6r^5}{-r^5} - \dfrac{r^4}{-r^5}$

$= 7r^2 - 6 + \dfrac{1}{r}$

19. $\dfrac{12x^5 - 9x^4 + 6x^3}{3x^2}$

$= \dfrac{12x^5}{3x^2} - \dfrac{9x^4}{3x^2} + \dfrac{6x^3}{3x^2}$

$= 4x^3 - 3x^2 + 2x$

21. $\dfrac{3x^2 + 15x^3 - 27x^4}{3x^2}$

$= \dfrac{3x^2}{3x^2} + \dfrac{15x^3}{3x^2} - \dfrac{27x^4}{3x^2}$

$= 1 + 5x - 9x^2$ or $-9x^2 + 5x + 1$

23. $\dfrac{36x + 24x^2 + 6x^3}{3x^2}$

$= \dfrac{36x}{3x^2} + \dfrac{24x^2}{3x^2} + \dfrac{6x^3}{3x^2}$

$= \dfrac{12}{x} + 8 + 2x$ or $2x + 8 + \dfrac{12}{x}$

25. $\dfrac{4x^4 + 3x^3 + 2x}{3x^2}$

$= \dfrac{4x^4}{3x^2} + \dfrac{3x^3}{3x^2} + \dfrac{2x}{3x^2}$

$= \dfrac{4x^2}{3} + x + \dfrac{2}{3x}$

27. $\dfrac{-81x^5 + 30x^4 + 12x^2}{3x^2}$

$= \dfrac{-81x^5}{3x^2} + \dfrac{30x^4}{3x^2} + \dfrac{12x^2}{3x^2}$

$= -27x^3 + 10x^2 + 4$

29. $\dfrac{-27r^4 + 36r^3 - 6r^2 - 26r + 2}{-3r}$

$= \dfrac{-27r^4}{-3r} + \dfrac{36r^3}{-3r} - \dfrac{6r^2}{-3r} - \dfrac{26r}{-3r} + \dfrac{2}{-3r}$

$= 9r^3 - 12r^2 + 2r + \dfrac{26}{3} - \dfrac{2}{3r}$

31. $\dfrac{2m^5 - 6m^4 + 8m^2}{-2m^3}$

$= \dfrac{2m^5}{-2m^3} - \dfrac{6m^4}{-2m^3} + \dfrac{8m^2}{-2m^3}$

$= -m^2 + 3m - \dfrac{4}{m}$

33. $(20a^4 - 15a^5 + 25a^3) \div (5a^4)$

$= \dfrac{20a^4 - 15a^5 + 25a^3}{5a^4}$

$= \dfrac{20a^4}{5a^4} - \dfrac{15a^5}{5a^4} + \dfrac{25a^3}{5a^4}$

$= 4 - 3a + \dfrac{5}{a}$ or $-3a + 4 + \dfrac{5}{a}$

35. $(120x^{11} - 60x^{10} + 140x^9 - 100x^8) \div (10x^{12})$

$= \dfrac{120x^{11} - 60x^{10} + 140x^9 - 100x^8}{10x^{12}}$

$= \dfrac{120x^{11}}{10x^{12}} - \dfrac{60x^{10}}{10x^{12}} + \dfrac{140x^9}{10x^{12}} - \dfrac{100x^8}{10x^{12}}$

$= \dfrac{12}{x} - \dfrac{6}{x^2} + \dfrac{14}{x^3} - \dfrac{10}{x^4}$

37. $(120x^5y^4 - 80x^2y^3 + 40x^2y^4 - 20x^5y^3)$

$\div (20xy^2)$

$= \dfrac{120x^5y^4 - 80x^2y^3 + 40x^2y^4 - 20x^5y^3}{20xy^2}$

$= \dfrac{120x^5y^4}{20xy^2} - \dfrac{80x^2y^3}{20xy^2} + \dfrac{40x^2y^4}{20xy^2} - \dfrac{20x^5y^3}{20xy^2}$

$= 6x^4y^2 - 4xy + 2xy^2 - x^4y$

39.
$$
2 \enclose{longdiv}{\overset{1423}{2846}}
$$

40. $1423 = (1 \times 10^3) + (4 \times 10^2)$

$\qquad + (2 \times 10^1) + (3 \times 10^0)$

41. $\dfrac{2x^3 + 8x^2 + 4x + 6}{2}$

$= \dfrac{2x^3}{2} + \dfrac{8x^2}{2} + \dfrac{4x}{2} + \dfrac{6}{2}$

$= x^3 + 4x^2 + 2x + 3$

42. They are similar in that the coefficients of powers of 10 are equal to the coefficients of the powers of x. They are different in that one is a constant while the other is a polynomial. They are equal if $x = 10$ (the base of our decimal system).

43. $\dfrac{x^2 - x - 6}{x - 3}$

$$
\begin{array}{r}
x + 2 \\
x - 3 \enclose{longdiv}{x^2 - x - 6} \\
\underline{x^2 - 3x} \\
2x - 6 \\
\underline{2x - 6} \\
0
\end{array}
$$

The remainder is 0. The answer is the quotient, $x + 2$.

45. $\dfrac{2y^2 + 9y - 35}{y + 7}$

$$
\begin{array}{r}
2y - 5 \\
y + 7 \enclose{longdiv}{2y^2 + 9y - 35} \\
\underline{2y^2 + 14y} \\
-5y - 35 \\
\underline{-5y - 35} \\
0
\end{array}
$$

The remainder is 0. The answer is the quotient, $2y - 5$.

47. $\dfrac{p^2 + 2p + 20}{p + 6}$

$$
\begin{array}{r}
p - 4 \\
p + 6 \enclose{longdiv}{p^2 + 2p + 20} \\
\underline{p^2 + 6p} \\
-4p + 20 \\
\underline{-4p - 24} \\
44
\end{array}
$$

The remainder is 44. Write the remainder as the numerator of a fraction that has the divisor $p + 6$ as its denominator. The answer is

$$
p - 4 + \frac{44}{p + 6}.
$$

49. $\dfrac{12m^2 - 20m + 3}{2m - 3}$

$$
\begin{array}{r}
6m - 1 \\
2m - 3 \enclose{longdiv}{12m^2 - 20m + 3} \\
\underline{12m^2 - 18m} \\
-2m + 3 \\
\underline{-2m + 3} \\
0
\end{array}
$$

The remainder is 0. The answer is the quotient, $6m - 1$.

51. $\dfrac{4a^2 - 22a + 32}{2a + 3}$

$$
\begin{array}{r}
2a - 14 \\
2a + 3 \enclose{longdiv}{4a^2 - 22a + 32} \\
\underline{4a^2 + 6a} \\
-28a + 32 \\
\underline{-28a - 42} \\
74
\end{array}
$$

The remainder is 74. The answer is

$$
2a - 14 + \frac{74}{2a + 3}.
$$

53. $\dfrac{8x^3 - 10x^2 - x + 3}{2x + 1}$

$$
\begin{array}{r}
4x^2 - 7x + 3 \\
2x + 1 \enclose{longdiv}{8x^3 - 10x^2 - x + 3} \\
\underline{8x^3 + 4x^2} \\
-14x^2 - x \\
\underline{-14x^2 - 7x} \\
6x + 3 \\
\underline{6x + 3} \\
0
\end{array}
$$

The remainder is 0. The answer is the quotient,

$$
4x^2 - 7x + 3.
$$

55. $\dfrac{8k^4 - 12k^3 - 2k^2 + 7k - 6}{2k - 3}$

$$
\begin{array}{r}
4k^3 \qquad\quad - k \; +2 \\
2k - 3 \;\overline{\big)\; 8k^4 \;-12k^3 \;-2k^2 \;+7k \;-6} \\
\underline{8k^4 \;-12k^3} \qquad\qquad\qquad \\
-2k^2 \;+7k \\
\underline{-2k^2 \;+3k} \\
+4k \;-6 \\
\underline{+4k \;-6} \\
0
\end{array}
$$

The remainder is 0. The answer is the quotient,

$4k^3 - k + 2.$

57. $\dfrac{5y^4 + 5y^3 + 2y^2 - y - 8}{y + 1}$

$$
\begin{array}{r}
5y^3 \qquad\quad + 2y \; -3 \\
y + 1 \;\overline{\big)\; 5y^4 \;+5y^3 \;+2y^2 \;- y \;-8} \\
\underline{5y^4 \;+5y^3} \qquad\qquad\qquad \\
2y^2 \;- y \\
\underline{2y^2 \;+2y} \\
-3y \;-8 \\
\underline{-3y \;-3} \\
-5
\end{array}
$$

The remainder is -5.

The quotient is $5y^3 + 2y - 3$.

The answer is $5y^3 + 2y - 3 + \dfrac{-5}{y + 1}$.

59. $\dfrac{3k^3 - 4k^2 - 6k + 10}{k - 2}$

$$
\begin{array}{r}
3k^2 \;+2k \;- 2 \\
k - 2 \;\overline{\big)\; 3k^3 \;-4k^2 \;-6k \;+10} \\
\underline{3k^3 \;-6k^2} \qquad\qquad\qquad \\
2k^2 \;-6k \\
\underline{2k^2 \;-4k} \\
-2k \;+10 \\
\underline{-2k \;+4} \\
6
\end{array}
$$

The remainder is 6.

The quotient is $3k^2 + 2k - 2$.

The answer is $3k^2 + 2k - 2 + \dfrac{6}{k - 2}$.

61. $\dfrac{6p^4 - 16p^3 + 15p^2 - 5p + 10}{3p + 1}$

$$
\begin{array}{r}
2p^3 \;- 6p^2 \;+7p \;- 4 \\
3p + 1 \;\overline{\big)\; 6p^4 \;-16p^3 \;+15p^2 \;-5p \;+10} \\
\underline{6p^4 \;+ 2p^3} \qquad\qquad\qquad\qquad \\
-18p^3 \;+15p^2 \\
\underline{-18p^3 \;- 6p^2} \\
21p^2 \;- 5p \\
\underline{21p^2 \;+ 7p} \\
-12p \;+10 \\
\underline{-12p \;- 4} \\
14
\end{array}
$$

The remainder is 14.

The quotient is $2p^3 - 6p^2 + 7p - 4$.

The answer is $2p^3 - 6p^2 + 7p - 4 + \dfrac{14}{3p + 1}$.

63. $(x^3 + 2x^2 - 3) \div (x - 1)$

Use 0 as the coefficient of the missing x-term.

$$
\begin{array}{r}
x^2 \;+ 3x \;+ 3 \\
x - 1 \;\overline{\big)\; x^3 \;+ 2x^2 \;+ 0x \;- 3} \\
\underline{x^3 \;- x^2} \qquad\qquad\qquad \\
3x^2 \;+ 0x \\
\underline{3x^2 \;- 3x} \\
3x \;- 3 \\
\underline{3x \;- 3} \\
0
\end{array}
$$

$(x^3 + 2x^2 - 3) \div (x - 1) = x^2 + 3x + 3$

65. $(2x^3 + x + 2) \div (x + 1)$

Use 0 as the coefficient of the missing x^2-term.

$$
\begin{array}{r}
2x^2 \;- 2x \;+ 3 \\
x + 1 \;\overline{\big)\; 2x^3 \;+ 0x^2 \;+ x \;+ 2} \\
\underline{2x^3 \;+ 2x^2} \qquad\qquad\qquad \\
-2x^2 \;+ x \\
\underline{-2x^2 \;- 2x} \\
3x \;+ 2 \\
\underline{3x \;+ 3} \\
-1
\end{array}
$$

$(2x^3 + x + 2) \div (x + 1) = 2x^2 - 2x + 3 + \dfrac{-1}{x + 1}$

67. $\dfrac{5 - 2r^2 + r^4}{r^2 - 1}$

Use 0 as the coefficient for the missing terms. Rearrange terms of the dividend in descending powers.

$$
\begin{array}{r}
r^2 \qquad\quad - 1 \\
r^2 + 0r - 1\ \overline{\smash{\big)}\ r^4 + 0r^3 - 2r^2 + 0r + 5} \\
\underline{r^4 + 0r^3 - r^2 \qquad\qquad} \\
-r^2 + 0r + 5 \\
\underline{-r^2 + 0r + 1} \\
4
\end{array}
$$

The remainder is 4.

The quotient is $r^2 - 1$.

The answer is $r^2 - 1 + \dfrac{4}{r^2 - 1}$.

69. $\dfrac{-4x + 3x^3 + 2}{x - 1}$

Use 0 as the coefficient for the missing terms. Rearrange terms of the dividend in descending powers.

$$
\begin{array}{r}
3x^2 + 3x - 1 \\
x - 1\ \overline{\smash{\big)}\ 3x^3 + 0x^2 - 4x + 2} \\
\underline{3x^3 - 3x^2 \qquad\qquad} \\
3x^2 - 4x \\
\underline{3x^2 - 3x} \\
-x + 2 \\
\underline{-x + 1} \\
1
\end{array}
$$

$$
\dfrac{-4x + 3x^3 + 2}{x - 1} = 3x^2 + 3x - 1 + \dfrac{1}{x - 1}
$$

71. $\dfrac{y^3 + 1}{y + 1}$

$$
\begin{array}{r}
y^2 - y + 1 \\
y + 1\ \overline{\smash{\big)}\ y^3 + 0y^2 + 0y + 1} \\
\underline{y^3 + y^2 \qquad\qquad} \\
-y^2 + 0y \\
\underline{-y^2 - y} \\
y + 1 \\
\underline{y + 1} \\
0
\end{array}
$$

The remainder is 0. The answer is the quotient,

$$y^2 - y + 1.$$

73. $\dfrac{a^4 - 1}{a^2 - 1}$

$$
\begin{array}{r}
a^2 \qquad\qquad + 1 \\
a^2 + 0a - 1\ \overline{\smash{\big)}\ a^4 + 0a^3 + 0a^2 + 0a - 1} \\
\underline{a^4 + 0a^3 - a^2 \qquad\qquad\qquad} \\
a^2 + 0a - 1 \\
\underline{a^2 + 0a - 1} \\
0
\end{array}
$$

The remainder is 0. The answer is the quotient,

$$a^2 + 1.$$

75. $\dfrac{x^4 - 4x^3 + 5x^2 - 3x + 2}{x^2 + 3}$

$$
\begin{array}{r}
x^2 - 4x + 2 \\
x^2 + 0x + 3\ \overline{\smash{\big)}\ x^4 - 4x^3 + 5x^2 - 3x + 2} \\
\underline{x^4 + 0x^3 + 3x^2 \qquad\qquad\qquad} \\
-4x^3 + 2x^2 - 3x \\
\underline{-4x^3 + 0x^2 - 12x} \\
2x^2 + 9x + 2 \\
\underline{2x^2 + 0x + 6} \\
9x - 4
\end{array}
$$

$$
\dfrac{x^4 - 4x^3 + 5x^2 - 3x + 2}{x^2 + 3}
$$

$$
= x^2 - 4x + 2 + \dfrac{9x - 4}{x^2 + 3}
$$

77. $\dfrac{2x^5 + 9x^4 + 8x^3 + 10x^2 + 14x + 5}{2x^2 + 3x + 1}$

$$
\begin{array}{r}
x^3 + 3x^2 - x + 5 \\
2x^2 + 3x + 1\ \overline{\smash{\big)}\ 2x^5 + 9x^4 + 8x^3 + 10x^2 + 14x + 5} \\
\underline{2x^5 + 3x^4 + x^3 \qquad\qquad\qquad\qquad} \\
6x^4 + 7x^3 + 10x^2 \\
\underline{6x^4 + 9x^3 + 3x^2} \\
-2x^3 + 7x^2 + 14x \\
\underline{-2x^3 - 3x^2 - x} \\
10x^2 + 15x + 5 \\
\underline{10x^2 + 15x + 5} \\
0
\end{array}
$$

The remainder is 0. The answer is the quotient,

$$x^3 + 3x^2 - x + 5.$$

79. $(3a^2 - 11a + 17) \div (2a + 6)$

$$
\begin{array}{r}
\tfrac{3}{2}a - 10 \\
2a + 6\ \overline{\smash{\big)}\ 3a^2 - 11a + 17} \\
\underline{3a^2 + 9a \qquad\quad} \\
-20a + 17 \\
\underline{-20a - 60} \\
77
\end{array}
$$

The remainder is 77.

The quotient is $\frac{3}{2}a - 10$.

The answer is $\frac{3}{2}a - 10 + \frac{77}{2a+6}$.

81. $\dfrac{3x^3 + 5x^2 - 9x + 5}{3x - 3}$

$$
\begin{array}{r}
x^2 + \frac{8}{3}x - \frac{1}{3} \\
3x - 3 \overline{\smash{\big)}\, 3x^3 + 5x^2 - 9x + 5} \\
\underline{3x^3 - 3x^2} \\
8x^2 - 9x \qquad \frac{8x^2}{3x} = \frac{8}{3}x \\
\underline{8x^2 - 8x} \\
-x + 5 \quad \frac{-x}{3x} = -\frac{1}{3} \\
\underline{-x + 1} \\
4
\end{array}
$$

$(3x^3 + 5x^2 - 9x + 5)$ divided by $(3x - 3)$ is

$$x^2 + \frac{8}{3}x - \frac{1}{3} + \frac{4}{3x - 3}.$$

83. Use $\mathcal{A} = LW$ with

$$\mathcal{A} = 5x^3 + 7x^2 - 13x - 6$$

and $W = 5x + 2$.

$$5x^3 + 7x^2 - 13x - 6 = L(5x + 2)$$
$$\frac{5x^3 + 7x^2 - 13x - 6}{5x + 2} = L$$

$$
\begin{array}{r}
x^2 + x - 3 \\
5x + 2 \overline{\smash{\big)}\, 5x^3 + 7x^2 - 13x - 6} \\
\underline{5x^3 + 2x^2} \\
5x^2 - 13x \\
\underline{5x^2 + 2x} \\
-15x - 6 \\
\underline{-15x - 6} \\
0
\end{array}
$$

The length L is $(x^2 + x - 3)$ units.

85. $\qquad\qquad \mathcal{A} = \frac{1}{2}bh$

$$24m^3 + 48m^2 + 12m = \frac{1}{2}(b)m$$

$$48m^3 + 96m^2 + 24m = bm \qquad \textit{Multiply by 2.}$$

$$\frac{48m^3 + 96m^2 + 24m}{m} = b \qquad \textit{Divide by m.}$$

$$\frac{48m^3}{m} + \frac{96m^2}{m} + \frac{24m}{m} = b$$

$$48m^2 + 96m + 24 = b$$

87. Use the distance formula, $d = rt$, with $d = (5x^3 - 6x^2 + 3x + 14)$ miles and $r = (x + 1)$ miles per hour.

$$(5x^3 - 6x^2 + 3x + 14) = (x + 1)t$$
$$\frac{(5x^3 - 6x^2 + 3x + 14)}{(x + 1)} = t$$

$$
\begin{array}{r}
5x^2 - 11x + 14 \\
x + 1 \overline{\smash{\big)}\, 5x^3 - 6x^2 + 3x + 14} \\
\underline{5x^3 + 5x^2} \\
-11x^2 + 3x \\
\underline{-11x^2 - 11x} \\
14x + 14 \\
\underline{14x + 14} \\
0
\end{array}
$$

The time t is $(5x^2 - 11x + 14)$ hours.

89. We can write 18 as

$$1 \cdot 18, \quad 2 \cdot 9, \quad \text{or} \quad 3 \cdot 6,$$

so the positive integer factors of 18 are

$$1, 2, 3, 6, 9, \quad \text{and} \quad 18.$$

91. We can write 48 as

$$1 \cdot 48, \, 2 \cdot 24, \, 3 \cdot 16, \, 4 \cdot 12, \quad \text{or} \quad 6 \cdot 8,$$

so the positive integer factors of 48 are

$$1, 2, 3, 4, 6, 8, 12, 16, 24, \quad \text{and} \quad 48.$$

Chapter 5 Review Exercises

1. $4^3 \cdot 4^8 = 4^{3+8} = 4^{11}$

2. $(-5)^6(-5)^5 = (-5)^{6+5} = (-5)^{11}$

3. $(-8x^4)(9x^3) = -8(9)(x^4)(x^3)$
$$= -72x^{4+3} = -72x^7$$

4. $(2x^2)(5x^3)(x^9) = 2(5)(x^2)(x^3)(x^9)$
$$= 10x^{2+3+9} = 10x^{14}$$

5. $(19x)^5 = 19^5x^5$

6. $(-4y)^7 = (-4)^7y^7$

7. $5(pt)^4 = 5p^4t^4$

8. $\left(\dfrac{7}{5}\right)^6 = \dfrac{7^6}{5^6}$

9. $(3x^2y^3)^3 = 3^3(x^2)^3(y^3)^3$
$$= 3^3x^{2 \cdot 3}y^{3 \cdot 3}$$
$$= 3^3x^6y^9$$

10. $(t^4)^8(t^2)^5 = t^{4 \cdot 8}t^{2 \cdot 5}$
$$= t^{32}t^{10}$$
$$= t^{32+10}$$
$$= t^{42}$$

11. $(6x^2z^4)^2(x^3yz^2)^4$
$= 6^2(x^2)^2(z^4)^2(x^3)^4(y)^4(z^2)^4$
$= 6^2x^4z^8x^{12}y^4z^8$
$= 6^2x^{4+12}y^4z^{8+8}$
$= 6^2x^{16}y^4z^{16}$

12. $\left(\dfrac{2m^3n}{p^2}\right)^3 = \dfrac{2^3(m^3)^3n^3}{(p^2)^3}$
$= \dfrac{2^3m^9n^3}{p^6}$

13. The product rule does not apply to $7^2 + 7^4$ because you are adding powers of 7, not multiplying them.

14. $6^0 + (-6)^0 = 1 + 1 = 2$

15. $-(-23)^0 = -1 \cdot (-23)^0 = -1 \cdot 1 = -1$

16. $-10^0 = -(10^0) = -(1) = -1$

17. $-7^{-2} = -\dfrac{1}{7^2} = -\dfrac{1}{49}$

18. $\left(\dfrac{5}{8}\right)^{-2} = \left(\dfrac{8}{5}\right)^2 = \dfrac{64}{25}$

19. $(5^{-2})^{-4} = 5^{(-2)(-4)}$ *Power rule*
$= 5^8$

20. $9^3 \cdot 9^{-5} = 9^{3+(-5)} = 9^{-2} = \dfrac{1}{9^2} = \dfrac{1}{81}$

21. $2^{-1} + 4^{-1} = \dfrac{1}{2^1} + \dfrac{1}{4^1}$
$= \dfrac{1}{2} + \dfrac{1}{4}$
$= \dfrac{2}{4} + \dfrac{1}{4} = \dfrac{3}{4}$

22. $\dfrac{6^{-5}}{6^{-3}} = \dfrac{6^3}{6^5} = \dfrac{1}{6^2} = \dfrac{1}{36}$

23. $\dfrac{x^{-7}}{x^{-9}} = \dfrac{x^9}{x^7} = x^{9-7} = x^2$ *Quotient rule*

24. $\dfrac{y^4 \cdot y^{-2}}{y^{-5}} = \dfrac{y^4 \cdot y^5}{y^2} = \dfrac{y^9}{y^2} = y^7$

25. $(3r^{-2})^{-4} = (3)^{-4}(r^{-2})^{-4}$
$= \left(3^{-4}\right)\left(r^{-2(-4)}\right)$
$= \dfrac{1}{3^4}r^8$
$= \dfrac{r^8}{81}$

26. $(3p)^4(3p^{-7}) = (3^4p^4)(3p^{-7})$
$= 3^{4+1} \cdot p^{4+(-7)}$
$= 3^5p^{-3}$
$= \dfrac{3^5}{p^3}$

27. $\dfrac{ab^{-3}}{a^4b^2} = \dfrac{a}{a^4b^2b^3} = \dfrac{1}{a^3b^5}$

28. $\dfrac{(6r^{-1})^2(2r^{-4})}{r^{-5}(r^2)^{-3}} = \dfrac{(6^2r^{-2})(2r^{-4})}{r^{-5}r^{-6}}$
$= \dfrac{72r^{-6}}{r^{-11}}$
$= \dfrac{72r^{11}}{r^6}$
$= 72r^5$

29. $48{,}000{,}000 = 4.8 \times 10^7$

Move the decimal point left 7 places so it is to the right of the first nonzero digit. $48{,}000{,}000$ is *greater* than 4.8, so the power is *positive*.

30. $28{,}988{,}000{,}000 = 2.8988 \times 10^{10}$

Move the decimal point left 10 places so it is to the right of the first nonzero digit. $28{,}988{,}000{,}000$ is *greater* than 2.8988, so the power is *positive*.

31. $0.000\,000\,082\,4 = 8.24 \times 10^{-8}$

Move the decimal point right 8 places so it is to the right of the first nonzero digit. $0.000\,000\,082\,4$ is *less* than 8.24, so the power is *negative*.

32. $2.4 \times 10^4 = 24{,}000$

Move the decimal point 4 places to the right.

33. $7.83 \times 10^7 = 78{,}300{,}000$

Move the decimal point 7 places to the right.

34. $8.97 \times 10^{-7} = 0.000\,000\,897$

Move the decimal point 7 places to the left.

35. $(2 \times 10^{-3}) \times (4 \times 10^5)$
$= (2 \times 4)(10^{-3} \times 10^5)$
$= 8 \times 10^{-3+5} = 8 \times 10^2 = 800$

36. $\dfrac{8 \times 10^4}{2 \times 10^{-2}} = \dfrac{8}{2} \times \dfrac{10^4}{10^{-2}} = 4 \times 10^{4-(-2)}$
$= 4 \times 10^6 = 4{,}000{,}000$

37. $\dfrac{12 \times 10^{-5} \times 5 \times 10^4}{4 \times 10^3 \times 6 \times 10^{-2}}$
$= \dfrac{12 \times 5}{4 \times 6} \times \dfrac{10^{-5} \times 10^4}{10^3 \times 10^{-2}}$
$= \dfrac{60}{24} \times \dfrac{10^{-1}}{10^1}$
$= \dfrac{5}{2} \times 10^{-1-1}$
$= 2.5 \times 10^{-2}$
$= 0.025$

38. $2 \times 10^{-6} = 0.000\,002$

39. $1.6 \times 10^{-12} = 0.000\,000\,000\,001\,6$

40. There are 41 zeros, so the number is 4.2×10^{42}.

41. $97{,}000 = 9.7 \times 10^4;\; 5000 = 5 \times 10^3$

42. There are 100 zeros, so the number is 1×10^{100}.

43. $1000 = 1 \times 10^3;\quad 2000 = 2 \times 10^3;$

$50{,}000 = 5 \times 10^4;\quad 100{,}000 = 1 \times 10^5$

44. $9m^2 + 11m^2 + 2m^2 = (9 + 11 + 2)m^2$
$$= 22m^2$$

The degree is 2.

To determine if the polynomial is a monomial, binomial, or trinomial, count the number of terms in the final expression.

There is one term, so this is a *monomial*.

45. $-4p + p^3 - p^2 + 8p + 2$
$$= p^3 - p^2 + (-4 + 8)p + 2$$
$$= p^3 - p^2 + 4p + 2$$

The degree is 3.

To determine if the polynomial is a monomial, binomial, or trinomial, count the number of terms in the final expression. Since there are four terms, it is none of these.

46. $12a^5 - 9a^4 + 8a^3 + 2a^2 - a + 3$ cannot be simplified further and is already written in descending powers of the variable.

The degree is 5.

This polynomial has 6 terms, so it is none of the names listed.

47. $-7y^5 - 8y^4 - y^5 + y^4 + 9y$
$$= -7y^5 - 1y^5 - 8y^4 + 1y^4 + 9y$$
$$= (-7 - 1)y^5 + (-8 + 1)y^4 + 9y$$
$$= -8y^5 - 7y^4 + 9y$$

The degree is 5.

There are three terms, so the polynomial is a *trinomial*.

48. $(12r^4 - 7r^3 + 2r^2) - (5r^4 - 3r^3 + 2r^2 - 1)$
$$= (12r^4 - 7r^3 + 2r^2)$$
$$\quad + (-5r^4 + 3r^3 - 2r^2 + 1)$$
Change signs in the second polynomial and add.
$$= 12r^4 - 7r^3 + 2r^2 - 5r^4 + 3r^3 - 2r^2 + 1$$
$$= 7r^4 - 4r^3 + 1$$

The degree is 4.

The polynomial is a *trinomial*.

49. $(5x^3y^2 - 3xy^5 + 12x^2)$
$$\quad - (-9x^2 - 8x^3y^2 + 2xy^5)$$
$$= (5x^3y^2 - 3xy^5 + 12x^2)$$
$$\quad + (9x^2 + 8x^3y^2 - 2xy^5)$$
$$= (5x^3y^2 + 8x^3y^2) + (-3xy^5 - 2xy^5)$$
$$\quad + (12x^2 + 9x^2)$$
$$= 13x^3y^2 - 5xy^5 + 21x^2$$

50. Add.
$$\begin{array}{r} -2a^3 + 5a^2 \\ 3a^3 - a^2 \\ \hline a^3 + 4a^2 \end{array}$$

51. Subtract.
$$\begin{array}{r} 6y^2 - 8y + 2 \\ 5y^2 + 2y - 7 \\ \hline \end{array}$$

Change all signs in the second row and then add.
$$\begin{array}{r} 6y^2 - 8y + 2 \\ -5y^2 - 2y + 7 \\ \hline y^2 - 10y + 9 \end{array}$$

52. Subtract.
$$\begin{array}{r} -12k^4 - 8k^2 + 7k \\ k^4 + 7k^2 - 11k \\ \hline \end{array}$$

Change all signs in the second row and then add.
$$\begin{array}{r} -12k^4 - 8k^2 + 7k \\ -k^4 - 7k^2 + 11k \\ \hline -13k^4 - 15k^2 + 18k \end{array}$$

53. $y = -x^2 + 5$

$x = -2 : y = -(-2)^2 + 5 = -4 + 5 = 1$
$x = -1 : y = -(-1)^2 + 5 = -1 + 5 = 4$
$x = 0 : y = -(0)^2 + 5 = 0 + 5 = 5$
$x = 1 : y = -(1)^2 + 5 = -1 + 5 = 4$
$x = 2 : y = -(2)^2 + 5 = -4 + 5 = 1$

x	-2	-1	0	1	2
y	1	4	5	4	1

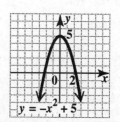

54. $y = 3x^2 - 2$

$$x = -2 : y = 3(-2)^2 - 2 = 3 \cdot 4 - 2 = 10$$
$$x = -1 : y = 3(-1)^2 - 2 = 3 \cdot 1 - 2 = 1$$
$$x = 0 : y = 3(0)^2 - 2 = 3 \cdot 0 - 2 = -2$$
$$x = 1 : y = 3(1)^2 - 2 = 3 \cdot 1 - 2 = 1$$
$$x = 2 : y = 3(2)^2 - 2 = 3 \cdot 4 - 2 = 10$$

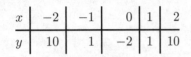

x	-2	-1	0	1	2
y	10	1	-2	1	10

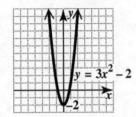

55. $(a + 2)(a^2 - 4a + 1)$

Multiply vertically.

$$
\begin{array}{rrrr}
a^2 & - & 4a & + & 1 \\
 & & a & + & 2 \\
\hline
2a^2 & - & 8a & + & 2 \\
a^3 & - 4a^2 & + & a & \\
\hline
a^3 & - 2a^2 & - & 7a & + & 2
\end{array}
$$

56. $(3r - 2)(2r^2 + 4r - 3)$

Multiply vertically.

$$
\begin{array}{rrrr}
2r^2 & + & 4r & - & 3 \\
 & & 3r & - & 2 \\
\hline
 & - 4r^2 & - & 8r & + & 6 \\
6r^3 & + 12r^2 & - & 9r & \\
\hline
6r^3 & + 8r^2 & - & 17r & + & 6
\end{array}
$$

57. $(5p^2 + 3p)(p^3 - p^2 + 5)$

$$= 5p^2(p^3) + 5p^2(-p^2) + 5p^2(5)$$
$$\quad + 3p(p^3) + 3p(-p^2) + 3p(5)$$
$$= 5p^5 - 5p^4 + 25p^2 + 3p^4 - 3p^3 + 15p$$
$$= 5p^5 - 2p^4 - 3p^3 + 25p^2 + 15p$$

58. $(m - 9)(m + 2)$

$$\qquad \textbf{F} \qquad \textbf{O} \qquad \textbf{I} \qquad \textbf{L}$$
$$= m(m) + m(2) + (-9)(m) + (-9)(2)$$
$$= m^2 + 2m - 9m - 18$$
$$= m^2 - 7m - 18$$

59. $(3k - 6)(2k + 1)$

$$\qquad \textbf{F} \qquad \textbf{O} \qquad \textbf{I} \qquad \textbf{L}$$
$$= 3k(2k) + 3k(1) + (-6)(2k) + (-6)(1)$$
$$= 6k^2 + 3k - 12k - 6$$
$$= 6k^2 - 9k - 6$$

60. $(a + 3b)(2a - b)$

$$\qquad \textbf{F} \qquad \textbf{O} \qquad \textbf{I} \qquad \textbf{L}$$
$$= a(2a) + a(-b) + 3b(2a) + 3b(-b)$$
$$= 2a^2 - ab + 6ab - 3b^2$$
$$= 2a^2 + 5ab - 3b^2$$

61. $(6k + 5q)(2k - 7q)$

$$\qquad \textbf{F} \qquad \textbf{O} \qquad \textbf{I} \qquad \textbf{L}$$
$$= 6k(2k) + 6k(-7q) + 5q(2k) + 5q(-7q)$$
$$= 12k^2 - 42kq + 10kq - 35q^2$$
$$= 12k^2 - 32kq - 35q^2$$

62. $(s - 1)^3 = (s - 1)^2(s - 1)$
$$= (s^2 - 2s + 1)(s - 1)$$

Now, use vertical multiplication.

$$
\begin{array}{rrrr}
s^2 & - & 2s & + & 1 \\
 & & s & - & 1 \\
\hline
- & s^2 & + & 2s & - & 1 \\
s^3 & - 2s^2 & + & s & \\
\hline
s^3 & - 3s^2 & + & 3s & - & 1
\end{array}
$$

$$(s - 1)^3 = s^3 - 3s^2 + 3s - 1$$

63. $(a + 4)^2 = a^2 + 2(a)(4) + 4^2$
$$= a^2 + 8a + 16$$

64. $(2r + 5t)^2 = (2r)^2 + 2(2r)(5t) + (5t)^2$
$$= 4r^2 + 20rt + 25t^2$$

65. $(6m - 5)(6m + 5) = (6m)^2 - 5^2$
$$= 36m^2 - 25$$

66. $(5a + 6b)(5a - 6b) = (5a)^2 - (6b)^2$
$$= 25a^2 - 36b^2$$

67. $(r + 2)^3 = (r + 2)^2(r + 2)$
$$= (r^2 + 4r + 4)(r + 2)$$
$$= r^3 + 4r^2 + 4r + 2r^2 + 8r + 8$$
$$= r^3 + 6r^2 + 12r + 8$$

68. $t(5t - 3)^2 = t(25t^2 - 30t + 9)$
$$= 25t^3 - 30t^2 + 9t$$

69. Answers will vary. One example is given here.

(a) $(x + y)^2 \neq x^2 + y^2$

Let $x = 2$ and $y = 3$.

$$(x + y)^2 = (2 + 3)^2 = 5^2 = 25$$
$$x^2 + y^2 = 2^2 + 3^2 = 4 + 9 = 13$$

Since $25 \neq 13$,

$$(x + y)^2 \neq x^2 + y^2.$$

(b) $(x+y)^3 \neq x^3 + y^3$

Let $x = 2$ and $y = 3$.

$$(x+y)^3 = (2+3)^3 = 5^3 = 125$$
$$x^3 + y^3 = 2^3 + 3^3 = 8 + 27 = 35$$

Since $125 \neq 35$,

$$(x+y)^3 \neq x^3 + y^3.$$

70. To find the third power of a binomial, such as $(a+b)^3$, first square the binomial and then multiply that result by the binomial:

$$\begin{aligned}(a+b)^3 &= (a+b)^2(a+b)\\ &= (a^2 + 2ab + b^2)(a+b)\\ &= (a^3 + 2a^2 b + ab^2)\\ &\quad + (a^2 b + 2ab^2 + b^3)\\ &= a^3 + 3a^2 b + 3ab^2 + b^3\end{aligned}$$

71. If we chose to let $x = 0$ and $y = 1$, we would get the true equation $1 = 1$ for both (a) and (b). These results would not be sufficient to illustrate the truth, in general, of the inequalities. The next step in working the exercise would be to use two other values instead of $x = 0$ and $y = 1$.

72. Use the formula for the volume of a cube, $V = e^3$, with $e = x^2 + 2$ centimeters.

$$\begin{aligned}V &= (x^2 + 2)^3\\ &= (x^2 + 2)^2(x^2 + 2)\\ &= (x^4 + 4x^2 + 4)(x^2 + 2)\end{aligned}$$

Now use vertical multiplication.

$$\begin{array}{r} x^4 + 4x^2 + 4 \\ x^2 + 2 \\ \hline 2x^4 + 8x^2 + 8 \\ x^6 + 4x^4 + 4x^2 \\ \hline x^6 + 6x^4 + 12x^2 + 8 \end{array}$$

The volume of the cube is

$$x^6 + 6x^4 + 12x^2 + 8$$

cubic centimeters.

73. Use the formula for the volume of a sphere, $V = \frac{4}{3}\pi r^3$, with $r = x + 1$ inches.

$$\begin{aligned}V &= \tfrac{4}{3}\pi(x+1)^3\\ &= \tfrac{4}{3}\pi(x+1)^2(x+1)\\ &= \tfrac{4}{3}\pi(x^2 + 2x + 1)(x+1)\end{aligned}$$

Now use vertical multiplication.

$$\begin{array}{r} x^2 + 2x + 1 \\ x + 1 \\ \hline x^2 + 2x + 1 \\ x^3 + 2x^2 + x \\ \hline x^3 + 3x^2 + 3x + 1 \end{array}$$

$$\begin{aligned}V &= \tfrac{4}{3}\pi(x^3 + 3x^2 + 3x + 1)\\ &= \tfrac{4}{3}\pi x^3 + 4\pi x^2 + 4\pi x + \tfrac{4}{3}\pi\end{aligned}$$

The volume of the sphere is

$$\tfrac{4}{3}\pi x^3 + 4\pi x^2 + 4\pi x + \tfrac{4}{3}\pi$$

cubic inches.

74. $\dfrac{-15y^4}{9y^2} = \dfrac{-15y^{4-2}}{9} = \dfrac{-5y^2}{3}$

75. $\dfrac{6y^4 - 12y^2 + 18y}{6y} = \dfrac{6y^4}{6y} - \dfrac{12y^2}{6y} + \dfrac{18y}{6y}$

$$= y^3 - 2y + 3$$

76. $(-10m^4 n^2 + 5m^3 n^2 + 6m^2 n^4) \div (5m^2 n)$

$$= \dfrac{-10m^4 n^2 + 5m^3 n^2 + 6m^2 n^4}{5m^2 n}$$

$$= \dfrac{-10m^4 n^2}{5m^2 n} + \dfrac{5m^3 n^2}{5m^2 n} + \dfrac{6m^2 n^4}{5m^2 n}$$

$$= -2m^2 n + mn + \dfrac{6n^3}{5}$$

77. Let P be the polynomial that when multiplied by $6m^2 n$ gives the product

$$12m^3 n^2 + 18m^6 n^3 - 24m^2 n^2.$$

$$(P)(6m^2 n) = 12m^3 n^2 + 18m^6 n^3 - 24m^2 n^2$$

$$P = \dfrac{12m^3 n^2 + 18m^6 n^3 - 24m^2 n^2}{6m^2 n}$$

$$= \dfrac{12m^3 n^2}{6m^2 n} + \dfrac{18m^6 n^3}{6m^2 n} - \dfrac{24m^2 n^2}{6m^2 n}$$

$$= 2mn + 3m^4 n^2 - 4n$$

78. The error made was not dividing both terms in the numerator by 6. The correct method is as follows:

$$\dfrac{6x^2 - 12x}{6} = \dfrac{6x^2}{6} - \dfrac{12x}{6} = x^2 - 2x.$$

79. $\dfrac{2r^2 + 3r - 14}{r - 2}$

$$\begin{array}{r} 2r + 7 \\ r-2\overline{\smash{\big)}\,2r^2 + 3r - 14} \\ \underline{2r^2 - 4r } \\ 7r - 14 \\ \underline{7r - 14} \\ 0 \end{array}$$

The remainder is 0.

The answer is the quotient, $2r + 7$.

80. $\dfrac{10a^3 + 9a^2 - 14a + 9}{5a - 3}$

$$
\begin{array}{r}
2a^2 + 3a - 1 \\
5a - 3\overline{\smash)10a^3 + 9a^2 - 14a + 9} \\
\underline{10a^3 - 6a^2} \\
15a^2 - 14a \\
\underline{15a^2 - 9a} \\
-5a + 9 \\
\underline{-5a + 3} \\
6
\end{array}
$$

The answer is

$$2a^2 + 3a - 1 + \frac{6}{5a - 3}.$$

81. $\dfrac{x^4 - 5x^2 + 3x^3 - 3x + 4}{x^2 - 1}$

Write the dividend in descending powers and use 0 as the coefficient of the missing term.

$$
\begin{array}{r}
x^2 + 3x - 4 \\
x^2 + 0x - 1\overline{\smash)x^4 + 3x^3 - 5x^2 - 3x + 4} \\
\underline{x^4 + 0x^3 - x^2} \\
3x^3 - 4x^2 - 3x \\
\underline{3x^3 + 0x^2 - 3x} \\
-4x^2 + 4 \\
\underline{-4x^2 + 4} \\
0
\end{array}
$$

The remainder is 0. The answer is the quotient,

$$x^2 + 3x - 4.$$

82. $\dfrac{m^4 + 4m^3 - 12m - 5m^2 + 6}{m^2 - 3}$

Write the dividend in descending powers and use 0 as the coefficient of the missing term.

$$
\begin{array}{r}
m^2 + 4m - 2 \\
m^2 + 0m - 3\overline{\smash)m^4 + 4m^3 - 5m^2 - 12m + 6} \\
\underline{m^4 + 0m^3 - 3m^2} \\
4m^3 - 2m^2 - 12m \\
\underline{4m^3 + 0m^2 - 12m} \\
-2m^2 + 0m + 6 \\
\underline{-2m^2 + 0m + 6} \\
0
\end{array}
$$

The remainder is 0. The answer is the quotient,

$$m^2 + 4m - 2.$$

83. $\dfrac{16x^2 - 25}{4x + 5}$

$$
\begin{array}{r}
4x - 5 \\
4x + 5\overline{\smash)16x^2 + 0x - 25} \\
\underline{16x^2 + 20x} \\
-20x - 25 \\
\underline{-20x - 25} \\
0
\end{array}
$$

The remainder is 0.
The answer is the quotient, $4x - 5$.

84. $\dfrac{25y^2 - 100}{5y + 10}$

$$
\begin{array}{r}
5y - 10 \\
5y + 10\overline{\smash)25y^2 + 0y - 100} \\
\underline{25y^2 + 50y} \\
-50y - 100 \\
\underline{-50y - 100} \\
0
\end{array}
$$

The remainder is 0.
The answer is the quotient, $5y - 10$.

85. $\dfrac{y^3 - 8}{y - 2}$

$$
\begin{array}{r}
y^2 + 2y + 4 \\
y - 2\overline{\smash)y^3 + 0y^2 + 0y - 8} \\
\underline{y^3 - 2y^2} \\
2y^2 + 0y \\
\underline{2y^2 - 4y} \\
4y - 8 \\
\underline{4y - 8} \\
0
\end{array}
$$

The remainder is 0.
The answer is the quotient, $y^2 + 2y + 4$.

86. $\dfrac{1000x^6 + 1}{10x^2 + 1}$

$$
\begin{array}{r}
100x^4 - 10x^2 + 1 \\
10x^2 + 1\overline{\smash)1000x^6 + 0x^4 + 0x^2 + 1} \\
\underline{1000x^6 + 100x^4} \\
-100x^4 + 0x^2 \\
\underline{-100x^4 - 10x^2} \\
10x^2 + 1 \\
\underline{10x^2 + 1} \\
0
\end{array}
$$

The remainder is 0.
The answer is the quotient, $100x^4 - 10x^2 + 1$.

87. $\dfrac{6y^4 - 15y^3 + 14y^2 - 5y - 1}{3y^2 + 1}$

$$\begin{array}{r} 2y^2 \ - \ 5y \ +4 \\ 3y^2+1\overline{)6y^4 \ - \ 15y^3 \ + \ 14y^2 \ - 5y \ - 1} \\ \underline{6y^4 \qquad\quad + \ 2y^2} \\ -15y^3 \ + \ 12y^2 \ - 5y \\ \underline{-15y^3 \qquad\qquad - 5y} \\ 12y^2 \qquad\quad - 1 \\ \underline{12y^2 \qquad\quad + 4} \\ - 5 \end{array}$$

The answer is $2y^2 - 5y + 4 + \dfrac{-5}{3y^2+1}$.

88. $\dfrac{4x^5 - 8x^4 - 3x^3 + 22x^2 - 15}{4x^2 - 3}$

$$\begin{array}{r} x^3 \ - \ 2x^2 \qquad + 4 \\ 4x^2-3\overline{)4x^5 \ - \ 8x^4 \ - 3x^3 \ + \ 22x^2 \ + 0x \ - 15} \\ \underline{4x^5 \qquad\qquad - 3x^3} \\ -8x^4 \qquad\quad + 22x^2 \\ \underline{-8x^4 \qquad\quad + \ 6x^2} \\ 16x^2 \qquad\quad - 15 \\ \underline{16x^2 \qquad\quad - 12} \\ -3 \end{array}$$

The answer is $x^3 - 2x^2 + 4 + \dfrac{-3}{4x^2-3}$.

89. **[5.2]** $5^0 + 7^0 = 1 + 1 = 2$

90. **[5.1]** $\left(\dfrac{6r^2p}{5}\right)^3 = \dfrac{6^3r^{2\cdot3}p^3}{5^3}$

$\qquad\qquad = \dfrac{6^3r^6p^3}{5^3}$

91. **[5.6]** $(12a+1)(12a-1) = (12a)^2 - 1^2$

$\qquad\qquad\qquad\qquad = 144a^2 - 1$

92. **[5.2]** $2^{-4} = \dfrac{1}{2^4} = \dfrac{1}{16}$

93. **[5.2]** $(8^{-3})^4 = 8^{(-3)(4)}$

$\qquad\qquad = 8^{-12}$

$\qquad\qquad = \dfrac{1}{8^{12}}$

94. **[5.7]** $\dfrac{2p^3 - 6p^2 + 5p}{2p^2}$

$\qquad = \dfrac{2p^3}{2p^2} - \dfrac{6p^2}{2p^2} + \dfrac{5p}{2p^2}$

$\qquad = p - 3 + \dfrac{5}{2p}$

95. **[5.2]** $\dfrac{(2m^{-5})(3m^2)^{-1}}{m^{-2}(m^{-1})^2}$

$\qquad = \dfrac{(2m^{-5})(3^{-1}m^{-2})}{m^{-2}(m^{-2})}$

$\qquad = \dfrac{2}{3} \cdot \dfrac{m^{-5+(-2)}}{m^{-2+(-2)}}$

$\qquad = \dfrac{2}{3} \cdot \dfrac{m^{-7}}{m^{-4}}$

$\qquad = \dfrac{2}{3} \cdot \dfrac{m^4}{m^7}$

$\qquad = \dfrac{2}{3m^3}$

96. **[5.5]** $(3k-6)(2k^2 + 4k + 1)$

Multiply vertically.

$$\begin{array}{r} 2k^2 \ + \ 4k \ + 1 \\ 3k \ - 6 \\ \hline -12k^2 \ - \ 24k \ - 6 \\ 6k^3 \ + \ 12k^2 \ + \ 3k \\ \hline 6k^3 \qquad\quad - \ 21k \ - 6 \end{array}$$

97. **[5.2]** $\dfrac{r^9 \cdot r^{-5}}{r^{-2} \cdot r^{-7}} = \dfrac{r^9 r^2 r^7}{r^5}$

$\qquad\qquad = \dfrac{r^{9+2+7}}{r^5}$

$\qquad\qquad = \dfrac{r^{18}}{r^5} = r^{13}$

98. **[5.6]** $(2r+5s)^2$

$\qquad = (2r)^2 + 2(2r)(5s) + (5s)^2$

$\qquad = 4r^2 + 20rs + 25s^2$

99. **[5.4]** $(-5y^2 + 3y - 11) + (4y^2 - 7y + 15)$

$\qquad = -5y^2 + 4y^2 + 3y - 7y - 11 + 15$

$\qquad = -y^2 - 4y + 4$

100. **[5.5]** $(2r+5)(5r-2)$

$\qquad\quad$ **F** $\qquad$ **O** $\qquad$ **I** $\qquad$ **L**

$\qquad = 2r(5r) + 2r(-2) + 5(5r) + 5(-2)$

$\qquad = 10r^2 - 4r + 25r - 10$

$\qquad = 10r^2 + 21r - 10$

101. **[5.7]** $\dfrac{2y^3 + 17y^2 + 37y + 7}{2y + 7}$

$$\begin{array}{r} y^2 \ + \ 5y \ +1 \\ 2y+7\overline{)2y^3 \ + 17y^2 \ + 37y \ + 7} \\ \underline{2y^3 \ + \ 7y^2} \\ 10y^2 \ + 37y \\ \underline{10y^2 \ + 35y} \\ 2y \ + 7 \\ \underline{2y \ + 7} \\ 0 \end{array}$$

The remainder is 0.
The answer is the quotient, $y^2 + 5y + 1$.

102. **[5.7]** $(25x^2y^3 - 8xy^2 + 15x^3y) \div (10x^2y^3)$

$$= \frac{25x^2y^3 - 8xy^2 + 15x^3y}{10x^2y^3}$$

$$= \frac{25x^2y^3}{10x^2y^3} - \frac{8xy^2}{10x^2y^3} + \frac{15x^3y}{10x^2y^3}$$

$$= \frac{5}{2} - \frac{4}{5xy} + \frac{3x}{2y^2}$$

103. **[5.4]** $(6p^2 - p - 8) - (-4p^2 + 2p - 3)$

$$= (6p^2 - p - 8) + (4p^2 - 2p + 3)$$

$$= 10p^2 - 3p - 5$$

104. **[5.7]** $\dfrac{3x^3 - 2x + 5}{x - 3}$

Use 0 as the coefficient for the missing term.

$$
\begin{array}{r}
3x^2 \quad + 9x \quad + 25 \\
x - 3 \overline{\smash{)}3x^3 \quad + 0x^2 \quad - 2x \quad + 5} \\
\underline{3x^3 \quad - 9x^2} \\
9x^2 \quad - 2x \\
\underline{9x^2 \quad - 27x} \\
25x \quad + 5 \\
\underline{25x \quad - 75} \\
80
\end{array}
$$

$$\frac{3x^3 - 2x + 5}{x - 3} = 3x^2 + 9x + 25 + \frac{80}{x - 3}$$

105. **[5.6]** $(-7 + 2k)^2$

$$= (-7)^2 + 2(-7)(2k) + (2k)^2$$

$$= 49 - 28k + 4k^2$$

106. **[5.2]** $\left(\dfrac{x}{y^{-3}}\right)^{-4} = \dfrac{x^{-4}}{(y^{-3})^{-4}}$

$$= \frac{x^{-4}}{y^{12}}$$

$$= \frac{1}{x^4y^{12}}$$

107. **[5.5] (a)** Use the formula for the perimeter of a rectangle, $P = 2L + 2W$, with $L = 2x - 3$ and $W = x + 2$.

$$P = 2(2x - 3) + 2(x + 2)$$

$$= 4x - 6 + 2x + 4$$

$$= 6x - 2$$

The perimeter of the rectangle is

$$(6x - 2) \text{ units.}$$

(b) Use the formula for the area of a rectangle, $\mathcal{A} = LW$, with $L = 2x - 3$ and $W = x + 2$.

$$\mathcal{A} = (2x - 3)(x + 2)$$

$$= (2x)(x) + (2x)(2) + (-3)(x) + (-3)(2)$$

$$= 2x^2 + 4x - 3x - 6$$

$$= 2x^2 + x - 6$$

The area of the rectangle is

$$(2x^2 + x - 6) \text{ square units.}$$

108. **[5.6] (a)** Use the formula for the perimeter of a square, $P = 4s$, with $s = 5x^4 + 2x^2$.

$$P = 4(5x^4 + 2x^2)$$

$$= 20x^4 + 8x^2$$

The perimeter of the square is

$$(20x^4 + 8x^2) \text{ units.}$$

(b) Use the formula for the area of a square, $\mathcal{A} = s^2$, with $s = 5x^4 + 2x^2$.

$$\mathcal{A} = (5x^4 + 2x^2)^2$$

$$= (5x^4)^2 + 2(5x^4)(2x^2) + (2x^2)^2$$

$$= 25x^8 + 20x^6 + 4x^4$$

The area of the square is

$$(25x^8 + 20x^6 + 4x^4) \text{ square units.}$$

Chapter 5 Test

1. $5^{-4} = \dfrac{1}{5^4} = \dfrac{1}{625}$

2. $(-3)^0 + 4^0 = 1 + 1 = 2$

3. $4^{-1} + 3^{-1} = \dfrac{1}{4^1} + \dfrac{1}{3^1} = \dfrac{3}{12} + \dfrac{4}{12} = \dfrac{7}{12}$

4. $\dfrac{(3x^2y)^2(xy^3)^2}{(xy)^3} = \dfrac{3^2(x^2)^2y^2x^2(y^3)^2}{x^3y^3}$

$$= \frac{9x^4y^2x^2y^6}{x^3y^3}$$

$$= \frac{9x^6y^8}{x^3y^3}$$

$$= 9x^3y^5$$

5. $\dfrac{8^{-1} \cdot 8^4}{8^{-2}} = \dfrac{8^{(-1)+4}}{8^{-2}} = \dfrac{8^3}{8^{-2}} = 8^{3-(-2)} = 8^5$

6.
$$\frac{(x^{-3})^{-2}(x^{-1}y)^2}{(xy^{-2})^2} = \frac{(x^{-3})^{-2}(x^{-1})^2(y)^2}{(x)^2(y^{-2})^2}$$
$$= \frac{x^6 x^{-2} y^2}{x^2 y^{-4}}$$
$$= \frac{x^4 y^2}{x^2 y^{-4}}$$
$$= x^{4-2} y^{2-(-4)}$$
$$= x^2 y^6$$

7. **(a)** $3^{-4} = \frac{1}{3^4} = \frac{1}{81}$, which is *positive*.
A negative exponent indicates a reciprocal, not a negative number.

(b) $(-3)^4 = 81$, which is *positive*.

(c) $-3^4 = -1 \cdot 3^4 = -81$, which is *negative*.

(d) $3^0 = 1$, which is *positive*.

(e) $(-3)^0 - 3^0 = 1 - 1 = 0$ (*zero*)

(f) $(-3)^{-3} = \frac{1}{(-3)^3} = \frac{1}{-27}$, which is *negative*.

8. **(a)** $45{,}000{,}000{,}000 = 4.5 \times 10^{10}$

Move the decimal point left 10 places so it is to the right of the first nonzero digit. $45{,}000{,}000{,}000$ is *greater* than 4.5, so the power is *positive*.

(b) $3.6 \times 10^{-6} = 0.000\,003\,6$

Move the decimal point 6 places to the left.

(c) $\frac{9.5 \times 10^{-1}}{5 \times 10^3} = \frac{9.5}{5} \times \frac{10^{-1}}{10^3}$
$$= 1.9 \times 10^{-1-3}$$
$$= 1.9 \times 10^{-4}$$
$$= 0.00019$$

9. **(a)** $1000 = 1 \times 10^3$

$5{,}890{,}000{,}000{,}000 = 5.89 \times 10^{12}$

(b) $\left(1 \times 10^3 \text{ light-years}\right)\left(5.89 \times 10^{12} \frac{\text{miles}}{\text{light-year}}\right)$
$$= 5.89 \times 10^{3+12} \text{ miles}$$
$$= 5.89 \times 10^{15} \text{ miles}$$

10. $5x^2 + 8x - 12x^2 = 5x^2 - 12x^2 + 8x$
$$= -7x^2 + 8x$$

degree 2; binomial (2 terms)

11. $13n^3 - n^2 + n^4 + 3n^4 - 9n^2$
$$= n^4 + 3n^4 + 13n^3 - n^2 - 9n^2$$
$$= 4n^4 + 13n^3 - 10n^2$$

degree 4; trinomial (3 terms)

12. $y = 2x^2 - 4$

$x = -2: y = 2(-2)^2 - 4 = 2\cdot4 - 4 = 4$
$x = -1: y = 2(-1)^2 - 4 = 2\cdot1 - 4 = -2$
$x = 0: y = 2(0)^2 - 4 = 2\cdot0 - 4 = -4$
$x = 1: y = 2(1)^2 - 4 = 2\cdot1 - 4 = -2$
$x = 2: y = 2(2)^2 - 4 = 2\cdot4 - 4 = 4$

x	-2	-1	0	1	2
y	4	-2	-4	-2	4

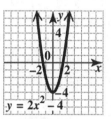

13. $(2y^2 - 8y + 8) + (-3y^2 + 2y + 3)$
$\quad - (y^2 + 3y - 6)$
$= (2y^2 - 8y + 8) + (-3y^2 + 2y + 3)$
$\quad + (-y^2 - 3y + 6)$
$= (2y^2 - 3y^2 - y^2) + (-8y + 2y - 3y)$
$\quad + (8 + 3 + 6)$
$= -2y^2 - 9y + 17$

14. $(-9a^3b^2 + 13ab^5 + 5a^2b^2)$
$\quad - (6ab^5 + 12a^3b^2 + 10a^2b^2)$
$= (-9a^3b^2 + 13ab^5 + 5a^2b^2)$
$\quad + (-6ab^5 - 12a^3b^2 - 10a^2b^2)$
$= (-9a^3b^2 - 12a^3b^2) + (13ab^5 - 6ab^5)$
$\quad + (5a^2b^2 - 10a^2b^2)$
$= -21a^3b^2 + 7ab^5 - 5a^2b^2$

15. Subtract.

$$9t^3 - 4t^2 + 2t + 2$$
$$\underline{9t^3 + 8t^2 - 3t - 6}$$

Change all signs in the second row and then add.

$$9t^3 - \quad 4t^2 + 2t + 2$$
$$\underline{-9t^3 - \quad 8t^2 + 3t + 6}$$
$$-12t^2 + 5t + 8$$

16. $3x^2(-9x^3 + 6x^2 - 2x + 1)$
$= 3x^2(-9x^3) + 3x^2(6x^2) + 3x^2(-2x) + 3x^2(1)$
$= -27x^5 + 18x^4 - 6x^3 + 3x^2$

17. $(t - 8)(t + 3)$
$\quad$ **F O I L**
$= t^2 + 3t - 8t - 24$
$= t^2 - 5t - 24$

18. $(4x + 3y)(2x - y)$

$$\text{F} \quad \text{O} \quad \text{I} \quad \text{L}$$
$$= 8x^2 - 4xy + 6xy - 3y^2$$
$$= 8x^2 + 2xy - 3y^2$$

19. $(5x - 2y)^2 = (5x)^2 - 2(5x)(2y) + (2y)^2$
$$= 25x^2 - 20xy + 4y^2$$

20. $(10v + 3w)(10v - 3w) = (10v)^2 - (3w)^2$
$$= 100v^2 - 9w^2$$

21. $(2r - 3)(r^2 + 2r - 5)$

Multiply vertically.

$$
\begin{array}{r}
r^2 + 2r - 5 \\
2r - 3 \\
\hline
-3r^2 - 6r + 15 \\
2r^3 + 4r^2 - 10r \\
\hline
2r^3 + r^2 - 16r + 15
\end{array}
$$

22. Use the formula for the perimeter of a square, $P = 4s$, with $s = 3x + 9$.

$$
\begin{aligned}
P &= 4s \\
&= 4(3x + 9) \\
&= 4(3x) + 4(9) \\
&= 12x + 36
\end{aligned}
$$

The perimeter of the square is $(12x + 26)$ units.

Use the formula for the area of a square, $A = s^2$, with $s = 3x + 9$.

$$
\begin{aligned}
A &= s^2 \\
&= (3x + 9)^2 \\
&= (3x)^2 + 2(3x)(9) + 9^2 \\
&= 9x^2 + 54x + 81
\end{aligned}
$$

The area of the square is $(9x^2 + 54x + 81)$ square units.

23. $\dfrac{8y^3 - 6y^2 + 4y + 10}{2y}$

$$= \frac{8y^3}{2y} - \frac{6y^2}{2y} + \frac{4y}{2y} + \frac{10}{2y}$$

$$= 4y^2 - 3y + 2 + \frac{5}{y}$$

24. $(-9x^2y^3 + 6x^4y^3 + 12xy^3) \div (3xy)$

$$= \frac{-9x^2y^3 + 6x^4y^3 + 12xy^3}{3xy}$$

$$= \frac{-9x^2y^3}{3xy} + \frac{6x^4y^3}{3xy} + \frac{12xy^3}{3xy}$$

$$= -3xy^2 + 2x^3y^2 + 4y^2$$

25. $\dfrac{5x^2 - x - 18}{5x + 9}$

$$
\begin{array}{r}
x - 2 \\
5x + 9 \overline{\smash{\big)}\, 5x^2 - x - 18} \\
\underline{5x^2 + 9x} \\
-10x - 18 \\
\underline{-10x - 18} \\
0
\end{array}
$$

The remainder is 0. The answer is the quotient, $x - 2$.

26. $(3x^3 - x + 4) \div (x - 2)$

$$
\begin{array}{r}
3x^2 + 6x + 11 \\
x - 2 \overline{\smash{\big)}\, 3x^3 + 0x^2 - x + 4} \\
\underline{3x^3 - 6x^2} \\
6x^2 - x \\
\underline{6x^2 - 12x} \\
11x + 4 \\
\underline{11x - 22} \\
26
\end{array}
$$

The answer is $3x^2 + 6x + 11 + \dfrac{26}{x - 2}$.

Cumulative Review Exercises (Chapters 1–5)

1. $\dfrac{28}{16} = \dfrac{7 \cdot 4}{4 \cdot 4} = \dfrac{7}{4}$

2. $\dfrac{55}{11} = \dfrac{5 \cdot 11}{1 \cdot 11} = \dfrac{5}{1} = 5$

3. Each shed requires $1\frac{1}{4}$ cubic yards of concrete, so the total amount of concrete needed for 25 sheds would be

$$
\begin{aligned}
25 \times 1\tfrac{1}{4} &= 25 \times \tfrac{5}{4} \\
&= \tfrac{125}{4} \\
&= 31\tfrac{1}{4} \text{ cubic yards.}
\end{aligned}
$$

4. Use the formula for simple interest, $I = Prt$, with $P = \$34,000$, $r = 5.4\%$, and $t = 1$.

$$
\begin{aligned}
I &= Prt \\
&= (34,000)(0.054)(1) \\
&= 1836
\end{aligned}
$$

She earned $1836 in interest.

5. The positive integer factors of 45 are

$$1, 3, 5, 9, 15, \text{ and } 45.$$

6. $\dfrac{4x - 2y}{x + y} = \dfrac{4(-2) - 2(4)}{(-2) + 4}$ *Let x = -2,*   *y = 4.*

$$= \dfrac{-8 - 8}{2} = \dfrac{-16}{2} = -8$$

7. $\dfrac{(-13 + 15) - (3 + 2)}{6 - 12} = \dfrac{2 - 5}{-6} = \dfrac{-3}{-6} = \dfrac{1}{2}$

8. $-7 - 3[2 + (5 - 8)] = -7 - 3[2 + (-3)]$
$$= -7 - 3[-1]$$
$$= -7 + 3 = -4$$

9. $(9 + 2) + 3 = 9 + (2 + 3)$

The numbers are in the same order but grouped differently, so this is an example of the associative property of addition.

10. $6(4 + 2) = 6(4) + 6(2)$

The number 6 outside the parentheses is "distributed" over the 4 and the 2. This is an example of the distributive property.

11. $-3(2x^2 - 8x + 9) - (4x^2 + 3x + 2)$
$$= -6x^2 + 24x - 27 - 4x^2 - 3x - 2$$
$$= -10x^2 + 21x - 29$$

12. $2 - 3(t - 5) = 4 + t$
$$2 - 3t + 15 = 4 + t$$
$$-3t + 17 = 4 + t$$
$$-4t + 17 = 4$$
$$-4t = -13$$
$$t = \dfrac{-13}{-4} = \dfrac{13}{4}$$

The solution set is $\left\{ \dfrac{13}{4} \right\}$.

13. $2(5x + 1) = 10x + 4$
$$10x + 2 = 10x + 4$$
$$2 = 4 \qquad \textit{False}$$

The false statement indicates that the equation has no solution, symbolized by $\emptyset$.

14. Solve $d = rt$ for r.

$$\dfrac{d}{t} = \dfrac{rt}{t} \qquad \textit{Divide by t.}$$

$$\dfrac{d}{t} = r$$

15. $\dfrac{x}{5} = \dfrac{x - 2}{7}$
$$7x = 5(x - 2) \quad \textit{Cross products are equal.}$$
$$7x = 5x - 10$$
$$2x = -10$$
$$x = -5$$

The solution set is $\{-5\}$.

16. $\frac{1}{3}p - \frac{1}{6}p = -2$

To clear fractions, multiply both sides of the equation by the least common denominator, which is 6.

$$6(\tfrac{1}{3}p - \tfrac{1}{6}p) = (6)(-2)$$
$$6(\tfrac{1}{3}p) - 6(\tfrac{1}{6}p) = -12$$
$$2p - p = -12$$
$$p = -12$$

The solution set is $\{-12\}$.

17. $0.05x + 0.15(50 - x) = 5.50$

To clear decimals, multiply both sides of the equation by 100.

$$100[0.05x + 0.15(50 - x)] = 100(5.50)$$
$$100(0.05x) + 100[0.15(50 - x)] = 100(5.50)$$
$$5x + 15(50 - x) = 550$$
$$5x + 750 - 15x = 550$$
$$-10x + 750 = 550$$
$$-10x = -200$$
$$x = 20$$

The solution set is $\{20\}$.

18. $4 - (3x + 12) = (2x - 9) - (5x - 1)$
$$4 - 3x - 12 = 2x - 9 - 5x + 1$$
$$-3x - 8 = -3x - 8 \qquad \textit{True}$$

The true statement indicates that the solution set is {all real numbers}.

19. Let $x =$ the number of calories burned in thermoregulation.

Then $5\frac{3}{8}x = \frac{43}{8}x =$ the number of calories burned in exertion.

$$x + \tfrac{43}{8}x = 11{,}200$$
$$8(x + \tfrac{43}{8}x) = 8(11{,}200)$$
$$8x + 43x = 89{,}600$$
$$51x = 89{,}600$$
$$x = \tfrac{89{,}600}{51} \approx 1757$$

A husky burns approximately 1757 calories for thermoregulation and $\frac{43}{8}\left(\frac{89{,}600}{51}\right) \approx 9443$ calories for exertion.

20. Let $x =$ one side of the triangle.
Then $2x =$ the other (unknown) side.

The third side is 17 feet. The perimeter of the triangle cannot be more than 50 feet. This is equivalent to stating that the sum of the lengths of the sides must be less than or equal to 50 feet. Write this statement as an inequality and solve.

$$x + 2x + 17 \le 50$$
$$3x + 17 \le 50$$
$$3x \le 33$$
$$x \le 11$$

One side cannot be more than 11 feet. The other side cannot be more than $2 \cdot 11 = 22$ feet.

21. $-2(x + 4) > 3x + 6$

$$-2x - 8 > 3x + 6$$

$$-2x > 3x + 14$$

$$-5x > 14$$

$$\frac{-5x}{-5} < \frac{14}{-5} \qquad \textit{Divide by –5;}$$
$$\qquad\qquad\qquad \textit{reverse the symbol.}$$

$$x < -\tfrac{14}{5}$$

The solution set is $\left(-\infty, -\tfrac{14}{5}\right)$.

22. $-3 \le 2x + 5 < 9$

$$-8 \le \quad 2x \quad < 4 \qquad \textit{Subtract 5.}$$

$$\frac{-8}{2} \le \quad \frac{2x}{2} \quad < \frac{4}{2} \qquad \textit{Divide by 2.}$$

$$-4 \le \quad x \quad < 2$$

The solution set is $[-4, 2)$.

23. We recognize $y = -3x + 6$ as the equation of a line with y-intercept $(0, 6)$ and slope -3. From the point $(0, 6)$, we can move right 1 unit and down 3 units to the point $(1, 3)$ to get another point on the graph.

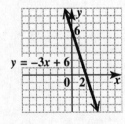

24. **(a)** Use the definition of slope with $(x_1, y_1) = (-1, 5)$ and $(x_2, y_2) = (2, 8)$.

$$m = \frac{\text{change in } y}{\text{change in } x} = \frac{y_2 - y_1}{x_2 - x_1}$$

$$= \frac{8 - 5}{2 - (-1)} = \frac{3}{3} = 1$$

(b) Use the point-slope form of the equation of a line with $m = 1$ and $(x_1, y_1) = (-1, 5)$.

$$y - y_1 = m(x - x_1)$$
$$y - 5 = 1[x - (-1)]$$
$$y - 5 = x + 1$$
$$y = x + 6$$

25. $y \ge x + 5$

$$3 \overset{?}{\ge} -1 + 5 \qquad \textit{Let y = 3; x = –1.}$$
$$3 \ge 4 \qquad\qquad \textit{False}$$

The false statement indicates that the point $(-1, 3)$ is not a solution of the inequality $y \ge x + 5$ and does not lie within the shaded region of the graph of the inequality.

26. $f(x) = x + 7$
$$f(-8) = -8 + 7 = -1$$

27.
$$y = 2x + 5 \qquad (1)$$
$$x + y = -4 \qquad (2)$$

To solve the system by the substitution method, let $y = 2x + 5$ in equation (2).

$$x + y = -4$$
$$x + (2x + 5) = -4$$
$$3x + 5 = -4$$
$$3x = -9$$
$$x = -3$$

From (1), $y = 2(-3) + 5 = -1$.
The solution set is $\{(-3, -1)\}$.

28. $3x + 2y = 2$ (1)
$2x + 3y = -7$ (2)

We solve this system by the elimination method. To eliminate x, multiply equation (1) by 2, equation (2) by -3, and then add.

$$\begin{array}{rcr} 6x + 4y &=& 4 \\ -6x - 9y &=& 21 \\ \hline -5y &=& 25 \\ y &=& -5 \end{array}$$

To eliminate y, multiply equation (1) by 3, equation (2) by -2, and then add.

$$\begin{array}{rcr} 9x + 6y &=& 6 \\ -4x - 6y &=& 14 \\ \hline 5x &=& 20 \\ x &=& 4 \end{array}$$

The solution set is $\{(4, -5)\}$.

29. $4^{-1} + 3^0 = \dfrac{1}{4^1} + 1 = 1\dfrac{1}{4}$, or $\dfrac{5}{4}$

30. $\dfrac{8^{-5} \cdot 8^7}{8^2} = \dfrac{8^{-5+7}}{8^2} = \dfrac{8^2}{8^2} = 1$

31. $\dfrac{(a^{-3}b^2)^2}{(2a^{-4}b^{-3})^{-1}} = \dfrac{(a^{-3})^2(b^2)^2}{2^{-1}(a^{-4})^{-1}(b^{-3})^{-1}}$

$= \dfrac{a^{-6}b^4}{2^{-1}a^4b^3}$

$= \dfrac{2b^4}{a^6a^4b^3} = \dfrac{2b}{a^{10}}$

32. $\left(3.6 \times 10^1 \text{ sec}\right)\left(3.0 \times 10^5 \dfrac{\text{km}}{\text{sec}}\right)$

$= 3.6 \times 3.0 \times 10^{1+5} \text{ km}$

$= 10.8 \times 10^6 \text{ km}$

Venus is about 10,800,000 km from the sun.

33. $y = (x + 4)^2$

$x = -6 : y = (-6 + 4)^2 = (-2)^2 = 4$
$x = -5 : y = (-5 + 4)^2 = (-1)^2 = 1$
$x = -4 : y = (-4 + 4)^2 = (0)^2 = 0$
$x = -3 : y = (-3 + 4)^2 = (1)^2 = 1$
$x = -2 : y = (-2 + 4)^2 = (2)^2 = 4$

x	-6	-5	-4	-3	-2
y	4	1	0	1	4

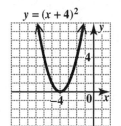

34. $(7x^3 - 12x^2 - 3x + 8) + (6x^2 + 4)$
$\quad - (-4x^3 + 8x^2 - 2x - 2)$
$= (7x^3 - 12x^2 - 3x + 8) + (6x^2 + 4)$
$\quad + (4x^3 - 8x^2 + 2x + 2)$
$= (7 + 4)x^3 + (-12 + 6 - 8)x^2$
$\quad + (-3 + 2)x + (8 + 4 + 2)$
$= 11x^3 - 14x^2 - x + 14$

35. $(7x + 4)(9x + 3)$
$= 63x^2 + 21x + 36x + 12$ *FOIL*
$= 63x^2 + 57x + 12$

36. $\dfrac{y^3 - 3y^2 + 8y - 6}{y - 1}$

$$\begin{array}{r} y^2 - 2y + 6 \\ y - 1 \overline{\smash{\big)}\, y^3 - 3y^2 + 8y - 6} \\ \underline{y^3 - y^2} \\ -2y^2 + 8y \\ \underline{-2y^2 + 2y} \\ 6y - 6 \\ \underline{6y - 6} \\ 0 \end{array}$$

The remainder is 0. The answer is the quotient,

$$y^2 - 2y + 6.$$

CHAPTER 6 FACTORING AND APPLICATIONS

6.1 The Greatest Common Factor; Factoring by Grouping

6.1 Now Try Exercises

N1. **(a)** $24, 36$

Write each number in prime factored form.

$$24 = 2 \cdot 2 \cdot 2 \cdot 3, \quad 36 = 2 \cdot 2 \cdot 3 \cdot 3$$

Use each prime the *least* number of times it appears in *all* the factored forms. The least number of times 2 appears in all the factored forms is 2, and the least number of times 3 appears is 1. The greatest common factor (**GCF**) is $2 \cdot 2 \cdot 3 = 12$.

(b) $54, 90, 108$

Write each number in prime factored form.

$$54 = 2 \cdot 3 \cdot 3 \cdot 3, \quad 90 = 2 \cdot 3 \cdot 3 \cdot 5,$$
$$108 = 2 \cdot 2 \cdot 3 \cdot 3 \cdot 3$$

Use each prime the least number of times it appears. The greatest common factor is $2 \cdot 3 \cdot 3 = 18$.

(c) $15, 19, 25$

Write each number in prime factored form.

$$15 = 3 \cdot 5, \quad 19 = 19, \quad 25 = 5 \cdot 5$$

Use each prime the least number of times it appears. Since no prime appears in all the factored forms, the GCF is 1.

N2. **(a)** $25k^3, 15k^2, 35k^5$

Write each number in prime factored form.

$25k^3 = 5^2 \cdot k^3, \quad 15k^2 = 3 \cdot 5 \cdot k^2,$
$35k^5 = 5 \cdot 7 \cdot k^5$

The greatest common factor of the coefficients 25, 15, and 35 is 5. The greatest common factor of the terms k^3, k^2, and k^5 is k^2 since 2 is the least exponent on k. Thus, the GCF of these terms is the product of 5 and k^2, that is, $5k^2$.

(b) m^3n^5, m^4n^4, m^5n^2

The least exponent on m is 3, and the least exponent on n is 2. Thus, the GCF is m^3n^2.

N3. **(a)** $7t^4 - 14t^3$

7 is the greatest common factor of 7 and 14, and 3 is the least exponent on t. Hence, $7t^3$ is the GCF.

$$7t^4 - 14t^3 = 7t^3(t) - 7t^3(2)$$
$$= 7t^3(t - 2)$$

(b) $8x^6 - 20x^5 + 28x^4$

4 is the greatest common factor of 8, 20, and 28, so 4 can be factored out. Since x occurs in every term and the least exponent on x is 4, x^4 can be factored out. Hence, $4x^4$ is the GCF.

$$8x^6 - 20x^5 + 28x^4$$
$$= 4x^4(2x^2) - 4x^4(5x) + 4x^4(7)$$
$$= 4x^4(2x^2 - 5x + 7)$$

(c) $30m^4n^3 - 42m^2n^2$

6 is the greatest common factor of 30 and 42. The least exponents on m and n are 2 and 2, respectively. Hence, $6m^2n^2$ is the GCF.

$$30m^4n^3 - 42m^2n^2$$
$$= (6m^2n^2)(5m^2n) - (6m^2n^2)(7)$$
$$= 6m^2n^2(5m^2n - 7)$$

N4. **(a)** $x(x + 2) + 5(x + 2)$

The binomial $x + 2$ is the greatest common factor here.

$$x(x + 2) + 5(x + 2) = (x + 2)(x + 5)$$

(b) $a(t + 10) - b(t + 10)$

The binomial $t + 10$ is the greatest common factor.

$$a(t + 10) - b(t + 10) = (t + 10)(a - b)$$

N5. **(a)** $ab + 3a + 5b + 15$
$= (ab + 3a) + (5b + 15)$ *Group terms.*
$= a(b + 3) + 5(b + 3)$ *Factor each group.*
$= (b + 3)(a + 5)$ *Factor out $b + 3$.*

(b) $12xy + 3x + 4y + 1$
$= (12xy + 3x) + (4y + 1)$ *Group terms.*
$= 3x(4y + 1) + 1(4y + 1)$ *Factor each group; remember the 1.*
$= (4y + 1)(3x + 1)$ *Factor out $4y + 1$.*

(c) $x^3 + 5x^2 - 8x - 40$
$= (x^3 + 5x^2) + (-8x - 40)$ *Group terms.*
$= x^2(x + 5) - 8(x + 5)$ *Factor each group.*
$= (x + 5)(x^2 - 8)$ *Factor out $x + 5$.*

N6. (a) $12p^2 - 28q - 16pq + 21p$

Factoring out the common factor 4 from the first two terms and the common factor p from the last two terms gives

$$12p^2 - 28q - 16pq + 21p$$
$$= 4(3p^2 - 7q) + p(-16q + 21).$$

This does not lead to a common factor, so we try rearranging the terms.

$$= (12p^2 - 16pq) + (21p - 28q)$$ *Rearrange.*

$$= 4p(3p - 4q) + 7(3p - 4q)$$ *Factor each group.*

$$= (3p - 4q)(4p + 7)$$ *Factor out $3p - 4q$.*

Here's another rearrangement.

$$= (12p^2 + 21p) + (-16pq - 28q)$$
$$= 3p(4p + 7) - 4q(4p + 7)$$
$$= (4p + 7)(3p - 4q)$$

This is an equivalent answer.

(b) $5xy - 6 - 15x + 2y$
$$= (5xy + 2y) + (-15x - 6)$$
$$= y(5x + 2) - 3(5x + 2)$$
$$= (5x + 2)(y - 3)$$

6.1 Section Exercises

1. Find the prime factored form of each number.

$$40 = 2 \cdot 2 \cdot 2 \cdot 5$$
$$20 = 2 \cdot 2 \cdot 5$$
$$4 = 2 \cdot 2$$

The least number of times 2 appears in all the factored forms is 2. There is no 5 in the prime factored form of 4, so the

$$\text{GCF} = 2^2 = 4.$$

3. Find the prime factored form of each number.

$$18 = 2 \cdot 3 \cdot 3$$
$$24 = 2 \cdot 2 \cdot 2 \cdot 3$$
$$36 = 2 \cdot 2 \cdot 3 \cdot 3$$
$$48 = 2 \cdot 2 \cdot 2 \cdot 2 \cdot 3$$

The least number of times the primes 2 and 3 appear in all four factored forms is once, so

$$\text{GCF} = 2 \cdot 3 = 6.$$

5. 6, 8, 9

Find the prime factored form of each number.

$$6 = 2 \cdot 3$$
$$8 = 2 \cdot 2 \cdot 2$$
$$9 = 3 \cdot 3$$

There are no primes common to all three numbers, so the GCF is 1.

7. Write each term in prime factored form.

$$16y = 2^4 \cdot y$$
$$24 = 2^3 \cdot 3$$

There is no y in the second term, so y will not appear in the GCF. Thus, the GCF of $16y$ and 24 is

$$2^3 = 8.$$

9. $30x^3 = 2 \cdot 3 \cdot 5 \cdot x^3$
$$40x^6 = 2^3 \cdot 5 \cdot x^6$$
$$50x^7 = 2 \cdot 5^2 \cdot x^7$$

The GCF of the coefficients, 30, 40, and 50, is $2^1 \cdot 5^1 = 10$. The smallest exponent on the variable x is 3. Thus, the GCF of the given terms is $10x^3$.

11. $x^4y^3 = x^4 \cdot y^3; \qquad xy^2 = x \cdot y^2$

The GCF is xy^2.

13. $12m^3n^2 = 2^2 \cdot 3 \cdot m^3 \cdot n^2$
$$18m^5n^4 = 2 \cdot 3^2 \cdot m^5 \cdot n^4$$
$$36m^8n^3 = 2^2 \cdot 3^2 \cdot m^8 \cdot n^3$$

The GCF is $2 \cdot 3 \cdot m^3 \cdot n^2 = 6m^3n^2$.

15. $2k^2(5k)$ is written as a product of $2k^2$ and $5k$ and hence, it is *factored*.

17. $2k^2 + (5k + 1)$ is written as a sum of $2k^2$ and $(5k + 1)$, and hence, it is *not factored*.

19. $9m^4 = 3m^2(3m^2)$

Factor out $3m^2$ from $9m^4$ to obtain $3m^2$.

21. $-8z^9 = -4z^5(2z^4)$

Factor out $-4z^5$ from $-8z^9$ to obtain $2z^4$.

23. $6m^4n^5 = 3m^3n(2mn^4)$

Factor out $3m^3n$ from $6m^4n^5$ to obtain $2mn^4$.

25. $12y + 24 = 12 \cdot y + 12 \cdot 2$
$$= 12(y + 2)$$

27. $10a^2 - 20a = 10a(a) - 10a(2)$
$$= 10a(a - 2)$$

29. $8x^2y + 12x^3y^2 = 4x^2y(2) + 4x^2y(3xy)$
$$= 4x^2y(2 + 3xy)$$

31. First, verify that you have factored completely. Then multiply the factors. The product should be the original polynomial.

33. The greatest common factor for $x^2 - 4x$ is x.
$$x^2 - 4x = x(x) + x(-4)$$
$$= x(x - 4)$$

35. The greatest common factor for $6t^2 + 15t$ is $3t$.
$$6t^2 + 15t = 3t(2t) + 3t(5)$$
$$= 3t(2t + 5)$$

37. $27m^3 - 9m$
The GCF is $9m$.
$$27m^3 - 9m = 9m(3m^2) + 9m(-1)$$
$$= 9m(3m^2 - 1)$$

39. $16z^4 + 24z^2$
The GCF is $8z^2$.
$$16z^4 + 24z^2 = 8z^2(2z^2) + 8z^2(3)$$
$$= 8z^2(2z^2 + 3)$$

41. $12x^3 + 6x^2$
The GCF is $6x^2$.
$$12x^3 + 6x^2 = 6x^2(2x) + 6x^2(1)$$
$$= 6x^2(2x + 1)$$

43. $65y^{10} + 35y^6$
The GCF is $5y^6$.
$$65y^{10} + 35y^6 = (5y^6)(13y^4) + (5y^6)(7)$$
$$= 5y^6(13y^4 + 7)$$

45. $11w^3 - 100$

The two terms of this expression have no common factor (except 1), so $11w^3 - 100$ is in factored form.

47. $8mn^3 + 24m^2n^3$
The GCF is $8mn^3$.
$$8mn^3 + 24m^2n^3$$
$$= (8mn^3)(1) + (8mn^3)(3m)$$
$$= 8mn^3(1 + 3m)$$

49. $13y^8 + 26y^4 - 39y^2$
The GCF is $13y^2$.
$$13y^8 + 26y^4 - 39y^2$$
$$= 13y^2(y^6) + 13y^2(2y^2) + 13y^2(-3)$$
$$= 13y^2(y^6 + 2y^2 - 3)$$

51. $36p^6q + 45p^5q^4 + 81p^3q^2$
The GCF is $9p^3q$.
$$36p^6q + 45p^5q^4 + 81p^3q^2$$
$$= 9p^3q(4p^3) + 9p^3q(5p^2q^3) + 9p^3q(9q)$$
$$= 9p^3q(4p^3 + 5p^2q^3 + 9q)$$

53. $a^5 + 2a^3b^2 - 3a^5b^2 + 4a^4b^3$
The GCF is a^3.

$$a^5 + 2a^3b^2 - 3a^5b^2 + 4a^4b^3$$
$$= a^3(a^2) + a^3(2b^2) + a^3(-3a^2b^2)$$
$$+ a^3(4ab^3)$$
$$= a^3(a^2 + 2b^2 - 3a^2b^2 + 4ab^3)$$

55. The GCF of the terms of $c(x + 2) - d(x + 2)$ is the binomial $x + 2$.
$$c(x + 2) - d(x + 2)$$
$$= (x + 2)(c) + (x + 2)(-d)$$
$$= (x + 2)(c - d)$$

57. The GCF of the terms of $m(m + 2n) + n(m + 2n)$ is the binomial $m + 2n$.
$$m(m + 2n) + n(m + 2n)$$
$$= (m + 2n)(m) + (m + 2n)(n)$$
$$= (m + 2n)(m + n)$$

59. The GCF for $q^2(p - 4) + 1(p - 4)$ is $p - 4$.
$$q^2(p - 4) + 1(p - 4) = (p - 4)(q^2 + 1)$$

61. $8(7t + 4) + x(7t + 4)$

This expression is the *sum* of two terms, $8(7t + 4)$ and $x(7t + 4)$, so it is not in factored form. We can factor out $7t + 4$.
$$8(7t + 4) + x(7t + 4)$$
$$= (7t + 4)(8) + (7t + 4)(x)$$
$$= (7t + 4)(8 + x)$$

63. $(8 + x)(7t + 4)$

This expression is the *product* of two factors, $8 + x$ and $7t + 4$, so it is in factored form.

65. $18x^2(y + 4) + 7(y - 4)$

This expression is the *sum* of two terms, $18x^2(y + 4)$ and $7(y - 4)$, so it is not in factored form.

67. It is not possible to factor the expression in Exercise 65 because the two terms, $18x^2(y + 4)$ and $7(y - 4)$, do not have a common factor.

69. $p^2 + 4p + pq + 4q$

The first two terms have a common factor of p, and the last two terms have a common factor of q. Thus,
$$p^2 + 4p + pq + 4q$$
$$= (p^2 + 4p) + (pq + 4q)$$
$$= p(p + 4) + q(p + 4).$$

Now we have two terms which have a common binomial factor of $p + 4$. Thus,
$$p^2 + 4p + pq + 4q$$
$$= p(p + 4) + q(p + 4)$$
$$= (p + 4)(p + q).$$

71. $a^2 - 2a + ab - 2b$

$= (a^2 - 2a) + (ab - 2b)$ *Group the terms.*

$= a(a - 2) + b(a - 2)$ *Factor each group.*

$= (a - 2)(a + b)$ *Factor out $a - 2$.*

73. $7z^2 + 14z - az - 2a$

$= (7z^2 + 14z) + (-az - 2a)$ *Group the terms.*

$= 7z(z + 2) - a(z + 2)$ *Factor each group.*

$= (z + 2)(7z - a)$ *Factor out $z + 2$.*

75. $18r^2 + 12ry - 3xr - 2xy$

$= (18r^2 + 12ry) + (-3xr - 2xy)$ *Group the terms.*

$= 6r(3r + 2y) - x(3r + 2y)$ *Factor each group.*

$= (3r + 2y)(6r - x)$ *Factor out $3r + 2y$.*

77. $3a^3 + 3ab^2 + 2a^2b + 2b^3$

$= (3a^3 + 3ab^2) + (2a^2b + 2b^3)$ *Group the terms.*

$= 3a(a^2 + b^2) + 2b(a^2 + b^2)$ *Factor each group.*

$= (a^2 + b^2)(3a + 2b)$ *Factor out $a^2 + b^2$.*

79. $12 - 4a - 3b + ab$

$= (12 - 4a) + (-3b + ab)$ *Group the terms.*

$= 4(3 - a) - b(3 - a)$ *Factor each group.*

$= (3 - a)(4 - b)$ *Factor out $3 - a$.*

81. $16m^3 - 4m^2p^2 - 4mp + p^3$

$= (16m^3 - 4m^2p^2) + (-4mp + p^3)$

$= 4m^2(4m - p^2) - p(4m - p^2)$

$= (4m - p^2)(4m^2 - p)$

83. $y^2 + 3x + 3y + xy$

$= y^2 + 3y + xy + 3x$ *Rearrange.*

$= (y^2 + 3y) + (xy + 3x)$

$= y(y + 3) + x(y + 3)$

$= (y + 3)(y + x)$

85. $5m - 6p - 2mp + 15$

We need to rearrange these terms to get two groups that each have a common factor. We could group $5m$ with either $-2mp$ or 15.

$5m + 15 - 2mp - 6p$ *Rearrange.*

$= (5m + 15) + (-2mp - 6p)$ *Group the terms.*

$= 5(m + 3) - 2p(m + 3)$ *Factor each group.*

$= (m + 3)(5 - 2p)$ *Factor out $m + 3$.*

87. $18r^2 - 2ty + 12ry - 3rt$

We'll rearrange the terms so that $18r^2$ is grouped with another term containing r.

$18r^2 + 12ry - 3rt - 2ty$ *Rearrange.*

$= (18r^2 + 12ry) + (-3rt - 2ty)$ *Group the terms.*

$= 6r(3r + 2y) - t(3r + 2y)$ *Factor each group.*

$= (3r + 2y)(6r - t)$ *Factor out $3r + 2y$.*

89. $a^5 - 3 + 2a^5b - 6b$

$= a^5 + 2a^5b - 3 - 6b$ *Rearrange.*

$= (a^5 + 2a^5b) + (-3 - 6b)$ *Group the terms.*

$= a^5(1 + 2b) - 3(1 + 2b)$ *Factor each group.*

$= (1 + 2b)(a^5 - 3)$ *Factor out $1 + 2b$.*

91. In order to rewrite

$$2xy + 12 - 3y - 8x$$

as $2xy - 8x - 3y + 12$,

we must change the order of the terms. The property that allows us to do this is the commutative property of addition.

92. After we group both pairs of terms in the rearranged polynomial, we have

$$(2xy - 8x) + (-3y + 12).$$

The greatest common factor for the first pair of terms is $2x$. The GCF for the second pair is -3. Factoring each group gives us

$$2x(y - 4) - 3(y - 4).$$

93. The expression obtained in Exercise 92 is the *difference* between two terms, $2x(y - 4)$ and $3(y - 4)$, so it is *not* in factored form.

94. $2x(y - 4) - 3(y - 4)$

$= (y - 4)(2x - 3)$

or $= (2x - 3)(y - 4)$

Yes, this is the same result as the one shown in Example 6(b), even though the terms were grouped in a different way.

95. $(x + 6)(x - 9) = x^2 - 9x + 6x - 54$

$= x^2 - 3x - 54$

97. $(x + 2)(x + 7) = x^2 + 7x + 2x + 14$

$= x^2 + 9x + 14$

99. $2x^2(x^2 + 3x + 5) = 2x^4 + 6x^3 + 10x^2$

6.2 Factoring Trinomials

6.2 Now Try Exercises

N1. $p^2 + 7p + 10$

Factors of 10	Sums of Factors
10, 1	$10 + 1 = 11$
5, 2	$5 + 2 = 7 \leftarrow$

The pair of integers whose product is 10 and whose sum is 7 is 5 and 2. Thus,

$$p^2 + 7p + 10 = (p + 2)(p + 5).$$

N2. $t^2 - 9t + 18$

Factors of 18	Sums of Factors
$-18, -1$	$-18 + (-1) = -19$
$-9, -2$	$-9 + (-2) = -11$
$-6, -3$	$-6 + (-3) = -9 \leftarrow$

The pair of integers whose product is 18 and whose sum is -9 is -6 and -3. Thus,

$$t^2 - 9t + 18 = (t - 3)(t - 6).$$

N3. $x^2 + x - 42$, or $x^2 + 1x - 42$

Find the two integers whose product is -42 and whose sum is 1. Because the last term is negative, the pair must include one positive and one negative integer.

Factors of -42	Sums of Factors
$42, -1$	$42 + (-1) = 41$
$21, -2$	$21 + (-2) = 19$
$14, -3$	$14 + (-3) = 11$
$7, -6$	$7 + (-6) = 1 \leftarrow$
$-42, 1$	$-42 + 1 = -41$
$-21, 2$	$-21 + 2 = -19$
$-14, 3$	$-14 + 3 = -11$
$-7, 6$	$-7 + 6 = -1$

The required integers are -6 and 7, so

$$x^2 + x - 42 = (x - 6)(x + 7).$$

N4. $x^2 - 4x - 21$

Find the two integers whose product is -21 and whose sum is -4. Because the last term is negative, the pair must include one positive and one negative integer.

Factors of -21	Sums of Factors
$21, -1$	$21 + (-1) = 20$
$7, -3$	$7 + (-3) = 4$
$-21, 1$	$-21 + 1 = -20$
$-7, 3$	$-7 + 3 = -4 \leftarrow$

The required integers are -7 and 3, so

$$x^2 - 4x - 21 = (x - 7)(x + 3).$$

N5. **(a)** $m^2 + 5m + 8$

There is no pair of integers whose product is 8 and whose sum is 5, so $m^2 + 5m + 8$ is a prime polynomial.

(b) $t^2 + 11t - 24$

There is no pair of integers whose product is -24 and whose sum is 11, so $t^2 + 11t - 24$ is a prime polynomial.

N6. $a^2 + 2ab - 15b^2$

Two expressions whose product is $-15b^2$ and whose sum is $2b$ are $5b$ and $-3b$, so

$$a^2 + 2ab - 15b^2 = (a + 5b)(a - 3b).$$

N7. $3y^4 - 27y^3 + 60y^2 = 3y^2(y^2 - 9y + 20)$

Factor $y^2 - 9y + 20$. The integers -5 and -4 have a product of 20 and a sum of -9, so

$$y^2 - 9y + 20 = (y - 5)(y - 4).$$

The completely factored form is

$$3y^4 - 27y^3 + 60y^2 = 3y^2(y - 5)(y - 4).$$

6.2 Section Exercises

1. Product: 48   Sum: -19

Factors of 48	Sums of Factors
1, 48	$1 + 48 = 49$
$-1, -48$	$-1 + (-48) = -49$
2, 24	$2 + 24 = 26$
$-2, -24$	$-2 + (-24) = -26$
3, 16	$3 + 16 = 19$
$-3, -16$	$-3 + (-16) = -19 \leftarrow$
4, 12	$4 + 12 = 16$
$-4, -12$	$-4 + (-12) = -16$
6, 8	$6 + 8 = 14$
$-6, -8$	$-6 + (-8) = -14$

The pair of integers whose product is 48 and whose sum is -19 is -3 and -16.

3. Product: -24 Sum: -5

Factors of -24	Sums of Factors
$1, -24$	$1 + (-24) = -23$
$-1, 24$	$-1 + 24 = 23$
$2, -12$	$2 + (-12) = -10$
$-2, 12$	$-2 + 12 = 10$
$3, -8$	$3 + (-8) = -5 \leftarrow$
$-3, 8$	$-3 + 8 = 5$
$4, -6$	$4 + (-6) = -2$
$-4, 6$	$-4 + 6 = 2$

The pair of integers whose product is -24 and whose sum is -5 is 3 and -8.

5. If the coefficient of the last term of the trinomial is negative, then a and b must have different signs, one positive and one negative.

7. A *prime polynomial* is one that cannot be factored using only integers in the factors.

9. $x^2 - 12x + 32$

Multiply each of the given pairs of factors to determine which one gives the required product.

A. $(x - 8)(x + 4) = x^2 - 4x - 32$

B. $(x + 8)(x - 4) = x^2 + 4x - 32$

C. $(x - 8)(x - 4) = x^2 - 12x + 32$

D. $(x + 8)(x + 4) = x^2 + 12x + 32$

Choice **C** is the correct factored form.

11. Multiply the factors using FOIL to determine the polynomial.

$$(a + 9)(a + 4)$$
$$\mathbf{F \quad O \quad I \quad L}$$
$$= a(a) + a(4) + 9(a) + 9(4)$$
$$= a^2 + 4a + 9a + 36$$
$$= a^2 + 13a + 36$$

13. $p^2 + 11p + 30 = (p + 5)(\underline{\quad})$

Look for an integer whose product with 5 is 30 and whose sum with 5 is 11. That integer is 6.

$$p^2 + 11p + 30 = (p + 5)(p + 6)$$

15. $x^2 + 15x + 44 = (x + 4)(\underline{\quad})$

Look for an integer whose product with 4 is 44 and whose sum with 4 is 15. That integer is 11.

$$x^2 + 15x + 44 = (x + 4)(x + 11)$$

17. $x^2 - 9x + 8 = (x - 1)(\underline{\quad})$

Look for an integer whose product with -1 is 8 and whose sum with -1 is -9. That integer is -8.

$$x^2 - 9x + 8 = (x - 1)(x - 8)$$

19. $y^2 - 2y - 15 = (y + 3)(\underline{\quad})$

Look for an integer whose product with 3 is -15 and whose sum with 3 is -2. That integer is -5.

$$y^2 - 2y - 15 = (y + 3)(y - 5)$$

21. $x^2 + 9x - 22 = (x - 2)(\underline{\quad})$

Look for an integer whose product with -2 is -22 and whose sum with -2 is 9. That integer is 11.

$$x^2 + 9x - 22 = (x - 2)(x + 11)$$

23. $y^2 - 7y - 18 = (y + 2)(\underline{\quad})$

Look for an integer whose product with 2 is -18 and whose sum with 2 is -7. That integer is -9.

$$y^2 - 7y - 18 = (y + 2)(y - 9)$$

25. $y^2 + 9y + 8$

Look for two integers whose product is 8 and whose sum is 9. Both integers must be positive because b and c are both positive.

Factors of 8	Sums of Factors
1, 8	9 ←
2, 4	6

Thus,

$$y^2 + 9y + 8 = (y + 8)(y + 1).$$

27. $b^2 + 8b + 15$

Look for two integers whose product is 15 and whose sum is 8. Both integers must be positive because b and c are both positive.

Factors of 15	Sums of Factors
1, 15	16
3, 5	8 ←

Thus,

$$b^2 + 8b + 15 = (b + 3)(b + 5).$$

29. $m^2 + m - 20$

Look for two integers whose product is -20 and whose sum is 1. Since c is negative, one integer must be positive and one must be negative.

Factors of -20	Sums of Factors
$-1, 20$	19
$1, -20$	-19
$-2, 10$	8
$2, -10$	-8
$-4, 5$	1 ←
$4, -5$	-1

Thus,

$$m^2 + m - 20 = (m - 4)(m + 5).$$

31. $y^2 - 8y + 15$

Find two integers whose product is 15 and whose sum is -8. Since c is positive and b is negative, both integers must be negative.

Factors of 15	Sums of Factors
$-1, -15$	-16
$-3, -5$	-8 ←

Thus,

$$y^2 - 8y + 15 = (y - 5)(y - 3).$$

33. $x^2 + 4x + 5$

Look for two integers whose product is 5 and whose sum is 4. Both integers must be positive since b and c are both positive.

Product	*Sum*
$5 \cdot 1 = 5$	$5 + 1 = 6$

There is no other pair of positive integers whose product is 5. Since there is no pair of integers whose product is 5 and whose sum is 4, $x^2 + 4x + 5$ is a *prime* polynomial.

35. $z^2 - 15z + 56$

Find two integers whose product is 56 and whose sum is -15. Since c is positive and b is negative, both integers must be negative.

Factors of 56	Sums of Factors
$-1, -56$	-57
$-2, -28$	-30
$-4, -14$	-18
$-7, -8$	-15 ←

Thus,
$$z^2 - 15z + 56 = (z - 7)(z - 8).$$

37. $r^2 - r - 30$

Look for two integers whose product is -30 and whose sum is -1. Because c is negative, one integer must be positive and the other must be negative.

Factors of -30	Sums of Factors
$-1, 30$	29
$1, -30$	-29
$-2, 15$	13
$2, -15$	-13
$-3, 10$	7
$3, -10$	-7
$-5, 6$	1
$5, -6$	-1 ←

Thus,
$$r^2 - r - 30 = (r + 5)(r - 6).$$

39. $a^2 - 8a - 48$

Find two integers whose product is -48 and whose sum is -8. Since c is negative, one integer must be positive and one must be negative.

Factors of -48	Sums of Factors
$-1, 48$	47
$1, -48$	-47
$-2, 24$	22
$2, -24$	-22
$-3, 16$	13
$3, -16$	-13
$-4, 12$	8
$4, -12$	-8 ←
$-6, 8$	2
$6, -8$	-2

Thus,
$$a^2 - 8a - 48 = (a + 4)(a - 12).$$

41. $x^2 + 3x - 39$

Look for two integers whose product is -39 and whose sum is 3. Because c is negative, one integer must be positive and one must be negative.

Factors of -39	Sums of Factors
$-1, 39$	38
$1, -39$	-38
$-3, 13$	10
$3, -13$	-10

This list does not produce the required integers, and there are no other possibilities to try. Therefore, $x^2 + 3x - 39$ is *prime*.

43. $-32 + 14x + x^2$, or $x^2 + 14x - 32$

Look for two integers whose product is -32 and whose sum is 14. Since c is negative, one integer must be positive and one must be negative.

Factors of -32	Sums of Factors
$-1, 32$	31
$1, -32$	-31
$-2, 16$	14 ←
$2, -16$	-14
$-4, 8$	4
$4, -8$	-4

Thus,
$$-32 + 14x + x^2 = (x - 2)(x + 16).$$

45. $r^2 + 3ra + 2a^2$

Look for two expressions whose product is $2a^2$ and whose sum is $3a$. They are $2a$ and a, so
$$r^2 + 3ra + 2a^2 = (r + 2a)(r + a).$$

47. $t^2 - tz - 6z^2$

Look for two expressions whose product is $-6z^2$ and whose sum is $-z$. They are $2z$ and $-3z$, so
$$t^2 - tz - 6z^2 = (t + 2z)(t - 3z).$$

49. $x^2 + 4xy + 3y^2$

Look for two expressions whose product is $3y^2$ and whose sum is $4y$. The expressions are $3y$ and y, so

$$x^2 + 4xy + 3y^2 = (x + 3y)(x + y).$$

51. $v^2 - 11vw + 30w^2$

Factors of $30w^2$	Sums of Factors
$-30w, -w$	$-31w$
$-15w, -2w$	$-17w$
$-10w, -3w$	$-13w$
$-5w, -6w$	$-11w \leftarrow$

The completely factored form is

$$v^2 - 11vw + 30w^2 = (v - 5w)(v - 6w).$$

53. $4x^2 + 12x - 40$

First, factor out the GCF, 4.

$$4x^2 + 12x - 40 = 4(x^2 + 3x - 10)$$

Now factor $x^2 + 3x - 10$.

Factors of -10	Sums of Factors
$-1, 10$	9
$1, -10$	-9
$2, -5$	-3
$-2, 5$	$3 \leftarrow$

Thus,

$$x^2 + 3x - 10 = (x - 2)(x + 5).$$

The completely factored form is

$$4x^2 + 12x - 40 = 4(x - 2)(x + 5).$$

55. $2t^3 + 8t^2 + 6t$

First, factor out the GCF, $2t$.

$$2t^3 + 8t^2 + 6t = 2t(t^2 + 4t + 3)$$

Then factor $t^2 + 4t + 3$.

$$t^2 + 4t + 3 = (t + 1)(t + 3)$$

The completely factored form is

$$2t^3 + 8t^2 + 6t = 2t(t + 1)(t + 3).$$

57. $2x^6 + 8x^5 - 42x^4$

First, factor out the GCF, $2x^4$.

$$2x^6 + 8x^5 - 42x^4 = 2x^4(x^2 + 4x - 21)$$

Now factor $x^2 + 4x - 21$.

Factors of -21	Sums of Factors
$1, -21$	-20
$-1, 21$	20
$3, -7$	-4
$-3, 7$	$4 \leftarrow$

Thus,

$$x^2 + 4x - 21 = (x - 3)(x + 7).$$

The completely factored form is

$$2x^6 + 8x^5 - 42x^4 = 2x^4(x - 3)(x + 7).$$

59. $5m^5 + 25m^4 - 40m^2$

Factor out the GCF, $5m^2$.

$$5m^5 + 25m^4 - 40m^2 = 5m^2(m^3 + 5m^2 - 8)$$

61. $m^3n - 10m^2n^2 + 24mn^3$

First, factor out the GCF, mn.

$$m^3n - 10m^2n^2 + 24mn^3$$
$$= mn(m^2 - 10mn + 24n^2)$$

The expressions $-6n$ and $-4n$ have a product of $24n^2$ and a sum of $-10n$. The completely factored form is

$$m^3n - 10m^2n^2 + 24mn^3$$
$$= mn(m - 6n)(m - 4n).$$

63. $a^5 + 3a^4b - 4a^3b^2$

The GCF is a^3, so

$$a^5 + 3a^4b - 4a^3b^2$$
$$= a^3(a^2 + 3ab - 4b^2).$$

Now factor $a^2 + 3ab - 4b^2$. The expressions $4b$ and $-b$ have a product of $-4b^2$ and a sum of $3b$. The completely factored form is

$$a^5 + 3a^4b - 4a^3b^2 = a^3(a + 4b)(a - b).$$

65. $y^3z + y^2z^2 - 6yz^3$

The GCF is yz, so

$$y^3z + y^2z^2 - 6yz^3 = yz(y^2 + yz - 6z^2).$$

Now factor $y^2 + yz - 6z^2$. The expressions $3z$ and $-2z$ have a product of $-6z^2$ and a sum of z. The completely factored form is

$$y^3z + y^2z^2 - 6yz^3 = yz(y + 3z)(y - 2z).$$

67. $z^{10} - 4z^9y - 21z^8y^2$
$$= z^8(z^2 - 4zy - 21y^2) \quad \text{GCF is } z^8.$$
$$= z^8(z - 7y)(z + 3y)$$

69. $(a+b)x^2 + (a+b)x - 12(a+b)$

The GCF is $(a+b)$, so
$$(a+b)x^2 + (a+b)x - 12(a+b)$$
$$= (a+b)(x^2 + x - 12).$$

Now factor $x^2 + x - 12$.
$$x^2 + x - 12 = (x+4)(x-3)$$

The completely factored form is
$$(a+b)x^2 + (a+b)x - 12(a+b)$$
$$= (a+b)(x+4)(x-3).$$

71. $(2p+q)r^2 - 12(2p+q)r + 27(2p+q)$

The GCF is $(2p+q)$, so
$$(2p+q)r^2 - 12(2p+q)r + 27(2p+q)$$
$$= (2p+q)(r^2 - 12r + 27).$$

Now factor $r^2 - 12r + 27$.
$$r^2 - 12r + 27 = (r-9)(r-3)$$

The completely factored form is
$$(2p+q)r^2 - 12(2p+q)r + 27(2p+q)$$
$$= (2p+q)(r-9)(r-3).$$

73. $(2y-7)(y+4) = 2y^2 + 8y - 7y - 28$
$$= 2y^2 + y - 28$$

75. $(5z+2)(3z-2) = 15z^2 - 10z + 6z - 4$
$$= 15z^2 - 4z - 4$$

6.3   More on Factoring Trinomials

6.3   Now Try Exercises

N1. **(a)** $2z^2 + 5z + 3$

Find the two integers whose product is $2(3) = 6$ and whose sum is 5. The integers are 2 and 3. Write the middle term, $5z$, as $2z + 3z$.
$$2z^2 + 5z + 3 = 2z^2 + 2z + 3z + 3$$
$$= (2z^2 + 2z) + (3z + 3)$$
$$= 2z(z+1) + 3(z+1)$$
$$= (z+1)(2z+3)$$

(b) $15m^2 + m - 2$

Find two integers whose product is $15(-2) = -30$ and whose sum is 1. The integers are 6 and -5.

$$15m^2 + m - 2 = 15m^2 + 6m - 5m - 2$$
$$= (15m^2 + 6m) + (-5m - 2)$$
$$= 3m(5m+2) - 1(5m+2)$$
$$= (5m+2)(3m-1)$$

Note that if we had written $-5m + 6m$ instead of $6m - 5m$, the order of the factors in the final expression would be reversed; that is, $(3m-1)(5m+2)$.

(c) $8x^2 - 2xy - 3y^2$

Find two integers whose product is $8(-3) = -24$ and whose sum is -2. The integers are -6 and 4.

$$8x^2 - 2xy - 3y^2$$
$$= 8x^2 - 6xy + 4xy - 3y^2$$
$$= (8x^2 - 6xy) + (4xy - 3y^2)$$
$$= 2x(4x - 3y) + y(4x - 3y)$$
$$= (4x - 3y)(2x + y)$$

N2. $15z^6 + 18z^5 - 24z^4$

First factor out the greatest common factor, $3z^4$.
$$15z^6 + 18z^5 - 24z^4 = 3z^4(5z^2 + 6z - 8)$$

Find two integers whose product is $5(-8) = -40$ and whose sum is 6. The integers are -4 and 10.

$$5z^2 + 6z - 8 = 5z^2 - 4z + 10z - 8$$
$$= (5z^2 - 4z) + (10z - 8)$$
$$= z(5z - 4) + 2(5z - 4)$$
$$= (5z - 4)(z + 2)$$

The completely factored form is
$$15z^6 + 18z^5 - 24z^4 = 3z^4(5z - 4)(z + 2).$$

N3. $8y^2 + 22y + 5$

The factors of $8y^2$ are $2y$ and $4y$, or $8y$ and y. The factors of 5 are 5 and 1. Try various combinations, checking to see if the middle term is $22y$ in each case.

$$(4y + 5)(2y + 1) = 8y^2 + 14y + 5 \quad \textit{Incorrect}$$
$$(4y + 1)(2y + 5) = 8y^2 + 22y + 5 \quad \textit{Correct}$$

Thus, $8y^2 + 22y + 5 = (4y + 1)(2y + 5)$.

N4. $10x^2 - 9x + 2$

The factors of $10x^2$ are $10x$ and x, or $5x$ and $2x$. Try $5x$ and $2x$. Since the last term is positive and the coefficient of the middle term is negative, only negative factors of 2 should be considered. The factors of 2 are -2 and -1.

$$(5x - 1)(2x - 2)$$
$$= 10x^2 - 12x + 2 \quad \textit{Incorrect}$$

$$(5x - 2)(2x - 1)$$
$$= 10x^2 - 9m + 2 \quad \textit{Correct}$$

Thus, $10x^2 - 9x + 2 = (5x - 2)(2x - 1)$.

N5. $10a^2 + 31a - 14$

The factors of $10a^2$ are $5a$ and $2a$, or $10a$ and a. Some factors of -14 are 14 and -1, or 7 and -2. Try various possibilities.

$(5a - 14)(2a + 1) = 10a^2 - 23a - 14$ *Incorrect*
$(5a - 2)(2a + 7) = 10a^2 + 31a - 14$ *Correct*

Thus, $10a^2 + 31a - 14 = (5a - 2)(2a + 7)$.

N6. $8z^2 + 2wz - 15w^2$

Try various possibilities.

$(4z + 5w)(2z - 3w)$
$= 8z^2 - 2wz - 15w^2$ *Incorrect*

The middle terms differ only in sign, so reverse the signs of the two factors.

$(4z - 5w)(2z + 3w)$
$= 8z^2 + 2wz - 15w^2$ *Correct*

Thus,

$8z^2 + 2wz - 15w^2 = (4z - 5w)(2z + 3w)$.

N7. $-10x^3 - 45x^2 + 90x$

The common factor could be $5x$ or $-5x$. If we factor out $-5x$, the first term of the trinomial factor will be positive, which makes it easier to factor.

$-10x^3 - 45x^2 + 90x$
$= -5x(2x^2 + 9x - 18)$

Factor $2x^2 + 9x - 18$ by trial and error to obtain

$2x^2 + 9x - 18 = (2x - 3)(x + 6)$.

The completely factored form is

$-10x^3 - 45x^2 + 90x$
$= -5x(2x - 3)(x + 6)$.

6.3 Section Exercises

1. $10t^2 + 5t + 4t + 2$
$= (10t^2 + 5t) + (4t + 2)$ *Group terms.*
$= 5t(2t + 1) + 2(2t + 1)$ *Factor each group.*
$= (2t + 1)(5t + 2)$ *Factor out $2t + 1$.*

3. $15z^2 - 10z - 9z + 6$
$= (15z^2 - 10z) + (-9z + 6)$ *Group terms.*
$= 5z(3z - 2) - 3(3z - 2)$ *Factor each group.*
$= (3z - 2)(5z - 3)$ *Factor out $3z - 2$.*

5. $8s^2 - 4st + 6st - 3t^2$
$= (8s^2 - 4st) + (6st - 3t^2)$ *Group terms.*
$= 4s(2s - t) + 3t(2s - t)$ *Factor each group.*
$= (2s - t)(4s + 3t)$ *Factor out $2s - t$.*

7. $2m^2 + 11m + 12$

(a) Find two integers whose product is $\underline{2} \cdot \underline{12} = \underline{24}$ and whose sum is $\underline{11}$.

(b) The required integers are $\underline{3}$ and $\underline{8}$. (Order is irrelevant.)

(c) Write the middle term, $11m$, as $\underline{3m} + \underline{8m}$.

(d) Rewrite the given trinomial as
$\underline{2m^2 + 3m + 8m + 12}$.

(e) $(2m^2 + 3m) + (8m + 12)$ *Group terms.*
$= m(2m + 3) + 4(2m + 3)$ *Factor each group.*
$= (2m + 3)(m + 4)$ *Factor out $2m + 3$.*

(f) $(2m + 3)(m + 4)$
       F   **O**   **I**   **L**
$= 2m(m) + 2m(4) + 3(m) + 3(4)$
$= 2m^2 + 8m + 3m + 12$
$= 2m^2 + 11m + 12$

9. To factor $12y^2 + 5y - 2$, we must find two integers with a product of $12(-2) = -24$ and a sum of 5. The only pair of integers satisfying those conditions is 8 and -3, choice **B**.

11. $2x^2 - x - 1$

Multiply the factors in the choices together to see which ones give the correct product. Since

$(2x - 1)(x + 1) = 2x^2 + x - 1$
and $(2x + 1)(x - 1) = 2x^2 - x - 1$,

the correct factored form is choice **B**,
$(2x + 1)(x - 1)$.

13. $4y^2 + 17y - 15$

Multiply the factors in the choices together to see which ones give the correct product. Since

$(y + 5)(4y - 3) = 4y^2 + 17y - 15$
and $(2y - 5)(2y + 3) = 4y^2 - 4y - 15$,

the correct factored form is choice **A**,
$(y + 5)(4y - 3)$.

15. $6a^2 + 7ab - 20b^2 = (3a - 4b)(\underline{\quad\quad})$

The first term in the missing expression must be $2a$ since

$$(3a)(2a) = 6a^2.$$

The second term in the missing expression must be $5b$ since

$$(-4b)(5b) = -20b^2.$$

Checking our answer by multiplying, we see that
$(3a - 4b)(\underline{2a + 5b}) = 6a^2 + 7ab - 20b^2$,
as desired.

17. $2x^2 + 6x - 8 = 2(x^2 + 3x - 4)$

To factor $x^2 + 3x - 4$, we look for two integers whose product is -4 and whose sum is 3. The integers are 4 and -1. Thus,

$$2x^2 + 6x - 8 = 2(x + 4)(x - 1).$$

19. $4z^3 - 10z^2 - 6z = 2z(2z^2 - 5z - 3)$
$$= 2z(2z + 1)(z - 3)$$

21. The binomial $2x - 6$ cannot be a factor of $12x^2 + 7x - 12$ because its terms have a common factor of 2, which the polynomial terms do not have.

In Exercises 23–78, either the trial and error method (which uses FOIL in reverse) or the grouping method can be used to factor each polynomial.

23. $3a^2 + 10a + 7$

Factor by the grouping method. Look for two integers whose product is $3(7) = 21$ and whose sum is 10. The integers are 3 and 7. Use these integers to rewrite the middle term, $10a$, as $3a + 7a$, and then factor the resulting four-term polynomial by grouping.

$3a^2 + 10a + 7$
$= 3a^2 + 3a + 7a + 7$ *$10a = 3a + 7a$*
$= (3a^2 + 3a) + (7a + 7)$ *Group the terms.*
$= 3a(a + 1) + 7(a + 1)$ *Factor each group.*
$= (a + 1)(3a + 7)$ *Factor out $a + 1$.*

25. $2y^2 + 7y + 6$

Factor by the grouping method. Look for two integers whose product is $2(6) = 12$ and whose sum is 7. The integers are 3 and 4.

$2y^2 + 7y + 6$
$= 2y^2 + 3y + 4y + 6$ *$7y = 3y + 4y$*
$= (2y^2 + 3y) + (4y + 6)$ *Group terms.*
$= y(2y + 3) + 2(2y + 3)$ *Factor each group.*
$= (2y + 3)(y + 2)$ *Factor out $2y + 3$.*

27. $15m^2 + m - 2$

Factor by the grouping method. Look for two integers whose product is $15(-2) = -30$ and whose sum is 1. The integers are 6 and -5.

$15m^2 + m - 2$
$= 15m^2 + 6m - 5m - 2$ *$m = 6m - 5m$*
$= (15m^2 + 6m) + (-5m - 2)$ *Group the terms.*
$= 3m(5m + 2) - 1(5m + 2)$ *Factor each group.*
$= (5m + 2)(3m - 1)$ *Factor out $5m + 2$.*

29. $12s^2 + 11s - 5$

Factor by trial and error.

Possible factors of $12s^2$ are s and $12s$, $2s$ and $6s$, or $3s$ and $4s$.

Factors of -5 are -1 and 5 or -5 and 1.

$(2s - 1)(6s + 5) = 12s^2 + 4s - 5$ *Incorrect*
$(2s + 1)(6s - 5) = 12s^2 - 4s - 5$ *Incorrect*
$(3s - 1)(4s + 5) = 12s^2 + 11s - 5$ *Correct*

31. $10m^2 - 23m + 12$

Factor by the grouping method. Look for two integers whose product is $10(12) = 120$ and whose sum is -23. The integers are -8 and -15.

$10m^2 - 23m + 12$
$= 10m^2 - 8m - 15m + 12$ *$-23m = -8m - 15m$*
$= (10m^2 - 8m) + (-15m + 12)$ *Group the terms.*
$= 2m(5m - 4) - 3(5m - 4)$ *Factor each group.*
$= (5m - 4)(2m - 3)$ *Factor out $5m - 4$.*

33. $8w^2 - 14w + 3$

Factor by trial and error. Possible factors of $8w^2$ are w and $8w$ or $2w$ and $4w$.
Factors of 3 are -1 and -3 (since $b = -14$ is negative).

$(4w - 3)(2w - 1) = 8w^2 - 10w + 3$ *Incorrect*
$(4w - 1)(2w - 3) = 8w^2 - 14w + 3$ *Correct*

35. $20y^2 - 39y - 11$

Factor by the grouping method. Look for two integers whose product is $20(-11) = -220$ and whose sum is -39. The integers are -44 and 5.

$20y^2 - 39y - 11$
$= 20y^2 - 44y + 5y - 11$ *$-39y = -44y + 5y$*
$= (20y^2 - 44y) + (5y - 11)$ *Group the terms.*
$= 4y(5y - 11) + 1(5y - 11)$ *Factor each group.*
$= (5y - 11)(4y + 1)$ *Factor out $5y - 11$.*

37. $3x^2 - 15x + 16$

Factor by the grouping method. Look for two integers whose product is $3(16) = 48$ and whose sum is -15. The negative factors and their sums are:

$$-1 + (-48) = -49$$
$$-2 + (-24) = -26$$
$$-3 + (-16) = -19$$
$$-4 + (-12) = -16$$
$$-6 + (-8) = -14$$

So there are no integers satisfying the conditions and the polynomial is *prime*.

39. First, factor out the greatest common factor, 2.

$$20x^2 + 22x + 6 = 2(10x^2 + 11x + 3)$$

Now factor $10x^2 + 11x + 3$ by trial and error to obtain

$$10x^2 + 11x + 3 = (5x + 3)(2x + 1).$$

The complete factorization is

$$20x^2 + 22x + 6 = 2(5x + 3)(2x + 1).$$

41. First, factor out the GCF, 3.

$$24x^2 - 42x + 9 = 3(8x^2 - 14x + 3)$$

Use the grouping method to factor $8x^2 - 14x + 3$. Look for two integers whose product is $8(3) = 24$ and whose sum is -14. The integers are -12 and -2.

$$24x^2 - 42x + 9$$
$$= 3(8x^2 - 12x - 2x + 3) \quad -14x = -12x - 2x$$
$$= 3[(8x^2 - 12x) + (-2x + 3)] \quad \textit{Group the terms.}$$
$$= 3[4x(2x - 3) - 1(2x - 3)] \quad \textit{Factor each group.}$$
$$= 3(2x - 3)(4x - 1) \quad \textit{Factor out } 2x - 3.$$

43. First, factor out the GCF, q.

$$40m^2q + mq - 6q = q(40m^2 + m - 6)$$

Now factor $40m^2 + m - 6$ by trial and error to obtain

$$40m^2 + m - 6 = (5m + 2)(8m - 3).$$

The complete factorization is

$$40m^2q + mq - 6q = q(5m + 2)(8m - 3).$$

45. First, factor out the GCF, $3n^2$.

$$15n^4 - 39n^3 + 18n^2 = 3n^2(5n^2 - 13n + 6)$$

Factor $5n^2 - 13n + 6$ by the trial and error method. Possible factors of $5n^2$ are $5n$ and n.

Possible factors of 6 are -6 and -1, or -3 and -2.

$$(5n - 6)(n - 1) = 5n^2 - 11n + 6 \quad \textit{Incorrect}$$
$$(5n - 3)(n - 2) = 5n^2 - 13n + 6 \quad \textit{Correct}$$

The completely factored form is

$$15n^4 - 39n^3 + 18n^2 = 3n^2(5n - 3)(n - 2).$$

47. First, factor out the GCF, y^2.

$$15x^2y^2 - 7xy^2 - 4y^2 = y^2(15x^2 - 7x - 4)$$

Factor $15x^2 - 7x - 4$ by the grouping method. Look for two integers whose product is $15(-4) = -60$ and whose sum is -7. The integers are -12 and 5.

$$15x^2y^2 - 7xy^2 - 4y^2$$
$$= y^2(15x^2 - 12x + 5x - 4)$$
$$= y^2[3x(5x - 4) + 1(5x - 4)]$$
$$= y^2(5x - 4)(3x + 1)$$

49. $5a^2 - 7ab - 6b^2$

Factor by the grouping method. Look for two integers whose product is $5(-6) = -30$ and whose sum is -7. The integers are -10 and 3.

$$5a^2 - 7ab - 6b^2$$
$$= 5a^2 - 10ab + 3ab - 6b^2$$
$$= (5a^2 - 10ab) + (3ab - 6b^2)$$
$$= 5a(a - 2b) + 3b(a - 2b)$$
$$= (a - 2b)(5a + 3b)$$

51. $12s^2 + 11st - 5t^2$

Factor by the grouping method. Look for two integers whose product is $12(-5) = -60$ and whose sum is 11. The integers are 15 and -4.

$$12s^2 + 11st - 5t^2$$
$$= 12s^2 + 15st - 4st - 5t^2$$
$$= (12s^2 + 15st) + (-4st - 5t^2)$$
$$= 3s(4s + 5t) - t(4s + 5t)$$
$$= (4s + 5t)(3s - t)$$

53. $6m^6n + 7m^5n^2 + 2m^4n^3$
$$= m^4n(6m^2 + 7mn + 2n^2) \quad GCF = m^4n$$

Now factor $6m^2 + 7mn + 2n^2$ by trial and error.

Possible factors of $6m^2$ are $6m$ and m or $3m$ and $2m$. Possible factors of $2n^2$ are $2n$ and n.

$$(3m + 2n)(2m + n) = 6m^2 + 7mn + 2n^2$$
$$\textit{Correct}$$

The completely factored form is

$$6m^6n + 7m^5n^2 + 2m^4n^3$$
$$= m^4n(3m + 2n)(2m + n).$$

55. $5 - 6x + x^2$

$= x^2 - 6x + 5$ — *Re-order.*

$= x^2 - 5x - x + 5$ — $-6x = -5x - x$

$= (x^2 - 5x) + (-x + 5)$ — *Group the terms.*

$= x(x - 5) - 1(x - 5)$ — *Factor each group.*

$= (x - 5)(x - 1)$ — *Factor out $x - 5$.*

57. $16 + 16x + 3x^2 = 3x^2 + 16x + 16$

Factor by the grouping method. Find two integers whose product is $(3)(16) = 48$ and whose sum is 16. The numbers are 4 and 12.

$3x^2 + 16x + 16$
$= 3x^2 + 4x + 12x + 16$
$= (3x^2 + 4x) + (12x + 16)$
$= x(3x + 4) + 4(3x + 4)$
$= (3x + 4)(x + 4)$

59. $-10x^3 + 5x^2 + 140x$

First, factor out $-5x$; then complete the factoring by trial and error, using FOIL to test various possibilities until the correct one is found.

$-10x^3 + 5x^2 + 140x$
$= -5x(2x^2 - x - 28)$
$= -5x(2x + 7)(x - 4)$

61. $12x^2 - 47x - 4$

Factor by the grouping method. Find two integers whose product is $12(-4) = -48$ and whose sum is -47. The numbers are 1 and -48.

$12x^2 - 47x - 4$
$= 12x^2 + x - 48x - 4$
$= (12x^2 + x) + (-48x - 4)$
$= x(12x + 1) - 4(12x + 1)$
$= (12x + 1)(x - 4)$

63. $24y^2 - 41xy - 14x^2$

Factor by using trial and error. Only positive factors of 24 should be considered: 1 and 24, 2 and 12, 3 and 8, or 4 and 6. The factors of -14 include 7 and -2. The correct factorization is

$24y^2 - 41xy - 14x^2 = (24y + 7x)(y - 2x)$.

65. $36x^4 - 64x^2y + 15y^2$

Factor by using trial and error. For $15y^2$, consider only $-y$ and $-15y$, and $-3y$ and $-5y$.

$36x^4 - 64x^2y + 15y^2 = (18x^2 - 5y)(2x^2 - 3y)$

67. $48a^2 - 94ab - 4b^2$

$= 2(24a^2 - 47ab - 2b^2)$ $GCF = 2$

Now factor $24a^2 - 47ab - 2b^2$ by the grouping method. Look for two integers whose product is $24(-2) = -48$ and whose sum is -47. The integers are 1 and -48.

$24a^2 - 47ab - 2b^2$
$= 24a^2 + ab - 48ab - 2b^2$
$= (24a^2 + ab) + (-48ab - 2b^2)$
$= a(24a + b) - 2b(24a + b)$
$= (24a + b)(a - 2b)$

The completely factored form is

$48a^2 - 94ab - 4b^2 = 2(24a + b)(a - 2b)$.

69. $10x^4y^5 + 39x^3y^5 - 4x^2y^5$

$= x^2y^5(10x^2 + 39x - 4)$ $GCF = x^2y^5$

Now factor $10x^2 + 39x - 4$ by the grouping method. Look for two integers whose product is $10(-4) = -40$ and whose sum is 39. The integers are -1 and 40.

$10x^2 + 39x - 4$
$= 10x^2 - x + 40x - 4$
$= (10x^2 - x) + (40x - 4)$
$= x(10x - 1) + 4(10x - 1)$
$= (10x - 1)(x + 4)$

The completely factored form is

$10x^4y^5 + 39x^3y^5 - 4x^2y^5$
$= x^2y^5(10x - 1)(x + 4)$.

71. $36a^3b^2 - 104a^2b^2 - 12ab^2$

$= 4ab^2(9a^2 - 26a - 3)$ $GCF = 4ab^2$

Now factor $9a^2 - 26a - 3$ by the grouping method. Look for two integers whose product is $9(-3) = -27$ and whose sum is -26. The integers are 1 and -27.

$9a^2 - 26a - 3$
$= 9a^2 + a - 27a - 3$
$= (9a^2 + a) + (-27a - 3)$
$= a(9a + 1) - 3(9a + 1)$
$= (9a + 1)(a - 3)$

The completely factored form is

$36a^3b^2 - 104a^2b^2 - 12ab^2$
$= 4ab^2(9a + 1)(a - 3)$.

73. $24x^2 - 46x + 15$

Factor by the grouping method. Look for two integers whose product is $24(15) = 360$ and whose sum is -46. The integers are -10 and -36.

$$24x^2 - 46x + 15$$
$$= 24x^2 - 10x - 36x + 15$$
$$= (24x^2 - 10x) + (-36x + 15)$$
$$= 2x(12x - 5) - 3(12x - 5)$$
$$= (12x - 5)(2x - 3)$$

75. $24x^4 + 55x^2 - 24$

Factor by the grouping method. Look for two integers whose product is $24(-24) = -576$ and whose sum is 55. The integers are -9 and 64. Note the x^2-term.

$$24x^4 + 55x^2 - 24$$
$$= 24x^4 - 9x^2 + 64x^2 - 24$$
$$= (24x^4 - 9x^2) + (64x^2 - 24)$$
$$= 3x^2(8x^2 - 3) + 8(8x^2 - 3)$$
$$= (8x^2 - 3)(3x^2 + 8)$$

77. $24x^2 + 38xy + 15y^2$

Factor by the grouping method. Look for two integers whose product is $24(15) = 360$ and whose sum is 38. The integers are 18 and 20. Note the xy-term.

$$24x^2 + 38xy + 15y^2$$
$$= 24x^2 + 18xy + 20xy + 15y^2$$
$$= (24x^2 + 18xy) + (20xy + 15y^2)$$
$$= 6x(4x + 3y) + 5y(4x + 3y)$$
$$= (4x + 3y)(6x + 5y)$$

79. $-x^2 - 4x + 21 = -1(x^2 + 4x - 21)$
$$= -1(x + 7)(x - 3)$$

81. $-3x^2 - x + 4 = -1(3x^2 + x - 4)$
$$= -1(3x + 4)(x - 1)$$

83. $-2a^2 - 5ab - 2b^2$
$$= -1(2a^2 + 5ab + 2b^2)$$
Factor out -1.
$$= -1(2a^2 + 4ab + ab + 2b^2)$$
$5ab = 4ab + ab$
$$= -1[(2a^2 + 4ab) + (ab + 2b^2)]$$
Group the terms.
$$= -1[2a(a + 2b) + b(a + 2b)]$$
Factor each group.
$$= -1(a + 2b)(2a + b)$$
Factor out $a + 2b$.

85. $25q^2(m + 1)^3 - 5q(m + 1)^3 - 2(m + 1)^3$

First, factor out the GCF, $(m + 1)^3$; then factor the resulting trinomial by trial and error.

$$25q^2(m + 1)^3 - 5q(m + 1)^3 - 2(m + 1)^3$$
$$= (m + 1)^3(25q^2 - 5q - 2)$$
$$= (m + 1)^3(5q - 2)(5q + 1)$$

87. $9x^2(r + 3)^3 + 12xy(r + 3)^3 + 4y^2(r + 3)^3$
$$= (r + 3)^3(9x^2 + 12xy + 4y^2)$$
$$= (r + 3)^3(3x + 2y)(3x + 2y)$$
$$= (r + 3)^3(3x + 2y)^2$$

89. $5x^2 + kx - 1$

Look for two integers whose product is $5(-1) = -5$ and whose sum is k.

Factors of -5	Sums of Factors
$-5, 1$	-4
$5, -1$	4

Thus, there are two possible integer values for k: -4 and 4.

91. $2m^2 + km + 5$

Look for two integers whose product is $2(5) = 10$ and whose sum is k.

Factors of 10	Sums of Factors
$-10, -1$	-11
$10, 1$	11
$-5, -2$	-7
$5, 2$	7

Thus, there are four possible integer values for k: $-11, -7, 7,$ and 11.

93. $(7p + 3)(7p - 3) = (7p)^2 - 3^2$
$$= 49p^2 - 9$$

95. $(x + 6)^2 = x^2 + 2(x)(6) + 6^2$
$$= x^2 + 12x + 36$$

6.4   Special Factoring Techniques

6.4 Now Try Exercises

N1. **(a)** $x^2 - 100 = x^2 - 10^2$
$$= (x + 10)(x - 10)$$

(b) $x^2 + 49$

This binomial is the sum of two squares and the terms have no common factor. Unlike the difference of two squares, it cannot be factored. It is a *prime* polynomial.

N2. **(a)** $9t^2 - 100 = (3t)^2 - 10^2$
$$= (3t + 10)(3t - 10)$$

(b) $36a^2 - 49b^2 = (6a)^2 - (7b)^2$
$$= (6a + 7b)(6a - 7b)$$

N3. (a) $16k^2 - 64$

$$= 16(k^2 - 4) \qquad \textit{Factor out 16.}$$
$$= 16(k^2 - 2^2)$$
$$= 16(k + 2)(k - 2) \qquad \begin{array}{l}\textit{Difference of}\\ \textit{squares}\end{array}$$

(b) $m^4 - 144 = (m^2)^2 - 12^2$

$$= (m^2 + 12)(m^2 - 12)$$

(c) $v^4 - 625 = (v^2)^2 - 25^2$

$$= (v^2 + 25)(v^2 - 25)$$
$$= (v^2 + 25)(v^2 - 5^2)$$
$$= (v^2 + 25)(v + 5)(v - 5)$$

N4. $y^2 + 14y + 49$

The term y^2 is a perfect square, and so is 49. Try to factor the trinomial as

$$y^2 + 14y + 49 = (y + 7)^2.$$

To check, take twice the product of the two terms in the squared binomial.

$$2 \cdot y \cdot 7 = 14y$$

Since $14y$ is the middle term of the trinomial, the trinomial is a perfect square and can be factored as $(y + 7)^2$. Thus,

$$y^2 + 14y + 49 = (y + 7)^2.$$

N5. (a) $t^2 - 18t + 81 \stackrel{?}{=} (t - 9)^2$

$2 \cdot t \cdot 9 = 18t$, so this is a perfect square trinomial, and

$$t^2 - 18t + 81 = (t - 9)^2.$$

(b) $4p^2 - 28p + 49 \stackrel{?}{=} (2p - 7)^2$

$2 \cdot 2p \cdot 7 = 28p$, so this is a perfect square trinomial, and

$$4p^2 - 28p + 49 = (2p - 7)^2.$$

(c) $9x^2 + 6x + 4 \stackrel{?}{=} (3x + 2)^2$

$2 \cdot 3x \cdot 2 = 12x \neq 6x$, so this is not a perfect square trinomial. The trinomial cannot be factored even with the methods of the previous sections. It is *prime*.

(d) $80x^3 + 120x^2 + 45x$

Factor out the greatest common factor, $5x$.

$$80x^3 + 120x^2 + 45x = 5x(16x^2 + 24x + 9)$$
$$\stackrel{?}{=} 5x(4x + 3)^2$$

$2 \cdot 4x \cdot 3 = 24x$, so this is a perfect square trinomial, and

$$5x(16x^2 + 24x + 9) = 5x(4x + 3)^2.$$

N6. (a) $a^3 - 27 = a^3 - 3^3$

Let $x = a$ and $y = 3$ in the pattern for the difference of cubes.

$$x^3 - y^3 = (x - y)(x^2 + xy + y^2)$$
$$a^3 - 3^3 = (a - 3)(a^2 + a \cdot 3 + 3^2)$$
$$= (a - 3)(a^2 + 3a + 9)$$

(b) $8t^3 - 125 = (2t)^3 - 5^3$

Let $x = 2t$ and $y = 5$ in the pattern for the difference of cubes.

$$x^3 - y^3 = (x - y)(x^2 + xy + y^2)$$
$$(2t)^3 - 5^3 = (2t - 5)[(2t)^2 + 2t \cdot 5 + 5^2]$$
$$= (2t - 5)(4t^2 + 10t + 25)$$

(c) $3k^3 - 192 = 3(k^3 - 64)$

$$= 3(k^3 - 4^3)$$
$$= 3(k - 4)(k^2 + k \cdot 4 + 4^2)$$
$$= 3(k - 4)(k^2 + 4k + 16)$$

(d) $125x^3 - 343y^6$

$$= (5x)^3 - (7y^2)^3$$
$$= (5x - 7y^2)[(5x)^2 + 5x(7y^2) + (7y^2)^2]$$
$$= (5x - 7y^2)(25x^2 + 35xy^2 + 49y^4)$$

N7. (a) $x^3 + 125 = x^3 + 5^3$

Let $x = x$ and $y = 5$ in the pattern for the sum of cubes.

$$x^3 + y^3 = (x + y)(x^2 - xy + y^2)$$
$$x^3 + 5^3 = (x + 5)(x^2 - x \cdot 5 + 5^2)$$
$$= (x + 5)(x^2 - 5x + 25)$$

(b) $27a^3 + 8b^3$

$$= (3a)^3 + (2b)^3$$
$$= (3a + 2b)[(3a)^2 - 3a(2b) + (2b)^2]$$
$$= (3a + 2b)(9a^2 - 6ab + 4b^2)$$

6.4 Section Exercises

1.

$1^2 = \underline{1}$	$2^2 = \underline{4}$	$3^2 = \underline{9}$
$4^2 = \underline{16}$	$5^2 = \underline{25}$	$6^2 = \underline{36}$
$7^2 = \underline{49}$	$8^2 = \underline{64}$	$9^2 = \underline{81}$
$10^2 = \underline{100}$	$11^2 = \underline{121}$	$12^2 = \underline{144}$
$13^2 = \underline{169}$	$14^2 = \underline{196}$	$15^2 = \underline{225}$
$16^2 = \underline{256}$	$17^2 = \underline{289}$	$18^2 = \underline{324}$
$19^2 = \underline{361}$	$20^2 = \underline{400}$	

3.

$1^3 = \underline{1}$	$2^3 = \underline{8}$	$3^3 = \underline{27}$
$4^3 = \underline{64}$	$5^3 = \underline{125}$	$6^3 = \underline{216}$
$7^3 = \underline{343}$	$8^3 = \underline{512}$	$9^3 = \underline{729}$
$10^3 = \underline{1000}$		

5. **(a)** $64x^6y^{12} = (8x^3y^6)^2$, so $64x^6y^{12}$ is a perfect square.

$64x^6y^{12} = (4x^2y^4)^3$, so $64x^6y^{12}$ is also a perfect cube.

Therefore, the answer is "both of these."

(b) $125t^6 = (5t^2)^3$, so $125t^6$ is a perfect cube. Since 125 is not a perfect square, $125t^6$ is not a perfect square.

(c) $49x^{12} = (7x^6)^2$, so $49x^{12}$ is a perfect square. Since 49 is not a perfect cube, $49x^{12}$ is not a perfect cube.

(d) $81r^{10} = (9r^5)^2$, so $81r^{10}$ is a perfect square. It is not a perfect cube.

7. $y^2 - 25$

To factor this binomial, use the rule for factoring a difference of squares.

$$a^2 - b^2 = (a+b)(a-b)$$
$$\downarrow \quad \downarrow \quad \downarrow \downarrow \downarrow \downarrow$$
$$y^2 - 25 = y^2 - 5^2 = (y+5)(y-5)$$

9. $x^2 - 144 = x^2 - 12^2$
$$= (x+12)(x-12)$$

11. $m^2 + 64$

This binomial is the *sum* of squares and the terms have no common factor. Unlike the *difference* of squares, it cannot be factored. It is a *prime* polynomial.

13. $4m^2 + 16$

Factor out the GCF, 4. The resulting sum of squares cannot be factored further.

$$4m^2 + 16 = 4(m^2 + 4)$$

15. $9r^2 - 4 = (3r)^2 - 2^2$
$$= (3r+2)(3r-2)$$

17. $36x^2 - 16$

First factor out the GCF, 4; then use the rule for factoring the difference of squares.

$$36x^2 - 16 = 4(9x^2 - 4)$$
$$= 4[(3x)^2 - 2^2]$$
$$= 4(3x+2)(3x-2)$$

19. $196p^2 - 225 = (14p)^2 - 15^2$
$$= (14p+15)(14p-15)$$

21. $16r^2 - 25a^2 = (4r)^2 - (5a)^2$
$$= (4r+5a)(4r-5a)$$

23. $100x^2 + 49$

This binomial is the *sum* of squares and the terms have no common factor. Unlike the *difference* of squares, it cannot be factored. It is a *prime* polynomial.

25. $p^4 - 49 = (p^2)^2 - 7^2$
$$= (p^2+7)(p^2-7)$$

27. $x^4 - 1$

To factor this binomial completely, factor the difference of squares twice.

$$x^4 - 1 = (x^2)^2 - 1^2$$
$$= (x^2+1)(x^2-1)$$
$$= (x^2+1)(x^2-1^2)$$
$$= (x^2+1)(x+1)(x-1)$$

29. $p^4 - 256$

To factor this binomial completely, factor the difference of squares twice.

$$p^4 - 256 = (p^2)^2 - 16^2$$
$$= (p^2+16)(p^2-16)$$
$$= (p^2+16)(p^2-4^2)$$
$$= (p^2+16)(p+4)(p-4)$$

31. The student's answer is not a complete factorization because $x^2 - 9$ can be factored further. The correct complete factorization is

$$k^4 - 81 = (k^2+9)(k+3)(k-3).$$

In Exercises 33–50, use the rules for factoring perfect square trinomials:

$$a^2 + 2ab + b^2 = (a+b)^2$$
$$a^2 - 2ab + b^2 = (a-b)^2.$$

33. Find b so that

$$x^2 + bx + 25 = (x+5)^2.$$

Since $(x+5)^2 = x^2 + 10x + 25$, $b = 10$.

35. Find a so that

$$ay^2 - 12y + 4 = (3y-2)^2.$$

Since $(3y-2)^2 = 9y^2 - 12y + 4$, $a = 9$.

37. $w^2 + 2w + 1$

The first and last terms are perfect squares, w^2 and 1^2. This trinomial is a perfect square, since the middle term is twice the product of w and 1, or

$$2 \cdot w \cdot 1 = 2w.$$

Therefore,

$$w^2 + 2w + 1 = (w+1)^2.$$

39. $x^2 - 8x + 16$

The first and last terms are perfect squares, x^2 and $(-4)^2$. This trinomial is a perfect square, since the middle term is twice the product of x and -4, or

$$2 \cdot x \cdot (-4) = -8x.$$

Therefore,

$$x^2 - 8x + 16 = (x - 4)^2.$$

41. $2x^2 + 24x + 72$

First, factor out the GCF, 2.

$$2x^2 + 24x + 72 = 2(x^2 + 12x + 36)$$

Now factor $x^2 + 12x + 36$ as a perfect square trinomial.

$$x^2 + 12x + 36 = (x + 6)^2$$

The final factored form is

$$2x^2 + 24x + 72 = 2(x + 6)^2.$$

43. $16x^2 - 40x + 25$

The first and last terms are perfect squares, $(4x)^2$ and $(-5)^2$. The middle term is

$$2(4x)(-5) = -40x.$$

Therefore,

$$16x^2 - 40x + 25 = (4x - 5)^2.$$

45. $49x^2 - 28xy + 4y^2$

The first and last terms are perfect squares, $(7x)^2$ and $(-2y)^2$. The middle term is

$$2(7x)(-2y) = -28xy.$$

Therefore,

$$49x^2 - 28xy + 4y^2 = (7x - 2y)^2.$$

47. $64x^2 + 48xy + 9y^2$
$$= (8x)^2 + 2(8x)(3y) + (3y)^2$$
$$= (8x + 3y)^2$$

49. $50h^2 - 40hy + 8y^2$
$$= 2(25h^2 - 20hy + 4y^2)$$
$$= 2[(5h)^2 - 2(5h)(2y) + (2y)^2]$$
$$= 2(5h - 2y)^2$$

51. $4k^3 - 4k^2 + 9k$

First, factor out the GCF, k.

$$4k^3 - 4k^2 + 9k = k(4k^2 - 4k + 9)$$

Since $4k^2 - 4k + 9$ cannot be factored, $k(4k^2 - 4k + 9)$ is the final factored form.

53. $25z^4 + 5z^3 + z^2$

First, factor out the GCF, z^2.

$$25z^4 + 5z^3 + z^2 = z^2(25z^2 + 5z + 1)$$

Since $25z^2 + 5z + 1$ cannot be factored, $z^2(25z^2 + 5z + 1)$ is the final factored form.

55. $a^3 - 1$

Let $x = a$ and $y = 1$ in the pattern for the difference of cubes.

$$x^3 - y^3 = (x - y)(x^2 + xy + y^2)$$
$$a^3 - 1 = a^3 - 1^3 = (a - 1)(a^2 + a \cdot 1 + 1^2)$$
$$= (a - 1)(a^2 + a + 1)$$

57. $m^3 + 8$

Let $x = m$ and $y = 2$ in the pattern for the sum of cubes.

$$x^3 + y^3 = (x + y)(x^2 - xy + y^2)$$
$$m^3 + 8 = m^3 + 2^3 = (m + 2)(m^2 - m \cdot 2 + 2^2)$$
$$= (m + 2)(m^2 - 2m + 4)$$

59. Factor $k^3 + 1000$ as the sum of cubes.

$$k^3 + 1000 = k^3 + 10^3$$
$$= (k + 10)(k^2 - 10k + 10^2)$$
$$= (k + 10)(k^2 - 10k + 100)$$

61. Factor $27x^3 - 64$ as the difference of cubes.

$$27x^3 - 64 = (3x)^3 - 4^3$$
$$= (3x - 4)[(3x)^2 + 3x \cdot 4 + 4^2]$$
$$= (3x - 4)(9x^2 + 12x + 16)$$

63. $6p^3 + 6 = 6(p^3 + 1)$ *GCF = 6*
$$= 6(p^3 + 1^3) \quad\quad\quad \textit{Sum of cubes}$$
$$= 6(p + 1)(p^2 - p \cdot 1 + 1^2)$$
$$= 6(p + 1)(p^2 - p + 1)$$

65. $5x^3 + 40 = 5(x^3 + 8)$ *GCF = 5*
$$= 5(x^3 + 2^3) \quad\quad\quad \textit{Sum of cubes}$$
$$= 5(x + 2)(x^2 - x \cdot 2 + 2^2)$$
$$= 5(x + 2)(x^2 - 2x + 4)$$

67. Factor $y^3 - 8x^3$ as the difference of cubes.

$$y^3 - 8x^3 = y^3 - (2x)^3$$
$$= (y - 2x)[y^2 + y(2x) + (2x)^2]$$
$$= (y - 2x)(y^2 + 2yx + 4x^2)$$

69. $2x^3 - 16y^3$
$$= 2(x^3 - 8y^3) \quad\quad\quad \textit{GCF = 2}$$
$$= 2[x^3 - (2y)^3] \quad\quad\quad \textit{Difference of cubes}$$
$$= 2(x - 2y)[x^2 + x \cdot 2y + (2y)^2]$$
$$= 2(x - 2y)(x^2 + 2xy + 4y^2)$$

71. Factor $8p^3 + 729q^3$ as the sum of cubes.

$$8p^3 + 729q^3$$
$$= (2p)^3 + (9q)^3$$
$$= (2p + 9q)[(2p)^2 - 2p \cdot 9q + (9q)^2]$$
$$= (2p + 9q)(4p^2 - 18pq + 81q^2)$$

73. Factor $27a^3 + 64b^3$ as the sum of cubes.

$$27a^3 + 64b^3$$
$$= (3a)^3 + (4b)^3$$
$$= (3a + 4b)[(3a)^2 - 3a \cdot 4b + (4b)^2]$$
$$= (3a + 4b)(9a^2 - 12ab + 16b^2)$$

75. Factor $125t^3 + 8s^3$ as the sum of cubes.

$$125t^3 + 8s^3$$
$$= (5t)^3 + (2s)^3$$
$$= (5t + 2s)[(5t)^2 - 5t \cdot 2s + (2s)^2]$$
$$= (5t + 2s)(25t^2 - 10ts + 4s^2)$$

77. Factor $8x^3 - 125y^6$ as the difference of cubes.

$$8x^3 - 125y^6$$
$$= (2x)^3 - (5y^2)^3$$
$$= (2x - 5y^2)[(2x)^2 + 2x(5y^2) + (5y^2)^2]$$
$$= (2x - 5y^2)(4x^2 + 10xy^2 + 25y^4)$$

79. Factor $27m^6 + 8n^3$ as the sum of cubes.

$$27m^6 + 8n^3$$
$$= (3m^2)^3 + (2n)^3$$
$$= (3m^2 + 2n)[(3m^2)^2 - 3m^2(2n) + (2n)^2]$$
$$= (3m^2 + 2n)(9m^4 - 6m^2n + 4n^2)$$

81. Factor $x^9 + y^9$ as the sum of cubes.

$$x^9 + y^9$$
$$= (x^3)^3 + (y^3)^3$$
$$= (x^3 + y^3)[(x^3)^2 - x^3(y^3) + (y^3)^2]$$
$$= (x + y)(x^2 - xy + y^2)(x^6 - x^3y^3 + y^6)$$

83. $p^2 - \frac{1}{9} = p^2 - \left(\frac{1}{3}\right)^2$
$$= \left(p + \frac{1}{3}\right)\left(p - \frac{1}{3}\right)$$

85. $36m^2 - \frac{16}{25} = (6m)^2 - \left(\frac{4}{5}\right)^2$
$$= \left(6m + \frac{4}{5}\right)\left(6m - \frac{4}{5}\right)$$

87. $x^2 - 0.64 = x^2 - (0.8)^2$
$$= (x + 0.8)(x - 0.8)$$

89. $t^2 + t + \frac{1}{4}$

t^2 is a perfect square, and $\frac{1}{4}$ is a perfect square since $\frac{1}{2} \cdot \frac{1}{2} = \frac{1}{4}$. The middle term is twice the product of t and $\frac{1}{2}$, or

$$t = 2(t)\left(\frac{1}{2}\right).$$

Therefore,

$$t^2 + t + \frac{1}{4} = \left(t + \frac{1}{2}\right)^2.$$

91. $x^2 - 1.0x + 0.25$

The first and last terms are perfect squares, x^2 and $(-0.5)^2$. The trinomial is a perfect square, since the middle term is

$$2 \cdot x \cdot (-0.5) = -1.0x.$$

Therefore,

$$x^2 - 1.0x + 0.25 = (x - 0.5)^2.$$

93. Factor $x^3 + \frac{1}{8}$ as the sum of cubes.

$$x^3 + \frac{1}{8} = x^3 + \left(\frac{1}{2}\right)^3$$
$$= \left(x + \frac{1}{2}\right)\left[x^2 - \frac{1}{2}x + \left(\frac{1}{2}\right)^2\right]$$
$$= \left(x + \frac{1}{2}\right)\left(x^2 - \frac{1}{2}x + \frac{1}{4}\right)$$

95. $(m + n)^2 - (m - n)^2$

Factor as the difference of squares. Substitute into the rule using $x = m + n$ and $y = m - n$.

$$(m + n)^2 - (m - n)^2$$
$$= [(m + n) + (m - n)] \cdot [(m + n) - (m - n)]$$
$$= (2m)(2n)$$
$$= 4mn$$

97. $m^2 - p^2 + 2m + 2p$

This expression can be factored by grouping.

$$= (m + p)(m - p) + 2(m + p) \quad \text{\textit{Factor each group.}}$$
$$= (m + p)(m - p + 2) \quad \text{\textit{Factor out } m + p.}$$

99. $m - 4 = 0$
$$m = 4 \quad \text{\textit{Add 4.}}$$

The solution set is $\{4\}$.

101. $2t + 10 = 0$
$$2t = -10 \quad \text{\textit{Subtract 10.}}$$
$$t = -5 \quad \text{\textit{Divide by 2.}}$$

The solution set is $\{-5\}$.

Summary Exercises on Factoring

1. $12x^2 + 20x + 8 = 4(3x^2 + 5x + 2)$
$$= 4(3x + 2)(x + 1)$$

Match with choice **G**. [Factor out the GCF; then factor a trinomial by grouping or trial and error.]

3. $16m^2n + 24mn - 40mn^2$
$$= 8mn(2m + 3 - 5n)$$

Match with choice **A**. [Factor out the GCF; no further factoring is possible.]

5. $36p^2 - 60pq + 25q^2$

The first and last terms are perfect squares, $(6p)^2$ and $(-5q)^2$. The middle term is

$$2(6p)(-5q) = -60pq.$$

Therefore,

$$36p^2 - 60pq + 25q^2 = (6p - 5q)^2.$$

Match with choice **E**. [Factor a perfect square trinomial.]

7. Factor $8r^3 - 125$ as the difference of cubes.

$$8r^3 - 125$$
$$= (2r)^3 - 5^3$$
$$= (2r - 5)[(2r)^2 + 2r(5) + 5^2]$$
$$= (2r - 5)(4r^2 + 10r + 25)$$

Match with choice **C**. [Factor a difference of cubes.]

9. $4w^2 + 49$

Match with choice **I**. [The polynomial is prime.]

11. $a^2 - 4a - 12 = (a - 6)(a + 2)$

13. $6y^2 - 6y - 12 = 6(y^2 - y - 2)$
$$= 6(y - 2)(y + 1)$$

15. $6a + 12b + 18c$
$$= 6(a + 2b + 3c)$$

17. $p^2 - 17p + 66 = (p - 11)(p - 6)$

19. $10z^2 - 7z - 6$

Use the grouping method.
Look for two integers whose product is $10(-6) = -60$ and whose sum is -7.
The integers are -12 and 5.

$$10z^2 - 7z - 6$$
$$= 10z^2 - 12z + 5z - 6$$
$$= 2z(5z - 6) + 1(5z - 6) \quad \textit{Factor each group.}$$
$$= (5z - 6)(2z + 1) \quad \textit{Factor out } 5z - 6.$$

21. $17x^3y^2 + 51xy = 17xy(x^2y + 3)$

23. $8a^5 - 8a^4 - 48a^3$
$$= 8a^3(a^2 - a - 6)$$
$$= 8a^3(a - 3)(a + 2)$$

25. $z^2 - 3za - 10a^2 = (z - 5a)(z + 2a)$

27. $x^2 - 4x - 5x + 20$
$$= x(x - 4) - 5(x - 4) \quad \textit{Factor each group.}$$
$$= (x - 4)(x - 5) \quad \textit{Factor out } x - 4.$$

29. $6n^2 - 19n + 10 = (3n - 2)(2n - 5)$

31. $16x + 20 = 4(4x + 5)$

33. $6y^2 - 5y - 4$

Factor by grouping. Find two integers whose product is $6(-4) = -24$ and whose sum is -5. The integers are -8 and 3.

$$6y^2 - 5y - 4$$
$$= 6y^2 - 8y + 3y - 4$$
$$= 2y(3y - 4) + 1(3y - 4)$$
$$= (3y - 4)(2y + 1)$$

35. $6z^2 + 31z + 5 = (6z + 1)(z + 5)$

37. $4k^2 - 12k + 9$
$$= (2k)^2 - 2 \cdot 2k \cdot 3 + 3^2$$
$$= (2k - 3)^2 \quad \textit{Perfect square trinomial}$$

39. $54m^2 - 24z^2$
$$= 6(9m^2 - 4z^2)$$
$$= 6[(3m)^2 - (2z)^2]$$
$$= 6(3m + 2z)(3m - 2z)$$

41. $3k^2 + 4k - 4 = (3k - 2)(k + 2)$

43. $14k^3 + 7k^2 - 70k$
$$= 7k(2k^2 + k - 10)$$
$$= 7k(2k + 5)(k - 2)$$

45. $y^4 - 16$
$$= (y^2)^2 - 4^2$$
$$= (y^2 + 4)(y^2 - 4) \quad \textit{Difference of squares}$$
$$= (y^2 + 4)(y + 2)(y - 2) \quad \textit{Difference of squares}$$

47. $8m - 16m^2 = 8m(1 - 2m)$

49. Factor $z^3 - 8$ as the difference of cubes.

$$z^3 - 8 = z^3 - 2^3$$
$$= (z - 2)(z^2 + z \cdot 2 + 2^2)$$
$$= (z - 2)(z^2 + 2z + 4)$$

51. $k^2 + 9$ cannot be factored because it is the sum of squares with no GCF. The expression is prime.

53. $32m^9 + 16m^5 + 24m^3$
$$= 8m^3(4m^6 + 2m^2 + 3)$$

55. $16r^2 + 24rm + 9m^2$
$$= (4r)^2 + 2 \cdot 4r \cdot 3m + (3m)^2$$
$$= (4r + 3m)^2 \quad \textit{Perfect square trinomial}$$

57. $15h^2 + 11hg - 14g^2$

Factor by grouping. Look for two integers whose product is $15(-14) = -210$ and whose sum is 11. The integers are 21 and -10.

$15h^2 + 11hg - 14g^2$

$= 15h^2 + 21hg - 10hg - 14g^2$

$= 3h(5h + 7g) - 2g(5h + 7g)$ *Factor each group.*

$= (5h + 7g)(3h - 2g)$ *Factor out $5h + 7g$.*

59. $k^2 - 11k + 30 = (k - 5)(k - 6)$

61. $3k^3 - 12k^2 - 15k$

$= 3k(k^2 - 4k - 5)$

$= 3k(k - 5)(k + 1)$

63. $1000p^3 + 27$

$= (10p)^3 + 3^3$ *Sum of Cubes*

$= (10p + 3)[(10p)^2 - 10p \cdot 3 + 3^2]$

$= (10p + 3)(100p^2 - 30p + 9)$

65. $6 + 3m + 2p + mp$

$= (6 + 3m) + (2p + mp)$

$= 3(2 + m) + p(2 + m)$

$= (2 + m)(3 + p)$

67. $16z^2 - 8z + 1$

$= (4z)^2 - 2 \cdot 4z \cdot 1 + 1^2$

$= (4z - 1)^2$ *Perfect square trinomial*

69. $108m^2 - 36m + 3$

$= 3(36m^2 - 12m + 1)$

$= 3(6m - 1)^2$ *Perfect square trinomial*

71. $x^2 - xy + y^2$ is *prime*. The middle term would have to be $+2xy$ or $-2xy$ in order to make this a perfect square trinomial.

73. $32z^3 + 56z^2 - 16z$

$= 8z(4z^2 + 7z - 2)$

$= 8z(4z - 1)(z + 2)$

75. $20 + 5m + 12n + 3mn$

$= 5(4 + m) + 3n(4 + m)$

$= (4 + m)(5 + 3n)$

77. $6a^2 + 10a - 4$

$= 2(3a^2 + 5a - 2)$

$= 2(3a - 1)(a + 2)$

79. $a^3 - b^3 + 2a - 2b$

Factor by grouping. The first two terms form a difference of cubes.

$a^3 - b^3 + 2a - 2b$

$= (a^3 - b^3) + (2a - 2b)$

$= (a - b)(a^2 + ab + b^2) + 2(a - b)$

$= (a - b)(a^2 + ab + b^2 + 2)$ *Factor out $(a - b)$.*

81. $64m^2 - 80mn + 25n^2$

$= (8m)^2 - 2(8m)(5n) + (5n)^2$

$= (8m - 5n)^2$ *Perfect square trinomial*

83. $8k^2 - 2kh - 3h^2$

$= 8k^2 - 6kh + 4kh - 3h^2$

$= 2k(4k - 3h) + h(4k - 3h)$ *Factor each group.*

$= (4k - 3h)(2k + h)$ *Factor out $4k - 3h$.*

85. $2x^3 + 128$

$= 2(x^3 + 64)$ *GCF = 2*

$= 2(x^3 + 4^3)$ *Sum of Cubes*

$= 2(x + 4)(x^2 - x \cdot 4 + 4^2)$

$= 2(x + 4)(x^2 - 4x + 16)$

87. $10y^2 - 7yz - 6z^2$

$= 10y^2 - 12yz + 5yz - 6z^2$

$= 2y(5y - 6z) + z(5y - 6z)$ *Factor each group.*

$= (5y - 6z)(2y + z)$ *Factor out $5y - 6z$.*

89. $8a^2 + 23ab - 3b^2$

$= 8a^2 + 24ab - ab - 3b^2$

$= 8a(a + 3b) - b(a + 3b)$ *Factor each group.*

$= (a + 3b)(8a - b)$ *Factor out $a + 3b$.*

91. $x^6 - 1 = (x^3)^2 - 1^2$

$= (x^3 + 1)(x^3 - 1)$

or $(x^3 - 1)(x^3 + 1)$

92. From Exercise 91, we have

$$x^6 - 1 = (x^3 - 1)(x^3 + 1).$$

Use the rules for the difference and sum of cubes to factor further.

Since

$$x^3 - 1 = (x - 1)(x^2 + x + 1)$$

and

$$x^3 + 1 = (x + 1)(x^2 - x + 1),$$

we obtain the factorization

$$x^6 - 1 = (x - 1)(x^2 + x + 1)$$
$$\cdot (x + 1)(x^2 - x + 1).$$

93. $x^6 - 1 = (x^2)^3 - 1^3$

$= (x^2 - 1)[(x^2)^2 + x^2 \cdot 1 + 1^2]$

$= (x^2 - 1)(x^4 + x^2 + 1)$

94. From Exercise 93, we have

$$x^6 - 1 = (x^2 - 1)(x^4 + x^2 + 1).$$

Use the rule for the difference of two squares to factor the binomial.

$$x^2 - 1 = (x - 1)(x + 1)$$

Thus, we obtain the factorization

$$x^6 - 1 = (x - 1)(x + 1)(x^4 + x^2 + 1).$$

95. The result in Exercise 92 is the completely factored form.

96. Multiply the trinomials from the factored form in Exercise 92 vertically.

$$
\begin{array}{rrrr}
x^2 & +x & +1 \\
x^2 & -x & +1 \\
\hline
x^2 & +x & +1 \\
-x^3 & -x^2 & -x \\
x^4 & +x^3 & +x^2 \\
\hline
x^4 & & +x^2 & +1 \\
\end{array}
$$

97. In general, if I must choose between factoring first using the method for difference of squares or the method for difference of cubes, I should choose the *difference of squares* method to eventually obtain the completely factored form.

98. $x^6 - 729 = (x^3)^2 - 27^2$

$$= (x^3 - 27)(x^3 + 27)$$
$$= (x^3 - 3^3)(x^3 + 3^3)$$
$$= (x - 3)(x^2 + 3x + 9)$$
$$\cdot (x + 3)(x^2 - 3x + 9)$$

6.5 Solving Quadratic Equations by Factoring

6.5 Now Try Exercises

N1. (a) $(x - 4)(3x + 1) = 0$

By the zero-factor property, either $x - 4 = 0$ or $3x + 1 = 0$. The equation $x - 4 = 0$ gives the solution $x = 4$; $3x + 1 = 0$ gives $x = -\frac{1}{3}$. Check both solutions by substituting first 4 and then $-\frac{1}{3}$ for x in the original equation. The solution set is $\left\{-\frac{1}{3}, 4\right\}$.

(b) $y(4y - 5) = 0$

By the zero-factor property, either $y = 0$ or $4y - 5 = 0$, so the solutions are $y = 0$ and $y = \frac{5}{4}$. Check each solution in the original equation. The solution set is $\left\{0, \frac{5}{4}\right\}$.

N2. $t^2 = -3t + 18$

Rewrite in standard form by adding $3t$ to each side and subtracting 18 from each side.

$$t^2 + 3t - 18 = 0$$
$$(t - 3)(t + 6) = 0$$

$$t - 3 = 0 \quad \text{or} \quad t + 6 = 0$$
$$t = 3 \quad \text{or} \quad t = -6$$

The solution set is $\{-6, 3\}$.

N3. $10p^2 + 65p = 35$

$$
\begin{array}{ll}
10p^2 + 65p - 35 = 0 & \textit{Standard form} \\
5(2p^2 + 13p - 7) = 0 & \textit{Factor out 5.} \\
2p^2 + 13p - 7 = 0 & \textit{Divide by 5.} \\
(2p - 1)(p + 7) = 0 & \textit{Factor.}
\end{array}
$$

$$2p - 1 = 0 \quad \text{or} \quad p + 7 = 0$$
$$p = \tfrac{1}{2} \quad \text{or} \quad p = -7$$

Check that the solution set is $\left\{-7, \frac{1}{2}\right\}$.

N4. (a) $9x^2 - 64 = 0$
$$(3x + 8)(3x - 8) = 0$$

$$3x + 8 = 0 \quad \text{or} \quad 3x - 8 = 0$$
$$3x = -8 \quad \text{or} \quad 3x = 8$$
$$x = -\tfrac{8}{3} \quad \text{or} \quad x = \tfrac{8}{3}$$

Check that the solution set is $\left\{-\frac{8}{3}, \frac{8}{3}\right\}$.

(b) $m^2 = 5m$

$$
\begin{array}{ll}
m^2 - 5m = 0 & \textit{Standard form} \\
m(m - 5) = 0 & \textit{Factor.} \\
m = 0 \ \text{or} \ m - 5 = 0 & \textit{Zero-factor prop.} \\
\quad\quad m = 5 & \textit{Solve.}
\end{array}
$$

Check that the solution set is $\{0, 5\}$.

(c) $p(6p - 1) = 2$

Write the equation in standard form.

$$6p^2 - p = 2$$
$$6p^2 - p - 2 = 0$$
$$(3p - 2)(2p + 1) = 0$$

$$3p - 2 = 0 \quad \text{or} \quad 2p + 1 = 0$$
$$3p = 2 \quad \text{or} \quad 2p = -1$$
$$p = \tfrac{2}{3} \quad \quad p = -\tfrac{1}{2}$$

Check that the solution set is $\left\{-\frac{1}{2}, \frac{2}{3}\right\}$.

N5. $4x^2 - 4x + 1 = 0$

Factor $4x^2 - 4x + 1$ as a perfect square trinomial.

$$(2x - 1)^2 = 0$$

Set the factor $2x - 1$ equal to 0 and solve.

$$2x - 1 = 0$$
$$2x = 1$$
$$x = \tfrac{1}{2}$$

Check this solution by substituting it in the original equation. The solution set is $\left\{\frac{1}{2}\right\}$.

N6. **(a)**
$$3x^3 - 27x = 0$$
$$3x(x^2 - 9) = 0$$
$$3x(x + 3)(x - 3) = 0$$

$3x = 0$ or $x + 3 = 0$ or $x - 3 = 0$
$x = 0$ or $x = -3$ or $x = 3$

Check that the solution set is $\{-3, 0, 3\}$.

(b) $(3a - 1)(2a^2 - 5a - 12) = 0$

We know that $a = \frac{1}{3}$ is the solution of $3a - 1 = 0$, so we need to find the solutions of $2a^2 - 5a - 12 = 0$.

$$2a^2 - 5a - 12 = 0$$
$$(2a + 3)(a - 4) = 0 \quad \textit{Factor.}$$

$2a + 3 = 0$ or $a - 4 = 0$ *Zero-factor property*
$a = -\frac{3}{2}$ or $a = 4$ *Solve.*

The solutions of the original equation are $-\frac{3}{2}$, $\frac{1}{3}$, and 4. Check each solution to verify that the solution set is $\left\{-\frac{3}{2}, \frac{1}{3}, 4\right\}$.

N7.
$$x(4x - 9) = (x - 2)^2 + 24$$
$$4x^2 - 9x = x^2 - 4x + 4 + 24 \quad \textit{Multiply.}$$
$$3x^2 - 5x - 28 = 0 \qquad \textit{St. form}$$
$$(3x + 7)(x - 4) = 0 \qquad \textit{Factor.}$$

$3x + 7 = 0$ or $x - 4 = 0$ *Zero-factor property*
$x = -\frac{7}{3}$ or $x = 4$ *Solve.*

Check that the solution set is $\left\{-\frac{7}{3}, 4\right\}$.

6.5   Section Exercises

For all equations in this section, answers should be checked by substituting into the original equation. These checks will be shown here for only a few of the exercises.

1. A quadratic equation is an equation that can be put in the form $\underline{ax^2 + bx + c} = 0$. $(a \neq 0)$

3. If a quadratic equation is in standard form, to solve the equation we should begin by attempting to *factor* the polynomial.

5. If a quadratic equation $ax^2 + bx + c = 0$ has $c = 0$, then $\underline{0}$ *must* be a solution because $\underline{x}$ is a factor of the polynomial. Note that if $c = 0$, the equation becomes $ax^2 + bx = 0$, which can be factored as $x(ax + b) = 0$, and has solutions $x = 0$ and $x = -b/a$.

7. $2x(3x - 4)$ is the product of two factors, $2x$ and $3x - 4$. Applying the zero-factor property yields two equations,

$$2x = 0 \quad \text{or} \quad 3x - 4 = 0.$$

Dividing each side of $2x = 0$ by 2 gives us $x = 0$, so we end up with the two solutions, $x = 0$ and $x = \frac{4}{3}$.

We conclude that multiplying a polynomial by a constant does not affect the solutions of the corresponding equation.

9. The variable x is another factor to set equal to 0, so the solution set is $\left\{0, \frac{1}{7}\right\}$.

Not including 0 as a solution is *WHAT WENT WRONG.*

For all equations in this section, answers should be checked by substituting into the original equation. These checks will be shown here for only a few of the exercises.

11. $(x + 5)(x - 2) = 0$

By the zero-factor property, the only way that the product of these two factors can be zero is if at least one of the factors is zero.

$$x + 5 = 0 \quad \text{or} \quad x - 2 = 0$$

Solve each of these linear equations.

$$x = -5 \quad \text{or} \quad x = 2$$

Check $x = -5$: $0(-7) = 0$ *True*
Check $x = 2$: $7(0) = 0$ *True*

The solution set is $\{-5, 2\}$.

13. $(2m - 7)(m - 3) = 0$

Set each factor equal to zero and solve the resulting linear equations.

$2m - 7 = 0$ or $m - 3 = 0$
$2m = 7$ or $m = 3$
$m = \frac{7}{2}$

The solution set is $\left\{3, \frac{7}{2}\right\}$.

15. $(2x + 1)(6x - 1) = 0$

Set each factor equal to zero and solve the resulting linear equations.

$2x + 1 = 0$ or $6x - 1 = 0$
$2x = -1$ or $6x = 1$
$x = -\frac{1}{2}$ $x = \frac{1}{6}$

The solution set is $\left\{-\frac{1}{2}, \frac{1}{6}\right\}$.

17. $t(6t + 5) = 0$

Set each factor equal to zero and solve the resulting linear equations.

$t = 0$ or $6t + 5 = 0$
$6t = -5$
$t = -\frac{5}{6}$

The solution set is $\left\{-\frac{5}{6}, 0\right\}$.

19. $2x(3x - 4) = 0$

Set each factor equal to zero and solve the resulting linear equations.

$$2x = 0 \quad \text{or} \quad 3x - 4 = 0$$
$$x = 0 \quad \text{or} \quad 3x = 4$$
$$x = \frac{4}{3}$$

The solution set is $\left\{0, \frac{4}{3}\right\}$.

21. $(x - 6)(x - 6) = 0$

Set each factor equal to zero and solve the resulting linear equations.

$$x - 6 = 0 \quad \text{or} \quad x - 6 = 0$$
$$x = 6 \quad \text{or} \quad x = 6$$

6 is called a *double solution* for $(x - 6)^2 = 0$ because it occurs twice when the equation is solved.

The solution set is $\{6\}$.

23. $y^2 + 3y + 2 = 0$

Factor the polynomial.

$$(y + 2)(y + 1) = 0$$

Set each factor equal to 0.

$$y + 2 = 0 \quad \text{or} \quad y + 1 = 0$$

Solve each equation.

$$y = -2 \quad \text{or} \quad y = -1$$

Check these solutions by substituting -2 for y and then -1 for y in the original equation.

$$y^2 + 3y + 2 = 0$$
$$(-2)^2 + 3(-2) + 2 \overset{?}{=} 0 \quad \textit{Let y = –2.}$$
$$4 - 6 + 2 \overset{?}{=} 0$$
$$-2 + 2 = 0 \quad \textit{True}$$

$$y^2 + 3y + 2 = 0$$
$$(-1)^2 + 3(-1) + 2 \overset{?}{=} 0 \quad \textit{Let y = –1.}$$
$$1 - 3 + 2 \overset{?}{=} 0$$
$$-2 + 2 = 0 \quad \textit{True}$$

The solution set is $\{-2, -1\}$.

25. $y^2 - 3y + 2 = 0$

Factor the polynomial.

$$(y - 1)(y - 2) = 0$$

Set each factor equal to 0.

$$y - 1 = 0 \quad \text{or} \quad y - 2 = 0$$

Solve each equation.

$$y = 1 \quad \text{or} \quad y = 2$$

The solution set is $\{1, 2\}$.

27. $x^2 = 24 - 5x$

Write the equation in standard form.

$$x^2 + 5x - 24 = 0$$

Factor the polynomial.

$$(x + 8)(x - 3) = 0$$

Set each factor equal to 0.

$$x + 8 = 0 \quad \text{or} \quad x - 3 = 0$$

Solve each equation.

$$x = -8 \quad \text{or} \quad x = 3$$

The solution set is $\{-8, 3\}$.

29. $x^2 = 3 + 2x$

Write the equation in standard form.

$$x^2 - 2x - 3 = 0$$
$$(x + 1)(x - 3) = 0$$

$$x + 1 = 0 \quad \text{or} \quad x - 3 = 0$$
$$x = -1 \quad \text{or} \quad x = 3$$

The solution set is $\{-1, 3\}$.

31. $z^2 + 3z = -2$

Write the equation in standard form.

$$z^2 + 3z + 2 = 0$$

Factor the polynomial.

$$(z + 2)(z + 1) = 0$$

Set each factor equal to 0.

$$z + 2 = 0 \quad \text{or} \quad z + 1 = 0$$
$$z = -2 \quad \text{or} \quad z = -1$$

The solution set is $\{-2, -1\}$.

33. $m^2 + 8m + 16 = 0$

Factor $m^2 + 8m + 16$ as a perfect square trinomial.

$$(m + 4)^2 = 0$$

Set the factor $m + 4$ equal to 0 and solve.

$$m + 4 = 0$$
$$m = -4$$

The solution set is $\{-4\}$.

35. $3x^2 + 5x - 2 = 0$

Factor the polynomial.

$$(3x - 1)(x + 2) = 0$$

Set each factor equal to 0.

$$3x - 1 = 0 \quad \text{or} \quad x + 2 = 0$$
$$3x = 1 \quad \text{or} \quad x = -2$$
$$x = \tfrac{1}{3}$$

The solution set is $\left\{-2, \tfrac{1}{3}\right\}$.

37.
$$12p^2 = 8 - 10p$$
$$12p^2 + 10p - 8 = 0 \qquad \textit{Standard form}$$
$$2(6p^2 + 5p - 4) = 0 \qquad \textit{Factor out 2.}$$
$$2(3p + 4)(2p - 1) = 0$$

$$3p + 4 = 0 \quad \text{or} \quad 2p - 1 = 0$$
$$3p = -4 \quad \text{or} \quad 2p = 1$$
$$p = -\tfrac{4}{3} \quad \text{or} \quad p = \tfrac{1}{2}$$

The solution set is $\left\{-\tfrac{4}{3}, \tfrac{1}{2}\right\}$.

39.
$$9s^2 + 12s = -4$$
$$9s^2 + 12s + 4 = 0$$
$$(3s + 2)^2 = 0$$

Set the factor $3s + 2$ equal to 0 and solve.

$$3s + 2 = 0$$
$$3s = -2$$
$$s = -\tfrac{2}{3}$$

The solution set is $\left\{-\tfrac{2}{3}\right\}$.

41.
$$y^2 - 9 = 0$$
$$(y + 3)(y - 3) = 0$$

$$y + 3 = 0 \quad \text{or} \quad y - 3 = 0$$
$$y = -3 \quad \text{or} \quad y = 3$$

The solution set is $\{-3, 3\}$.

43.
$$16x^2 - 49 = 0$$
$$(4x + 7)(4x - 7) = 0$$

$$4x + 7 = 0 \quad \text{or} \quad 4x - 7 = 0$$
$$4x = -7 \quad \text{or} \quad 4x = 7$$
$$x = -\tfrac{7}{4} \quad \text{or} \quad x = \tfrac{7}{4}$$

The solution set is $\left\{-\tfrac{7}{4}, \tfrac{7}{4}\right\}$.

45.
$$n^2 = 121$$
$$n^2 - 121 = 0$$
$$(n + 11)(n - 11) = 0$$

$$n + 11 = 0 \quad \text{or} \quad n - 11 = 0$$
$$n = -11 \quad \text{or} \quad n = 11$$

The solution set is $\{-11, 11\}$.

47.
$$x^2 = 7x$$
$$x^2 - 7x = 0$$
$$x(x - 7) = 0$$

$$x = 0 \quad \text{or} \quad x - 7 = 0$$
$$x = 0 \quad \text{or} \quad x = 7$$

Check $x = 0$: $0 = 0$ *True*
Check $x = 7$: $49 = 49$ *True*

The solution set is $\{0, 7\}$.

49.
$$6r^2 = 3r$$
$$6r^2 - 3r = 0$$
$$3r(2r - 1) = 0$$

$$3r = 0 \quad \text{or} \quad 2r - 1 = 0$$
$$r = 0 \quad \text{or} \quad 2r = 1$$
$$r = \tfrac{1}{2}$$

The solution set is $\left\{0, \tfrac{1}{2}\right\}$.

51.
$$x(x - 7) = -10$$
$$x^2 - 7x = -10$$
$$x^2 - 7x + 10 = 0$$
$$(x - 2)(x - 5) = 0$$

$$x - 2 = 0 \quad \text{or} \quad x - 5 = 0$$
$$x = 2 \quad \text{or} \quad x = 5$$

The solution set is $\{2, 5\}$.

53.
$$3z(2z + 7) = 12$$
$$z(2z + 7) = 4 \qquad \textit{Divide by 3.}$$
$$2z^2 + 7z = 4$$
$$2z^2 + 7z - 4 = 0$$
$$(2z - 1)(z + 4) = 0$$

$$2z - 1 = 0 \quad \text{or} \quad z + 4 = 0$$
$$2z = 1 \quad \text{or} \quad z = -4$$
$$z = \tfrac{1}{2}$$

The solution set is $\left\{-4, \tfrac{1}{2}\right\}$.

55.
$$2y(y + 13) = 136$$
$$y(y + 13) = 68 \qquad \textit{Divide by 2.}$$
$$y^2 + 13y = 68 \qquad \textit{Multiply.}$$
$$y^2 + 13y - 68 = 0 \qquad \textit{Standard form}$$
$$(y + 17)(y - 4) = 0 \qquad \textit{Factor.}$$

$x + 17 = 0 \quad \text{or} \quad y - 4 = 0$ *Zero-factor property*
$$y = -17 \text{ or} \quad y = 4 \quad \textit{Solve.}$$

Check $y = -17$:

$$2y(y + 13) = 136$$
$$2(-17)(-17 + 13) \overset{?}{=} 136 \quad \textit{Let y = -17.}$$
$$(-34)(-4) = 136 \quad \textit{True}$$

Check $y = 4$:

$$2(4)(4 + 13) \overset{?}{=} 136 \quad \textit{Let y = 4.}$$
$$8(17) = 136 \quad \textit{True}$$

The solution set is $\{-17, 4\}$.

57. $(2r + 5)(3r^2 - 16r + 5) = 0$

Begin by factoring $3r^2 - 16r + 5$.

$(2r + 5)(3r - 1)(r - 5) = 0$

Set each of the three factors equal to 0 and solve the resulting equations.

$$2r + 5 = 0 \quad \text{or} \quad 3r - 1 = 0 \quad \text{or} \quad r - 5 = 0$$
$$2r = -5 \qquad\qquad 3r = 1$$
$$r = -\tfrac{5}{2} \quad \text{or} \qquad r = \tfrac{1}{3} \quad \text{or} \qquad r = 5$$

The solution set is $\left\{ -\tfrac{5}{2}, \tfrac{1}{3}, 5 \right\}$.

59. $(2x + 7)(x^2 + 2x - 3) = 0$
$(2x + 7)(x + 3)(x - 1) = 0$

$$2x + 7 = 0 \quad \text{or} \quad x + 3 = 0 \quad \text{or} \quad x - 1 = 0$$
$$2x = -7$$
$$x = -\tfrac{7}{2} \quad \text{or} \qquad x = -3 \quad \text{or} \qquad x = 1$$

The solution set is $\left\{ -\tfrac{7}{2}, -3, 1 \right\}$.

61. $9y^3 - 49y = 0$

To factor the polynomial, begin by factoring out the greatest common factor.

$$y(9y^2 - 49) = 0$$

Now factor $9y^2 - 49$ as the difference of squares.

$$y(3y + 7)(3y - 7) = 0$$

Set each of the three factors equal to 0 and solve.

$$y = 0 \quad \text{or} \quad 3y + 7 = 0 \quad \text{or} \quad 3y - 7 = 0$$
$$3y = -7 \qquad\qquad 3y = 7$$
$$y = 0 \quad \text{or} \qquad y = -\tfrac{7}{3} \quad \text{or} \qquad y = \tfrac{7}{3}$$

The solution set is $\left\{ -\tfrac{7}{3}, 0, \tfrac{7}{3} \right\}$.

63. $r^3 - 2r^2 - 8r = 0$
$r(r^2 - 2r - 8) = 0$ *Factor out r.*
$r(r - 4)(r + 2) = 0$ *Factor.*

Set each factor equal to zero and solve.

$$r = 0 \quad \text{or} \quad r - 4 = 0 \quad \text{or} \quad r + 2 = 0$$
$$r = 0 \quad \text{or} \qquad r = 4 \quad \text{or} \qquad r = -2$$

The solution set is $\{-2, 0, 4\}$.

65. $x^3 + x^2 - 20x = 0$
$x(x^2 + x - 20) = 0$ *Factor out x.*
$x(x + 5)(x - 4) = 0$ *Factor.*

Set each factor equal to zero and solve.

$$x = 0 \quad \text{or} \quad x + 5 = 0 \quad \text{or} \quad x - 4 = 0$$
$$x = 0 \quad \text{or} \qquad x = -5 \quad \text{or} \qquad x = 4$$

The solution set is $\{-5, 0, 4\}$.

67. $r^4 = 2r^3 + 15r^2$

Rewrite with all terms on the left side.

$$r^4 - 2r^3 - 15r^2 = 0$$
$$r^2(r^2 - 2r - 15) = 0 \quad \textit{Factor out } r^2.$$
$$r^2(r - 5)(r + 3) = 0 \quad \textit{Factor.}$$

Set each factor equal to zero and solve.

$$r^2 = 0 \quad \text{or} \quad r - 5 = 0 \quad \text{or} \quad r + 3 = 0$$
$$r = 0 \quad \text{or} \qquad r = 5 \quad \text{or} \qquad r = -3$$

The solution set is $\{-3, 0, 5\}$.

69.
$$3x(x + 1) = (2x + 3)(x + 1)$$
$$3x^2 + 3x = 2x^2 + 5x + 3$$
$$x^2 - 2x - 3 = 0$$
$$(x + 1)(x - 3) = 0$$

$$x + 1 = 0 \quad \text{or} \quad x - 3 = 0$$
$$x = -1 \quad \text{or} \qquad x = 3$$

The solution set is $\{-1, 3\}$.

Alternatively, we could begin by moving all the terms to the left side and then factoring out $x + 1$.

$$3x(x + 1) - (2x + 3)(x + 1) = 0$$
$$(x + 1)[3x - (2x + 3)] = 0$$
$$(x + 1)(x - 3) = 0$$

The rest of the solution is the same.

71.
$$x^2 + (x + 1)^2 = (x + 2)^2$$
$$x^2 + x^2 + 2x + 1 = x^2 + 4x + 4$$
$$x^2 - 2x - 3 = 0$$
$$(x + 1)(x - 3) = 0$$

$$x + 1 = 0 \quad \text{or} \quad x - 3 = 0$$
$$x = -1 \quad \text{or} \qquad x = 3$$

The solution set is $\{-1, 3\}$.

73. $(2x)^2 = (2x + 4)^2 - (x + 5)^2$
$$4x^2 = 4x^2 + 16x + 16 - (x^2 + 10x + 25)$$
$$4x^2 = 4x^2 + 16x + 16 - x^2 - 10x - 25$$
$$4x^2 = 3x^2 + 6x - 9$$

$$x^2 - 6x + 9 = 0$$
$$(x - 3)(x - 3) = 0$$

Set the factor $x - 3$ equal to 0 and solve.

$$x - 3 = 0$$
$$x = 3$$

The solution set is $\{3\}$.

75.
$$(x+3)^2 - (2x-1)^2 = 0$$
$$(x^2 + 6x + 9) - (4x^2 - 4x + 1) = 0$$
Square the binomials.
$$x^2 + 6x + 9 - 4x^2 + 4x - 1 = 0$$
$$-3x^2 + 10x + 8 = 0$$
Combine like terms.
$$-1(3x^2 - 10x - 8) = 0$$
Factor out –1.
$$-1(3x + 2)(x - 4) = 0$$
Factor.

Set each factor equal to zero and solve.

$$3x + 2 = 0 \quad \text{or} \quad x - 4 = 0$$
$$3x = -2$$
$$x = -\tfrac{2}{3} \quad \text{or} \quad x = 4$$

The solution set is $\left\{-\tfrac{2}{3}, 4\right\}$.

Alternatively we could begin by factoring the left side as the difference of two squares.

$$(x+3)^2 - (2x-1)^2 = 0$$
$$[(x+3) + (2x-1)][(x+3) - (2x-1)] = 0$$
$$[3x+2][-x+4] = 0$$

The same solution set is obtained.

77. $6p^2(p+1) = 4(p+1) - 5p(p+1)$
$$6p^2(p+1) + 5p(p+1) - 4(p+1) = 0$$
Rewrite with all terms on the left.
$$(p+1)(6p^2 + 5p - 4) = 0$$
Factor out p + 1.
$$(p+1)(2p-1)(3p+4) = 0$$
Factor.

Set each factor equal to zero and solve.

$$p+1 = 0 \quad \text{or} \quad 2p - 1 = 0 \quad \text{or} \quad 3p + 4 = 0$$
$$2p = 1 \qquad 3p = -4$$
$$p = -1 \quad \text{or} \quad p = \tfrac{1}{2} \quad \text{or} \quad p = -\tfrac{4}{3}$$

The solution set is $\left\{-\tfrac{4}{3}, -1, \tfrac{1}{2}\right\}$.

79. (a) $d = 16t^2$

$$t = 2: \ d = 16(2)^2 = 16(4) = 64$$
$$t = 3: \ d = 16(3)^2 = 16(9) = 144$$
$$d = 256: \ 256 = 16t^2; \ 16 = t^2; \ t = 4$$
$$d = 576: \ 576 = 16t^2; \ 36 = t^2; \ t = 6$$

t in seconds	0	1	2	3	4	6
d in feet	0	16	64	144	256	576

(b) When $t = 0$, $d = 0$, no time has elapsed, so the object hasn't fallen (been released) yet.

81. From the calculator screens, we see that the solution set of
$$x^2 + 0.4x - 0.05 = 0$$
is $\{-0.5, 0.1\}$. Substituting -0.5 for x gives us
$$(-0.5)^2 + 0.4(-0.5) - 0.05 \overset{?}{=} 0$$
$$0.25 - 0.20 - 0.05 \overset{?}{=} 0$$
$$0.25 - 0.25 = 0 \quad \textit{True}$$

This shows that -0.5 is a solution of the given quadratic equation.

A similar check shows that 0.1 is also a solution.

The solution set is $\{-0.5, 0.1\}$.

83. Let $x =$ the number.
"If a number is doubled and 6 is subtracted from this result, the answer is 3684" translates to
$$2x - 6 = 3684.$$
$$2x = 3690$$
$$x = 1845$$

The number is 1845, so that is when Texas was admitted to the Union.

85. Let $x =$ the first consecutive integer. Then $x + 1 =$ the second consecutive integer. "Twice the sum of two consecutive integers is 28 more than the greater integer" is translated as
$$2[x + (x+1)] = 28 + (x+1).$$

Solve this equation.

$$2(2x + 1) = 28 + x + 1$$
$$4x + 2 = x + 29$$
$$3x + 2 = 29$$
$$3x = 27$$
$$x = 9$$

Since $x = 9$, $x + 1 = 10$, so the integers are 9 and 10.

6.6 Applications of Quadratic Equations

6.6 Now Try Exercises

N1. *Step 2*
Let $x =$ the length of one leg.
Then $x - 4 =$ the length of the other leg.

Step 3 $\qquad \mathcal{A} = \tfrac{1}{2}bh$
$$6 = \tfrac{1}{2}x(x-4)$$

Step 4
Solve the equation.

$$12 = x(x-4) \quad \textit{Multiply by 2.}$$
$$12 = x^2 - 4x \quad \textit{Distributive prop.}$$
$$x^2 - 4x - 12 = 0 \qquad \textit{Standard form}$$
$$(x+2)(x-6) = 0 \qquad \textit{Factor.}$$

$x + 2 = 0$ or $x - 6 = 0$ *Zero-factor prop.*
    $x = -2$ or $x = 6$

Step 5
Because a triangle's side cannot be negative, $x = 6$ and $x - 4 = 2$. The lengths of the legs are 2 feet and 6 feet.

Step 6
The length of one leg is 4 meters less than the length of the other leg, and the area is $\frac{1}{2}(6)(2) = 6$ m^2, as required.

N2. Let $x =$ the first integer.
Then $x + 1 =$ the second integer
and $x + 2 =$ the third integer.

From the given information, we write the equation

$$x(x + 1) = 8(x + 2) + 2.$$

Solve this equation.

$$x^2 + x = 8x + 16 + 2$$
$$x^2 - 7x - 18 = 0$$
$$(x + 2)(x - 9) = 0$$

$x + 2 = 0$ or $x - 9 = 0$
    $x = -2$ or $x = 9$

If $x = -2$, then $x + 1 = -1$ and $x + 2 = 0$.
If $x = 9$, then $x + 1 = 10$ and $x + 2 = 11$. There are two sets of integers that satisfy the statement of the problem: $-2, -1, 0$, and $9, 10, 11$.

N3. Let $x =$ the length of the shorter leg of the right triangle.
Then $x + 7 =$ the length of the longer leg
and $x + 8 =$ the length of the hypotenuse.

Use the Pythagorean theorem, substituting x for a, $x + 7$ for b, and $x + 8$ for c.

$$a^2 + b^2 = c^2$$
$$x^2 + (x + 7)^2 = (x + 8)^2$$
$$x^2 + x^2 + 14x + 49 = x^2 + 16x + 64$$
$$2x^2 + 14x + 49 = x^2 + 16x + 64$$
$$x^2 - 2x - 15 = 0$$
$$(x + 3)(x - 5) = 0$$

$x + 3 = 0$ or $x - 5 = 0$
    $x = -3$ or $x = 5$

Discard -3 since the length of a side of a triangle cannot be negative. The length of the shorter leg is 5 feet, the length of the longer leg is $x + 7 = 5 + 7 = 12$ feet, and the length of the hypotenuse is $x + 8 = 5 + 8 = 13$ feet.

N4. $h = -16t^2 + 180t + 6$

To find how long it will take for the ball to reach a height of 50 feet, let $h = 50$ and solve the resulting equation. (For convenience, we reverse the sides of the equation.)

$$-16t^2 + 180t + 6 = 50$$
$$-16t^2 + 180t - 44 = 0 \quad \textit{Standard form}$$
$$4t^2 - 45t + 11 = 0 \quad \textit{Divide by } -4.$$
$$(4t - 1)(t - 11) = 0 \quad \textit{Factor.}$$

$4t - 1 = 0$ or $t - 11 = 0$
    $t = \frac{1}{4}$ or $t = 11$

The ball will reach a height of 50 feet after $\frac{1}{4}$ second (on the way up) and 11 seconds (on the way down).

N5. For 2000, $x = 2000 - 1930 = 70$.

$$y = 0.01048x^2 - 0.5400x + 15.43$$
$$y = 0.01048(70)^2 - 0.5400(70) + 15.43 \quad \textit{Let x = 70.}$$
$$= 28.982 \approx 29.0$$

According to the model, the foreign-born population of the United States in 2000, to the nearest tenth of a million, was about 29.0 million. The actual value from the table is 28.4 million, so the answer using the model is slightly high.

6.6   Section Exercises

1.   Read; variable; equation; Solve; answer; Check, original

3. $A = bh$; $A = 45$, $b = 2x + 1$, $h = x + 1$

Step 3 $A = bh$
        $45 = (2x + 1)(x + 1)$

Step 4   Solve the equation.
        $45 = 2x^2 + 3x + 1$
        $0 = 2x^2 + 3x - 44$
        $0 = (2x + 11)(x - 4)$

$2x + 11 = 0$ or $x - 4 = 0$
    $2x = -11$
    $x = -\frac{11}{2}$ or $x = 4$

Step 5
The solution cannot be $x = -\frac{11}{2}$ since, when substituted, $2(-\frac{11}{2}) + 1$ and $-\frac{11}{2} + 1$ are negative numbers and base and height cannot be negative. Thus, $x = 4$ and

$$b = 2x + 1 = 8 + 1 = 9$$
$$h = x + 1 = 4 + 1 = 5.$$

The base is 9 units and the height is 5 units.

Step 6
$bh = 9 \cdot 5 = 45$, the desired value of A.

5. $\mathcal{A} = LW; \mathcal{A} = 80, L = x + 8, W = x - 8$

Step 3 $\mathcal{A} = LW$
$$80 = (x + 8)(x - 8)$$

Step 4
Solve the equation.

$$80 = x^2 - 64$$
$$0 = x^2 - 144$$
$$0 = (x + 12)(x - 12)$$

$$x + 12 = 0 \qquad \text{or} \qquad x - 12 = 0$$
$$x = -12 \quad \text{or} \qquad x = 12$$

Step 5
The solution cannot be $x = -12$ since, when substituted, $-12 + 8$ and $-12 - 8$ are negative numbers and length and width cannot be negative. Thus, $x = 12$ and

$$L = x + 8 = 12 + 8 = 20$$
$$W = x - 8 = 12 - 8 = 4.$$

The length is 20 units, and the width is 4 units.

Step 6
$LW = 20 \cdot 4 = 80$, the desired value of $\mathcal{A}$.

7. *Step 2*
Let $x = $ the width of the case.
Then $x + 2 = $ the length of the case.

Step 3

$$\mathcal{A} = LW$$
$$168 = (x + 2)x \quad \textit{Substitute.}$$

Step 4
$$168 = x^2 + 2x \quad \textit{Multiply.}$$
$$x^2 + 2x - 168 = 0 \qquad \textit{Standard form}$$
$$(x + 14)(x - 12) = 0 \qquad \textit{Factor.}$$

$$x + 14 = 0 \qquad \text{or} \quad x - 12 = 0 \; \textit{Zero-factor prop.}$$
$$x = -14 \quad \text{or} \qquad x = 12 \; \textit{Solve.}$$

Step 5
Because a width cannot be negative, $x = 12$ and $x + 2 = 14$. The width is 12 cm and the length is 14 cm.

Step 6
The length is 2 cm more than the width and the area is $14(12) = 168$ cm^2, as required.

9. Let $h = $ the height of the triangle.
Then $2h + 2 = $ the base of the triangle.

The area of the triangle is 30 square inches.

$$\mathcal{A} = \tfrac{1}{2}bh$$
$$30 = \tfrac{1}{2}(2h + 2) \cdot h$$
$$60 = (2h + 2)h$$
$$60 = 2h^2 + 2h$$
$$0 = 2h^2 + 2h - 60$$
$$0 = 2(h^2 + h - 30)$$
$$0 = 2(h + 6)(h - 5)$$

$$h + 6 = 0 \qquad \text{or} \qquad h - 5 = 0$$
$$h = -6 \quad \text{or} \qquad h = 5$$

The solution $h = -6$ must be discarded since a triangle cannot have a negative height. Thus,

$$h = 5 \text{ and } 2h + 2 = 2(5) + 2 = 12.$$

The height is 5 inches, and the base is 12 inches.

11. Let $x = $ the width of the aquarium.
Then $x + 3 = $ the height of the aquarium.

Use the formula for the volume of a rectangular box.

$$V = LWH$$
$$2730 = 21x(x + 3)$$
$$130 = x(x + 3) \qquad \textit{Divide by 21.}$$
$$130 = x^2 + 3x$$
$$0 = x^2 + 3x - 130$$
$$0 = (x + 13)(x - 10)$$

$$x + 13 = 0 \qquad \text{or} \quad x - 10 = 0$$
$$x = -13 \quad \text{or} \qquad x = 10$$

We discard -13 because the width cannot be negative. The width is 10 inches. The height is $10 + 3 = 13$ inches.

13. *Step 2*
Let $x = $ the width of the monitor.
Then $x + 3 = $ the length of the monitor.
The area is $x(x + 3)$.

Step 3
If the length were doubled $[2(x + 3)]$ and if the width were decreased by 1 in. $[x - 1]$, the area would be increased by 150 in.^{2} $[x(x + 3) + 150]$. Write an equation.

$$LW = \mathcal{A}$$
$$[2(x + 3)](x - 1) = x(x + 3) + 150$$

Step 4

$$(2x + 6)(x - 1) = x^2 + 3x + 150$$
$$2x^2 + 4x - 6 = x^2 + 3x + 150$$
$$x^2 + x - 156 = 0$$
$$(x + 13)(x - 12) = 0$$

$$x + 13 = 0 \qquad \text{or} \qquad x - 12 = 0$$
$$x = -13 \quad \text{or} \qquad x = 12$$

Step 5
Reject -13, so the width is 12 inches and the length is $12 + 3 = 15$ inches.

Step 6
The length is 3 inches more than the width. The area of the monitor is $15(12) = 180$ in.^{2}. Doubling the length and decreasing the width by 1 inch gives us an area of $30(11) = 330$ in.^{2}, which is 150 in.^{2} more than the area of the original monitor, as required.

15. Let $x =$ the length of a side of the square painting.
Then $x - 2 =$ the length of a side of the square mirror.

Since the formula for the area of a square is $A = s^2$, the area of the painting is x^2, and the area of the mirror is $(x - 2)^2$. The difference between their areas is 32, so

$$x^2 - (x - 2)^2 = 32$$
$$x^2 - (x^2 - 4x + 4) = 32$$
$$x^2 - x^2 + 4x - 4 = 32$$
$$4x - 4 = 32$$
$$4x = 36$$
$$x = 9.$$

The length of a side of the painting is 9 feet. The length of a side of the mirror is $9 - 2 = 7$ feet.

Check: $9^2 - 7^2 = 81 - 49 = 32$

17. Let $x =$ the first volume number.
Then $x + 1 =$ the second volume number.
The product of the numbers is 420.

$$x(x + 1) = 420$$
$$x^2 + x - 420 = 0$$
$$(x - 20)(x + 21) = 0$$

$$x - 20 = 0 \quad \text{or} \quad x + 21 = 0$$
$$x = 20 \quad \text{or} \quad x = -21$$

The volume number cannot be negative, so we reject -21. The volume numbers are 20 and $x + 1 = 20 + 1 = 21$.

19. Let $x =$ the first integer.
Then $x + 1 =$ the second integer,
and $x + 2 =$ the third integer.

The product of the second and third	is	2	more than	10 times the first
↓	↓	↓	↓	↓
$(x+1)(x+2)$	$=$	2	$+$	$10x$

$$x^2 + 3x + 2 = 2 + 10x$$
$$x^2 - 7x = 0$$
$$x(x - 7) = 0$$

$$x = 0 \quad \text{or} \quad x - 7 = 0$$
$$x = 7$$

If $x = 0$, then $x + 1 = 1$, and $x + 2 = 2$.
If $x = 7$, then $x + 1 = 8$, and $x + 2 = 9$.
So there are two sets of consecutive integers that satisfy the condition: $0, 1, 2$ and $7, 8, 9$.

Check 0, 1, 2: $1(2) \stackrel{?}{=} 2 + 10(0)$
$$2 = 2 + 0 \qquad \textit{True}$$

Check 7, 8, 9: $8(9) \stackrel{?}{=} 2 + 10(7)$
$$72 = 2 + 70 \qquad \textit{True}$$

21. Let $x =$ the first odd integer.
Then $x + 2 =$ the second odd integer and $x + 4 =$ the third odd integer.

$$3[x + (x + 2) + (x + 4)] = x(x + 2) + 18$$
$$3(3x + 6) = x^2 + 2x + 18$$
$$9x + 18 = x^2 + 2x + 18$$
$$0 = x^2 - 7x$$
$$0 = x(x - 7)$$

$$x = 0 \quad \text{or} \quad x - 7 = 0$$
$$x = 7$$

We must discard 0 because it is even and the problem requires the integers to be odd. If $x = 7$, $x + 2 = 9$, and $x + 4 = 11$. The three integers are $7, 9,$ and 11.

23. Let $x =$ the first even integer. Then $x + 2$ and $x + 4$ are the next two even integers.

$$x^2 + (x + 2)^2 = (x + 4)^2$$
$$x^2 + x^2 + 4x + 4 = x^2 + 8x + 16$$
$$x^2 - 4x - 12 = 0$$
$$(x - 6)(x + 2) = 0$$

$$x - 6 = 0 \quad \text{or} \quad x + 2 = 0$$
$$x = 6 \quad \text{or} \quad x = -2$$

If $x = 6$, $x + 2 = 8$, and $x + 4 = 10$.
If $x = -2$, $x + 2 = 0$, and $x + 4 = 2$.

The three integers are $6, 8,$ and 10 or $-2, 0,$ and 2.

25. Let $x =$ the length of the longer leg of the right triangle.
Then $x + 1 =$ the length of the hypotenuse and $x - 7 =$ the length of the shorter leg.

Refer to the figure in the text.
Use the Pythagorean theorem with $a = x$, $b = x - 7$, and $c = x + 1$.

$$a^2 + b^2 = c^2$$
$$x^2 + (x - 7)^2 = (x + 1)^2$$
$$x^2 + (x^2 - 14x + 49) = x^2 + 2x + 1$$
$$2x^2 - 14x + 49 = x^2 + 2x + 1$$
$$x^2 - 16x + 48 = 0$$
$$(x - 12)(x - 4) = 0$$

$$x - 12 = 0 \quad \text{or} \quad x - 4 = 0$$
$$x = 12 \quad \text{or} \quad x = 4$$

Discard 4 because if the length of the longer leg is 4 centimeters, by the conditions of the problem, the length of the shorter leg would be $4 - 7 = -3$ centimeters, which is impossible. The length of the longer leg is 12 centimeters.

Check: $12^2 + 5^2 = 13^2$; $169 = 169$ *True*

27. Let $x =$ Alan's distance from home. Then $x + 1 =$ the distance between Tram and Alan.

Refer to the diagram in the textbook.
Use the Pythagorean theorem.

$$a^2 + b^2 = c^2$$
$$x^2 + 5^2 = (x + 1)^2$$
$$x^2 + 25 = x^2 + 2x + 1$$
$$24 = 2x$$
$$12 = x$$

Alan is 12 miles from home.

Check: $12^2 + 5^2 = 13^2$; $169 = 169$ *True*

29. Let $x =$ the length of the ladder. Then
$x - 4 =$ the distance from the bottom of the ladder to the building and
$x - 2 =$ the distance on the side of the building to the top of the ladder.

Substitute into the Pythagorean theorem.

$$a^2 + b^2 = c^2$$
$$(x - 2)^2 + (x - 4)^2 = x^2$$
$$x^2 - 4x + 4 + x^2 - 8x + 16 = x^2$$
$$x^2 - 12x + 20 = 0$$
$$(x - 10)(x - 2) = 0$$

$$x - 10 = 0 \quad \text{or} \quad x - 2 = 0$$
$$x = 10 \quad \text{or} \quad x = 2$$

The solution cannot be 2 because then a negative distance results. Thus, $x = 10$ and the top of the ladder reaches $x - 2 = 10 - 2 = 8$ feet up the side of the building.

Check: $8^2 + 6^2 = 10^2$; $100 = 100$ *True*

31. $h = -16t^2 + 128t$
$h = -16(1)^2 + 128(1)$　*Let t = 1.*
$ = -16 + 128$
$ = 112$

After 1 second, the height is 112 feet.

33. $h = -16t^2 + 128t$
$h = -16(4)^2 + 128(4)$　*Let t = 4.*
$ = -16(16) + 512$
$ = -256 + 512$
$ = 256$

After 4 seconds, the height is 256 feet.

35. **(a)** Let $h = 64$ in the given formula and solve for t.

$$h = -16t^2 + 32t + 48$$
$$64 = -16t^2 + 32t + 48$$
$$16t^2 - 32t + 16 = 0$$
$$16(t^2 - 2t + 1) = 0$$
$$16(t - 1)^2 = 0$$
$$t - 1 = 0$$
$$t = 1$$

The height of the object will be 64 feet after 1 second.

(b) To find the time when the height is 60 feet, let $h = 60$ in the given equation and solve for t.

$$h = -16t^2 + 32t + 48$$
$$60 = -16t^2 + 32t + 48$$
$$16t^2 - 32t + 12 = 0$$
$$4(4t^2 - 8t + 3) = 0$$
$$4(2t - 1)(2t - 3) = 0$$

$$2t - 1 = 0 \quad \text{or} \quad 2t - 3 = 0$$
$$2t = 1 \quad \text{or} \quad 2t = 3$$
$$t = \tfrac{1}{2} \quad \text{or} \quad t = \tfrac{3}{2}$$

The height of the object is 60 feet after $\frac{1}{2}$ second (on the way up) and after $\frac{3}{2}$ or $1\frac{1}{2}$ seconds (on the way down).

(c) To find the time when the object hits the ground, let $h = 0$ and solve for t.

$$h = -16t^2 + 32t + 48$$
$$0 = -16t^2 + 32t + 48$$
$$16t^2 - 32t - 48 = 0$$
$$16(t^2 - 2t - 3) = 0$$
$$16(t + 1)(t - 3) = 0$$

$$t + 1 = 0 \quad \text{or} \quad t - 3 = 0$$
$$t = -1 \quad \text{or} \quad t = 3$$

We discard -1 because time cannot be negative. The object will hit the ground after 3 seconds.

(d) The negative solution, -1, does not make sense, since t represents time, which cannot be negative.

37. (a) $x = 10$ in 2000.
$$y = 0.590x^2 + 4.523x + 0.136$$
$$y = 0.590(10)^2 + 4.523(10) + 0.136$$
$$y = 104.366 \approx 104.4$$

The model indicates there were about 104.4 million cellular phone subscribers in 2000. The result using the model is less than 109 million, the actual number for 2000.

(b) $x = 2008 - 1990 = 18$
$x = 18$ corresponds to 2008.

(c) $y = 0.590(18)^2 + 4.523(18) + 0.136$
$$y = 272.71 \approx 272.7$$

The model indicates there were about 272.7 million cellular phone subscribers in 2008. The result using the model is more than 263 million, the actual number for 2008.

(d) $x = 2010 - 1990 = 20$
$$y = 0.590(20)^2 + 4.523(20) + 0.136$$
$$y = 326.596 \approx 326.6$$

The model gives an estimate of 326.6 million cellular phone subscribers in 2010.

39. $\dfrac{50}{72} = \dfrac{25 \cdot 2}{36 \cdot 2} = \dfrac{25}{36}$

41. $\dfrac{48}{-27} = \dfrac{16 \cdot 3}{-9 \cdot 3} = \dfrac{16}{-9}$, or $-\dfrac{16}{9}$

Chapter 6 Review Exercises

1. $7t + 14 = 7 \cdot t + 7 \cdot 2 = 7(t + 2)$

2. $60z^3 + 30z = 30z \cdot 2z^2 + 30z \cdot 1$
$$= 30z(2z^2 + 1)$$

3. $2xy - 8y + 3x - 12$
$= (2xy - 8y) + (3x - 12)$ *Group terms.*
$= 2y(x - 4) + 3(x - 4)$ *Factor each group.*
$= (x - 4)(2y + 3)$ *Factor out x − 4.*

4. $6y^2 + 9y + 4xy + 6x$
$= (6y^2 + 9y) + (4xy + 6x)$
$= 3y(2y + 3) + 2x(2y + 3)$
$= (2y + 3)(3y + 2x)$

5. $x^2 + 5x + 6$

Find two integers whose product is 6 and whose sum is 5. The integers are 3 and 2. Thus,
$$x^2 + 5x + 6 = (x + 3)(x + 2).$$

6. $y^2 - 13y + 40$

Find two integers whose product is 40 and whose sum is -13.

Factors of 40	Sums of Factors
$-1, -40$	-41
$-2, -20$	-22
$-4, -10$	-14
$-5, -8$	-13 ←

The integers are -5 and -8, so
$$y^2 - 13y + 40 = (y - 5)(y - 8).$$

7. $q^2 + 6q - 27$

Find two integers whose product is -27 and whose sum is 6. The integers are -3 and 9, so
$$q^2 + 6q - 27 = (q - 3)(q + 9).$$

8. $r^2 - r - 56$

Find two integers whose product is -56 and whose sum is -1. The integers are 7 and -8, so
$$r^2 - r - 56 = (r + 7)(r - 8).$$

9. $r^2 - 4rs - 96s^2$

Find two expressions whose product is $-96s^2$ and whose sum is $-4s$. The expressions are $8s$ and $-12s$, so
$$r^2 - 4rs - 96s^2 = (r + 8s)(r - 12s).$$

10. $p^2 + 2pq - 120q^2$

Find two expressions whose product is $-120q^2$ and whose sum is $2q$. The expressions are $12q$ and $-10q$, so
$$p^2 + 2pq - 120q^2 = (p + 12q)(p - 10q).$$

11. $8p^3 - 24p^2 - 80p$

First, factor out the GCF, $8p$.
$$8p^3 - 24p^2 - 80p = 8p(p^2 - 3p - 10)$$
Now factor $p^2 - 3p - 10$.
$$p^2 - 3p - 10 = (p + 2)(p - 5)$$
The completely factored form is
$$8p^3 - 24p^2 - 80p = 8p(p + 2)(p - 5).$$

12. $3x^4 + 30x^3 + 48x^2$
$= 3x^2(x^2 + 10x + 16)$
$= 3x^2(x + 2)(x + 8)$

13. $p^7 - p^6q - 2p^5q^2$
$= p^5(p^2 - pq - 2q^2)$
$= p^5(p + q)(p - 2q)$

14. $3r^5 - 6r^4 s - 45r^3 s^2$
$$= 3r^3(r^2 - 2rs - 15s^2)$$
$$= 3r^3(r + 3s)(r - 5s)$$

15. $9x^4 y - 9x^3 y - 54x^2 y$
$$= 9x^2 y(x^2 - x - 6)$$
$$= 9x^2 y(x + 2)(x - 3)$$

16. $2x^7 + 2x^6 y - 12x^5 y^2$
$$= 2x^5(x^2 + xy - 6y^2)$$
$$= 2x^5(x - 2y)(x + 3y)$$

17. To begin factoring $6r^2 - 5r - 6$, the possible first terms of the two binomial factors are r and $6r$, or $2r$ and $3r$, if we consider only positive integer coefficients.

18. When factoring $2z^3 + 9z^2 - 5z$, the first step is to factor out the GCF, z.

In Exercises 19–27, either the trial and error method or the grouping method can be used to factor each polynomial.

19. Factor $2k^2 - 5k + 2$ by trial and error.
$$2k^2 - 5k + 2 = (2k - 1)(k - 2)$$

20. Factor $3r^2 + 11r - 4$ by grouping. Look for two integers whose product is $3(-4) = -12$ and whose sum is 11. The integers are 12 and -1.
$$3r^2 + 11r - 4 = 3r^2 + 12r - r - 4$$
$$= (3r^2 + 12r) + (-r - 4)$$
$$= 3r(r + 4) - 1(r + 4)$$
$$= (r + 4)(3r - 1)$$

21. Factor $6r^2 - 5r - 6$ by grouping. Find two integers whose product is $6(-6) = -36$ and whose sum is -5. The integers are -9 and 4.
$$6r^2 - 5r - 6 = 6r^2 - 9r + 4r - 6$$
$$= (6r^2 - 9r) + (4r - 6)$$
$$= 3r(2r - 3) + 2(2r - 3)$$
$$= (2r - 3)(3r + 2)$$

22. Factor $10z^2 - 3z - 1$ by trial and error.
$$10z^2 - 3z - 1 = (5z + 1)(2z - 1)$$

23. Factor $8v^2 + 17v - 21$ by grouping. Look for two integers whose product is $8(-21) = -168$ and whose sum is 17. The integers are 24 and -7.
$$8v^2 + 17v - 21 = 8v^2 + 24v - 7v - 21$$
$$= (8v^2 + 24v) + (-7v - 21)$$
$$= 8v(v + 3) - 7(v + 3)$$
$$= (v + 3)(8v - 7)$$

24. $24x^5 - 20x^4 + 4x^3$

Factor out the GCF, $4x^3$. Then complete the factoring by trial and error.
$$24x^5 - 20x^4 + 4x^3$$
$$= 4x^3(6x^2 - 5x + 1)$$
$$= 4x^3(3x - 1)(2x - 1)$$

25. $-6x^2 + 3x + 30$
$$= -3(2x^2 - x - 10)$$
$$= -3(2x - 5)(x + 2)$$

26. $10r^3 s + 17r^2 s^2 + 6rs^3$
$$= rs(10r^2 + 17rs + 6s^2)$$
$$= rs(5r + 6s)(2r + s)$$

27. $48x^4 y + 4x^3 y^2 - 4x^2 y^3$
$$= 4x^2 y(12x^2 + xy - y^2)$$
$$= 4x^2 y(3x + y)(4x - y)$$

28. The student stopped too soon. He needs to factor out the common factor $4x - 1$ to get $(4x - 1)(4x - 5)$ as the correct answer.

29. Only choice **B**, $4x^2 y^2 - 25z^2$, is the difference of squares. In **A**, 32 is not a perfect square. In **C**, we have a sum, not a difference. In **D**, y^3 is not a square. The correct choice is **B**.

30. Only choice **D**, $x^2 - 20x + 100$, is a perfect square trinomial because $x^2 = x \cdot x$, $100 = 10 \cdot 10$, and $-20x = -2(x)(10)$.

In Exercises 31–34, use the rule for factoring a difference of squares.

31. $n^2 - 49 = n^2 - 7^2 = (n + 7)(n - 7)$

32. $25b^2 - 121 = (5b)^2 - 11^2$
$$= (5b + 11)(5b - 11)$$

33. $49y^2 - 25w^2 = (7y)^2 - (5w)^2$
$$= (7y + 5w)(7y - 5w)$$

34. $144p^2 - 36q^2 = 36(4p^2 - q^2)$
$$= 36[(2p)^2 - q^2]$$
$$= 36(2p + q)(2p - q)$$

35. $x^2 + 100$

This polynomial is *prime* because it is the sum of squares and the two terms have no common factor.

In Exercises 36–37, use the rules for factoring a perfect square trinomial.

36. $r^2 - 12r + 36 = r^2 - 2(6)(r) + 6^2$
$$= (r - 6)^2$$

37. $9t^2 - 42t + 49 = (3t)^2 - 2(3t)(7) + 7^2$
$$= (3t - 7)^2$$

In Exercises 38–39, use the rule for factoring a sum of cubes.

38. $m^3 + 1000$
$= m^3 + 10^3$
$= (m + 10)(m^2 - 10 \cdot m + 10^2)$
$= (m + 10)(m^2 - 10m + 100)$

39. $125k^3 + 64x^3$
$= (5k)^3 + (4x)^3$
$= (5k + 4x)[(5k)^2 - 5k \cdot 4x + (4x)^2]$
$= (5k + 4x)(25k^2 - 20kx + 16x^2)$

In Exercises 40–41, use the rule for factoring a difference of cubes.

40. $343x^3 - 64$
$= (7x)^3 - 4^3$
$= (7x - 4)[(7x)^2 + 7x \cdot 4 + 4^2]$
$= (7x - 4)(49x^2 + 28x + 16)$

41. $1000 - 27x^6$
$= 10^3 - (3x^2)^3$
$= (10 - 3x^2)[10^2 + 10(3x^2) + (3x^2)^2]$
$= (10 - 3x^2)(100 + 30x^2 + 9x^4)$

42. $x^6 - y^6$
$= (x^3)^2 - (y^3)^2$ *Difference of squares*
$= (x^3 + y^3)(x^3 - y^3)$
Now factor as the sum and difference of cubes.
$= (x + y)(x^2 - xy + y^2)(x - y)(x^2 + xy + y^2)$

In Exercises 43–56, all solutions should be checked by substituting in the original equations. The checks will not be shown here.

43. $(4t + 3)(t - 1) = 0$
$4t + 3 = 0$ or $t - 1 = 0$
$4t = -3$ or $t = 1$
$t = -\frac{3}{4}$
The solution set is $\left\{-\frac{3}{4}, 1\right\}$.

44. $(x + 7)(x - 4)(x + 3) = 0$
$x + 7 = 0$ or $x - 4 = 0$ or $x + 3 = 0$
$x = -7$ or $x = 4$ or $x = -3$
The solution set is $\{-7, -3, 4\}$.

45. $x(2x - 5) = 0$
$x = 0$ or $2x - 5 = 0$
$2x = 5$
$x = \frac{5}{2}$
The solution set is $\left\{0, \frac{5}{2}\right\}$.

46. $z^2 + 4z + 3 = 0$
$(z + 3)(z + 1) = 0$
$z + 3 = 0$ or $z + 1 = 0$
$z = -3$ or $z = -1$
The solution set is $\{-3, -1\}$.

47. $m^2 - 5m + 4 = 0$
$(m - 1)(m - 4) = 0$
$m - 1 = 0$ or $m - 4 = 0$
$m = 1$ or $m = 4$
The solution set is $\{1, 4\}$.

48. $x^2 = -15 + 8x$
$x^2 - 8x + 15 = 0$
$(x - 3)(x - 5) = 0$
$x - 3 = 0$ or $x - 5 = 0$
$x = 3$ or $x = 5$
The solution set is $\{3, 5\}$.

49. $3z^2 - 11z - 20 = 0$
$(3z + 4)(z - 5) = 0$
$3z + 4 = 0$ or $z - 5 = 0$
$3z = -4$ or $z = 5$
$z = -\frac{4}{3}$
The solution set is $\left\{-\frac{4}{3}, 5\right\}$.

50. $81t^2 - 64 = 0$
$(9t + 8)(9t - 8) = 0$
$9t + 8 = 0$ or $9t - 8 = 0$
$9t = -8$ or $9t = 8$
$t = -\frac{8}{9}$ or $t = \frac{8}{9}$
The solution set is $\left\{-\frac{8}{9}, \frac{8}{9}\right\}$.

51. $y^2 = 8y$
$y^2 - 8y = 0$
$y(y - 8) = 0$
$y = 0$ or $y - 8 = 0$
$y = 8$
The solution set is $\{0, 8\}$.

52. $n(n - 5) = 6$
$n^2 - 5n = 6$
$n^2 - 5n - 6 = 0$
$(n + 1)(n - 6) = 0$
$n + 1 = 0$ or $n - 6 = 0$
$n = -1$ or $n = 6$
The solution set is $\{-1, 6\}$.

53. $t^2 - 14t + 49 = 0$
$$(t - 7)^2 = 0$$
$$t - 7 = 0$$
$$t = 7$$

The solution set is $\{7\}$.

54. $$t^2 = 12(t - 3)$$
$$t^2 = 12t - 36$$
$$t^2 - 12t + 36 = 0$$
$$(t - 6)^2 = 0$$
$$t - 6 = 0$$
$$t = 6$$

The solution set is $\{6\}$.

55. $(5z + 2)(z^2 + 3z + 2) = 0$
$$(5z + 2)(z + 2)(z + 1) = 0$$

$5z + 2 = 0$ or $z + 2 = 0$ or $z + 1 = 0$
$5z = -2$

$z = -\frac{2}{5}$ or $z = -2$ or $z = -1$

The solution set is $\left\{-2, -1, -\frac{2}{5}\right\}$.

56. $$x^2 = 9$$
$$x^2 - 9 = 0$$
$$(x + 3)(x - 3) = 0$$

$x + 3 = 0$ or $x - 3 = 0$
$x = -3$ or $x = 3$

The solution set is $\{-3, 3\}$.

57. Let $x =$ the width of the rug.
Then $x + 6 =$ the length of the rug.

$$\mathcal{A} = LW$$
$$40 = (x + 6)x$$
$$40 = x^2 + 6x$$
$$0 = x^2 + 6x - 40$$
$$0 = (x + 10)(x - 4)$$

$x + 10 = 0$ or $x - 4 = 0$
$x = -10$ or $x = 4$

Reject -10 since the width cannot be negative. The width of the rug is 4 feet and the length is $4 + 6$ or 10 feet.

58. From the figure, we have $L = 20$, $W = x$, and $H = x + 4$.

$$S = 2WH + 2WL + 2LH$$
$$650 = 2x(x + 4) + 2x(20) + 2(20)(x + 4)$$
$$650 = 2x^2 + 8x + 40x + 40(x + 4)$$
$$650 = 2x^2 + 48x + 40x + 160$$
$$0 = 2x^2 + 88x - 490$$
$$0 = 2(x^2 + 44x - 245)$$
$$0 = 2(x + 49)(x - 5)$$

$x + 49 = 0$ or $x - 5 = 0$
$x = -49$ or $x = 5$

Reject -49 because the width cannot be negative. The width of the chest is 5 feet.

59. Let $x =$ the first integer.
Then $x + 1 =$ the next integer.

The product of the integers is 29 more than their sum, so

$$x(x + 1) = 29 + [x + (x + 1)].$$

Solve this equation.

$$x^2 + x = 29 + 2x + 1$$
$$x^2 - x - 30 = 0$$
$$(x - 6)(x + 5) = 0$$

$x - 6 = 0$ or $x + 5 = 0$
$x = 6$ or $x = -5$

If $x = 6$, $x + 1 = 6 + 1 = 7$.
If $x = -5$, $x + 1 = -5 + 1 = -4$.

The consecutive integers are 6 and 7 or -5 and -4.

60. Let $x =$ the distance traveled west.
Then $x - 14 =$ the distance traveled south,
and $(x - 14) + 16 = x + 2 =$ the distance between the cars.

These three distances form a right triangle with x and $x - 14$ representing the lengths of the legs and $x + 2$ representing the length of the hypotenuse. Use the Pythagorean theorem.

$$a^2 + b^2 = c^2$$
$$x^2 + (x - 14)^2 = (x + 2)^2$$
$$x^2 + x^2 - 28x + 196 = x^2 + 4x + 4$$
$$x^2 - 32x + 192 = 0$$
$$(x - 8)(x - 24) = 0$$

$x - 8 = 0$ or $x - 24 = 0$
$x = 8$ or $x = 24$

If $x = 8$, then $x - 14 = -6$, which is not possible because a distance cannot be negative.

If $x = 24$, then $x - 14 = 10$ and $x + 2 = 26$. The cars were 26 miles apart.

61. (a) $d = 16t^2$

$$t = 4: \ d = 16(4)^2 = 256$$

In 4 seconds, the object would fall 256 feet.

(b) $t = 8: \ d = 16(8)^2 = 1024$

In 8 seconds, the object would fall 1024 feet.

62. (a) In 2005, $x = 5$

$$y = -2.84x^2 + 61.1x + 366$$
$$y = -2.84(5)^2 + 61.1(5) + 366$$
$$y = 600.5 \approx 601$$

If $x = 5$, then $y = 601$, so the prediction is 601,000 vehicles. The result is slightly higher than the actual number for 2005.

(b) In 2007, $x = 7$

$$y = -2.84x^2 + 61.1x + 366$$
$$y = -2.84(7)^2 + 61.1(7) + 366$$
$$y = 654.54 \approx 655$$

If $x = 7$, then $y = 655$, so the prediction is 655,000 vehicles.

(c) The estimate may be unreliable because the conditions that prevailed in the years 2001–2006 may have changed, causing either a greater increase or a decrease predicted by the model for the number of alternative-fueled vehicles.

63. [6.1] D is not factored completely.
$$3(7t + 4) + x(7t + 4) = (7t + 4)(3 + x)$$

64. [6.1] The factor $2x + 8$ has a factor of 2. The completely factored form is $2(x + 4)(3x - 4)$.

65. [6.3] $3k^2 + 11k + 10$

Two integers with product $3(10) = 30$ and sum 11 are 5 and 6.

$$3k^2 + 11k + 10$$
$$= 3k^2 + 5k + 6k + 10$$
$$= (3k^2 + 5k) + (6k + 10)$$
$$= k(3k + 5) + 2(3k + 5)$$
$$= (3k + 5)(k + 2)$$

66. [6.2] $z^2 - 11zx + 10x^2 = (z - x)(z - 10x)$

67. [6.4] $y^4 - 625$

$$= (y^2)^2 - 25^2$$
$$= (y^2 + 25)(y^2 - 25) \quad \textit{Difference of squares}$$
$$= (y^2 + 25)(y + 5)(y - 5) \quad \textit{Difference of squares}$$

68. [6.1] $15m^2 + 20m - 12mp - 16p$
$$= 5m(3m + 4) - 4p(3m + 4)$$
$$\qquad\qquad \textit{Factor by grouping.}$$
$$= (3m + 4)(5m - 4p)$$

69. [6.1] $24ab^3c^2 - 56a^2bc^3 + 72a^2b^2c$
$$= 8abc(3b^2c - 7ac^2 + 9ab)$$

70. [6.3] $6m^3 - 21m^2 - 45m$
$$= 3m(2m^2 - 7m - 15)$$
$$= 3m[(2m^2 - 10m) + (3m - 15)]$$
$$\qquad\qquad \textit{Factor by grouping.}$$
$$= 3m[2m(m - 5) + 3(m - 5)]$$
$$= 3m(m - 5)(2m + 3)$$

71. [6.1] $12x^2yz^3 + 12xy^2z - 30x^3y^2z^4$
$$= 6xyz(2xz^2 + 2y - 5x^2yz^3)$$

72. [6.3] $25a^2 + 15ab + 9b^2$ is a *prime* polynomial.

73. [6.1] $12r^2 + 18rq - 10r - 15q$
$$= 6r(2r + 3q) - 5(2r + 3q) \quad \textit{Factor by grouping.}$$
$$= (2r + 3q)(6r - 5)$$

74. [6.2] $2a^5 - 8a^4 - 24a^3$
$$= 2a^3(a^2 - 4a - 12)$$
$$= 2a^3(a - 6)(a + 2)$$

75. [6.4] $49t^2 + 56t + 16$
$$= (7t)^2 + 2(7t)(4) + 4^2$$
$$= (7t + 4)^2$$

76. [6.4] $1000a^3 + 27$
$$= (10a)^3 + 3^3$$
$$= (10a + 3)[(10a)^2 - 10a \cdot 3 + 3^2]$$
$$= (10a + 3)(100a^2 - 30a + 9)$$

77. [6.5] $t(t - 7) = 0$

$$t = 0 \quad \text{or} \quad t - 7 = 0$$
$$t = 7$$

The solution set is $\{0, 7\}$.

78. [6.5]
$$x^2 + 3x = 10$$
$$x^2 + 3x - 10 = 0$$
$$(x + 5)(x - 2) = 0$$

$$x + 5 = 0 \quad \text{or} \quad x - 2 = 0$$
$$x = -5 \quad \text{or} \quad x = 2$$

The solution set is $\{-5, 2\}$.

79. [6.5]
$$25x^2 + 20x + 4 = 0$$
$$(5x)^2 + 2(5x)(2) + 2^2 = 0$$
$$(5x + 2)^2 = 0$$
$$5x + 2 = 0$$
$$5x = -2$$
$$x = -\tfrac{2}{5}$$

The solution set is $\left\{-\tfrac{2}{5}\right\}$.

80. [6.6] Let $x =$ the first integer. Then $x + 1$ and $x + 2$ are the next two integers.

The product of the first two of three consecutive integers is equal to 23 plus the third.

$$x(x+1) = 23 + (x+2)$$
$$x^2 + x = 23 + x + 2$$
$$x^2 - 25 = 0$$
$$(x+5)(x-5) = 0$$

$$x + 5 = 0 \quad \text{or} \quad x - 5 = 0$$
$$x = -5 \quad \text{or} \quad x = 5$$

If $x = -5$, then $x + 1 = -4$ and $x + 2 = -3$.
If $x = 5$, then $x + 1 = 6$ and $x + 2 = 7$.
The integers are $-5, -4$, and -3, or $5, 6$, and 7.

81. **[6.6]** Let x = the width of the base.
Then $x + 2$ = the length of the base.

The area of the base, B, is given by LW, so

$$B = x(x+2).$$

Use the formula for the volume of a pyramid,

$$V = \tfrac{1}{3} \cdot B \cdot h.$$
$$48 = \tfrac{1}{3} x(x+2)(6)$$
$$48 = 2x(x+2)$$
$$24 = x^2 + 2x$$
$$x^2 + 2x - 24 = 0$$
$$(x+6)(x-4) = 0$$

$$x + 6 = 0 \quad \text{or} \quad x - 4 = 0$$
$$x = -6 \quad \text{or} \quad x = 4$$

Reject -6. The width of the base is 4 meters and the length is $4 + 2$ or 6 meters.

82. **[6.6]** Let x = the length of the shorter leg.
Then $2x + 6$ = the length of the longer leg and $(2x + 6) + 3 = 2x + 9$ = the length of the hypotenuse.

Use the Pythagorean theorem, $a^2 + b^2 = c^2$.

$$x^2 + (2x+6)^2 = (2x+9)^2$$
$$x^2 + 4x^2 + 24x + 36 = 4x^2 + 36x + 81$$
$$x^2 - 12x - 45 = 0$$
$$(x-15)(x+3) = 0$$

$$x - 15 = 0 \quad \text{or} \quad x + 3 = 0$$
$$x = 15 \quad \text{or} \quad x = -3$$

Reject -3 because a length cannot be negative. The sides of the lot are 15 meters, $2(15) + 6 = 36$ meters, and $36 + 3 = 39$ meters.

83. **[6.6]** Let b = the base of the sail.
Then $b + 4$ = the height of the sail.

Use the formula for the area of a triangle.

$$\mathcal{A} = \tfrac{1}{2}bh$$
$$30 = \tfrac{1}{2}(b)(b+4) \qquad \text{Let } A = 30.$$
$$60 = b^2 + 4b$$
$$0 = b^2 + 4b - 60$$
$$0 = (b+10)(b-6)$$

$$b + 10 = 0 \quad \text{or} \quad b - 6 = 0$$
$$b = -10 \quad \text{or} \quad b = 6$$

Discard -10 since the base of a triangle cannot be negative. The base of the triangular sail is 6 meters.

84. **[6.6]** Let x = the width of the house.
Then $x + 7$ = the length of the house.

Use $\mathcal{A} = LW$ with 170 for $\mathcal{A}$, $x + 7$ for L, and x for W.

$$170 = (x+7)(x)$$
$$170 = x^2 + 7x$$
$$0 = x^2 + 7x - 170$$
$$0 = (x+17)(x-10)$$

$$x + 17 = 0 \quad \text{or} \quad x - 10 = 0$$
$$x = -17 \quad \text{or} \quad x = 10$$

Discard -17 because the width cannot be negative. If $x = 10$, $x + 7 = 10 + 7 = 17$.

The width is 10 meters and the length is 17 meters.

Chapter 6 Test

1.
$$2x^2 - 2x - 24 = 2(x^2 - x - 12)$$
$$= 2(x+3)(x-4)$$

The correct completely factored form is choice **D**. Note that the factored forms **A**, $(2x+6)(x-4)$, and **B**, $(x+3)(2x-8)$, also can be multiplied to give a product of $2x^2 - 2x - 24$, but neither of these is completely factored because $2x + 6$ and $2x - 8$ both contain a common factor of 2.

2. $12x^2 - 30x = 6x(2x - 5)$

3. $2m^3n^2 + 3m^3n - 5m^2n^2$
$$= m^2n(2mn + 3m - 5n)$$

4. $2ax - 2bx + ay - by$
$$= 2x(a-b) + y(a-b)$$
$$= (a-b)(2x+y)$$

5. $x^2 - 5x - 24$
Find two integers whose product is -24 and whose sum is -5. The integers are 3 and -8.
$$x^2 - 5x - 24 = (x+3)(x-8)$$

6. Factor $2x^2 + x - 3$ by trial and error.
$$2x^2 + x - 3 = (2x+3)(x-1)$$

7. Factor $10z^2 - 17z + 3$ by trial and error.
$$10z^2 - 17z + 3 = (2z-3)(5z-1)$$

8. $t^2 + 2t + 3$

We cannot find two integers whose product is 3 and whose sum is 2. This polynomial is prime.

9. $x^2 + 36$

This polynomial is *prime* because the sum of squares cannot be factored and the two terms have no common factor.

10. $12 - 6a + 2b - ab$
$= (12 - 6a) + (2b - ab)$
$= 6(2 - a) + b(2 - a)$
$= (2 - a)(6 + b)$

11. $9y^2 - 64$
$= (3y)^2 - 8^2$
$= (3y + 8)(3y - 8)$

12. $4x^2 - 28xy + 49y^2$
$= (2x)^2 - 2(2x)(7y) + (7y)^2$
$= (2x - 7y)^2$

13. $-2x^2 - 4x - 2$
$= -2(x^2 + 2x + 1)$
$= -2(x^2 + 2 \cdot x \cdot 1 + 1^2)$
$= -2(x + 1)^2$

14. $6t^4 + 3t^3 - 108t^2$
$= 3t^2(2t^2 + t - 36)$
$= 3t^2(2t + 9)(t - 4)$

15. $r^3 - 125 = r^3 - 5^3$
$= (r - 5)(r^2 + 5 \cdot r + 5^2)$
$= (r - 5)(r^2 + 5r + 25)$

16. $8k^3 + 64 = 8(k^3 + 8)$
$= 8(k^3 + 2^3)$
$= 8(k + 2)(k^2 - 2 \cdot k + 2^2)$
$= 8(k + 2)(k^2 - 2k + 4)$

17. $x^4 - 81 = (x^2)^2 - 9^2$
$= (x^2 + 9)(x^2 - 9)$
$= (x^2 + 9)(x + 3)(x - 3)$

18. $81x^4 - 16y^4 = (9x^2)^2 - (4y^2)^2$
$= (9x^2 + 4y^2)(9x^2 - 4y^2)$
$= (9x^2 + 4y^2)[(3x)^2 - (2y)^2]$
$= (9x^2 + 4y^2)[(3x + 2y)(3x - 2y)]$
$= (3x + 2y)(3x - 2y)(9x^2 + 4y^2)$

19. $9x^6y^4 + 12x^3y^2 + 4$
$= (3x^3y^2)^2 + 2(3x^3y^2)(2) + (2)^2$
$= (3x^3y^2 + 2)^2$

20. $2r^2 - 13r + 6 = 0$
$(2r - 1)(r - 6) = 0$

$2r - 1 = 0$ or $r - 6 = 0$
$2r = 1$ or $r = 6$
$r = \frac{1}{2}$

The solution set is $\left\{ \frac{1}{2}, 6 \right\}$.

21. $25x^2 - 4 = 0$
$(5x + 2)(5x - 2) = 0$

$5x + 2 = 0$ or $5x - 2 = 0$
$5x = -2$ or $5x = 2$
$x = -\frac{2}{5}$ or $x = \frac{2}{5}$

The solution set is $\left\{ -\frac{2}{5}, \frac{2}{5} \right\}$.

22. $t^2 = 9t$
$t^2 - 9t = 0$
$t(t - 9) = 0$

$t = 0$ or $t - 9 = 0$
$t = 0$ or $t = 9$

The solution set is $\{0, 9\}$.

23. $x(x - 20) = -100$
$x^2 - 20x = -100$
$x^2 - 20x + 100 = 0$
$(x - 10)^2 = 0$
$x - 10 = 0$
$x = 10$

The solution set is $\{10\}$.

24. $(s + 8)(6s^2 + 13s - 5) = 0$

If $s + 8 = 0$, then $s = -8$.

$6s^2 + 13s - 5 = 0$
$(2s + 5)(3s - 1) = 0$

$2s + 5 = 0$ or $3s - 1 = 0$
$2s = -5$ or $3s = 1$
$s = -\frac{5}{2}$ or $s = \frac{1}{3}$

The solution set is $\left\{ -8, -\frac{5}{2}, \frac{1}{3} \right\}$.

25. Let $x =$ the width of the flower bed.
Then $2x - 3 =$ the length of the flower bed.

Use the formula $A = LW$.

$x(2x - 3) = 54$
$2x^2 - 3x = 54$
$2x^2 - 3x - 54 = 0$
$(2x + 9)(x - 6) = 0$

$2x + 9 = 0$ or $x - 6 = 0$
$2x = -9$ or $x = 6$
$x = -\frac{9}{2}$

Reject $-\frac{9}{2}$. If $x = 6$, $2x - 3 = 2(6) - 3 = 9$.
The dimensions of the flower bed are 6 feet by 9 feet.

26. Let $x =$ the first integer.
Then $x + 1 =$ the second integer.

The square of the sum of the two integers is 11 more than the first integer.

$$[x + (x + 1)]^2 = x + 11$$
$$(2x + 1)^2 = x + 11$$
$$4x^2 + 4x + 1 = x + 11$$
$$4x^2 + 3x - 10 = 0$$
$$(4x - 5)(x + 2) = 0$$

$$4x - 5 = 0 \quad \text{or} \quad x + 2 = 0$$
$$4x = 5 \quad \text{or} \qquad x = -2$$
$$x = \tfrac{5}{4}$$

Reject $\frac{5}{4}$ because it is not an integer. If $x = -2$, $x + 1 = -1$. The integers are -2 and -1.

27. Let $x = $ the length of the stud.
Then $3x - 7 = $ the length of the brace.

The figure shows that a right triangle is formed with the brace as the hypotenuse. Use the Pythagorean theorem, $a^2 + b^2 = c^2$.

$$x^2 + 15^2 = (3x - 7)^2$$
$$x^2 + 225 = 9x^2 - 42x + 49$$
$$0 = 8x^2 - 42x - 176$$
$$0 = 2(4x^2 - 21x - 88)$$
$$0 = 2(4x + 11)(x - 8)$$

$$4x + 11 = 0 \qquad \text{or} \qquad x - 8 = 0$$
$$4x = -11 \quad \text{or} \qquad x = 8$$
$$x = -\tfrac{11}{4}$$

Reject $-\frac{11}{4}$. If $x = 8$, $3x - 7 = 24 - 7 = 17$, so the brace should be 17 feet long.

28. For 2006, $x = 2006 - 2000 = 6$.

$$y = 29.92x^2 + 305.8x + 5581$$
$$y = 29.92(6)^2 + 305.8(6) + 5581$$
$$= 8492.92 \approx 8493$$

In 2006, the model estimates that the public debt of the United States was $8493 billion.

Cumulative Review Exercises (Chapters 1–6)

1.
$$3x + 2(x - 4) = 4(x - 2)$$
$$3x + 2x - 8 = 4x - 8$$
$$5x - 8 = 4x - 8$$
$$x - 8 = -8$$
$$x = 0$$

The solution set is $\{0\}$.

2. $0.3x + 0.9x = 0.06$

Multiply both sides by 100 to clear decimals.

$$100(0.3x + 0.9x) = 100(0.06)$$
$$30x + 90x = 6$$
$$120x = 6$$
$$x = \tfrac{6}{120} = \tfrac{1}{20} = 0.05$$

The solution set is $\{0.05\}$.

3. $\frac{2}{3}m - \frac{1}{2}(m - 4) = 3$

To clear fractions, multiply both sides by the least common denominator, which is 6.

$$6[\tfrac{2}{3}m - \tfrac{1}{2}(m - 4)] = 6(3)$$
$$4m - 3(m - 4) = 18$$
$$4m - 3m + 12 = 18$$
$$m + 12 = 18$$
$$m = 6$$

The solution set is $\{6\}$.

4.
$$A = P + Prt$$
$$A = P(1 + rt) \qquad \textit{Factor out P.}$$
$$\frac{A}{1 + rt} = \frac{P(1 + rt)}{1 + rt} \qquad \textit{Divide by 1 + rt.}$$
$$\frac{A}{1 + rt} = P \quad \text{or} \qquad P = \frac{A}{1 + rt}$$

5. The angles are supplementary, so the sum of the angles is $180°$.

$$(2x + 16) + (x + 23) = 180$$
$$3x + 39 = 180$$
$$3x = 141$$
$$x = 47$$

Since $x = 47$, $2x + 16 = 2(47) + 16 = 110$ and $x + 23 = 47 + 23 = 70$.
The angles are $110°$ and $70°$.

6. Let $x = $ number of bronze medals.
Then $x + 5 = $ number of gold medals,
and $(x + 5) + 1 = x + 6 = $ the number of silver medals.

The total number of medals was 29.

$$x + (x + 5) + (x + 6) = 29$$
$$3x + 11 = 29$$
$$3x = 18$$
$$x = 6$$

Since $x = 6$, $x + 5 = 11$, and $x + 6 = 12$.
Germany won 11 gold medals, 12 silver medals, and 6 bronze medals.

7. 46% of 500 is what number?

$$\frac{\text{part}}{\text{whole}} = \frac{p}{100}$$
$$\frac{a}{500} = \frac{46}{100}$$
$$100a = 46(500)$$
$$a = 230$$

41% of 500 is what number?

$$\frac{\text{part}}{\text{whole}} = \frac{p}{100}$$
$$\frac{a}{500} = \frac{41}{100}$$
$$100a = 500(41)$$
$$a = 205$$

What percent of 500 is 190?

$$\frac{\text{part}}{\text{whole}} = \frac{p}{100}$$

$$\frac{190}{500} = \frac{p}{100}$$

$$500p = 190(100)$$

$$p = 38$$

What percent of 500 is 60?

$$\frac{\text{part}}{\text{whole}} = \frac{p}{100}$$

$$\frac{60}{500} = \frac{p}{100}$$

$$500p = 60(100)$$

$$p = 12$$

Item	Percent	Number
Personal computer	46%	230
Cell phone	41%	205
High-speed internet	38%	190
MP3 player	12%	60

8. The point with coordinates (a, b) is in

(a) quadrant II if a is *negative* and b is *positive*.

(b) quadrant III if a is *negative* and b is *negative*.

9. **(a)** The equation $y = 12x + 3$ is in slope-intercept form, so the y-intercept is $(0, 3)$.

Let $y = 0$ to find the x-intercept.

$$0 = 12x + 3$$
$$-3 = 12x$$
$$-\tfrac{1}{4} = x$$

The x-intercept is $(-\tfrac{1}{4}, 0)$.

(b) The equation $y = 12x + 3$ is in slope-intercept form, so the slope is the coefficient of x, that is, 12.

(c)

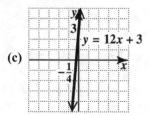

10. **(a)** $(2001, 161), (2007, 259)$

$$m = \frac{y_2 - y_1}{x_2 - x_1}$$

$$= \frac{259 - 161}{2007 - 2001}$$

$$= \frac{98}{6} \approx 16 \text{ (to the nearest whole number)}$$

A slope of (approximately) 16 means that the retail sales of prescription drugs increased by about $16 billion per year.

(b) The graph of the line will fall about 16 units from 2007 to 2006 and also from 2006 to 2005, so the y-value is $259 - 16 - 16 = 227$. The ordered pair is $(2005, 227)$, or about $(2005, 230)$.

11.
$$4x - y = -6 \quad (1)$$
$$2x + 3y = 4 \quad (2)$$

$$
\begin{array}{rl}
12x - 3y = -18 & (3) \quad 3 \times \text{Eq. (1)} \\
2x + 3y = 4 & (2) \\
\hline
14x = -14 & \quad Add\ (3)\ and\ (2). \\
x = -1 &
\end{array}
$$

To find y, substitute -1 for x in equation (1).

$$4(-1) - y = -6$$
$$-4 - y = -6$$
$$-y = -2$$
$$y = 2$$

The solution set is $\{(-1, 2)\}$.

12.
$$5x + 3y = 10 \quad (1)$$
$$2x + \frac{6}{5}y = 5 \quad (2)$$

$$
\begin{array}{rl}
-10x - 6y = -20 & (3) \quad -2 \times \text{Eq. (1)} \\
10x + 6y = 25 & (4) \quad 5 \times \text{Eq. (2)} \\
\hline
0 = 5 & \quad Add\ (3)\ and\ (4).
\end{array}
$$

The system of equations has no solution, symbolized by $\emptyset$.

13. $\left(\frac{3}{4}\right)^{-2} = \left(\frac{4}{3}\right)^2 = \frac{16}{9}$

14. $\left(\frac{4^{-3} \cdot 4^4}{4^5}\right)^{-1} = \left(\frac{4^5}{4^{-3} \cdot 4^4}\right)^1$

$$= \frac{4^5}{4^1} = 4^4 = 256$$

15. $\dfrac{(p^2)^3 p^{-4}}{(p^{-3})^{-1} p} = \dfrac{p^{2 \cdot 3} p^{-4}}{p^{(-3)(-1)} p}$

$$= \frac{p^6 p^{-4}}{p^3 p^1}$$

$$= \frac{p^{6-4}}{p^{3+1}}$$

$$= \frac{p^2}{p^4} = \frac{1}{p^2}$$

16. $\dfrac{(m^{-2})^3 m}{m^5 m^{-4}} = \dfrac{m^{-2(3)} m^1}{m^{5+(-4)}}$

$$= \frac{m^{-6+1}}{m^1}$$

$$= \frac{m^{-5}}{m^1} = \frac{1}{m^6}$$

17. $(2k^2 + 4k) - (5k^2 - 2) - (k^2 + 8k - 6)$
$= (2k^2 + 4k) + (-5k^2 + 2)$
$\quad + (-k^2 - 8k + 6)$
$= 2k^2 + 4k - 5k^2 + 2 - k^2 - 8k + 6$
$= -4k^2 - 4k + 8$

18. $(9x + 6)(5x - 3)$

$\qquad$ **F** $\qquad$ **O** $\qquad$ **I** $\qquad$ **L**
$= 9x(5x) + 9x(-3) + 6(5x) + 6(-3)$
$= 45x^2 - 27x + 30x - 18$
$= 45x^2 + 3x - 18$

19. $(3p + 2)^2 = (3p)^2 + 2 \cdot 3p \cdot 2 + 2^2$
$\qquad\qquad = 9p^2 + 12p + 4$

20. $\dfrac{8x^4 + 12x^3 - 6x^2 + 20x}{2x}$

$= \dfrac{8x^4}{2x} + \dfrac{12x^3}{2x} - \dfrac{6x^2}{2x} + \dfrac{20x}{2x}$

$= 4x^3 + 6x^2 - 3x + 10$

21. $55{,}000 = 5.5 \times 10^4$

Move the decimal point left 4 places so it is to the right of the first nonzero digit. $55{,}000$ is *greater* than 5.5, so the power is *positive*.

$2{,}000{,}000 = 2.0 \times 10^6$

Move the decimal point left 6 places so it is to the right of the first nonzero digit. $2{,}000{,}000$ is *greater* than 2, so the power is *positive*.

22. Factor $2a^2 + 7a - 4$ by trial and error.

$\qquad 2a^2 + 7a - 4 = (a + 4)(2a - 1)$

23. $10m^2 + 19m + 6$

To factor by grouping, find two integers whose product is $10(6) = 60$ and whose sum is 19. The integers are 15 and 4.

$10m^2 + 19m + 6 = 10m^2 + 15m + 4m + 6$
$\qquad\qquad\qquad = 5m(2m + 3) + 2(2m + 3)$
$\qquad\qquad\qquad = (2m + 3)(5m + 2)$

24. Factor $8t^2 + 10tv + 3v^2$ by trial and error.

$\qquad 8t^2 + 10tv + 3v^2 = (4t + 3v)(2t + v)$

25. $4p^2 - 12p + 9 = (2p - 3)(2p - 3)$
$\qquad\qquad\qquad = (2p - 3)^2$

26. $25r^2 - 81t^2 = (5r)^2 - (9t)^2$
$\qquad\qquad\qquad = (5r + 9t)(5r - 9t)$

27. $2pq + 6p^3q + 8p^2q$
$\qquad = 2pq(1 + 3p^2 + 4p)$
$\qquad = 2pq(3p^2 + 4p + 1)$
$\qquad = 2pq(3p + 1)(p + 1)$

28. $\qquad 6m^2 + m - 2 = 0$
$\qquad (3m + 2)(2m - 1) = 0$

$\quad 3m + 2 = 0 \qquad$ or $\qquad 2m - 1 = 0$
$\qquad 3m = -2 \qquad$ or $\qquad\quad 2m = 1$
$\qquad\quad m = -\tfrac{2}{3} \qquad$ or $\qquad\qquad m = \tfrac{1}{2}$

The solution set is $\left\{ -\tfrac{2}{3}, \tfrac{1}{2} \right\}$.

29. $\qquad\quad 8x^2 = 64x$
$\qquad 8x^2 - 64x = 0$
$\qquad 8x(x - 8) = 0$

$\qquad 8x = 0 \quad$ or $\quad x - 8 = 0$
$\qquad\; x = 0 \quad$ or $\qquad\; x = 8$

The solution set is $\{0, 8\}$.

30. Let $x =$ the length of the shorter leg. Then $x + 7 =$ the length of the longer leg, and $2x + 3 =$ the length of the hypotenuse.

Use the Pythagorean theorem.

$$x^2 + (x + 7)^2 = (2x + 3)^2$$
$$x^2 + (x^2 + 14x + 49) = 4x^2 + 12x + 9$$
$$2x^2 + 14x + 49 = 4x^2 + 12x + 9$$
$$0 = 2x^2 - 2x - 40$$
$$0 = 2(x^2 - x - 20)$$
$$0 = (x - 5)(x + 4)$$

$\quad x - 5 = 0 \quad$ or $\quad x + 4 = 0$
$\qquad x = 5 \quad$ or $\qquad\; x = -4$

Reject -4 because the length of a leg cannot be negative. Since $x = 5$, $x + 7 = 12$, and $2x + 3 = 2(5) + 3 = 13$. The length of the sides are 5 meters, 12 meters, and 13 meters.

CHAPTER 7   RATIONAL EXPRESSIONS AND APPLICATIONS

7.1 The Fundamental Property of Rational Expressions

7.1 Now Try Exercises

N1. $\dfrac{2x-1}{x+4} = \dfrac{2(-3)-1}{-3+4}$ *Let x = −3.*

$= \dfrac{-7}{1} = -7$

N2. (a) $\dfrac{k-4}{2k-1}$

Solve $2k - 1 = 0$.
Since $k = \frac{1}{2}$ will make the denominator zero, the expression is undefined for $k = \frac{1}{2}$. We write the answer as $k \neq \frac{1}{2}$.

(b) $\dfrac{2x}{x^2+5x-14}$

Solve $x^2 + 5x - 14 = 0$.
$(x+7)(x-2) = 0$, so the denominator is zero when either $x + 7 = 0$ or $x - 2 = 0$, that is, when $x = -7$ or $x = 2$. The expression is undefined for $x = -7$ and $x = 2$. We write the answer as $x \neq -7, x \neq 2$.

(c) $\dfrac{y+10}{y^2+10}$

This denominator will not equal 0 for any value of y, because y^2 is always greater than or equal to 0, and adding 10 makes the sum greater than 0. Thus, there are no values for which this rational expression is undefined.

N3. $\dfrac{21y^5}{7y^2} = \dfrac{3 \cdot 7 \cdot y \cdot y \cdot y \cdot y \cdot y}{7 \cdot y \cdot y}$ *Factor.*

$= \dfrac{3 \cdot y \cdot y \cdot y (7 \cdot y \cdot y)}{1(7 \cdot y \cdot y)}$ *Group.*

$= \dfrac{3 \cdot y \cdot y \cdot y}{1}$ *Fundamental property*

$= 3y^3$

N4. (a) $\dfrac{3x+15}{5x+25} = \dfrac{3(x+5)}{5(x+5)}$ *Factor.*

$= \dfrac{3}{5}$ *Fundamental property*

(b) $\dfrac{k^2-36}{k^2+8k+12}$

$= \dfrac{(k+6)(k-6)}{(k+6)(k+2)}$ *Factor.*

$= \dfrac{k-6}{k+2}$ *Fundamental property*

N5. $\dfrac{10-a^2}{a^2-10} = \dfrac{1(10-a^2)}{-1(10-a^2)} = \dfrac{1}{-1} = -1$

N6. (a) $\dfrac{p-4}{4-p}$

Since $p - 4$ and $4 - p$ are opposites, this expression equals −1.

(b) $\dfrac{4m^2-n^2}{2n-4m} = \dfrac{(2m)^2-n^2}{2(n-2m)}$

$= \dfrac{(2m+n)(2m-n)}{2(-1)(2m-n)}$

$= \dfrac{2m+n}{2(-1)}$

$= \dfrac{2m+n}{-2}$, or $-\dfrac{2m+n}{2}$

(c) $\dfrac{x+y}{x-y}$

$x - y = -1(-x+y) \neq -1(x+y)$

The expressions $x + y$ and $x - y$ are not opposites of each other. They do not have any common factors (other than 1), so the rational expression is already in lowest terms.

N7. To write four equivalent expressions for $-\dfrac{4k-9}{k+3}$, we will follow the outline in Example 7. Applying the negative sign to the numerator we have

$$\dfrac{-(4k-9)}{k+3}.$$

Distributing the negative sign gives us

$$\dfrac{-4k+9}{k+3}.$$

Applying the negative sign to the denominator yields

$$\dfrac{4k-9}{-(k+3)}.$$

Again, we distribute to get

$$\dfrac{4k-9}{-k-3}.$$

7.1 Section Exercises

1. (a) $\dfrac{3x+1}{5x} = \dfrac{3(2)+1}{5(2)}$ *Let x = 2.*

$= \dfrac{7}{10}$

(b) $\dfrac{3x+1}{5x} = \dfrac{3(-3)+1}{5(-3)}$ *Let x = −3.*

$= \dfrac{-8}{-15} = \dfrac{8}{15}$

3.   **(a)** $\dfrac{x^2 - 4}{2x + 1} = \dfrac{(2)^2 - 4}{2(2) + 1}$ *Let x = 2.*

$$= \dfrac{0}{5} = 0$$

    (b) $\dfrac{x^2 - 4}{2x + 1} = \dfrac{(-3)^2 - 4}{2(-3) + 1}$ *Let x = –3.*

$$= \dfrac{5}{-5} = -1$$

5.   **(a)** $\dfrac{(-2x)^3}{3x + 9} = \dfrac{(-2 \cdot 2)^3}{3 \cdot 2 + 9}$ *Let x = 2.*

$$= \dfrac{-64}{15} = -\dfrac{64}{15}$$

    (b) $\dfrac{(-2x)^3}{3x + 9} = \dfrac{[-2(-3)]^3}{3(-3) + 9}$ *Let x = –3.*

$$= \dfrac{216}{0}$$

Since substituting -3 for x makes the denominator zero, the given rational expression is undefined when $x = -3$.

7.   **(a)** $\dfrac{7 - 3x}{3x^2 - 7x + 2}$

$$= \dfrac{7 - 3(2)}{3(2)^2 - 7(2) + 2}$$ *Let x = 2.*

$$= \dfrac{7 - 6}{12 - 14 + 2} = \dfrac{1}{0}$$

Since substituting 2 for x makes the denominator zero, the given rational expression is undefined when $x = 2$.

    (b) $\dfrac{7 - 3x}{3x^2 - 7x + 2}$

$$= \dfrac{7 - 3(-3)}{3(-3)^2 - 7(-3) + 2}$$ *Let x = –3.*

$$= \dfrac{7 + 9}{27 + 21 + 2} = \dfrac{16}{50} = \dfrac{8}{25}$$

9.   **(a)** $\dfrac{(x + 3)(x - 2)}{500x}$

$$= \dfrac{(2 + 3)(2 - 2)}{500(2)}$$ *Let x = 2.*

$$= \dfrac{5(0)}{1000} = \dfrac{0}{1000} = 0$$

    (b) $\dfrac{(x + 3)(x - 2)}{500x}$

$$= \dfrac{(-3 + 3)(-3 - 2)}{500(-3)}$$ *Let x = –3.*

$$= \dfrac{0(-5)}{-1500} = \dfrac{0}{-1500} = 0$$

11.   **(a)** $\dfrac{x^2 - 4}{x^2 - 9} = \dfrac{2^2 - 4}{2^2 - 9}$ *Let x = 2.*

$$= \dfrac{4 - 4}{4 - 9} = \dfrac{0}{-5} = 0$$

    (b) $\dfrac{x^2 - 4}{x^2 - 9} = \dfrac{(-3)^2 - 4}{(-3)^2 - 9}$ *Let x = –3.*

$$= \dfrac{9 - 4}{9 - 9} = \dfrac{5}{0}$$

Since substituting -3 for x makes the denominator zero, the given rational expression is undefined when $x = -3$.

13. A rational expression is a quotient of two polynomials, such as $\dfrac{x^2 + 3x - 6}{x + 4}$. One can think of this as an algebraic fraction.

15. Division by 0 is undefined. If the denominator of a rational expression equals 0, the expression is undefined.

17. $\dfrac{12}{5y}$

The denominator $5y$ will be zero when $y = 0$, so the given expression is undefined for $y = 0$. We write the answer as $y \neq 0$.

19. $\dfrac{x + 1}{x - 6}$

To find the values for which this expression is undefined, set the denominator equal to zero and solve for x.

$$x - 6 = 0$$
$$x = 6$$

Because $x = 6$ will make the denominator zero, the given expression is undefined for 6. We write the answer as $x \neq 6$.

21. $\dfrac{4x^2}{3x + 5}$

To find the values for which this expression is undefined, set the denominator equal to zero and solve for x.

$$3x + 5 = 0$$
$$3x = -5$$
$$x = -\dfrac{5}{3}$$

Because $x = -\dfrac{5}{3}$ will make the denominator zero, the given expression is undefined for $-\dfrac{5}{3}$. We write the answer as $x \neq -\dfrac{5}{3}$.

23. $\dfrac{5m + 2}{m^2 + m - 6}$

To find the numbers that make the denominator 0, we must solve

$$m^2 + m - 6 = 0$$
$$(m + 3)(m - 2) = 0$$

$$m + 3 = 0 \quad \text{or} \quad m - 2 = 0$$
$$m = -3 \quad \text{or} \quad m = 2$$

The given expression is undefined for $m = -3$ and for $m = 2$. We write the answer as $m \neq -3$ and $m \neq 2$.

25. $\dfrac{x^2 + 3x}{4}$ is never undefined since the denominator is never zero.

27. $\dfrac{3x - 1}{x^2 + 2}$

This denominator cannot equal zero for any value of x because x^2 is always greater than or equal to zero, and adding 2 makes the sum greater than zero. Thus, the given rational expression is never undefined.

29. (a) $\dfrac{x^2 + 4x}{x + 4}$

The two terms in the numerator are x^2 and $4x$. The two terms in the denominator are x and 4.

(b) To express the rational expression in lowest terms, factor the numerator and denominator and divide both by the common factor $x + 4$ to get x.

$$\frac{x^2 + 4x}{x + 4} = \frac{x(x + 4)}{x + 4} = \frac{x}{1} = x$$

31. $\dfrac{18r^3}{6r} = \dfrac{3r^2(6r)}{1(6r)}$ *Factor.*

$\qquad = 3r^2$ *Fundamental property*

33. $\dfrac{4(y - 2)}{10(y - 2)} = \dfrac{2 \cdot 2(y - 2)}{5 \cdot 2(y - 2)}$ *Factor.*

$\qquad = \dfrac{2}{5}$ *Fundamental property*

35. $\dfrac{(x + 1)(x - 1)}{(x + 1)^2} = \dfrac{(x + 1)(x - 1)}{(x + 1)(x + 1)}$

$\qquad = \dfrac{x - 1}{x + 1}$ *Fundamental property*

37. $\dfrac{7m + 14}{5m + 10} = \dfrac{7(m + 2)}{5(m + 2)}$ *Factor.*

$\qquad = \dfrac{7}{5}$ *Fundamental property*

39. $\dfrac{6m - 18}{7m - 21} = \dfrac{6(m - 3)}{7(m - 3)}$ *Factor.*

$\qquad = \dfrac{6}{7}$ *Fundamental property*

41. $\dfrac{m^2 - n^2}{m + n} = \dfrac{(m + n)(m - n)}{m + n}$

$\qquad = m - n$

43. $\dfrac{2t + 6}{t^2 - 9} = \dfrac{2(t + 3)}{(t + 3)(t - 3)}$

$\qquad = \dfrac{2}{t - 3}$

45. $\dfrac{12m^2 - 3}{8m - 4} = \dfrac{3(4m^2 - 1)}{4(2m - 1)}$

$\qquad = \dfrac{3(2m + 1)(2m - 1)}{4(2m - 1)}$

$\qquad = \dfrac{3(2m + 1)}{4}$

47. $\dfrac{3m^2 - 3m}{5m - 5} = \dfrac{3m(m - 1)}{5(m - 1)}$

$\qquad = \dfrac{3m}{5}$

49. $\dfrac{9r^2 - 4s^2}{9r + 6s} = \dfrac{(3r + 2s)(3r - 2s)}{3(3r + 2s)}$

$\qquad = \dfrac{3r - 2s}{3}$

51. $\dfrac{5k^2 - 13k - 6}{5k + 2} = \dfrac{(5k + 2)(k - 3)}{5k + 2}$

$\qquad = k - 3$

53. $\dfrac{x^2 + 2x - 15}{x^2 + 6x + 5} = \dfrac{(x + 5)(x - 3)}{(x + 5)(x + 1)}$

$\qquad = \dfrac{x - 3}{x + 1}$

55. $\dfrac{2x^2 - 3x - 5}{2x^2 - 7x + 5} = \dfrac{(2x - 5)(x + 1)}{(2x - 5)(x - 1)}$

$\qquad = \dfrac{x + 1}{x - 1}$

57. $\dfrac{3x^3 + 13x^2 + 14x}{3x^3 - 5x^2 - 28x}$

$\qquad = \dfrac{x(3x^2 + 13x + 14)}{x(3x^2 - 5x - 28)}$

$\qquad = \dfrac{x(3x + 7)(x + 2)}{x(3x + 7)(x - 4)}$

$\qquad = \dfrac{x + 2}{x - 4}$ *Fundamental property*

59. $\dfrac{-3t + 6t^2 - 3t^3}{7t^2 - 14t^3 + 7t^4}$

$\qquad = \dfrac{-3t(1 - 2t + t^2)}{7t^2(1 - 2t + t^2)}$

$\qquad = -\dfrac{3}{7t}$ *Fundamental property*

61. $\dfrac{zw + 4z - 3w - 12}{zw + 4z + 5w + 20}$

$= \dfrac{z(w + 4) - 3(w + 4)}{z(w + 4) + 5(w + 4)}$ *Factor by grouping.*

$= \dfrac{(w + 4)(z - 3)}{(w + 4)(z + 5)}$

$= \dfrac{z - 3}{z + 5}$ *Fundamental property*

63. $\dfrac{pr + qr + ps + qs}{pr + qr - ps - qs}$

$= \dfrac{r(p + q) + s(p + q)}{r(p + q) - s(p + q)}$ *Factor by grouping.*

$= \dfrac{(p + q)(r + s)}{(p + q)(r - s)}$

$= \dfrac{r + s}{r - s}$ *Fundamental property*

65. $\dfrac{ac - ad + bc - bd}{ac - ad - bc + bd}$

$= \dfrac{a(c - d) + b(c - d)}{a(c - d) - b(c - d)}$ *Factor by grouping.*

$= \dfrac{(c - d)(a + b)}{(c - d)(a - b)}$

$= \dfrac{a + b}{a - b}$ *Fundamental property*

67. $\dfrac{m^2 - n^2 - 4m - 4n}{2m - 2n - 8}$

$= \dfrac{(m + n)(m - n) - 4(m + n)}{2(m - n - 4)}$ *Factor by grouping.*

$= \dfrac{(m + n)(m - n - 4)}{2(m - n - 4)}$

$= \dfrac{m + n}{2}$ *Fundamental property*

69. $\dfrac{x^2 y + y + x^2 z + z}{xy + xz}$

$= \dfrac{y(x^2 + 1) + z(x^2 + 1)}{x(y + z)}$ *Factor by grouping.*

$= \dfrac{(x^2 + 1)(y + z)}{x(y + z)}$

$= \dfrac{x^2 + 1}{x}$ *Fundamental property*

71. The numerator is the sum of cubes.

$$\frac{1 + p^3}{1 + p} = \frac{1^3 + p^3}{1 + p}$$

$$= \frac{(1 + p)(1 - p + p^2)}{1 + p}$$

$$= 1 - p + p^2$$

73. The numerator is the difference of cubes.

$$\frac{x^3 - 27}{x - 3} = \frac{x^3 - 3^3}{x - 3}$$

$$= \frac{(x - 3)(x^2 + 3x + 9)}{x - 3}$$

$$= x^2 + 3x + 9$$

75. The numerator is the difference of cubes and the denominator is the difference of squares.

$\dfrac{b^3 - a^3}{a^2 - b^2}$

$= \dfrac{(b - a)(b^2 + ba + a^2)}{(a - b)(a + b)}$

$= (-1) \cdot \dfrac{(b^2 + ba + a^2)}{(a + b)} \qquad \dfrac{(b - a)}{(a - b)} = -1$

$= -\dfrac{b^2 + ba + a^2}{a + b}$

77. The numerator is the sum of cubes and the denominator is the difference of squares.

$\dfrac{k^3 + 8}{k^2 - 4} = \dfrac{k^3 + 2^3}{(k + 2)(k - 2)}$

$= \dfrac{(k + 2)(k^2 - 2k + 4)}{(k + 2)(k - 2)}$

$= \dfrac{k^2 - 2k + 4}{k - 2}$

79. The numerator is the sum of cubes. The denominator has a common factor of z.

$\dfrac{z^3 + 27}{z^3 - 3z^2 + 9z} = \dfrac{z^3 + 3^3}{z(z^2 - 3z + 9)}$

$= \dfrac{(z + 3)(z^2 - 3z + 9)}{z(z^2 - 3z + 9)}$

$= \dfrac{z + 3}{z}$

81. The numerator is the difference of cubes. The denominator has a common factor of 2.

$\dfrac{1 - 8r^3}{8r^2 + 4r + 2} = \dfrac{1^3 - (2r)^3}{2(4r^2 + 2r + 1)}$

$= \dfrac{(1 - 2r)(1 + 2r + 4r^2)}{2(4r^2 + 2r + 1)}$

$= \dfrac{1 - 2r}{2}$

83. **A.** $\dfrac{2x+3}{2x-3} \neq -1$

B. $\dfrac{2x-3}{3-2x} = \dfrac{-1(3-2x)}{3-2x} = -1$

C. $\dfrac{2x+3}{3+2x} = 1 \neq -1$

D. $\dfrac{2x+3}{-2x-3} = \dfrac{2x+3}{-1(2x+3)} = -1$

B and **D** are equal to -1.

85. $\dfrac{6-t}{t-6} = \dfrac{-1(t-6)}{1(t-6)} = \dfrac{-1}{1} = -1$

Note that $6-t$ and $t-6$ are opposites, so we know that their quotient will be -1.

87. $\dfrac{m^2-1}{1-m} = \dfrac{(m+1)(m-1)}{-1(m-1)}$

$= \dfrac{m+1}{-1}$

$= -(m+1)$ or $-m-1$

89. $\dfrac{q^2-4q}{4q-q^2} = \dfrac{q(q-4)}{q(4-q)}$

$= \dfrac{q-4}{4-q} = -1$

$q-4$ and $4-q$ are opposites.

91. In the expression $\dfrac{p+6}{p-6}$, neither numerator nor denominator can be factored. It is already in lowest terms. *Note:* $(p+6)$ and $(p-6)$ are not opposites.

93. **A.** $\dfrac{3-x}{x-4} = \dfrac{-1(3-x)}{-1(x-4)} = \dfrac{-3+x}{-x+4} = \dfrac{x-3}{4-x}$

C. $-\dfrac{3-x}{4-x} = \dfrac{-1(3-x)}{4-x} = \dfrac{-3+x}{4-x} = \dfrac{x-3}{4-x}$

D. $-\dfrac{x-3}{x-4} = \dfrac{x-3}{-1(x-4)} = \dfrac{x-3}{-x+4} = \dfrac{x-3}{4-x}$

Since **A**, **C**, and **D** are equivalent to $\dfrac{x-3}{4-x}$, **B** is the one that is not.

There are many possible answers for Exercises 95–100.

95. To write four equivalent expressions for $-\dfrac{x+4}{x-3}$, we will follow the outline in Example 7. Applying the negative sign to the numerator we have

$\dfrac{-(x+4)}{x-3}.$

Distributing the negative sign gives us

$\dfrac{-x-4}{x-3}.$

Applying the negative sign to the denominator yields

$\dfrac{x+4}{-(x-3)}.$

Again, we distribute to get

$\dfrac{x+4}{-x+3}.$

97. $-\dfrac{2x-3}{x+3}$ is equivalent to each of the following:

$\dfrac{-(2x-3)}{x+3},\quad \dfrac{-2x+3}{x+3},$

$\dfrac{2x-3}{-(x+3)},\quad \dfrac{2x-3}{-x-3}$

99. $-\dfrac{3x-1}{5x-6}$ is equivalent to each of the following:

$\dfrac{-(3x-1)}{5x-6},\quad \dfrac{-3x+1}{5x-6},$

$\dfrac{3x-1}{-(5x-6)},\quad \dfrac{3x-1}{-5x+6}$

101. $L \cdot W = \mathcal{A}$

$W = \dfrac{\mathcal{A}}{L}$

$W = \dfrac{x^4+10x^2+21}{x^2+7}$

$= \dfrac{(x^2+7)(x^2+3)}{x^2+7}$

$= x^2+3$

Note: If it is not apparent that we can factor $\mathcal{A}$ as $x^4+10x^2+21 = (x^2+7)(x^2+3)$, we may use "long division" to find the quotient $\dfrac{\mathcal{A}}{L}$. Remember to insert zeros for the coefficients of the missing terms in the dividend and divisor.

$$\begin{array}{r} x^2 \qquad\quad +3 \\ x^2+0x+7\overline{)x^4+0x^3+10x^2+0x+21} \\ \underline{x^4+0x^3+7x^2} \\ 3x^2+0x+21 \\ \underline{3x^2+0x+21} \\ 0 \end{array}$$

The width of the rectangle is x^2+3.

103. Let $w = \dfrac{x^2}{2(1-x)}$.

   (a) $w = \dfrac{(0.1)^2}{2(1-0.1)}$ *Let x = 0.1.*

        $= \dfrac{0.01}{2(0.9)} \approx 0.006$

To the nearest tenth, w is 0.

   (b) $w = \dfrac{(0.8)^2}{2(1-0.8)}$ *Let x = 0.8.*

        $= \dfrac{0.64}{2(0.2)} = 1.6$

   (c) $w = \dfrac{(0.9)^2}{2(1-0.9)}$ *Let x = 0.9.*

        $= \dfrac{0.81}{2(0.1)} = 4.05 \approx 4.1$

   (d) Based on the answers in (a), (b), and (c), we see that as the traffic intensity increases, the waiting time also increases.

105. $\dfrac{2}{3} \cdot \dfrac{5}{6} = \dfrac{2 \cdot 5}{3 \cdot 2 \cdot 3} = \dfrac{5}{3 \cdot 3} = \dfrac{5}{9}$

107. $\dfrac{10}{3} \div \dfrac{5}{6} = \dfrac{10}{3} \cdot \dfrac{6}{5}$

        $= \dfrac{2 \cdot 5 \cdot 2 \cdot 3}{3 \cdot 5}$

        $= \dfrac{2 \cdot 2}{1} = 4$

7.2   Multiplying and Dividing Rational Expressions

7.2   Now Try Exercises

N1. $\dfrac{4k^2}{7} \cdot \dfrac{14}{11k} = \dfrac{2 \cdot 2 \cdot k \cdot k \cdot 2 \cdot 7}{7 \cdot 11 \cdot k}$

        $= \dfrac{2 \cdot 2 \cdot k \cdot 2}{11} = \dfrac{8k}{11}$

N2. $\dfrac{m-3}{3m} \cdot \dfrac{9m^2}{8(m-3)^2}$

     $= \dfrac{(m-3) \cdot 9m^2}{3m \cdot 8(m-3)^2}$

     $= \dfrac{(m-3) \cdot 3 \cdot 3 \cdot m \cdot m}{3 \cdot m \cdot 2 \cdot 2 \cdot 2(m-3)^2} = \dfrac{3m}{8(m-3)}$

N3. $\dfrac{y^2 - 3y - 28}{y^2 - 9y + 14} \cdot \dfrac{y^2 - 7y + 10}{y^2 + 4y}$

     $= \dfrac{(y-7)(y+4)}{(y-7)(y-2)} \cdot \dfrac{(y-2)(y-5)}{y(y+4)}$ *Factor.*

     $= \dfrac{(y-7)(y+4)(y-2)(y-5)}{(y-7)(y-2)y(y+4)}$ *Multiply.*

     $= \dfrac{y-5}{y}$ *Lowest terms*

N4. $\dfrac{2x-5}{3x^2} \div \dfrac{2x-5}{12x}$

     $= \dfrac{2x-5}{3x^2} \cdot \dfrac{12x}{2x-5}$ *Multiply by reciprocal.*

     $= \dfrac{2x-5}{3 \cdot x \cdot x} \cdot \dfrac{2 \cdot 2 \cdot 3 \cdot x}{2x-5}$ *Factor.*

     $= \dfrac{2 \cdot 2 \cdot 3 \cdot x \cdot (2x-5)}{3 \cdot x \cdot x \cdot (2x-5)}$ *Multiply.*

     $= \dfrac{4}{x}$ *Lowest terms*

N5. $\dfrac{(3k)^3}{2j^4} \div \dfrac{9k^2}{6j}$

     $= \dfrac{(3k)^3}{2j^4} \cdot \dfrac{6j}{9k^2}$ *Multiply by reciprocal.*

     $= \dfrac{27k^3 \cdot 6j}{2j^4 \cdot 9k^2}$ *Multiply.*

     $= \dfrac{9k}{j^3}$ *Lowest terms*

N6. $\dfrac{(t+2)(t-5)}{-4t} \div \dfrac{t^2-25}{(t+5)(t+2)}$

     $= \dfrac{(t+2)(t-5)}{-4t} \cdot \dfrac{(t+5)(t+2)}{t^2-25}$ *Multiply by reciprocal.*

     $= \dfrac{(t+2)(t-5)}{-4t} \cdot \dfrac{(t+5)(t+2)}{(t+5)(t-5)}$ *Factor.*

     $= \dfrac{(t+2)(t-5)(t+5)(t+2)}{-4t(t+5)(t-5)}$ *Multiply.*

     $= \dfrac{(t+2)^2}{-4t}$, or $-\dfrac{(t+2)^2}{4t}$ *Lowest terms*

N7. $\dfrac{7-x}{2x+6} \div \dfrac{x^2-49}{x^2+6x+9}$

     $= \dfrac{7-x}{2x+6} \cdot \dfrac{x^2+6x+9}{x^2-49}$

     $= \dfrac{(7-x)(x^2+6x+9)}{(2x+6)(x^2-49)}$

     $= \dfrac{(7-x)(x+3)(x+3)}{2(x+3)(x+7)(x-7)}$

     $= \dfrac{(-1)(x+3)}{2(x+7)}$

     $= \dfrac{-x-3}{2(x+7)}$, or $-\dfrac{x+3}{2(x+7)}$

7.2   Section Exercises

1.   **(a)** $\dfrac{5x^3}{10x^4} \cdot \dfrac{10x^7}{4x} = \dfrac{5 \cdot 10 \cdot x^3 \cdot x^7}{10 \cdot 4 \cdot x^4 \cdot x}$

             $= \dfrac{5x^{10}}{4x^5}$

             $= \dfrac{5x^5}{4}$ **(B)**

(b) $\dfrac{10x^4}{5x^3} \cdot \dfrac{10x^7}{4x} = \dfrac{5 \cdot 2 \cdot 5 \cdot 2 \cdot x^4 \cdot x^7}{5 \cdot 2 \cdot 2 \cdot x^3 \cdot x}$

$\qquad\qquad = \dfrac{5x^{11}}{1x^4}$

$\qquad\qquad = 5x^7$ **(D)**

(c) $\dfrac{5x^3}{10x^4} \cdot \dfrac{4x}{10x^7} = \dfrac{5 \cdot 2 \cdot 2 \cdot x^3 \cdot x}{5 \cdot 2 \cdot 5 \cdot 2 \cdot x^4 \cdot x^7}$

$\qquad\qquad = \dfrac{1x^4}{5x^{11}}$

$\qquad\qquad = \dfrac{1}{5x^7}$ **(C)**

(d) $\dfrac{10x^4}{5x^3} \cdot \dfrac{4x}{10x^7} = \dfrac{10 \cdot 4 \cdot x^4 \cdot x}{5 \cdot 10 \cdot x^3 \cdot x^7}$

$\qquad\qquad = \dfrac{4x^5}{5x^{10}}$

$\qquad\qquad = \dfrac{4}{5x^5}$ **(A)**

3. $\dfrac{15a^2}{14} \cdot \dfrac{7}{5a} = \dfrac{3 \cdot 5 \cdot a \cdot a \cdot 7}{2 \cdot 7 \cdot 5 \cdot a}$ *Multiply and factor.*

$\qquad = \dfrac{3 \cdot a(5 \cdot 7 \cdot a)}{2(5 \cdot 7 \cdot a)}$

$\qquad = \dfrac{3a}{2}$ *Lowest terms*

5. $\dfrac{12x^4}{18x^3} \cdot \dfrac{-8x^5}{4x^2} = \dfrac{-96x^9}{72x^5}$ *Multiply numerators; multiply denominators.*

$\qquad = \dfrac{-4x^4(24x^5)}{3(24x^5)}$ *Group common factors.*

$\qquad = -\dfrac{4x^4}{3}$ *Lowest terms*

7. $\dfrac{2(c+d)}{3} \cdot \dfrac{18}{6(c+d)^2}$

$\qquad = \dfrac{3 \cdot 3 \cdot 2 \cdot 2(c+d)}{3 \cdot 3 \cdot 2(c+d)(c+d)}$ *Multiply and factor.*

$\qquad = \dfrac{2}{c+d}$ *Lowest terms*

9. $\dfrac{(x-y)^2}{2} \cdot \dfrac{24}{3(x-y)}$

$\qquad = \dfrac{6 \cdot 4(x-y)(x-y)}{6(x-y)}$

$\qquad = 4(x-y)$

11. $\dfrac{t-4}{8} \cdot \dfrac{4t^2}{t-4}$

$\qquad = \dfrac{4t^2(t-4)}{2 \cdot 4(t-4)}$

$\qquad = \dfrac{t^2}{2}$

13. $\dfrac{3x}{x+3} \cdot \dfrac{(x+3)^2}{6x^2}$

$\qquad = \dfrac{3x(x+3)(x+3)}{2 \cdot 3 \cdot x \cdot x(x+3)}$

$\qquad = \dfrac{x+3}{2x}$

15. $\dfrac{9z^4}{3z^5} \div \dfrac{3z^2}{5z^3} = \dfrac{9z^4}{3z^5} \cdot \dfrac{5z^3}{3z^2}$

$\qquad\qquad = \dfrac{9 \cdot 5z^7}{3 \cdot 3z^7}$

$\qquad\qquad = 5$

17. $\dfrac{4t^4}{2t^5} \div \dfrac{(2t)^3}{-6} = \dfrac{4t^4}{2t^5} \cdot \dfrac{-6}{(2t)^3}$

$\qquad\qquad = \dfrac{4t^4}{2t^5} \cdot \dfrac{-6}{8t^3}$

$\qquad\qquad = \dfrac{-24t^4}{16t^8}$

$\qquad\qquad = \dfrac{-3(8t^4)}{2t^4(8t^4)}$

$\qquad\qquad = \dfrac{-3}{2t^4} = -\dfrac{3}{2t^4}$

19. $\dfrac{3}{2y-6} \div \dfrac{6}{y-3} = \dfrac{3}{2y-6} \cdot \dfrac{y-3}{6}$

$\qquad\qquad = \dfrac{3}{2(y-3)} \cdot \dfrac{y-3}{6}$

$\qquad\qquad = \dfrac{3(y-3)}{2 \cdot 2 \cdot 3(y-3)}$

$\qquad\qquad = \dfrac{1}{2 \cdot 2} = \dfrac{1}{4}$

21. $\dfrac{7t+7}{-6} \div \dfrac{4t+4}{15}$

$\qquad = \dfrac{7t+7}{-6} \cdot \dfrac{15}{4t+4}$

$\qquad = \dfrac{7(t+1)}{-2 \cdot 3} \cdot \dfrac{3 \cdot 5}{4(t+1)}$

$\qquad = \dfrac{3 \cdot 5 \cdot 7(t+1)}{-2 \cdot 3 \cdot 4(t+1)} = -\dfrac{35}{8}$

23. $\dfrac{2x}{x-1} \div \dfrac{x^2}{x+2}$

$= \dfrac{2x}{x-1} \cdot \dfrac{x+2}{x^2}$

$= \dfrac{2x(x+2)}{x \cdot x(x-1)} = \dfrac{2(x+2)}{x(x-1)}$

25. $\dfrac{(x-3)^2}{6x} \div \dfrac{x-3}{x^2}$

$= \dfrac{(x-3)^2}{6x} \cdot \dfrac{x^2}{x-3}$

$= \dfrac{x \cdot x(x-3)(x-3)}{6x(x-3)} = \dfrac{x(x-3)}{6}$

27. $\dfrac{5x-15}{3x+9} \cdot \dfrac{4x+12}{6x-18}$

$= \dfrac{5(x-3)}{3(x+3)} \cdot \dfrac{4(x+3)}{6(x-3)}$

$= \dfrac{5 \cdot 4 \cdot (x-3)(x+3)}{3 \cdot 6 \cdot (x-3)(x+3)}$

$= \dfrac{10}{9}$

29. $\dfrac{2-t}{8} \div \dfrac{t-2}{6}$

$= \dfrac{2-t}{8} \cdot \dfrac{6}{t-2}$ *Multiply by reciprocal.*

$= \dfrac{6(2-t)}{8(t-2)}$ *Multiply numerators; multiply denominators.*

$= \dfrac{6(-1)}{8}$ $\quad \dfrac{2-t}{t-2} = -1$

$= -\dfrac{3}{4}$ *Lowest terms*

31. $\dfrac{27-3z}{4} \cdot \dfrac{12}{2z-18}$

$= \dfrac{3(9-z)}{4} \cdot \dfrac{3 \cdot 4}{2(z-9)}$

$= \dfrac{3 \cdot 3 \cdot 4(9-z)}{4 \cdot 2(z-9)}$

$= \dfrac{3 \cdot 3 \cdot (-1)}{2} = -\dfrac{9}{2}$

33. $\dfrac{p^2+4p-5}{p^2+7p+10} \div \dfrac{p-1}{p+4}$

$= \dfrac{p^2+4p-5}{p^2+7p+10} \cdot \dfrac{p+4}{p-1}$

$= \dfrac{(p+5)(p-1) \cdot (p+4)}{(p+5)(p+2) \cdot (p-1)}$

$= \dfrac{p+4}{p+2}$

35. $\dfrac{m^2-4}{16-8m} \div \dfrac{m+2}{8}$

$= \dfrac{(m+2)(m-2)}{8(2-m)} \cdot \dfrac{8}{m+2}$

$= \dfrac{8(m+2)(m-2)}{8(m+2)(2-m)}$

$= -1$

37. $\dfrac{2x^2-7x+3}{x-3} \cdot \dfrac{x+2}{x-1}$

$= \dfrac{(2x-1)(x-3)}{x-3} \cdot \dfrac{x+2}{x-1}$

$= \dfrac{(2x-1)(x+2)}{x-1}$

39. $\dfrac{2k^2-k-1}{2k^2+5k+3} \div \dfrac{4k^2-1}{2k^2+k-3}$

$= \dfrac{2k^2-k-1}{2k^2+5k+3} \cdot \dfrac{2k^2+k-3}{4k^2-1}$

$= \dfrac{(2k+1)(k-1)(2k+3)(k-1)}{(2k+3)(k+1)(2k+1)(2k-1)}$

$= \dfrac{(k-1)(k-1)}{(k+1)(2k-1)}$

$= \dfrac{(k-1)^2}{(k+1)(2k-1)}$

41. $\dfrac{2k^2+3k-2}{6k^2-7k+2} \cdot \dfrac{4k^2-5k+1}{k^2+k-2}$

$= \dfrac{(2k-1)(k+2)}{(3k-2)(2k-1)} \cdot \dfrac{(4k-1)(k-1)}{(k+2)(k-1)}$

$= \dfrac{(2k-1)(k+2)(4k-1)(k-1)}{(3k-2)(2k-1)(k+2)(k-1)}$

$= \dfrac{4k-1}{3k-2}$

43. $\dfrac{m^2+2mp-3p^2}{m^2-3mp+2p^2} \div \dfrac{m^2+4mp+3p^2}{m^2+2mp-8p^2}$

$= \dfrac{m^2+2mp-3p^2}{m^2-3mp+2p^2} \cdot \dfrac{m^2+2mp-8p^2}{m^2+4mp+3p^2}$

$= \dfrac{(m+3p)(m-p)(m+4p)(m-2p)}{(m-2p)(m-p)(m+3p)(m+p)}$

$= \dfrac{m+4p}{m+p}$

45. $\dfrac{m^2+3m+2}{m^2+5m+4} \cdot \dfrac{m^2+10m+24}{m^2+5m+6}$

$= \dfrac{(m+2)(m+1)}{(m+4)(m+1)} \cdot \dfrac{(m+6)(m+4)}{(m+3)(m+2)}$

$= \dfrac{m+6}{m+3}$ *Multiply and use fundamental property.*

47. $\dfrac{y^2 + y - 2}{y^2 + 3y - 4} \div \dfrac{y + 2}{y + 3}$

$= \dfrac{y^2 + y - 2}{y^2 + 3y - 4} \cdot \dfrac{y + 3}{y + 2}$

$= \dfrac{(y + 2)(y - 1)}{(y + 4)(y - 1)} \cdot \dfrac{y + 3}{y + 2}$

$= \dfrac{y + 3}{y + 4}$ *Multiply and use*
fundamental property.

49. $\dfrac{2m^2 + 7m + 3}{m^2 - 9} \cdot \dfrac{m^2 - 3m}{2m^2 + 11m + 5}$

$= \dfrac{(2m + 1)(m + 3)}{(m - 3)(m + 3)} \cdot \dfrac{m(m - 3)}{(2m + 1)(m + 5)}$

$= \dfrac{(2m + 1)(m + 3)m(m - 3)}{(m - 3)(m + 3)(2m + 1)(m + 5)}$

$= \dfrac{m}{m + 5}$

51. $\dfrac{r^2 + rs - 12s^2}{r^2 - rs - 20s^2} \div \dfrac{r^2 - 2rs - 3s^2}{r^2 + rs - 30s^2}$

$= \dfrac{r^2 + rs - 12s^2}{r^2 - rs - 20s^2} \cdot \dfrac{r^2 + rs - 30s^2}{r^2 - 2rs - 3s^2}$

$= \dfrac{(r - 3s)(r + 4s)(r + 6s)(r - 5s)}{(r - 5s)(r + 4s)(r - 3s)(r + s)}$

$= \dfrac{r + 6s}{r + s}$

53. $\dfrac{(q - 3)^4(q + 2)}{q^2 + 3q + 2} \div \dfrac{q^2 - 6q + 9}{q^2 + 4q + 4}$

$= \dfrac{(q - 3)^4(q + 2)}{q^2 + 3q + 2} \cdot \dfrac{q^2 + 4q + 4}{q^2 - 6q + 9}$

$= \dfrac{(q - 3)^4(q + 2)(q + 2)^2}{(q + 2)(q + 1)(q - 3)^2}$

$= \dfrac{(q - 3)^2(q + 2)^2}{q + 1}$

55. $\dfrac{x + 5}{x + 10} \div \left(\dfrac{x^2 + 10x + 25}{x^2 + 10x} \cdot \dfrac{10x}{x^2 + 15x + 50} \right)$

$= \dfrac{x + 5}{x + 10} \div \left[\dfrac{(x + 5)^2 \cdot 10x}{x(x + 10)(x + 5)(x + 10)} \right]$

$= \dfrac{x + 5}{x + 10} \div \left[\dfrac{10(x + 5)}{(x + 10)^2} \right]$

$= \dfrac{x + 5}{x + 10} \cdot \dfrac{(x + 10)^2}{10(x + 5)}$

$= \dfrac{x + 10}{10}$

57. $\dfrac{3a - 3b - a^2 + b^2}{4a^2 - 4ab + b^2} \cdot \dfrac{4a^2 - b^2}{2a^2 - ab - b^2}$

Factor $3a - 3b - a^2 + b^2$ by grouping.

$3a - 3b - a^2 + b^2$
$= 3(a - b) - (a^2 - b^2)$
$= 3(a - b) - (a - b)(a + b)$
$= (a - b)[3 - (a + b)]$
$= (a - b)(3 - a - b)$

Thus,

$\dfrac{3a - 3b - a^2 + b^2}{4a^2 - 4ab + b^2} \cdot \dfrac{4a^2 - b^2}{2a^2 - ab - b^2}$

$= \dfrac{(a - b)(3 - a - b)}{(2a - b)(2a - b)} \cdot \dfrac{(2a - b)(2a + b)}{(2a + b)(a - b)}$

$= \dfrac{(a - b)(3 - a - b)(2a - b)(2a + b)}{(2a - b)(2a - b)(2a + b)(a - b)}$

$= \dfrac{3 - a - b}{2a - b}.$

59. $\dfrac{-x^3 - y^3}{x^2 - 2xy + y^2} \div \dfrac{3y^2 - 3xy}{x^2 - y^2}$

$= \dfrac{-1(x^3 + y^3)}{x^2 - 2xy + y^2} \cdot \dfrac{x^2 - y^2}{3y^2 - 3xy}$

$= \dfrac{-1(x + y)(x^2 - xy + y^2)}{(x - y)(x - y)}$

$\cdot \dfrac{(x - y)(x + y)}{3y(y - x)}$

$= \dfrac{-1(x + y)(x^2 - xy + y^2)(x - y)(x + y)}{-1(x - y)(x - y)(3y)(x - y)}$

$= \dfrac{(x + y)^2(x^2 - xy + y^2)}{3y(x - y)^2}$

If we had not changed $y - x$ to $-1(x - y)$ in the denominator, we would have obtained an alternate form of the answer,

$-\dfrac{(x + y)^2(x^2 - xy + y^2)}{3y(y - x)(x - y)}.$

61. Use the formula for the area of a rectangle with $\mathcal{A} = \dfrac{5x^2y^3}{2pq}$ and $L = \dfrac{2xy}{p}$ to solve for W.

$\mathcal{A} = L \cdot W$

$\dfrac{5x^2y^3}{2pq} = \dfrac{2xy}{p} \cdot W$

$W = \dfrac{5x^2y^3}{2pq} \div \dfrac{2xy}{p}$

$= \dfrac{5x^2y^3}{2pq} \cdot \dfrac{p}{2xy}$

$= \dfrac{5x^2y^3p}{4pqxy} = \dfrac{5xy^2}{4q}$

Thus, the rational expression $\dfrac{5xy^2}{4q}$ represents the width of the rectangle.

63. $18 = 2 \cdot 9$
$ = 2 \cdot 3 \cdot 3$

The prime factored form of 18 is $2 \cdot 3^2$.

65. $108 = 2 \cdot 54$
$ = 2 \cdot 2 \cdot 27$
$ = 2 \cdot 2 \cdot 3 \cdot 9$
$ = 2 \cdot 2 \cdot 3 \cdot 3 \cdot 3$

The prime factored form of 108 is $2^2 \cdot 3^3$.

67. $24m = 2^3 \cdot 3 \cdot m$
$18m^2 = 2 \cdot 3^2 \cdot m^2$
$6 = 2 \cdot 3$

The GCF is $2 \cdot 3 = 6$.

69. $84q^3 = 2^2 \cdot 3 \cdot 7 \cdot q^3$
$90q^6 = 2 \cdot 3^2 \cdot 5 \cdot q^6$

The GCF is $2 \cdot 3 \cdot q^3 = 6q^3$.

7.3   Least Common Denominators

7.3   Now Try Exercises

N1. (a) $\dfrac{5}{48}, \dfrac{1}{30}$

Factor each denominator.

$$48 = 2 \cdot 2 \cdot 2 \cdot 2 \cdot 3, \qquad 30 = 2 \cdot 3 \cdot 5$$

Take each different factor the *greatest* number of times it appears as a factor in any of the denominators, and use it to form the **least common denominator (LCD).**

$$\text{LCD} = 2 \cdot 2 \cdot 2 \cdot 2 \cdot 3 \cdot 5 = 240$$

(b) $\dfrac{3}{10y}, \dfrac{1}{6y}$

Factor each denominator.

$$10y = 2 \cdot 5 \cdot y$$
$$6y = 2 \cdot 3 \cdot y$$

Take each factor the greatest number of times it appears in any denominator; then multiply.

$$\text{LCD} = 2 \cdot 3 \cdot 5 \cdot y = 30y$$

N2. $\dfrac{5}{6x^4}, \dfrac{7}{8x^3}$

Factor each denominator.

$$6x^4 = 2 \cdot 3 \cdot x^4$$
$$8x^3 = 2 \cdot 2 \cdot 2 \cdot x^3$$

Take each factor the greatest number of times it appears in any denominator; then multiply.

$$\text{LCD} = 2 \cdot 2 \cdot 2 \cdot 3 \cdot x^4 = 24x^4$$

N3. (a) $\dfrac{3t}{2t^2 - 10t}, \dfrac{t+4}{t^2 - 25}$

Factor each denominator.

$$2t^2 - 10t = 2t(t - 5)$$
$$t^2 - 25 = (t + 5)(t - 5)$$

Take each factor the greatest number of times it appears in any denominator; then multiply.

$$\text{LCD} = 2t(t - 5)(t + 5)$$

(b) $\dfrac{1}{x^2 + 7x + 12}, \dfrac{2}{x^2 + 6x + 9}, \dfrac{5}{x^2 + 2x - 8}$

Factor each denominator.

$$x^2 + 7x + 12 = (x + 3)(x + 4)$$
$$x^2 + 6x + 9 = (x + 3)(x + 3)$$
$$x^2 + 2x - 8 = (x + 4)(x - 2)$$

Take each factor the greatest number of times it appears in any denominator; then multiply.

$$\text{LCD} = (x + 3)^2(x + 4)(x - 2)$$

(c) $\dfrac{2}{a - 4}, \dfrac{1}{4 - a}$

The expressions $a - 4$ and $4 - a$ are opposites of each other because

$$-(a - 4) = -a + 4 = 4 - a.$$

Therefore, either $a - 4$ or $4 - a$ can be used as the LCD.

N4. (a) $\dfrac{2}{9} = \dfrac{?}{27}$

First factor the denominator on the right. Then compare the denominator on the left with the one on the right to decide what factors are missing.

$$\dfrac{2}{9} = \dfrac{?}{3 \cdot 9}$$

A factor of 3 is missing, so multiply $\frac{2}{9}$ by $\frac{3}{3}$, which is equal to 1.

$$\dfrac{2}{9} = \dfrac{2}{9} \cdot \dfrac{3}{3} = \dfrac{6}{27}$$

(b) $\dfrac{4t}{11} = \dfrac{?}{33t}$

Factor the denominator on the right; then compare it to the denominator on the left.

$$\dfrac{4t}{11} = \dfrac{?}{3 \cdot 11 \cdot t}$$

The factors 3 and t are missing on the left, so multiply by $\frac{3t}{3t}$.

$$\dfrac{4t}{11} = \dfrac{4t}{11} \cdot \dfrac{3t}{3t} = \dfrac{12t^2}{33t}$$

N5. (a) $\dfrac{8k}{5k-2} = \dfrac{?}{25k-10}$

Factor the denominator on the right.

$$\frac{8k}{5k-2} = \frac{?}{5(5k-2)}$$

The factor 5 is missing on the left, so multiply by $\frac{5}{5}$.

$$\frac{8k}{5k-2} = \frac{8k}{5k-2} \cdot \frac{5}{5} = \frac{40k}{25k-10}$$

(b) $\dfrac{2t-1}{t^2+4t} = \dfrac{?}{t^3+12t^2+32t}$

Factor and compare the denominators.

$$\frac{2t-1}{t(t+4)} = \frac{?}{t(t+4)(t+8)}$$

The factor $t+8$ is missing on the left, so multiply by $\frac{t+8}{t+8}$.

$$\frac{2t-1}{t^2+4t} = \frac{2t-1}{t(t+4)} \cdot \frac{t+8}{t+8}$$
$$= \frac{(2t-1)(t+8)}{t(t+4)(t+8)}$$

7.3 Section Exercises

1. The factor x appears at most one time in any denominator as does the factor y. Thus, the LCD is the product of the two factors, xy. The correct response is **C**.

3. Since $20 = 2^2 \cdot 5$, the LCD of $\frac{9}{20}$ and $\frac{1}{2}$ must have 5 as a factor and 2^2 as a factor. Because 2 appears twice in $2^2 \cdot 5$, we don't have to include another 2 in the LCD for the number $\frac{1}{2}$. Thus, the LCD is just $2^2 \cdot 5 = 20$. Note that this is a specific case of Exercise 2 since 2 is a factor of 20. The correct response is **C**.

5. $\dfrac{7}{15}, \dfrac{21}{20}$

Factor each denominator.

$$15 = 3 \cdot 5$$
$$20 = 2 \cdot 2 \cdot 5 = 2^2 \cdot 5$$

Take each factor the greatest number of times it appears as a factor in any one of the denominators.

$$\text{LCD} = 2^2 \cdot 3 \cdot 5 = 60$$

7. $\dfrac{17}{100}, \dfrac{23}{120}, \dfrac{43}{180}$

Factor each denominator.

$$100 = 2^2 \cdot 5^2$$
$$120 = 2^3 \cdot 3 \cdot 5$$
$$180 = 2^2 \cdot 3^2 \cdot 5$$

Take each factor the greatest number of times it appears as a factor in any one of the denominators.

$$\text{LCD} = 2^3 \cdot 3^2 \cdot 5^2 = 1800$$

9. $\dfrac{9}{x^2}, \dfrac{8}{x^5}$

The greatest number of times x appears as a factor in any denominator is the greatest exponent on x, which is 5.

$$\text{LCD} = x^5$$

11. $\dfrac{-2}{5p}, \dfrac{13}{6p}$

Factor each denominator.

$$5p = 5 \cdot p$$
$$6p = 2 \cdot 3 \cdot p$$

Take each factor the greatest number of times it appears; then multiply.

$$\text{LCD} = 2 \cdot 3 \cdot 5 \cdot p = 30p$$

13. $\dfrac{17}{15y^2}, \dfrac{55}{36y^4}$

Factor each denominator.

$$15y^2 = 3 \cdot 5 \cdot y^2$$
$$36y^4 = 2^2 \cdot 3^2 \cdot y^4$$

Take each factor the greatest number of times it appears; then multiply.

$$\text{LCD} = 2^2 \cdot 3^2 \cdot 5 \cdot y^4 = 180y^4$$

15. $\dfrac{5}{21r^3}, \dfrac{7}{12r^5}$

Factor each denominator.

$$21r^3 = 3 \cdot 7 \cdot r^3$$
$$12r^5 = 2^2 \cdot 3 \cdot r^5$$

Take each factor the greatest number of times it appears; then multiply.

$$\text{LCD} = 2^2 \cdot 3 \cdot 7 \cdot r^5 = 84r^5$$

17. $\dfrac{13}{5a^2b^3}, \dfrac{29}{15a^5b}$

Factor each denominator.

$$5a^2b^3 = 5 \cdot a^2 \cdot b^3$$
$$15a^5b = 3 \cdot 5 \cdot a^5 \cdot b$$

Take each factor the greatest number of times it appears; then multiply.

$$\text{LCD} = 3 \cdot 5 \cdot a^5 \cdot b^3 = 15a^5b^3$$

19. $\dfrac{7}{6p}, \dfrac{15}{4p-8}$

Factor each denominator.

$$6p = 2 \cdot 3 \cdot p$$
$$4p - 8 = 4(p-2) = 2^2(p-2)$$

Take each factor the greatest number of times it appears; then multiply.

$$\text{LCD} = 2^2 \cdot 3 \cdot p(p-2) = 12p(p-2)$$

21. $\dfrac{9}{28m^2}, \dfrac{3}{12m-20}$

Factor each denominator.

$$28m^2 = 2^2 \cdot 7 \cdot m^2$$
$$12m - 20 = 4(3m-5) = 2^2(3m-5)$$

Take each factor the greatest number of times it appears; then multiply.

$$\text{LCD} = 2^2 \cdot 7m^2(3m-5) = 28m^2(3m-5)$$

23. $\dfrac{7}{5b-10}, \dfrac{11}{6b-12}$

Factor each denominator.

$$5b - 10 = 5(b-2)$$
$$6b - 12 = 6(b-2) = 2 \cdot 3(b-2)$$

Take each factor the greatest number of times it appears; then multiply.

$$\text{LCD} = 2 \cdot 3 \cdot 5(b-2) = 30(b-2)$$

25. $\dfrac{37}{6r-12}, \dfrac{25}{9r-18}$

Factor each denominator.

$$6r - 12 = 6(r-2) = 2 \cdot 3(r-2)$$
$$9r - 18 = 9(r-2) = 3^2(r-2)$$

Take each factor the greatest number of times it appears; then multiply.

$$\text{LCD} = 2 \cdot 3^2(r-2) = 18(r-2)$$

27. $\dfrac{5}{12p+60}, \dfrac{-17}{p^2+5p}, \dfrac{-16}{p^2+10p+25}$

Factor each denominator.

$$12p + 60 = 12(p+5) = 2^2 \cdot 3(p+5)$$
$$p^2 + 5p = p(p+5)$$
$$p^2 + 10p + 25 = (p+5)(p+5)$$

$$\text{LCD} = 2^2 \cdot 3 \cdot p(p+5)^2 = 12p(p+5)^2$$

29. $\dfrac{-3}{8y+16}, \dfrac{-22}{y^2+3y+2}$

Factor each denominator.

$$8y + 16 = 8(y+2) = 2^3(y+2)$$
$$y^2 + 3y + 2 = (y+2)(y+1)$$

$$\text{LCD} = 8(y+2)(y+1)$$

31. $\dfrac{5}{c-d}, \dfrac{8}{d-c}$

The denominators, $c-d$ and $d-c$, are opposites of each other since

$$-(c-d) = -c + d = d - c.$$

Therefore, either $c-d$ or $d-c$ can be used as the LCD.

33. $\dfrac{12}{m-3}, \dfrac{-4}{3-m}$

The expression $3-m$ can be written as $-1(m-3)$, since

$$-1(m-3) = -m + 3 = 3 - m.$$

Because of this, either $m-3$ or $3-m$ can be used as the LCD.

35. $\dfrac{29}{p-q}, \dfrac{18}{q-p}$

The expression $q-p$ can be written as $-1(p-q)$, since

$$-1(p-q) = -p + q = q - p.$$

Because of this, either $p-q$ or $q-p$ can be used as the LCD.

37. $\dfrac{3}{k^2+5k}, \dfrac{2}{k^2+3k-10}$

Factor each denominator.

$$k^2 + 5k = k(k+5)$$
$$k^2 + 3k - 10 = (k+5)(k-2)$$

$$\text{LCD} = k(k+5)(k-2)$$

39. $\dfrac{6}{a^2+6a}, \dfrac{-5}{a^2+3a-18}$

Factor each denominator.

$$a^2 + 6a = a(a+6)$$
$$a^2 + 3a - 18 = (a+6)(a-3)$$

$$\text{LCD} = a(a+6)(a-3)$$

41. $\dfrac{5}{p^2+8p+15}, \dfrac{3}{p^2-3p-18}, \dfrac{12}{p^2-p-30}$

Factor each denominator.

$$p^2 + 8p + 15 = (p+5)(p+3)$$
$$p^2 - 3p - 18 = (p-6)(p+3)$$
$$p^2 - p - 30 = (p-6)(p+5)$$

$$\text{LCD} = (p+3)(p+5)(p-6)$$

43. $\dfrac{-5}{k^2 + 2k - 35}, \dfrac{-8}{k^2 + 3k - 40}, \dfrac{19}{k^2 - 2k - 15}$

Factor each denominator.

$$k^2 + 2k - 35 = (k + 7)(k - 5)$$
$$k^2 + 3k - 40 = (k + 8)(k - 5)$$
$$k^2 - 2k - 15 = (k - 5)(k + 3)$$

$$\text{LCD} = (k + 7)(k - 5)(k + 8)(k + 3)$$

45. $\dfrac{3}{4} = \dfrac{?}{28}$

To change 4 into 28, multiply by 7. If you multiply the denominator by 7, you must multiply the numerator by 7.

46. $\dfrac{3}{4} = \dfrac{3}{4} \cdot \dfrac{7}{7} = \dfrac{21}{28}$

Note that numerator and denominator are being multiplied by 7, so $\frac{3}{4}$ is being multiplied by the fraction $\frac{7}{7}$, which is equal to 1.

47. Since $\frac{7}{7}$ has a value of 1, the multiplier is 1. The *identity property of multiplication* is being used when we write a common fraction as an equivalent one with a larger denominator.

48. $\dfrac{2x + 5}{x - 4} = \dfrac{?}{7x - 28} = \dfrac{?}{7(x - 4)}$

The expression $7x - 28$ is factored as $7(x - 4)$, so the multiplier is 7.

49. $\dfrac{2x + 5}{x - 4} = \dfrac{?}{7x - 28} = \dfrac{?}{7(x - 4)}$

To form the new denominator, 7 must be used as the multiplier for the denominator. To form an equivalent fraction, the same multiplier must be used for numerator and denominator. Thus, the multiplier is $\frac{7}{7}$, which is equal to 1.

50. The *identity property of multiplication* is being used when we write an algebraic fraction as an equivalent one with a larger denominator.

51. $\dfrac{4}{11} = \dfrac{?}{55}$

First factor the denominator on the right. Then compare the denominator on the left with the one on the right to decide what factors are missing.

$$\dfrac{4}{11} = \dfrac{?}{11 \cdot 5}$$

A factor of 5 is missing, so multiply $\frac{4}{11}$ by $\frac{5}{5}$, which is equal to 1.

$$\dfrac{4}{11} \cdot \dfrac{5}{5} = \dfrac{20}{55}$$

53. $\dfrac{-5}{k} = \dfrac{?}{9k}$

A factor of 9 is missing in the first fraction, so multiply numerator and denominator by 9.

$$\dfrac{-5}{k} \cdot \dfrac{9}{9} = \dfrac{-45}{9k}$$

55. $\dfrac{15m^2}{8k} = \dfrac{?}{32k^4}$

$32k^4 = (8k)(4k^3)$, so we must multiply the numerator and the denominator by $4k^3$.

$$\dfrac{15m^2}{8k} = \dfrac{15m^2}{8k} \cdot \dfrac{4k^3}{4k^3} \quad \textit{Multiplicative identity property}$$

$$= \dfrac{60m^2k^3}{32k^4}$$

57. $\dfrac{19z}{2z - 6} = \dfrac{?}{6z - 18}$

Begin by factoring each denominator.

$$2z - 6 = 2(z - 3)$$
$$6z - 18 = 6(z - 3)$$

The fractions may now be written as follows.

$$\dfrac{19z}{2(z - 3)} = \dfrac{?}{6(z - 3)}$$

Comparing the two factored forms, we see that the denominator of the fraction on the left side must be multiplied by 3; the numerator must also be multiplied by 3.

$$\dfrac{19z}{2z - 6}$$

$$= \dfrac{19z}{2(z - 3)} \cdot \dfrac{3}{3} \quad \textit{Multiplicative identity property}$$

$$= \dfrac{19z(3)}{2(z - 3)(3)} \quad \textit{Multiplication of rational expressions}$$

$$= \dfrac{57z}{6z - 18} \quad \textit{Multiply the factors}$$

59. $\dfrac{-2a}{9a - 18} = \dfrac{?}{18a - 36}$

$$\dfrac{-2a}{9(a - 2)} = \dfrac{?}{18(a - 2)} \quad \textit{Factor each denominator}$$

$$\dfrac{-2a}{9a - 18} = \dfrac{-2a}{9(a - 2)} \cdot \dfrac{2}{2} \quad \textit{Multiplicative identity property}$$

$$= \dfrac{-4a}{18a - 36} \quad \textit{Multiply}$$

61. $\dfrac{6}{k^2 - 4k} = \dfrac{?}{k(k-4)(k+1)}$

$\dfrac{6}{k(k-4)} = \dfrac{?}{k(k-4)(k+1)}$ *Factor first denominator*

$\dfrac{6}{k^2 - 4k} = \dfrac{6}{k(k-4)} \cdot \dfrac{(k+1)}{(k+1)}$ *Multiplicative identity property*

$= \dfrac{6(k+1)}{k(k-4)(k+1)}$ *Multiply*

63. $\dfrac{36r}{r^2 - r - 6} = \dfrac{?}{(r-3)(r+2)(r+1)}$

$\dfrac{36r}{(r-3)(r+2)} = \dfrac{?}{(r-3)(r+2)(r+1)}$

Factor first denominator

$\dfrac{36r}{r^2 - r - 6} = \dfrac{36r}{(r-3)(r+2)} \cdot \dfrac{(r+1)}{(r+1)}$

Multiplicative identity property

$= \dfrac{36r(r+1)}{(r-3)(r+2)(r+1)}$

65. $\dfrac{a + 2b}{2a^2 + ab - b^2} = \dfrac{?}{2a^3b + a^2b^2 - ab^3}$

$\dfrac{a + 2b}{2a^2 + ab - b^2} = \dfrac{?}{ab(2a^2 + ab - b^2)}$

Factor second denominator

$\dfrac{a + 2b}{2a^2 + ab - b^2} = \dfrac{(a + 2b)}{(2a^2 + ab - b^2)} \cdot \dfrac{ab}{ab}$

Multiplicative identity property

$= \dfrac{ab(a + 2b)}{2a^3b + a^2b^2 - ab^3}$

67. $\dfrac{4r - t}{r^2 + rt + t^2} = \dfrac{?}{t^3 - r^3}$

Factor the second denominator as the difference of cubes.

$t^3 - r^3 = (t - r)(t^2 + rt + r^2)$

$\dfrac{4r - t}{r^2 + rt + t^2} = \dfrac{(4r - t)}{(r^2 + rt + t^2)} \cdot \dfrac{(t - r)}{(t - r)}$

Multiplicative identity property

$= \dfrac{(4r - t)(t - r)}{t^3 - r^3}$

Multiply the factors

69. $\dfrac{2(z - y)}{y^2 + yz + z^2} = \dfrac{?}{y^4 - z^3y}$

Factor the second denominator.

$y^4 - z^3y = y(y^3 - z^3)$ *GCF = y*

$= y(y - z)(y^2 + yz + z^2)$

Difference of cubes

$\dfrac{2(z - y)}{y^2 + yz + z^2} = \dfrac{?}{y(y - z)(y^2 + yz + z^2)}$

$\dfrac{2(z - y)}{y^2 + yz + z^2} = \dfrac{2(z - y)}{(y^2 + yz + z^2)} \cdot \dfrac{y(y - z)}{y(y - z)}$

$= \dfrac{2y(z - y)(y - z)}{y(y - z)(y^2 + yz + z^2)}$

Multiplicative identity property

$= \dfrac{2y(z - y)(y - z)}{y(y^3 - z^3)}$

$= \dfrac{2y(z - y)(y - z)}{y^4 - z^3y}$

or $\dfrac{-2y(y - z)^2}{y^4 - z^3y}$, since $z - y = -1(y - z)$

71. $\dfrac{1}{2} + \dfrac{7}{8} = \dfrac{4}{8} + \dfrac{7}{8} = \dfrac{4 + 7}{8} = \dfrac{11}{8}$

73. $\dfrac{7}{5} - \dfrac{3}{4} = \dfrac{28}{20} - \dfrac{15}{20} = \dfrac{28 - 15}{20} = \dfrac{13}{20}$

7.4 Adding and Subtracting Rational Expressions

7.4 Now Try Exercises

N1. (a) $\dfrac{2}{7k} + \dfrac{4}{7k} = \dfrac{2 + 4}{7k}$

$= \dfrac{6}{7k}$

(b) $\dfrac{4y}{y + 3} + \dfrac{12}{y + 3} = \dfrac{4y + 12}{y + 3}$

$= \dfrac{4(y + 3)}{y + 3}$

$= 4$

N2. (a) $\dfrac{5}{12} + \dfrac{3}{20}$

Step 1

$\text{LCD} = 2 \cdot 2 \cdot 3 \cdot 5 = 60$

Step 2

$\dfrac{5}{12} = \dfrac{5(5)}{12(5)} = \dfrac{25}{60}$

$\dfrac{3}{20} = \dfrac{3(3)}{20(3)} = \dfrac{9}{60}$

Step 3

$\dfrac{25}{60} + \dfrac{9}{60} = \dfrac{25 + 9}{60}$

$= \dfrac{34}{60}$

Step 4

$\dfrac{34}{60} = \dfrac{17(2)}{30(2)} = \dfrac{17}{30}$

(b) $\dfrac{3}{5x} + \dfrac{2}{7x}$

Step 1

$LCD = 5 \cdot 7 \cdot x = 35x$

Step 2

$$\frac{3}{5x} = \frac{3(7)}{5x(7)} = \frac{21}{35x}$$

$$\frac{2}{7x} = \frac{2(5)}{7x(5)} = \frac{10}{35x}$$

Step 3

$$\frac{21}{35x} + \frac{10}{35x} = \frac{31}{35x}$$

N3. $\dfrac{6t}{t^2 - 9} + \dfrac{-3}{t + 3}$

$= \dfrac{6t}{(t+3)(t-3)} + \dfrac{-3}{t+3}$ *Factor denominator.*

$= \dfrac{6t}{(t+3)(t-3)} + \dfrac{-3(t-3)}{(t+3)(t-3)}$

$\qquad\qquad LCD = (t+3)(t-3)$

$= \dfrac{6t + [-3(t-3)]}{(t+3)(t-3)}$ *Add numerators.*

$= \dfrac{6t - 3t + 9}{(t+3)(t-3)}$ *Dist. property*

$= \dfrac{3t + 9}{(t+3)(t-3)}$ *Combine terms.*

$= \dfrac{3(t+3)}{(t+3)(t-3)}$ *Factor numerator.*

$= \dfrac{3}{t-3}$ *Lowest terms*

N4. $\dfrac{x-1}{x^2 + 6x + 8} + \dfrac{4x}{x^2 + x - 12}$

$= \dfrac{x-1}{(x+2)(x+4)} + \dfrac{4x}{(x+4)(x-3)}$ *Factor.*

$= \dfrac{(x-1)(x-3)}{(x+2)(x+4)(x-3)}$

$\quad + \dfrac{4x(x+2)}{(x+4)(x-3)(x+2)}$

$\qquad\qquad LCD = (x+4)(x+2)(x-3)$

$= \dfrac{(x-1)(x-3) + 4x(x+2)}{(x+4)(x+2)(x-3)}$ *Add numerators.*

$= \dfrac{x^2 - 4x + 3 + 4x^2 + 8x}{(x+4)(x+2)(x-3)}$ *Multiply.*

$= \dfrac{5x^2 + 4x + 3}{(x+4)(x+2)(x-3)}$ *Combine terms.*

N5. $\dfrac{2k}{k-7} + \dfrac{5}{7-k}$

$= \dfrac{2k}{k-7} + \dfrac{5(-1)}{(7-k)(-1)}$

$= \dfrac{2k}{k-7} + \dfrac{-5}{k-7}$

$= \dfrac{2k-5}{k-7}, \quad \text{or} \quad \dfrac{5-2k}{7-k}$

N6. $\dfrac{2x}{x+5} - \dfrac{x+1}{x+5}$

$= \dfrac{2x - (x+1)}{x+5}$ *Use parentheses.*

$= \dfrac{2x - x - 1}{x+5}$

$= \dfrac{x-1}{x+5}$

N7. $\dfrac{6}{y-6} - \dfrac{2}{y}$

$= \dfrac{6(y)}{(y-6)(y)} - \dfrac{2(y-6)}{y(y-6)}$ *$LCD = y(y-6)$*

$= \dfrac{6y - 2(y-6)}{y(y-6)}$

$= \dfrac{6y - 2y + 12}{y(y-6)}$

$= \dfrac{4y + 12}{y(y-6)} = \dfrac{4(y+3)}{y(y-6)}$

N8. $\dfrac{2m}{m-4} - \dfrac{m-12}{4-m}$

The denominators are opposites, so either may be used as the common denominator. We will choose $m - 4$.

$\qquad \dfrac{2m}{m-4} - \dfrac{m-12}{4-m}$

$= \dfrac{2m}{m-4} - \dfrac{(m-12)(-1)}{(4-m)(-1)}$

$= \dfrac{2m}{m-4} - \dfrac{-m+12}{m-4}$

$= \dfrac{2m - (-m+12)}{m-4}$

$= \dfrac{2m + m - 12}{m-4}$

$= \dfrac{3m - 12}{m-4}$

$= \dfrac{3(m-4)}{m-4} = 3$

N9. $\dfrac{5}{t^2 - 6t + 9} - \dfrac{2t}{t^2 - 9}$

$= \dfrac{5}{(t-3)^2} - \dfrac{2t}{(t+3)(t-3)}$ *Factor.*

$= \dfrac{5(t+3)}{(t-3)^2(t+3)} - \dfrac{2t(t-3)}{(t-3)^2(t+3)}$

$\qquad\qquad LCD = (t-3)^2(t+3)$

$= \dfrac{5(t+3) - 2t(t-3)}{(t-3)^2(t+3)}$

$= \dfrac{5t + 15 - 2t^2 + 6t}{(t-3)^2(t+3)}$

$= \dfrac{-2t^2 + 11t + 15}{(t-3)^2(t+3)}$

7.4   Section Exercises

1. $\dfrac{x}{x+8} + \dfrac{8}{x+8}$

The denominators are the same, so the sum is found by adding the two numerators and keeping the same (common) denominator.

$$\frac{x}{x+8} + \frac{8}{x+8} = \frac{x+8}{x+8} = 1$$

Choice **E** is correct.

3. $\dfrac{8}{x-8} - \dfrac{x}{x-8}$

The denominators are the same, so the difference is found by subtracting the two numerators and keeping the same (common) denominator.

$$\frac{8}{x-8} - \frac{x}{x-8} = \frac{8-x}{x-8}$$
$$= \frac{-1(x-8)}{x-8} = -1$$

Choice **C** is correct.

5. $\dfrac{x}{x+8} - \dfrac{8}{x+8}$

The denominators are the same, so the difference is found by subtracting the two numerators and keeping the same (common) denominator.

$$\frac{x}{x+8} - \frac{8}{x+8} = \frac{x-8}{x+8}$$

Choice **B** is correct.

7. $\dfrac{1}{8} - \dfrac{1}{x}$

The LCD is $8x$. Now rewrite each rational expression as a fraction with the LCD as its denominator.

$$\frac{1}{8} \cdot \frac{x}{x} = \frac{x}{8x}$$
$$\frac{1}{x} \cdot \frac{8}{8} = \frac{8}{8x}$$

Since the fractions now have a common denominator, subtract the numerators and use the LCD as the denominator of the sum.

$$\frac{1}{8} - \frac{1}{x} = \frac{x}{8x} - \frac{8}{8x} = \frac{x-8}{8x}$$

Choice **G** is correct.

9. $\dfrac{4}{m} + \dfrac{7}{m}$

The denominators are the same, so the sum is found by adding the two numerators and keeping the same (common) denominator.

$$\frac{4}{m} + \frac{7}{m} = \frac{4+7}{m} = \frac{11}{m}$$

11. $\dfrac{5}{y+4} - \dfrac{1}{y+4}$

The denominators are the same, so the difference is found by subtracting the two numerators and keeping the same (common) denominator.

$$\frac{5}{y+4} - \frac{1}{y+4} = \frac{5-1}{y+4} = \frac{4}{y+4}$$

13. $\dfrac{x}{x+y} + \dfrac{y}{x+y}$

The denominators are the same, so the sum is found by adding the two numerators and keeping the same (common) denominator.

$$\frac{x}{x+y} + \frac{y}{x+y} = \frac{x+y}{x+y} = 1$$

15. $\dfrac{5m}{m+1} - \dfrac{1+4m}{m+1}$

The denominators are the same, so the difference is found by subtracting the two numerators and keeping the same (common) denominator. Don't forget the parentheses on the second numerator.

$$\frac{5m}{m+1} - \frac{1+4m}{m+1} = \frac{5m-(1+4m)}{m+1}$$
$$= \frac{5m-1-4m}{m+1}$$
$$= \frac{m-1}{m+1}$$

17. $\dfrac{a+b}{2} - \dfrac{a-b}{2}$

The denominators are the same, so the difference is found by subtracting the two numerators and keeping the same (common) denominator. Don't forget the parentheses on the second numerator.

$$\frac{a+b}{2} - \frac{a-b}{2} = \frac{(a+b)-(a-b)}{2}$$
$$= \frac{a+b-a+b}{2}$$
$$= \frac{2b}{2} = b$$

19. $\dfrac{x^2}{x+5} + \dfrac{5x}{x+5} = \dfrac{x^2+5x}{x+5}$ *Add numerators.*

$\qquad\qquad\qquad = \dfrac{x(x+5)}{x+5}$ *Factor numerator.*

$\qquad\qquad\qquad = x$ *Lowest terms*

21. $\dfrac{y^2-3y}{y+3} + \dfrac{-18}{y+3} = \dfrac{y^2-3y-18}{y+3}$

$\qquad\qquad\qquad = \dfrac{(y-6)(y+3)}{y+3}$

$\qquad\qquad\qquad = y - 6$

23. $\dfrac{x}{x^2-9} - \dfrac{-3}{x^2-9}$

The denominators are the same, so the sum is found by adding the two numerators and keeping the same (common) denominator.

$\dfrac{x}{x^2-9} - \dfrac{-3}{x^2-9} = \dfrac{x+3}{x^2-9}$

$\qquad\qquad\qquad = \dfrac{x+3}{(x+3)(x-3)}$

$\qquad\qquad\qquad = \dfrac{1}{x-3}$

25. $\dfrac{z}{5} + \dfrac{1}{3}$

The LCD is 15. Now rewrite each rational expression as a fraction with the LCD as its denominator.

$\dfrac{z}{5} \cdot \dfrac{3}{3} = \dfrac{3z}{15}$

$\dfrac{1}{3} \cdot \dfrac{5}{5} = \dfrac{5}{15}$

Since the fractions now have a common denominator, add the numerators and use the LCD as the denominator of the sum.

$\dfrac{z}{5} + \dfrac{1}{3} = \dfrac{3z}{15} + \dfrac{5}{15} = \dfrac{3z+5}{15}$

27. $\dfrac{5}{7} - \dfrac{r}{2} = \dfrac{5}{7} \cdot \dfrac{2}{2} - \dfrac{r}{2} \cdot \dfrac{7}{7}$ *LCD = 14*

$\qquad = \dfrac{10}{14} - \dfrac{7r}{14}$

$\qquad = \dfrac{10-7r}{14}$

29. $-\dfrac{3}{4} - \dfrac{1}{2x} = -\dfrac{3 \cdot x}{4 \cdot x} - \dfrac{1 \cdot 2}{2x \cdot 2}$ *LCD = 4x*

$\qquad = \dfrac{-3x-2}{4x}$

31. $\dfrac{6}{5x} + \dfrac{9}{2x} = \dfrac{6}{5x} \cdot \dfrac{2}{2} + \dfrac{9}{2x} \cdot \dfrac{5}{5}$ *LCD = 10x*

$\qquad = \dfrac{12}{10x} + \dfrac{45}{10x}$

$\qquad = \dfrac{12+45}{10x} = \dfrac{57}{10x}$

33. $\dfrac{x+1}{6} + \dfrac{3x+3}{9}$

First reduce the second fraction.

$\dfrac{3x+3}{9} = \dfrac{3(x+1)}{9} = \dfrac{x+1}{3}$

Now the LCD of $\dfrac{x+1}{6}$ and $\dfrac{x+1}{3}$ is 6. Thus,

$\dfrac{x+1}{6} + \dfrac{x+1}{3} = \dfrac{x+1}{6} + \dfrac{x+1}{3} \cdot \dfrac{2}{2}$

$\qquad = \dfrac{x+1+2x+2}{6}$

$\qquad = \dfrac{3x+3}{6}$

$\qquad = \dfrac{3(x+1)}{6} = \dfrac{x+1}{2}.$

35. $\dfrac{x+3}{3x} + \dfrac{2x+2}{4x} = \dfrac{x+3}{3x} + \dfrac{2(x+1)}{4x}$

$\qquad = \dfrac{x+3}{3x} + \dfrac{x+1}{2x}$ *Reduce.*

$\qquad = \dfrac{x+3}{3x} \cdot \dfrac{2}{2} + \dfrac{x+1}{2x} \cdot \dfrac{3}{3}$

$\qquad\qquad\qquad\qquad$ *LCD = 6x*

$\qquad = \dfrac{2x+6+3x+3}{6x}$

$\qquad = \dfrac{5x+9}{6x}$

37. $\dfrac{7}{3p^2} - \dfrac{2}{p} = \dfrac{7}{3p^2} - \dfrac{2}{p} \cdot \dfrac{3p}{3p}$ *LCD = 3p²*

$\qquad = \dfrac{7-6p}{3p^2}$

39. $\dfrac{1}{k+4} - \dfrac{2}{k} = \dfrac{1}{k+4} \cdot \dfrac{k}{k} - \dfrac{2}{k} \cdot \dfrac{k+4}{k+4}$ *LCD = k(k+4)*

$\qquad = \dfrac{k}{k(k+4)} - \dfrac{2(k+4)}{k(k+4)}$

$\qquad = \dfrac{k-2k-8}{k(k+4)}$

$\qquad = \dfrac{-k-8}{k(k+4)}$

41. $\dfrac{x}{x-2} + \dfrac{-8}{x^2-4}$

$$= \dfrac{x}{x-2} + \dfrac{-8}{(x+2)(x-2)}$$

$$= \dfrac{x}{x-2} \cdot \dfrac{x+2}{x+2} + \dfrac{-8}{(x+2)(x-2)}$$

$$LCD = (x+2)(x-2)$$

$$= \dfrac{x(x+2)-8}{(x+2)(x-2)}$$

$$= \dfrac{x^2+2x-8}{(x+2)(x-2)}$$

$$= \dfrac{(x+4)(x-2)}{(x+2)(x-2)} = \dfrac{x+4}{x+2}$$

43. $\dfrac{4m}{m^2+3m+2} + \dfrac{2m-1}{m^2+6m+5}$

$$= \dfrac{4m}{(m+2)(m+1)} + \dfrac{2m-1}{(m+1)(m+5)}$$

$$= \dfrac{4m(m+5)}{(m+2)(m+1)(m+5)}$$

$$\quad + \dfrac{(2m-1)(m+2)}{(m+1)(m+5)(m+2)}$$

$$LCD = (m+2)(m+1)(m+5)$$

$$= \dfrac{(4m^2+20m)+(2m^2+3m-2)}{(m+2)(m+1)(m+5)}$$

$$= \dfrac{6m^2+23m-2}{(m+2)(m+1)(m+5)}$$

45. $\dfrac{4y}{y^2-1} - \dfrac{5}{y^2+2y+1}$

$$= \dfrac{4y}{(y+1)(y-1)} - \dfrac{5}{(y+1)(y+1)}$$

$$= \dfrac{4y(y+1)}{(y+1)^2(y-1)} - \dfrac{5(y-1)}{(y+1)^2(y-1)}$$

$$LCD = (y+1)^2(y-1)$$

$$= \dfrac{(4y^2+4y)-(5y-5)}{(y+1)^2(y-1)}$$

$$= \dfrac{4y^2-y+5}{(y+1)^2(y-1)}$$

47. $\dfrac{t}{t+2} + \dfrac{5-t}{t} - \dfrac{4}{t^2+2t}$

$$= \dfrac{t}{t+2} + \dfrac{5-t}{t} - \dfrac{4}{t(t+2)}$$

$$= \dfrac{t}{t+2} \cdot \dfrac{t}{t} + \dfrac{5-t}{t} \cdot \dfrac{t+2}{t+2}$$

$$\quad - \dfrac{4}{t(t+2)} \quad LCD = t(t+2)$$

$$= \dfrac{t \cdot t + (5-t)(t+2)-4}{t(t+2)}$$

$$= \dfrac{t^2+5t+10-t^2-2t-4}{t(t+2)}$$

$$= \dfrac{3t+6}{t(t+2)}$$

$$= \dfrac{3(t+2)}{t(t+2)} = \dfrac{3}{t}$$

49. $\dfrac{10}{m-2} + \dfrac{5}{2-m}$

Since

$$2-m = -1(m-2),$$

either $m-2$ or $2-m$ could be used as the LCD.

51. $\dfrac{4}{x-5} + \dfrac{6}{5-x}$

The two denominators, $x-5$ and $5-x$, are opposites of each other, so either one may be used as the common denominator. We will work the exercise both ways and compare the answers.

$$\dfrac{4}{x-5} + \dfrac{6}{5-x} = \dfrac{4}{x-5} + \dfrac{6(-1)}{(5-x)(-1)}$$

$$LCD = x-5$$

$$= \dfrac{4}{x-5} + \dfrac{-6}{x-5}$$

$$= \dfrac{-2}{x-5}$$

$$\dfrac{4}{x-5} + \dfrac{6}{5-x} = \dfrac{4(-1)}{(x-5)(-1)} + \dfrac{6}{5-x}$$

$$LCD = 5-x$$

$$= \dfrac{-4}{5-x} + \dfrac{6}{5-x}$$

$$= \dfrac{2}{5-x}$$

The two answers are equivalent, since

$$\dfrac{-2}{x-5} \cdot \dfrac{-1}{-1} = \dfrac{2}{5-x}.$$

53. $\dfrac{-1}{1-y} - \dfrac{4y-3}{y-1}$

The LCD is either $1-y$ or $y-1$. We'll use $y-1$.

$$\dfrac{-1}{1-y} - \dfrac{4y-3}{y-1} = \dfrac{-1 \cdot -1}{-1 \cdot (1-y)} - \dfrac{4y-3}{y-1}$$

$$= \dfrac{1-(4y-3)}{y-1}$$

$$= \dfrac{1-4y+3}{y-1}$$

$$= \frac{4 - 4y}{y - 1}$$

$$= \frac{4(1 - y)}{y - 1} = -4$$

55. $\dfrac{2}{x - y^2} + \dfrac{7}{y^2 - x}$

LCD $= x - y^2$ or $y^2 - x$

We will use $x - y^2$.

$$\frac{2}{x - y^2} + \frac{7}{y^2 - x}$$

$$= \frac{2}{x - y^2} + \frac{-1(7)}{-1(y^2 - x)}$$

$$= \frac{2}{x - y^2} + \frac{-7}{-y^2 + x}$$

$$= \frac{2}{x - y^2} + \frac{-7}{x - y^2}$$

$$= \frac{2 + (-7)}{x - y^2} = \frac{-5}{x - y^2}$$

If $y^2 - x$ is used as the LCD, we will obtain the equivalent answer

$$\frac{5}{y^2 - x}.$$

57. $\dfrac{x}{5x - 3y} - \dfrac{y}{3y - 5x}$

LCD $= 5x - 3y$ or $3y - 5x$

We will use $5x - 3y$.

$$\frac{x}{5x - 3y} - \frac{y}{3y - 5x}$$

$$= \frac{x}{5x - 3y} - \frac{-1(y)}{-1(3y - 5x)}$$

$$= \frac{x}{5x - 3y} - \frac{-y}{-3y + 5x}$$

$$= \frac{x}{5x - 3y} - \frac{-y}{5x - 3y}$$

$$= \frac{x - (-y)}{5x - 3y} = \frac{x + y}{5x - 3y}$$

If $3y - 5x$ is used as the LCD, we will obtain the equivalent answer

$$\frac{-x - y}{3y - 5x}.$$

59. $\dfrac{3}{4p - 5} + \dfrac{9}{5 - 4p}$

LCD $= 4p - 5$ or $5 - 4p$

We will use $4p - 5$.

$$\frac{3}{4p - 5} + \frac{9}{5 - 4p}$$

$$= \frac{3}{4p - 5} + \frac{-1(9)}{-1(5 - 4p)}$$

$$= \frac{3}{4p - 5} + \frac{-9}{-5 + 4p}$$

$$= \frac{3}{4p - 5} + \frac{-9}{4p - 5}$$

$$= \frac{3 + (-9)}{4p - 5} = \frac{-6}{4p - 5}$$

If $5 - 4p$ is used as the LCD, we will obtain the equivalent answer

$$\frac{6}{5 - 4p}.$$

61. $\dfrac{2m}{m - n} - \dfrac{5m + n}{2m - 2n}$

$$= \frac{2m}{m - n} - \frac{5m + n}{2(m - n)} \quad \begin{array}{l} \textit{Factor second} \\ \textit{denominator.} \end{array}$$

$$= \frac{2m}{m - n} \cdot \frac{2}{2} - \frac{5m + n}{2(m - n)} \quad LCD = 2(m - n)$$

$$= \frac{4m - (5m + n)}{2(m - n)}$$

$$= \frac{4m - 5m - n}{2(m - n)}$$

$$= \frac{-m - n}{2(m - n)}$$

63. $\dfrac{5}{x^2 - 9} - \dfrac{x + 2}{x^2 + 4x + 3}$

To find the LCD, factor the denominators.

$$x^2 - 9 = (x + 3)(x - 3)$$

$$x^2 + 4x + 3 = (x + 3)(x + 1)$$

The LCD is $(x + 3)(x - 3)(x + 1)$.

$$\frac{5}{x^2 - 9} - \frac{x + 2}{x^2 + 4x + 3}$$

$$= \frac{5 \cdot (x + 1)}{(x + 3)(x - 3) \cdot (x + 1)}$$

$$\quad - \frac{(x + 2) \cdot (x - 3)}{(x + 3)(x + 1) \cdot (x - 3)}$$

$$= \frac{5x + 5}{(x + 3)(x - 3)(x + 1)}$$

$$\quad - \frac{x^2 - x - 6}{(x + 3)(x + 1)(x - 3)}$$

$$= \frac{(5x + 5) - (x^2 - x - 6)}{(x + 3)(x - 3)(x + 1)}$$

$$= \frac{5x + 5 - x^2 + x + 6}{(x + 3)(x - 3)(x + 1)}$$

$$= \frac{-x^2 + 6x + 11}{(x + 3)(x - 3)(x + 1)}$$

65. $\dfrac{2q+1}{3q^2+10q-8} - \dfrac{3q+5}{2q^2+5q-12}$

$= \dfrac{2q+1}{(3q-2)(q+4)} - \dfrac{3q+5}{(2q-3)(q+4)}$

$= \dfrac{(2q+1)\cdot(2q-3)}{(3q-2)(q+4)\cdot(2q-3)}$

$\quad - \dfrac{(3q+5)\cdot(3q-2)}{(2q-3)(q+4)\cdot(3q-2)}$

$\quad\quad LCD = (3q-2)(q+4)(2q-3)$

$= \dfrac{(4q^2-4q-3)-(9q^2+9q-10)}{(3q-2)(q+4)(2q-3)}$

$= \dfrac{4q^2-4q-3-9q^2-9q+10}{(3q-2)(q+4)(2q-3)}$

$= \dfrac{-5q^2-13q+7}{(3q-2)(q+4)(2q-3)}$

67. $\dfrac{4}{r^2-r} + \dfrac{6}{r^2+2r} - \dfrac{1}{r^2+r-2}$

$= \dfrac{4}{r(r-1)} + \dfrac{6}{r(r+2)} - \dfrac{1}{(r+2)(r-1)}$

$= \dfrac{4\cdot(r+2)}{r(r-1)\cdot(r+2)} + \dfrac{6\cdot(r-1)}{r(r+2)\cdot(r-1)}$

$\quad - \dfrac{1\cdot r}{r\cdot(r+2)(r-1)}$

$\quad\quad LCD = r(r+2)(r-1)$

$= \dfrac{4r+8+6r-6-r}{r(r+2)(r-1)}$

$= \dfrac{9r+2}{r(r+2)(r-1)}$

69. $\dfrac{x+3y}{x^2+2xy+y^2} + \dfrac{x-y}{x^2+4xy+3y^2}$

$= \dfrac{x+3y}{(x+y)(x+y)} + \dfrac{x-y}{(x+3y)(x+y)}$

$= \dfrac{(x+3y)\cdot(x+3y)}{(x+y)(x+y)\cdot(x+3y)}$

$\quad + \dfrac{(x-y)\cdot(x+y)}{(x+3y)(x+y)\cdot(x+y)}$

$\quad\quad LCD = (x+y)(x+y)(x+3y)$

$= \dfrac{(x^2+6xy+9y^2)+(x^2-y^2)}{(x+y)(x+y)(x+3y)}$

$= \dfrac{2x^2+6xy+8y^2}{(x+y)(x+y)(x+3y)}$

$= \dfrac{2(x^2+3xy+4y^2)}{(x+y)(x+y)(x+3y)},$

or $\dfrac{2(x^2+3xy+4y^2)}{(x+y)^2(x+3y)}$

71. $\dfrac{r+y}{18r^2+9ry-2y^2} + \dfrac{3r-y}{36r^2-y^2}$

Factor the first denominator by grouping. Find integers whose product is $18(-2)=-36$ and whose sum is 9. The integers are 12 and -3.

$18r^2+9ry-2y^2$
$= 18r^2+12ry-3ry-2y^2$
$= 6r(3r+2y)-y(3r+2y)$
$= (3r+2y)(6r-y)$

Factor the second denominator.

$\dfrac{r+y}{(3r+2y)(6r-y)} + \dfrac{3r-y}{(6r-y)(6r+y)}$

Rewrite fractions with the LCD, $(3r+2y)(6r-y)(6r+y)$.

$= \dfrac{(r+y)\cdot(6r+y)}{(3r+2y)(6r-y)\cdot(6r+y)}$

$\quad + \dfrac{(3r-y)\cdot(3r+2y)}{(6r-y)(6r+y)\cdot(3r+2y)}$

$= \dfrac{6r^2+7ry+y^2}{(3r+2y)(6r-y)(6r+y)}$

$\quad + \dfrac{9r^2+3ry-2y^2}{(3r+2y)(6r-y)(6r+y)}$

$= \dfrac{6r^2+7ry+y^2+9r^2+3ry-2y^2}{(3r+2y)(6r-y)(6r+y)}$

$= \dfrac{15r^2+10ry-y^2}{(3r+2y)(6r-y)(6r+y)}$

73. (a) $P = 2L+2W$

$= 2\left(\dfrac{3k+1}{10}\right) + 2\left(\dfrac{5}{6k+2}\right)$

$= 2\left(\dfrac{3k+1}{2\cdot5}\right) + 2\left(\dfrac{5}{2(3k+1)}\right)$

$= \dfrac{3k+1}{5} + \dfrac{5}{3k+1}$

To add the two fractions on the right, use $5(3k+1)$ as the LCD.

$P = \dfrac{(3k+1)(3k+1)}{5(3k+1)} + \dfrac{(5)(5)}{5(3k+1)}$

$= \dfrac{(3k+1)(3k+1)+(5)(5)}{5(3k+1)}$

$= \dfrac{9k^2+6k+1+25}{5(3k+1)}$

$= \dfrac{9k^2+6k+26}{5(3k+1)}$

(b) $A = L\cdot W$

$A = \dfrac{3k+1}{10}\cdot\dfrac{5}{6k+2}$

$$= \frac{3k+1}{5 \cdot 2} \cdot \frac{5}{2(3k+1)}$$

$$= \frac{1}{2 \cdot 2} = \frac{1}{4}$$

75. $\dfrac{1010}{49(101-x)} - \dfrac{10}{49}$

$$= \frac{1010}{49(101-x)} - \frac{10(101-x)}{49(101-x)}$$

$$= \frac{1010 - 1010 + 10x}{49(101-x)}$$

$$= \frac{10x}{49(101-x)}$$

77. $\dfrac{\frac{5}{6}}{\frac{2}{3}} = \dfrac{5}{6} \div \dfrac{2}{3} = \dfrac{5}{6} \cdot \dfrac{3}{2} = \dfrac{5}{\cancel{6}} \cdot \dfrac{\cancel{3}^{1}}{2} = \dfrac{5}{4}$

79. $\dfrac{\frac{3}{2}}{\frac{7}{4}} = \dfrac{3}{2} \div \dfrac{7}{4} = \dfrac{3}{2} \cdot \dfrac{4}{7} = \dfrac{3}{\cancel{2}} \cdot \dfrac{\cancel{4}^{2}}{7} = \dfrac{6}{7}$

7.5 Complex Fractions

7.5 Now Try Exercises

N1. (a) $\dfrac{\frac{2}{5} + \frac{1}{4}}{\frac{1}{6} + \frac{3}{8}} = \dfrac{\frac{2(4)}{5(4)} + \frac{1(5)}{4(5)}}{\frac{1(4)}{6(4)} + \frac{3(3)}{8(3)}}$

$$= \frac{\frac{8+5}{20}}{\frac{4+9}{24}}$$

$$= \frac{13}{20} \div \frac{13}{24}$$

$$= \frac{13}{20} \cdot \frac{24}{13} \qquad \textit{Multiply by reciprocal.}$$

$$= \frac{13 \cdot 4 \cdot 6}{4 \cdot 5 \cdot 13} = \frac{6}{5}$$

(b) $\dfrac{2 + \frac{4}{x}}{\frac{5}{6} + \frac{5x}{12}} = \dfrac{\frac{2}{1} + \frac{4}{x}}{\frac{5}{6} + \frac{5x}{12}} = \dfrac{\frac{2(x)}{1(x)} + \frac{4}{x}}{\frac{5(2)}{6(2)} + \frac{5x}{12}}$

$$= \frac{\frac{2x+4}{x}}{\frac{10+5x}{12}} = \frac{2x+4}{x} \div \frac{10+5x}{12}$$

$$= \frac{2x+4}{x} \cdot \frac{12}{10+5x}$$

$$= \frac{2(x+2) \cdot 12}{x \cdot 5(2+x)}$$

$$= \frac{24}{5x}$$

N2. $\dfrac{\frac{a^2 b}{c}}{\frac{ab^2}{c^3}} = \dfrac{a^2 b}{c} \div \dfrac{ab^2}{c^3}$

$$= \frac{a^2 b}{c} \cdot \frac{c^3}{ab^2} = \frac{ac^2}{b}$$

N3. $\dfrac{5 + \dfrac{2}{a-3}}{\dfrac{1}{a-3} - 2}$

$$= \frac{\dfrac{5(a-3)}{1(a-3)} + \dfrac{2}{a-3}}{\dfrac{1}{a-3} - \dfrac{2(a-3)}{1(a-3)}}$$

$$= \frac{\dfrac{5(a-3) + 2}{a-3}}{\dfrac{1 - 2(a-3)}{a-3}}$$

$$= \frac{\dfrac{5a - 15 + 2}{a-3}}{\dfrac{1 - 2a + 6}{a-3}}$$

$$= \frac{\dfrac{5a - 13}{a-3}}{\dfrac{7 - 2a}{a-3}}$$

$$= \frac{5a - 13}{a-3} \div \frac{7 - 2a}{a-3}$$

$$= \frac{5a - 13}{a-3} \cdot \frac{a-3}{7 - 2a}$$

$$= \frac{5a - 13}{7 - 2a}$$

N4. (a) $\dfrac{\frac{3}{5} - \frac{1}{4}}{\frac{1}{8} + \frac{3}{20}}$

$$= \frac{40 \left(\dfrac{3}{5} - \dfrac{1}{4} \right)}{40 \left(\dfrac{1}{8} + \dfrac{3}{20} \right)} \qquad LCD = 2^3 \cdot 5 = 40$$

$$= \frac{40 \left(\dfrac{3}{5} \right) - 40 \left(\dfrac{1}{4} \right)}{40 \left(\dfrac{1}{8} \right) + 40 \left(\dfrac{3}{20} \right)}$$

$$= \frac{24 - 10}{5 + 6}$$

$$= \frac{14}{11}$$

(b) $\dfrac{\dfrac{2}{x} - 3}{7 + \dfrac{x}{5}}$ $LCD = 5x$

$= \dfrac{5x\left(\dfrac{2}{x} - 3\right)}{5x\left(7 + \dfrac{x}{5}\right)}$ *Multiply numerator and denominator by 5x.*

$= \dfrac{5x\left(\dfrac{2}{x}\right) - 5x(3)}{5x(7) + 5x\left(\dfrac{x}{5}\right)}$

$= \dfrac{10 - 15x}{35x + x^2}$, or $\dfrac{10 - 15x}{x^2 + 35x}$

N5. $\dfrac{\dfrac{1}{y} + \dfrac{2}{3y^2}}{\dfrac{5}{4y^2} - \dfrac{3}{2y^3}}$ $LCD = 12y^3$

$= \dfrac{12y^3\left(\dfrac{1}{y} + \dfrac{2}{3y^2}\right)}{12y^3\left(\dfrac{5}{4y^2} - \dfrac{3}{2y^3}\right)}$

$= \dfrac{12y^3\left(\dfrac{1}{y}\right) + 12y^3\left(\dfrac{2}{3y^2}\right)}{12y^3\left(\dfrac{5}{4y^2}\right) - 12y^3\left(\dfrac{3}{2y^3}\right)}$

$= \dfrac{12y^2 + 8y}{15y - 18}$

N6. **(a)** $\dfrac{1 - \dfrac{2}{x} - \dfrac{15}{x^2}}{1 + \dfrac{5}{x} + \dfrac{6}{x^2}}$ $LCD = x^2$

$= \dfrac{x^2\left(1 - \dfrac{2}{x} - \dfrac{15}{x^2}\right)}{x^2\left(1 + \dfrac{5}{x} + \dfrac{6}{x^2}\right)}$ *Use Method 2.*

$= \dfrac{x^2(1) - x^2\left(\dfrac{2}{x}\right) - x^2\left(\dfrac{15}{x^2}\right)}{x^2(1) + x^2\left(\dfrac{5}{x}\right) + x^2\left(\dfrac{6}{x^2}\right)}$

$= \dfrac{x^2 - 2x - 15}{x^2 + 5x + 6}$

$= \dfrac{(x - 5)(x + 3)}{(x + 2)(x + 3)} = \dfrac{x - 5}{x + 2}$

(b) $\dfrac{\dfrac{9y^2 - 16}{y^2 - 100}}{\dfrac{3y - 4}{y + 10}} = \dfrac{9y^2 - 16}{y^2 - 100} \div \dfrac{3y - 4}{y + 10}$

$= \dfrac{9y^2 - 16}{y^2 - 100} \cdot \dfrac{y + 10}{3y - 4}$

$= \dfrac{(3y + 4)(3y - 4)}{(y + 10)(y - 10)} \cdot \dfrac{y + 10}{3y - 4}$

$= \dfrac{3y + 4}{y - 10}$

7.5   Section Exercises

1. **(a)** The LCD of $\dfrac{3}{2}$ and $\dfrac{4}{3}$ is $2 \cdot 3 = 6$. The simplified form of the numerator is

$$\dfrac{3}{2} - \dfrac{4}{3} = \dfrac{9}{6} - \dfrac{8}{6} = \dfrac{1}{6}.$$

(b) The LCD of $\dfrac{1}{6}$ and $\dfrac{5}{12}$ is 12 since 12 is a multiple of 6. The simplified form of the denominator is

$$\dfrac{1}{6} - \dfrac{5}{12} = \dfrac{2}{12} - \dfrac{5}{12} = -\dfrac{3}{12} = -\dfrac{1}{4}.$$

(c) $\dfrac{\frac{1}{6}}{-\frac{1}{4}} = \dfrac{1}{6} \div \left(-\dfrac{1}{4}\right)$

(d) $\dfrac{1}{6} \div \left(-\dfrac{1}{4}\right) = \dfrac{1}{6} \cdot \left(-\dfrac{4}{1}\right)$

$= -\dfrac{2 \cdot 2}{2 \cdot 3} = -\dfrac{2}{3}$

3. $\dfrac{2 - \frac{1}{4}}{3 - \frac{1}{2}} = \dfrac{-2 + \frac{1}{4}}{-3 + \frac{1}{2}}$

Choice **D** is equivalent to the given fraction. Each term of the numerator and denominator has been multiplied by -1. Since $\dfrac{-1}{-1} = 1$, the fraction has been multiplied by the identity element, so its value is unchanged.

In Exercises 5–40, either Method 1 or Method 2 can be used to simplify each complex fraction. Only one method will be shown for each exercise.

5. To use Method 1, divide the numerator of the complex fraction by the denominator.

$$\dfrac{-\frac{4}{3}}{\frac{2}{9}} = -\dfrac{4}{3} \div \dfrac{2}{9}$$

$$= -\dfrac{4}{3} \cdot \dfrac{9}{2} = -\dfrac{36}{6} = -6$$

7. To use Method 2, multiply the numerator and denominator of the complex fraction by the LCD, y^2.

$$\dfrac{\dfrac{x}{y^2}}{\dfrac{x^2}{y}} = \dfrac{y^2\left(\dfrac{x}{y^2}\right)}{y^2\left(\dfrac{x^2}{y}\right)}$$

$$= \dfrac{x}{yx^2} = \dfrac{1}{xy}$$

9. $\dfrac{\dfrac{4a^4b^3}{3a}}{\dfrac{2ab^4}{b^2}} = \dfrac{4a^4b^3}{3a} \div \dfrac{2ab^4}{b^2}$ *Method 1*

$= \dfrac{4a^4b^3}{3a} \cdot \dfrac{b^2}{2ab^4}$

$= \dfrac{4a^4b^3 \cdot b^2}{3a \cdot 2ab^4}$

$= \dfrac{4a^4b^5}{6a^2b^4}$

$= \dfrac{2a^2b}{3}$

11. To use Method **2**, multiply the numerator and denominator of the complex fraction by the LCD, $3m$.

$$\dfrac{\dfrac{m+2}{3}}{\dfrac{m-4}{m}} = \dfrac{3m\left(\dfrac{m+2}{3}\right)}{3m\left(\dfrac{m-4}{m}\right)}$$

$$= \dfrac{m(m+2)}{3(m-4)}$$

13. $\dfrac{\dfrac{2}{x}-3}{\dfrac{2-3x}{2}} = \dfrac{2x\left(\dfrac{2}{x}-3\right)}{2x\left(\dfrac{2-3x}{2}\right)}$ *Method 2;*
LCD = 2x

$= \dfrac{2x\left(\frac{2}{x}\right) - 2x(3)}{x(2-3x)}$

$= \dfrac{4-6x}{x(2-3x)}$

$= \dfrac{2(2-3x)}{x(2-3x)}$ *Factor.*

$= \dfrac{2}{x}$ *Lowest terms*

15. $\dfrac{\dfrac{1}{x}+x}{\dfrac{x^2+1}{8}} = \dfrac{8x\left(\dfrac{1}{x}+x\right)}{8x\left(\dfrac{x^2+1}{8}\right)}$ *Method 2;*
LCD = 8x

$= \dfrac{8+8x^2}{x(x^2+1)}$ *Distributive property*

$= \dfrac{8(1+x^2)}{x(x^2+1)}$ *Factor.*

$= \dfrac{8}{x}$ *Lowest terms*

17. $\dfrac{a-\dfrac{5}{a}}{a+\dfrac{1}{a}} = \dfrac{a\left(a-\dfrac{5}{a}\right)}{a\left(a+\dfrac{1}{a}\right)}$ *Method 2;*
LCD = a

$= \dfrac{a^2-5}{a^2+1}$

19. $\dfrac{\dfrac{5}{8}+\dfrac{2}{3}}{\dfrac{7}{3}-\dfrac{1}{4}} = \dfrac{24\left(\dfrac{5}{8}+\dfrac{2}{3}\right)}{24\left(\dfrac{7}{3}-\dfrac{1}{4}\right)}$ *Method 2;*
LCD = 24

$= \dfrac{24\left(\frac{5}{8}\right)+24\left(\frac{2}{3}\right)}{24\left(\frac{7}{3}\right)-24\left(\frac{1}{4}\right)}$

$= \dfrac{15+16}{56-6} = \dfrac{31}{50}$

21. $\dfrac{\dfrac{1}{x^2}+\dfrac{1}{y^2}}{\dfrac{1}{x}-\dfrac{1}{y}}$

$= \dfrac{x^2y^2\left(\dfrac{1}{x^2}+\dfrac{1}{y^2}\right)}{x^2y^2\left(\dfrac{1}{x}-\dfrac{1}{y}\right)}$ *Method 2;*
LCD = x²y²

$= \dfrac{x^2y^2\left(\dfrac{1}{x^2}\right)+x^2y^2\left(\dfrac{1}{y^2}\right)}{x^2y^2\left(\dfrac{1}{x}\right)-x^2y^2\left(\dfrac{1}{y}\right)}$

$= \dfrac{y^2+x^2}{xy^2-x^2y} = \dfrac{y^2+x^2}{xy(y-x)}$

23. $\dfrac{\dfrac{2}{p^2}-\dfrac{3}{5p}}{\dfrac{4}{p}+\dfrac{1}{4p}} = \dfrac{20p^2\left(\dfrac{2}{p^2}-\dfrac{3}{5p}\right)}{20p^2\left(\dfrac{4}{p}+\dfrac{1}{4p}\right)}$ *Method 2;*
LCD = 20p²

$= \dfrac{20p^2\left(\dfrac{2}{p^2}\right)-20p^2\left(\dfrac{3}{5p}\right)}{20p^2\left(\dfrac{4}{p}\right)+20p^2\left(\dfrac{1}{4p}\right)}$

$= \dfrac{40-12p}{80p+5p}$

$= \dfrac{40-12p}{85p}$

25. $\dfrac{\dfrac{5}{x^2y} - \dfrac{2}{xy^2}}{\dfrac{3}{x^2y^2} + \dfrac{4}{xy}}$

$= \dfrac{x^2y^2\left(\dfrac{5}{x^2y} - \dfrac{2}{xy^2}\right)}{x^2y^2\left(\dfrac{3}{x^2y^2} + \dfrac{4}{xy}\right)}$ *Method 2;*
  LCD = x^2y^2

$= \dfrac{x^2y^2\left(\dfrac{5}{x^2y}\right) - x^2y^2\left(\dfrac{2}{xy^2}\right)}{x^2y^2\left(\dfrac{3}{x^2y^2}\right) + x^2y^2\left(\dfrac{4}{xy}\right)}$

$= \dfrac{5y - 2x}{3 + 4xy}$

27. $\dfrac{\dfrac{1}{4} - \dfrac{1}{a^2}}{\dfrac{1}{2} + \dfrac{1}{a}}$

$= \dfrac{4a^2\left(\dfrac{1}{4} - \dfrac{1}{a^2}\right)}{4a^2\left(\dfrac{1}{2} + \dfrac{1}{a}\right)}$ *Method 2;*
  LCD = $4a^2$

$= \dfrac{a^2 - 4}{2a^2 + 4a}$ *Distributive property*

$= \dfrac{(a-2)(a+2)}{2a(a+2)}$ *Factor numerator and denominator.*

$= \dfrac{a-2}{2a}$ *Fundamental property*

29. $\dfrac{\dfrac{1}{z+5}}{\dfrac{4}{z^2-25}}$

$= \dfrac{1}{z+5} \div \dfrac{4}{z^2-25}$ *Method 1*

$= \dfrac{1}{z+5} \cdot \dfrac{z^2-25}{4}$ *Multiply by reciprocal.*

$= \dfrac{1 \cdot (z^2-25)}{(z+5) \cdot 4}$ *Multiply.*

$= \dfrac{(z+5)(z-5)}{(z+5) \cdot 4}$ *Factor numerator.*

$= \dfrac{z-5}{4}$ *Fundamental property*

31. $\dfrac{\dfrac{1}{m+1} - 1}{\dfrac{1}{m+1} + 1}$

$= \dfrac{(m+1)\left(\dfrac{1}{m+1} - 1\right)}{(m+1)\left(\dfrac{1}{m+1} + 1\right)}$ *Method 2;*
  LCD = m + 1

$= \dfrac{1 - 1(m+1)}{1 + 1(m+1)}$ *Distributive property*

$= \dfrac{1 - m - 1}{1 + m + 1}$ *Distributive property*

$= \dfrac{-m}{m+2}$

33. $\dfrac{\dfrac{1}{m-1} + \dfrac{2}{m+2}}{\dfrac{2}{m+2} - \dfrac{1}{m-3}}$

$= \dfrac{(m-1)(m+2)(m-3)\left(\frac{1}{m-1} + \frac{2}{m+2}\right)}{(m-1)(m+2)(m-3)\left(\frac{2}{m+2} - \frac{1}{m-3}\right)}$

Method 2;
LCD = (m − 1)(m + 2)(m − 3)

$= \dfrac{(m+2)(m-3) + 2(m-1)(m-3)}{2(m-1)(m-3) - (m-1)(m+2)}$

Distributive property

$= \dfrac{(m-3)[(m+2) + 2(m-1)]}{(m-1)[2(m-3) - (m+2)]}$

Factor out m − 3 in numerator and m − 1 in denominator.

$= \dfrac{(m-3)[m+2+2m-2]}{(m-1)[2m-6-m-2]}$

Distributive property

$= \dfrac{3m(m-3)}{(m-1)(m-8)}$ *Combine like terms.*

35. $\dfrac{2 + \dfrac{1}{x} - \dfrac{28}{x^2}}{3 + \dfrac{13}{x} + \dfrac{4}{x^2}}$ *LCD = x^2*

$= \dfrac{x^2\left(2 + \dfrac{1}{x} - \dfrac{28}{x^2}\right)}{x^2\left(3 + \dfrac{13}{x} + \dfrac{4}{x^2}\right)}$ *Method 2*

$= \dfrac{2x^2 + x - 28}{3x^2 + 13x + 4}$

$= \dfrac{(2x-7)(x+4)}{(3x+1)(x+4)}$

$= \dfrac{2x-7}{3x+1}$

37. $\dfrac{\dfrac{y+8}{y-4}}{\dfrac{y^2-64}{y^2-16}}$

$= \dfrac{y+8}{y-4} \div \dfrac{y^2-64}{y^2-16}$ 　　*Method 1*

$= \dfrac{y+8}{y-4} \cdot \dfrac{y^2-16}{y^2-64}$ 　　*Multiply by reciprocal.*

$= \dfrac{(y+8) \cdot (y^2-16)}{(y-4) \cdot (y^2-64)}$ 　　*Multiply.*

$= \dfrac{(y+8)(y+4)(y-4)}{(y-4)(y+8)(y-8)}$ 　　*Factor.*

$= \dfrac{y+4}{y-8}$ 　　*Fundamental property*

39. $\dfrac{1+x^{-1}-12x^{-2}}{1-x^{-1}-20x^{-2}}$

$= \dfrac{1+\dfrac{1}{x}-\dfrac{12}{x^2}}{1-\dfrac{1}{x}-\dfrac{20}{x^2}}$ 　　$LCD = x^2$

$= \dfrac{x^2\left(1+\dfrac{1}{x}-\dfrac{12}{x^2}\right)}{x^2\left(1-\dfrac{1}{x}-\dfrac{20}{x^2}\right)}$ 　　*Method 2*

$= \dfrac{x^2+x-12}{x^2-x-20}$ 　　*Multiply.*

$= \dfrac{(x+4)(x-3)}{(x-5)(x+4)}$ 　　*Factor.*

$= \dfrac{x-3}{x-5}$ 　　*Fundamental property*

41. In a fraction, the fraction bar represents division. For example, $\frac{3}{5}$ can be read "3 divided by 5."

43. "The sum of $\frac{3}{8}$ and $\frac{5}{6}$, divided by 2" is written

$$\dfrac{\frac{3}{8}+\frac{5}{6}}{2}.$$

44. $\dfrac{\frac{3}{8}+\frac{5}{6}}{2} = \dfrac{\frac{9}{24}+\frac{20}{24}}{2}$ 　　*Method 1*

$= \dfrac{\frac{29}{24}}{\frac{2}{1}} = \dfrac{29}{24} \cdot \dfrac{1}{2} = \dfrac{29}{48}$

45. $\dfrac{\frac{3}{8}+\frac{5}{6}}{2} = \dfrac{24\left(\frac{3}{8}+\frac{5}{6}\right)}{24(2)}$ 　　*Method 2; LCD = 24*

$= \dfrac{24\left(\frac{3}{8}\right)+24\left(\frac{5}{6}\right)}{24(2)}$

$= \dfrac{9+20}{48} = \dfrac{29}{48}$

46. Method 2 is usually shorter for more complex problems because the problem can be worked without adding and subtracting rational

expressions, which can be complicated and time-consuming.

47. $1+\dfrac{1}{1+\dfrac{1}{1+1}} = 1+\dfrac{1}{1+\dfrac{1}{2}}$

$= 1+\dfrac{1}{\frac{2}{2}+\frac{1}{2}}$

$= 1+\dfrac{1}{\frac{3}{2}}$

$= 1+1\cdot\dfrac{2}{3}$

$= 1+\dfrac{2}{3}$

$= \dfrac{3}{3}+\dfrac{2}{3} = \dfrac{5}{3}$

49. $7-\dfrac{3}{5+\dfrac{2}{4-2}} = 7-\dfrac{3}{5+\dfrac{2}{2}}$

$= 7-\dfrac{3}{5+1}$

$= 7-\dfrac{3}{6}$

$= 7-\dfrac{1}{2}$

$= \dfrac{14}{2}-\dfrac{1}{2} = \dfrac{13}{2}$

51. $r+\dfrac{r}{4-\dfrac{2}{6+2}} = r+\dfrac{r}{4-\dfrac{2}{8}}$

$= r+\dfrac{r}{4-\dfrac{1}{4}}$

$= r+\dfrac{r}{\frac{16}{4}-\dfrac{1}{4}}$

$= r+\dfrac{r}{\frac{15}{4}}$

$= r+r\cdot\dfrac{4}{15}$

$= r+\dfrac{4r}{15}$

$= \dfrac{15r}{15}+\dfrac{4r}{15}$

$= \dfrac{19r}{15}$

53. $9\left(\dfrac{4x}{3}+\dfrac{2}{9}\right) = 9\left(\dfrac{4x}{3}\right)+9\left(\dfrac{2}{9}\right)$

$= 3(4x)+2$

$= 12x+2$

55. $-12\left(\dfrac{11p^2}{3}-\dfrac{9p}{4}\right)$

$= -12\left(\dfrac{11p^2}{3}\right)-12\left(-\dfrac{9p}{4}\right)$

$= -4(11p^2)+3(9p)$

$= -44p^2+27p$

57. $3x + 5 = 7x + 3$

$\quad\quad 5 = 4x + 3$ *Subtract 3x.*

$\quad\quad 2 = 4x$ *Subtract 3.*

$\quad\quad x = \frac{2}{4} = \frac{1}{2}$ *Divide by 4.*

The solution set is $\left\{\frac{1}{2}\right\}$.

59. $6(z - 3) + 5 = 8z - 3$

$\quad 6z - 18 + 5 = 8z - 3$

$\quad\quad 6z - 13 = 8z - 3$

$\quad\quad\quad -13 = 2z - 3$

$\quad\quad\quad -10 = 2z$

$\quad\quad\quad\quad z = \frac{-10}{2} = -5$

The solution set is $\{-5\}$.

7.6 Solving Equations with Rational Expressions

7.6 Now Try Exercises

N1. (a) $\frac{3}{2}t - \frac{5}{7}t = \frac{11}{7}$ has an equals sign, so this is an *equation* to be solved. Use the multiplication property of equality to clear fractions. The LCD is 14.

$$\frac{3}{2}t - \frac{5}{7}t = \frac{11}{7}$$

$14\left(\frac{3}{2}t - \frac{5}{7}t\right) = 14\left(\frac{11}{7}\right)$ *Multiply by 14.*

$14\left(\frac{3}{2}t\right) - 14\left(\frac{5}{7}t\right) = 14\left(\frac{11}{7}\right)$ *Distributive property*

$\quad\quad\quad 21t - 10t = 22$ *Multiply.*

$\quad\quad\quad\quad 11t = 22$ *Combine terms.*

$\quad\quad\quad\quad\quad t = 2$ *Divide by 11.*

Check $t = 2$: $\frac{21}{7} - \frac{10}{7} = \frac{11}{7}$ *True*

The solution set is $\{2\}$.

(b) $\frac{3}{2}t - \frac{5}{7}t$ is the difference of two terms, so it is an *expression* to be simplified. Simplify by finding the LCD, writing each coefficient with this LCD, and combining like terms.

$\frac{3}{2}t - \frac{5}{7}t = \frac{3 \cdot 7}{2 \cdot 7}t - \frac{5 \cdot 2}{7 \cdot 2}t$ *LCD = 14*

$\quad\quad\quad = \frac{21}{14}t - \frac{10}{14}t$ *Multiply.*

$\quad\quad\quad = \frac{11}{14}t$ *Combine terms.*

N2. $\frac{x + 5}{5} - \frac{x}{7} = \frac{3}{7}$

$35\left(\frac{x + 5}{5} - \frac{x}{7}\right) = 35\left(\frac{3}{7}\right)$

         Multiply by the LCD, 35.

$\quad 7(x + 5) - 5x = 5(3)$

$\quad 7x + 35 - 5x = 15$

$2x + 35 = 15$ *Combine like terms.*

$\quad\quad 2x = -20$ *Subtract 35.*

$\quad\quad\quad x = -10$ *Divide by 2.*

Check $x = -10$: $-1 + \frac{10}{7} = \frac{3}{7}$ *True*

The solution set is $\{-10\}$.

N3. $\quad\quad 4 + \frac{6}{x - 3} = \frac{2x}{x - 3}$

$(x - 3)\left(4 + \frac{6}{x - 3}\right) = (x - 3)\frac{2x}{x - 3}$

        Multiply by the LCD, x − 3.

$\quad 4(x - 3) + 6 = 2x$

$\quad 4x - 12 + 6 = 2x$

$\quad\quad 4x - 6 = 2x$

$\quad\quad\quad 2x = 6$

$\quad\quad\quad\quad x = 3$

Check $x = 3$: $4 + \frac{6}{0} = \frac{6}{0}$

The fractions are undefined, so the equation has no solution. The solution set is $\emptyset$.

N4. $\quad\quad \frac{3}{2x^2 - 8x} = \frac{1}{x^2 - 16}$

$\quad\quad \frac{3}{2x(x - 4)} = \frac{1}{(x + 4)(x - 4)}$ *Factor.*

   Multiply by the LCD, 2x(x + 4)(x − 4).

$2x(x + 4)(x - 4)\left(\frac{3}{2x(x - 4)}\right)$

$\quad\quad = 2x(x + 4)(x - 4)\left(\frac{1}{(x + 4)(x - 4)}\right)$

$\quad\quad 3(x + 4) = 2x(1)$

$\quad\quad 3x + 12 = 2x$

$\quad\quad\quad\quad x = -12$

Check $x = -12$: $\frac{3}{384} = \frac{1}{128}$ *True*

The solution set is $\{-12\}$.

N5. $\quad\quad \frac{2y}{y^2 - 25} = \frac{8}{y + 5} - \frac{1}{y - 5}$

$\quad\quad \frac{2y}{(y + 5)(y - 5)} = \frac{8}{y + 5} - \frac{1}{y - 5}$ *Factor.*

   Multiply by the LCD, (y + 5)(y − 5).

$(y + 5)(y - 5)\left(\frac{2y}{(y + 5)(y - 5)}\right)$

$\quad = (y + 5)(y - 5)\left(\frac{8}{y + 5} - \frac{1}{y - 5}\right)$

$\quad 2y = 8(y - 5) - 1(y + 5)$

$\quad 2y = 8y - 40 - y - 5$

$\quad 2y = 7y - 45$

$\quad 45 = 5y$

$\quad\quad 9 = y$

Check $y = 9$: $\frac{18}{56} = \frac{8}{14} - \frac{1}{4}$ *True*

The solution set is $\{9\}$.

N6.
$$\frac{3}{m^2 - 9} = \frac{1}{2(m - 3)} - \frac{1}{4}$$

$$\frac{3}{(m + 3)(m - 3)} = \frac{1}{2(m - 3)} - \frac{1}{4} \quad \textit{Factor.}$$

Multiply by the LCD, 4(m − 3)(m + 3).

$$4(m - 3)(m + 3)\left(\frac{3}{(m + 3)(m - 3)}\right)$$

$$= 4(m - 3)(m + 3)\left(\frac{1}{2(m - 3)} - \frac{1}{4}\right)$$

$$4(3) = 2(m + 3) - (m - 3)(m + 3)$$

$$12 = 2m + 6 - (m^2 - 9)$$

$$12 = 2m + 6 - m^2 + 9$$

$$0 = -m^2 + 2m + 3$$

$$0 = m^2 - 2m - 3$$

$$0 = (m - 3)(m + 1)$$

$$m - 3 = 0 \quad \text{or} \quad m + 1 = 0$$

$$m = 3 \quad \text{or} \quad m = -1$$

Check $m = 3$: $\frac{3}{0} = \frac{1}{0} - \frac{1}{4}$ (undefined)

Check $m = -1$: $-\frac{3}{8} = -\frac{1}{8} - \frac{1}{4}$ *True*

The solution set is $\{-1\}$.

N7.
$$\frac{5}{k^2 + k - 2} = \frac{1}{3k - 3} - \frac{1}{k + 2}$$

$$\frac{5}{(k + 2)(k - 1)} = \frac{1}{3(k - 1)} - \frac{1}{k + 2} \quad \textit{Factor.}$$

Multiply by the LCD, 3(k + 2)(k − 1).

$$3(k + 2)(k - 1)\left(\frac{5}{(k + 2)(k - 1)}\right)$$

$$= 3(k + 2)(k - 1)\left(\frac{1}{3(k - 1)} - \frac{1}{k + 2}\right)$$

$$3(5) = k + 2 - 3(k - 1)$$

$$15 = k + 2 - 3k + 3$$

$$15 = -2k + 5$$

$$10 = -2k$$

$$-5 = k$$

Check $k = -5$: $\frac{5}{18} = -\frac{1}{18} + \frac{1}{3}$ *True*

The solution set is $\{-5\}$.

N8. (a) Solve $p = \frac{x - y}{z}$ for x.

$$pz = \left(\frac{x - y}{z}\right)(z) \quad \textit{Multiply by the LCD, z.}$$

$$pz = x - y$$

$$pz + y = x \quad \textit{Isolate x.}$$

(b) Solve $a = \frac{b}{c + d}$ for d.

Multiply by the LCD, $c + d$.

$$a(c + d) = \left(\frac{b}{c + d}\right)(c + d)$$

$$ac + ad = b$$

$$ad = b - ac \quad \textit{Isolate ad.}$$

$$d = \frac{b - ac}{a} \quad \textit{Divide by a.}$$

Another solution method is to multiply by $c + d$, divide by a, and then subtract c to get $y = \frac{b}{a} - c$.

N9. Solve the formula $\frac{2}{w} = \frac{1}{x} - \frac{3}{y}$ for x.

Multiply by the LCD, wxy.

$$wxy\left(\frac{2}{w}\right) = wxy\left(\frac{1}{x} - \frac{3}{y}\right)$$

$$wxy\left(\frac{2}{w}\right) = wxy\left(\frac{1}{x}\right) - wxy\left(\frac{3}{y}\right)$$

$$2xy = wy - 3wx$$

Get the x-terms on one side.

$$2xy + 3wx = wy$$

$$x(2y + 3w) = wy \quad \textit{Factor.}$$

$$x = \frac{wy}{2y + 3w}$$

7.6 Section Exercises

1. $\frac{7}{8}x + \frac{1}{5}x$ is the sum of two terms, so it is an *expression* to be simplified. Simplify by finding the LCD, writing each coefficient with this LCD, and combining like terms.

$$\frac{7}{8}x + \frac{1}{5}x = \frac{35}{40}x + \frac{8}{40}x \quad \textit{LCD = 40}$$

$$= \frac{43}{40}x \quad \begin{array}{l}\textit{Combine}\\\textit{like terms.}\end{array}$$

3. $\frac{7}{8}x + \frac{1}{5}x = 1$ has an equals sign, so this is an *equation* to be solved. Use the multiplication property of equality to clear fractions. The LCD is 40.

$$\frac{7}{8}x + \frac{1}{5}x = 1$$

$$40\left(\frac{7}{8}x + \frac{1}{5}x\right) = 40 \cdot 1 \quad \textit{Multiply by 40.}$$

$$40\left(\frac{7}{8}x\right) + 40\left(\frac{1}{5}x\right) = 40 \cdot 1 \quad \begin{array}{l}\textit{Distributive}\\\textit{property}\end{array}$$

$$35x + 8x = 40 \quad \textit{Multiply.}$$

$$43x = 40 \quad \begin{array}{l}\textit{Combine}\\\textit{like terms.}\end{array}$$

$$x = \frac{40}{43} \quad \textit{Divide by 43.}$$

The solution set is $\left\{\frac{40}{43}\right\}$.

5. $\frac{3}{5}x - \frac{7}{10}x$ is the difference of two terms, so it is an *expression* to be simplified.

$$\frac{3}{5}x - \frac{7}{10}x = \frac{6}{10}x - \frac{7}{10}x \quad LCD = 10$$

$$= -\frac{1}{10}x \quad \begin{array}{l} \textit{Combine} \\ \textit{like terms.} \end{array}$$

7. $\frac{3}{5}x - \frac{7}{10}x = 1$ has an equals sign, so it is an *equation* to be solved.

$$\frac{3}{5}x - \frac{7}{10}x = 1$$

$$10\left(\frac{3}{5}x - \frac{7}{10}x\right) = 10 \cdot 1 \quad LCD = 10$$

$$10\left(\frac{3}{5}x\right) - 10\left(\frac{7}{10}x\right) = 10 \cdot 1 \quad \begin{array}{l} \textit{Distributive} \\ \textit{property} \end{array}$$

$$6x - 7x = 10 \quad \textit{Multiply.}$$

$$-x = 10 \quad \begin{array}{l} \textit{Combine} \\ \textit{like terms.} \end{array}$$

$$x = -10 \quad \textit{Divide by } -1.$$

The solution set is $\{-10\}$.

9. $\frac{3}{4}x - \frac{1}{2}x = 0$ has an equals sign, so it is an *equation* to be solved.

$$\frac{3}{4}x - \frac{1}{2}x = 0$$

$$4\left(\frac{3}{4}x - \frac{1}{2}x\right) = 4(0) \quad LCD = 4$$

$$4\left(\frac{3}{4}x\right) - 4\left(\frac{1}{2}x\right) = 4(0) \quad \begin{array}{l} \textit{Distributive} \\ \textit{property} \end{array}$$

$$3x - 2x = 0 \quad \textit{Multiply.}$$

$$x = 0 \quad \begin{array}{l} \textit{Combine} \\ \textit{like terms.} \end{array}$$

The solution set is $\{0\}$.

11. $\frac{3}{x+2} - \frac{5}{x} = 1$

The denominators, $x + 2$ and x, are equal to 0 for the values -2 and 0, respectively. Thus, $x \neq -2, 0$.

13. $\frac{-1}{(x+3)(x-4)} = \frac{1}{2x+1}$

The denominators, $(x+3)(x-4)$ and $2x+1$, are equal to 0 for the values -3, 4, and $-\frac{1}{2}$, respectively. Thus, $x \neq -3, 4, -\frac{1}{2}$.

15. $\frac{4}{x^2 + 8x - 9} + \frac{1}{x^2 - 4} = 0$

The denominators, $x^2 + 8x - 9 = (x+9)(x-1)$ and $x^2 - 4 = (x+2)(x-2)$, are equal to 0 for the values $-9, 1, -2,$ and 2, respectively. Thus, $x \neq -9, 1, -2, 2$.

17. We cannot solve

$$\frac{2}{3x} + \frac{1}{5x}$$

because it is an expression, not an equation. An expression cannot be solved. Only equations and inequalities are "solved."

Note: In Exercises 19–76, all proposed solutions should be checked by substituting in the original equation. It is essential to determine whether a proposed solution will make any denominator in the original equation equal to zero. Full checks will be shown here for only a few of the exercises.

19. $\frac{5}{m} - \frac{3}{m} = 8$

Multiply each side by the LCD, m.

$$m\left(\frac{5}{m} - \frac{3}{m}\right) = m \cdot 8$$

Use the distributive property to remove parentheses; then solve.

$$m\left(\frac{5}{m}\right) - m\left(\frac{3}{m}\right) = 8m$$

$$5 - 3 = 8m$$

$$2 = 8m$$

$$m = \frac{2}{8} = \frac{1}{4}$$

Check this proposed solution by replacing m with $\frac{1}{4}$ in the original equation.

$$\frac{5}{\frac{1}{4}} - \frac{3}{\frac{1}{4}} \stackrel{?}{=} 8 \quad \textit{Let } m = \frac{1}{4}.$$

$$5 \cdot 4 - 3 \cdot 4 \stackrel{?}{=} 8 \quad \begin{array}{l} \textit{Multiply by} \\ \textit{reciprocals.} \end{array}$$

$$20 - 12 \stackrel{?}{=} 8$$

$$8 = 8 \quad \textit{True}$$

Thus, the solution set is $\left\{\frac{1}{4}\right\}$.

21. $\frac{5}{y} + 4 = \frac{2}{y}$

$$y\left(\frac{5}{y} + 4\right) = y\left(\frac{2}{y}\right) \quad \begin{array}{l} \textit{Multiply by} \\ \textit{LCD, } y. \end{array}$$

$$y\left(\frac{5}{y}\right) + y(4) = y\left(\frac{2}{y}\right) \quad \begin{array}{l} \textit{Distributive} \\ \textit{property} \end{array}$$

$$5 + 4y = 2$$

$$4y = -3$$

$$y = -\frac{3}{4}$$

Check $y = -\frac{3}{4}$: $-\frac{8}{3} = -\frac{8}{3}$ *True*

Thus, the solution set is $\left\{-\frac{3}{4}\right\}$.

23.
$$\frac{3x}{5} - 6 = x$$

$$5\left(\frac{3x}{5} - 6\right) = 5(x) \quad \text{\textit{Multiply by LCD, 5.}}$$

$$5\left(\frac{3x}{5}\right) - 5(6) = 5x \quad \text{\textit{Distributive property}}$$

$$3x - 30 = 5x$$
$$-30 = 2x$$
$$-15 = x$$

Check $x = -15$: $-15 = -15$ *True*

Thus, the solution set is $\{-15\}$.

25.
$$\frac{4m}{7} + m = 11$$

$$7\left(\frac{4m}{7} + m\right) = 7(11) \quad \text{\textit{Multiply by LCD, 7.}}$$

$$7\left(\frac{4m}{7}\right) + 7(m) = 77 \quad \text{\textit{Distributive property}}$$

$$4m + 7m = 77$$
$$11m = 77$$
$$m = 7$$

Check $m = 7$: $11 = 11$ *True*

Thus, the solution set is $\{7\}$.

27.
$$\frac{z-1}{4} = \frac{z+3}{3}$$

$$12\left(\frac{z-1}{4}\right) = 12\left(\frac{z+3}{3}\right) \quad \text{\textit{Multiply by LCD, 12.}}$$

$$3(z-1) = 4(z+3)$$
$$3z - 3 = 4z + 12 \quad \text{\textit{Dist. prop.}}$$
$$-15 = z$$

Check $z = -15$: $-4 = -4$ *True*

Thus, the solution set is $\{-15\}$.

29.
$$\frac{3p+6}{8} = \frac{3p-3}{16}$$

$$16\left(\frac{3p+6}{8}\right) = 16\left(\frac{3p-3}{16}\right) \quad \text{\textit{Multiply by LCD, 16.}}$$

$$2(3p+6) = 3p - 3$$
$$6p + 12 = 3p - 3 \quad \text{\textit{Dist. prop.}}$$
$$3p = -15$$
$$p = -5$$

Check $p = -5$: $-\frac{9}{8} = -\frac{9}{8}$ *True*

Thus, the solution set is $\{-5\}$.

31.
$$\frac{2x+3}{x} = \frac{3}{2}$$

$$2x\left(\frac{2x+3}{x}\right) = 2x\left(\frac{3}{2}\right) \quad \text{\textit{Multiply by LCD, 2x.}}$$

$$2(2x+3) = 3x$$
$$4x + 6 = 3x \quad \text{\textit{Dist. prop.}}$$
$$x = -6$$

Check $x = -6$: $\frac{3}{2} = \frac{3}{2}$ *True*

Thus, the solution set is $\{-6\}$.

33.
$$\frac{k}{k-4} - 5 = \frac{4}{k-4}$$

$$(k-4)\left(\frac{k}{k-4} - 5\right) = (k-4)\left(\frac{4}{k-4}\right)$$
$$\text{\textit{Multiply by LCD, k - 4.}}$$

$$(k-4)\left(\frac{k}{k-4}\right) - 5(k-4) = 4 \quad \text{\textit{Distributive property}}$$

$$k - 5k + 20 = 4$$
$$-4k = -16$$
$$k = 4$$

The proposed solution is 4. However, 4 cannot be a solution because it makes the denominator $k - 4$ equal 0. Therefore, the solution set is $\emptyset$.

35.
$$\frac{q+2}{3} + \frac{q-5}{5} = \frac{7}{3}$$

$$15\left(\frac{q+2}{3} + \frac{q-5}{5}\right) = 15\left(\frac{7}{3}\right) \quad \text{\textit{Mult. by LCD, 15.}}$$

$$15\left(\frac{q+2}{3}\right) + 15\left(\frac{q-5}{5}\right) = 5 \cdot 7$$

$$5(q+2) + 3(q-5) = 35$$
$$5q + 10 + 3q - 15 = 35$$
$$8q - 5 = 35$$
$$8q = 40$$
$$q = 5$$

Check $q = 5$: $\frac{7}{3} = \frac{7}{3}$ *True*

Thus, the solution set is $\{5\}$.

37.
$$\frac{x}{2} = \frac{5}{4} + \frac{x-1}{4}$$

$$4\left(\frac{x}{2}\right) = 4\left(\frac{5}{4} + \frac{x-1}{4}\right) \quad \text{\textit{Multiply by LCD, 4.}}$$

$$2(x) = 4\left(\frac{5}{4}\right) + 4\left(\frac{x-1}{4}\right)$$

$$2x = 5 + x - 1$$
$$x = 4$$

Check $x = 4$: $2 = 2$ *True*

Thus, the solution set is $\{4\}$.

39.
$$x + \frac{17}{2} = \frac{x}{2} + x + 6$$

$$2\left(x + \frac{17}{2}\right) = 2\left(\frac{x}{2} + x + 6\right) \quad \text{\textit{Multiply by LCD, 2.}}$$

$$2(x) + 2\left(\frac{17}{2}\right) = 2\left(\frac{x}{2}\right) + 2(x) + 2(6)$$

$$2x + 17 = x + 2x + 12$$
$$5 = x$$

Check $x = 5$: $\frac{27}{2} = \frac{27}{2}$ *True*

Thus, the solution set is $\{5\}$.

41. $\dfrac{9}{3x+4} = \dfrac{36-27x}{16-9x^2}$

$\dfrac{9}{3x+4} = \dfrac{9(4-3x)}{(4+3x)(4-3x)}$ *Factor.*

$\dfrac{9}{3x+4} = \dfrac{9}{4+3x}$ *Reduce.*

                       $x \ne 4/3$

This is an identity, which is true for all real numbers for which the original equation is defined. We must exclude $\pm \frac{4}{3}$. Thus, the solution set is $\left\{ x \mid x \ne \pm \frac{4}{3} \right\}$.

43. $\dfrac{a+7}{8} - \dfrac{a-2}{3} = \dfrac{4}{3}$

$24\left(\dfrac{a+7}{8} - \dfrac{a-2}{3} \right) = 24\left(\dfrac{4}{3} \right)$

                     Multiply by LCD, 24.

$24\left(\dfrac{a+7}{8} \right) - 24\left(\dfrac{a-2}{3} \right) = 8(4)$

$3(a+7) - 8(a-2) = 32$

$3a + 21 - 8a + 16 = 32$

$-5a + 37 = 32$

$-5a = -5$

$a = 1$

Check $a = 1$: $\frac{4}{3} = \frac{4}{3}$ *True*

Thus, the solution set is $\{1\}$.

45. $\dfrac{p}{2} - \dfrac{p-1}{4} = \dfrac{5}{4}$

$4\left(\dfrac{p}{2} - \dfrac{p-1}{4} \right) = 4\left(\dfrac{5}{4} \right)$ *Multiply by LCD, 4.*

$4\left(\dfrac{p}{2} \right) - 4\left(\dfrac{p-1}{4} \right) = 5$

$2p - 1(p-1) = 5$

$2p - p + 1 = 5$

$p = 4$

Check $p = 4$: $\frac{5}{4} = \frac{5}{4}$ *True*

Thus, the solution set is $\{4\}$.

47. $\dfrac{3x}{5} - \dfrac{x-5}{7} = 3$

$35\left(\dfrac{3x}{5} - \dfrac{x-5}{7} \right) = 35(3)$ *Multiply by LCD, 35.*

$35\left(\dfrac{3x}{5} \right) - 35\left(\dfrac{x-5}{7} \right) = 105$

$7(3x) - 5(x-5) = 105$

$21x - 5x + 25 = 105$

$16x = 80$

$x = 5$

Check $x = 5$: $3 = 3$ *True*

Thus, the solution set is $\{5\}$.

49. $\dfrac{4}{x^2 - 3x} = \dfrac{1}{x^2 - 9}$

$\dfrac{4}{x(x-3)} = \dfrac{1}{(x+3)(x-3)}$ *Factor denominators.*

$x(x+3)(x-3) \cdot \dfrac{4}{x(x-3)}$

$= x(x+3)(x-3) \cdot \dfrac{1}{(x+3)(x-3)}$

           Multiply by LCD, x(x + 3)(x − 3).

$4(x+3) = x \cdot 1$

$4x + 12 = x$

$3x = -12$

$x = -4$

Check $x = -4$: $\frac{1}{7} = \frac{1}{7}$ *True*

Thus, the solution set is $\{-4\}$.

51. $\dfrac{2}{m} = \dfrac{m}{5m+12}$

$m(5m+12)\left(\dfrac{2}{m} \right) = m(5m+12)\left(\dfrac{m}{5m+12} \right)$

           Multiply by LCD, m(5m + 12).

$(5m+12)(2) = m(m)$

$10m + 24 = m^2$

$-m^2 + 10m + 24 = 0$

$m^2 - 10m - 24 = 0$ *Multiply by −1.*

$(m-12)(m+2) = 0$

$m - 12 = 0$ or $m + 2 = 0$

$m = 12$ or $m = -2$

Check $m = 12$: $\frac{1}{6} = \frac{1}{6}$ *True*

Check $m = -2$: $-1 = -1$ *True*

Thus, the solution set is $\{-2, 12\}$.

53. $\dfrac{-2}{z+5} + \dfrac{3}{z-5} = \dfrac{20}{z^2 - 25}$

$\dfrac{-2}{z+5} + \dfrac{3}{z-5} = \dfrac{20}{(z+5)(z-5)}$

$(z+5)(z-5)\left(\dfrac{-2}{z+5} + \dfrac{3}{z-5} \right)$

$= (z+5)(z-5)\left(\dfrac{20}{(z+5)(z-5)} \right)$

          Multiply by LCD, (z + 5)(z − 5).

$(z+5)(z-5)\left(\dfrac{-2}{z+5} \right)$

$+ (z+5)(z-5)\left(\dfrac{3}{z-5} \right) = 20$

$-2(z-5) + 3(z+5) = 20$

$-2z + 10 + 3z + 15 = 20$

$z + 25 = 20$

$z = -5$

The proposed solution, −5, cannot be a solution because it would make the denominators $z + 5$

and $z^2 - 25$ equal 0 and the corresponding fractions undefined. Since -5 cannot be a solution, the solution set is $\emptyset$.

55.

$$\frac{3}{x-1} + \frac{2}{4x-4} = \frac{7}{4}$$

$$\frac{3}{x-1} + \frac{2}{4(x-1)} = \frac{7}{4}$$

$$4(x-1)\left(\frac{3}{x-1} + \frac{2}{4(x-1)}\right) = 4(x-1)\left(\frac{7}{4}\right)$$

Multiply by LCD, 4(x − 1).

$$4(3) + 2 = (x-1)(7)$$
$$14 = 7x - 7$$
$$21 = 7x$$
$$3 = x$$

Check $x = 3$: $\frac{7}{4} = \frac{7}{4}$ *True*

Thus, the solution set is $\{3\}$.

57.

$$\frac{x}{3x+3} = \frac{2x-3}{x+1} - \frac{2x}{3x+3}$$

$$\frac{x}{3(x+1)} = \frac{2x-3}{x+1} - \frac{2x}{3(x+1)}$$

$$3(x+1)\left(\frac{x}{3(x+1)}\right) =$$

$$3(x+1)\left[\frac{2x-3}{x+1} - \frac{2x}{3(x+1)}\right]$$

Multiply by LCD, 3(x + 1).

$$x = 3(x+1)\left(\frac{2x-3}{x+1}\right)$$

$$- 3(x+1)\left(\frac{2x}{3(x+1)}\right)$$

$$x = 3(2x-3) - 2x$$
$$x = 6x - 9 - 2x$$
$$x = 4x - 9$$
$$-3x = -9$$
$$x = 3$$

Check $x = 3$: $\frac{1}{4} = \frac{1}{4}$ *True*

Thus, the solution set is $\{3\}$.

59.

$$\frac{2p}{p^2-1} = \frac{2}{p+1} - \frac{1}{p-1}$$

$$\frac{2p}{(p+1)(p-1)} = \frac{2}{p+1} - \frac{1}{p-1}$$

$$(p+1)(p-1)\left[\frac{2p}{(p+1)(p-1)}\right]$$

$$= (p+1)(p-1)\left(\frac{2}{p+1}\right)$$

$$- (p+1)(p-1)\left(\frac{1}{p-1}\right)$$

Multiply by LCD, (p + 1)(p − 1).

$$2p = 2(p-1) - 1(p+1)$$
$$2p = 2p - 2 - p - 1$$
$$p = -3$$

Check $p = -3$: $-\frac{3}{4} = -1 + \frac{1}{4}$ *True*

Thus, the solution set is $\{-3\}$.

61.

$$\frac{5x}{14x+3} = \frac{1}{x}$$

$$x(14x+3)\left(\frac{5x}{14x+3}\right) = x(14x+3)\left(\frac{1}{x}\right)$$

Multiply by LCD, x(14x + 3).

$$x(5x) = (14x+3)(1)$$
$$5x^2 = 14x + 3$$
$$5x^2 - 14x - 3 = 0$$
$$(5x+1)(x-3) = 0$$

Note to reader: We may skip writing out the zero-factor property since this step can be easily performed mentally.

$$x = -\frac{1}{5} \quad \text{or} \quad x = 3$$

Check $x = -\frac{1}{5}$: $-5 = -5$ *True*
Check $x = 3$: $\frac{1}{3} = \frac{1}{3}$ *True*

Thus, the solution set is $\left\{-\frac{1}{5}, 3\right\}$.

63.

$$\frac{2}{x-1} - \frac{2}{3} = \frac{-1}{x+1}$$

$$3(x-1)(x+1)\left(\frac{2}{x-1} - \frac{2}{3}\right)$$

$$= 3(x-1)(x+1)\left(\frac{-1}{x+1}\right)$$

Multiply by LCD, 3(x − 1)(x + 1).

$$3(x-1)(x+1)\left(\frac{2}{x-1}\right)$$

$$- 3(x-1)(x+1)\left(\frac{2}{3}\right)$$

$$= 3(x-1)(x+1)\left(\frac{-1}{x+1}\right)$$

$$3(x+1)(2) - (x-1)(x+1)(2)$$
$$= 3(x-1)(-1)$$
$$6(x+1) - 2(x^2 - 1) = -3(x-1)$$
$$6x + 6 - 2x^2 + 2 = -3x + 3$$
$$-2x^2 + 9x + 5 = 0$$
$$2x^2 - 9x - 5 = 0$$
$$(2x+1)(x-5) = 0$$

$$x = -\frac{1}{2} \quad \text{or} \quad x = 5$$

Check $x = -\frac{1}{2}$: $-2 = -2$ *True*
Check $x = 5$: $-\frac{1}{6} = -\frac{1}{6}$ *True*

Thus, the solution set is $\left\{-\frac{1}{2}, 5\right\}$.

65.
$$\frac{x}{2x+2} = \frac{-2x}{4x+4} + \frac{2x-3}{x+1}$$

$$\frac{x}{2(x+1)} = \frac{-2x}{4(x+1)} + \frac{2x-3}{x+1}$$

$$4(x+1)\left(\frac{x}{2(x+1)}\right) = 4(x+1)\left(\frac{-2x}{4(x+1)}\right)$$
$$+ 4(x+1)\left(\frac{2x-3}{x+1}\right)$$

Multiply by LCD, 4(x + 1).

$$2(x) = -2x + 4(2x-3)$$
$$2x = -2x + 8x - 12$$
$$-4x = -12$$
$$x = 3$$

Check $x = 3$: $\frac{3}{8} = \frac{3}{8}$ *True*

Thus, the solution set is $\{3\}$.

67.
$$\frac{8x+3}{x} = 3x$$

$$x\left(\frac{8x+3}{x}\right) = x(3x) \quad \begin{array}{l}\textit{Multiply by} \\ \textit{LCD, x.}\end{array}$$

$$8x + 3 = 3x^2$$
$$0 = 3x^2 - 8x - 3$$
$$0 = (3x+1)(x-3)$$
$$x = -\tfrac{1}{3} \quad \text{or} \quad x = 3$$

Check $x = -\frac{1}{3}$: $-1 = -1$ *True*
Check $x = 3$: $9 = 9$ *True*

Thus, the solution set is $\left\{-\frac{1}{3}, 3\right\}$.

69.
$$\frac{1}{x+4} + \frac{x}{x-4} = \frac{-8}{x^2-16}$$

$$(x+4)(x-4)\left(\frac{1}{x+4}\right) + (x+4)(x-4)\left(\frac{x}{x-4}\right)$$
$$= (x+4)(x-4)\left(\frac{-8}{x^2-16}\right)$$

Multiply by LCD, (x + 4)(x − 4).

$$1(x-4) + x(x+4) = -8$$
$$x - 4 + x^2 + 4x = -8$$
$$x^2 + 5x + 4 = 0$$
$$(x+4)(x+1) = 0$$
$$x = -4 \quad \text{or} \quad x = -1$$

$x = -4$ cannot be a solution because it would make the denominators $x + 4$ and $x^2 - 16$ equal 0 and the corresponding fractions undefined.

Check $x = -1$: $\frac{1}{3} + \frac{1}{5} = \frac{8}{15}$ *True*

Thus, the solution set is $\{-1\}$.

71.
$$\frac{4}{3x+6} - \frac{3}{x+3} = \frac{8}{x^2+5x+6}$$

$$\frac{4}{3(x+2)} - \frac{3}{x+3} = \frac{8}{(x+2)(x+3)}$$

$$3(x+2)(x+3) \cdot \frac{4}{3(x+2)} - 3(x+2)(x+3) \cdot \frac{3}{x+3}$$

$$= 3(x+2)(x+3) \cdot \frac{8}{(x+2)(x+3)}$$

Multiply by LCD, 3(x + 2)(x + 3).

$$4(x+3) - 3(x+2)(3) = 3(8)$$
$$4x + 12 - 9x - 18 = 24$$
$$-5x = 30$$
$$x = -6$$

Check $x = -6$: $-\frac{1}{3} - (-1) = \frac{2}{3}$ *True*

Thus, the solution set is $\{-6\}$.

73.
$$\frac{3x}{x^2+5x+6}$$

$$= \frac{5x}{x^2+2x-3} - \frac{2}{x^2+x-2}$$

$$\frac{3x}{(x+2)(x+3)}$$

$$= \frac{5x}{(x+3)(x-1)} - \frac{2}{(x-1)(x+2)}$$

$$(x+2)(x+3)(x-1) \cdot \left[\frac{3x}{(x+2)(x+3)}\right]$$

$$= (x+2)(x+3)(x-1) \cdot \left[\frac{5x}{(x+3)(x-1)}\right]$$

$$- (x+2)(x+3)(x-1) \cdot \left[\frac{2}{(x-1)(x+2)}\right]$$

Multiply by LCD, (x + 2)(x + 3)(x − 1).

$$3x(x-1) = 5x(x+2) - 2(x+3)$$
$$3x^2 - 3x = 5x^2 + 10x - 2x - 6$$
$$0 = 2x^2 + 11x - 6$$
$$0 = (2x-1)(x+6)$$
$$x = \tfrac{1}{2} \quad \text{or} \quad x = -6$$

Check $x = \frac{1}{2}$: $\frac{6}{35} = -\frac{10}{7} - \left(-\frac{8}{5}\right)$ *True*
Check $x = -6$: $-\frac{3}{2} = -\frac{10}{7} - \frac{1}{14}$ *True*

Thus, the solution set is $\left\{-6, \frac{1}{2}\right\}$.

75.
$$\frac{x+4}{x^2-3x+2} - \frac{5}{x^2-4x+3}$$

$$= \frac{x-4}{x^2-5x+6}$$

$$\frac{x+4}{(x-2)(x-1)} - \frac{5}{(x-3)(x-1)}$$

$$= \frac{x-4}{(x-3)(x-2)}$$

$(x-2)(x-1)(x-3)$

$$\cdot \left[\frac{x+4}{(x-2)(x-1)} - \frac{5}{(x-3)(x-1)} \right]$$

$$= (x-2)(x-1)(x-3)\left[\frac{x-4}{(x-3)(x-2)} \right]$$

Multiply by LCD, $(x-2)(x-1)(x-3)$.

$(x+4)(x-3) - 5(x-2) = (x-1)(x-4)$

$x^2 + x - 12 - 5x + 10 = x^2 - 5x + 4$

$-4x - 2 = -5x + 4$

$x = 6$

Check $x = 6$: $\frac{1}{2} - \frac{1}{3} = \frac{1}{6}$ *True*

Thus, the solution set is $\{6\}$.

77. $kr - mr = km$

If you are solving for k, put both terms with k on one side and the remaining term on the other side.

$$kr - km = mr$$

79. $m = \dfrac{kF}{a}$ for F

We need to isolate F on one side of the equation.

$$m \cdot a = \left(\frac{kF}{a} \right)(a) \quad \text{Multiply by } a.$$

$$ma = kF$$

$$\frac{ma}{k} = \frac{kF}{k} \qquad \text{Divide by } k.$$

$$\frac{ma}{k} = F$$

81. $m = \dfrac{kF}{a}$ for a

$$m \cdot a = \left(\frac{kF}{a} \right)(a) \quad \text{Multiply by } a.$$

$$ma = kF$$

$$\frac{ma}{m} = \frac{kF}{m} \qquad \text{Divide by } m.$$

$$a = \frac{kF}{m}$$

83. $I = \dfrac{E}{R+r}$ for R

We need to isolate R on one side of the equation.

$$I(R+r) = \left(\frac{E}{R+r} \right)(R+r) \quad \begin{array}{l} \textit{Multiply} \\ \textit{by } R+r. \end{array}$$

$$IR + Ir = E \qquad \begin{array}{l} \textit{Distributive} \\ \textit{property} \end{array}$$

$$IR = E - Ir \qquad \textit{Subtract Ir.}$$

$$R = \frac{E - Ir}{I}, \qquad \textit{Divide by I.}$$

$$\text{or } R = \frac{E}{I} - r$$

85. $h = \dfrac{2A}{B+b}$ for A

$$(B+b)h = (B+b) \cdot \frac{2A}{B+b}$$

Multiply by $B + b$.

$$h(B+b) = 2A$$

$$\frac{h(B+b)}{2} = A \qquad \textit{Divide by 2.}$$

87. $d = \dfrac{2S}{n(a+L)}$ for a

We need to isolate a on one side of the equation.

$$d \cdot n(a+L) = \frac{2S}{n(a+L)} \cdot n(a+L)$$

Multiply by $n(a+L)$.

$$nd(a+L) = 2S$$

$$and + ndL = 2S$$

$$and = 2S - ndL \qquad \textit{Subtract ndL.}$$

$$a = \frac{2S - ndL}{nd}, \qquad \textit{Divide by nd.}$$

$$\text{or } a = \frac{2S}{nd} - L$$

89. $\dfrac{1}{x} = \dfrac{1}{y} - \dfrac{1}{z}$ for y

The LCD of all the fractions in the equation is xyz, so multiply both sides by xyz.

$$xyz\left(\frac{1}{x} \right) = xyz\left(\frac{1}{y} - \frac{1}{z} \right)$$

$$xyz\left(\frac{1}{x} \right) = xyz\left(\frac{1}{y} \right) - xyz\left(\frac{1}{z} \right)$$

Distributive property

$$yz = xz - xy$$

Since we are solving for y, get all terms with y on one side of the equation.

$$xy + yz = xz \quad \textit{Add xy.}$$

Factor out the common factor y on the left.

$$y(x+z) = xz$$

Finally, divide both sides by the coefficient of y, which is $x + z$.

$$y = \frac{xz}{x+z}$$

91. $\dfrac{2}{r} + \dfrac{3}{s} + \dfrac{1}{t} = 1$ for t

The LCD of all the fractions in the equation is rst, so multiply both sides by rst.

$$rst\left(\dfrac{2}{r} + \dfrac{3}{s} + \dfrac{1}{t}\right) = rst(1)$$

$$rst\left(\dfrac{2}{r}\right) + rst\left(\dfrac{3}{s}\right) + rst\left(\dfrac{1}{t}\right) = rst$$

$$2st + 3rt + rs = rst$$

Since we are solving for t, get all terms with t on one side of the equation.

$$2st + 3rt - rst = -rs$$

Factor out the common factor t on the left.

$$t(2s + 3r - rs) = -rs$$

Finally, divide both sides by the coefficient of t, which is $2s + 3r - rs$.

$$t = \dfrac{-rs}{2s + 3r - rs}, \quad \text{or} \quad t = \dfrac{rs}{-2s - 3r + rs}$$

93. $\qquad 9x + \dfrac{3}{z} = \dfrac{5}{y} \quad$ for z

$$yz\left(9x + \dfrac{3}{z}\right) = yz\left(\dfrac{5}{y}\right) \quad \begin{array}{l}\textit{Multiply by}\\ \textit{LCD, yz.}\end{array}$$

$$yz(9x) + yz\left(\dfrac{3}{z}\right) = yz\left(\dfrac{5}{y}\right) \quad \begin{array}{l}\textit{Distributive}\\ \textit{property}\end{array}$$

$$9xyz + 3y = 5z$$

$$9xyz - 5z = -3y \quad \begin{array}{l}\textit{Get the z terms}\\ \textit{on one side.}\end{array}$$

$$z(9xy - 5) = -3y \quad \textit{Factor out z.}$$

$$z = \dfrac{-3y}{9xy - 5}, \textit{Divide by 9xy - 5.}$$

$$\text{or} \quad z = \dfrac{3y}{5 - 9xy}$$

95. $\qquad \dfrac{t}{x-1} - \dfrac{2}{x+1} = \dfrac{1}{x^2-1} \quad$ for t

$$(x+1)(x-1)\left(\dfrac{t}{x-1} - \dfrac{2}{x+1}\right)$$

$$= (x+1)(x-1)\left(\dfrac{1}{x^2-1}\right)$$

$$\textit{Multiply by LCD, (x + 1)(x - 1).}$$

$$t(x+1) - 2(x-1) = 1$$

$$t(x+1) = 1 + 2(x-1)$$

$$t(x+1) = 1 + 2x - 2$$

$$t = \dfrac{2x-1}{x+1}, \text{or } t = \dfrac{-2x+1}{-x-1}$$

97. Using $d = rt$, we get

$$r = \dfrac{d}{t} = \dfrac{288}{t}.$$

His rate is $\dfrac{288}{t}$ mph.

99. Using $d = rt$, we get

$$t = \dfrac{d}{r} = \dfrac{289}{z}.$$

His time is $\dfrac{289}{z}$ hr.

Summary Exercises on Rational Expressions and Equations

1. No equals sign appears so this is an *expression*.

$$\dfrac{4}{p} + \dfrac{6}{p} = \dfrac{4+6}{p} = \dfrac{10}{p}$$

3. No equals sign appears so this is an *expression*.

$$\dfrac{1}{x^2+x-2} \div \dfrac{4x^2}{2x-2}$$

$$= \dfrac{1}{x^2+x-2} \cdot \dfrac{2x-2}{4x^2}$$

$$= \dfrac{1}{(x+2)(x-1)} \cdot \dfrac{2(x-1)}{2 \cdot 2x^2}$$

$$= \dfrac{1}{2x^2(x+2)}$$

5. No equals sign appears so this is an *expression*.

$$\dfrac{2y^2+y-6}{2y^2-9y+9} \cdot \dfrac{y^2-2y-3}{y^2-1}$$

$$= \dfrac{(2y-3)(y+2)(y-3)(y+1)}{(2y-3)(y-3)(y+1)(y-1)}$$

$$= \dfrac{y+2}{y-1}$$

7. $\dfrac{x-4}{5} = \dfrac{x+3}{6}$

There is an equals sign, so this is an *equation*.

$$30\left(\dfrac{x-4}{5}\right) = 30\left(\dfrac{x+3}{6}\right) \quad \begin{array}{l}\textit{Multiply by}\\ \textit{LCD, 30.}\end{array}$$

$$6(x-4) = 5(x+3)$$

$$6x - 24 = 5x + 15$$

$$x = 39$$

Check $x = 39$: $7 = 7$ *True*

The solution is set is $\{39\}$.

9. No equals sign appears so this is an *expression*.

$$\dfrac{4}{p+2} + \dfrac{1}{3p+6}$$

$$= \dfrac{4}{p+2} + \dfrac{1}{3(p+2)}$$

$$= \dfrac{3 \cdot 4}{3(p+2)} + \dfrac{1}{3(p+2)} \quad LCD = 3(p+2)$$

$$= \dfrac{12+1}{3(p+2)}$$

$$= \dfrac{13}{3(p+2)}$$

11. $\dfrac{3}{t-1} + \dfrac{1}{t} = \dfrac{7}{2}$

There is an equals sign, so this is an *equation*.

$$2t(t-1)\left(\dfrac{3}{t-1} + \dfrac{1}{t}\right) = 2t(t-1)\left(\dfrac{7}{2}\right)$$

Multiply by LCD, 2t(t – 1).

$$2t(t-1)\left(\dfrac{3}{t-1}\right) + 2t(t-1)\left(\dfrac{1}{t}\right) = 7t(t-1)$$

$$2t(3) + 2(t-1) = 7t(t-1)$$

$$6t + 2t - 2 = 7t^2 - 7t$$

$$0 = 7t^2 - 15t + 2$$

$$0 = (7t-1)(t-2)$$

$$t = \tfrac{1}{7} \quad \text{or} \quad t = 2$$

Check $t = \tfrac{1}{7}$: $-\tfrac{7}{2} + 7 = \tfrac{7}{2}$ *True*

Check $t = 2$: $3 + \tfrac{1}{2} = \tfrac{7}{2}$ *True*

The solution set is $\left\{\tfrac{1}{7}, 2\right\}$.

13. No equals sign appears so this is an *expression*.

$$\dfrac{5}{4z} - \dfrac{2}{3z} = \dfrac{3 \cdot 5}{3 \cdot 4z} - \dfrac{4 \cdot 2}{4 \cdot 3z} \quad LCD = 12z$$

$$= \dfrac{15}{12z} - \dfrac{8}{12z}$$

$$= \dfrac{15 - 8}{12z} = \dfrac{7}{12z}$$

15. No equals sign appears so this is an *expression*.

$$\dfrac{1}{m^2 + 5m + 6} + \dfrac{2}{m^2 + 4m + 3}$$

$$= \dfrac{1}{(m+2)(m+3)} + \dfrac{2}{(m+1)(m+3)}$$

$$= \dfrac{1(m+1)}{(m+2)(m+3)(m+1)}$$

$$+ \dfrac{2(m+2)}{(m+1)(m+3)(m+2)}$$

$$LCD = (m+1)(m+2)(m+3)$$

$$= \dfrac{(m+1) + (2m+4)}{(m+1)(m+2)(m+3)}$$

$$= \dfrac{3m+5}{(m+1)(m+2)(m+3)}$$

17. $\dfrac{2}{x+1} + \dfrac{5}{x-1} = \dfrac{10}{x^2 - 1}$

There is an equals sign, so this is an *equation*.

$$\dfrac{2}{x+1} + \dfrac{5}{x-1} = \dfrac{10}{(x+1)(x-1)}$$

$$(x+1)(x-1)\left(\dfrac{2}{x+1} + \dfrac{5}{x-1}\right)$$

$$= (x+1)(x-1)\left[\dfrac{10}{(x+1)(x-1)}\right]$$

Multiply by LCD, (x + 1)(x – 1).

$$(x+1)(x-1)\left(\dfrac{2}{x+1}\right)$$

$$+ (x+1)(x-1)\left(\dfrac{5}{x-1}\right) = 10$$

Distributive property

$$2(x-1) + 5(x+1) = 10$$

$$2x - 2 + 5x + 5 = 10$$

$$3 + 7x = 10$$

$$7x = 7$$

$$x = 1$$

Replacing x by 1 in the original equation makes the denominators $x - 1$ and $x^2 - 1$ equal to 0, so the solution set is $\emptyset$.

19. No equals sign appears so this is an *expression*.

$$\dfrac{4t^2 - t}{6t^2 + 10t} \div \dfrac{8t^2 + 2t - 1}{3t^2 + 11t + 10}$$

$$= \dfrac{4t^2 - t}{6t^2 + 10t} \cdot \dfrac{3t^2 + 11t + 10}{8t^2 + 2t - 1}$$

Multiply by reciprocal.

$$= \dfrac{t(4t-1)}{2t(3t+5)} \cdot \dfrac{(3t+5)(t+2)}{(4t-1)(2t+1)}$$

Factor numerators and denominators.

$$= \dfrac{t+2}{2(2t+1)}$$

7.7 Applications of Rational Expressions

7.7 Now Try Exercises

N1. *Step 2*

Let $x =$ the denominator of the original fraction, so that the numerator is $x - 4$.

Step 3

If 7 is added to both the numerator and denominator, the resulting fraction is equivalent to $\tfrac{7}{8}$ translates to

$$\dfrac{(x-4) + 7}{x+7} = \dfrac{7}{8}.$$

Step 4

Multiply by the LCD, $8(x+7)$.

$$8(x+7)\dfrac{(x-4)+7}{x+7} = 8(x+7)\dfrac{7}{8}$$

$$8(x+3) = 7(x+7)$$

$$8x + 24 = 7x + 49$$

$$x = 25$$

Step 5

The denominator is 25 and the numerator is $25 - 4 = 21$, so the original fraction is $\tfrac{21}{25}$.

Step 6

If 7 is added to the numerator and the denominator of $\tfrac{21}{25}$, the result is $\tfrac{28}{32}$, which is equal to $\tfrac{7}{8}$.

N2. Let x = the rate of the boat with no current. Complete a table.

	d	r	t
With the Current	12	$x + 2$	$\dfrac{12}{x + 2}$
Against the Current	4	$x - 2$	$\dfrac{4}{x - 2}$

Since the times are equal, we get the following equation.

$$\frac{12}{x + 2} = \frac{4}{x - 2}$$

Multiply by the LCD, $(x + 2)(x - 2)$.

$$12(x - 2) = 4(x + 2)$$
$$12x - 24 = 4x + 8$$
$$8x - 24 = 8$$
$$8x = 32$$
$$x = 4$$

The rate of the boat with no current is 4 miles per hour.

N3.

	Rate	Time Working Together	Fractional Part of the Job Done when Working Together
Sarah	$\frac{1}{10}$	x	$\frac{1}{10}x$
Joyce	$\frac{1}{12}$	x	$\frac{1}{12}x$

Since together Sarah and Joyce complete 1 whole job, we must add their individual parts and set the sum equal to 1.

$$\frac{1}{10}x + \frac{1}{12}x = 1$$
$$60\left(\frac{1}{10}x + \frac{1}{12}x\right) = 60(1) \quad \text{\textit{Multiply by the LCD, 60.}}$$
$$60\left(\frac{1}{10}x\right) + 60\left(\frac{1}{12}x\right) = 60$$
$$6x + 5x = 60$$
$$11x = 60$$
$$x = \frac{60}{11}, \quad \text{or} \quad 5\frac{5}{11}$$

Working together, Sarah and Joyce can proofread the manuscript in $5\frac{5}{11}$ hours.

7.7   Section Exercises

1. (a) Let x = <u>the amount</u>.

(b) An expression for "the numerator of the fraction $\frac{5}{6}$ is increased by an amount" is <u>$5 + x$</u>. We could also use $\dfrac{5 + x}{6}$.

(c) An equation that can be used to solve the problem is

$$\frac{5 + x}{6} = \frac{13}{3}.$$

3. *Step 2*

Let x = the numerator of the original fraction. Then $x + 6$ = the denominator of the original fraction.

Step 3

If 3 is added to both the numerator and denominator, the resulting fraction is equivalent to $\frac{5}{7}$ translates to

$$\frac{x + 3}{(x + 6) + 3} = \frac{5}{7}.$$

Step 4

Since we have a fraction equal to another fraction, we can use cross multiplication.

$$7(x + 3) = 5[(x + 6) + 3]$$
$$7x + 21 = 5x + 45$$
$$2x = 24$$
$$x = 12$$

Step 5

The original fraction is

$$\frac{x}{x + 6} = \frac{12}{12 + 6} = \frac{12}{18}.$$

Step 6

Adding 3 to both the numerator and the denominator gives us

$$\frac{12 + 3}{18 + 3} = \frac{15}{21},$$

which is equivalent to $\frac{5}{7}$.

5. *Step 2*

Let x = the denominator of the original fraction. Then $4x$ = the numerator of the original fraction.

Step 3

If 6 is added to both the numerator and the denominator, the resulting fraction is equivalent to 2 translates to

$$\frac{4x + 6}{x + 6} = 2.$$

Step 4
$$4x + 6 = 2(x + 6)$$
$$4x + 6 = 2x + 12$$
$$2x = 6$$
$$x = 3$$

Step 5

The original fraction is

$$\frac{4x}{x} = \frac{4(3)}{3} = \frac{12}{3}.$$

Step 6
Adding 6 to both the numerator and the denominator gives us

$$\frac{12+6}{3+6} = \frac{18}{9} = 2.$$

7. *Step 2*
Let x = the number.

Step 3
One-third of a number is 2 greater than one-sixth of the same number translates to

$$\tfrac{1}{3}x = 2 + \tfrac{1}{6}x.$$

Step 4
Multiply both sides by the LCD, 6.

$$6(\tfrac{1}{3}x) = 6(2 + \tfrac{1}{6}x)$$
$$2x = 12 + x$$
$$x = 12$$

Step 5   The number is 12.

Step 6
One-third of 12 is 4 and one-sixth of 12 is 2. So $x = 12$ checks since 4 is 2 more than 2.

9. *Step 2*
Let x = the quantity.
Then $\tfrac{2}{3}$ of it, $\tfrac{1}{2}$ of it, and $\tfrac{1}{7}$ of it are

$$\tfrac{2}{3}x, \tfrac{1}{2}x, \text{ and } \tfrac{1}{7}x.$$

Step 3
Added together equals 33 translates to

$$x + \tfrac{2}{3}x + \tfrac{1}{2}x + \tfrac{1}{7}x = 33.$$

Step 4
Multiply both sides by the LCD of 3, 2, and 7, which is 42.

$$42(x + \tfrac{2}{3}x + \tfrac{1}{2}x + \tfrac{1}{7}x) = 42(33)$$
$$42x + 42(\tfrac{2}{3}x) + 42(\tfrac{1}{2}x) + 42(\tfrac{1}{7}x) = 42(33)$$
$$42x + 28x + 21x + 6x = 1386$$
$$97x = 1386$$
$$x = \tfrac{1386}{97}$$

Step 5
The quantity is $\tfrac{1386}{97}$. (Note that this fraction is already in lowest terms since 97 is a prime number and is not a factor of 1386.)

Step 6
Check $\tfrac{1386}{97}$ in the original problem.

$$x = \tfrac{1386}{97}, \tfrac{2}{3}x = \tfrac{924}{97}, \tfrac{1}{2}x = \tfrac{693}{97}, \tfrac{1}{7}x = \tfrac{198}{97}$$

Adding gives us

$$\frac{1386 + 924 + 693 + 198}{97} = \frac{3201}{97} = 33,$$

as desired.

11. We are asked to find the *time*, so we'll use the distance, rate, and time relationship $t = d/r$.

$$t = \frac{d}{r} = \frac{0.6 \text{ miles}}{0.0319 \text{ miles per minute}}$$
$$\approx 18.809 \text{ minutes}$$

13. We are asked to find the average *rate*, so we'll use the distance, rate, and time relationship $r = d/t$.

$$r = \frac{d}{t} = \frac{5000 \text{ meters}}{15.911 \text{ minutes}}$$
$$\approx 314.248 \text{ meters per minute}$$

15. We are asked to find the *time*, so we'll use the distance, rate, and time relationship $t = d/r$.

$$t = \frac{d}{r} = \frac{500 \text{ miles}}{152.672 \text{ miles per hour}}$$
$$\approx 3.275 \text{ hours}$$

17. Use time $= \frac{\text{distance}}{\text{rate}}$. Since we know that the times for Stephanie and Wally are the same,

$$\text{time}_{\text{Stephanie}} = \text{time}_{\text{Wally}}$$

$$\text{or} \qquad \frac{D}{R} = \frac{d}{r}$$

19. Let x = rate of the plane in still air. Then the rate against the wind is $x - 10$ and the rate with the wind is $x + 10$. The time flying against the wind is

$$t = \frac{d}{r} = \frac{500}{x - 10},$$

and the time flying with the wind is

$$t = \frac{d}{r} = \frac{600}{x + 10}.$$

Now complete the chart.

	d	r	t
Against the Wind	500	$x - 10$	$\dfrac{500}{x - 10}$
With the Wind	600	$x + 10$	$\dfrac{600}{x + 10}$

Since the problem states that the two times are equal, we have

$$\frac{500}{x - 10} = \frac{600}{x + 10}.$$

We would use this equation to solve the problem.

21. Let x represent the rate of the boat in still water. Then $x - 4$ is the rate against the current and $x + 4$ is the rate with the current. We fill in the chart as follows, realizing that the time column is filled in by using the formula $t = d/r$.

	d	r	t
Against the Current	20	$x - 4$	$\dfrac{20}{x - 4}$
With the Current	60	$x + 4$	$\dfrac{60}{x + 4}$

Since the times are equal, we get the following equation.

$$\frac{20}{x - 4} = \frac{60}{x + 4}$$

$$(x + 4)(x - 4)\frac{20}{x - 4} = (x + 4)(x - 4)\frac{60}{x + 4}$$

Multiply by LCD, (x + 4)(x − 4)

$$20(x + 4) = 60(x - 4)$$
$$20x + 80 = 60x - 240$$
$$320 = 40x$$
$$8 = x$$

The rate of the boat in still water is 8 miles per hour.

23. Let x = rate of the bird in still air. Then the rate against the wind is $x - 8$ and the rate with the wind is $x + 8$. Use $t = d/r$ to complete the chart.

	d	r	t
Against the Wind	18	$x - 8$	$\dfrac{18}{x - 8}$
With the Wind	30	$x + 8$	$\dfrac{30}{x + 8}$

Since the problem states that the two times are equal, we get the following equation.

$$\frac{18}{x - 8} = \frac{30}{x + 8}$$

$$(x + 8)(x - 8)\frac{18}{x - 8} = (x + 8)(x - 8)\frac{30}{x + 8}$$

Multiply by LCD, (x + 8)(x − 8)

$$18(x + 8) = 30(x - 8)$$
$$18x + 144 = 30x - 240$$
$$384 = 12x$$
$$32 = x$$

The rate of the bird in still air is 32 miles per hour.

25. Let x = rate of the plane in still air. Then the rate against the wind is $x - 15$ and the rate with the wind is $x + 15$. Use $t = d/r$ to complete the chart.

	d	r	t
Against the Wind	375	$x - 15$	$\dfrac{375}{x - 15}$
With the Wind	450	$x + 15$	$\dfrac{450}{x + 15}$

Since the problem states that the two times are equal, we get the following equation.

$$\frac{375}{x - 15} = \frac{450}{x + 15}$$

$$(x + 15)(x - 15)\frac{375}{x - 15} = (x + 15)(x - 15)\frac{450}{x + 15}$$

Multiply by LCD, (x + 15)(x − 15)

$$375(x + 15) = 450(x - 15)$$
$$375x + 5625 = 450x - 6750$$
$$12{,}375 = 75x$$
$$165 = x$$

The rate of the plane in still air is 165 miles per hour.

27. Let x represent the rate of the current of the river. Then $12 - x$ is the rate upstream (against the current) and $12 + x$ is the rate downstream (with the current). Use $t = d/r$ to complete the table.

	d	r	t
Upstream	6	$12 - x$	$\dfrac{6}{12 - x}$
Downstream	10	$12 + x$	$\dfrac{10}{12 + x}$

Since the times are equal, we get the following equation.

$$\frac{6}{12 - x} = \frac{10}{12 + x}$$

$$(12 + x)(12 - x)\frac{6}{12 - x} = (12 + x)(12 - x)\frac{10}{12 + x}$$

Multiply by LCD, (12 + x)(12 − x)

$$6(12 + x) = 10(12 - x)$$
$$72 + 6x = 120 - 10x$$
$$16x = 48$$
$$x = 3$$

The rate of the current of the river is 3 miles per hour.

29. Let x = the average rate of the ferry.

Use the formula $t = d/r$ to make a chart.

	d	r	t
Seattle-Victoria	148	x	$\dfrac{148}{x}$
Victoria-Vancouver	74	x	$\dfrac{74}{x}$

Since the time for the Victoria-Vancouver trip is 4 hours less than the time for the Seattle-Victoria trip, solve the equation

$$\frac{74}{x} = \frac{148}{x} - 4.$$

$$x\left(\frac{74}{x}\right) = x\left(\frac{148}{x} - 4\right) \quad \begin{array}{l}\textit{Multiply by}\\ \textit{LCD, } x.\end{array}$$
$$74 = 148 - 4x$$
$$4x = 74$$
$$x = \frac{74}{4} = \frac{37}{2}, \text{ or } 18\tfrac{1}{2}$$

The average rate of the ferry is $18\frac{1}{2}$ miles per hour.

31. If it takes Elayn 10 hours to do a job, her rate is $\frac{1}{10}$ job per hour.

33. Let t = the number of hours it will take Edward and Abdalla to paint the room working together.

	Rate	Time Working Together	Fractional Part of the Job Done when Working Together
Edward	$\frac{1}{8}$	t	$\frac{1}{8}t$
Abdalla	$\frac{1}{6}$	t	$\frac{1}{6}t$

Working together, they complete 1 whole job, so add their individual fractional parts and set the sum equal to 1.

$$\begin{array}{ccccc}\text{part done} & & \text{part done} & & \text{1 whole} \\ \text{by Edward} & + & \text{by Abdalla} & = & \text{job} \\ \downarrow & \downarrow & \downarrow & \downarrow & \downarrow \\ \frac{1}{8}t & + & \frac{1}{6}t & = & 1 \end{array}$$

An equation that can be used to solve this problem is

$$\tfrac{1}{8}t + \tfrac{1}{6}t = 1.$$

Alternatively, we can compare the hourly rates of completion. In one hour, Edward will complete $\frac{1}{8}$ of the job, Abdalla will complete $\frac{1}{6}$ of the job, and together they will complete $\frac{1}{t}$ of the job. So another equation that can be used to solve this problem is

$$\tfrac{1}{8} + \tfrac{1}{6} = \tfrac{1}{t}.$$

35. Let x represent the number of hours it will take for Heather and Courtney to grade the tests working together. Since Heather can grade the test in 4 hours, her rate alone is $\frac{1}{4}$ job per hour. Also, since Courtney can do the job alone in 6 hours, her rate is $\frac{1}{6}$ job per hour.

	Rate	Time Working Together	Fractional Part of the Job Done when Working Together
Heather	$\frac{1}{4}$	x	$\frac{1}{4}x$
Courtney	$\frac{1}{6}$	x	$\frac{1}{6}x$

Since together Heather and Courtney complete 1 whole job, we must add their individual fractional parts and set the sum equal to 1.

$$\tfrac{1}{4}x + \tfrac{1}{6}x = 1$$
$$12(\tfrac{1}{4}x + \tfrac{1}{6}x) = 12(1) \quad LCD = 12$$
$$12(\tfrac{1}{4}x) + 12(\tfrac{1}{6}x) = 12$$
$$3x + 2x = 12$$
$$5x = 12$$
$$x = \tfrac{12}{5}, \text{ or } 2\tfrac{2}{5}$$

It will take Heather and Courtney $2\frac{2}{5}$ hours to grade the tests if they work together.

37. Let x = the number of hours to pump the water using both pumps.

	Rate	Time Working Together	Fractional Part of the Job Done when Working Together
Pump 1	$\frac{1}{10}$	x	$\frac{1}{10}x$
Pump 2	$\frac{1}{12}$	x	$\frac{1}{12}x$

Since together the two pumps complete 1 whole job, we must add their individual fractional parts and set the sum equal to 1.

$$\tfrac{1}{10}x + \tfrac{1}{12}x = 1$$
$$60(\tfrac{1}{10}x + \tfrac{1}{12}x) = 60(1) \quad LCD = 60$$
$$60(\tfrac{1}{10}x) + 60(\tfrac{1}{12}x) = 60$$
$$6x + 5x = 60$$
$$11x = 60$$
$$x = \tfrac{60}{11}, \text{ or } 5\tfrac{5}{11}$$

It would take $5\frac{5}{11}$ hours to pump out the basement if both pumps were used.

39. Let x represent the number of hours it will take the experienced employee to enter the data. Then $2x$ represents the number of hours it will take the new employee (the experienced employee takes less time). The experienced employee's rate is $\dfrac{1}{x}$ job per hour and the new employee's rate is $\dfrac{1}{2x}$ job per hour.

	Rate	Time Working Together	Fractional Part of the Job Done when Working Together
Experienced employee	$\dfrac{1}{x}$	2	$\dfrac{1}{x} \cdot 2 = \dfrac{2}{x}$
New employee	$\dfrac{1}{2x}$	2	$\dfrac{1}{2x} \cdot 2 = \dfrac{1}{x}$

Since together the two employees complete the whole job, we must add their individual fractional parts and set the sum equal to 1.

$$\frac{2}{x} + \frac{1}{x} = 1$$

$$x\left(\frac{2}{x} + \frac{1}{x}\right) = x(1) \qquad \text{\textit{Multiply by the LCD, x.}}$$

$$x\left(\frac{2}{x}\right) + x\left(\frac{1}{x}\right) = x$$

$$2 + 1 = x$$

$$3 = x$$

Working alone, it will take the experienced employee 3 hours to enter the data.

41. Let x = the number of hours to fill the pool $\frac{3}{4}$ full with both pipes working together.

	Rate	Time Working Together	Fractional Part of the Job Done when Working Together
First pipe	$\dfrac{1}{6}$	x	$\dfrac{1}{6}x$
Second pipe	$\dfrac{1}{9}$	x	$\dfrac{1}{9}x$

$$
\begin{array}{ccccc}
\text{Part done} & & \text{Part done by} & & \frac{3}{4} \text{ full} \\
\text{by first pipe} & + & \text{second pipe} & = & \\
\downarrow & \downarrow & \downarrow & \downarrow & \downarrow \\
\frac{1}{6}x & + & \frac{1}{9}x & = & \frac{3}{4}
\end{array}
$$

$$36\left(\tfrac{1}{6}x + \tfrac{1}{9}x\right) = 36\left(\tfrac{3}{4}\right) \quad \text{\textit{LCD = 36}}$$

$$36\left(\tfrac{1}{6}x\right) + 36\left(\tfrac{1}{9}x\right) = 36\left(\tfrac{3}{4}\right)$$

$$6x + 4x = 27$$

$$10x = 27$$

$$x = \tfrac{27}{10}, \text{ or } 2\tfrac{7}{10}$$

It takes $2\frac{7}{10}$ hours to fill the pool $\frac{3}{4}$ full using both pipes.

Alternatively, we could solve $\frac{1}{6}x + \frac{1}{9}x = 1$ (filling the whole pool) and then multiply that answer by $\frac{3}{4}$. In this case, x would represent the number of hours to fill the pool with both pipes working together.

43. Let x = the number of minutes it takes to fill the sink.

In 1 minute, the cold water faucet (alone) can fill $\frac{1}{12}$ of the sink. In the same time, the hot water faucet (alone) can fill $\frac{1}{15}$ of the sink. In 1 minute, the drain (alone) empties $\frac{1}{25}$ of the sink. Together, they fill $\frac{1}{x}$ of the sink in one minute, so solve the equation

$$\tfrac{1}{12} + \tfrac{1}{15} - \tfrac{1}{25} = \tfrac{1}{x}.$$

$$300x\left(\tfrac{1}{12} + \tfrac{1}{15} - \tfrac{1}{25}\right) = 300x\left(\tfrac{1}{x}\right)$$

$$\qquad \text{\textit{Multiply by the LCD, 300x.}}$$

$$25x + 20x - 12x = 300$$

$$33x = 300$$

$$x = \tfrac{300}{33} = \tfrac{100}{11}, \text{ or } 9\tfrac{1}{11}$$

It will take $9\frac{1}{11}$ minutes to fill the sink.

45. $200 = 15k$

$\quad k = \frac{200}{15} \quad$ *Divide by 15.*

$\quad k = \frac{40}{3} \quad$ *Reduce.*

The solution set is $\left\{ \frac{40}{3} \right\}$.

47. $180 = \dfrac{k}{20}$

$\quad k = 180 \cdot 20 \quad$ *Multiply by 20.*

$\quad k = 3600$

The solution set is $\{3600\}$.

49. $y = kx$

$\quad \dfrac{y}{x} = k \text{ or } k = \dfrac{y}{x} \quad$ *Divide by x.*

51. $y = \dfrac{k}{x}$

$\quad xy = k \text{ or } k = xy \quad$ *Multiply by x.*

7.8 Variation

7.8 Now Try Exercises

N1. $W = kr \quad$ *W varies directly as r.*

$\quad 40 = 5k \quad$ *Substitute W = 40 and r = 5.*

$\quad 8 = k \quad$ *Solve for k.*

Since $W = kr$ and $k = 8$,

$\quad W = 8r. \qquad$ *Substitute k = 8 in W = kr.*

$\quad W = 8(10) \quad$ *Substitute r = 10.*

$\quad W = 80 \qquad$ *Multiply.*

N2. $V = kr^2$ *Volume varies directly as the square of the radius.*

$80 = k(4)^2$ *Substitute V = 80 and r = 4.*

$5 = k$ *Solve for k.*

Since $V = kr^2$ and $k = 5$,

$$V = 5r^2.$$
$$V = 5(5)^2 \quad \text{Substitute } r = 5.$$
$$V = 125$$

The volume is 125 cubic feet when the radius is 5 feet.

N3. $t = \dfrac{k}{r}$ *t varies inversely as r.*

$12 = \dfrac{k}{3}$ *Substitute t = 12 and r = 3.*

$36 = k$ *Solve for k.*

Since $t = \dfrac{k}{r}$ and $k = 36$,

$$t = \frac{36}{r}. \quad \text{Substitute } k = 36 \text{ in } t = \frac{k}{r}.$$
$$t = \frac{36}{6} \quad \text{Substitute } r = 6.$$
$$t = 6 \quad \text{Reduce.}$$

N4. $h = \dfrac{k}{b}$ *Height varies inversely as base.*

$6 = \dfrac{k}{4}$ *Substitute h = 6 and b = 4.*

$24 = k$ *Solve for k.*

Since $h = \dfrac{k}{b}$ and $k = 24$,

$$h = \frac{24}{b}. \quad \text{Substitute } k = 24 \text{ in } h = \frac{k}{b}.$$
$$h = \frac{24}{12} \quad \text{Substitute } b = 12.$$
$$h = 2 \quad \text{Reduce.}$$

The height of the triangle is 2 feet when the base is 12 feet.

7.8 Section Exercises

1. As the number of candy bars you buy *increases*, the total price for the candy *increases*. Thus, the variation between the quantities is *direct*.

3. As the amount of pressure put on the accelerator of a truck *increases*, the rate of the truck *increases*. Thus, the variation between the quantities is *direct*.

5. As the number of days until the end of the baseball season *decreases*, the number of home runs that Evan Longoria has *increases*. Thus, the variation between the quantities is *inverse*.

7. As the number of days from now until December 25 *decreases*, the magnitude of the frenzy of Christmas shopping *increases*. Thus, the variation between the quantities is *inverse*.

9. $y = \dfrac{3}{x}$ represents *inverse* variation since it is of the form $y = \dfrac{k}{x}$.

11. $y = 10x^2$ represents *direct* variation since it is of the form $y = kx^n$.

13. $y = 50x$ represents *direct* variation since it is of the form $y = kx$.

15. $y = \dfrac{12}{x^2}$ represents *inverse* variation since it is of the form $y = \dfrac{k}{x^n}$.

17. **(a)** If the constant of variation is positive and y varies directly as x, then as x increases, y *increases*.

(b) If the constant of variation is positive and y varies inversely as x, then as x increases, y *decreases*.

19. Since x varies directly as y, there is a constant k such that $x = ky$. First find the value of k.

$$27 = k(6) \quad \text{Let } x = 27,\ y = 6.$$
$$k = \tfrac{27}{6} = \tfrac{9}{2}$$

When $k = \tfrac{9}{2}$, $x = ky$ becomes

$$x = \tfrac{9}{2}y.$$

Now find x when $y = 2$.

$$x = \tfrac{9}{2}(2) \quad \text{Let } y = 2.$$
$$= 9$$

21. Since d varies directly as t, there is a constant k such that $d = kt$. First find the value of k.

$$150 = k(3) \quad \text{Let } d = 150,\ t = 3.$$
$$k = \tfrac{150}{3} = 50$$

When $k = 50$, $d = kt$ becomes

$$d = 50t.$$

Now find d when $t = 5$.

$$d = 50(5) \quad \text{Let } t = 5.$$
$$= 250$$

23. Since x varies inversely as y, there is a constant k such that $x = \frac{k}{y}$. First find the value of k.

$$3 = \frac{k}{8} \quad \textit{Let x = 3, y = 8.}$$
$$k = 3(8) = 24$$

When $k = 24$, $x = \frac{k}{y}$ becomes

$$x = \frac{24}{y}.$$

Now find y when $x = 4$.

$$4 = \frac{24}{y} \quad \textit{Let x = 4.}$$
$$4y = 24$$
$$y = 6$$

25. Since p varies inversely as q, there is a constant k such that $p = \frac{k}{q}$. First find the value of k.

$$7 = \frac{k}{6} \quad \textit{Let p = 7, q = 6.}$$
$$k = 7(6) = 42$$

When $k = 42$, $p = \frac{k}{q}$ becomes

$$p = \frac{42}{q}.$$

Now find p when $q = 2$.

$$p = \frac{42}{2} = 21$$

27. Since m varies inversely as p^2, there is a constant k such that $m = \frac{k}{p^2}$. First find the value of k.

$$20 = \frac{k}{2^2} \quad \textit{Let m = 20, p = 2.}$$
$$k = 20(4) = 80$$

When $k = 80$, $m = \frac{k}{p^2}$ becomes

$$m = \frac{80}{p^2}.$$

Now find m when $p = 5$.

$$m = \frac{80}{5^2} \quad \textit{Let p = 5.}$$
$$m = \frac{80}{25} = \frac{16}{5}, \quad \text{or} \quad 3\frac{1}{5}$$

29. Since p varies inversely as q^2, there is a constant k such that $p = \frac{k}{q^2}$. First find the value of k.

$$4 = \frac{k}{(\frac{1}{2})^2} \quad \textit{Let p = 4, q = }\frac{1}{2}.$$
$$k = 4(\tfrac{1}{4}) = 1$$

When $k = 1$, $p = \frac{k}{q^2}$ becomes

$$p = \frac{1}{q^2}.$$

Now find p when $q = \frac{3}{2}$.

$$p = \frac{1}{(\frac{3}{2})^2} \quad \textit{Let q = }\frac{3}{2}.$$
$$p = \frac{1}{\frac{9}{4}} = 1 \cdot \frac{4}{9} = \frac{4}{9}$$

31. The interest I on an investment varies directly as the rate of interest r, so there is a constant k such that $I = kr$. Find the value of k.

$$48 = k(0.05) \quad \textit{Let I = 48, r = 5\% = 0.05.}$$
$$k = \frac{48}{0.05} = 960$$

When $k = 960$, $I = kr$ becomes

$$I = 960r.$$

Now find I when $r = 4.2\% = 0.042$.

$$I = 960(0.042) = 40.32$$

The interest on the investment when the rate is 4.2% is $40.32.

33. The distance d that a spring stretches varies directly with the force F applied, so

$$d = kF.$$
$$16 = k(75) \quad \textit{Let d = 16, F = 75.}$$
$$\tfrac{16}{75} = k$$

So $d = \frac{16}{75}F$ and when $F = 200$,

$$d = \tfrac{16}{75}(200) = \tfrac{16(8)}{3} = \tfrac{128}{3}.$$

A force of 200 pounds stretches the spring $42\frac{2}{3}$ inches.

35. The rate r varies inversely with time t, so there is a constant k such that $r = \frac{k}{t}$. Find the value of k.

$$160 = \frac{k}{\frac{1}{2}} \quad \textit{Let r = 160, t = }\frac{1}{2}.$$
$$k = \tfrac{1}{2} \cdot 160 = 80$$

When $k = 80$, $r = \frac{k}{t}$ becomes

$$r = \frac{80}{t}.$$

Now find r when $t = \frac{3}{4}$.

$$r = \frac{80}{\frac{3}{4}} = 80 \cdot \tfrac{4}{3} = \tfrac{320}{3}, \quad \text{or} \quad 106\tfrac{2}{3}$$

A rate of $106\frac{2}{3}$ miles per hour is needed to go the same distance in three-fourths of a minute.

37. The current c in a simple electrical circuit varies inversely as the resistance r, so there is a constant k such that $c = \frac{k}{r}$. Find the value of k.

$$20 = \frac{k}{5} \quad \textit{Let c = 20, r = 5.}$$
$$k = 5 \cdot 20 = 100$$

When $k = 100$, $c = \frac{k}{r}$ becomes

$$c = \frac{100}{r}.$$

Now find c when $r = 8$.

$$c = \frac{100}{8} = \frac{25}{2} = 12\frac{1}{2}$$

When the resistance is 8 ohms, the current is $12\frac{1}{2}$ amps.

39. The force F required to compress a spring varies directly as the change C in the length of the spring, so $F = kC$.

$$12 = k(3) \quad \textit{Let F = 12, C = 3.}$$
$$k = \frac{12}{3} = 4$$

So $F = 4C$ and when $C = 5$,

$$F = 4(5) = 20.$$

The force required to compress the spring 5 inches is 20 pounds.

41. The area A of a circle varies directly as the square of its radius r, so $A = kr^2$.

$$28.278 = k(3)^2 \quad \textit{Let A = 28.278, r = 3.}$$
$$k = \frac{28.278}{9} = 3.142$$

So $A = 3.142r^2$ and when $r = 4.1$,

$$A = 3.142(4.1)^2 = 52.81702.$$

With $k = 3.142$, the area of a circle with radius 4.1 inches is 52.817 square inches (to the nearest thousandth).

43. The amount of light A produced by a light source varies inversely as the square of the distance d from the source, so

$$A = \frac{k}{d^2}.$$

$$75 = \frac{k}{4^2} \quad \textit{Let A = 75, d = 4.}$$

$$k = 75(4^2) = 1200$$

So $A = \frac{1200}{d^2}$ and when $d = 9$,

$$A = \frac{1200}{9^2} = \frac{400}{27}, \quad \text{or} \quad 14\frac{22}{27}.$$

The amount of light is $14\frac{22}{27}$ footcandles at a distance of 9 feet.

45. $8^2 = 8 \cdot 8 = 64$

47. $-12^2 = -1 \cdot 12 \cdot 12 = -144$

49. $a^2 + b^2 = 5^2 + 12^2$
$$= 25 + 144$$
$$= 169$$

Chapter 7 Review Exercises

1. **(a)** $\dfrac{4x - 3}{5x + 2} = \dfrac{4(-2) - 3}{5(-2) + 2} \quad \textit{Let x = -2.}$

$$= \frac{-8 - 3}{-10 + 2} = \frac{-11}{-8} = \frac{11}{8}$$

(b) $\dfrac{4x - 3}{5x + 2} = \dfrac{4(4) - 3}{5(4) + 2} \quad \textit{Let x = 4.}$

$$= \frac{16 - 3}{20 + 2} = \frac{13}{22}$$

2. **(a)** $\dfrac{3x}{x^2 - 4} = \dfrac{3(-2)}{(-2)^2 - 4} \quad \textit{Let x = -2.}$

$$= \frac{-6}{4 - 4} = \frac{-6}{0}$$

Substituting -2 for x makes the denominator zero, so the given expression is undefined when $x = -2$.

(b) $\dfrac{3x}{x^2 - 4} = \dfrac{3(4)}{(4)^2 - 4} \quad \textit{Let x = 4.}$

$$= \frac{12}{16 - 4} = \frac{12}{12} = 1$$

3. $\dfrac{4}{x - 3}$

To find the values for which this expression is undefined, set the denominator equal to zero and solve for x.

$$x - 3 = 0$$
$$x = 3$$

Because $x = 3$ will make the denominator zero, the given expression is undefined for 3. Thus, $x \neq 3$.

4. $\dfrac{y + 3}{2y}$

Set the denominator equal to zero and solve for y.

$$2y = 0$$
$$y = 0$$

The given expression is undefined for 0. Thus, $y \neq 0$.

5. $\dfrac{2k+1}{3k^2+17k+10}$

Set the denominator equal to zero and solve for k.

$$3k^2 + 17k + 10 = 0$$
$$(3k+2)(k+5) = 0$$
$$k = -\tfrac{2}{3} \text{ or } k = -5$$

The given expression is undefined for -5 and $-\tfrac{2}{3}$. Thus, $k \neq -5, -\tfrac{2}{3}$.

6. Set the denominator equal to 0 and solve the equation. Any solutions are values for which the rational expression is undefined.

7. $\dfrac{5a^3b^3}{15a^4b^2} = \dfrac{b \cdot 5a^3b^2}{3a \cdot 5a^3b^2} = \dfrac{b}{3a}$

8. $\dfrac{m-4}{4-m} = \dfrac{-1(4-m)}{4-m} = -1$

9. $\dfrac{4x^2-9}{6-4x} = \dfrac{(2x+3)(2x-3)}{-2(2x-3)}$

$$= \dfrac{2x+3}{-2} = \dfrac{-1(2x+3)}{2}$$

$$= \dfrac{-(2x+3)}{2}$$

10. $\dfrac{4p^2+8pq-5q^2}{10p^2-3pq-q^2} = \dfrac{(2p-q)(2p+5q)}{(5p+q)(2p-q)}$

$$= \dfrac{2p+5q}{5p+q}$$

11. $-\dfrac{4x-9}{2x+3}$

Apply the negative sign to the numerator:

$$\dfrac{-(4x-9)}{2x+3}$$

Now distribute the negative sign:

$$\dfrac{-4x+9}{2x+3}$$

Apply the negative sign to the denominator:

$$\dfrac{4x-9}{-(2x+3)}$$

Again, distribute the negative sign:

$$\dfrac{4x-9}{-2x-3}$$

12. $-\dfrac{8-3x}{3-6x}$

Four equivalent forms are:

$$\dfrac{-(8-3x)}{3-6x}, \quad \dfrac{-8+3x}{3-6x},$$

$$\dfrac{8-3x}{-(3-6x)}, \quad \dfrac{8-3x}{-3+6x}$$

13. $\dfrac{18p^3}{6} \cdot \dfrac{24}{p^4} = \dfrac{6 \cdot 3 \cdot 24p^3}{6p^4} = \dfrac{72}{p}$

14. $\dfrac{8x^2}{12x^5} \cdot \dfrac{6x^4}{2x} = \dfrac{2 \cdot 4}{3 \cdot 4x^3} \cdot \dfrac{3x^3}{1} = 2$

15. $\dfrac{x-3}{4} \cdot \dfrac{5}{2x-6} = \dfrac{x-3}{4} \cdot \dfrac{5}{2(x-3)} = \dfrac{5}{8}$

16. $\dfrac{2r+3}{r-4} \cdot \dfrac{r^2-16}{6r+9}$

$$= \dfrac{2r+3}{r-4} \cdot \dfrac{(r+4)(r-4)}{3(2r+3)}$$

$$= \dfrac{r+4}{3}$$

17. $\dfrac{6a^2+7a-3}{2a^2-a-6} \div \dfrac{a+5}{a-2}$

$$= \dfrac{6a^2+7a-3}{2a^2-a-6} \cdot \dfrac{a-2}{a+5}$$

$$= \dfrac{(3a-1)(2a+3)}{(2a+3)(a-2)} \cdot \dfrac{a-2}{a+5}$$

$$= \dfrac{3a-1}{a+5}$$

18. $\dfrac{y^2-6y+8}{y^2+3y-18} \div \dfrac{y-4}{y+6}$

$$= \dfrac{y^2-6y+8}{y^2+3y-18} \cdot \dfrac{y+6}{y-4}$$

$$= \dfrac{(y-4)(y-2)}{(y+6)(y-3)} \cdot \dfrac{y+6}{y-4}$$

$$= \dfrac{y-2}{y-3}$$

19. $\dfrac{2p^2+13p+20}{p^2+p-12} \cdot \dfrac{p^2+2p-15}{2p^2+7p+5}$

$$= \dfrac{(2p+5)(p+4)}{(p+4)(p-3)} \cdot \dfrac{(p+5)(p-3)}{(2p+5)(p+1)}$$

$$= \dfrac{p+5}{p+1}$$

20. $\dfrac{3z^2+5z-2}{9z^2-1} \cdot \dfrac{9z^2+6z+1}{z^2+5z+6}$

$$= \dfrac{(3z-1)(z+2)}{(3z-1)(3z+1)} \cdot \dfrac{(3z+1)^2}{(z+3)(z+2)}$$

$$= \dfrac{3z+1}{z+3}$$

21. $\dfrac{4}{9y}, \dfrac{7}{12y^2}, \dfrac{5}{27y^4}$

Factor each denominator.

$$9y = 3^2 y$$
$$12y^2 = 2^2 \cdot 3 \cdot y^2$$
$$27y^4 = 3^3 \cdot y^4$$

$$LCD = 2^2 \cdot 3^3 \cdot y^4 = 108y^4$$

22. $\dfrac{3}{x^2 + 4x + 3}, \dfrac{5}{x^2 + 5x + 4}$

Factor each denominator.

$$x^2 + 4x + 3 = (x+3)(x+1)$$
$$x^2 + 5x + 4 = (x+1)(x+4)$$

$$LCD = (x+3)(x+1)(x+4)$$

23. $\dfrac{3}{2a^3} = \dfrac{?}{10a^4}$

$$\dfrac{3}{2a^3} = \dfrac{3}{2a^3} \cdot \dfrac{5a}{5a} = \dfrac{15a}{10a^4}$$

24. $\dfrac{9}{x-3} = \dfrac{?}{18 - 6x} = \dfrac{?}{-6(x-3)}$

$$\dfrac{9}{x-3} = \dfrac{9}{x-3} \cdot \dfrac{-6}{-6}$$

$$= \dfrac{-54}{-6x + 18}$$

$$= \dfrac{-54}{18 - 6x}$$

25. $\dfrac{-3y}{2y-10} = \dfrac{?}{50 - 10y} = \dfrac{?}{-5(2y-10)}$

$$\dfrac{-3y}{2y-10} = \dfrac{-3y}{2y-10} \cdot \dfrac{-5}{-5}$$

$$= \dfrac{15y}{-10y + 50}$$

$$= \dfrac{15y}{50 - 10y}$$

26. $\dfrac{4b}{b^2 + 2b - 3} = \dfrac{?}{(b+3)(b-1)(b+2)}$

$$\dfrac{4b}{b^2 + 2b - 3} = \dfrac{4b}{(b+3)(b-1)}$$

$$= \dfrac{4b}{(b+3)(b-1)} \cdot \dfrac{b+2}{b+2}$$

$$= \dfrac{4b(b+2)}{(b+3)(b-1)(b+2)}$$

27. $\dfrac{10}{x} + \dfrac{5}{x} = \dfrac{10+5}{x} = \dfrac{15}{x}$

28. $\dfrac{6}{3p} - \dfrac{12}{3p} = \dfrac{6-12}{3p} = \dfrac{-6}{3p} = -\dfrac{2}{p}$

29. $\dfrac{9}{k} - \dfrac{5}{k-5}$

$$= \dfrac{9(k-5)}{k(k-5)} - \dfrac{5 \cdot k}{(k-5)k} \quad LCD = k(k-5)$$

$$= \dfrac{9(k-5) - 5k}{k(k-5)}$$

$$= \dfrac{9k - 45 - 5k}{k(k-5)}$$

$$= \dfrac{4k - 45}{k(k-5)}$$

30. $\dfrac{4}{y} + \dfrac{7}{7+y}$

$$= \dfrac{4(7+y)}{y(7+y)} + \dfrac{7 \cdot y}{(7+y)y} \quad LCD = y(7+y)$$

$$= \dfrac{28 + 4y + 7y}{y(7+y)}$$

$$= \dfrac{28 + 11y}{y(7+y)}$$

31. $\dfrac{m}{3} - \dfrac{2 + 5m}{6}$

$$= \dfrac{m \cdot 2}{3 \cdot 2} - \dfrac{2 + 5m}{6} \quad LCD = 6$$

$$= \dfrac{2m - (2 + 5m)}{6}$$

$$= \dfrac{2m - 2 - 5m}{6}$$

$$= \dfrac{-2 - 3m}{6}$$

32. $\dfrac{12}{x^2} - \dfrac{3}{4x}$

$$= \dfrac{12 \cdot 4}{x^2 \cdot 4} - \dfrac{3 \cdot x}{4x \cdot x} \quad LCD = 4x^2$$

$$= \dfrac{48 - 3x}{4x^2}$$

$$= \dfrac{3(16 - x)}{4x^2}$$

33. $\dfrac{5}{a - 2b} + \dfrac{2}{a + 2b}$

$$= \dfrac{5(a + 2b)}{(a-2b)(a+2b)} + \dfrac{2(a-2b)}{(a+2b)(a-2b)}$$

$$LCD = (a - 2b)(a + 2b)$$

$$= \dfrac{5(a+2b) + 2(a-2b)}{(a-2b)(a+2b)}$$

$$= \dfrac{5a + 10b + 2a - 4b}{(a-2b)(a+2b)}$$

$$= \dfrac{7a + 6b}{(a-2b)(a+2b)}$$

34. $\dfrac{4}{k^2 - 9} - \dfrac{k + 3}{3k - 9}$

$= \dfrac{4}{(k + 3)(k - 3)} - \dfrac{k + 3}{3(k - 3)}$

$\qquad\qquad LCD = 3(k + 3)(k - 3)$

$= \dfrac{4 \cdot 3}{(k + 3)(k - 3) \cdot 3} - \dfrac{(k + 3)(k + 3)}{3(k - 3)(k + 3)}$

$= \dfrac{12 - (k + 3)(k + 3)}{3(k + 3)(k - 3)}$

$= \dfrac{12 - (k^2 + 6k + 9)}{3(k + 3)(k - 3)}$

$= \dfrac{12 - k^2 - 6k - 9}{3(k + 3)(k - 3)}$

$= \dfrac{-k^2 - 6k + 3}{3(k + 3)(k - 3)}$

35. $\dfrac{8}{z^2 + 6z} - \dfrac{3}{z^2 + 4z - 12}$

$= \dfrac{8}{z(z + 6)} - \dfrac{3}{(z + 6)(z - 2)}$

$\qquad\qquad LCD = z(z + 6)(z - 2)$

$= \dfrac{8(z - 2)}{z(z + 6)(z - 2)} - \dfrac{3 \cdot z}{(z + 6)(z - 2) \cdot z}$

$= \dfrac{8(z - 2) - 3z}{z(z + 6)(z - 2)}$

$= \dfrac{8z - 16 - 3z}{z(z + 6)(z - 2)}$

$= \dfrac{5z - 16}{z(z + 6)(z - 2)}$

36. $\dfrac{11}{2p - p^2} - \dfrac{2}{p^2 - 5p + 6}$

$= \dfrac{11}{p(2 - p)} - \dfrac{2}{(p - 3)(p - 2)}$

$\qquad\qquad LCD = p(p - 3)(p - 2)$

$= \dfrac{11(-1)(p - 3)}{p(2 - p)(-1)(p - 3)}$

$\quad - \dfrac{2 \cdot p}{(p - 3)(p - 2)p}$

$= \dfrac{-11(p - 3) - 2p}{p(p - 2)(p - 3)}$

$= \dfrac{-11p + 33 - 2p}{p(p - 2)(p - 3)}$

$= \dfrac{-13p + 33}{p(p - 2)(p - 3)}$

37. $\dfrac{\dfrac{y - 3}{y}}{\dfrac{y + 3}{4y}} = \dfrac{y - 3}{y} \cdot \dfrac{4y}{y + 3} = \dfrac{4(y - 3)}{y + 3}$

38. $\dfrac{\frac{2}{3} - \frac{1}{6}}{\frac{1}{4} + \frac{2}{5}} = \dfrac{60\left(\frac{2}{3} - \frac{1}{6}\right)}{60\left(\frac{1}{4} + \frac{2}{5}\right)}$ *Multiply by LCD, 60*

$= \dfrac{60 \cdot \frac{2}{3} - 60 \cdot \frac{1}{6}}{60 \cdot \frac{1}{4} + 60 \cdot \frac{2}{5}}$

$= \dfrac{40 - 10}{15 + 24} = \dfrac{30}{39} = \dfrac{10}{13}$

39. $\dfrac{x + \frac{1}{w}}{x - \frac{1}{w}}$

$= \dfrac{\left(x + \frac{1}{w}\right) \cdot w}{\left(x - \frac{1}{w}\right) \cdot w}$ *Multiply by LCD, w.*

$= \dfrac{xw + \left(\frac{1}{w}\right)w}{xw - \left(\frac{1}{w}\right)w}$

$= \dfrac{xw + 1}{xw - 1}$

40. $\dfrac{\dfrac{1}{p} - \dfrac{1}{q}}{\dfrac{1}{q - p}}$

$= \dfrac{\left(\dfrac{1}{p} - \dfrac{1}{q}\right)pq(q - p)}{\left(\dfrac{1}{q - p}\right)pq(q - p)}$ *Multiply by LCD, pq(q - p).*

$= \dfrac{\frac{1}{p}[pq(q - p)] - \frac{1}{q}[pq(q - p)]}{pq}$

$= \dfrac{q(q - p) - p(q - p)}{pq}$

$= \dfrac{q^2 - pq - pq + p^2}{pq}$

$= \dfrac{q^2 - 2pq + p^2}{pq}$

$= \dfrac{(q - p)^2}{pq}$

41. $\dfrac{\dfrac{x^2 - 25}{x + 3}}{\dfrac{x + 5}{x^2 - 9}}$

$= \dfrac{x^2 - 25}{x + 3} \cdot \dfrac{x^2 - 9}{x + 5}$

$= \dfrac{(x + 5)(x - 5)}{x + 3} \cdot \dfrac{(x + 3)(x - 3)}{x + 5}$

$= \dfrac{x - 5}{1} \cdot \dfrac{x - 3}{1}$

$= (x - 5)(x - 3), \text{ or } x^2 - 8x + 15$

42.
$$\frac{\dfrac{1}{r+t} - 1}{\dfrac{1}{r+t} + 1}$$

$$= \frac{\left(\dfrac{1}{r+t} - 1\right)(r+t)}{\left(\dfrac{1}{r+t} + 1\right)(r+t)} \quad \begin{array}{l}\textit{Multiply by} \\ \textit{LCD, } r + t.\end{array}$$

$$= \frac{\dfrac{1}{r+t}(r+t) - 1(r+t)}{\dfrac{1}{r+t}(r+t) + 1(r+t)}$$

$$= \frac{1 - r - t}{1 + r + t}$$

43.
$$\frac{3x-1}{x-2} = \frac{5}{x-2} + 1$$

$$(x-2)\left(\frac{3x-1}{x-2}\right) = (x-2)\left(\frac{5}{x-2} + 1\right)$$
$$\textit{Multiply by LCD, } x - 2.$$

$$(x-2)\left(\frac{3x-1}{x-2}\right) = (x-2)\left(\frac{5}{x-2}\right)$$
$$\qquad + (x-2)(1)$$
$$\textit{Distributive property}$$

$$3x - 1 = 5 + x - 2$$
$$3x - 1 = 3 + x$$
$$2x = 4$$
$$x = 2$$

The solution set is $\emptyset$ because $x = 2$ makes the original denominators equal to zero.

44.
$$\frac{4-z}{z} + \frac{3}{2} = \frac{-4}{z}$$

Multiply each side by the LCD, $2z$.

$$2z\left(\frac{4-z}{z} + \frac{3}{2}\right) = 2z\left(-\frac{4}{z}\right)$$

$$2z\left(\frac{4-z}{z}\right) + 2z\left(\frac{3}{2}\right) = -8$$

$$2(4-z) + 3z = -8$$
$$8 - 2z + 3z = -8$$
$$8 + z = -8$$
$$z = -16$$

Check $z = -16$: $\quad -\frac{5}{4} + \frac{3}{2} = \frac{1}{4} \quad$ *True*

Thus, the solution set is $\{-16\}$.

45.
$$\frac{3}{x+4} - \frac{2x}{5} = \frac{3}{x+4}$$

$$-\frac{2x}{5} = 0 \quad \textit{Subtract } \frac{3}{x+4}.$$

$$x = 0 \quad \textit{Multiply by } -\frac{5}{2}.$$

Check $x = 0$: $\quad \frac{3}{4} - 0 = \frac{3}{4} \quad$ *True*

Thus, the solution set is $\{0\}$.

46.
$$\frac{3}{m-2} + \frac{1}{m-1} = \frac{7}{m^2 - 3m + 2}$$

$$\frac{3}{m-2} + \frac{1}{m-1} = \frac{7}{(m-2)(m-1)}$$

$$(m-2)(m-1)\left(\frac{3}{m-2} + \frac{1}{m-1}\right)$$
$$= (m-2)(m-1)\cdot\frac{7}{(m-2)(m-1)}$$
$$\textit{Multiply by LCD, } (m-2)(m-1).$$

$$3(m-1) + 1(m-2) = 7$$
$$3m - 3 + m - 2 = 7$$
$$4m - 5 = 7$$
$$4m = 12$$
$$m = 3$$

Check $m = 3$: $\quad 3 + \frac{1}{2} = \frac{7}{2} \quad$ *True*

Thus, the solution set is $\{3\}$.

47.
$$m = \frac{Ry}{t} \text{ for } t$$

$$t \cdot m = t\left(\frac{Ry}{t}\right) \quad \textit{Multiply by } t.$$

$$tm = Ry$$

$$t = \frac{Ry}{m} \qquad \textit{Divide by } m.$$

48.
$$x = \frac{3y-5}{4} \text{ for } y$$

$$4x = 4\left(\frac{3y-5}{4}\right)$$

$$4x = 3y - 5$$

$$4x + 5 = 3y$$

$$\frac{4x+5}{3} = y$$

49.
$$p^2 = \frac{4}{3m-q} \text{ for } m$$

$$(3m-q)p^2 = (3m-q)\left(\frac{4}{3m-q}\right)$$

$$3mp^2 - p^2 q = 4$$

$$3mp^2 = 4 + p^2 q$$

$$m = \frac{4 + p^2 q}{3p^2}$$

50. Let x = the numerator. Then $x - 5$ = the denominator. Adding 5 to both the numerator and the denominator gives us a fraction that is equivalent to $\frac{5}{4}$.

$$\frac{x+5}{x-5+5} = \frac{5}{4}$$
$$\frac{x+5}{x} = \frac{5}{4}$$
$$4x\left(\frac{x+5}{x}\right) = 4x\left(\frac{5}{4}\right)$$
$$4(x+5) = x(5)$$
$$4x + 20 = 5x$$
$$20 = x$$

The numerator is 20 and the denominator is $20 - 5 = 15$, so the original fraction is $\frac{20}{15}$.

51. Let x = the numerator. Then $6x$ = the denominator. Adding 3 to the numerator and subtracting 3 from the denominator gives us a fraction equivalent to $\frac{2}{5}$.

$$\frac{x+3}{6x-3} = \frac{2}{5}$$
$$5(6x-3)\left(\frac{x+3}{6x-3}\right) = 5(6x-3)\left(\frac{2}{5}\right)$$
$$5(x+3) = 2(6x-3)$$
$$5x + 15 = 12x - 6$$
$$21 = 7x$$
$$3 = x$$

The numerator is 3 and the denominator is $6 \cdot 3 = 18$, so the original fraction is $\frac{3}{18}$.

52. Let x = the rate of the wind. Then the rate against the wind is $165 - x$ and the rate with the wind is $165 + x$. Complete the chart using $t = d/r$.

	d	r	t
Against the Wind	310	$165 - x$	$\dfrac{310}{165-x}$
With the Wind	350	$165 + x$	$\dfrac{350}{165+x}$

Since the times are equal, we get the following equation.

$$\frac{310}{165-x} = \frac{350}{165+x}$$
$$(165+x)(165-x)\frac{310}{165-x} = (165+x)(165-x)\frac{350}{165+x}$$
$$310(165+x) = 350(165-x)$$
$$51{,}150 + 310x = 57{,}750 - 350x$$
$$660x = 6600$$
$$x = 10$$

The rate of the wind is 10 miles per hour.

53. *Step 2*
Let x = the number of hours it takes them to do the job working together.

	Rate	Time Working Together	Fractional Part of the Job Done when Working Together
Susan	$\frac{1}{5}$	x	$\frac{1}{5}x$
Friend	$\frac{1}{8}$	x	$\frac{1}{8}x$

Step 3
Working together, they do 1 whole job, so

$$\tfrac{1}{5}x + \tfrac{1}{8}x = 1.$$

Step 4
Solve this equation by multiplying both sides by the LCD, 40.

$$40(\tfrac{1}{5}x + \tfrac{1}{8}x) = 40(1)$$
$$8x + 5x = 40$$
$$13x = 40$$
$$x = \tfrac{40}{13}, \text{ or } 3\tfrac{1}{13}$$

Step 5
Working together, it takes them $3\frac{1}{13}$ hours.

Step 6
Susan does $\frac{1}{5}$ of the job per hour for $\frac{40}{13}$ hours:

$$\tfrac{1}{5} \cdot \tfrac{40}{13} = \tfrac{8}{13} \text{ of the job}$$

Her friend does $\frac{1}{8}$ of the job per hour for $\frac{40}{13}$ hours:

$$\tfrac{1}{8} \cdot \tfrac{40}{13} = \tfrac{5}{13} \text{ of the job}$$

Together, they have done

$$\tfrac{8}{13} + \tfrac{5}{13} = \tfrac{13}{13} = 1 \text{ total job.}$$

54. *Step 2*
Let x = the time needed by the head gardener to mow the lawns.
Then $2x$ = the time needed by the assistant to mow the lawns.

	Rate	Time Working Together	Fractional Part of the Job Done when Working Together
Head gardener	$\dfrac{1}{x}$	$1\dfrac{1}{3} = \dfrac{4}{3}$	$\dfrac{1}{x} \cdot \dfrac{4}{3} = \dfrac{4}{3x}$
Assistant	$\dfrac{1}{2x}$	$1\dfrac{1}{3} = \dfrac{4}{3}$	$\dfrac{1}{2x} \cdot \dfrac{4}{3} = \dfrac{2}{3x}$

Step 3 $$\frac{4}{3x} + \frac{2}{3x} = 1$$

Step 4
$$\frac{4+2}{3x} = 1$$
$$\frac{6}{3x} = 1$$
$$3x\left(\frac{6}{3x}\right) = 3x(1)$$
$$6 = 3x$$
$$2 = x$$

Step 5
It takes the head gardener 2 hours to mow the lawns.

Step 6
The head gardener does $\frac{1}{2}$ of the job per hour for $\frac{4}{3}$ hours:
$$\tfrac{1}{2}\cdot\tfrac{4}{3} = \tfrac{2}{3} \text{ of the job}$$

The assistant does $\frac{1}{4}$ of the job per hour for $\frac{4}{3}$ hours:
$$\tfrac{1}{4}\cdot\tfrac{4}{3} = \tfrac{1}{3} \text{ of the job}$$

Together, they have done
$$\tfrac{2}{3} + \tfrac{1}{3} = \tfrac{3}{3} = 1 \text{ total job.}$$

55. Since a longer term is related to a lower rate per year, this is an *inverse* variation.

56. Let h = the height of the parallelogram and b = the length of the base of the parallelogram.

The height varies inversely as the base, so
$$h = \tfrac{k}{b}.$$

Find k by replacing h with 8 and b with 12.
$$8 = \frac{k}{12}$$
$$k = 8\cdot12 = 96$$

So $h = \frac{96}{b}$ and when $b = 24$,
$$h = \tfrac{96}{24} = 4.$$

The height of the parallelogram is 4 centimeters.

57. Since y varies directly as x, there is a constant k such that $y = kx$. First find the value of k.
$$5 = k(12) \quad \textit{Let x = 12, y = 5.}$$
$$k = \tfrac{5}{12}$$

When $k = \frac{5}{12}$, $y = kx$ becomes
$$y = \tfrac{5}{12}x.$$

Now find x when $y = 3$.
$$3 = \tfrac{5}{12}x \quad \textit{Let y = 3.}$$
$$x = 3\cdot\tfrac{12}{5} = \tfrac{36}{5}$$

58. **[7.4]** $\dfrac{4}{m-1} - \dfrac{3}{m+1}$

To perform the indicated subtraction, use $(m-1)(m+1)$ as the LCD.
$$\frac{4}{m-1} - \frac{3}{m+1}$$
$$= \frac{4(m+1)}{(m-1)(m+1)} - \frac{3(m-1)}{(m+1)(m-1)}$$
$$= \frac{4(m+1) - 3(m-1)}{(m-1)(m+1)}$$
$$= \frac{4m+4-3m+3}{(m-1)(m+1)}$$
$$= \frac{m+7}{(m-1)(m+1)}$$

59. **[7.2]** $\dfrac{8p^5}{5} \div \dfrac{2p^3}{10}$

To perform the indicated division, multiply the first rational expression by the reciprocal of the second.
$$\frac{8p^5}{5} \div \frac{2p^3}{10} = \frac{8p^5}{5}\cdot\frac{10}{2p^3}$$
$$= \frac{80p^5}{10p^3}$$
$$= 8p^2$$

60. **[7.2]** $\dfrac{r-3}{8} \div \dfrac{3r-9}{4} = \dfrac{r-3}{8}\cdot\dfrac{4}{3r-9}$
$$= \frac{r-3}{8}\cdot\frac{4}{3(r-3)}$$
$$= \frac{4}{24} = \frac{1}{6}$$

61. **[7.5]** $\dfrac{\frac{5}{x}-1}{\frac{5-x}{3x}} = \dfrac{\left(\frac{5}{x}-1\right)3x}{\left(\frac{5-x}{3x}\right)3x}$ *Multiply by LCD, 3x.*
$$= \frac{\frac{5}{x}(3x) - 1(3x)}{5-x}$$
$$= \frac{15-3x}{5-x}$$
$$= \frac{3(5-x)}{5-x} = 3$$

62. **[7.4]** $\dfrac{4}{z^2 - 2z + 1} - \dfrac{3}{z^2 - 1}$

$= \dfrac{4}{(z-1)^2} - \dfrac{3}{(z+1)(z-1)}$

$LCD = (z+1)(z-1)^2$

$= \dfrac{4(z+1)}{(z-1)^2(z+1)}$

$- \dfrac{3(z-1)}{(z+1)(z-1)(z-1)}$

$= \dfrac{4(z+1) - 3(z-1)}{(z+1)(z-1)^2}$

$= \dfrac{4z+4 - 3z+3}{(z+1)(z-1)^2}$

$= \dfrac{z+7}{(z+1)(z-1)^2}$

63. **[7.4]** $\dfrac{1}{t^2 - 4} + \dfrac{1}{2 - t}$

$= \dfrac{1}{(t+2)(t-2)} + \dfrac{1}{2-t}$

$LCD = (t+2)(t-2)$

$= \dfrac{1}{(t+2)(t-2)} - \dfrac{1(t+2)}{(t-2)(t+2)}$

$= \dfrac{1 - t - 2}{(t+2)(t-2)}$

$= \dfrac{-t-1}{(t+2)(t-2)}$, or $\dfrac{t+1}{(t+2)(2-t)}$

64. **[7.6]** $\dfrac{2}{z} - \dfrac{z}{z+3} = \dfrac{1}{z+3}$

Multiply each side of the equation by the LCD, $z(z+3)$.

$z(z+3)\left(\dfrac{2}{z} - \dfrac{z}{z+3}\right) = z(z+3)\left(\dfrac{1}{z+3}\right)$

$z(z+3)\left(\dfrac{2}{z}\right) - z(z+3)\left(\dfrac{z}{z+3}\right) = z(1)$

$2(z+3) - z^2 = z$

$2z + 6 - z^2 = z$

$0 = z^2 - z - 6$

$0 = (z-3)(z+2)$

$z - 3 = 0$ or $z + 2 = 0$

$z = 3$ or $z = -2$

Check $z = -2$: $-1 - (-2) = 1$ *True*
Check $z = 3$: $\frac{2}{3} - \frac{1}{2} = \frac{1}{6}$ *True*
Thus, the solution set is $\{-2, 3\}$.

65. **[7.6]** Solve $a = \dfrac{v - w}{t}$ for v.

$t \cdot a = v - w$ *Multiply by t.*
$at + w = v$ *Add w.*

66. **[7.7]** Let $x =$ the rate of the plane in still air. Then the rate of the plane with the wind is $x + 50$, and the rate of the plane against the wind is $x - 50$. Use $t = d/r$ to complete the chart.

	d	r	t
With the Wind	400	$x+50$	$\dfrac{400}{x+50}$
Against the Wind	200	$x-50$	$\dfrac{200}{x-50}$

The times are the same, so

$$\dfrac{400}{x+50} = \dfrac{200}{x-50}.$$

To solve this equation, multiply both sides by the LCD, $(x+50)(x-50)$.

$(x+50)(x-50) \cdot \dfrac{400}{x+50}$

$= (x+50)(x-50) \cdot \dfrac{200}{x-50}$

$400(x-50) = 200(x+50)$
$400x - 20{,}000 = 200x + 10{,}000$
$200x = 30{,}000$
$x = 150$

The rate of the plane is 150 kilometers per hour.

67. **[7.7]** Let $x =$ the number of hours it takes them to do the job working together.

	Rate	Time Working Together	Fractional Part of the Job Done when Working Together
Lizette	$\frac{1}{8}$	x	$\frac{1}{8}x$
Seyed	$\frac{1}{14}$	x	$\frac{1}{14}x$

Working together, they do 1 whole job, so

$$\tfrac{1}{8}x + \tfrac{1}{14}x = 1.$$

To clear fractions, multiply both sides by the LCD, 56.

$$56(\tfrac{1}{8}x + \tfrac{1}{14}x) = 56(1)$$
$$7x + 4x = 56$$
$$11x = 56$$
$$x = \tfrac{56}{11}, \text{ or } 5\tfrac{1}{11}$$

Working together, they can paint the woodwork in $5\tfrac{1}{11}$ hours.

68. **[7.8]** Since w varies inversely as z, there is a constant k such that $w = \frac{k}{z}$. First find the value of k.

$$16 = \frac{k}{3} \quad \textit{Let w = 16, z = 3.}$$
$$k = 16(3) = 48$$

When $k = 48$, $w = \frac{k}{z}$ becomes

$$w = \frac{48}{z}.$$

Now find w when $z = 2$.

$$w = \frac{48}{2} = 24$$

69. **[7.8]** For a constant area, the length L of a rectangle varies inversely as the width W, so

$$L = \frac{k}{W}.$$
$$24 = \frac{k}{2} \quad \textit{Let L = 24, W = 2.}$$
$$k = 24 \cdot 2 = 48$$

So $L = \frac{48}{W}$ and when $L = 12$,

$$12 = \frac{48}{W}$$
$$12W = 48$$
$$W = \frac{48}{12} = 4.$$

When the length is 12, the width is 4.

70. **[7.8]** $x = ky^3$ *x varies directly as the cube of y.*

$$54 = k(3)^3 \quad \textit{Substitute x = 54 and y = 3.}$$
$$2 = k \quad \textit{Solve for k.}$$

Since $x = ky^3$ and $k = 2$,

$$x = 2y^3.$$
$$x = 2(-2)^3 \quad \textit{Substitute y = -2.}$$
$$x = -16$$

71. **(a)** $x + 3 = 0$
$$x = -3$$

This value of x makes the value of the denominator zero, so P will be undefined when $x = -3$.

(b) $x + 1 = 0$
$$x = -1$$

This value of x makes the value of the denominator zero, so Q will be undefined when $x = -1$.

(c) $x^2 + 4x + 3 = 0$
$$(x + 3)(x + 1) = 0$$
$$x = -3 \quad \text{or} \quad x = -1$$

These values of x make the value of the denominator zero, so the values for which R is undefined are -3 and -1.

72. $(P \cdot Q) \div R$

$$= \left(\frac{6}{x+3} \cdot \frac{5}{x+1} \right) \div \frac{4x}{x^2 + 4x + 3}$$

$$= \frac{30}{(x+3)(x+1)} \cdot \frac{x^2 + 4x + 3}{4x}$$

$$= \frac{30}{(x+3)(x+1)} \cdot \frac{(x+3)(x+1)}{4x}$$

$$= \frac{30}{4x}$$

$$= \frac{15}{2x}$$

73. If $x = 0$, the divisor R is equal to 0, and division by 0 is undefined.

74. List the three denominators and factor if possible.

$$x + 3, \, x + 1, \, x^2 + 4x + 3 = (x+3)(x+1)$$

The LCD for P, Q, and R is $(x+3)(x+1)$.

75. $P + Q - R$

$$= \frac{6}{x+3} + \frac{5}{x+1} - \frac{4x}{x^2 + 4x + 3}$$

$$= \frac{6}{x+3} + \frac{5}{x+1} - \frac{4x}{(x+3)(x+1)}$$

$$LCD = (x+3)(x+1)$$

$$= \frac{6(x+1)}{(x+3)(x+1)} + \frac{5(x+3)}{(x+3)(x+1)}$$

$$\quad - \frac{4x}{(x+3)(x+1)}$$

$$= \frac{6(x+1) + 5(x+3) - 4x}{(x+3)(x+1)}$$

$$= \frac{6x + 6 + 5x + 15 - 4x}{(x+3)(x+1)}$$

$$= \frac{7x + 21}{(x+3)(x+1)}$$

$$= \frac{7(x+3)}{(x+3)(x+1)}$$

$$= \frac{7}{x+1}$$

76. $\dfrac{P+Q}{R} = \dfrac{\dfrac{6}{x+3} + \dfrac{5}{x+1}}{\dfrac{4x}{x^2+4x+3}} = \dfrac{\dfrac{6}{x+3} + \dfrac{5}{x+1}}{\dfrac{4x}{(x+3)(x+1)}}$

To simplify this complex fraction, use Method 2. Multiply numerator and denominator by the LCD for all the fractions, $(x+3)(x+1)$.

$\dfrac{(x+3)(x+1)\left(\dfrac{6}{x+3} + \dfrac{5}{x+1}\right)}{(x+3)(x+1)\left(\dfrac{4x}{(x+3)(x+1)}\right)}$

$= \dfrac{(x+3)(x+1)\left(\dfrac{6}{x+3}\right) + (x+3)(x+1)\left(\dfrac{5}{x+1}\right)}{4x}$

$= \dfrac{6(x+1) + 5(x+3)}{4x}$

$= \dfrac{6x+6+5x+15}{4x}$

$= \dfrac{11x+21}{4x}$

77. $\qquad\qquad P + Q = R$

$\dfrac{6}{x+3} + \dfrac{5}{x+1} = \dfrac{4x}{x^2+4x+3}$

$\dfrac{6}{x+3} + \dfrac{5}{x+1} = \dfrac{4x}{(x+3)(x+1)}$

$\qquad\qquad$ *Multiply by LCD, $(x+3)(x+1)$.*

$(x+3)(x+1)\left(\dfrac{6}{x+3}\right) + (x+3)(x+1)\left(\dfrac{5}{x+1}\right)$

$\qquad\qquad = (x+3)(x+1)\left(\dfrac{4x}{(x+3)(x+1)}\right)$

$6(x+1) + 5(x+3) = 4x$

$6x+6+5x+15 = 4x$

$7x = -21$

$x = -3$

To check, substitute -3 for x. The denominators of the first and third fractions become zero. Reject -3 as a solution. There is no solution to the equation, so the solution set is $\emptyset$.

78. We know that -3 is not allowed because P and R are undefined for $x = -3$.

79. Solving $d = rt$ for r gives us $r = d/t$. If $d = 6$ miles and $t = (x+3)$ minutes, then $r = \dfrac{d}{t} = \dfrac{6}{x+3}$ miles per minute. Thus,

$$P = \dfrac{6}{x+3}$$

represents the rate of the car (in miles per minute).

80. $\qquad\qquad R = \dfrac{4x}{x^2+4x+3}$

$\dfrac{40}{77} = \dfrac{4x}{x^2+4x+3}$

$40(x^2+4x+3) = (4x)(77)$ $\quad$ *Cross products are equal.*

$40x^2 + 160x + 120 = 308x$

$40x^2 - 148x + 120 = 0$

$10x^2 - 37x + 30 = 0$

$(5x-6)(2x-5) = 0$

$x = \tfrac{6}{5}$ or $x = \tfrac{5}{2}$

Chapter 7 Test

1. (a) $\dfrac{6r+1}{2r^2-3r-20}$

$= \dfrac{6(-2)+1}{2(-2)^2 - 3(-2) - 20}$ $\quad$ *Let $r = -2$.*

$= \dfrac{-12+1}{2\cdot 4 + 6 - 20}$

$= \dfrac{-11}{8+6-20}$

$= \dfrac{-11}{-6} = \dfrac{11}{6}$

(b) $\dfrac{6r+1}{2r^2-3r-20}$

$= \dfrac{6(4)+1}{2(4)^2 - 3(4) - 20}$ $\quad$ *Let $r = 4$.*

$= \dfrac{24+1}{2\cdot 16 - 12 - 20}$

$= \dfrac{25}{32-12-20}$

$= \dfrac{25}{20-20} = \dfrac{25}{0}$

The expression is undefined when $r = 4$ because the denominator is 0.

2. $\dfrac{3x-1}{x^2-2x-8}$

Set the denominator equal to zero and solve for x.

$x^2 - 2x - 8 = 0$

$(x+2)(x-4) = 0$

$x+2 = 0$ $\quad$ or $\quad$ $x-4 = 0$

$x = -2$ $\quad$ or $\quad\quad$ $x = 4$

The expression is undefined for -2 and 4, so $x \neq -2, 4$.

3. $-\dfrac{6x-5}{2x+3}$

Apply the negative sign to the numerator:

$$\dfrac{-(6x-5)}{2x+3}$$

Now distribute the negative sign:

$$\dfrac{-6x+5}{2x+3}$$

Apply the negative sign to the denominator:

$$\dfrac{6x-5}{-(2x+3)}$$

Again, distribute the negative sign:

$$\dfrac{6x-5}{-2x-3}$$

4. $\dfrac{-15x^6y^4}{5x^4y} = \dfrac{(5x^4y)(-3x^2y^3)}{(5x^4y)(1)}$

$$= \dfrac{5x^4y}{5x^4y} \cdot \dfrac{-3x^2y^3}{1} = -3x^2y^3$$

5. $\dfrac{6a^2+a-2}{2a^2-3a+1} = \dfrac{(3a+2)(2a-1)}{(2a-1)(a-1)}$

$$= \dfrac{3a+2}{a-1}$$

6. $\dfrac{5(d-2)}{9} \div \dfrac{3(d-2)}{5}$

$$= \dfrac{5(d-2)}{9} \cdot \dfrac{5}{3(d-2)}$$

$$= \dfrac{5 \cdot 5}{9 \cdot 3} = \dfrac{25}{27}$$

7. $\dfrac{6k^2-k-2}{8k^2+10k+3} \cdot \dfrac{4k^2+7k+3}{3k^2+5k+2}$

$$= \dfrac{(3k-2)(2k+1)}{(4k+3)(2k+1)} \cdot \dfrac{(4k+3)(k+1)}{(3k+2)(k+1)}$$

$$= \dfrac{3k-2}{3k+2}$$

8. $\dfrac{4a^2+9a+2}{3a^2+11a+10} \div \dfrac{4a^2+17a+4}{3a^2+2a-5}$

$$= \dfrac{4a^2+9a+2}{3a^2+11a+10} \cdot \dfrac{3a^2+2a-5}{4a^2+17a+4}$$

$$= \dfrac{(4a+1)(a+2)}{(3a+5)(a+2)} \cdot \dfrac{(3a+5)(a-1)}{(4a+1)(a+4)}$$

$$= \dfrac{a-1}{a+4}$$

9. $\dfrac{x^2-10x+25}{9-6x+x^2} \cdot \dfrac{x-3}{5-x}$

$$= \dfrac{(x-5)(x-5)}{(3-x)(3-x)} \cdot \dfrac{x-3}{5-x} \quad \text{Factor.}$$

$$= \dfrac{(x-5)(x-5)(x-3)}{(3-x)(3-x)(5-x)} \quad \text{Multiply.}$$

$$= \dfrac{-1(x-5)}{-1(3-x)} \quad \text{Lowest terms}$$

$$= \dfrac{x-5}{3-x}$$

10. $\dfrac{-3}{10p^2}, \dfrac{21}{25p^3}, \dfrac{-7}{30p^5}$

Factor each denominator.

$$10p^2 = 2 \cdot 5 \cdot p^2$$
$$25p^3 = 5^2 \cdot p^3$$
$$30p^5 = 2 \cdot 3 \cdot 5 \cdot p^5$$

$$\text{LCD} = 2 \cdot 3 \cdot 5^2 \cdot p^5 = 150p^5$$

11. $\dfrac{r+1}{2r^2+7r+6}, \dfrac{-2r+1}{2r^2-7r-15}$

Factor each denominator.

$$2r^2+7r+6 = (2r+3)(r+2)$$
$$2r^2-7r-15 = (2r+3)(r-5)$$

$$\text{LCD} = (2r+3)(r+2)(r-5)$$

12. $\dfrac{15}{4p} = \dfrac{?}{64p^3} = \dfrac{?}{4p \cdot 16p^2}$

$$\dfrac{15}{4p} = \dfrac{15 \cdot 16p^2}{4p \cdot 16p^2} = \dfrac{240p^2}{64p^3}$$

13. $\dfrac{3}{6m-12} = \dfrac{?}{42m-84} = \dfrac{?}{7(6m-12)}$

$$\dfrac{3}{6m-12} = \dfrac{3 \cdot 7}{(6m-12)7} = \dfrac{21}{42m-84}$$

14. $\dfrac{4x+2}{x+5} + \dfrac{-2x+8}{x+5}$

$$= \dfrac{(4x+2)+(-2x+8)}{x+5}$$

$$= \dfrac{2x+10}{x+5}$$

$$= \dfrac{2(x+5)}{x+5} = 2$$

15. $\dfrac{-4}{y+2} + \dfrac{6}{5y+10}$

$$= \dfrac{-4}{y+2} + \dfrac{6}{5(y+2)} \quad \text{LCD} = 5(y+2)$$

$$= \dfrac{-4 \cdot 5}{(y+2) \cdot 5} + \dfrac{6}{5(y+2)}$$

$$= \dfrac{-20+6}{5(y+2)} = \dfrac{-14}{5(y+2)}$$

16. Using LCD $= 3 - x$,

$$\frac{x+1}{3-x} + \frac{x^2}{x-3} = \frac{x+1}{3-x} + \frac{-1(x^2)}{-1(x-3)}$$

$$= \frac{x+1}{3-x} + \frac{-x^2}{-x+3}$$

$$= \frac{x+1}{3-x} + \frac{-x^2}{3-x}$$

$$= \frac{(x+1) + (-x^2)}{3-x}$$

$$= \frac{-x^2 + x + 1}{3-x}.$$

If we use $x - 3$ for the LCD, we obtain the equivalent answer

$$\frac{x^2 - x - 1}{x - 3}.$$

17. $\dfrac{3}{2m^2 - 9m - 5} - \dfrac{m+1}{2m^2 - m - 1}$

$$= \frac{3}{(2m+1)(m-5)} - \frac{m+1}{(2m+1)(m-1)}$$

$$LCD = (2m + 1)(m - 5)(m - 1)$$

$$= \frac{3(m-1)}{(2m+1)(m-5)(m-1)}$$

$$- \frac{(m+1)(m-5)}{(2m+1)(m-1)(m-5)}$$

$$= \frac{3(m-1) - (m+1)(m-5)}{(2m+1)(m-5)(m-1)}$$

$$= \frac{(3m-3) - (m^2 - 4m - 5)}{(2m+1)(m-5)(m-1)}$$

$$= \frac{3m - 3 - m^2 + 4m + 5}{(2m+1)(m-5)(m-1)}$$

$$= \frac{-m^2 + 7m + 2}{(2m+1)(m-5)(m-1)}$$

18. $\dfrac{\dfrac{2p}{k^2}}{\dfrac{3p^2}{k^3}} = \dfrac{2p}{k^2} \div \dfrac{3p^2}{k^3}$

$$= \frac{2p}{k^2} \cdot \frac{k^3}{3p^2}$$

$$= \frac{2k^3 p}{3k^2 p^2} = \frac{2k}{3p}$$

19. $\dfrac{\dfrac{1}{x+3} - 1}{1 + \dfrac{1}{x+3}}$

Start by multiplying the numerator and the denominator by the LCD, $x + 3$.

$$= \frac{(x+3)\left(\dfrac{1}{x+3} - 1\right)}{(x+3)\left(1 + \dfrac{1}{x+3}\right)}$$

$$= \frac{(x+3)\left(\dfrac{1}{x+3}\right) - (x+3)(1)}{(x+3)(1) + (x+3)\left(\dfrac{1}{x+3}\right)}$$

$$= \frac{1 - (x+3)}{(x+3) + 1}$$

$$= \frac{1 - x - 3}{x + 4}$$

$$= \frac{-2 - x}{x + 4}$$

20. $$\frac{3x}{x+1} = \frac{3}{2x}$$

$$2x(x+1)\frac{3x}{x+1} = 2x(x+1)\frac{3}{2x}$$

Multiply by LCD, 2x(x + 1).

$$2x(3x) = 3(x+1)$$

$$6x^2 = 3x + 3$$

$$6x^2 - 3x - 3 = 0$$

$$3(2x^2 - x - 1) = 0$$

$$3(2x+1)(x-1) = 0$$

$$x = -\tfrac{1}{2} \quad \text{or} \quad x = 1$$

Check $x = -\tfrac{1}{2}$: $-3 = -3$ *True*
Check $x = 1$: $\tfrac{3}{2} = \tfrac{3}{2}$ *True*

Thus, the solution set is $\left\{-\tfrac{1}{2}, 1\right\}$.

21. $$\frac{2x}{x-3} + \frac{1}{x+3} = \frac{-6}{x^2 - 9}$$

$$\frac{2x}{x-3} + \frac{1}{x+3} = \frac{-6}{(x+3)(x-3)}$$

$$(x+3)(x-3)\left(\frac{2x}{x-3} + \frac{1}{x+3}\right)$$

$$= (x+3)(x-3)\left(\frac{-6}{(x+3)(x-3)}\right)$$

Multiply by LCD, (x + 3)(x − 3).

$$2x(x+3) + 1(x-3) = -6$$

$$2x^2 + 6x + x - 3 = -6$$

$$2x^2 + 7x + 3 = 0$$

$$(2x+1)(x+3) = 0$$

$$x = -\tfrac{1}{2} \quad \text{or} \quad x = -3$$

x cannot equal -3 because the denominator $x + 3$ would equal 0.

Check $x = -\tfrac{1}{2}$: $\tfrac{2}{7} + \tfrac{2}{5} = \tfrac{24}{35}$ *True*

Thus, the solution set is $\left\{-\tfrac{1}{2}\right\}$.

22.   Solve $F = \dfrac{k}{d-D}$ for D.

$$(d-D)(F) = (d-D)\left(\dfrac{k}{d-D}\right) \quad \begin{array}{l}\textit{Multiply by}\\ \textit{LCD, } d-D.\end{array}$$

$$(d-D)(F) = k$$
$$dF - DF = k$$
$$-DF = k - dF$$
$$DF = dF - k$$
$$D = \dfrac{dF-k}{F}, \quad \text{or} \quad D = d - \dfrac{k}{F}$$

23.   Let $x =$ the rate of the current.

	d	r	t
Upstream	20	$7-x$	$\dfrac{20}{7-x}$
Downstream	50	$7+x$	$\dfrac{50}{7+x}$

The times are equal, so

$$\dfrac{20}{7-x} = \dfrac{50}{7+x}.$$

$$(7-x)(7+x)\left(\dfrac{20}{7-x}\right) = (7-x)(7+x)\left(\dfrac{50}{7+x}\right)$$

$$\textit{Multiply by LCD, } (7-x)(7+x).$$

$$20(7+x) = 50(7-x)$$
$$140 + 20x = 350 - 50x$$
$$70x = 210$$
$$x = 3$$

The rate of the current is 3 miles per hour.

24.   Let $x =$ the time required for Sanford and his neighbor to paint the room working together.

	Rate	Time Working Together	Fractional Part of the Job Done when Working Together
Sanford	$\frac{1}{5}$	x	$\frac{1}{5}x$
Neighbor	$\frac{1}{4}$	x	$\frac{1}{4}x$

Working together, they do 1 whole job, so

$$\tfrac{1}{5}x + \tfrac{1}{4}x = 1.$$

$$20\left(\tfrac{1}{5}x + \tfrac{1}{4}x\right) = 20(1) \quad \begin{array}{l}\textit{Multiply by}\\ \textit{LCD, 20.}\end{array}$$

$$20\left(\tfrac{1}{5}x\right) + 20\left(\tfrac{1}{4}x\right) = 20$$
$$4x + 5x = 20$$
$$9x = 20$$
$$x = \tfrac{20}{9}, \quad \text{or} \quad 2\tfrac{2}{9}$$

They can paint the room in $2\frac{2}{9}$ hours.

25.   Since x varies directly as y, there is a constant k such that $x = ky$. First find the value of k.

$$x = ky$$
$$12 = k \cdot 4 \quad \textit{Let x = 12, y = 4.}$$
$$k = \tfrac{12}{4} = 3$$

Since $x = ky$ and $k = 3$,

$$x = 3y.$$

Now find x when $y = 9$.

$$x = 3(9) = 27$$

26.   The length of time L that it takes for fruit to ripen during the growing season varies inversely as the average maximum temperature T during the season, so

$$L = \dfrac{k}{T}.$$

$$25 = \dfrac{k}{80} \quad \textit{Let L = 25, T = 80.}$$

$$k = 25 \cdot 80 = 2000$$

So $L = \frac{2000}{T}$ and when $T = 75$,

$$L = \dfrac{2000}{75} = \dfrac{80}{3}, \quad \text{or} \quad 26\tfrac{2}{3}.$$

It would take 27 days (to the nearest whole number) for fruit to ripen with an average maximum temperature of $75°$.

Cumulative Review Exercises (Chapters 1–7)

1. $3 + 4\left(\tfrac{1}{2} - \tfrac{3}{4}\right)$
$= 3 + 4\left(\tfrac{2}{4} - \tfrac{3}{4}\right)$
$= 3 + 4\left(-\tfrac{1}{4}\right) \quad \textit{Parentheses}$
$= 3 + (-1) \quad \textit{Multiplication}$
$= 2 \quad \textit{Addition}$

2. $3(2y - 5) = 2 + 5y$
$6y - 15 = 2 + 5y$
$y = 17$

The solution set is $\{17\}$.

3.   Solve $A = \tfrac{1}{2}bh$ for b.

$$2 \cdot A = 2 \cdot \tfrac{1}{2}bh \quad \textit{Multiply by 2.}$$
$$2A = bh$$
$$\dfrac{2A}{h} = b \quad \textit{Divide by h.}$$

4. $\dfrac{2+m}{2-m} = \dfrac{3}{4}$
$4(2+m) = 3(2-m) \quad \textit{Cross multiply.}$
$8 + 4m = 6 - 3m$
$7m = -2$
$m = -\tfrac{2}{7}$

The solution set is $\left\{-\tfrac{2}{7}\right\}$.

5. $5y \leq 6y + 8$

$-y \leq 8$

$y \geq -8$ *Reverse the inequality symbol.*

The solution set is $[-8, \infty)$.

6. $4x + 3y = -12$

(a) Let $y = 0$ to find the x-intercept.

$$4x + 3(0) = -12$$
$$4x = -12$$
$$x = -3$$

The x-intercept is $(-3, 0)$.

(b) Let $x = 0$ to find the y-intercept.

$$4(0) + 3y = -12$$
$$3y = -12$$
$$y = -4$$

The y-intercept is $(0, -4)$.

7. $y = -3x + 2$

This is an equation of a line.

If $x = 0$, $y = 2$; so the y-intercept is $(0, 2)$.
If $x = 1$, $y = -1$; and if $x = 2$, $y = -4$.

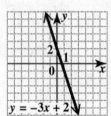

8. $y = -x^2 + 1$

This is the equation of a parabola opening downward with a y-intercept $(0, 1)$.
If $x = +2$ or -2, $y = -3$.

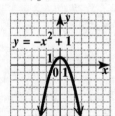

9. $4x - y = -7$ (1)
$5x + 2y = 1$ (2)

$$\begin{array}{rll} 8x - 2y &= -14 & (3)\quad 2 \times \text{Eq.}(1) \\ 5x + 2y &= 1 & (2) \\ \hline 13x &= -13 & \textit{Add} \text{ (3) } \textit{and} \text{ (2).} \\ x &= -1 \end{array}$$

Substitute -1 for x in equation (1).

$$4x - y = -7$$
$$4(-1) - y = -7$$

$$-4 - y = -7$$
$$-y = -3$$
$$y = 3$$

The solution set is $\{(-1, 3)\}$.

10. $5x + 2y = 7$ (1)
$10x + 4y = 12$ (2)

Multiply equation (2) by $\frac{1}{2}$.

$$5x + 2y = 6 \quad (3) \quad \tfrac{1}{2} \times \text{Eq.}(2)$$

The left sides of (1) and (3) are equal, so they can't be equal to different numbers (7 and 6). There are no solutions, so the solution set is $\emptyset$.

11. $\dfrac{(2x^3)^{-1} \cdot x}{2^3 x^5} = \dfrac{2^{-1}(x^3)^{-1} \cdot x}{2^3 x^5} = \dfrac{2^{-1} x^{-3} x}{2^3 x^5}$

$$= \dfrac{2^{-1} x^{-2}}{2^3 x^5} = \dfrac{1}{2^1 \cdot 2^3 \cdot x^2 \cdot x^5}$$

$$= \dfrac{1}{2^4 x^7}$$

12. $\dfrac{(m^{-2})^3 m}{m^5 m^{-4}} = \dfrac{m^{-6} m}{m^5 m^{-4}} = \dfrac{m \cdot m^4}{m^5 \cdot m^6}$

$$= \dfrac{m^5}{m^{11}} = \dfrac{1}{m^6}$$

13. $(2k^2 + 3k) - (k^2 + k - 1)$
$= 2k^2 + 3k - k^2 - k + 1$
$= k^2 + 2k + 1$

14. $(2a - b)^2 = (2a)^2 - 2(2a)(b) + (b)^2$
$= 4a^2 - 4ab + b^2$

15. $(y^2 + 3y + 5)(3y - 1)$

Multiply vertically.

$$\begin{array}{r} y^2 + 3y + 5 \\ 3y - 1 \\ \hline -y^2 - 3y - 5 \\ 3y^3 + 9y^2 + 15y \\ \hline 3y^3 + 8y^2 + 12y - 5 \end{array}$$

16. $\dfrac{12p^3 + 2p^2 - 12p + 4}{2p - 2}$

$$\begin{array}{r} 6p^2 + 7p + 1 \\ 2p - 2 \overline{\smash{)}12p^3 + 2p^2 - 12p + 4} \\ \underline{12p^3 - 12p^2} \\ 14p^2 - 12p \\ \underline{14p^2 - 14p} \\ 2p + 4 \\ \underline{2p - 2} \\ 6 \end{array}$$

The result is

$$6p^2 + 7p + 1 + \frac{6}{2p-2}$$

$$= 6p^2 + 7p + 1 + \frac{2 \cdot 3}{2(p-1)}$$

$$= 6p^2 + 7p + 1 + \frac{3}{p-1}.$$

17. $8t^2 + 10tv + 3v^2$

$$= 8t^2 + 6tv + 4tv + 3v^2 \quad \begin{array}{l} 6 \cdot 4 = 24; \\ 6 + 4 = 10 \end{array}$$

$$= (8t^2 + 6tv) + (4tv + 3v^2)$$

$$= 2t(4t + 3v) + v(4t + 3v)$$

$$= (4t + 3v)(2t + v)$$

18. $8r^2 - 9rs + 12s^2$

To factor this polynomial by the grouping method, we must find two integers whose product is $(8)(12) = 96$ and whose sum is -9. There is no pair of integers that satisfies both of these conditions, so the polynomial is *prime*.

19. $16x^4 - 1$

$$= (4x^2)^2 - (1)^2$$

$$= (4x^2 + 1)(4x^2 - 1)$$

$$= (4x^2 + 1)[(2x)^2 - (1)^2]$$

$$= (4x^2 + 1)(2x + 1)(2x - 1)$$

20.
$$r^2 = 2r + 15$$
$$r^2 - 2r - 15 = 0$$
$$(r + 3)(r - 5) = 0$$

$$r + 3 = 0 \quad \text{or} \quad r - 5 = 0$$
$$r = -3 \quad \text{or} \quad r = 5$$

The solution set is $\{-3, 5\}$.

21. $(r - 5)(2r + 1)(3r - 2) = 0$

$$r - 5 = 0 \quad \text{or} \quad 2r + 1 = 0 \quad \text{or} \quad 3r - 2 = 0$$
$$2r = -1 \quad \text{or} \quad 3r = 2$$
$$r = 5 \quad \text{or} \quad r = -\tfrac{1}{2} \quad \text{or} \quad r = \tfrac{2}{3}$$

The solution set is $\left\{5, -\tfrac{1}{2}, \tfrac{2}{3}\right\}$.

22. Let $x =$ the lesser number
Then $x + 4 =$ the greater number.

The product of the numbers is 2 less than the lesser number translates to

$$x(x + 4) = x - 2.$$
$$x^2 + 4x = x - 2$$
$$x^2 + 3x + 2 = 0$$
$$(x + 2)(x + 1) = 0$$
$$x = -2 \quad \text{or} \quad x = -1$$

The lesser number can be either -2 or -1.

23. Let $w =$ the width of the rectangle.
Then $2w - 2 =$ the length of the rectangle.

Use the formula $A = LW$ with the area $= 60$.

$$60 = (2w - 2)w$$
$$60 = 2w^2 - 2w$$
$$0 = 2w^2 - 2w - 60$$
$$0 = 2(w^2 - w - 30)$$
$$0 = 2(w - 6)(w + 5)$$
$$w = 6 \quad \text{or} \quad w = -5$$

Discard -5 because the width cannot be negative. The width of the rectangle is 6 meters.

24. All of the given expressions are equal to 1 for all real numbers for which they are defined. However, expressions **B**, **C**, and **D** all have one or more values for which the expression is undefined and therefore cannot be equal to 1 at these values. Since $k^2 + 2$ is *always* positive, the denominator in expression **A** is never equal to zero. This expression is defined and equal to 1 for all real numbers, so the correct choice is **A**.

25. The appropriate choice is **D** since

$$\frac{-(3x + 4)}{7} = \frac{-3x - 4}{7}$$

$$\neq \frac{4 - 3x}{7}.$$

26. $\dfrac{5}{q} - \dfrac{1}{q} = \dfrac{5 - 1}{q} = \dfrac{4}{q}$

27. $\dfrac{3}{7} + \dfrac{4}{r} = \dfrac{3 \cdot r}{7 \cdot r} + \dfrac{4 \cdot 7}{r \cdot 7} \quad LCD = 7r$

$$= \frac{3r + 28}{7r}$$

28.
$$\frac{4}{5q - 20} - \frac{1}{3q - 12}$$

$$= \frac{4}{5(q - 4)} - \frac{1}{3(q - 4)}$$

$$= \frac{4 \cdot 3}{5(q - 4) \cdot 3} - \frac{1 \cdot 5}{3(q - 4) \cdot 5}$$

$$LCD = 5 \cdot 3 \cdot (q - 4) = 15(q - 4)$$

$$= \frac{12 - 5}{15(q - 4)} = \frac{7}{15(q - 4)}$$

29. $\dfrac{2}{k^2 + k} - \dfrac{3}{k^2 - k}$

$= \dfrac{2}{k(k+1)} - \dfrac{3}{k(k-1)}$

$= \dfrac{2(k-1)}{k(k+1)(k-1)} - \dfrac{3(k+1)}{k(k-1)(k+1)}$

$\qquad\qquad\qquad LCD = k(k+1)(k-1)$

$= \dfrac{2(k-1) - 3(k+1)}{k(k+1)(k-1)}$

$= \dfrac{2k - 2 - 3k - 3}{k(k+1)(k-1)}$

$= \dfrac{-k - 5}{k(k+1)(k-1)}$

30. $\dfrac{7z^2 + 49z + 70}{16z^2 + 72z - 40} \div \dfrac{3z + 6}{4z^2 - 1}$

$= \dfrac{7z^2 + 49z + 70}{16z^2 + 72z - 40} \cdot \dfrac{4z^2 - 1}{3z + 6}$

$= \dfrac{7(z^2 + 7z + 10)}{8(2z^2 + 9z - 5)} \cdot \dfrac{(2z+1)(2z-1)}{3(z+2)}$

$= \dfrac{7(z+5)(z+2)}{8(2z-1)(z+5)} \cdot \dfrac{(2z+1)(2z-1)}{3(z+2)}$

$= \dfrac{7(2z+1)}{8 \cdot 3} = \dfrac{7(2z+1)}{24}$

31. $\dfrac{\dfrac{4}{a} + \dfrac{5}{2a}}{\dfrac{7}{6a} - \dfrac{1}{5a}}$

$= \dfrac{\left(\dfrac{4}{a} + \dfrac{5}{2a}\right) \cdot 30a}{\left(\dfrac{7}{6a} - \dfrac{1}{5a}\right) \cdot 30a}$ *Multiply by LCD, 30a.*

$= \dfrac{\dfrac{4}{a}(30a) + \dfrac{5}{2a}(30a)}{\dfrac{7}{6a}(30a) - \dfrac{1}{5a}(30a)}$

$= \dfrac{4 \cdot 30 + 5 \cdot 15}{7 \cdot 5 - 1 \cdot 6}$

$= \dfrac{120 + 75}{35 - 6} = \dfrac{195}{29}$

32. $\dfrac{r+2}{5} = \dfrac{r-3}{3}$

$15\left(\dfrac{r+2}{5}\right) = 15\left(\dfrac{r-3}{3}\right)$ *Multiply by LCD, 15.*

$3(r+2) = 5(r-3)$

$3r + 6 = 5r - 15$

$21 = 2r$

$\dfrac{21}{2} = r$

Check $r = \dfrac{21}{2}$: $\dfrac{5}{2} = \dfrac{5}{2}$ *True*

The solution set is $\left\{ \dfrac{21}{2} \right\}$.

33. $\dfrac{1}{x} = \dfrac{1}{x+1} + \dfrac{1}{2}$

$2x(x+1)\left(\dfrac{1}{x}\right) = 2x(x+1)\left(\dfrac{1}{x+1} + \dfrac{1}{2}\right)$

Multiply by LCD, 2x(x + 1).

$2(x+1) = 2x(x+1)\left(\dfrac{1}{x+1}\right)$
$\qquad\qquad + 2x(x+1)\left(\dfrac{1}{2}\right)$

$2(x+1) = 2x + x(x+1)$

$2x + 2 = 2x + x^2 + x$

$0 = x^2 + x - 2$

$0 = (x+2)(x-1)$

$x = -2 \quad \text{or} \quad x = 1$

Check $x = -2$: $-\dfrac{1}{2} = -1 + \dfrac{1}{2}$ *True*

Check $x = 1$: $1 = \dfrac{1}{2} + \dfrac{1}{2}$ *True*

Thus, the solution set is $\{-2, 1\}$.

34. Let $x =$ the number of hours it will take Jody and Pat to weed the yard working together.

	Rate	Time Working Together	Fractional Part of the Job Done when Working Together
Jody	$\dfrac{1}{3}$	x	$\dfrac{1}{3}x$
Pat	$\dfrac{1}{2}$	x	$\dfrac{1}{2}x$

Working together, they can do 1 whole job, so

$\dfrac{1}{3}x + \dfrac{1}{2}x = 1.$

$6\left(\dfrac{1}{3}x + \dfrac{1}{2}x\right) = 6 \cdot 1 \quad LCD = 6$

$6\left(\dfrac{1}{3}x\right) + 6\left(\dfrac{1}{2}x\right) = 6$

$2x + 3x = 6$

$5x = 6$

$x = \dfrac{6}{5}, \quad \text{or} \quad 1\dfrac{1}{5}$

If Jody and Pat worked together, it would take them $1\dfrac{1}{5}$ hours to weed the yard.

35. The circumference C of a circle varies directly as its radius r, so $C = kr$.

$9.42 = k(1.5)$ *Let C = 9.42, r = 1.5.*

$k = \dfrac{9.42}{1.5} = 6.28$

So $C = 6.28r$ and when $r = 5.25$,

$C = 6.28(5.25) = 32.97.$

Using $k = 6.28$, a circle with radius 5.25 inches has circumference 32.97 inches.

CHAPTER 8 ROOTS AND RADICALS

8.1 Evaluating Roots

8.1 Now Try Exercises

N1. The square roots of 81 are 9 and -9 because

$$9 \cdot 9 = 81 \quad \text{and} \quad (-9)(-9) = 81.$$

N2. **(a)** $\sqrt{400}$ is the positive square root of 400.

$$\sqrt{400} = 20$$

(b) $-\sqrt{169}$ is the negative square root of 169.

$$-\sqrt{169} = -13$$

(c) $\sqrt{\frac{100}{121}}$ is the positive square root of $\frac{100}{121}$.

$$\sqrt{\frac{100}{121}} = \frac{10}{11}$$

N3. **(a)** The square of $\sqrt{15}$ is

$$\left(\sqrt{15}\right)^2 = 15.$$

(b) The square of $-\sqrt{23}$ is

$$\left(-\sqrt{23}\right)^2 = \left(\sqrt{23}\right)^2 = 23.$$

(c) The square of $\sqrt{2k^2 + 5}$ is

$$\left(\sqrt{2k^2 + 5}\right)^2 = 2k^2 + 5.$$

N4. **(a)** 31 is not a perfect square, so $\sqrt{31}$ is irrational.

(b) 900 is a perfect square, so $\sqrt{900} = 30$ is rational.

(c) There is no real number whose square is -16, so $\sqrt{-16}$ is not a real number.

N5. Use the square root key of a calculator to find a decimal approximation for each root. Answers are given to the nearest thousandth.

(a) $\sqrt{51} \approx 7.141$

(b) $-\sqrt{360} \approx -18.974$

N6. Substitute the given values in the Pythagorean theorem, $c^2 = a^2 + b^2$. Then solve for the variable that is not given.

(a) $c^2 = a^2 + b^2$
$\quad c^2 = 5^2 + 12^2 \quad$ *Let a = 5, b = 12.*
$\quad c^2 = 25 + 144 \quad$ *Square.*
$\quad c^2 = 169 \quad\quad$ *Add.*
$\quad\; c = \sqrt{169} \quad$ *Take square root.*
$\quad\; c = 13 \quad\quad 13^2 = 169$

(b) $c^2 = a^2 + b^2$
$\quad 14^2 = 9^2 + b^2 \quad$ *Let c = 14, a = 9.*
$\quad\; 196 = 81 + b^2$
$\quad\; 115 = b^2 \quad\quad$ *Subtract 81.*
$\quad\quad\; b = \sqrt{115} \quad$ *Take square root.*
$\quad\quad\; b \approx 10.724 \quad$ *Use a calculator.*

N7. Let c = the length of the diagonal. Use the Pythagorean theorem, since the width, length, and diagonal of a rectangle form a right triangle.

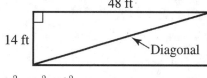

$c^2 = a^2 + b^2$
$c^2 = 14^2 + 48^2 \quad$ *Let a = 14, b = 48.*
$c^2 = 196 + 2304$
$c^2 = 2500$
$\; c = \sqrt{2500}$
$\; c = 50$

The diagonal is 50 feet long.

N8. Let $(x_1, y_1) = (-4, 3)$ and $(x_2, y_2) = (-7, 1)$.

Use the distance formula.

$$\begin{aligned} d &= \sqrt{(x_2 - x_1)^2 + (y_2 - y_1)^2} \\ &= \sqrt{[-7 - (-4)]^2 + (1 - 3)^2} \\ &= \sqrt{(-3)^2 + (-2)^2} \\ &= \sqrt{9 + 4} \\ &= \sqrt{13} \end{aligned}$$

N9. **(a)** $\sqrt[3]{343} = 7$ because $7^3 = 343$.

(b) $\sqrt[3]{-1000} = -10$ because $(-10)^3 = -1000$.

(c) $\sqrt[3]{27} = 3$ because $3^3 = 27$.

N10. **(a)** $\sqrt[4]{625} = 5$ because $5^4 = 625$.

(b) $\sqrt[4]{-625}$ is not a real number because a fourth power of a real number cannot be negative.

(c) $\sqrt[5]{3125} = 5$ because $5^5 = 3125$.

(d) $\sqrt[5]{-3125} = -5$ because $(-5)^5 = -3125$.

8.1 Section Exercises

1. Every positive number has two real square roots. This statement is *true*. One of the real square roots is a positive number and the other is its opposite.

3. Every nonnegative number has two real square roots. This statement is *false* since zero is a nonnegative number that has only one square root, namely 0.

5. The cube root of every nonzero real number has the same sign as the number itself. This statement is *true*. The cube root of a positive real number is positive and the cube root of a negative real number is negative. The cube root of 0 is 0.

7. The square roots of 9 are -3 and 3 because $(-3)(-3) = 9$ and $3 \cdot 3 = 9$.

9. The square roots of 64 are -8 and 8 because $(-8)(-8) = 64$ and $8 \cdot 8 = 64$.

11. The square roots of 169 are -13 and 13 because $(-13)(-13) = 169$ and $13 \cdot 13 = 169$.

13. The square roots of $\frac{25}{196}$ are $-\frac{5}{14}$ and $\frac{5}{14}$ because $(-\frac{5}{14})(-\frac{5}{14}) = \frac{25}{196}$ and $\frac{5}{14} \cdot \frac{5}{14} = \frac{25}{196}$.

15. The square roots of 900 are -30 and 30 because $(-30)(-30) = 900$ and $30 \cdot 30 = 900$.

17. $\sqrt{1}$ represents the positive square root of 1. Since $1 \cdot 1 = 1$, $\sqrt{1} = 1$.

19. $\sqrt{49}$ represents the positive square root of 49. Since $7 \cdot 7 = 49$, $\sqrt{49} = 7$.

21. $-\sqrt{256}$ represents the negative square root of 256. Since $16 \cdot 16 = 256$, $-\sqrt{256} = -16$.

23. $-\sqrt{\frac{144}{121}}$ represents the negative square root of $\frac{144}{121}$. Since $\frac{12}{11} \cdot \frac{12}{11} = \frac{144}{121}$, $-\sqrt{\frac{144}{121}} = -\frac{12}{11}$.

25. $\sqrt{0.64}$ represents the positive square root of 0.64. Since $0.8 \cdot 0.8 = 0.64$, $\sqrt{0.64} = 0.8$.

27. $\sqrt{-121}$ is not a real number because there is no real number whose square is -121.

29. $-\sqrt{-49}$ is not a real number because there is no real number whose square is -49. The leading negative sign is irrelevant.

31. The square of $\sqrt{19}$ is

$$\left(\sqrt{19}\right)^2 = 19,$$

by the definition of square root.

33. The square of $-\sqrt{19}$ is

$$\left(-\sqrt{19}\right)^2 = 19,$$

by the definition of square root and since the square of a negative number is positive.

35. The square of $\sqrt{\frac{2}{3}}$ is

$$\left(\sqrt{\frac{2}{3}}\right)^2 = \frac{2}{3},$$

by the definition of square root.

37. The square of $\sqrt{3x^2 + 4}$ is

$$\left(\sqrt{3x^2 + 4}\right)^2 = 3x^2 + 4.$$

39. For the statement "$\sqrt{a}$ represents a positive number" to be true, a must be positive because the square root of a negative number is not a real number and $\sqrt{0} = 0$.

41. For the statement "$\sqrt{a}$ is not a real number" to be true, a must be negative.

43. $\sqrt{25}$
The number 25 is a perfect square, 5^2, so $\sqrt{25}$ is a *rational* number.

$$\sqrt{25} = 5$$

45. $\sqrt{29}$
Because 29 is not a perfect square, $\sqrt{29}$ is *irrational*. Using a calculator, we obtain

$$\sqrt{29} \approx 5.385.$$

47. $-\sqrt{64}$
The number 64 is a perfect square, 8^2, so $-\sqrt{64}$ is *rational*.

$$-\sqrt{64} = -8$$

49. $-\sqrt{300}$
The number 300 is not a perfect square, so $-\sqrt{300}$ is *irrational*. Using a calculator, we obtain

$$-\sqrt{300} \approx -17.321.$$

51. $\sqrt{-29}$
There is no real number whose square is -29. Therefore, $\sqrt{-29}$ is *not a real number*.

53. $\sqrt{1200}$
Because 1200 is not a perfect square, $\sqrt{1200}$ is *irrational*. Using a calculator, we obtain

$$\sqrt{1200} \approx 34.641.$$

55. Since $81 < 94 < 100$, $\sqrt{81} = 9$, and $\sqrt{100} = 10$, we conclude that $\sqrt{94}$ is between 9 and 10.

57. Since $49 < 51 < 64$, $\sqrt{49} = 7$, and $\sqrt{64} = 8$, we conclude that $\sqrt{51}$ is between 7 and 8.

59. Since $36 < 40 < 49$, $\sqrt{36} = 6$, and $\sqrt{49} = 7$, we conclude that $-\sqrt{40}$ is between -7 and -6.

61. Since $16 < 23.2 < 25$, $\sqrt{16} = 4$, and $\sqrt{25} = 5$, we conclude that $\sqrt{23.2}$ is between 4 and 5.

63. $\sqrt{103} \approx \sqrt{100} = 10$

$\sqrt{48} \approx \sqrt{49} = 7$

The best estimate for the length and width of the rectangle is 10 by 7, choice **C**.

65. $a = 8, b = 15$

Substitute the given values in the Pythagorean theorem and then solve for c^2.

$$c^2 = a^2 + b^2$$
$$c^2 = 8^2 + 15^2$$
$$= 64 + 225$$
$$= 289$$

Now find the positive square root of 289 to obtain the length of the hypotenuse, c.

$$c = \sqrt{289} = 17$$

67. $a = 6, c = 10$

Substitute the given values in the Pythagorean theorem and then solve for b^2.

$$c^2 = a^2 + b^2$$
$$10^2 = 6^2 + b^2$$
$$100 = 36 + b^2$$
$$64 = b^2$$

Now find the positive square root of 64 to obtain the length of the leg b.

$$b = \sqrt{64} = 8$$

69. $a = 11, b = 4$

$$c^2 = a^2 + b^2$$
$$c^2 = 11^2 + 4^2$$
$$= 121 + 16$$
$$= 137$$
$$c = \sqrt{137} \approx 11.705$$

71. The given information involves a right triangle with hypotenuse 25 centimeters and a leg of length 7 centimeters. Let a represent the length of the other leg, and use the Pythagorean theorem.

$$c^2 = a^2 + b^2$$
$$25^2 = a^2 + 7^2$$
$$625 = a^2 + 49$$
$$576 = a^2$$
$$a = \sqrt{576} = 24$$

The length of the rectangle is 24 centimeters.

73. *Step 2*

Let x represent the vertical distance of the kite above Tyler's hand. The kite string forms the hypotenuse of a right triangle.

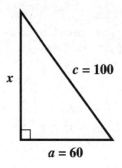

Step 3

Use the Pythagorean theorem.

$$a^2 + x^2 = c^2$$
$$60^2 + x^2 = 100^2$$

Step 4

$$3600 + x^2 = 10,000$$
$$x^2 = 6400$$
$$x = \sqrt{6400} = 80$$

Step 5

The kite is 80 feet above his hand.

Step 6

From the figure, we see that we must have

$$60^2 + 80^2 \stackrel{?}{=} 100^2$$
$$3600 + 6400 = 10,000. \quad \textit{True}$$

75. *Step 2*

Let x represent the distance from R to S.

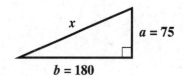

Step 3

Triangle RST is a right triangle, so we can use the Pythagorean theorem.

$$x^2 = a^2 + b^2$$
$$x^2 = 75^2 + 180^2$$
$$= 5625 + 32,400$$

Step 4

$$x^2 = 38,025$$
$$x = \sqrt{38,025} = 195$$

Step 5

The distance across the lake is 195 feet.

Step 6

From the figure, we see that we must have

$$75^2 + 180^2 \stackrel{?}{=} 195^2$$
$$5625 + 32,400 = 38,025. \quad \textit{True}$$

77. Let $a = 4.5$ and $c = 12$. Use the Pythagorean theorem to find b, the distance along the ground as described in the problem.

$$c^2 = a^2 + b^2$$
$$12^2 = (4.5)^2 + b^2$$
$$144 = 20.25 + b^2$$
$$123.75 = b^2$$
$$b = \sqrt{123.75} \approx 11.1$$

The distance from the base of the tree to the point where the broken part touches the ground is 11.1 feet (to the nearest tenth).

79. Refer to the right triangle shown in the figure in the textbook. Note that the given distances are the lengths of the hypotenuse (193.0 feet) and one of the legs (110.0 feet) of the triangle. Use the Pythagorean theorem with $a = 110.0$, $c = 193.0$, and $b = $ the height of the building.

$$a^2 + b^2 = c^2$$
$$(110.0)^2 + b^2 = (193.0)^2$$
$$12{,}100 + b^2 = 37{,}249$$
$$b^2 = 25{,}149$$
$$b = \sqrt{25{,}149} \approx 158.6$$

The height of the building, to the nearest tenth, is 158.6 feet.

81. Use the Pythagorean theorem with $a = 5$, $b = 8$, and $c = x$.

$$c^2 = a^2 + b^2$$
$$x^2 = 5^2 + 8^2$$
$$= 25 + 64$$
$$= 89$$
$$x = \sqrt{89} \approx 9.434$$

83. The area of the large square on the left is $c \cdot c = c^2$. The side of the small square inside that figure has length $b - a$, so the area of that square is $(b - a)^2$.

The area of one of the rectangles on the right is $a \cdot b$, so the sum of the areas of the two rectangles in the figure on the right is $2ab$. Since the areas of the two figures are the same, we have

$$c^2 = 2ab + (b - a)^2$$
$$c^2 = 2ab + b^2 - 2ab + a^2$$
$$c^2 = a^2 + b^2,$$

which is the Pythagorean theorem.

85. Let $(x_1, y_1) = (5, 7)$ and $(x_2, y_2) = (1, 4)$. Use the distance formula.

$$d = \sqrt{(x_2 - x_1)^2 + (y_2 - y_1)^2}$$
$$= \sqrt{(1 - 5)^2 + (4 - 7)^2}$$

$$= \sqrt{(-4)^2 + (-3)^2}$$
$$= \sqrt{16 + 9}$$
$$= \sqrt{25} = 5$$

87. Let $(x_1, y_1) = (2, 9)$ and $(x_2, y_2) = (-3, -3)$. Use the distance formula.

$$d = \sqrt{(x_2 - x_1)^2 + (y_2 - y_1)^2}$$
$$= \sqrt{(-3 - 2)^2 + (-3 - 9)^2}$$
$$= \sqrt{(-5)^2 + (-12)^2}$$
$$= \sqrt{25 + 144}$$
$$= \sqrt{169} = 13$$

89. Let $(x_1, y_1) = (-1, -2)$ and $(x_2, y_2) = (-3, 1)$. Use the distance formula.

$$d = \sqrt{(x_2 - x_1)^2 + (y_2 - y_1)^2}$$
$$= \sqrt{[-3 - (-1)]^2 + [1 - (-2)]^2}$$
$$= \sqrt{(-2)^2 + 3^2}$$
$$= \sqrt{4 + 9} = \sqrt{13}$$

91. Let $(x_1, y_1) = \left(-\frac{1}{4}, \frac{2}{3}\right)$ and $(x_2, y_2) = \left(\frac{3}{4}, -\frac{1}{3}\right)$.

$$d = \sqrt{(x_2 - x_1)^2 + (y_2 - y_1)^2}$$
$$= \sqrt{\left[\frac{3}{4} - \left(-\frac{1}{4}\right)\right]^2 + \left(-\frac{1}{3} - \frac{2}{3}\right)^2}$$
$$= \sqrt{1^2 + (-1)^2}$$
$$= \sqrt{1 + 1} = \sqrt{2}$$

93.

$1^3 = \underline{1}$	$6^3 = \underline{216}$
$2^3 = \underline{8}$	$7^3 = \underline{343}$
$3^3 = \underline{27}$	$8^3 = \underline{512}$
$4^3 = \underline{64}$	$9^3 = \underline{729}$
$5^3 = \underline{125}$	$10^3 = \underline{1000}$

95. $\sqrt[3]{1} = 1$ because $1^3 = 1$.

97. $\sqrt[3]{125} = 5$ because $5^3 = 125$.

99. $\sqrt[3]{-27} = -3$ because $(-3)^3 = -27$.

101. $\sqrt[3]{-216} = -6$ because $(-6)^3 = -216$.

103. $\sqrt[3]{-8} = -2$ because $(-2)^3 = -8$. Thus, $-\sqrt[3]{-8} = -(-2) = 2$.

105. $\sqrt[4]{625} = 5$ because 5 is positive and $5^4 = 625$.

107. $\sqrt[4]{1296} = 6$ because 6 is positive and $6^4 = 1296$.

109. $\sqrt[4]{-1}$ is not a real number because the fourth power of a real number cannot be negative.

111. $\sqrt[4]{81} = 3$ because 3 is positive and $3^4 = 81$. Thus, $-\sqrt[4]{81} = -3$.

113. $\sqrt[5]{-1024} = -4$ because $(-4)^5 = -1024$.

115. $\sqrt[3]{12} \approx 2.289$

117. $\sqrt[3]{130.6} \approx 5.074$

119. $\sqrt[3]{-87} = -\sqrt[3]{87} \approx -4.431$

121. $72 = 2 \cdot 36$
$\qquad = 2 \cdot 2 \cdot 18$
$\qquad = 2 \cdot 2 \cdot 2 \cdot 9$
$\qquad = 2 \cdot 2 \cdot 2 \cdot 3 \cdot 3$

The prime factored form of 72 is $2^3 \cdot 3^2$.

123. $40 = 2 \cdot 20$
$\qquad = 2 \cdot 2 \cdot 10$
$\qquad = 2 \cdot 2 \cdot 2 \cdot 5$

The prime factored form of 40 is $2^3 \cdot 5$.

125. 23 is a prime number.

8.2 Multiplying, Dividing, and Simplifying Radicals

8.2 Now Try Exercises

N1. (a) $\sqrt{5} \cdot \sqrt{11} = \sqrt{5 \cdot 11} = \sqrt{55}$ *Product rule*

(b) $\sqrt{11} \cdot \sqrt{11} = 11$

(c) $\sqrt{7} \cdot \sqrt{k} = \sqrt{7 \cdot k} = \sqrt{7k}$ *Product rule*
Assume $k \geq 0$.

N2. (a) $\sqrt{28} = \sqrt{4 \cdot 7}$ *4 is a perfect square.*
$\qquad = \sqrt{4} \cdot \sqrt{7}$ *Product rule*
$\qquad = 2\sqrt{7}$

(b) $\sqrt{99} = \sqrt{9 \cdot 11}$ *9 is a perfect square.*
$\qquad = \sqrt{9} \cdot \sqrt{11}$ *Product rule*
$\qquad = 3\sqrt{11}$

(c) $\sqrt{85}$ cannot be simplified further because 85 has no perfect square factors (except 1).

N3. (a) $\sqrt{16} \cdot \sqrt{50} = 4\sqrt{50}$ $\sqrt{16} = 4$
$\qquad = 4\sqrt{25 \cdot 2}$ *Factor; 25 is a perfect square.*
$\qquad = 4\sqrt{25} \cdot \sqrt{2}$ *Product rule*
$\qquad = 4 \cdot 5\sqrt{2}$ $\sqrt{25} = 5$
$\qquad = 20\sqrt{2}$ *Multiply.*

(b) $\sqrt{6} \cdot \sqrt{30} = \sqrt{6 \cdot 30}$ *Product rule*
$\qquad = \sqrt{6 \cdot 6 \cdot 5}$ *Factor.*
$\qquad = \sqrt{36} \cdot \sqrt{5}$ *36 is a perfect square.*
$\qquad = 6\sqrt{5}$

N4. (a) $\sqrt{\dfrac{81}{100}} = \dfrac{\sqrt{81}}{\sqrt{100}} = \dfrac{9}{10}$ *Quotient rule*

(b) $\dfrac{\sqrt{245}}{\sqrt{5}} = \sqrt{\dfrac{245}{5}} = \sqrt{49} = 7$

(c) $\sqrt{\dfrac{11}{49}} = \dfrac{\sqrt{11}}{\sqrt{49}} = \dfrac{\sqrt{11}}{7}$

N5. $\dfrac{24\sqrt{39}}{4\sqrt{13}} = \dfrac{24}{4} \cdot \dfrac{\sqrt{39}}{\sqrt{13}}$
$\qquad = 6 \cdot \sqrt{\dfrac{39}{13}}$
$\qquad = 6\sqrt{3}$

N6. $\sqrt{\dfrac{1}{2}} \cdot \sqrt{\dfrac{5}{18}} = \sqrt{\dfrac{1}{2} \cdot \dfrac{5}{18}}$ *Product rule*
$\qquad = \sqrt{\dfrac{5}{36}}$ *Multiply fractions.*
$\qquad = \dfrac{\sqrt{5}}{\sqrt{36}}$ *Quotient rule*
$\qquad = \dfrac{\sqrt{5}}{6}$ $\sqrt{36} = 6$

N7. (a) $\sqrt{16y^8} = 4y^4$ since $(4y^4)^2 = 16y^8$.

(b) $\sqrt{x^5} = \sqrt{x^4} \cdot \sqrt{x} = x^2\sqrt{x}$

(c) $\sqrt{\dfrac{13}{t^2}} = \dfrac{\sqrt{13}}{\sqrt{t^2}} = \dfrac{\sqrt{13}}{t}$ Assume $t \neq 0$.

N8. (a) $\sqrt[3]{250} = \sqrt[3]{125 \cdot 2}$ *125 is a perfect cube.*
$\qquad = \sqrt[3]{125} \cdot \sqrt[3]{2}$ *Product rule*
$\qquad = 5\sqrt[3]{2}$

(b) $\sqrt[4]{48} = \sqrt[4]{16 \cdot 3}$ *16 is a perfect fourth power.*
$\qquad = \sqrt[4]{16} \cdot \sqrt[4]{3}$ *Product rule*
$\qquad = 2\sqrt[4]{3}$

(c) $\sqrt[3]{\dfrac{1}{125}} = \dfrac{\sqrt[3]{1}}{\sqrt[3]{125}}$ *Quotient rule*
$\qquad = \dfrac{1}{5}$

N9. (a) $\sqrt[3]{x^{12}} = x^4$ since $(x^4)^3 = x^{12}$.

(b) $\sqrt[3]{64t^3} = \sqrt[3]{64} \cdot \sqrt[3]{t^3}$ *Product rule*
$\qquad = 4t$ $4^3 = 64$

(c) $\sqrt[3]{40a^7} = \sqrt[3]{8a^6 \cdot 5a}$ *$8a^6$ is a cube.*
$\qquad = \sqrt[3]{8a^6} \cdot \sqrt[3]{5a}$ *Product rule*
$\qquad = 2a^2\sqrt[3]{5a}$ $(2a^2)^3 = 8a^6$

(d) $\sqrt[3]{\dfrac{x^{15}}{1000}} = \dfrac{\sqrt[3]{x^{15}}}{\sqrt[3]{1000}}$ *Quotient rule*
$\qquad = \dfrac{x^5}{10}$

8.2 Section Exercises

1. $\sqrt{3} \cdot \sqrt{5}$

Since 3 and 5 are nonnegative real numbers, the *Product Rule for Radicals* applies. Thus,

$$\sqrt{3} \cdot \sqrt{5} = \sqrt{3 \cdot 5} = \sqrt{15}.$$

3. $\sqrt{2} \cdot \sqrt{11} = \sqrt{2 \cdot 11} = \sqrt{22}$

5. $\sqrt{6} \cdot \sqrt{7} = \sqrt{6 \cdot 7} = \sqrt{42}$

7. $\sqrt{3} \cdot \sqrt{27} = \sqrt{3 \cdot 27} = \sqrt{81}$, or 9

9. $\sqrt{13} \cdot \sqrt{13} = 13$

11. $\sqrt{13} \cdot \sqrt{r} \quad (r \geq 0) = \sqrt{13r}$

13. $\sqrt{47}$ is in simplified form since 47 has no perfect square factor (other than 1).

The other three choices could be simplified as follows.

$$\sqrt{45} = \sqrt{9 \cdot 5} = 3\sqrt{5}$$
$$\sqrt{48} = \sqrt{16 \cdot 3} = 4\sqrt{3}$$
$$\sqrt{44} = \sqrt{4 \cdot 11} = 2\sqrt{11}$$

The correct choice is **A**.

15. $\sqrt{45} = \sqrt{9 \cdot 5} = \sqrt{9} \cdot \sqrt{5} = 3\sqrt{5}$

17. $\sqrt{24} = \sqrt{4 \cdot 6} = \sqrt{4} \cdot \sqrt{6} = 2\sqrt{6}$

19. $\sqrt{90} = \sqrt{9 \cdot 10} = \sqrt{9} \cdot \sqrt{10} = 3\sqrt{10}$

21. $\sqrt{75} = \sqrt{25 \cdot 3} = \sqrt{25} \cdot \sqrt{3} = 5\sqrt{3}$

23. $\sqrt{125} = \sqrt{25 \cdot 5} = \sqrt{25} \cdot \sqrt{5} = 5\sqrt{5}$

25. $145 = 5 \cdot 29$, so 145 has no perfect square factors (except 1) and $\sqrt{145}$ cannot be simplified further.

27. $\sqrt{160} = \sqrt{16 \cdot 10} = \sqrt{16} \cdot \sqrt{10} = 4\sqrt{10}$

29. $-\sqrt{700} = -\sqrt{100 \cdot 7}$
$$= -\sqrt{100} \cdot \sqrt{7} = -10\sqrt{7}$$

31. $3\sqrt{27} = 3\sqrt{9} \cdot \sqrt{3}$
$$= 3 \cdot 3 \cdot \sqrt{3}$$
$$= 9\sqrt{3}$$

33. $5\sqrt{50} = 5\sqrt{25} \cdot \sqrt{2}$
$$= 5 \cdot 5 \cdot \sqrt{2}$$
$$= 25\sqrt{2}$$

35. $a = 13, b = 9$

$$c^2 = a^2 + b^2 \qquad c = \sqrt{250}$$
$$c^2 = 13^2 + 9^2 \qquad = \sqrt{25} \cdot \sqrt{10}$$
$$= 169 + 81 \qquad = 5\sqrt{10}$$
$$= 250$$

37. $a = 7, c = 11$

$$c^2 = a^2 + b^2 \qquad b = \sqrt{72}$$
$$11^2 = 7^2 + b^2 \qquad = \sqrt{36} \cdot \sqrt{2}$$
$$121 = 49 + b^2 \qquad = 6\sqrt{2}$$
$$72 = b^2$$

39. Let $(x_1, y_1) = (-6, 5)$ and $(x_2, y_2) = (3, -4)$.

$$d = \sqrt{(x_2 - x_1)^2 + (y_2 - y_1)^2}$$
$$= \sqrt{[3 - (-6)]^2 + (-4 - 5)^2}$$
$$= \sqrt{9^2 + (-9)^2}$$
$$= \sqrt{81 + 81} = \sqrt{162}$$
$$= \sqrt{81} \cdot \sqrt{2} = 9\sqrt{2}$$

41. Let $(x_1, y_1) = (-5, -1)$ and $(x_2, y_2) = (3, 1)$.

$$d = \sqrt{(x_2 - x_1)^2 + (y_2 - y_1)^2}$$
$$= \sqrt{[3 - (-5)]^2 + [1 - (-1)]^2}$$
$$= \sqrt{8^2 + 2^2}$$
$$= \sqrt{64 + 4} = \sqrt{68}$$
$$= \sqrt{4} \cdot \sqrt{17} = 2\sqrt{17}$$

43. $\sqrt{9} \cdot \sqrt{32} = 3 \cdot \sqrt{16} \cdot \sqrt{2}$
$$= 3 \cdot 4 \cdot \sqrt{2}$$
$$= 12\sqrt{2}$$

45. $\sqrt{3} \cdot \sqrt{18} = \sqrt{3 \cdot 18} = \sqrt{54} = \sqrt{9 \cdot 6}$
$$= \sqrt{9} \cdot \sqrt{6} = 3\sqrt{6}$$

47. $\sqrt{12} \cdot \sqrt{48} = \sqrt{12 \cdot 48}$
$$= \sqrt{12 \cdot 12 \cdot 4}$$
$$= \sqrt{12 \cdot 12} \cdot \sqrt{4}$$
$$= 12 \cdot 2$$
$$= 24$$

49. $\sqrt{12} \cdot \sqrt{30} = \sqrt{12 \cdot 30}$
$$= \sqrt{360} = \sqrt{36 \cdot 10}$$
$$= \sqrt{36} \cdot \sqrt{10} = 6\sqrt{10}$$

51. $2\sqrt{10} \cdot 3\sqrt{2} = 2 \cdot 3 \cdot \sqrt{10 \cdot 2}$ *Product rule*
$$= 6\sqrt{20} \qquad\qquad \textit{Multiply.}$$
$$= 6\sqrt{4 \cdot 5} \qquad\quad \begin{array}{l}\textit{Factor; 4 is a}\\ \textit{perfect square.}\end{array}$$
$$= 6\sqrt{4} \cdot \sqrt{5} \qquad \textit{Product rule}$$
$$= 6 \cdot 2 \cdot \sqrt{5} \qquad\; \sqrt{4} = 2$$
$$= 12\sqrt{5} \qquad\qquad \textit{Multiply.}$$

53. $5\sqrt{3} \cdot 2\sqrt{15} = 5 \cdot 2 \cdot \sqrt{3 \cdot 15}$ *Product rule*

$= 10\sqrt{45}$ *Multiply.*

$= 10\sqrt{9 \cdot 5}$ *Factor; 9 is a perfect square.*

$= 10\sqrt{9} \cdot \sqrt{5}$ *Product rule*

$= 10 \cdot 3 \cdot \sqrt{5}$ $\sqrt{9} = 3$

$= 30\sqrt{5}$ *Multiply.*

55. $\sqrt{8} \cdot \sqrt{32} = \sqrt{8 \cdot 32} = \sqrt{256} = 16$.

Also, $\sqrt{8} = 2\sqrt{2}$ and $\sqrt{32} = 4\sqrt{2}$, so

$\sqrt{8} \cdot \sqrt{32} = 2\sqrt{2} \cdot 4\sqrt{2} = 8 \cdot 2 = 16$.

Both methods give the same answer, and the correct answer can always be obtained using either method.

57. $\sqrt{\dfrac{16}{225}} = \dfrac{\sqrt{16}}{\sqrt{225}} = \dfrac{4}{15}$

59. $\sqrt{\dfrac{7}{16}} = \dfrac{\sqrt{7}}{\sqrt{16}} = \dfrac{\sqrt{7}}{4}$

61. $\sqrt{\dfrac{4}{50}} = \sqrt{\dfrac{2 \cdot 2}{25 \cdot 2}} = \sqrt{\dfrac{2}{25}} = \dfrac{\sqrt{2}}{\sqrt{25}} = \dfrac{\sqrt{2}}{5}$

63. $\dfrac{\sqrt{75}}{\sqrt{3}} = \sqrt{\dfrac{75}{3}} = \sqrt{25} = 5$

65. $\dfrac{30\sqrt{10}}{5\sqrt{2}} = \dfrac{30}{5}\sqrt{\dfrac{10}{2}} = 6\sqrt{5}$

67. $\sqrt{\dfrac{5}{2}} \cdot \sqrt{\dfrac{125}{8}} = \sqrt{\dfrac{5}{2} \cdot \dfrac{125}{8}}$

$= \sqrt{\dfrac{625}{16}}$

$= \dfrac{\sqrt{625}}{\sqrt{16}} = \dfrac{25}{4}$

69. $\sqrt{m^2} = m \quad (m \geq 0)$

71. $\sqrt{y^4} = \sqrt{(y^2)^2} = y^2$

73. $\sqrt{36z^2} = \sqrt{36} \cdot \sqrt{z^2} = 6z$

75. $\sqrt{400x^6} = \sqrt{20 \cdot 20 \cdot x^3 \cdot x^3} = 20x^3$

77. $\sqrt{18x^8} = \sqrt{9 \cdot 2 \cdot x^8}$

$= \sqrt{9} \cdot \sqrt{2} \cdot \sqrt{x^8}$

$= 3 \cdot \sqrt{2} \cdot x^4$

$= 3x^4\sqrt{2}$

79. $\sqrt{45c^{14}} = \sqrt{9 \cdot 5 \cdot c^{14}}$

$= \sqrt{9} \cdot \sqrt{5} \cdot \sqrt{c^{14}}$

$= 3 \cdot \sqrt{5} \cdot c^7$

$= 3c^7\sqrt{5}$

81. $\sqrt{z^5} = \sqrt{z^4 \cdot z} = \sqrt{z^4} \cdot \sqrt{z} = z^2\sqrt{z}$

83. $\sqrt{a^{13}} = \sqrt{a^{12}} \cdot \sqrt{a}$

$= \sqrt{(a^6)^2} \cdot \sqrt{a} = a^6\sqrt{a}$

85. $\sqrt{64x^7} = \sqrt{64} \cdot \sqrt{x^6}\sqrt{x}$

$= 8x^3\sqrt{x}$

87. $\sqrt{x^6 y^{12}} = \sqrt{(x^3)^2 \cdot (y^6)^2} = x^3 y^6$

89. $\sqrt{81m^4 n^2} = \sqrt{81} \cdot \sqrt{m^4} \cdot \sqrt{n^2}$

$= 9m^2 n$

91. $\sqrt{\dfrac{7}{x^{10}}} = \dfrac{\sqrt{7}}{\sqrt{x^{10}}} = \dfrac{\sqrt{7}}{x^5} \quad (x \neq 0)$

93. $\sqrt{\dfrac{y^4}{100}} = \dfrac{\sqrt{y^4}}{\sqrt{100}} = \dfrac{y^2}{10}$

95. $\sqrt{\dfrac{x^4 y^6}{169}} = \dfrac{\sqrt{x^4 y^6}}{\sqrt{169}} = \dfrac{\sqrt{x^4}\sqrt{y^6}}{13}$

$= \dfrac{x^2 y^3}{13}$

97. $\sqrt[3]{40}$

8 is a perfect cube that is a factor of 40.

$\sqrt[3]{40} = \sqrt[3]{8 \cdot 5}$

$= \sqrt[3]{8} \cdot \sqrt[3]{5} = 2\sqrt[3]{5}$

99. $\sqrt[3]{54}$

27 is a perfect cube that is a factor of 54.

$\sqrt[3]{54} = \sqrt[3]{27 \cdot 2}$

$= \sqrt[3]{27} \cdot \sqrt[3]{2} = 3\sqrt[3]{2}$

101. $\sqrt[3]{128}$

64 is a perfect cube that is a factor of 128.

$\sqrt[3]{128} = \sqrt[3]{64 \cdot 2}$

$= \sqrt[3]{64} \cdot \sqrt[3]{2} = 4\sqrt[3]{2}$

103. $\sqrt[4]{80}$

16 is a perfect fourth power that is a factor of 80.

$\sqrt[4]{80} = \sqrt[4]{16 \cdot 5}$

$= \sqrt[4]{16} \cdot \sqrt[4]{5} = 2\sqrt[4]{5}$

105. $\sqrt[3]{\dfrac{8}{27}}$

8 and 27 are both perfect cubes.

$\sqrt[3]{\dfrac{8}{27}} = \dfrac{\sqrt[3]{8}}{\sqrt[3]{27}} = \dfrac{2}{3}$

107. $\sqrt[3]{-\dfrac{216}{125}} = \sqrt[3]{\left(-\dfrac{6}{5}\right)^3} = -\dfrac{6}{5}$

109. $\sqrt[3]{p^3} = p$ because $(p)^3 = p^3$.

111. $\sqrt[3]{x^9} = \sqrt[3]{(x^3)^3} = x^3$

113. $\sqrt[3]{64z^6} = \sqrt[3]{64} \cdot \sqrt[3]{(z^2)^3} = 4z^2$

115. $\sqrt[3]{343a^9b^3} = \sqrt[3]{343} \cdot \sqrt[3]{a^9} \cdot \sqrt[3]{b^3} = 7a^3b$

117. $\sqrt[3]{16t^5} = \sqrt[3]{8t^3} \cdot \sqrt[3]{2t^2} = 2t\sqrt[3]{2t^2}$

119. $\sqrt[3]{\dfrac{m^{12}}{8}} = \sqrt[3]{\dfrac{(m^4)^3}{2^3}} = \dfrac{m^4}{2}$

121. Use the formula for the volume of a cube.

$$V = s^3$$
$$216 = s^3 \quad \textit{Let V = 216.}$$
$$\sqrt[3]{216} = s$$
$$6 = s$$

The length of each side of the container is 6 centimeters.

123. Use the formula for the volume of a sphere, $V = \frac{4}{3}\pi r^3$. Let $V = 288\pi$ and solve for r.

$$\frac{4}{3}\pi r^3 = 288\pi$$
$$\frac{3}{4}\left(\frac{4}{3}\pi r^3\right) = \frac{3}{4}(288\pi)$$
$$\pi r^3 = 216\pi$$
$$r^3 = 216$$
$$r = \sqrt[3]{216} = 6$$

The radius is 6 inches.

125. $2\sqrt{26} \approx 2\sqrt{25} = 2 \cdot 5 = 10$
$\sqrt{83} \approx \sqrt{81} = 9$

Using 10 and 9 as estimates for the length and the width of the rectangle gives us $10 \cdot 9 = 90$ as an estimate for the area. Thus, choice **D** is the best estimate.

127. $4x + 7 - 9x + 12 = 4x - 9x + 7 + 12$
$\qquad\qquad\qquad\qquad = -5x + 19$

129. $2xy + 3x^2y - 9xy + 8x^2y$
$\quad = 3x^2y + 8x^2y + 2xy - 9xy$
$\quad = 11x^2y - 7xy$

8.3 Adding and Subtracting Radicals

8.3 Now Try Exercises

N1. (a) $4\sqrt{3} + \sqrt{3}$
$\qquad = (4+1)\sqrt{3} \quad \textit{Distributive property}$
$\qquad = 5\sqrt{3}$

(b) $2\sqrt{11} - 6\sqrt{11} = (2-6)\sqrt{11}$
$\qquad\qquad\qquad\quad = -4\sqrt{11}$

(c) $\sqrt{5} + \sqrt{14}$ cannot be added using the distributive property.

N2. (a) $\sqrt{2} + \sqrt{18} = \sqrt{2} + \sqrt{9 \cdot 2}$
$\qquad\qquad\qquad = \sqrt{2} + \sqrt{9} \cdot \sqrt{2}$
$\qquad\qquad\qquad = \sqrt{2} + 3\sqrt{2}$
$\qquad\qquad\qquad = 4\sqrt{2}$

(b) $3\sqrt{48} - 2\sqrt{75}$
$\qquad = 3\left(\sqrt{16} \cdot \sqrt{3}\right) - 2\left(\sqrt{25} \cdot \sqrt{3}\right)$
$\qquad = 3\left(4\sqrt{3}\right) - 2\left(5\sqrt{3}\right)$
$\qquad = 12\sqrt{3} - 10\sqrt{3}$
$\qquad = 2\sqrt{3}$

(c) $8\sqrt[3]{5} + 10\sqrt[3]{40} = 8\sqrt[3]{5} + 10\sqrt[3]{8 \cdot 5}$
$\qquad\qquad\qquad\qquad = 8\sqrt[3]{5} + 10\sqrt[3]{8} \cdot \sqrt[3]{5}$
$\qquad\qquad\qquad\qquad = 8\sqrt[3]{5} + 10 \cdot 2 \cdot \sqrt[3]{5}$
$\qquad\qquad\qquad\qquad = 8\sqrt[3]{5} + 20\sqrt[3]{5}$
$\qquad\qquad\qquad\qquad = 28\sqrt[3]{5}$

N3. (a) $\sqrt{7} \cdot \sqrt{14} + 5\sqrt{2}$
$\qquad = \sqrt{7} \cdot \sqrt{7} \cdot \sqrt{2} + 5\sqrt{2}$
$\qquad = \sqrt{49} \cdot \sqrt{2} + 5\sqrt{2}$
$\qquad = 7\sqrt{2} + 5\sqrt{2}$
$\qquad = 12\sqrt{2}$

(b) $\sqrt{150x} + 2\sqrt{24x}$
$\qquad = \sqrt{25 \cdot 6x} + 2\sqrt{4 \cdot 6x}$
$\qquad = \sqrt{25} \cdot \sqrt{6x} + 2\sqrt{4} \cdot \sqrt{6x}$
$\qquad = 5\sqrt{6x} + 2 \cdot 2\sqrt{6x}$
$\qquad = 5\sqrt{6x} + 4\sqrt{6x}$
$\qquad = 9\sqrt{6x}$

(c) $5k^2\sqrt{12} - 4\sqrt{27k^4}$
$\qquad = 5k^2\sqrt{4 \cdot 3} - 4\sqrt{9k^4 \cdot 3}$
$\qquad = 5k^2\sqrt{4} \cdot \sqrt{3} - 4\sqrt{9k^4} \cdot \sqrt{3}$
$\qquad = 5k^2 \cdot 2\sqrt{3} - 4 \cdot 3k^2\sqrt{3}$
$\qquad = (10k^2 - 12k^2)\sqrt{3}$
$\qquad = -2k^2\sqrt{3}$

(d) $\sqrt[3]{128y^5} + 5y\sqrt[3]{16y^2}$
$\qquad = \sqrt[3]{64y^3 \cdot 2y^2} + 5y\sqrt[3]{8 \cdot 2y^2}$
$\qquad = \sqrt[3]{64y^3} \cdot \sqrt[3]{2y^2} + 5y\left(\sqrt[3]{8} \cdot \sqrt[3]{2y^2}\right)$
$\qquad = 4y\sqrt[3]{2y^2} + 5y\left(2\sqrt[3]{2y^2}\right)$
$\qquad = 4y\sqrt[3]{2y^2} + 10y\sqrt[3]{2y^2}$
$\qquad = (4y + 10y)\sqrt[3]{2y^2}$
$\qquad = 14y\sqrt[3]{2y^2}$

8.3 Section Exercises

1. Simplifying $5\sqrt{2} + 6\sqrt{2}$ as $(5+6)\sqrt{2}$, or $11\sqrt{2}$, is an application of the <u>distributive</u> property.

3. $\sqrt{2} - 2\sqrt{3}$ cannot be simplified because the <u>radicands</u> are different.

5. $2\sqrt{3} + 5\sqrt{3}$
$= (2+5)\sqrt{3}$ *Distributive property*
$= 7\sqrt{3}$

7. $4\sqrt{7} - 9\sqrt{7}$
$= (4-9)\sqrt{7}$ *Distributive property*
$= -5\sqrt{7}$

9. $\sqrt{6} + \sqrt{6} = 1\sqrt{6} + 1\sqrt{6}$
$= (1+1)\sqrt{6}$
$= 2\sqrt{6}$

11. $\sqrt{17} + 2\sqrt{17} = 1\sqrt{17} + 2\sqrt{17}$
$= (1+2)\sqrt{17}$
$= 3\sqrt{17}$

13. $5\sqrt{3} + \sqrt{12} = 5\sqrt{3} + \sqrt{4 \cdot 3}$
$= 5\sqrt{3} + \sqrt{4} \cdot \sqrt{3}$
$= 5\sqrt{3} + 2\sqrt{3}$
$= 7\sqrt{3}$

15. $\sqrt{6} + \sqrt{7}$ cannot be added using the distributive property.

17. $2\sqrt{75} - \sqrt{12} = 2\sqrt{25 \cdot 3} - \sqrt{4 \cdot 3}$
$= 2 \cdot \sqrt{25} \cdot \sqrt{3} - \sqrt{4} \cdot \sqrt{3}$
$= 2 \cdot 5 \cdot \sqrt{3} - 2\sqrt{3}$
$= 10\sqrt{3} - 2\sqrt{3}$
$= 8\sqrt{3}$

19. $2\sqrt{50} - 5\sqrt{72}$
$= 2\sqrt{25 \cdot 2} - 5\sqrt{36 \cdot 2}$
$= 2 \cdot \sqrt{25} \cdot \sqrt{2} - 5 \cdot \sqrt{36} \cdot \sqrt{2}$
$= 2 \cdot 5 \cdot \sqrt{2} - 5 \cdot 6 \cdot \sqrt{2}$
$= 10\sqrt{2} - 30\sqrt{2}$
$= -20\sqrt{2}$

21. $5\sqrt{7} - 2\sqrt{28} + 6\sqrt{63}$
$= 5\sqrt{7} - 2\sqrt{4 \cdot 7} + 6\sqrt{9 \cdot 7}$
$= 5\sqrt{7} - 2 \cdot \sqrt{4} \cdot \sqrt{7} + 6 \cdot \sqrt{9} \cdot \sqrt{7}$
$= 5\sqrt{7} - 2 \cdot 2 \cdot \sqrt{7} + 6 \cdot 3 \cdot \sqrt{7}$
$= 5\sqrt{7} - 4\sqrt{7} + 18\sqrt{7}$
$= (5 - 4 + 18)\sqrt{7}$
$= 19\sqrt{7}$

23. $9\sqrt{24} - 2\sqrt{54} + 3\sqrt{20}$
$= 9\sqrt{4 \cdot 6} - 2\sqrt{9 \cdot 6} + 3\sqrt{4 \cdot 5}$
$= 9\sqrt{4} \cdot \sqrt{6} - 2\sqrt{9} \cdot \sqrt{6} + 3\sqrt{4} \cdot \sqrt{5}$
$= 9 \cdot 2\sqrt{6} - 2 \cdot 3\sqrt{6} + 3 \cdot 2\sqrt{5}$
$= 18\sqrt{6} - 6\sqrt{6} + 6\sqrt{5}$
$= 12\sqrt{6} + 6\sqrt{5}$

(Because $\sqrt{6}$ and $\sqrt{5}$ are unlike radicals, this expression cannot be simplified further.)

25. $5\sqrt{72} - 3\sqrt{48} - 4\sqrt{128}$
$= 5\sqrt{36} \cdot \sqrt{2} - 3\sqrt{16} \cdot \sqrt{3} - 4\sqrt{64} \cdot \sqrt{2}$
$= 5\left(6\sqrt{2}\right) - 3\left(4\sqrt{3}\right) - 4\left(8\sqrt{2}\right)$
$= 30\sqrt{2} - 12\sqrt{3} - 32\sqrt{2}$
$= -2\sqrt{2} - 12\sqrt{3}$

27. $\frac{1}{4}\sqrt{288} + \frac{1}{6}\sqrt{72}$
$= \frac{1}{4}\left(\sqrt{144} \cdot \sqrt{2}\right) + \frac{1}{6}\left(\sqrt{36} \cdot \sqrt{2}\right)$
$= \frac{1}{4}\left(12\sqrt{2}\right) + \frac{1}{6}\left(6\sqrt{2}\right)$
$= 3\sqrt{2} + 1\sqrt{2} = 4\sqrt{2}$

29. $\frac{3}{5}\sqrt{75} - \frac{2}{3}\sqrt{45}$
$= \frac{3}{5}\left(\sqrt{25} \cdot \sqrt{3}\right) - \frac{2}{3}\left(\sqrt{9} \cdot \sqrt{5}\right)$
$= \frac{3}{5}\left(5\sqrt{3}\right) - \frac{2}{3}\left(3\sqrt{5}\right)$
$= 3\sqrt{3} - 2\sqrt{5}$

31. $\sqrt{3} \cdot \sqrt{7} + 2\sqrt{21} = \sqrt{3 \cdot 7} + 2\sqrt{21}$
$= 1\sqrt{21} + 2\sqrt{21}$
$= 3\sqrt{21}$

33. $\sqrt{6} \cdot \sqrt{2} + 3\sqrt{3}$
$= \sqrt{6 \cdot 2} + 3\sqrt{3}$
$= \sqrt{12} + 3\sqrt{3}$
$= \sqrt{4 \cdot 3} + 3\sqrt{3}$
$= \sqrt{4} \cdot \sqrt{3} + 3\sqrt{3}$
$= 2\sqrt{3} + 3\sqrt{3}$
$= 5\sqrt{3}$

35. $4\sqrt[3]{16} - 3\sqrt[3]{54}$

Recall that 8 and 27 are perfect cubes.
$4\sqrt[3]{16} - 3\sqrt[3]{54}$
$= 4\left(\sqrt[3]{8 \cdot 2}\right) - 3\left(\sqrt[3]{27 \cdot 2}\right)$
$= 4\left(\sqrt[3]{8} \cdot \sqrt[3]{2}\right) - 3\left(\sqrt[3]{27} \cdot \sqrt[3]{2}\right)$
$= 4\left(2\sqrt[3]{2}\right) - 3\left(3\sqrt[3]{2}\right)$
$= 8\sqrt[3]{2} - 9\sqrt[3]{2}$
$= (8 - 9)\sqrt[3]{2} = -1\sqrt[3]{2} = -\sqrt[3]{2}$

37. $3\sqrt[3]{24} + 6\sqrt[3]{81}$

$\qquad = 3\left(\sqrt[3]{8}\cdot\sqrt[3]{3}\right) + 6\left(\sqrt[3]{27}\cdot\sqrt[3]{3}\right)$

$\qquad = 3\left(2\sqrt[3]{3}\right) + 6\left(3\sqrt[3]{3}\right)$

$\qquad = 6\sqrt[3]{3} + 18\sqrt[3]{3}$

$\qquad = 24\sqrt[3]{3}$

39. $5\sqrt[4]{32} + 2\sqrt[4]{32}\cdot\sqrt[4]{4}$

$\qquad = 5\left(\sqrt[4]{16}\cdot\sqrt[4]{2}\right) + 2\sqrt[4]{16}\cdot\sqrt[4]{2}\cdot\sqrt[4]{4}$

$\qquad = 5\left(2\sqrt[4]{2}\right) + 2\cdot2\sqrt[4]{2\cdot4}$

$\qquad = 10\sqrt[4]{2} + 4\sqrt[4]{8}$

41. $\sqrt{32x} - \sqrt{18x} = \sqrt{16\cdot2x} - \sqrt{9\cdot2x}$

$\qquad = \sqrt{16}\cdot\sqrt{2x} - \sqrt{9}\cdot\sqrt{2x}$

$\qquad = 4\sqrt{2x} - 3\sqrt{2x}$

$\qquad = (4-3)\sqrt{2x}$

$\qquad = 1\sqrt{2x} = \sqrt{2x}$

43. $\sqrt{27r} + \sqrt{48r}$

$\qquad = \sqrt{9}\cdot\sqrt{3r} + \sqrt{16}\cdot\sqrt{3r}$

$\qquad = 3\sqrt{3r} + 4\sqrt{3r}$

$\qquad = 7\sqrt{3r}$

45. $\sqrt{75x^2} + x\sqrt{300}$

$\qquad = \sqrt{25x^2}\sqrt{3} + x\sqrt{100}\cdot\sqrt{3}$

$\qquad = 5x\sqrt{3} + 10x\sqrt{3}$

$\qquad = (5x + 10x)\sqrt{3} = 15x\sqrt{3}$

47. $3\sqrt{8x^2} - 4x\sqrt{2}$

$\qquad = 3\sqrt{4x^2}\sqrt{2} - 4x\sqrt{2}$

$\qquad = 3(2x)\sqrt{2} - 4x\sqrt{2}$

$\qquad = 6x\sqrt{2} - 4x\sqrt{2}$

$\qquad = (6x - 4x)\sqrt{2} = 2x\sqrt{2}$

49. $5\sqrt{75p^2} - 4\sqrt{27p^2}$

$\qquad = 5\sqrt{25p^2}\sqrt{3} - 4\sqrt{9p^2}\sqrt{3}$

$\qquad = 5(5p)\sqrt{3} - 4(3p)\sqrt{3}$

$\qquad = 25p\sqrt{3} - 12p\sqrt{3}$

$\qquad = (25p - 12p)\sqrt{3} = 13p\sqrt{3}$

51. $2\sqrt{125x^2z} + 8x\sqrt{80z}$

$\qquad = 2\sqrt{25x^2}\sqrt{5z} + 8x\sqrt{16}\sqrt{5z}$

$\qquad = 2\cdot5x\cdot\sqrt{5z} + 8x\cdot4\cdot\sqrt{5z}$

$\qquad = 10x\sqrt{5z} + 32x\sqrt{5z}$

$\qquad = (10x + 32x)\sqrt{5z} = 42x\sqrt{5z}$

53. $3k\sqrt{24k^2h^2} + 9h\sqrt{54k^3}$

$\qquad = 3k\sqrt{4k^2h^2}\sqrt{6} + 9h\sqrt{9k^2}\sqrt{6k}$

$\qquad = 3k(2kh)\sqrt{6} + 9h(3k)\sqrt{6k}$

$\qquad = 6k^2h\sqrt{6} + 27hk\sqrt{6k}$

55. $6\sqrt[3]{8p^2} - 2\sqrt[3]{27p^2}$

$\qquad = 6\cdot\sqrt[3]{8}\cdot\sqrt[3]{p^2} - 2\cdot\sqrt[3]{27}\cdot\sqrt[3]{p^2}$

$\qquad = 6\cdot2\cdot\sqrt[3]{p^2} - 2\cdot3\cdot\sqrt[3]{p^2}$

$\qquad = 12\sqrt[3]{p^2} - 6\sqrt[3]{p^2}$

$\qquad = 6\sqrt[3]{p^2}$

57. $5\sqrt[4]{m^3} + 8\sqrt[4]{16m^3}$

$\qquad = 5\sqrt[4]{m^3} + 8\sqrt[4]{16}\sqrt[4]{m^3}$

$\qquad = 5\sqrt[4]{m^3} + 8\cdot2\cdot\sqrt[4]{m^3}$

$\qquad = 5\sqrt[4]{m^3} + 16\sqrt[4]{m^3}$

$\qquad = 21\sqrt[4]{m^3}$

59. $2\sqrt[4]{p^5} - 5p\sqrt[4]{16p}$

$\qquad = 2\sqrt[4]{p^4}\sqrt[4]{p} - 5p\sqrt[4]{16}\sqrt[4]{p}$

$\qquad = 2\cdot p\cdot\sqrt[4]{p} - 5p\cdot2\cdot\sqrt[4]{p}$

$\qquad = 2p\sqrt[4]{p} - 10p\sqrt[4]{p}$

$\qquad = (2p - 10p)\sqrt[4]{p} = -8p\sqrt[4]{p}$

61. $-5\sqrt[3]{256z^4} - 2z\sqrt[3]{32z}$

$\qquad = -5\sqrt[3]{64z^3}\sqrt[3]{4z} - 2z\sqrt[3]{8}\sqrt[3]{4z}$

$\qquad = -5\cdot4z\cdot\sqrt[3]{4z} - 2z\cdot2\cdot\sqrt[3]{4z}$

$\qquad = -20z\sqrt[3]{4z} - 4z\sqrt[3]{4z}$

$\qquad = (-20z - 4z)\sqrt[3]{4z} = -24z\sqrt[3]{4z}$

63. $2\sqrt[4]{6k^7} - k\sqrt[4]{96k^3}$

$\qquad = 2\sqrt[4]{k^4\cdot6k^3} - k\sqrt[4]{16\cdot6k^3}$

$\qquad = 2\sqrt[4]{k^4}\cdot\sqrt[4]{6k^3} - k\cdot\sqrt[4]{16}\cdot\sqrt[4]{6k^3}$

$\qquad = 2k\sqrt[4]{6k^3} - k\cdot2\sqrt[4]{6k^3}$

$\qquad = 0\cdot\sqrt[4]{6k^3}$

$\qquad = 0$

65. Use the formula for the perimeter of a rectangle.

$\qquad P = 2L + 2W$

$\qquad\quad = 2\left(7\sqrt{2}\right) + 2\left(4\sqrt{2}\right)$

$\qquad\quad = 14\sqrt{2} + 8\sqrt{2}$

$\qquad\quad = 22\sqrt{2}$

67. $\sqrt{(-3-1)^2 + (1-4)^2}$

$\qquad = \sqrt{(-4)^2 + (-3)^2}$

$\qquad = \sqrt{16 + 9}$

$\qquad = \sqrt{25} = 5$

69. $\sqrt{(2-1)^2 + (6-(-3))^2}$
$$= \sqrt{(1)^2 + (9)^2}$$
$$= \sqrt{1+81}$$
$$= \sqrt{82}$$

71. $\sqrt{(-5)^2 - 4(1)(-6)} = \sqrt{25 - 4(1)(-6)}$
$$= \sqrt{25+24}$$
$$= \sqrt{49}$$
$$= 7$$

73. $\sqrt{(-10)^2 - 4(3)(-8)} = \sqrt{100+96}$
$$= \sqrt{196}$$
$$= 14$$

75. $\sqrt{(-4)^2 - 4(2)(1)} = \sqrt{16-8}$
$$= \sqrt{8} = \sqrt{4 \cdot 2}$$
$$= \sqrt{4} \cdot \sqrt{2} = 2\sqrt{2}$$

77. $\left(\sqrt{6}\right)^2 = \sqrt{6} \cdot \sqrt{6} = 6$

79. $\sqrt[3]{2} \cdot \sqrt[3]{4} = \sqrt[3]{2 \cdot 4} = \sqrt[3]{8} = 2$

81. $\sqrt{7500} = \sqrt{2500 \cdot 3} = \sqrt{2500} \cdot \sqrt{3} = 50\sqrt{3}$

8.4 Rationalizing the Denominator

8.4 Now Try Exercises

N1. (a) $\dfrac{15}{\sqrt{5}} = \dfrac{15 \cdot \sqrt{5}}{\sqrt{5} \cdot \sqrt{5}}$
$$= \dfrac{15\sqrt{5}}{5}$$
$$= 3\sqrt{5}$$

(b) $\dfrac{3}{\sqrt{24}} = \dfrac{3}{\sqrt{4 \cdot 6}} = \dfrac{3}{\sqrt{4} \cdot \sqrt{6}}$
$$= \dfrac{3}{2 \cdot \sqrt{6}} = \dfrac{3 \cdot \sqrt{6}}{2\sqrt{6} \cdot \sqrt{6}}$$
$$= \dfrac{3\sqrt{6}}{2 \cdot 6} = \dfrac{3\sqrt{6}}{12} = \dfrac{\sqrt{6}}{4}$$

N2. $\sqrt{\dfrac{27}{7}} = \dfrac{\sqrt{27}}{\sqrt{7}} = \dfrac{\sqrt{9} \cdot \sqrt{3}}{\sqrt{7}}$
$$= \dfrac{3 \cdot \sqrt{3}}{\sqrt{7}} = \dfrac{3 \cdot \sqrt{3} \cdot \sqrt{7}}{\sqrt{7} \cdot \sqrt{7}}$$
$$= \dfrac{3\sqrt{21}}{7}$$

N3. $\sqrt{\dfrac{1}{6}} \cdot \sqrt{\dfrac{3}{10}} = \sqrt{\dfrac{1}{6} \cdot \dfrac{3}{10}} = \sqrt{\dfrac{3}{60}}$
$$= \sqrt{\dfrac{1}{20}} = \dfrac{\sqrt{1}}{\sqrt{20}}$$
$$= \dfrac{1}{\sqrt{4} \cdot \sqrt{5}}$$

N4. (a) $\dfrac{\sqrt{9m}}{\sqrt{n}} = \dfrac{\sqrt{9m} \cdot \sqrt{n}}{\sqrt{n} \cdot \sqrt{n}} = \dfrac{\sqrt{9mn}}{n}$
$$= \dfrac{\sqrt{9} \cdot \sqrt{mn}}{n} = \dfrac{3\sqrt{mn}}{n}$$

(b) $\sqrt{\dfrac{16m^2n}{5}} = \dfrac{\sqrt{16m^2n}}{\sqrt{5}} = \dfrac{\sqrt{16m^2n} \cdot \sqrt{5}}{\sqrt{5} \cdot \sqrt{5}}$
$$= \dfrac{\sqrt{80m^2n}}{5} = \dfrac{\sqrt{16m^2} \cdot \sqrt{5n}}{5}$$
$$= \dfrac{4m\sqrt{5n}}{5}$$

N5. (a) $\sqrt[3]{\dfrac{2}{7}} = \dfrac{\sqrt[3]{2}}{\sqrt[3]{7}} = \dfrac{\sqrt[3]{2} \cdot \sqrt[3]{7^2}}{\sqrt[3]{7} \cdot \sqrt[3]{7^2}}$
$$= \dfrac{\sqrt[3]{2 \cdot 7^2}}{\sqrt[3]{7^3}} = \dfrac{\sqrt[3]{98}}{7}$$

(b) $\dfrac{\sqrt[3]{2}}{\sqrt[3]{5}} = \dfrac{\sqrt[3]{2} \cdot \sqrt[3]{5^2}}{\sqrt[3]{5} \cdot \sqrt[3]{5^2}} = \dfrac{\sqrt[3]{2 \cdot 5^2}}{\sqrt[3]{5^3}} = \dfrac{\sqrt[3]{50}}{5}$

(c) $\dfrac{\sqrt[3]{4}}{\sqrt[3]{9t}} = \dfrac{\sqrt[3]{4} \cdot \sqrt[3]{3t^2}}{\sqrt[3]{3^2t} \cdot \sqrt[3]{3t^2}} = \dfrac{\sqrt[3]{4 \cdot 3t^2}}{\sqrt[3]{3^3t^3}}$
$$= \dfrac{\sqrt[3]{12t^2}}{\sqrt[3]{(3t)^3}} = \dfrac{\sqrt[3]{12t^2}}{3t}, t \neq 0$$

8.4 Section Exercises

1. $\dfrac{6}{\sqrt{5}}$

To rationalize the denominator, multiply the numerator and denominator by $\sqrt{5}$.

$$\dfrac{6}{\sqrt{5}} = \dfrac{6 \cdot \sqrt{5}}{\sqrt{5} \cdot \sqrt{5}} = \dfrac{6\sqrt{5}}{5}$$

3. $\dfrac{5}{\sqrt{5}} = \dfrac{5 \cdot \sqrt{5}}{\sqrt{5} \cdot \sqrt{5}} = \dfrac{5\sqrt{5}}{5} = \sqrt{5}$

5. $\dfrac{4}{\sqrt{6}} = \dfrac{4 \cdot \sqrt{6}}{\sqrt{6} \cdot \sqrt{6}} = \dfrac{4\sqrt{6}}{6} = \dfrac{2\sqrt{6}}{3}$

7. $\dfrac{8\sqrt{3}}{\sqrt{5}} = \dfrac{8\sqrt{3} \cdot \sqrt{5}}{\sqrt{5} \cdot \sqrt{5}} = \dfrac{8\sqrt{15}}{5}$

9. $\dfrac{12\sqrt{10}}{8\sqrt{3}} = \dfrac{12\sqrt{10} \cdot \sqrt{3}}{8\sqrt{3} \cdot \sqrt{3}}$
$$= \dfrac{12\sqrt{30}}{8 \cdot 3}$$
$$= \dfrac{12\sqrt{30}}{24} = \dfrac{\sqrt{30}}{2}$$

11. $\dfrac{8}{\sqrt{27}} = \dfrac{8}{\sqrt{9 \cdot 3}} = \dfrac{8}{\sqrt{9} \cdot \sqrt{3}} = \dfrac{8}{3\sqrt{3}}$

$\qquad = \dfrac{8 \cdot \sqrt{3}}{3\sqrt{3} \cdot \sqrt{3}} = \dfrac{8\sqrt{3}}{9}$

13. $\dfrac{6}{\sqrt{200}} = \dfrac{6}{\sqrt{100 \cdot 2}}$

$\qquad = \dfrac{6}{10\sqrt{2}} = \dfrac{3}{5\sqrt{2}} \cdot \dfrac{\sqrt{2}}{\sqrt{2}}$

$\qquad = \dfrac{3 \cdot \sqrt{2}}{5\sqrt{2} \cdot \sqrt{2}}$

$\qquad = \dfrac{3\sqrt{2}}{5 \cdot 2} = \dfrac{3\sqrt{2}}{10}$

15. $\dfrac{12}{\sqrt{72}} = \dfrac{12}{\sqrt{36} \cdot \sqrt{2}}$

$\qquad = \dfrac{12 \cdot \sqrt{2}}{6 \cdot \sqrt{2} \cdot \sqrt{2}}$

$\qquad = \dfrac{2 \cdot 6 \cdot \sqrt{2}}{6 \cdot 2}$

$\qquad = \sqrt{2}$

17. $\dfrac{\sqrt{10}}{\sqrt{5}} = \sqrt{\dfrac{10}{5}}$ *Quotient Rule*

$\qquad = \sqrt{2}$

19. $\sqrt{\dfrac{40}{3}} = \dfrac{\sqrt{40}}{\sqrt{3}} = \dfrac{\sqrt{4} \cdot \sqrt{10} \cdot \sqrt{3}}{\sqrt{3} \cdot \sqrt{3}}$

$\qquad = \dfrac{2\sqrt{30}}{3}$

21. $\sqrt{\dfrac{1}{32}} = \dfrac{\sqrt{1}}{\sqrt{32}} = \dfrac{1 \cdot \sqrt{2}}{\sqrt{16} \cdot \sqrt{2} \cdot \sqrt{2}}$

$\qquad = \dfrac{\sqrt{2}}{4 \cdot 2}$

$\qquad = \dfrac{\sqrt{2}}{8}$

23. $\sqrt{\dfrac{9}{5}} = \dfrac{\sqrt{9}}{\sqrt{5}} = \dfrac{3 \cdot \sqrt{5}}{\sqrt{5} \cdot \sqrt{5}}$

$\qquad = \dfrac{3\sqrt{5}}{5}$

25. $\dfrac{-3}{\sqrt{50}} = \dfrac{-3}{\sqrt{25 \cdot 2}}$

$\qquad = \dfrac{-3}{5\sqrt{2}}$

$\qquad = \dfrac{-3 \cdot \sqrt{2}}{5\sqrt{2} \cdot \sqrt{2}}$

$\qquad = \dfrac{-3\sqrt{2}}{5 \cdot 2} = \dfrac{-3\sqrt{2}}{10}$

27. $\dfrac{63}{\sqrt{45}} = \dfrac{63}{\sqrt{9} \cdot \sqrt{5}} = \dfrac{63}{3\sqrt{5}} = \dfrac{21}{\sqrt{5}}$

$\qquad = \dfrac{21 \cdot \sqrt{5}}{\sqrt{5} \cdot \sqrt{5}} = \dfrac{21\sqrt{5}}{5}$

29. $\dfrac{\sqrt{8}}{\sqrt{24}} = \dfrac{\sqrt{8}}{\sqrt{8} \cdot \sqrt{3}}$

$\qquad = \dfrac{1 \cdot \sqrt{3}}{\sqrt{3} \cdot \sqrt{3}} = \dfrac{\sqrt{3}}{3}$

31. $-\sqrt{\dfrac{1}{5}} = -\dfrac{\sqrt{1}}{\sqrt{5}}$ *Quotient rule*

$\qquad = -\dfrac{1 \cdot \sqrt{5}}{\sqrt{5} \cdot \sqrt{5}} = -\dfrac{\sqrt{5}}{5}$

33. $\sqrt{\dfrac{13}{5}} = \dfrac{\sqrt{13}}{\sqrt{5}} = \dfrac{\sqrt{13} \cdot \sqrt{5}}{\sqrt{5} \cdot \sqrt{5}} = \dfrac{\sqrt{65}}{5}$

35. The given expression is being multiplied by $\frac{\sqrt{3}}{\sqrt{3}}$, which is 1. According to the identity property for multiplication, multiplying an expression by 1 does not change the value of the expression.

37. $\sqrt{\dfrac{7}{13}} \cdot \sqrt{\dfrac{13}{3}} = \sqrt{\dfrac{7}{13} \cdot \dfrac{13}{3}}$ *Product rule*

$\qquad = \sqrt{\dfrac{7}{3}} = \dfrac{\sqrt{7}}{\sqrt{3}}$

$\qquad = \dfrac{\sqrt{7} \cdot \sqrt{3}}{\sqrt{3} \cdot \sqrt{3}} = \dfrac{\sqrt{21}}{3}$

39. $\sqrt{\dfrac{21}{7}} \cdot \sqrt{\dfrac{21}{8}} = \dfrac{\sqrt{21}}{\sqrt{7}} \cdot \dfrac{\sqrt{21}}{\sqrt{8}} = \dfrac{21}{\sqrt{7 \cdot 2 \cdot 4}}$

$\qquad = \dfrac{21}{2\sqrt{14}} = \dfrac{21 \cdot \sqrt{14}}{2 \cdot \sqrt{14} \cdot \sqrt{14}}$

$\qquad = \dfrac{21\sqrt{14}}{2 \cdot 14} = \dfrac{3\sqrt{14}}{4}$

41. $\sqrt{\dfrac{1}{12}} \cdot \sqrt{\dfrac{1}{3}} = \sqrt{\dfrac{1}{12} \cdot \dfrac{1}{3}}$

$\qquad = \sqrt{\dfrac{1}{36}} = \dfrac{\sqrt{1}}{\sqrt{36}} = \dfrac{1}{6}$

43. $\sqrt{\dfrac{2}{9}} \cdot \sqrt{\dfrac{9}{2}} = \sqrt{\dfrac{2}{9} \cdot \dfrac{9}{2}} = \sqrt{1} = 1$

45. $\sqrt{\dfrac{3}{4}} \cdot \sqrt{\dfrac{1}{5}} = \dfrac{\sqrt{3}}{\sqrt{4}} \cdot \dfrac{\sqrt{1}}{\sqrt{5}}$

$\qquad = \dfrac{\sqrt{3} \cdot \sqrt{5}}{2 \cdot \sqrt{5} \cdot \sqrt{5}}$

$\qquad = \dfrac{\sqrt{15}}{2 \cdot 5} = \dfrac{\sqrt{15}}{10}$

47. $\sqrt{\dfrac{17}{3}} \cdot \sqrt{\dfrac{17}{6}} = \dfrac{\sqrt{17} \cdot \sqrt{17}}{\sqrt{3} \cdot \sqrt{6}}$

$= \dfrac{17}{\sqrt{18}}$

$= \dfrac{17 \cdot \sqrt{2}}{\sqrt{9} \cdot \sqrt{2} \cdot \sqrt{2}}$

$= \dfrac{17\sqrt{2}}{3 \cdot 2}$

$= \dfrac{17\sqrt{2}}{6}$

49. $\sqrt{\dfrac{2}{5}} \cdot \sqrt{\dfrac{3}{10}} = \sqrt{\dfrac{2}{5} \cdot \dfrac{3}{10}}$

$= \sqrt{\dfrac{3}{25}}$

$= \dfrac{\sqrt{3}}{\sqrt{25}} = \dfrac{\sqrt{3}}{5}$

51. $\sqrt{\dfrac{16}{27}} \cdot \sqrt{\dfrac{1}{9}} = \dfrac{\sqrt{16} \cdot \sqrt{1}}{\sqrt{27} \cdot \sqrt{9}}$

$= \dfrac{4 \cdot 1}{\sqrt{9} \cdot \sqrt{3 \cdot 3}}$

$= \dfrac{4 \cdot 1 \cdot \sqrt{3}}{3 \cdot \sqrt{3 \cdot 3} \cdot \sqrt{3}}$

$= \dfrac{4\sqrt{3}}{3 \cdot 3 \cdot 3} = \dfrac{4\sqrt{3}}{27}$

53. $\sqrt{\dfrac{6}{p}} = \dfrac{\sqrt{6}}{\sqrt{p}} \cdot \dfrac{\sqrt{p}}{\sqrt{p}}$

$= \dfrac{\sqrt{6p}}{p}$

55. $\sqrt{\dfrac{3}{y}} = \dfrac{\sqrt{3}}{\sqrt{y}} \cdot \dfrac{\sqrt{y}}{\sqrt{y}}$

$= \dfrac{\sqrt{3y}}{y}$

57. $\sqrt{\dfrac{16}{m}} = \dfrac{\sqrt{16}}{\sqrt{m}} \cdot \dfrac{\sqrt{m}}{\sqrt{m}}$

$= \dfrac{4\sqrt{m}}{m}$

59. $\dfrac{\sqrt{3p^2}}{\sqrt{q}} = \dfrac{\sqrt{3}\sqrt{p^2}}{\sqrt{q}} \cdot \dfrac{\sqrt{q}}{\sqrt{q}}$

$= \dfrac{p\sqrt{3q}}{q}$

61. $\dfrac{\sqrt{7x^3}}{\sqrt{y}} = \dfrac{\sqrt{7}\sqrt{x^2}\sqrt{x}}{\sqrt{y}} \cdot \dfrac{\sqrt{y}}{\sqrt{y}}$

$= \dfrac{x\sqrt{7xy}}{y}$

63. $\sqrt{\dfrac{6p^3}{3m}} = \sqrt{\dfrac{2p^3}{m}} = \dfrac{\sqrt{2}\sqrt{p^2}\sqrt{p}}{\sqrt{m}} \cdot \dfrac{\sqrt{m}}{\sqrt{m}}$

$= \dfrac{p\sqrt{2pm}}{m}$

65. $\sqrt{\dfrac{x^2}{4y}} = \dfrac{\sqrt{x^2}}{\sqrt{4}\sqrt{y}} \cdot \dfrac{\sqrt{y}}{\sqrt{y}}$

$= \dfrac{x\sqrt{y}}{2y}$

67. $\sqrt{\dfrac{9a^2r}{5}} = \dfrac{\sqrt{9a^2}\sqrt{r}}{\sqrt{5}} \cdot \dfrac{\sqrt{5}}{\sqrt{5}}$

$= \dfrac{3a\sqrt{5r}}{5}$

69. We need to multiply the numerator and denominator of $\dfrac{\sqrt[3]{2}}{\sqrt[3]{5}}$ by enough factors of 5 to make the radicand in the denominator a perfect cube. In this case we have one factor of 5, so we need to multiply by two more factors of 5 to make three factors of 5. Thus, the correct choice for a rationalizing factor in this problem is $\sqrt[3]{5^2} = \sqrt[3]{25}$, which corresponds to choice **B**.

71. $\sqrt[3]{\dfrac{1}{2}}$

Multiply the numerator and the denominator by enough factors of 2 to make the radicand in the denominator a perfect cube. This will eliminate the radical in the denominator. Here, we multiply by $\sqrt[3]{2^2}$ or $\sqrt[3]{4}$.

$\sqrt[3]{\dfrac{1}{2}} = \dfrac{\sqrt[3]{1}}{\sqrt[3]{2}} = \dfrac{1 \cdot \sqrt[3]{2^2}}{\sqrt[3]{2} \cdot \sqrt[3]{2^2}}$

$= \dfrac{\sqrt[3]{4}}{\sqrt[3]{2 \cdot 2^2}} = \dfrac{\sqrt[3]{4}}{\sqrt[3]{2^3}} = \dfrac{\sqrt[3]{4}}{2}$

73. $\sqrt[3]{\dfrac{1}{32}} = \dfrac{\sqrt[3]{1}}{\sqrt[3]{32}} = \dfrac{1}{\sqrt[3]{8}\sqrt[3]{4}} \cdot \dfrac{\sqrt[3]{2}}{\sqrt[3]{2}}$

$= \dfrac{\sqrt[3]{2}}{2 \cdot \sqrt[3]{8}}$

$= \dfrac{\sqrt[3]{2}}{2 \cdot 2} = \dfrac{\sqrt[3]{2}}{4}$

75. $\sqrt[3]{\dfrac{1}{11}} = \dfrac{\sqrt[3]{1}}{\sqrt[3]{11}} = \dfrac{1 \cdot \sqrt[3]{11^2}}{\sqrt[3]{11} \cdot \sqrt[3]{11^2}}$

$= \dfrac{\sqrt[3]{121}}{\sqrt[3]{11 \cdot 11^2}} = \dfrac{\sqrt[3]{121}}{\sqrt[3]{11^3}} = \dfrac{\sqrt[3]{121}}{11}$

77. $\sqrt[3]{\dfrac{2}{5}}$

Multiply the numerator and the denominator by enough factors of 5 to make the radicand in the denominator a perfect cube. This will eliminate the radical in the denominator. Here, we multiply by $\sqrt[3]{5^2}$ or $\sqrt[3]{25}$.

$$\sqrt[3]{\frac{2}{5}} = \frac{\sqrt[3]{2}}{\sqrt[3]{5}} = \frac{\sqrt[3]{2} \cdot \sqrt[3]{5^2}}{\sqrt[3]{5} \cdot \sqrt[3]{5^2}}$$

$$= \frac{\sqrt[3]{2 \cdot 5^2}}{\sqrt[3]{5^3}} = \frac{\sqrt[3]{50}}{5}$$

79. $\dfrac{\sqrt[3]{4}}{\sqrt[3]{7}} = \dfrac{\sqrt[3]{4} \cdot \sqrt[3]{7^2}}{\sqrt[3]{7} \cdot \sqrt[3]{7^2}}$

$$= \frac{\sqrt[3]{4} \cdot \sqrt[3]{49}}{\sqrt[3]{7^3}} = \frac{\sqrt[3]{196}}{7}$$

81. To make the radicand in the denominator, $4y^2$, into a perfect cube, we must multiply 4 by 2 to get the perfect cube 8 and y^2 by y to get the perfect cube y^3. So we multiply the numerator and denominator by $\sqrt[3]{2y}$.

$$\sqrt[3]{\frac{3}{4y^2}} = \frac{\sqrt[3]{3}}{\sqrt[3]{4y^2}} = \frac{\sqrt[3]{3} \cdot \sqrt[3]{2y}}{\sqrt[3]{4y^2} \cdot \sqrt[3]{2y}}$$

$$= \frac{\sqrt[3]{6y}}{\sqrt[3]{8y^3}} = \frac{\sqrt[3]{6y}}{2y}$$

83. $\dfrac{\sqrt[3]{7m}}{\sqrt[3]{36n}} = \dfrac{\sqrt[3]{7m}}{\sqrt[3]{6^2 n}}$

$$= \frac{\sqrt[3]{7m} \cdot \sqrt[3]{6n^2}}{\sqrt[3]{6^2 n} \cdot \sqrt[3]{6n^2}}$$

$$= \frac{\sqrt[3]{42mn^2}}{\sqrt[3]{6^3 n^3}} = \frac{\sqrt[3]{42mn^2}}{6n}$$

85. $\sqrt[4]{\dfrac{1}{8}} = \dfrac{\sqrt[4]{1}}{\sqrt[4]{8}} = \dfrac{1 \cdot \sqrt[4]{2}}{\sqrt[4]{8} \cdot \sqrt[4]{2}} = \dfrac{\sqrt[4]{2}}{\sqrt[4]{16}} = \dfrac{\sqrt[4]{2}}{2}$

87. (a) $p = k \cdot \sqrt{\dfrac{L}{g}}$

$$p = 6 \cdot \sqrt{\frac{9}{32}} \quad \textit{Let k = 6, L = 9, g = 32.}$$

$$= \frac{6\sqrt{9}}{\sqrt{32}} = \frac{6 \cdot 3}{\sqrt{16 \cdot 2}}$$

$$= \frac{18}{4\sqrt{2}} = \frac{9}{2\sqrt{2}}$$

$$= \frac{9 \cdot \sqrt{2}}{2\sqrt{2} \cdot \sqrt{2}} \quad \begin{array}{l}\textit{Rationalize} \\ \textit{the denominator.}\end{array}$$

$$= \frac{9\sqrt{2}}{4}$$

The period of the pendulum is $\dfrac{9\sqrt{2}}{4}$ seconds.

(b) Using a calculator, we obtain

$$\frac{9\sqrt{2}}{4} \approx 3.182 \text{ seconds.}$$

89. $(4x + 7)(8x - 3)$

$$= 32x^2 - 12x + 56x - 21 \quad \textit{FOIL}$$

$$= 32x^2 + 44x - 21$$

91. $(6x - 1)(6x + 1)$

$$= (6x)^2 - 1^2 \quad \textit{Special product}$$

$$= 36x^2 - 1$$

93. $(p + q)(a - m)$

$$= pa - pm + qa - qm \quad \textit{FOIL}$$

8.5 More Simplifying and Operations with Radicals

8.5 Now Try Exercises

N1. (a) $\sqrt{3}\left(\sqrt{45} - \sqrt{20}\right)$

$$= \sqrt{3}\left(\sqrt{9} \cdot \sqrt{5} - \sqrt{4} \cdot \sqrt{5}\right)$$

$$= \sqrt{3}\left(3\sqrt{5} - 2\sqrt{5}\right)$$

$$= \sqrt{3}\left(\sqrt{5}\right)$$

$$= \sqrt{15}$$

(b) $\left(2\sqrt{3} + \sqrt{7}\right)\left(\sqrt{3} + 3\sqrt{7}\right)$

$$= 2\sqrt{3} \cdot \sqrt{3} + 2\sqrt{3} \cdot 3\sqrt{7}$$

$$\qquad + \sqrt{7} \cdot \sqrt{3} + \sqrt{7} \cdot 3\sqrt{7} \quad \textit{FOIL}$$

$$= 2 \cdot 3 + 6\sqrt{21} + \sqrt{21} + 3 \cdot 7$$

$$= 6 + 6\sqrt{21} + \sqrt{21} + 21$$

$$= 27 + 7\sqrt{21}$$

(c) $\left(\sqrt{10} - 8\right)\left(2\sqrt{10} + 3\sqrt{2}\right)$

$$= 2\sqrt{10} \cdot \sqrt{10} + \sqrt{10} \cdot 3\sqrt{2}$$

$$\qquad - 8 \cdot 2\sqrt{10} - 8 \cdot 3\sqrt{2}$$

$$= 2 \cdot 10 + 3\sqrt{20} - 16\sqrt{10} - 24\sqrt{2}$$

$$= 2 \cdot 10 + 3\sqrt{4 \cdot 5} - 16\sqrt{10} - 24\sqrt{2}$$

$$= 20 + 3 \cdot 2\sqrt{5} - 16\sqrt{10} - 24\sqrt{2}$$

$$= 20 + 6\sqrt{5} - 16\sqrt{10} - 24\sqrt{2}$$

N2. (a) Use the special product formula,

$$(a - b)^2 = a^2 - 2ab + b^2.$$

$$\left(\sqrt{7}-4\right)^2 = \left(\sqrt{7}\right)^2 - 2\left(\sqrt{7}\right)(4) + 4^2$$
$$= 7 - 8\sqrt{7} + 16$$
$$= 23 - 8\sqrt{7}$$

(b) Use the special product formula,

$$(a+b)^2 = a^2 + 2ab + b^2.$$

$$\left(3+\sqrt{y}\right)^2 = 3^2 + 2(3)\left(\sqrt{y}\right) + \left(\sqrt{y}\right)^2$$
$$= 9 + 6\sqrt{y} + y$$

N3. (a) Use the rule for the product of the sum and the difference of two terms,

$$(a+b)(a-b) = a^2 - b^2.$$

$$\left(8+\sqrt{10}\right)\left(8-\sqrt{10}\right) = 8^2 - \left(\sqrt{10}\right)^2$$
$$= 64 - 10$$
$$= 54$$

(b) $\left(\sqrt{x}+2\sqrt{3}\right)\left(\sqrt{x}-2\sqrt{3}\right)$

$$= \left(\sqrt{x}\right)^2 - \left(2\sqrt{3}\right)^2$$
$$= x - 2^2 \cdot 3$$
$$= x - 12$$

N4. (a) $\dfrac{6}{4+\sqrt{3}}$

We can eliminate the radical in the denominator by multiplying both the numerator and denominator by $4 - \sqrt{3}$, the conjugate of the denominator.

$$\frac{6}{4+\sqrt{3}}$$

$$= \frac{6\left(4-\sqrt{3}\right)}{\left(4+\sqrt{3}\right)\left(4-\sqrt{3}\right)}$$

$$= \frac{6\left(4-\sqrt{3}\right)}{4^2 - \left(\sqrt{3}\right)^2} \qquad \begin{array}{l}(a+b)(a-b)\\ = a^2 - b^2\end{array}$$

$$= \frac{6\left(4-\sqrt{3}\right)}{16-3}$$

$$= \frac{6\left(4-\sqrt{3}\right)}{13}$$

(b) $\dfrac{5+\sqrt{7}}{\sqrt{7}-2} = \dfrac{\left(5+\sqrt{7}\right)\left(\sqrt{7}+2\right)}{\left(\sqrt{7}-2\right)\left(\sqrt{7}+2\right)}$

$$= \frac{5\sqrt{7}+10+7+2\sqrt{7}}{\left(\sqrt{7}\right)^2 - 2^2}$$

$$= \frac{17+7\sqrt{7}}{7-4}$$
$$= \frac{17+7\sqrt{7}}{3}$$

(c) $\dfrac{9}{\sqrt{k}-6} = \dfrac{9\left(\sqrt{k}+6\right)}{\left(\sqrt{k}-6\right)\left(\sqrt{k}+6\right)}$

$$= \frac{9\left(\sqrt{k}+6\right)}{k-36}, \, k \neq 36$$

N5. $\dfrac{12\sqrt{6}+28}{20} = \dfrac{4\left(3\sqrt{6}+7\right)}{4(5)}$

$$= \frac{3\sqrt{6}+7}{5}$$

8.5 Section Exercises

1. $\sqrt{25} + \sqrt{64} = 13$
$$\left[\sqrt{25} + \sqrt{64} = 5 + 8\right]$$

3. $\sqrt{8} \cdot \sqrt{2} = 4$
$$\left[\sqrt{8} \cdot \sqrt{2} = \sqrt{16}\right]$$

5. $\sqrt{5}\left(\sqrt{3}-\sqrt{7}\right) = \sqrt{5} \cdot \sqrt{3} - \sqrt{5} \cdot \sqrt{7}$
$$= \sqrt{15} - \sqrt{35}$$

7. $2\sqrt{5}\left(3\sqrt{5}+\sqrt{2}\right)$
$$= 2\sqrt{5} \cdot 3\sqrt{5} + 2\sqrt{5} \cdot \sqrt{2}$$
$$= 2 \cdot 3 \cdot \sqrt{5} \cdot \sqrt{5} + 2\sqrt{10}$$
$$= 6 \cdot 5 + 2\sqrt{10}$$
$$= 30 + 2\sqrt{10}$$

9. $3\sqrt{14} \cdot \sqrt{2} - \sqrt{28} = 3\sqrt{14 \cdot 2} - \sqrt{28}$
$$= 3\sqrt{28} - 1\sqrt{28}$$
$$= 2\sqrt{28}$$
$$= 2\sqrt{4 \cdot 7}$$
$$= 2 \cdot \sqrt{4} \cdot \sqrt{7}$$
$$= 2 \cdot 2 \cdot \sqrt{7}$$
$$= 4\sqrt{7}$$

11. $\left(2\sqrt{6}+3\right)\left(3\sqrt{6}+7\right)$
$$= 2\sqrt{6} \cdot 3\sqrt{6} + 7 \cdot 2\sqrt{6} + 3 \cdot 3\sqrt{6}$$
$$\qquad + 3 \cdot 7 \qquad\qquad\qquad\qquad \textit{FOIL}$$
$$= 2 \cdot 3 \cdot \sqrt{6} \cdot \sqrt{6} + 14\sqrt{6} + 9\sqrt{6} + 21$$
$$= 6 \cdot 6 + 23\sqrt{6} + 21$$
$$= 36 + 23\sqrt{6} + 21$$
$$= 57 + 23\sqrt{6}$$

13. $\left(5\sqrt{7}-2\sqrt{3}\right)\left(3\sqrt{7}+4\sqrt{3}\right)$

$= 5\sqrt{7}\left(3\sqrt{7}\right)+5\sqrt{7}\left(4\sqrt{3}\right)$

$\quad -2\sqrt{3}\left(3\sqrt{7}\right)-2\sqrt{3}\left(4\sqrt{3}\right)$ *FOIL*

$= 15\cdot 7+20\sqrt{21}-6\sqrt{21}-8\cdot 3$

$= 105+14\sqrt{21}-24$

$= 81+14\sqrt{21}$

15. $\left(8-\sqrt{7}\right)^2$

$= (8)^2-2(8)\left(\sqrt{7}\right)+\left(\sqrt{7}\right)^2$

Square of a binomial

$= 64-16\sqrt{7}+7$

$= 71-16\sqrt{7}$

17. $\left(2\sqrt{7}+3\right)^2$

$= \left(2\sqrt{7}\right)^2+2\left(2\sqrt{7}\right)(3)+(3)^2$

Square of a binomial

$= 4\cdot 7+12\sqrt{7}+9$

$= 28+12\sqrt{7}+9$

$= 37+12\sqrt{7}$

19. $\left(\sqrt{6}+1\right)^2$

$= \left(\sqrt{6}\right)^2+2\left(\sqrt{6}\right)(1)+(1)^2$

Square of a binomial

$= 6+2\sqrt{6}+1$

$= 7+2\sqrt{6}$

21. $\left(5-\sqrt{2}\right)\left(5+\sqrt{2}\right)=(5)^2-\left(\sqrt{2}\right)^2$

Product of the sum and difference of two terms

$= 25-2 = 23$

23. $\left(\sqrt{8}-\sqrt{7}\right)\left(\sqrt{8}+\sqrt{7}\right)$

$= \left(\sqrt{8}\right)^2-\left(\sqrt{7}\right)^2$

Product of the sum and difference of two terms

$= 8-7 = 1$

25. $\left(\sqrt{78}-\sqrt{76}\right)\left(\sqrt{78}+\sqrt{76}\right)$

$= \left(\sqrt{78}\right)^2-\left(\sqrt{76}\right)^2$

$= 78-76 = 2$

27. $\left(\sqrt{2}+\sqrt{3}\right)\left(\sqrt{6}-\sqrt{2}\right)$

$= \sqrt{2}\left(\sqrt{6}\right)-\sqrt{2}\left(\sqrt{2}\right)+\sqrt{3}\left(\sqrt{6}\right)$

$\quad -\sqrt{3}\left(\sqrt{2}\right)$ *FOIL*

$= \sqrt{12}-2+\sqrt{18}-\sqrt{6}$ *Product rule*

$= \sqrt{4}\cdot\sqrt{3}-2+\sqrt{9}\cdot\sqrt{2}-\sqrt{6}$

$= 2\sqrt{3}-2+3\sqrt{2}-\sqrt{6}$

29. $\left(\sqrt{10}-\sqrt{5}\right)\left(\sqrt{5}+\sqrt{20}\right)$

$= \sqrt{10}\cdot\sqrt{5}+\sqrt{10}\cdot\sqrt{20}-\sqrt{5}\cdot\sqrt{5}$

$\quad -\sqrt{5}\cdot\sqrt{20}$ *FOIL*

$= \sqrt{50}+\sqrt{200}-5-\sqrt{100}$

$= \sqrt{25\cdot 2}+\sqrt{100\cdot 2}-5-10$

$= 5\sqrt{2}+10\sqrt{2}-15$

$= 15\sqrt{2}-15$

31. $\left(\sqrt{5}+\sqrt{30}\right)\left(\sqrt{6}+\sqrt{3}\right)$

$= \sqrt{5}\cdot\sqrt{6}+\sqrt{5}\cdot\sqrt{3}+\sqrt{30}\cdot\sqrt{6}$

$\quad +\sqrt{30}\cdot\sqrt{3}$ *FOIL*

$= \sqrt{30}+\sqrt{15}+\sqrt{180}+\sqrt{90}$

$= \sqrt{30}+\sqrt{15}+\sqrt{36\cdot 5}+\sqrt{9\cdot 10}$

$= \sqrt{30}+\sqrt{15}+6\sqrt{5}+3\sqrt{10}$

33. $\left(5\sqrt{7}-2\sqrt{3}\right)^2$

$= \left(5\sqrt{7}\right)^2-2\left(5\sqrt{7}\right)\left(2\sqrt{3}\right)+\left(2\sqrt{3}\right)^2$

$= 5^2\left(\sqrt{7}\right)^2-20\sqrt{21}+2^2\left(\sqrt{3}\right)^2$

$= 25\cdot 7-20\sqrt{21}+4\cdot 3$

$= 175-20\sqrt{21}+12$

$= 187-20\sqrt{21}$

35. Because multiplication must be performed before addition, it is incorrect to add -37 and -2. Since $-2\sqrt{15}$ cannot be simplified, the expression cannot be written in a simpler form, and the final answer is $-37-2\sqrt{15}$.

37. $\left(7+\sqrt{x}\right)^2$

$= (7)^2+2(7)\left(\sqrt{x}\right)+\left(\sqrt{x}\right)^2$

Square of a binomial

$= 49+14\sqrt{x}+x$

39. $\left(3\sqrt{t}+\sqrt{7}\right)\left(2\sqrt{t}-\sqrt{14}\right)$

$= 3\sqrt{t}\cdot 2\sqrt{t}-3\sqrt{t}\cdot\sqrt{14}$

$\quad +2\sqrt{t}\cdot\sqrt{7}-\sqrt{7}\cdot\sqrt{14}$ *FOIL*

$= 3\cdot 2\cdot\sqrt{t}\cdot\sqrt{t}-3\sqrt{14t}+2\sqrt{7t}-\sqrt{98}$

$= 6t-3\sqrt{14t}+2\sqrt{7t}-\sqrt{49}\cdot\sqrt{2}$

$= 6t-3\sqrt{14t}+2\sqrt{7t}-7\sqrt{2}$

41. $\left(\sqrt{3m} + \sqrt{2n}\right)\left(\sqrt{3m} - \sqrt{2n}\right)$

$= \left(\sqrt{3m}\right)^2 - \left(\sqrt{2n}\right)^2$

Product of the sum and difference of two terms

$= 3m - 2n$

43. **(a)** The denominator is $\sqrt{5} + \sqrt{3}$, so to rationalize the denominator, we should multiply the numerator and denominator by its conjugate, $\sqrt{5} - \sqrt{3}$.

(b) The denominator is $\sqrt{6} - \sqrt{5}$, so to rationalize the denominator, we should multiply the numerator and denominator by its conjugate, $\sqrt{6} + \sqrt{5}$.

45. $\dfrac{1}{2 + \sqrt{5}} = \dfrac{1\left(2 - \sqrt{5}\right)}{\left(2 + \sqrt{5}\right)\left(2 - \sqrt{5}\right)}$

Multiply numerator and denominator by the conjugate of the denominator.

$= \dfrac{2 - \sqrt{5}}{2^2 - \left(\sqrt{5}\right)^2}$

$(a + b)(a - b) = a^2 - b^2$

$= \dfrac{2 - \sqrt{5}}{4 - 5} = \dfrac{2 - \sqrt{5}}{-1}$

$= -1\left(2 - \sqrt{5}\right) = -2 + \sqrt{5}$

47. $\dfrac{7}{2 - \sqrt{11}} = \dfrac{7\left(2 + \sqrt{11}\right)}{\left(2 - \sqrt{11}\right)\left(2 + \sqrt{11}\right)}$

Multiply numerator and denominator by the conjugate of the denominator.

$= \dfrac{7\left(2 + \sqrt{11}\right)}{(2)^2 - \left(\sqrt{11}\right)^2}$

$(a + b)(a - b) = a^2 - b^2$

$= \dfrac{7\left(2 + \sqrt{11}\right)}{4 - 11}$

$= \dfrac{7\left(2 + \sqrt{11}\right)}{-7}$

$= -1\left(2 + \sqrt{11}\right)$

$= -2 - \sqrt{11}$

49. $\dfrac{\sqrt{12}}{\sqrt{3} + 1} = \dfrac{\sqrt{4}\sqrt{3}\left(\sqrt{3} - 1\right)}{\left(\sqrt{3} + 1\right)\left(\sqrt{3} - 1\right)}$ *Multiply by the conjugate.*

$= \dfrac{2\left(3 - \sqrt{3}\right)}{\left(\sqrt{3}\right)^2 - 1^2}$

$= \dfrac{2\left(3 - \sqrt{3}\right)}{3 - 1}$

$= \dfrac{2\left(3 - \sqrt{3}\right)}{2} = 3 - \sqrt{3}$

51. $\dfrac{2\sqrt{3}}{\sqrt{3} + 5} = \dfrac{2\sqrt{3}\left(\sqrt{3} - 5\right)}{\left(\sqrt{3} + 5\right)\left(\sqrt{3} - 5\right)}$ *Multiply by the conjugate.*

$= \dfrac{2 \cdot 3 - 2 \cdot 5\sqrt{3}}{\left(\sqrt{3}\right)^2 - 5^2}$

$= \dfrac{6 - 10\sqrt{3}}{3 - 25}$

$= \dfrac{6 - 10\sqrt{3}}{-22}$

$= \dfrac{2\left(3 - 5\sqrt{3}\right)}{2(-11)}$ *Factor.*

$= \dfrac{3 - 5\sqrt{3}}{-11}$ or $\dfrac{-3 + 5\sqrt{3}}{11}$

53. $\dfrac{\sqrt{2} + 3}{\sqrt{3} - 1} = \dfrac{\left(\sqrt{2} + 3\right)\left(\sqrt{3} + 1\right)}{\left(\sqrt{3} - 1\right)\left(\sqrt{3} + 1\right)}$ *Multiply by the conjugate.*

$= \dfrac{\sqrt{2} \cdot \sqrt{3} + \sqrt{2} + 3\sqrt{3} + 3}{\left(\sqrt{3}\right)^2 - 1^2}$

$= \dfrac{\sqrt{6} + \sqrt{2} + 3\sqrt{3} + 3}{3 - 1}$

$= \dfrac{\sqrt{6} + \sqrt{2} + 3\sqrt{3} + 3}{2}$

55. $\dfrac{6 - \sqrt{5}}{\sqrt{2} + 2} = \dfrac{\left(6 - \sqrt{5}\right)\left(\sqrt{2} - 2\right)}{\left(\sqrt{2} + 2\right)\left(\sqrt{2} - 2\right)}$

$= \dfrac{6\sqrt{2} - 12 - \sqrt{5} \cdot \sqrt{2} + 2\sqrt{5}}{\left(\sqrt{2}\right)^2 - 2^2}$

$= \dfrac{6\sqrt{2} - 12 - \sqrt{10} + 2\sqrt{5}}{2 - 4}$

$= \dfrac{6\sqrt{2} - 12 - \sqrt{10} + 2\sqrt{5}}{-2}$

or $\dfrac{-6\sqrt{2} + 12 + \sqrt{10} - 2\sqrt{5}}{2}$

57. $\dfrac{2\sqrt{6}+1}{\sqrt{2}+5} = \dfrac{\left(2\sqrt{6}+1\right)\left(\sqrt{2}-5\right)}{\left(\sqrt{2}+5\right)\left(\sqrt{2}-5\right)}$

$= \dfrac{2\sqrt{12}-10\sqrt{6}+\sqrt{2}-5}{\left(\sqrt{2}\right)^2-5^2}$

$= \dfrac{2\sqrt{4\cdot 3}-10\sqrt{6}+\sqrt{2}-5}{2-25}$

$= \dfrac{2\cdot 2\sqrt{3}-10\sqrt{6}+\sqrt{2}-5}{-23}$

$= \dfrac{-4\sqrt{3}+10\sqrt{6}-\sqrt{2}+5}{23}$

59. $\dfrac{\sqrt{7}+\sqrt{2}}{\sqrt{3}-\sqrt{2}}$

$= \dfrac{\left(\sqrt{7}+\sqrt{2}\right)\left(\sqrt{3}+\sqrt{2}\right)}{\left(\sqrt{3}-\sqrt{2}\right)\left(\sqrt{3}+\sqrt{2}\right)}$

$= \dfrac{\sqrt{7}\cdot\sqrt{3}+\sqrt{7}\cdot\sqrt{2}+\sqrt{2}\cdot\sqrt{3}+\sqrt{2}\cdot\sqrt{2}}{\left(\sqrt{3}\right)^2-\left(\sqrt{2}\right)^2}$

$= \dfrac{\sqrt{21}+\sqrt{14}+\sqrt{6}+2}{3-2}$

$= \dfrac{\sqrt{21}+\sqrt{14}+\sqrt{6}+2}{1}$

$= \sqrt{21}+\sqrt{14}+\sqrt{6}+2$

61. $\dfrac{\sqrt{5}}{\sqrt{2}+\sqrt{3}} = \dfrac{\sqrt{5}\left(\sqrt{2}-\sqrt{3}\right)}{\left(\sqrt{2}+\sqrt{3}\right)\left(\sqrt{2}-\sqrt{3}\right)}$

Multiply by the conjugate.

$= \dfrac{\sqrt{5}\cdot\sqrt{2}-\sqrt{5}\cdot\sqrt{3}}{\left(\sqrt{2}\right)^2-\left(\sqrt{3}\right)^2}$

$= \dfrac{\sqrt{10}-\sqrt{15}}{2-3}$

$= \dfrac{\sqrt{10}-\sqrt{15}}{-1} = -\sqrt{10}+\sqrt{15}$

63. $\dfrac{\sqrt{108}}{3+3\sqrt{3}} = \dfrac{\sqrt{36}\cdot\sqrt{3}}{3\left(1+\sqrt{3}\right)}$

$= \dfrac{6\sqrt{3}}{3\left(1+\sqrt{3}\right)} = \dfrac{2\sqrt{3}}{1+\sqrt{3}}$

$= \dfrac{2\sqrt{3}}{1+\sqrt{3}}\cdot\dfrac{1-\sqrt{3}}{1-\sqrt{3}}$

$= \dfrac{2\sqrt{3}\left(1-\sqrt{3}\right)}{1^2-\left(\sqrt{3}\right)^2}$

$= \dfrac{2\sqrt{3}\left(1-\sqrt{3}\right)}{1-3} = \dfrac{2\sqrt{3}\left(1-\sqrt{3}\right)}{-2}$

$= -\sqrt{3}\left(1-\sqrt{3}\right) = -\sqrt{3}+3 \text{ or } 3-\sqrt{3}$

65. $\dfrac{8}{4-\sqrt{x}} = \dfrac{8\left(4+\sqrt{x}\right)}{\left(4-\sqrt{x}\right)\left(4+\sqrt{x}\right)}$

$= \dfrac{8\left(4+\sqrt{x}\right)}{4^2-\left(\sqrt{x}\right)^2}$

$= \dfrac{8\left(4+\sqrt{x}\right)}{16-x}$

67. $\dfrac{1}{\sqrt{x}+\sqrt{y}} = \dfrac{1\left(\sqrt{x}-\sqrt{y}\right)}{\left(\sqrt{x}+\sqrt{y}\right)\left(\sqrt{x}-\sqrt{y}\right)}$

$= \dfrac{\sqrt{x}-\sqrt{y}}{\left(\sqrt{x}\right)^2-\left(\sqrt{y}\right)^2}$

$= \dfrac{\sqrt{x}-\sqrt{y}}{x-y}$

69. $\dfrac{5\sqrt{7}-10}{5} = \dfrac{5\left(\sqrt{7}-2\right)}{5}$ *Factor numerator.*

$= \sqrt{7}-2$ *Lowest terms*

71. $\dfrac{2\sqrt{3}+10}{8} = \dfrac{2\left(\sqrt{3}+5\right)}{2\cdot 4}$ *Factor numerator and denominator.*

$= \dfrac{\sqrt{3}+5}{4}$ *Lowest terms*

73. $\dfrac{12-2\sqrt{10}}{4} = \dfrac{2\left(6-\sqrt{10}\right)}{2\cdot 2} = \dfrac{6-\sqrt{10}}{2}$

75. $\dfrac{16+\sqrt{128}}{24} = \dfrac{16+\sqrt{64}\sqrt{2}}{24}$

$= \dfrac{16+8\sqrt{2}}{24}$

$= \dfrac{8\left(2+\sqrt{2}\right)}{8\cdot 3} = \dfrac{2+\sqrt{2}}{3}$

77. $\sqrt[3]{4}\left(\sqrt[3]{2}-3\right)$

$= \sqrt[3]{4}\left(\sqrt[3]{2}\right)-\sqrt[3]{4}(3)$ *Distributive property*

$= \sqrt[3]{8}-3\sqrt[3]{4}$ *Product rule*

$= 2-3\sqrt[3]{4}$ $\sqrt[3]{8}=2$

79. $2\sqrt[4]{2}\left(3\sqrt[4]{8}+5\sqrt[4]{4}\right)$

$= 2\cdot 3\cdot\sqrt[4]{2}\cdot\sqrt[4]{8}+2\cdot 5\cdot\sqrt[4]{2}\cdot\sqrt[4]{4}$

Distributive property

$= 6\sqrt[4]{16}+10\sqrt[4]{8}$ *Product rule*

$= 6\cdot 2+10\sqrt[4]{8}$ $\sqrt[4]{16}=2$

$= 12+10\sqrt[4]{8}$

81. $\left(\sqrt[3]{2}-1\right)\left(\sqrt[3]{4}+3\right)$

$= \sqrt[3]{8}+3\sqrt[3]{2}-\sqrt[3]{4}-3$ *FOIL*

$= 2+3\sqrt[3]{2}-\sqrt[3]{4}-3$

$= -1+3\sqrt[3]{2}-\sqrt[3]{4}$

83. $\left(\sqrt[3]{5}-\sqrt[3]{4}\right)\left(\sqrt[3]{25}+\sqrt[3]{20}+\sqrt[3]{16}\right)$

$= \sqrt[3]{5}\left(\sqrt[3]{25}+\sqrt[3]{20}+\sqrt[3]{16}\right)$

$\quad - \sqrt[3]{4}\left(\sqrt[3]{25}+\sqrt[3]{20}+\sqrt[3]{16}\right)$

Distributive property

$= \sqrt[3]{5}\cdot\sqrt[3]{25}+\sqrt[3]{5}\cdot\sqrt[3]{20}+\sqrt[3]{5}\cdot\sqrt[3]{16}$

$\quad - \sqrt[3]{4}\cdot\sqrt[3]{25}-\sqrt[3]{4}\cdot\sqrt[3]{20}-\sqrt[3]{4}\cdot\sqrt[3]{16}$

Distributive property

$= \sqrt[3]{125}+\sqrt[3]{100}+\sqrt[3]{80}-\sqrt[3]{100}$

$\quad - \sqrt[3]{80}-\sqrt[3]{64}$ *Product rule*

$= \sqrt[3]{125}-\sqrt[3]{64}$

$= 5-4=1$

85. $r = \dfrac{-h+\sqrt{h^2+0.64S}}{2}$

Substitute 12 for h and 400 for S.

$r = \dfrac{-12+\sqrt{12^2+0.64(400)}}{2}$

$= \dfrac{-12+\sqrt{144+256}}{2}$

$= \dfrac{-12+\sqrt{400}}{2}$

$= \dfrac{-12+20}{2}=\dfrac{8}{2}=4$

The radius should be 4 inches.

87. $6(5+3x) = (6)(5)+(6)(3x)$

$\quad\quad\quad\quad = 30+18x$

88. 30 and $18x$ cannot be combined because they are not like terms.

89. $\left(2\sqrt{10}+5\sqrt{2}\right)\left(3\sqrt{10}-3\sqrt{2}\right)$

$= 2\sqrt{10}\left(3\sqrt{10}\right)+2\sqrt{10}\left(-3\sqrt{2}\right)$

$\quad + 5\sqrt{2}\left(3\sqrt{10}\right)+5\sqrt{2}\left(-3\sqrt{2}\right)$ *FOIL*

$= 6\cdot 10-6\sqrt{20}+15\sqrt{20}-15\cdot 2$

$= 60+9\sqrt{20}-30$

$= 30+9\sqrt{4}\cdot\sqrt{5}$

$= 30+9\left(2\sqrt{5}\right)$

$= 30+18\sqrt{5}$

90. 30 and $18\sqrt{5}$ cannot be combined because they are not like radicals.

91. In the expression $30+18x$, make the first term $30x$, so that

$$30x+18x = 48x.$$

In the expression $30+18\sqrt{5}$, make the first term $30\sqrt{5}$, so that

$$30\sqrt{5}+18\sqrt{5} = 48\sqrt{5}.$$

92. When combining like terms, we add (or subtract) the coefficients of the common factors of the terms: $2xy+5xy=7xy$. When combining like radicals, we add (or subtract) the coefficients of the common radical terms:

$$2\sqrt{ab}+5\sqrt{ab}=7\sqrt{ab}.$$

93. $(2x-1)(4x-3)=0$

$2x-1=0$ or $4x-3=0$

$2x=1$ or $4x=3$

$x=\frac{1}{2}$ or $x=\frac{3}{4}$

The solution set is $\left\{\frac{1}{2},\frac{3}{4}\right\}$.

95. $x^2+4x+3=0$

$(x+3)(x+1)=0$

$x+3=0$ or $x+1=0$

$x=-3$ or $x=-1$

The solution set is $\{-3,-1\}$.

97. $x(x+2)=3$

$x^2+2x-3=0$

$(x+3)(x-1)=0$

$x+3=0$ or $x-1=0$

$x=-3$ or $x=1$

The solution set is $\{-3,1\}$.

Summary Exercises on Operations with Radicals

1. $5\sqrt{10}-8\sqrt{10}=(5-8)\sqrt{10}$

$\quad\quad\quad\quad\quad = -3\sqrt{10}$

3. $\left(1+\sqrt{3}\right)\left(2-\sqrt{6}\right)$

$= 1\cdot 2-1\cdot\sqrt{6}+2\cdot\sqrt{3}-\sqrt{3}\cdot\sqrt{6}$

$= 2-\sqrt{6}+2\sqrt{3}-\sqrt{18}$

$= 2-\sqrt{6}+2\sqrt{3}-\sqrt{9\cdot 2}$

$= 2-\sqrt{6}+2\sqrt{3}-3\sqrt{2}$

5. $\left(3\sqrt{5}-2\sqrt{7}\right)^2$

$= \left(3\sqrt{5}\right)^2-2\left(3\sqrt{5}\right)\left(2\sqrt{7}\right)+\left(2\sqrt{7}\right)^2$

$= 3^2\left(\sqrt{5}\right)^2-2\cdot 3\cdot 2\cdot\sqrt{5}\cdot\sqrt{7}+2^2\left(\sqrt{7}\right)^2$

$= 9\cdot 5-12\sqrt{35}+4\cdot 7$

$= 45-12\sqrt{35}+28$

$= 73-12\sqrt{35}$

7. $\dfrac{1+\sqrt{2}}{1-\sqrt{2}} = \dfrac{1+\sqrt{2}}{1-\sqrt{2}} \cdot \dfrac{1+\sqrt{2}}{1+\sqrt{2}}$

$= \dfrac{1+\sqrt{2}+\sqrt{2}+\sqrt{2}\cdot\sqrt{2}}{1^2 - \left(\sqrt{2}\right)^2}$

$= \dfrac{1+2\sqrt{2}+2}{1-2}$

$= \dfrac{3+2\sqrt{2}}{-1} = -3 - 2\sqrt{2}$

9. $\left(\sqrt{3}+6\right)\left(\sqrt{3}-6\right) = \left(\sqrt{3}\right)^2 - 6^2$

$= 3 - 36$

$= -33$

11. $\sqrt[3]{8x^3y^5z^6} = \sqrt[3]{8x^3y^3z^6} \cdot \sqrt[3]{y^2}$

$= 2xyz^2\sqrt[3]{y^2}$

13. $\dfrac{5}{\sqrt{6}-1} = \dfrac{5}{\sqrt{6}-1} \cdot \dfrac{\sqrt{6}+1}{\sqrt{6}+1}$

$= \dfrac{5\left(\sqrt{6}+1\right)}{\left(\sqrt{6}\right)^2 - 1^2}$

$= \dfrac{5\left(\sqrt{6}+1\right)}{6-1}$

$= \dfrac{5\left(\sqrt{6}+1\right)}{5} = \sqrt{6}+1$

15. $\dfrac{6\sqrt{3}}{5\sqrt{12}} = \dfrac{6\sqrt{3}}{5\sqrt{4}\cdot\sqrt{3}} = \dfrac{6}{5\cdot 2} = \dfrac{3}{5}$

17. $\dfrac{-4}{\sqrt[3]{4}} = \dfrac{-4\cdot\sqrt[3]{2}}{\sqrt[3]{4}\cdot\sqrt[3]{2}}$

$= \dfrac{-4\sqrt[3]{2}}{\sqrt[3]{8}}$

$= \dfrac{-4\sqrt[3]{2}}{2} = -2\sqrt[3]{2}$

19. $\sqrt{75x} - \sqrt{12x} = \sqrt{25\cdot 3x} - \sqrt{4\cdot 3x}$

$= 5\sqrt{3x} - 2\sqrt{3x}$

$= (5-2)\sqrt{3x}$

$= 3\sqrt{3x}$

21. $\sqrt[3]{\dfrac{16}{81}} = \dfrac{\sqrt[3]{16}}{\sqrt[3]{81}} = \dfrac{\sqrt[3]{8}\sqrt[3]{2}}{\sqrt[3]{27}\sqrt[3]{3}}$

$= \dfrac{2\sqrt[3]{2}}{3\sqrt[3]{3}} \cdot \dfrac{\sqrt[3]{9}}{\sqrt[3]{9}}$

$= \dfrac{2\sqrt[3]{18}}{3\sqrt[3]{27}}$

$= \dfrac{2\sqrt[3]{18}}{3\cdot 3} = \dfrac{2\sqrt[3]{18}}{9}$

23. $x\sqrt[4]{x^5} - 3\sqrt[4]{x^9} + x^2\sqrt[4]{x}$

$= x\sqrt[4]{x^4}\cdot\sqrt[4]{x} - 3\sqrt[4]{x^8}\cdot\sqrt[4]{x} + x^2\sqrt[4]{x}$

$= x\cdot x\cdot\sqrt[4]{x} - 3\cdot x^2\cdot\sqrt[4]{x} + x^2\sqrt[4]{x}$

$= (x^2 - 3x^2 + x^2)\sqrt[4]{x}$

$= -x^2\sqrt[4]{x}$

25. $\left(1+\sqrt[3]{3}\right)\left(1-\sqrt[3]{3}+\sqrt[3]{9}\right)$

$= 1\left(1-\sqrt[3]{3}+\sqrt[3]{9}\right) + \sqrt[3]{3}\left(1-\sqrt[3]{3}+\sqrt[3]{9}\right)$

$= 1 - \sqrt[3]{3} + \sqrt[3]{9} + \sqrt[3]{3} - \sqrt[3]{9} + \sqrt[3]{27}$

$= 1 + \sqrt[3]{27}$

$= 1 + 3 = 4$

27. **(a)** $\sqrt{81} = 9$

(b) $x^2 = 81$

$x = -\sqrt{81}$ or $x = \sqrt{81}$
$x = -9$ or $x = 9$

The solution set is $\{-9, 9\}$.

29. **(a)** $x^2 = 9$

$x = -\sqrt{9}$ or $x = \sqrt{9}$
$x = -3$ or $x = 3$

The solution set is $\{-3, 3\}$.

(b) $-\sqrt{9} = -\left(\sqrt{9}\right) = -3$

31. **(a)** $x^2 = \frac{1}{49}$

$x = -\sqrt{\frac{1}{49}}$ or $x = \sqrt{\frac{1}{49}}$
$x = -\frac{1}{7}$ or $x = \frac{1}{7}$

The solution set is $\left\{-\frac{1}{7}, \frac{1}{7}\right\}$.

(b) $\sqrt{\frac{1}{49}} = \frac{1}{7}$ because $\left(\frac{1}{7}\right)^2 = \frac{1}{49}$.

33. **(a)** Since $\left(\frac{7}{10}\right)^2 = \frac{49}{100}$, $-\sqrt{\frac{49}{100}} = -\frac{7}{10}$.

(b) $x^2 = \frac{49}{100}$

$x = -\sqrt{\frac{49}{100}}$ or $x = \sqrt{\frac{49}{100}}$
$x = -\frac{7}{10}$ or $x = \frac{7}{10}$

The solution set is $\left\{-\frac{7}{10}, \frac{7}{10}\right\}$.

35. **(a)** $x^2 = 0.16$

$x = -\sqrt{0.16}$ or $x = \sqrt{0.16}$
$x = -0.4$ or $x = 0.4$

The solution set is $\{-0.4, 0.4\}$.

(b) Since $(0.4)^2 = 0.16$, $\sqrt{0.16} = 0.4$.

8.6   Solving Equations with Radicals

8.6   Now Try Exercises

N1.
$$\sqrt{x-5}=6$$
$$\left(\sqrt{x-5}\right)^2=6^2 \quad \textit{Square each side.}$$
$$x-5=36$$
$$x=41 \quad \textit{Add 5.}$$

Check $x=41$: $\sqrt{x-5}=6$
$$\sqrt{41-5}\overset{?}{=}6 \quad \textit{Let x = 41.}$$
$$\sqrt{36}\overset{?}{=}6$$
$$6=6 \quad \textit{True}$$

The solution set is $\{41\}$.

N2.
$$4\sqrt{x}=\sqrt{10x+12}$$
$$\left(4\sqrt{x}\right)^2=\left(\sqrt{10x+12}\right)^2 \quad \textit{Square each side.}$$
$$16x=10x+12$$
$$6x=12 \quad\quad\quad\quad \textit{Subtract 10x.}$$
$$x=2 \quad\quad\quad\quad \textit{Divide by 6.}$$

Check $x=2$:
$$4\sqrt{x}=\sqrt{10x+12}$$
$$4\sqrt{2}\overset{?}{=}\sqrt{10(2)+12} \quad \textit{Let x = 2.}$$
$$4\sqrt{2}\overset{?}{=}\sqrt{32}$$
$$4\sqrt{2}\overset{?}{=}\sqrt{16\cdot 2}$$
$$4\sqrt{2}=4\sqrt{2} \quad\quad \textit{True}$$

The solution set is $\{2\}$.

N3.
$$\sqrt{x}=-6$$
$$\left(\sqrt{x}\right)^2=(-6)^2 \quad \textit{Square each side.}$$
$$x=36$$

Check $x=36$: $\sqrt{x}=-6$
$$\sqrt{36}\overset{?}{=}-6 \quad \textit{Let x = 36.}$$
$$6=-6 \quad \textit{False}$$

Because the statement $6=-6$ is false, the number 36 is *not* a solution (36 is called an extraneous solution). The equation has no solution. The solution set is $\emptyset$.

Another approach: Because $\sqrt{x}$ represents the *principal* or *nonnegative* square root of x, it cannot equal -6.

N4.
$$t=\sqrt{t^2+3t+9}$$
$$t^2=\left(\sqrt{t^2+3t+9}\right)^2 \quad \textit{Square each side.}$$
$$t^2=t^2+3t+9$$
$$0=3t+9 \quad\quad \textit{Subtract }t^2.$$
$$-3t=9 \quad\quad \textit{Subtract 3t.}$$
$$t=-3 \quad\quad \textit{Divide by }-3.$$

Check $t=-3$:
$$t=\sqrt{t^2+3t+9}$$
$$-3\overset{?}{=}\sqrt{(-3)^2+3(-3)+9} \quad \textit{Let t = -3.}$$
$$-3\overset{?}{=}\sqrt{9-9+9}$$
$$-3\overset{?}{=}\sqrt{9}$$
$$-3=3 \quad\quad\quad \textit{False}$$

The only potential solution does not check, so -3 is an extraneous solution, and the equation has no solution. The solution set is $\emptyset$.

N5.
$$\sqrt{4x+1}=x-5$$
$$\left(\sqrt{4x+1}\right)^2=(x-5)^2 \quad \textit{Square each side.}$$
$$4x+1=x^2-10x+25$$
$$0=x^2-14x+24 \quad \textit{Subtract } 4x+1.$$
$$0=(x-2)(x-12) \quad \textit{Factor.}$$
$$x-2=0 \quad \text{or} \quad x-12=0 \quad \textit{Zero-factor property}$$
$$x=2 \quad \text{or} \quad x=12 \quad \textit{Solve.}$$

Check $x=2$:
$$\sqrt{4x+1}=x-5$$
$$\sqrt{4(2)+1}\overset{?}{=}2-5 \quad \textit{Let x = 2.}$$
$$\sqrt{9}\overset{?}{=}-3$$
$$3=-3 \quad \textit{False}$$

Check $x=12$:
$$\sqrt{4x+1}=x-5$$
$$\sqrt{4(12)+1}\overset{?}{=}12-5 \quad \textit{Let x = 12.}$$
$$\sqrt{49}\overset{?}{=}7$$
$$7=7 \quad \textit{True}$$

The number 2 does not satisfy the original equation, so it is extraneous. The only solution is 12. The solution set is $\{12\}$.

N6.
$$\sqrt{27x}-3=2x$$
$$\sqrt{27x}=2x+3 \quad \textit{Isolate the radical.}$$
$$\left(\sqrt{27x}\right)^2=(2x+3)^2 \quad \textit{Square both sides.}$$
$$27x=4x^2+12x+9$$
$$0=4x^2-15x+9 \quad \textit{Standard form}$$
$$0=(4x-3)(x-3) \quad \textit{Factor.}$$
$$4x-3=0 \quad \text{or} \quad x-3=0 \quad \textit{Zero-factor property}$$
$$x=\tfrac{3}{4} \quad \text{or} \quad x=3 \quad \textit{Solve.}$$

Check $x=\tfrac{3}{4}$: $\sqrt{\tfrac{81}{4}}-3\overset{?}{=}2(\tfrac{3}{4})$
$$\tfrac{9}{2}-3=\tfrac{3}{2} \quad \textit{True}$$

Check $x=3$: $\sqrt{81}-3\overset{?}{=}2(3)$
$$9-3=6 \quad \textit{True}$$

The solution set is $\{\tfrac{3}{4},3\}$.

N7. $\sqrt{x} + 2 = \sqrt{x+8}$

$$\left(\sqrt{x}+2\right)^2 = \left(\sqrt{x+8}\right)^2$$

$$x + 4\sqrt{x} + 4 = x + 8$$

$$4\sqrt{x} = 4$$

$$\sqrt{x} = 1$$

$$\left(\sqrt{x}\right)^2 = 1^2$$

$$x = 1$$

Check $x = 1$:

$$\sqrt{x} + 2 = \sqrt{x+8}$$

$$\sqrt{1} + 2 \stackrel{?}{=} \sqrt{1+8} \quad \textit{Let x = 1.}$$

$$1 + 2 \stackrel{?}{=} \sqrt{9}$$

$$3 = 3 \qquad \textit{True}$$

The solution set is $\{1\}$.

N8. (a) $\sqrt[3]{8x-3} = \sqrt[3]{4x}$

$$\left(\sqrt[3]{8x-3}\right)^3 = \left(\sqrt[3]{4x}\right)^3 \quad \textit{Cube each side.}$$

$$8x - 3 = 4x \qquad \left(\sqrt[3]{a}\right)^3 = a$$

$$4x = 3$$

$$x = \tfrac{3}{4} \qquad\qquad \textit{Divide by 4.}$$

Check $x = \tfrac{3}{4}$: $\sqrt[3]{8\left(\tfrac{3}{4}\right)-3} \stackrel{?}{=} \sqrt[3]{4\left(\tfrac{3}{4}\right)}$

$$\sqrt[3]{3} = \sqrt[3]{3} \qquad\qquad \textit{True}$$

The solution set is $\left\{\tfrac{3}{4}\right\}$.

(b) $\sqrt[3]{2x^2} = \sqrt[3]{10x-12}$

$$\left(\sqrt[3]{2x^2}\right)^3 = \left(\sqrt[3]{10x-12}\right)^3 \quad \begin{array}{l}\textit{Cube each}\\ \textit{side.}\end{array}$$

$$2x^2 = 10x - 12 \qquad \left(\sqrt[3]{a}\right)^3 = a$$

$$2x^2 - 10x + 12 = 0 \qquad \begin{array}{l}\textit{Standard}\\ \textit{form}\end{array}$$

$$2(x^2 - 5x + 6) = 0 \qquad \textit{Factor.}$$

$$(x-2)(x-3) = 0 \qquad \textit{Factor.}$$

$$x - 2 = 0 \quad \text{or} \quad x - 3 = 0 \qquad \begin{array}{l}\textit{Zero-factor}\\ \textit{property}\end{array}$$

$$x = 2 \quad \text{or} \qquad x = 3 \qquad \textit{Solve.}$$

Check $x = 2$:

$$\sqrt[3]{2(2)^2} \stackrel{?}{=} \sqrt[3]{10(2)-12}$$

$$\sqrt[3]{8} = \sqrt[3]{8} \qquad\qquad \textit{True}$$

Check $x = 3$:

$$\sqrt[3]{2(3)^2} \stackrel{?}{=} \sqrt[3]{10(3)-12}$$

$$\sqrt[3]{18} = \sqrt[3]{18} \qquad\qquad \textit{True}$$

The solution set is $\{2, 3\}$.

8.6   Section Exercises

1. $\sqrt{x} = 7$

Use the *squaring property of equality* to square each side of the equation.

$$\left(\sqrt{x}\right)^2 = 7^2$$

$$x = 49$$

Now check this proposed solution in the original equation.

Check $x = 49$: $\sqrt{x} = 7$

$$\sqrt{49} \stackrel{?}{=} 7 \quad \textit{Let x = 49.}$$

$$7 = 7 \quad \textit{True}$$

Since this statement is true, the solution set of the original equation is $\{49\}$.

3. $\sqrt{x+2} = 3$

$$\left(\sqrt{x+2}\right)^2 = 3^2 \quad \textit{Square each side.}$$

$$x + 2 = 9$$

$$x = 7$$

Check $x = 7$:

$$\sqrt{x+2} = 3$$

$$\sqrt{7+2} \stackrel{?}{=} 3 \quad \textit{Let x = 7.}$$

$$\sqrt{9} \stackrel{?}{=} 3$$

$$3 = 3 \quad \textit{True}$$

Since this statement is true, the solution set of the original equation is $\{7\}$.

5. $\sqrt{r-4} = 9$

$$\left(\sqrt{r-4}\right)^2 = 9^2 \quad \textit{Square each side.}$$

$$r - 4 = 81$$

$$r = 85$$

Check $r = 85$:

$$\sqrt{r-4} = 9$$

$$\sqrt{85-4} \stackrel{?}{=} 9 \quad \textit{Let r = 85.}$$

$$\sqrt{81} \stackrel{?}{=} 9$$

$$9 = 9 \quad \textit{True}$$

Since this statement is true, the solution set of the original equation is $\{85\}$.

7. $\sqrt{4-t} = 7$

$$\left(\sqrt{4-t}\right)^2 = 7^2 \quad \textit{Square each side.}$$

$$4 - t = 49$$

$$-t = 45$$

$$t = -45$$

Check $t = -45$:

$$\sqrt{4-t} = 7$$
$$\sqrt{4-(-45)} \stackrel{?}{=} 7 \quad \textit{Let } t = -45.$$
$$\sqrt{49} \stackrel{?}{=} 7$$
$$7 = 7 \quad \textit{True}$$

Since this statement is true, the solution set of the original equation is $\{-45\}$.

9.
$$\sqrt{2t+3} = 0$$
$$\left(\sqrt{2t+3}\right)^2 = 0^2 \quad \textit{Square each side.}$$
$$2t + 3 = 0$$
$$2t = -3$$
$$t = -\frac{3}{2}$$

Check $t = -\frac{3}{2}$:
$$\sqrt{2t+3} = 0$$
$$\sqrt{2\left(-\frac{3}{2}\right)+3} \stackrel{?}{=} 0 \quad \textit{Let } t = -\frac{3}{2}.$$
$$\sqrt{-3+3} \stackrel{?}{=} 0$$
$$\sqrt{0} \stackrel{?}{=} 0$$
$$0 = 0 \quad \textit{True}$$

Since this statement is true, the solution set of the original equation is $\left\{-\frac{3}{2}\right\}$.

11. $\sqrt{t} = -5$

Because $\sqrt{t}$ represents the *principal* or *nonnegative* square root of t, it cannot equal -5. Thus, the solution set is $\emptyset$.

13. $\sqrt{w} - 4 = 7$

Add 4 to both sides of the equation before squaring.
$$\sqrt{w} = 11$$
$$\left(\sqrt{w}\right)^2 = (11)^2$$
$$w = 121$$

Check $w = 121$:
$$\sqrt{w} - 4 = 7$$
$$\sqrt{121} - 4 \stackrel{?}{=} 7 \quad \textit{Let } w = 121.$$
$$11 - 4 \stackrel{?}{=} 7$$
$$7 = 7 \quad \textit{True}$$

Since this statement is true, the solution set of the original equation is $\{121\}$.

15.
$$\sqrt{10x-8} = 3\sqrt{x}$$
$$\left(\sqrt{10x-8}\right)^2 = \left(3\sqrt{x}\right)^2 \quad \textit{Square sides.}$$
$$10x - 8 = (3)^2\left(\sqrt{x}\right)^2 \quad \textit{(ab)}^2 = a^2 b^2$$
$$10x - 8 = 9x$$
$$x = 8$$

Check $x = 8$:

$$\sqrt{10x-8} = 3\sqrt{x}$$
$$\sqrt{10(8)-8} \stackrel{?}{=} 3\sqrt{8} \quad \textit{Let } x = 8.$$
$$\sqrt{72} \stackrel{?}{=} 3\sqrt{8}$$
$$\sqrt{36 \cdot 2} \stackrel{?}{=} 3 \cdot 2\sqrt{2}$$
$$6\sqrt{2} = 6\sqrt{2} \quad \textit{True}$$

Since this statement is true, the solution set of the original equation is $\{8\}$.

17.
$$5\sqrt{x} = \sqrt{10x+15}$$
$$\left(5\sqrt{x}\right)^2 = \left(\sqrt{10x+15}\right)^2$$
$$25x = 10x + 15$$
$$15x = 15$$
$$x = 1$$

Check $x = 1$:
$$5\sqrt{x} = \sqrt{10x+15}$$
$$5\sqrt{1} \stackrel{?}{=} \sqrt{10 \cdot 1 + 15} \quad \textit{Let } x = 1.$$
$$5 \cdot 1 \stackrel{?}{=} \sqrt{25}$$
$$5 = 5 \quad \textit{True}$$

Since this statement is true, the solution set of the original equation is $\{1\}$.

19.
$$\sqrt{3x-5} = \sqrt{2x+1}$$
$$\left(\sqrt{3x-5}\right)^2 = \left(\sqrt{2x+1}\right)^2$$
$$3x - 5 = 2x + 1$$
$$x = 6$$

Check $x = 6$:
$$\sqrt{3x-5} = \sqrt{2x+1}$$
$$\sqrt{3(6)-5} \stackrel{?}{=} \sqrt{2(6)+1} \quad \textit{Let } x = 6.$$
$$\sqrt{13} = \sqrt{13} \quad \textit{True}$$

Since this statement is true, the solution set of the original equation is $\{6\}$.

21.
$$k = \sqrt{k^2 - 5k - 15}$$
$$(k)^2 = \left(\sqrt{k^2 - 5k - 15}\right)^2$$
$$k^2 = k^2 - 5k - 15$$
$$0 = -5k - 15$$
$$5k = -15$$
$$k = -3$$

Check $k = -3$:
$$k = \sqrt{k^2 - 5k - 15}$$
$$-3 \stackrel{?}{=} \sqrt{(-3)^2 - 5(-3) - 15} \quad \textit{Let } k = -3.$$
$$-3 \stackrel{?}{=} \sqrt{9 + 15 - 15}$$
$$-3 \stackrel{?}{=} \sqrt{9}$$
$$-3 = 3 \quad \textit{False}$$

Since this statement is false, the solution set of the original equation is $\emptyset$.

23.
$$7x = \sqrt{49x^2 + 2x - 10}$$
$$(7x)^2 = \left(\sqrt{49x^2 + 2x - 10}\right)^2$$
$$49x^2 = 49x^2 + 2x - 10$$
$$0 = 2x - 10$$
$$10 = 2x$$
$$5 = x$$

Check $x = 5$:

$$7x = \sqrt{49x^2 + 2x - 10}$$
$$7(5) \stackrel{?}{=} \sqrt{49(5)^2 + 2(5) - 10} \quad \textit{Let x = 5.}$$
$$35 \stackrel{?}{=} \sqrt{1225 + 10 - 10}$$
$$35 \stackrel{?}{=} \sqrt{1225}$$
$$35 = 35 \qquad \textit{True}$$

Since this statement is true, the solution set of the original equation is $\{5\}$.

25.
$$\sqrt{2x + 2} = \sqrt{3x - 5}$$
$$\left(\sqrt{2x + 2}\right)^2 = \left(\sqrt{3x - 5}\right)^2$$
$$2x + 2 = 3x - 5$$
$$7 = x$$

Check $x = 7$: $4 = 4$ *True*

The solution set is $\{7\}$.

27.
$$\sqrt{5x - 5} = \sqrt{4x + 1}$$
$$\left(\sqrt{5x - 5}\right)^2 = \left(\sqrt{4x + 1}\right)^2$$
$$5x - 5 = 4x + 1$$
$$x = 6$$

Check $x = 6$: $5 = 5$ *True*

The solution set is $\{6\}$.

29.
$$\sqrt{3x - 8} = -2$$
$$\left(\sqrt{3x - 8}\right)^2 = (-2)^2 \quad \textit{Square each side.}$$
$$3x - 8 = 4$$
$$3x = 12$$
$$x = 4$$

Check $x = 4$:

$$\sqrt{3x - 8} = -2$$
$$\sqrt{3(4) - 8} \stackrel{?}{=} -2 \quad \textit{Let x = 4.}$$
$$\sqrt{12 - 8} \stackrel{?}{=} -2$$
$$\sqrt{4} \stackrel{?}{=} -2$$
$$2 = -2 \quad \textit{False}$$

Since this statement is false, the solution set of the original equation is $\emptyset$. (Note that, in the original equation, the result of a square root cannot equal a negative number.)

31. The error occurs in the first step, when both sides are squared. When the left side is squared, the result should be $x - 1$, not $-(x - 1)$. Thus, we have:

$$x - 1 = 16$$
$$x = 17$$

The correct solution set is $\{17\}$.

33.
$$\sqrt{5x + 11} = x + 3$$
$$\left(\sqrt{5x + 11}\right)^2 = (x + 3)^2$$
$$5x + 11 = x^2 + 6x + 9$$
$$0 = x^2 + x - 2$$
$$0 = (x + 2)(x - 1)$$
$$x = -2 \quad \text{or} \quad x = 1$$

Check $x = -2$: $1 = 1$ *True*
Check $x = 1$: $4 = 4$ *True*

The solution set is $\{-2, 1\}$.

35.
$$\sqrt{2x + 1} = x - 7$$
$$\left(\sqrt{2x + 1}\right)^2 = (x - 7)^2$$
$$2x + 1 = x^2 - 14x + 49$$
$$0 = x^2 - 16x + 48$$
$$0 = (x - 4)(x - 12)$$
$$x = 4 \quad \text{or} \quad x = 12$$

Check $x = 4$: $3 = -3$ *False*
Check $x = 12$: $5 = 5$ *True*

The solution set is $\{12\}$.

37.
$$\sqrt{x + 2} - 2 = x$$
$$\sqrt{x + 2} = x + 2 \qquad \textit{Isolate the radical.}$$
$$\left(\sqrt{x + 2}\right)^2 = (x + 2)^2 \qquad \textit{Square each side.}$$
$$x + 2 = x^2 + 4x + 4$$
$$0 = x^2 + 3x + 2 \qquad \textit{Standard form}$$
$$0 = (x + 2)(x + 1) \qquad \textit{Factor.}$$

$$x + 2 = 0 \qquad \text{or} \qquad x + 1 = 0$$
$$x = -2 \qquad \text{or} \qquad x = -1$$

Check $x = -2$: $\sqrt{0} - 2 = -2$ *True*
Check $x = -1$: $\sqrt{1} - 2 = -1$ *True*

The solution set is $\{-2, -1\}$.

39. $\sqrt{12x+12}+10=2x$

$$\sqrt{12x+12}=2x-10$$

$$\left(\sqrt{12x+12}\right)^2=(2x-10)^2$$

$$12x+12=4x^2-40x+100$$

$$0=4x^2-52x+88$$

$$0=4(x-2)(x-11)$$

$$x=2 \quad\text{or}\quad x=11$$

Check $x=2$: $\quad 6+10=4 \quad$ *False*
Check $x=11$: $12+10=22 \quad$ *True*

The solution set is $\{11\}$.

Alternative solution: Since
$\sqrt{12x+12}=\sqrt{4}\sqrt{3x+3}=2\sqrt{3x+3}$,
all the terms in the original equation are divisible by 2. Dividing by 2 gives us $\sqrt{3x+3}+5=x$, which is easier to solve and gives us the same solution set.

41. $\sqrt{6x+7}-1=x+1$

$$\sqrt{6x+7}=x+2$$

$$\left(\sqrt{6x+7}\right)^2=(x+2)^2$$

$$6x+7=x^2+4x+4$$

$$0=x^2-2x-3$$

$$0=(x+1)(x-3)$$

$$x=-1 \quad\text{or}\quad x=3$$

Check $x=-1$: $\quad 0=0 \quad$ *True*
Check $x=3$: $\qquad 4=4 \quad$ *True*

The solution set is $\{-1,3\}$.

43. $2\sqrt{x+7}=x-1$

$$\left(2\sqrt{x+7}\right)^2=(x-1)^2$$

$$4(x+7)=x^2-2x+1$$

$$4x+28=x^2-2x+1$$

$$0=x^2-6x-27$$

$$0=(x-9)(x+3)$$

$$x=-3 \quad\text{or}\quad x=9$$

Check $x=-3$: $\quad 4=-4 \quad$ *False*
Check $x=9$: $\qquad 8=8 \qquad$ *True*

The solution set is $\{9\}$.

45. $\sqrt{2x}+4=x$

$$\sqrt{2x}=x-4$$

$$\left(\sqrt{2x}\right)^2=(x-4)^2$$

$$2x=x^2-8x+16$$

$$0=x^2-10x+16$$

$$0=(x-2)(x-8)$$

$$x=2 \quad\text{or}\quad x=8$$

Check $x=2$: $\quad 6=2 \quad$ *False*
Check $x=8$: $\quad 8=8 \quad$ *True*

The solution set is $\{8\}$.

47. $\sqrt{x}+9=x+3$

$$\sqrt{x}=x-6$$

$$\left(\sqrt{x}\right)^2=(x-6)^2$$

$$x=x^2-12x+36$$

$$0=x^2-13x+36$$

$$0=(x-4)(x-9)$$

$$x=4 \quad\text{or}\quad x=9$$

Check $x=4$: $\quad 11=7 \quad$ *False*
Check $x=9$: $\quad 12=12 \quad$ *True*

The solution set is $\{9\}$.

49. $3\sqrt{x-2}=x-2$

$$\left(3\sqrt{x-2}\right)^2=(x-2)^2$$

$$9(x-2)=x^2-4x+4$$

$$9x-18=x^2-4x+4$$

$$0=x^2-13x+22$$

$$0=(x-2)(x-11)$$

$$x=2 \quad\text{or}\quad x=11$$

Check $x=2$: $\quad 0=0 \quad$ *True*
Check $x=11$: $\quad 9=9 \quad$ *True*

The solution set is $\{2,11\}$.

51. The error occurs in the first step. We cannot square term by term. The left side must be squared as a binomial, using the formula

$$(a+b)^2=a^2+2ab+b^2.$$

To find the correct solution set, we'll isolate one radical on the left side.

$$\sqrt{3x+4}+\sqrt{x+5}=7$$

$$\sqrt{3x+4}=7-\sqrt{x+5}$$

$$\left(\sqrt{3x+4}\right)^2=\left(7-\sqrt{x+5}\right)^2$$

$$3x+4=49-14\sqrt{x+5}+(x+5)$$

$$2x-50=-14\sqrt{x+5}$$

Isolate the remaining radical.

$$x-25=-7\sqrt{x+5} \quad \text{Divide by 2.}$$

$$(x-25)^2=\left(-7\sqrt{x+5}\right)^2$$

$$x^2-50x+625=49(x+5)$$

$$x^2-50x+625=49x+245$$

$$x^2-99x+380=0$$

$$(x-4)(x-95)=0$$

$$x=4 \quad\text{or}\quad x=95$$

Check $x=4$: $\quad 4+3=7 \quad$ *True*
Check $x=95$: $\quad 17+10=7 \quad$ *False*

The solution set is $\{4\}$.

53. $\sqrt{3x+3} + \sqrt{x+2} = 5$

Rewrite the equation so that there is one radical on each side.

$$\sqrt{3x+3} = 5 - \sqrt{x+2}$$

Square both sides. On the right-hand side, use the formula for the square of a binomial.

$$\left(\sqrt{3x+3}\right)^2 = \left(5 - \sqrt{x+2}\right)^2$$

$$3x+3 = 5^2 - 2 \cdot 5 \cdot \sqrt{x+2} + \left(\sqrt{x+2}\right)^2$$

$$3x+3 = 25 - 10\sqrt{x+2} + x + 2$$

$$3x+3 = 27 + x - 10\sqrt{x+2}$$

$$2x - 24 = -10\sqrt{x+2}$$

$$x - 12 = -5\sqrt{x+2} \qquad \textit{Divide by 2.}$$

We still have a radical on the right, so we must square both sides again.

$$(x-12)^2 = \left(-5\sqrt{x+2}\right)^2$$

$$x^2 - 24x + 144 = 25(x+2)$$

$$x^2 - 24x + 144 = 25x + 50$$

$$x^2 - 49x + 94 = 0$$

$$(x-2)(x-47) = 0$$

$$x = 2 \quad \text{or} \quad x = 47$$

Check $x = 2$: $\quad 3 + 2 = 5 \quad \textit{True}$
Check $x = 47$: $12 + 7 = 5 \quad \textit{False}$

The solution set is $\{2\}$.

55. $\sqrt{x} + 6 = \sqrt{x+72}$

$$\left(\sqrt{x}+6\right)^2 = \left(\sqrt{x+72}\right)^2$$

$$\left(\sqrt{x}\right)^2 + 2 \cdot 6 \cdot \sqrt{x} + 6^2 = x + 72$$

$$x + 12\sqrt{x} + 36 = x + 72$$

$$12\sqrt{x} = 36$$

$$\sqrt{x} = 3$$

$$\left(\sqrt{x}\right)^2 = 3^2$$

$$x = 9$$

Check $x = 9$: $\quad 3 + 6 = 9 \quad \textit{True}$

The solution set is $\{9\}$.

57. $\sqrt{3x+4} - \sqrt{2x-4} = 2$

$$\sqrt{3x+4} = 2 + \sqrt{2x-4}$$

$$\left(\sqrt{3x+4}\right)^2 = \left(2 + \sqrt{2x-4}\right)^2$$

$$3x+4 = 4 + 4\sqrt{2x-4} + 2x - 4$$

$$x + 4 = 4\sqrt{2x-4}$$

$$(x+4)^2 = \left(4\sqrt{2x-4}\right)^2$$

$$x^2 + 8x + 16 = 16(2x-4)$$

$$x^2 + 8x + 16 = 32x - 64$$

$$x^2 - 24x + 80 = 0$$

$$(x-4)(x-20) = 0$$

$$x = 4 \quad \text{or} \quad x = 20$$

Check $x = 4$: $\quad 4 - 2 = 2 \quad \textit{True}$
Check $x = 20$: $\quad 8 - 6 = 2 \quad \textit{True}$

The solution set is $\{4, 20\}$.

59. $\sqrt{2x+11} + \sqrt{x+6} = 2$

$$\sqrt{2x+11} = 2 - \sqrt{x+6}$$

$$\left(\sqrt{2x+11}\right)^2 = \left(2 - \sqrt{x+6}\right)^2$$

$$2x+11 = 4 - 4\sqrt{x+6} + x + 6$$

$$x + 1 = -4\sqrt{x+6}$$

$$(x+1)^2 = \left(-4\sqrt{x+6}\right)^2$$

$$x^2 + 2x + 1 = 16(x+6)$$

$$x^2 + 2x + 1 = 16x + 96$$

$$x^2 - 14x - 95 = 0$$

$$(x+5)(x-19) = 0$$

$$x = -5 \quad \text{or} \quad x = 19$$

Check $x = -5$: $\quad 1 + 1 = 2 \quad \textit{True}$
Check $x = 19$: $\quad 7 + 5 = 2 \quad \textit{False}$

The solution set is $\{-5\}$.

61. $\sqrt[3]{2x} = \sqrt[3]{5x+2}$

$$\left(\sqrt[3]{2x}\right)^3 = \left(\sqrt[3]{5x+2}\right)^3$$

$$2x = 5x + 2$$

$$-3x = 2$$

$$x = -\frac{2}{3}$$

Check $x = -\frac{2}{3}$: $\sqrt[3]{-\frac{4}{3}} = \sqrt[3]{-\frac{10}{3} + 2} \quad \textit{True}$

The solution set is $\left\{-\frac{2}{3}\right\}$.

63. $\sqrt[3]{x^2} = \sqrt[3]{8+7x}$

$$\left(\sqrt[3]{x^2}\right)^3 = \left(\sqrt[3]{8+7x}\right)^3$$

$$x^2 = 8 + 7x$$

$$x^2 - 7x - 8 = 0$$

$$(x+1)(x-8) = 0$$

$$x = -1 \quad \text{or} \quad x = 8$$

Check $x = -1$: $\sqrt[3]{1} = \sqrt[3]{8-7} \quad \textit{True}$
Check $x = 8$: $\sqrt[3]{64} = \sqrt[3]{8+56} \quad \textit{True}$

The solution set is $\{-1, 8\}$.

65. $\sqrt[3]{3x^2 - 9x + 8} = \sqrt[3]{x}$

$$\left(\sqrt[3]{3x^2 - 9x + 8}\right)^3 = \left(\sqrt[3]{x}\right)^3$$

$$3x^2 - 9x + 8 = x$$

$$3x^2 - 10x + 8 = 0$$

$$(3x-4)(x-2) = 0$$

$$x = \frac{4}{3} \quad \text{or} \quad x = 2$$

Check $x = \frac{4}{3}$: $\sqrt[3]{\frac{16}{3} - 12 + 8} \stackrel{?}{=} \sqrt[3]{\frac{4}{3}}$

$\sqrt[3]{\frac{16}{3} - \frac{12}{3}} = \sqrt[3]{\frac{4}{3}}$ *True*

Check $x = 2$: $\sqrt[3]{12 - 18 + 8} = \sqrt[3]{2}$ *True*

The solution set is $\left\{\frac{4}{3}, 2\right\}$.

67. $\sqrt[4]{x^2 + 24x} = 3$

$\left(\sqrt[4]{x^2 + 24x}\right)^4 = 3^4$

$x^2 + 24x = 81$

$x^2 + 24x - 81 = 0$

$(x + 27)(x - 3) = 0$

$x = -27 \quad \text{or} \quad x = 3$

Check $x = -27$: $\sqrt[4]{729 - 648} = \sqrt[4]{81} = 3$ *True*

Check $x = 3$: $\sqrt[4]{9 + 72} = \sqrt[4]{81} = 3$ *True*

The solution set is $\{-27, 3\}$.

69. Let $x =$ the number.

"The square root of the sum of a number and 4 is 5" translates to

$$\sqrt{x + 4} = 5.$$

$$\left(\sqrt{x + 4}\right)^2 = 5^2$$

$$x + 4 = 25$$

$$x = 21$$

Check $x = 21$: $\sqrt{25} = 5$ *True*

The number is 21.

71. Let $x =$ the number.

"Three times the square root of 2 equals the square root of the sum of some number and 10" translates to

$$3\sqrt{2} = \sqrt{x + 10}.$$

$$\left(3\sqrt{2}\right)^2 = \left(\sqrt{x + 10}\right)^2$$

$$9 \cdot 2 = x + 10$$

$$18 = x + 10$$

$$8 = x$$

Check $x = 8$: $3\sqrt{2} = \sqrt{18}$ *True*

The number is 8.

73. $s = 30\sqrt{\dfrac{a}{p}}$

Use a calculator and round answers to the nearest tenth.

(a) $s = 30\sqrt{\frac{862}{156}}$ *Let a = 862 and p = 156.*

≈ 70.5 miles per hour

(b) $s = 30\sqrt{\frac{382}{96}}$ *Let a = 382 and p = 96.*

≈ 59.8 miles per hour

(c) $s = 30\sqrt{\frac{84}{26}}$ *Let a = 84 and p = 26.*

≈ 53.9 miles per hour

75. Let $S =$ sight distance (in kilometers) and $h =$ height of the structure (in kilometers). The equation given is then

$$S = 111.7\sqrt{h}.$$

The height of the London Eye is 135 meters, or 0.135 kilometer.

$$S = 111.7\sqrt{0.135}$$

$$\approx 41.041201 \text{ km.}$$

To convert to miles, we multiply by 0.621371 to get 25.502 miles. So yes, the passengers on the London Eye can see Windsor Castle, which is 25 miles away.

77. $1483 \text{ ft} \approx 1483(0.3048)$

$= 452.0184 \text{ m}$

So $h = 0.4520184$ kilometer and

$$S = 111.7\sqrt{0.4520184}$$

$$\approx 75.098494 \text{ km.}$$

Converting to miles gives us

$$(75.098494)(0.621371) \approx 46.7,$$

or about 47 miles.

79. $s = \frac{1}{2}(a + b + c)$

$= \frac{1}{2}(7 + 7 + 12)$

$= \frac{1}{2}(26) = 13$ units

80. $\mathcal{A} = \sqrt{s(s - a)(s - b)(s - c)}$

$= \sqrt{13(13 - 7)(13 - 7)(13 - 12)}$

$= \sqrt{13(6)(6)(1)}$

$= 6\sqrt{13}$ square units

81. $c^2 = a^2 + b^2$

$7^2 = 6^2 + h^2$

$49 = 36 + h^2$

$h^2 = 13$

$h = \sqrt{13}$ units

82. $\mathcal{A} = \frac{1}{2}bh$

$= \frac{1}{2}(6)(\sqrt{13})$

$= 3\sqrt{13}$ square units

83. $2(3\sqrt{13}) = 6\sqrt{13}$ square units

84. They are both $6\sqrt{13}$.

85. $(5^2)^3 = 5^{2 \cdot 3} = 5^6$

87. $\dfrac{a^{-2}a^3}{a^4} = \dfrac{a^1}{a^4} = a^{1-4} = a^{-3} = \dfrac{1}{a^3}$

89. $\left(\dfrac{p}{3}\right)^{-2} = \left(\dfrac{3}{p}\right)^2 = \dfrac{3^2}{p^2}$

91. $\dfrac{(c^3)^2 c^4}{(c^{-1})^3} = \dfrac{c^6 c^4}{c^{-3}} = \dfrac{c^{10}}{c^{-3}} = c^{10-(-3)} = c^{13}$

8.7 Using Rational Numbers as Exponents

8.7 Now Try Exercises

N1. (a) $144^{1/2} = \sqrt{144} = 12$

(b) $729^{1/3} = \sqrt[3]{729} = 9$

(c) $256^{1/4} = \sqrt[4]{256} = 4$

N2. (a) $27^{4/3} = (27^{1/3})^4 = \left(\sqrt[3]{27}\right)^4 = 3^4 = 81$

(b) $16^{3/2} = (16^{1/2})^3 = \left(\sqrt{16}\right)^3 = 4^3 = 64$

(c) $-16^{7/4} = -(16^{1/4})^7 = -\left(\sqrt[4]{16}\right)^7$
$= -(2)^7 = -128$

N3. (a) $16^{-3/4} = \dfrac{1}{16^{3/4}} = \dfrac{1}{(16^{1/4})^3} = \dfrac{1}{2^3} = \dfrac{1}{8}$

(b) $8^{-2/3} = \dfrac{1}{8^{2/3}} = \dfrac{1}{(8^{1/3})^2} = \dfrac{1}{2^2} = \dfrac{1}{4}$

N4. (a) $7^{1/4} \cdot 7^{5/4} = 7^{1/4+5/4} = 7^{6/4} = 7^{3/2}$

(b) $\dfrac{11^{1/3}}{11^{2/3}} = 11^{1/3-2/3} = 11^{-1/3} = \dfrac{1}{11^{1/3}}$

(c) $\left(\dfrac{64}{125}\right)^{2/3} = \dfrac{64^{2/3}}{125^{2/3}} = \dfrac{(64^{1/3})^2}{(125^{1/3})^2}$
$= \dfrac{\left(\sqrt[3]{64}\right)^2}{\left(\sqrt[3]{125}\right)^2} = \dfrac{4^2}{5^2} = \dfrac{16}{25}$

OR $\left(\dfrac{64}{125}\right)^{2/3} = \left[\left(\dfrac{64}{125}\right)^{1/3}\right]^2$
$= \left(\sqrt[3]{\dfrac{64}{125}}\right)^2 = \left(\dfrac{4}{5}\right)^2 = \dfrac{16}{25}$

(d) $\dfrac{5^{1/3} \cdot 5^{-2/3}}{5^{-4/3}} = \dfrac{5^{1/3+(-2/3)}}{5^{-4/3}} = \dfrac{5^{-1/3}}{5^{-4/3}}$
$= 5^{-1/3-(-4/3)} = 5^{3/3} = 5^1 = 5$

N5. (a) $(x^{2/3} y^{3/2})^6 = (x^{2/3})^6 (y^{3/2})^6$
$= x^{6(2/3)} y^{6(3/2)} = x^4 y^9$

(b) $\left(\dfrac{p^{1/5}}{q^{3/4}}\right)^3 = \dfrac{(p^{1/5})^3}{(q^{3/4})^3} = \dfrac{p^{3/5}}{q^{9/4}}$

(c) $\dfrac{t^{-1} \cdot t^{3/4}}{t^{7/4}} = \dfrac{t^{-1+3/4}}{t^{7/4}} = \dfrac{t^{-1/4}}{t^{7/4}}$
$= t^{-1/4-7/4} = t^{-2} = \dfrac{1}{t^2}$

N6. (a) $\sqrt[6]{64^2} = (64^2)^{1/6} = 64^{2/6} = 64^{1/3}$
$= \sqrt[3]{64} = 4$

(b) $\left(\sqrt[10]{y}\right)^5 = (y^{1/10})^5 = y^{5/10} = y^{1/2}$
$= \sqrt{y}, y \geq 0.$

8.7 Section Exercises

1. $49^{1/2}$ can be written as $49^{0.5}$ or $\sqrt{49}$, which is 7 and not -7. Thus, all of the choices are equal to 7 except **A**, which is the answer.

3. $-64^{1/3}$ can be written as $-\sqrt[3]{64}$, which is -4. Also, $-\sqrt{16} = -4$. Thus, all of the choices are equal to -4 except **C**, which is the answer.

5. $25^{1/2} = \sqrt{25} = \sqrt{5^2} = 5$

7. $64^{1/3} = \sqrt[3]{64} = \sqrt[3]{4^3} = 4$

9. $16^{1/4} = \sqrt[4]{16} = \sqrt[4]{2^4} = 2$

11. $32^{1/5} = \sqrt[5]{32} = \sqrt[5]{2^5} = 2$

13. $4^{3/2} = (4^{1/2})^3$
$= \left(\sqrt{4}\right)^3 = 2^3 = 8$

15. $27^{2/3} = (27^{1/3})^2$
$= \left(\sqrt[3]{27}\right)^2 = 3^2 = 9$

17. $16^{3/4} = (16^{1/4})^3$
$= \left(\sqrt[4]{16}\right)^3 = 2^3 = 8$

19. $32^{2/5} = (32^{1/5})^2$
$= \left(\sqrt[5]{32}\right)^2 = 2^2 = 4$

21. $-8^{2/3} = -(8^{1/3})^2$
$= -\left(\sqrt[3]{8}\right)^2 = -2^2 = -4$

23. $-64^{1/3} = -\sqrt[3]{64} = -4$

25. $49^{-3/2} = \dfrac{1}{49^{3/2}} = \dfrac{1}{(49^{1/2})^3} = \dfrac{1}{\left(\sqrt{49}\right)^3}$
$= \dfrac{1}{7^3} = \dfrac{1}{343}$

27. $216^{-2/3} = \dfrac{1}{216^{2/3}} = \dfrac{1}{(216^{1/3})^2} = \dfrac{1}{\left(\sqrt[3]{216}\right)^2}$
$= \dfrac{1}{6^2} = \dfrac{1}{36}$

29. $-16^{-5/4} = -\dfrac{1}{16^{5/4}} = -\dfrac{1}{(16^{1/4})^5}$
$= -\dfrac{1}{\left(\sqrt[4]{16}\right)^5} = -\dfrac{1}{2^5} = -\dfrac{1}{32}$

31. $2^{1/3} \cdot 2^{7/3} = 2^{1/3+7/3}$ *Product rule*

$\qquad\qquad\quad = 2^{8/3}$

33. $6^{1/4} \cdot 6^{-3/4} = 6^{1/4+(-3/4)}$ *Product rule*

$\qquad\qquad\quad = 6^{-2/4}$

$\qquad\qquad\quad = 6^{-1/2} = \dfrac{1}{6^{1/2}}$

35. $\dfrac{15^{3/4}}{15^{5/4}} = 15^{3/4-5/4}$ *Quotient rule*

$\qquad\quad = 15^{-2/4}$

$\qquad\quad = 15^{-1/2} = \dfrac{1}{15^{1/2}}$

37. $\dfrac{11^{-2/7}}{11^{-3/7}} = 11^{-2/7-(-3/7)}$ *Quotient rule*

$\qquad\qquad = 11^{1/7}$

39. $\left(8^{3/2}\right)^2 = 8^{(3/2)(2)}$ *Power rule*

$\qquad\qquad = 8^3$

41. $\left(6^{1/3}\right)^{3/2} = 6^{(1/3)(3/2)}$ *Power rule*

$\qquad\qquad = 6^{1/2}$

43. $\left(\dfrac{25}{4}\right)^{3/2} = \dfrac{25^{3/2}}{4^{3/2}} = \dfrac{(25^{1/2})^3}{(4^{1/2})^3}$

$\qquad\qquad\qquad = \dfrac{\left(\sqrt{25}\right)^3}{\left(\sqrt{4}\right)^3}$

$\qquad\qquad\qquad = \dfrac{5^3}{2^3}$

45. $\dfrac{2^{2/5} \cdot 2^{-3/5}}{2^{7/5}} = \dfrac{2^{2/5}}{2^{3/5} \cdot 2^{7/5}}$

$\qquad\qquad = \dfrac{2^{2/5}}{2^{10/5}}$

$\qquad\qquad = \dfrac{1}{2^{8/5}}$

47. $\dfrac{6^{-2/9}}{6^{1/9} \cdot 6^{-5/9}} = \dfrac{6^{5/9}}{6^{1/9} \cdot 6^{2/9}}$

$\qquad\qquad\quad = \dfrac{6^{5/9}}{6^{3/9}}$

$\qquad\qquad\quad = 6^{2/9}$

49. $x^{2/5} \cdot x^{7/5} = x^{2/5+7/5}$

$\qquad\qquad\quad = x^{9/5}$

51. $\dfrac{r^{4/9}}{r^{3/9}} = r^{4/9-3/9}$

$\qquad\quad = r^{1/9}$

53. $\left(m^3 n^{1/4}\right)^{2/3} = (m^3)^{2/3}(n^{1/4})^{2/3}$

$\qquad\qquad\qquad = m^{3 \cdot (2/3)} n^{(1/4) \cdot (2/3)}$

$\qquad\qquad\qquad = m^2 n^{1/6}$

55. $\left(\dfrac{a^{2/3}}{b^{1/4}}\right)^6 = \dfrac{(a^{2/3})^6}{(b^{1/4})^6}$

$\qquad\qquad\quad = \dfrac{a^{(2/3) \cdot 6}}{b^{(1/4) \cdot 6}}$

$\qquad\qquad\quad = \dfrac{a^4}{b^{3/2}}$

57. $\dfrac{m^{3/4} \cdot m^{-1/4}}{m^{1/3}} = \dfrac{m^{3/4+(-1/4)}}{m^{1/3}}$

$\qquad\qquad\qquad = \dfrac{m^{2/4}}{m^{1/3}}$

$\qquad\qquad\qquad = m^{(1/2)-(1/3)}$

$\qquad\qquad\qquad = m^{(3/6)-(2/6)}$

$\qquad\qquad\qquad = m^{1/6}$

59. $\sqrt[6]{4^3} = 4^{3/6} = 4^{1/2} = \sqrt{4} = 2$

61. $\sqrt[8]{16^2} = 16^{2/8} = 16^{1/4} = \sqrt[4]{16} = 2$

63. $\sqrt[4]{a^2} = a^{2/4} = a^{1/2} = \sqrt{a}$

65. $\sqrt[6]{k^4} = k^{4/6} = k^{2/3} = \sqrt[3]{k^2}$

67. The real square roots of 121 are -11 and 11 because $(-11)(-11) = 121$ and $11 \cdot 11 = 121$.

69. The real square roots of $\frac{1}{4}$ are $-\frac{1}{2}$ and $\frac{1}{2}$ because

$\qquad \left(-\frac{1}{2}\right)\left(-\frac{1}{2}\right) = \frac{1}{4}$ and $\frac{1}{2} \cdot \frac{1}{2} = \frac{1}{4}$.

71. $\sqrt{236} = \sqrt{4 \cdot 59} = 2\sqrt{59}$

73. $\sqrt{147} = \sqrt{49 \cdot 3} = 7\sqrt{3}$

Chapter 8 Review Exercises

1. The square roots of 49 are -7 and 7 because $(-7)^2 = 49$ and $7^2 = 49$.

2. The square roots of 81 are -9 and 9 because $(-9)^2 = 81$ and $9^2 = 81$.

3. The square roots of 196 are -14 and 14 because $(-14)^2 = 196$ and $14^2 = 196$.

4. The square roots of 121 are -11 and 11 because $(-11)^2 = 121$ and $11^2 = 121$.

5. The square roots of 225 are -15 and 15 because $(-15)^2 = 225$ and $15^2 = 225$.

6. The square roots of 729 are -27 and 27 because $(-27)^2 = 729$ and $27^2 = 729$.

7. $\sqrt{16} = 4$ because $4^2 = 16$.

8. $-\sqrt{36}$ represents the negative square root of 36. Since $6 \cdot 6 = 36, -\sqrt{36} = -6$.

9. $\sqrt[3]{1000} = 10$ because $10^3 = 1000$.

10. $\sqrt[4]{81} = 3$ because 3 is positive and $3^4 = 81$.

11. $\sqrt{-8100}$ is not a real number.

12. $-\sqrt{4225}$ represents the negative square root of 4225. Since $65 \cdot 65 = 4225$, $-\sqrt{4225} = -65$.

13. $\sqrt{\dfrac{49}{36}} = \dfrac{\sqrt{49}}{\sqrt{36}} = \dfrac{7}{6}$

14. $\sqrt{\dfrac{100}{81}} = \dfrac{\sqrt{100}}{\sqrt{81}} = \dfrac{10}{9}$

15. Let $(x_1, y_1) = (-3, -5)$ and $(x_2, y_2) = (4, -3)$. Use the distance formula.

$$\begin{aligned} d &= \sqrt{(x_2 - x_1)^2 + (y_2 - y_1)^2} \\ &= \sqrt{[4 - (-3)]^2 + [-3 - (-5)]^2} \\ &= \sqrt{7^2 + 2^2} \\ &= \sqrt{49 + 4} = \sqrt{53} \end{aligned}$$

16. Use the Pythagorean theorem with $a = 15$, $b = x$, and $c = 17$.

$$\begin{aligned} c^2 &= a^2 + b^2 \\ 17^2 &= 15^2 + x^2 \\ 289 &= 225 + x^2 \\ 64 &= x^2 \\ x &= \sqrt{64} = 8 \end{aligned}$$

17. Use the Pythagorean theorem with $a = 30.4$ cm and $b = 37.5$ cm.

$$\begin{aligned} c^2 &= a^2 + b^2 \\ &= (30.4)^2 + (37.5)^2 \\ &= 924.16 + 1406.25 \\ &= 2330.41 \\ c &= \sqrt{2330.41} \approx 48.3 \text{ cm} \end{aligned}$$

18. $\sqrt{111}$

This number is *irrational* because 111 is not a perfect square.

$$\sqrt{111} \approx 10.536$$

19. $-\sqrt{25}$

This number is *rational* because 25 is a perfect square.

$$-\sqrt{25} = -5$$

20. $\sqrt{-4}$

This is not a real number.

21. $\sqrt{5} \cdot \sqrt{15} = \sqrt{5} \cdot \sqrt{5} \cdot \sqrt{3}$
$= \sqrt{25} \cdot \sqrt{3} = 5\sqrt{3}$

22. $-\sqrt{27} = -\sqrt{9 \cdot 3} = -\sqrt{9} \cdot \sqrt{3} = -3\sqrt{3}$

23. $\sqrt{160} = \sqrt{16 \cdot 10} = \sqrt{16} \cdot \sqrt{10} = 4\sqrt{10}$

24. $\sqrt[3]{-1331} = -11$ because $(-11)^3 = -1331$.

25. $\sqrt[3]{1728} = 12$ because $12^3 = 1728$.

26. $\sqrt{12} \cdot \sqrt{27} = \sqrt{4 \cdot 3} \cdot \sqrt{9 \cdot 3}$
$= 2\sqrt{3} \cdot 3\sqrt{3}$
$= 2 \cdot 3 \cdot \left(\sqrt{3}\right)^2$
$= 2 \cdot 3 \cdot 3 = 18$

27. $\sqrt{32} \cdot \sqrt{48} = \sqrt{16 \cdot 2} \cdot \sqrt{16 \cdot 3}$
$= 4\sqrt{2} \cdot 4\sqrt{3}$
$= 4 \cdot 4 \cdot \sqrt{2 \cdot 3}$
$= 16\sqrt{6}$

28. $\sqrt{50} \cdot \sqrt{125} = \sqrt{25 \cdot 2} \cdot \sqrt{25 \cdot 5}$
$= 5\sqrt{2} \cdot 5\sqrt{5}$
$= 5 \cdot 5 \cdot \sqrt{2 \cdot 5}$
$= 25\sqrt{10}$

29. $-\sqrt{\dfrac{121}{400}} = -\dfrac{\sqrt{121}}{\sqrt{400}} = -\dfrac{11}{20}$

30. $\sqrt{\dfrac{3}{49}} = \dfrac{\sqrt{3}}{\sqrt{49}} = \dfrac{\sqrt{3}}{7}$

31. $\sqrt{\dfrac{7}{169}} = \dfrac{\sqrt{7}}{\sqrt{169}} = \dfrac{\sqrt{7}}{13}$

32. $\sqrt{\dfrac{1}{6}} \cdot \sqrt{\dfrac{5}{6}} = \sqrt{\dfrac{1}{6} \cdot \dfrac{5}{6}}$
$= \sqrt{\dfrac{5}{36}}$
$= \dfrac{\sqrt{5}}{\sqrt{36}} = \dfrac{\sqrt{5}}{6}$

33. $\sqrt{\dfrac{2}{5}} \cdot \sqrt{\dfrac{2}{45}} = \sqrt{\dfrac{2}{5} \cdot \dfrac{2}{45}}$
$= \sqrt{\dfrac{4}{225}}$
$= \dfrac{\sqrt{4}}{\sqrt{225}} = \dfrac{2}{15}$

34. $\dfrac{3\sqrt{10}}{\sqrt{5}} = \dfrac{3 \cdot \sqrt{5} \cdot \sqrt{2}}{\sqrt{5}}$
$= 3\sqrt{2}$

35. $\dfrac{24\sqrt{12}}{6\sqrt{3}} = \dfrac{24 \cdot \sqrt{4} \cdot \sqrt{3}}{6\sqrt{3}}$
$= 4\sqrt{4} = 4 \cdot 2 = 8$

36. $\dfrac{8\sqrt{150}}{4\sqrt{75}} = \dfrac{8 \cdot \sqrt{75} \cdot \sqrt{2}}{4\sqrt{75}}$
$= 2\sqrt{2}$

37. $\sqrt{p} \cdot \sqrt{p} = p$

38. $\sqrt{k} \cdot \sqrt{m} = \sqrt{km}$

39. $\sqrt{r^{18}} = r^9$ because $(r^9)^2 = r^{18}$.

40. $\sqrt{x^{10}y^{16}} = x^5y^8$ because $(x^5y^8)^2 = x^{10}y^{16}$.

41. $\sqrt{a^{15}b^{21}} = \sqrt{a^{14}b^{20} \cdot ab}$
$$= \sqrt{a^{14}b^{20}} \cdot \sqrt{ab}$$
$$= a^7b^{10}\sqrt{ab}$$

42. $\sqrt{121x^6y^{10}} = 11x^3y^5$ because
$$(11x^3y^5)^2 = 121x^6y^{10}.$$

43. $7\sqrt{11} + \sqrt{11} = (7+1)\sqrt{11} = 8\sqrt{11}$

44. $3\sqrt{2} + 6\sqrt{2} = (3+6)\sqrt{2} = 9\sqrt{2}$

45. $3\sqrt{75} + 2\sqrt{27}$
$$= 3\left(\sqrt{25} \cdot \sqrt{3}\right) + 2\left(\sqrt{9} \cdot \sqrt{3}\right)$$
$$= 3\left(5\sqrt{3}\right) + 2\left(3\sqrt{3}\right)$$
$$= 15\sqrt{3} + 6\sqrt{3} = 21\sqrt{3}$$

46. $4\sqrt{12} + \sqrt{48}$
$$= 4\left(\sqrt{4} \cdot \sqrt{3}\right) + \sqrt{16} \cdot \sqrt{3}$$
$$= 4\left(2\sqrt{3}\right) + 4\sqrt{3}$$
$$= 8\sqrt{3} + 4\sqrt{3} = 12\sqrt{3}$$

47. $4\sqrt{24} - 3\sqrt{54} + \sqrt{6}$
$$= 4\left(\sqrt{4} \cdot \sqrt{6}\right) - 3\left(\sqrt{9} \cdot \sqrt{6}\right) + \sqrt{6}$$
$$= 4\left(2\sqrt{6}\right) - 3\left(3\sqrt{6}\right) + \sqrt{6}$$
$$= 8\sqrt{6} - 9\sqrt{6} + 1\sqrt{6}$$
$$= 0\sqrt{6} = 0$$

48. $2\sqrt{7} - 4\sqrt{28} + 3\sqrt{63}$
$$= 2\sqrt{7} - 4\left(\sqrt{4} \cdot \sqrt{7}\right) + 3\left(\sqrt{9} \cdot \sqrt{7}\right)$$
$$= 2\sqrt{7} - 4\left(2\sqrt{7}\right) + 3\left(3\sqrt{7}\right)$$
$$= 2\sqrt{7} - 8\sqrt{7} + 9\sqrt{7} = 3\sqrt{7}$$

49. $\frac{2}{5}\sqrt{75} + \frac{3}{4}\sqrt{160}$
$$= \frac{2}{5}\left(\sqrt{25} \cdot \sqrt{3}\right) + \frac{3}{4}\left(\sqrt{16} \cdot \sqrt{10}\right)$$
$$= \frac{2}{5}\left(5\sqrt{3}\right) + \frac{3}{4}\left(4\sqrt{10}\right)$$
$$= 2\sqrt{3} + 3\sqrt{10}$$

50. $\frac{1}{3}\sqrt{18} + \frac{1}{4}\sqrt{32}$
$$= \frac{1}{3}\left(\sqrt{9} \cdot \sqrt{2}\right) + \frac{1}{4}\left(\sqrt{16} \cdot \sqrt{2}\right)$$
$$= \frac{1}{3}\left(3\sqrt{2}\right) + \frac{1}{4}\left(4\sqrt{2}\right)$$
$$= 1\sqrt{2} + 1\sqrt{2} = 2\sqrt{2}$$

51. $\sqrt{15} \cdot \sqrt{2} + 5\sqrt{30} = \sqrt{30} + 5\sqrt{30}$
$$= 1\sqrt{30} + 5\sqrt{30}$$
$$= 6\sqrt{30}$$

52. $\sqrt{4x} + \sqrt{36x} - \sqrt{9x}$
$$= \sqrt{4}\sqrt{x} + \sqrt{36}\sqrt{x} - \sqrt{9}\sqrt{x}$$
$$= 2\sqrt{x} + 6\sqrt{x} - 3\sqrt{x} = 5\sqrt{x}$$

53. $\sqrt{20m^2} - m\sqrt{45}$
$$= \sqrt{4m^2 \cdot 5} - m\left(\sqrt{9} \cdot \sqrt{5}\right)$$
$$= \sqrt{4m^2} \cdot \sqrt{5} - m\left(3\sqrt{5}\right)$$
$$= 2m\sqrt{5} - 3m\sqrt{5} = -m\sqrt{5}$$

54. $3k\sqrt{8k^2n} + 5k^2\sqrt{2n}$
$$= 3k\left(\sqrt{4k^2} \cdot \sqrt{2n}\right) + 5k^2\sqrt{2n}$$
$$= 3k\left(2k\sqrt{2n}\right) + 5k^2\sqrt{2n}$$
$$= 6k^2\sqrt{2n} + 5k^2\sqrt{2n}$$
$$= (6k^2 + 5k^2)\sqrt{2n}$$
$$= 11k^2\sqrt{2n}$$

55. $\frac{8\sqrt{2}}{\sqrt{5}} = \frac{8\sqrt{2} \cdot \sqrt{5}}{\sqrt{5} \cdot \sqrt{5}} = \frac{8\sqrt{10}}{5}$

56. $\frac{5}{\sqrt{5}} = \frac{5 \cdot \sqrt{5}}{\sqrt{5} \cdot \sqrt{5}} = \frac{5\sqrt{5}}{5} = \sqrt{5}$

57. $\frac{12}{\sqrt{24}} = \frac{12}{\sqrt{4 \cdot 6}} = \frac{12}{2\sqrt{6}}$
$$= \frac{12 \cdot \sqrt{6}}{2\sqrt{6} \cdot \sqrt{6}} = \frac{12\sqrt{6}}{2 \cdot 6}$$
$$= \frac{12\sqrt{6}}{12} = \sqrt{6}$$

58. $\frac{\sqrt{2}}{\sqrt{15}} = \frac{\sqrt{2} \cdot \sqrt{15}}{\sqrt{15} \cdot \sqrt{15}} = \frac{\sqrt{30}}{15}$

59. $\sqrt{\frac{2}{5}} = \frac{\sqrt{2}}{\sqrt{5}} = \frac{\sqrt{2} \cdot \sqrt{5}}{\sqrt{5} \cdot \sqrt{5}} = \frac{\sqrt{10}}{5}$

60. $\sqrt{\frac{5}{14}} \cdot \sqrt{28} = \sqrt{\frac{5}{14} \cdot 28}$
$$= \sqrt{5 \cdot 2} = \sqrt{10}$$

61. $\sqrt{\frac{2}{7}} \cdot \sqrt{\frac{1}{3}} = \sqrt{\frac{2}{7} \cdot \frac{1}{3}}$
$$= \sqrt{\frac{2}{21}} = \frac{\sqrt{2}}{\sqrt{21}}$$
$$= \frac{\sqrt{2} \cdot \sqrt{21}}{\sqrt{21} \cdot \sqrt{21}} = \frac{\sqrt{42}}{21}$$

62. $\sqrt{\dfrac{r^2}{16x}} = \dfrac{\sqrt{r^2}}{\sqrt{16x}}$

$= \dfrac{r \cdot \sqrt{x}}{\sqrt{16x} \cdot \sqrt{x}}$

$= \dfrac{r\sqrt{x}}{\sqrt{16x^2}} = \dfrac{r\sqrt{x}}{4x}$

63. $\sqrt[3]{\dfrac{1}{3}} = \dfrac{\sqrt[3]{1}}{\sqrt[3]{3}} = \dfrac{1 \cdot \sqrt[3]{3^2}}{\sqrt[3]{3} \cdot \sqrt[3]{3^2}}$

$= \dfrac{\sqrt[3]{3^2}}{\sqrt[3]{3^3}} = \dfrac{\sqrt[3]{9}}{3}$

64. $\sqrt[3]{\dfrac{2}{7}} = \dfrac{\sqrt[3]{2}}{\sqrt[3]{7}} = \dfrac{\sqrt[3]{2} \cdot \sqrt[3]{7^2}}{\sqrt[3]{7} \cdot \sqrt[3]{7^2}}$

$= \dfrac{\sqrt[3]{2 \cdot 7^2}}{\sqrt[3]{7^3}} = \dfrac{\sqrt[3]{98}}{7}$

65. $-\sqrt{3}\left(\sqrt{5} + \sqrt{27}\right)$

$= -\sqrt{3}\left(\sqrt{5}\right) + \left(-\sqrt{3}\right)\left(\sqrt{27}\right)$

$= -\sqrt{3 \cdot 5} - \sqrt{3 \cdot 27}$

$= -\sqrt{15} - \sqrt{81}$

$= -\sqrt{15} - 9$

66. $3\sqrt{2}\left(\sqrt{3} + 2\sqrt{2}\right)$

$= 3\sqrt{2}\left(\sqrt{3}\right) + 3\sqrt{2}\left(2\sqrt{2}\right)$

$= 3\sqrt{6} + 6 \cdot 2$

$= 3\sqrt{6} + 12$

67. $\left(2\sqrt{3} - 4\right)\left(5\sqrt{3} + 2\right)$

$= 2\sqrt{3}\left(5\sqrt{3}\right) + \left(2\sqrt{3}\right)(2) - 4\left(5\sqrt{3}\right)$

$\quad - 4(2)$ *FOIL*

$= 10 \cdot 3 + 4\sqrt{3} - 20\sqrt{3} - 8$

$= 30 - 16\sqrt{3} - 8$

$= 22 - 16\sqrt{3}$

68. $\left(5\sqrt{7} + 2\right)^2$

$= \left(5\sqrt{7}\right)^2 + 2\left(5\sqrt{7}\right)(2) + 2^2$

     Square of a binomial

$= 25 \cdot 7 + 20\sqrt{7} + 4$

$= 175 + 20\sqrt{7} + 4$

$= 179 + 20\sqrt{7}$

69. $\left(\sqrt{5} - \sqrt{7}\right)\left(\sqrt{5} + \sqrt{7}\right)$

$= \left(\sqrt{5}\right)^2 - \left(\sqrt{7}\right)^2$

$= 5 - 7 = -2$

70. $\left(2\sqrt{3} + 5\right)\left(2\sqrt{3} - 5\right)$

$= \left(2\sqrt{3}\right)^2 - (5)^2$

$= 4 \cdot 3 - 25$

$= 12 - 25 = -13$

71. $\dfrac{1}{2 + \sqrt{5}}$

$= \dfrac{1\left(2 - \sqrt{5}\right)}{\left(2 + \sqrt{5}\right)\left(2 - \sqrt{5}\right)}$ *Multiply by the conjugate.*

$= \dfrac{2 - \sqrt{5}}{(2)^2 - \left(\sqrt{5}\right)^2}$

$= \dfrac{2 - \sqrt{5}}{4 - 5}$

$= \dfrac{2 - \sqrt{5}}{-1} = -2 + \sqrt{5}$

72. $\dfrac{\sqrt{8}}{\sqrt{2} + 6}$

$= \dfrac{\sqrt{8}\left(\sqrt{2} - 6\right)}{\left(\sqrt{2} + 6\right)\left(\sqrt{2} - 6\right)}$ *Multiply by the conjugate.*

$= \dfrac{\sqrt{16} - 6\sqrt{8}}{\left(\sqrt{2}\right)^2 - 6^2}$

$= \dfrac{4 - 6 \cdot \sqrt{4 \cdot 2}}{2 - 36}$

$= \dfrac{4 - 6 \cdot 2\sqrt{2}}{-34}$

$= \dfrac{4 - 12\sqrt{2}}{-34}$

$= \dfrac{-2\left(-2 + 6\sqrt{2}\right)}{-2(17)}$ *Factor numerator and denominator.*

$= \dfrac{-2 + 6\sqrt{2}}{17}$ *Lowest terms*

73. $\dfrac{2 + \sqrt{6}}{\sqrt{3} - 1} = \dfrac{\left(2 + \sqrt{6}\right)\left(\sqrt{3} + 1\right)}{\left(\sqrt{3} - 1\right)\left(\sqrt{3} + 1\right)}$

$= \dfrac{2\sqrt{3} + 2 + \sqrt{18} + \sqrt{6}}{3 - 1}$

$= \dfrac{2\sqrt{3} + 2 + \sqrt{9 \cdot 2} + \sqrt{6}}{2}$

$= \dfrac{2\sqrt{3} + 2 + 3\sqrt{2} + \sqrt{6}}{2}$

74. $\dfrac{15 + 10\sqrt{6}}{15} = \dfrac{5\left(3 + 2\sqrt{6}\right)}{5(3)}$ *Factor.*

$= \dfrac{3 + 2\sqrt{6}}{3}$ *Lowest terms*

75. $\dfrac{3 + 9\sqrt{7}}{12} = \dfrac{3\left(1 + 3\sqrt{7}\right)}{3(4)}$ *Factor.*

$= \dfrac{1 + 3\sqrt{7}}{4}$ *Lowest terms*

76. $\dfrac{6 + \sqrt{192}}{2} = \dfrac{6 + \sqrt{64 \cdot 3}}{2}$

$= \dfrac{6 + 8\sqrt{3}}{2}$

$= \dfrac{2\left(3 + 4\sqrt{3}\right)}{2}$

$= 3 + 4\sqrt{3}$

77. $\sqrt{m} - 5 = 0$

$\sqrt{m} = 5$ *Isolate the radical.*

$\left(\sqrt{m}\right)^2 = 5^2$ *Square both sides.*

$m = 25$

Check $m = 25$: $5 - 5 = 0$ *True*

The solution set is $\{25\}$.

78. $\sqrt{p} + 4 = 0$

$\sqrt{p} = -4$

Since a square root cannot equal a negative number, there is no solution and the solution set is $\emptyset$.

79. $\sqrt{x + 1} = 7$

$\left(\sqrt{x + 1}\right)^2 = 7^2$

$x + 1 = 49$

$x = 48$

Check $x = 48$: $\sqrt{49} = 7$ *True*

The solution set is $\{48\}$.

80. $\sqrt{5m + 4} = 3\sqrt{m}$

$\left(\sqrt{5m + 4}\right)^2 = \left(3\sqrt{m}\right)^2$

$5m + 4 = 9m$

$4 = 4m$

$1 = m$

Check $m = 1$: $\sqrt{9} = 3\sqrt{1}$ *True*

The solution set is $\{1\}$.

81. $\sqrt{2p + 3} = \sqrt{5p - 3}$

$\left(\sqrt{2p + 3}\right)^2 = \left(\sqrt{5p - 3}\right)^2$

$2p + 3 = 5p - 3$

$6 = 3p$

$2 = p$

Check $p = 2$: $\sqrt{7} = \sqrt{7}$ *True*

The solution set is $\{2\}$.

82. $\sqrt{-2t - 4} = t + 2$

$\left(\sqrt{-2t - 4}\right)^2 = (t + 2)^2$

$-2t - 4 = t^2 + 4t + 4$

$0 = t^2 + 6t + 8$

$0 = (t + 2)(t + 4)$

$t = -2$ or $t = -4$

Check $t = -2$: $\sqrt{0} = 0$ *True*
Check $t = -4$: $\sqrt{4} = -2$ *False*

Of the two potential solutions, -2 checks in the original equation, but -4 does not. Thus, the solution set is $\{-2\}$.

83. $\sqrt{13 + 4t} = t + 4$

$\left(\sqrt{13 + 4t}\right)^2 = (t + 4)^2$

$13 + 4t = t^2 + 8t + 16$

$0 = t^2 + 4t + 3$

$0 = (t + 3)(t + 1)$

$t = -3$ or $t = -1$

Check $t = -3$: $1 = 1$ *True*
Check $t = -1$: $3 = 3$ *True*

The solution set is $\{-3, -1\}$.

84. $\sqrt{2 - x} + 3 = x + 7$

$\sqrt{2 - x} = x + 4$ *Isolate the radical.*

$\left(\sqrt{2 - x}\right)^2 = (x + 4)^2$

$2 - x = x^2 + 8x + 16$

$0 = x^2 + 9x + 14$

$0 = (x + 2)(x + 7)$

$x = -2$ or $x = -7$

Check $x = -2$: $2 + 3 = 5$ *True*
Check $x = -7$: $3 + 3 = 0$ *False*

Of the two potential solutions, -2 checks in the original equation, but -7 does not. Thus, the solution set is $\{-2\}$.

85. $\sqrt[3]{x+4} = \sqrt[3]{16-2x}$

$\left(\sqrt[3]{x+4}\right)^3 = \left(\sqrt[3]{16-2x}\right)^3$ *Cube both sides.*

$x + 4 = 16 - 2x$

$3x = 12$

$x = 4$

Check $x = 4$: $\sqrt[3]{8} = \sqrt[3]{16-8}$ *True*

The solution set is $\{4\}$.

86. $\sqrt{5x+6} + \sqrt{3x+4} = 2$

$\sqrt{5x+6} = 2 - \sqrt{3x+4}$

$\left(\sqrt{5x+6}\right)^2 = \left(2 - \sqrt{3x+4}\right)^2$

$5x + 6 = 4 - 4\sqrt{3x+4} + 3x + 4$

$2x - 2 = -4\sqrt{3x+4}$

$x - 1 = -2\sqrt{3x+4}$

$(x-1)^2 = \left(-2\sqrt{3x+4}\right)^2$

$x^2 - 2x + 1 = 4(3x+4)$

$x^2 - 2x + 1 = 12x + 16$

$x^2 - 14x - 15 = 0$

$(x+1)(x-15) = 0$

$x = -1$ or $x = 15$

Check $x = -1$: $1 + 1 = 2$ *True*
Check $x = 15$: $9 + 7 = 2$ *False*

The solution set is $\{-1\}$.

87. $81^{1/2} = \sqrt{81} = 9$

88. $-125^{1/3} = -\sqrt[3]{125} = -5$

89. $7^{2/3} \cdot 7^{7/3} = 7^{2/3 + 7/3}$

$= 7^{9/3} = 7^3$,   or 343

90. $\dfrac{13^{4/5}}{13^{-3/5}} = 13^{4/5 - (-3/5)} = 13^{7/5}$

91. $\dfrac{x^{1/4} \cdot x^{5/4}}{x^{3/4}} = \dfrac{x^{1/4 + 5/4}}{x^{3/4}} = \dfrac{x^{6/4}}{x^{3/4}}$

$= x^{6/4 - 3/4} = x^{3/4}$

92. $\sqrt[8]{49^4} = 49^{4/8} = 49^{1/2} = \sqrt{49} = 7$

93.   **[8.7]** $64^{2/3} = \left(\sqrt[3]{64}\right)^2 = 4^2 = 16$

94.   **[8.4]** $\sqrt{\dfrac{1}{3}} \cdot \sqrt{\dfrac{24}{5}} = \sqrt{\dfrac{1}{3} \cdot \dfrac{24}{5}} = \sqrt{\dfrac{8}{5}} = \dfrac{\sqrt{8}}{\sqrt{5}}$

$= \dfrac{\sqrt{8} \cdot \sqrt{5}}{\sqrt{5} \cdot \sqrt{5}} = \dfrac{\sqrt{40}}{5}$

$= \dfrac{\sqrt{4 \cdot 10}}{5} = \dfrac{2\sqrt{10}}{5}$

95.   **[8.5]** $\dfrac{1}{5 + \sqrt{2}} = \dfrac{1\left(5 - \sqrt{2}\right)}{\left(5 + \sqrt{2}\right)\left(5 - \sqrt{2}\right)}$

$= \dfrac{5 - \sqrt{2}}{(5)^2 - \left(\sqrt{2}\right)^2}$

$= \dfrac{5 - \sqrt{2}}{25 - 2}$

$= \dfrac{5 - \sqrt{2}}{23}$

96.   **[8.1]** $\sqrt[3]{-125} = -5$ because $(-5)^3 = -125$.

97.   **[8.2]** $\sqrt{50y^2} = \sqrt{25y^2 \cdot 2}$

$= \sqrt{25y^2} \cdot \sqrt{2}$

$= 5y\sqrt{2}$

98.   **[8.4]** $\sqrt{\dfrac{16r^3}{3s}} = \dfrac{\sqrt{16r^3}}{\sqrt{3s}} = \dfrac{\sqrt{16r^2} \cdot \sqrt{r}}{\sqrt{3s}}$

$= \dfrac{4r\sqrt{r}}{\sqrt{3s}} = \dfrac{4r\sqrt{r} \cdot \sqrt{3s}}{\sqrt{3s} \cdot \sqrt{3s}}$

$= \dfrac{4r\sqrt{3rs}}{3s}$

99.   **[8.3]** $-\sqrt{5}\left(\sqrt{2} + \sqrt{75}\right)$

$= -\sqrt{5}\left(\sqrt{2}\right) + \left(-\sqrt{5}\right)\left(\sqrt{75}\right)$

$= -\sqrt{10} - \sqrt{375}$

$= -\sqrt{10} - \sqrt{25 \cdot 15}$

$= -\sqrt{10} - 5\sqrt{15}$

100.   **[8.3]** $-\sqrt{162} + \sqrt{8} = -\sqrt{81 \cdot 2} + \sqrt{4 \cdot 2}$

$= -9\sqrt{2} + 2\sqrt{2}$

$= -7\sqrt{2}$

101.   **[8.5]** $\dfrac{12 + 6\sqrt{13}}{12} = \dfrac{6\left(2 + \sqrt{13}\right)}{6(2)}$

$= \dfrac{2 + \sqrt{13}}{2}$

102.   **[8.5]** $\left(6\sqrt{7} + 2\right)\left(4\sqrt{7} - 1\right)$

$= 6\sqrt{7}\left(4\sqrt{7}\right) - 1\left(6\sqrt{7}\right)$

$\quad + 2\left(4\sqrt{7}\right) + 2(-1)$ *FOIL*

$= 24 \cdot 7 - 6\sqrt{7} + 8\sqrt{7} - 2$

$= 168 - 2 + 2\sqrt{7}$

$= 166 + 2\sqrt{7}$

103. **[8.5]** $\left(\sqrt{5}-\sqrt{2}\right)^2$

$= \left(\sqrt{5}\right)^2 - 2\sqrt{5}\sqrt{2} + \left(\sqrt{2}\right)^2$

Square of a binomial

$= 5 - 2\sqrt{10} + 2$

$= 7 - 2\sqrt{10}$

104. **[8.7]** $\dfrac{x^{8/3}}{x^{2/3}} = x^{8/3-2/3} = x^{6/3} = x^2$

105. **[8.1]** $-\sqrt{121} = -11$

106. **[8.3]** $2\sqrt{27} + 3\sqrt{75} - \sqrt{300}$

$= 2\sqrt{9\cdot 3} + 3\sqrt{25\cdot 3} - \sqrt{100\cdot 3}$

$= 2\cdot 3\sqrt{3} + 3\cdot 5\sqrt{3} - 10\sqrt{3}$

$= 6\sqrt{3} + 15\sqrt{3} - 10\sqrt{3}$

$= 11\sqrt{3}$

107. **[8.6]** $\sqrt{x+2} = x - 4$

$\left(\sqrt{x+2}\right)^2 = (x-4)^2$

$x + 2 = x^2 - 8x + 16$

$0 = x^2 - 9x + 14$

$0 = (x-2)(x-7)$

$x = 2 \quad \text{or} \quad x = 7$

Check $x = 2$: $\sqrt{4} = -2$ *False*
Check $x = 7$: $\sqrt{9} = 3$ *True*

The solution set is $\{7\}$.

108. **[8.6]** $\sqrt{x} + 3 = 0$

$\sqrt{x} = -3$

Since a square root cannot equal a negative number, there is no solution and the solution set is $\emptyset$.

109. **[8.6]** $\sqrt{1+3t} - t = -3$

$\sqrt{1+3t} = t - 3$

$\left(\sqrt{1+3t}\right)^2 = (t-3)^2$

$1 + 3t = t^2 - 6t + 9$

$0 = t^2 - 9t + 8$

$0 = (t-1)(t-8)$

$t = 1 \quad \text{or} \quad t = 8$

Check $t = 1$: $2 - 1 = -3$ *False*
Check $t = 8$: $5 - 8 = -3$ *True*

The solution set is $\{8\}$.

110. **[8.1]** **(a)** $S = 28.6\sqrt[3]{A}$

$S = 28.6\sqrt[3]{8}$ *Let $A = 8$.*

$= 28.6(2)$

$= 57.2 \approx 57$

There would be 57 species.

(b) $S = 28.6\sqrt[3]{A}$

$S = 28.6\sqrt[3]{1790}$ *Let $A = 1790$.*

≈ 347.3

There would be 347 species.

Chapter 8 Test

1. The square roots of 196 are -14 and 14 because $(-14)^2 = 196$ and $14^2 = 196$.

2. **(a)** $\sqrt{142}$ is *irrational* because 142 is not a perfect square.

(b) $\sqrt{142} \approx 11.916$

3. **(a)** $\sqrt{64} = 8$; **B**

(b) $-\sqrt{64} = -8$; **F**

(c) $\sqrt{-64}$ is not a real number; **D**

(d) $\sqrt[3]{64} = 4$; **A**

(e) $\sqrt[3]{-64} = -4$; **C**

(f) $-\sqrt[3]{-64} = -(-4) = 4$; **A**

4. $\sqrt{\dfrac{128}{25}} = \dfrac{\sqrt{128}}{\sqrt{25}} = \dfrac{\sqrt{64\cdot 2}}{5} = \dfrac{8\sqrt{2}}{5}$

5. $\sqrt[3]{32} = \sqrt[3]{8\cdot 4} = \sqrt[3]{8}\cdot\sqrt[3]{4} = 2\sqrt[3]{4}$

6. $\dfrac{20\sqrt{18}}{5\sqrt{3}} = \dfrac{4\sqrt{9\cdot 2}}{\sqrt{3}}$

$= \dfrac{4\cdot 3\sqrt{2}}{\sqrt{3}}$

$= \dfrac{12\sqrt{2}\cdot\sqrt{3}}{\sqrt{3}\cdot\sqrt{3}}$

$= \dfrac{12\sqrt{6}}{3} = 4\sqrt{6}$

7. $3\sqrt{28} + \sqrt{63} = 3\left(\sqrt{4\cdot 7}\right) + \sqrt{9\cdot 7}$

$= 3\left(2\sqrt{7}\right) + 3\sqrt{7}$

$= 6\sqrt{7} + 3\sqrt{7} = 9\sqrt{7}$

8. $3\sqrt{27x} - 4\sqrt{48x} + 2\sqrt{3x}$

$= 3\left(\sqrt{9\cdot 3x}\right) - 4\left(\sqrt{16\cdot 3x}\right) + 2\sqrt{3x}$

$= 3\left(3\sqrt{3x}\right) - 4\left(4\sqrt{3x}\right) + 2\sqrt{3x}$

$= 9\sqrt{3x} - 16\sqrt{3x} + 2\sqrt{3x} = -5\sqrt{3x}$

9. $\sqrt[3]{32x^2y^3} = \sqrt[3]{8y^3\cdot 4x^2}$

$= \sqrt[3]{8y^3}\cdot\sqrt[3]{4x^2}$

$= 2y\sqrt[3]{4x^2}$

10. $\left(6 - \sqrt{5}\right)\left(6 + \sqrt{5}\right)$
$$= (6)^2 - \left(\sqrt{5}\right)^2$$
$$= 36 - 5 = 31$$

11. $\left(2 - \sqrt{7}\right)\left(3\sqrt{2} + 1\right)$
$$= 2\left(3\sqrt{2}\right) + 2(1) - \sqrt{7}\left(3\sqrt{2}\right) - \sqrt{7}(1)$$
$$= 6\sqrt{2} + 2 - 3\sqrt{14} - \sqrt{7}$$

12. $\left(\sqrt{5} + \sqrt{6}\right)^2$
$$= \left(\sqrt{5}\right)^2 + 2\left(\sqrt{5}\right)\left(\sqrt{6}\right) + \left(\sqrt{6}\right)^2$$
$$= 5 + 2\sqrt{30} + 6$$
$$= 11 + 2\sqrt{30}$$

13. $\sqrt[3]{16x^4} - 2\sqrt[3]{128x^4}$
$$= \sqrt[3]{8x^3} \cdot \sqrt[3]{2x} - 2 \cdot \sqrt[3]{64x^3} \cdot \sqrt[3]{2x}$$
$$= 2x \cdot \sqrt[3]{2x} - 2 \cdot 4x \cdot \sqrt[3]{2x}$$
$$= (2x - 8x)\sqrt[3]{2x} = -6x\sqrt[3]{2x}$$

14. $\sqrt[3]{\dfrac{2}{3}}$

Multiply the numerator and the denominator by enough factors of 3 to make the radicand in the denominator a perfect cube. This will eliminate the radical in the denominator. Here, we multiply by $\sqrt[3]{3^2}$ or $\sqrt[3]{9}$.

$$\sqrt[3]{\frac{2}{3}} = \frac{\sqrt[3]{2}}{\sqrt[3]{3}} = \frac{\sqrt[3]{2} \cdot \sqrt[3]{3^2}}{\sqrt[3]{3} \cdot \sqrt[3]{3^2}}$$
$$= \frac{\sqrt[3]{2 \cdot 3^2}}{\sqrt[3]{3 \cdot 3^2}} = \frac{\sqrt[3]{18}}{\sqrt[3]{3^3}} = \frac{\sqrt[3]{18}}{3}$$

15. Use the Pythagorean theorem with $c = 9$ and $b = 3$.
$$c^2 = a^2 + b^2$$
$$9^2 = a^2 + 3^2$$
$$81 = a^2 + 9$$
$$72 = a^2$$
$$\sqrt{72} = a$$

   (a) $a = \sqrt{72} = \sqrt{36 \cdot 2} = 6\sqrt{2}$ inches

   (b) $a = \sqrt{72} \approx 8.485$ inches

16. $Z = \sqrt{R^2 + X^2}$
$$= \sqrt{40^2 + 30^2} \qquad \textit{Let } R = 40,\, X = 30.$$
$$= \sqrt{1600 + 900}$$
$$= \sqrt{2500} = 50 \text{ ohms}$$

17. $\dfrac{5\sqrt{2}}{\sqrt{7}} = \dfrac{5\sqrt{2} \cdot \sqrt{7}}{\sqrt{7} \cdot \sqrt{7}} = \dfrac{5\sqrt{14}}{7}$

18. $\sqrt{\dfrac{2}{3x}} = \dfrac{\sqrt{2}}{\sqrt{3x}} = \dfrac{\sqrt{2} \cdot \sqrt{3x}}{\sqrt{3x} \cdot \sqrt{3x}} = \dfrac{\sqrt{6x}}{3x}$

19. $\dfrac{-2}{\sqrt[3]{4}} = \dfrac{-2 \cdot \sqrt[3]{2}}{\sqrt[3]{4} \cdot \sqrt[3]{2}} = \dfrac{-2\sqrt[3]{2}}{\sqrt[3]{8}}$
$$= \dfrac{-2\sqrt[3]{2}}{2} = -\sqrt[3]{2}$$

20. $\dfrac{-3}{4 - \sqrt{3}} = \dfrac{-3\left(4 + \sqrt{3}\right)}{\left(4 - \sqrt{3}\right)\left(4 + \sqrt{3}\right)}$
$$= \dfrac{-12 - 3\sqrt{3}}{(4)^2 - \left(\sqrt{3}\right)^2}$$
$$= \dfrac{-12 - 3\sqrt{3}}{16 - 3}$$
$$= \dfrac{-12 - 3\sqrt{3}}{13}$$

21. $\dfrac{2 + \sqrt{8}}{4} = \dfrac{2 + \sqrt{4} \cdot \sqrt{2}}{4}$
$$= \dfrac{2 + 2\sqrt{2}}{4}$$
$$= \dfrac{2\left(1 + \sqrt{2}\right)}{2(2)}$$
$$= \dfrac{1 + \sqrt{2}}{2}$$

22. $\sqrt{2x + 6} + 4 = 2$
$$\sqrt{2x + 6} = -2 \qquad \textit{Isolate the radical.}$$
$$\left(\sqrt{2x + 6}\right)^2 = (-2)^2$$
$$2x + 6 = 4$$
$$2x = -2$$
$$x = -1$$

Check $x = -1$: $\quad \sqrt{4} + 4 \overset{?}{=} 2$
$$6 = 2 \qquad \textit{False}$$

The solution set is $\emptyset$. Note that in the second line we have a square root equal to a negative number, but the square root of a real number is nonnegative.

23. $\sqrt{x + 1} = 5 - x$
$$\left(\sqrt{x + 1}\right)^2 = (5 - x)^2$$
$$x + 1 = 25 - 10x + x^2$$
$$0 = x^2 - 11x + 24$$
$$0 = (x - 3)(x - 8)$$
$$x = 3 \quad \text{or} \quad x = 8$$

Check $x = 3$: $\quad \sqrt{4} = 2 \quad \textit{True}$
Check $x = 8$: $\quad \sqrt{9} = -3 \quad \textit{False}$

The solution set is $\{3\}$.

24. $3\sqrt{x} - 1 = 2x$

$\qquad 3\sqrt{x} = 2x + 1$ *Isolate the radical.*

$\qquad \left(3\sqrt{x}\right)^2 = (2x + 1)^2$

$\qquad\qquad 9x = 4x^2 + 4x + 1$

$\qquad\qquad\; 0 = 4x^2 - 5x + 1$

$\qquad\qquad\; 0 = (4x - 1)(x - 1)$

$\qquad\qquad\; x = \tfrac{1}{4} \;\; \text{or} \;\; x = 1$

Check $x = \tfrac{1}{4}$: $\tfrac{1}{2} = \tfrac{1}{2}$ *True*
Check $x = 1$: $2 = 2$ *True*

The solution set is $\left\{\tfrac{1}{4}, 1\right\}$.

25. $\sqrt{2x + 9} + \sqrt{x + 5} = 2$

$\qquad\qquad \sqrt{2x + 9} = 2 - \sqrt{x + 5}$

$\qquad \left(\sqrt{2x + 9}\right)^2 = \left(2 - \sqrt{x + 5}\right)^2$

$\qquad\quad 2x + 9 = 4 - 4\sqrt{x + 5} + x + 5$

$\qquad\qquad\quad x = -4\sqrt{x + 5}$

$\qquad\qquad (x)^2 = \left(-4\sqrt{x + 5}\right)^2$

$\qquad\qquad\; x^2 = 16(x + 5)$

$\qquad\qquad\; x^2 = 16x + 80$

$\quad x^2 - 16x - 80 = 0$

$\;(x + 4)(x - 20) = 0$

$\qquad\qquad\; x = -4 \;\; \text{or} \;\; x = 20$

Check $x = -4$: $1 + 1 = 2$ *True*
Check $x = 20$: $7 + 5 = 2$ *False*

The solution set is $\{-4\}$.

26.   Nothing is wrong with the steps taken so far, but the potential solution must be checked.

Let $x = 12$ in the original equation.

$$\sqrt{2x + 1} + 5 = 0$$
$$\sqrt{2(12) + 1} + 5 \overset{?}{=} 0 \quad \text{\textit{Let x = 12.}}$$
$$\sqrt{25} + 5 \overset{?}{=} 0$$
$$5 + 5 \overset{?}{=} 0$$
$$10 = 0 \quad \textit{False}$$

12 is not a solution because it does not satisfy the original equation. The equation has no solution, so the solution set is $\emptyset$.

27. $8^{4/3} = \left(\sqrt[3]{8}\right)^4 = 2^4 = 16$

28. $-125^{2/3} = -\left(\sqrt[3]{125}\right)^2 = -(5)^2 = -25$

29. $5^{3/4} \cdot 5^{1/4} = 5^{3/4 + 1/4} = 5^{4/4} = 5^1 = 5$

30. $\dfrac{(3^{1/4})^3}{3^{7/4}} = \dfrac{3^{(1/4)\cdot 3}}{3^{7/4}} = \dfrac{3^{3/4}}{3^{7/4}} = 3^{3/4 - 7/4}$

$\qquad\qquad = 3^{-4/4} = 3^{-1} = \dfrac{1}{3^1} = \dfrac{1}{3}$

Cumulative Review Exercises (Chapters 1–8)

1. $3(6 + 7) + 6 \cdot 4 - 3^2$

$\quad = 3(13) + 24 - 9$

$\quad = 39 + 24 - 9$

$\quad = 63 - 9 = 54$

2. $\dfrac{3(6 + 7) + 3}{2(4) - 1} = \dfrac{3(13) + 3}{8 - 1}$

$\qquad\qquad\qquad = \dfrac{39 + 3}{7}$

$\qquad\qquad\qquad = \tfrac{42}{7} = 6$

3. $|-6| - |-3| = 6 - 3 = 3$

4. $5(k - 4) - k = k - 11$

$\qquad 5k - 20 - k = k - 11$

$\qquad\quad 4k - 20 = k - 11$

$\qquad\qquad\quad 3k = 9$

$\qquad\qquad\quad\; k = 3$

The solution set is $\{3\}$.

5. $-\tfrac{3}{4}x \le 12$

$\quad -\tfrac{4}{3}\left(-\tfrac{3}{4}x\right) \ge -\tfrac{4}{3}(12)$

$\qquad\qquad\quad x \ge -16$

The solution set is $[-16, \infty)$.

6. $5z + 3 - 4 > 2z + 9 + z$

$\qquad\; 5z - 1 > 3z + 9$

$\qquad\qquad 2z > 10$

$\qquad\qquad\; z > 5$

The solution set is $(5, \infty)$.

7.   Let $x =$ the amount earned in 2006.
Then $x + \$95{,}191 =$ the amount earned in 2007.
The amounts total $\$755{,}039$, so

$$x + (x + 95{,}191) = 755{,}039$$
$$2x + 95{,}191 = 755{,}039$$
$$2x = 659{,}848$$
$$x = 329{,}924$$

Brazile earned $\$329{,}924$ in 2006 and
$\$329{,}924 + \$95{,}191 = \$425{,}115$ in 2007.

8. $-4x + 5y = -20$

Find the intercepts.

If $y = 0$, $x = 5$, so the x-intercept is $(5, 0)$.
If $x = 0$, $y = -4$, so the y-intercept is $(0, -4)$.

Draw the line that passes through the points $(5, 0)$ and $(0, -4)$.

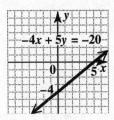

9. $x = 2$

For any value of y, the value of x is 2, so this is a vertical line through $(2, 0)$.

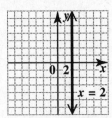

10. $2x - 5y > 10$

The boundary, $2x - 5y = 10$, is the line that passes through $(5, 0)$ and $(0, -2)$; draw it as a dashed line because of the $>$ symbol. Use $(0, 0)$ as a test point. Because

$$2(0) - 5(0) > 10$$

is a false statement, shade the side of the dashed boundary that does not include the origin, $(0, 0)$. The dashed line shows that the boundary is not part of the graph.

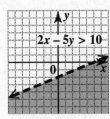

11. **(a)** The slope of the line through the points $(0, 140.8)$ and $(6, 262.7)$ is

$$m = \frac{y_2 - y_1}{x_2 - x_1}$$

$$= \frac{262.7 - 140.8}{6 - 0}$$

$$= \frac{121.9}{6} \approx 20.3$$

An interpretation of the slope is that the number of cell phone subscribers increased by an average of $20.3 million per year.

(b) Using the slope-intercept form of a line with $m = 20.3$ and $b = 140.8$, we get

$$y = 20.3x + 140.8.$$

(c) For 2010, $x = 2010 - 2002 = 8$.

$$y = 20.3(8) + 140.8$$
$$= 162.4 + 140.8$$
$$= 303.2$$

The estimated number of cell phone subscribers for 2010 is about 303.2 million.

12. $4x - y = 19$ (1)
    $3x + 2y = -5$ (2)

We will solve this system by the elimination method. Multiply both sides of equation (1) by 2, and then add the result to equation (2).

$$
\begin{array}{rrrrr}
8x & - & 2y & = & 38 \\
3x & + & 2y & = & -5 \\
\hline
11x & & & = & 33 \\
& & x & = & 3
\end{array}
$$

Let $x = 3$ in equation (1).

$$4(3) - y = 19$$
$$12 - y = 19$$
$$-y = 7$$
$$y = -7$$

The solution set is $\{(3, -7)\}$.

13. $2x - y = 6$ (1)
    $3y = 6x - 18$ (2)

We will solve this system by the substitution method. Solve equation (2) for y by dividing both sides by 3.

$$y = 2x - 6$$

Substitute $2x - 6$ for y in (1).

$$2x - (2x - 6) = 6$$
$$2x - 2x + 6 = 6$$
$$6 = 6$$

This true statement indicates that the two original equations both describe the same line. This system has an infinite number of solutions. The solution set is $\{(x, y) \mid 2x - y = 6\}$.

14.   Let $x =$ the average rate of the slower car (departing from Des Moines).

Then $x + 7 =$ the average rate of the faster car (departing from Chicago).

In 3 hours, the slower car travels $3x$ miles and the faster car travels $3(x + 7)$ miles. The total distance traveled is 345 miles, so

$$3x + 3(x + 7) = 345$$
$$3x + 3x + 21 = 345$$
$$6x = 324$$
$$x = 54.$$

The car departing from Des Moines averaged 54 miles per hour and traveled $3(54) = 162$ miles. The car departing from Chicago averaged 61 miles per hour and traveled $3(61) = 183$ miles.

15. $(3x^6)(2x^2y)^2$
$$= (3x^6)(2)^2(x^2)^2(y)^2$$
$$= (3x^6) \cdot 4x^4y^2$$
$$= 12x^{10}y^2$$

16. $\left(\dfrac{3^2y^{-2}}{2^{-1}y^3}\right)^{-3} = \left(\dfrac{2^{-1}y^3}{3^2y^{-2}}\right)^3$
$$= \left(\dfrac{y^3 \cdot y^2}{2^1 \cdot 3^2}\right)^3$$
$$= \dfrac{(y^5)^3}{(2 \cdot 3^2)^3}$$
$$= \dfrac{y^{15}}{2^3 \cdot 3^6}, \text{ or } \dfrac{y^{15}}{5832}$$

17. $(10x^3 + 3x^2 - 9) - (7x^3 - 8x^2 + 4)$
$$= 10x^3 + 3x^2 - 9 - 7x^3 + 8x^2 - 4$$
$$= 3x^3 + 11x^2 - 13$$

18.

$$\begin{array}{r} 4t^2 - 8t + 5 \\ 2t+3\overline{)8t^3 - 4t^2 - 14t + 15} \\ \underline{8t^3 + 12t^2} \\ -16t^2 - 14t \\ \underline{-16t^2 - 24t} \\ 10t + 15 \\ \underline{10t + 15} \\ 0 \end{array}$$

The remainder is 0, so the answer is the quotient, $4t^2 - 8t + 5$.

19. $m^2 + 12m + 32$

Find two integers whose product is 32 and whose sum is 12. The required integers are 8 and 4. Thus,

$$m^2 + 12m + 32 = (m + 8)(m + 4).$$

20. $12a^2 + 4ab - 5b^2$

Factor by the grouping method. Look for two integers whose product is $12(-5) = -60$ and whose sum is 4. The integers are 10 and -6.

$$12a^2 + 4ab - 5b^2$$
$$= 12a^2 + 10ab - 6ab - 5b^2$$
$$= (12a^2 + 10ab) + (-6ab - 5b^2)$$
$$= 2a(6a + 5b) - b(6a + 5b)$$
$$= (6a + 5b)(2a - b)$$

21. $81z^2 + 72z + 16$
$$= (9z)^2 + 2(9z)(4) + 4^2$$
$$= (9z + 4)^2$$

22. $\dfrac{x^2 - 3x - 4}{x^2 + 3x} \cdot \dfrac{x^2 + 2x - 3}{x^2 - 5x + 4}$
$$= \dfrac{(x-4)(x+1)}{x(x+3)} \cdot \dfrac{(x-1)(x+3)}{(x-4)(x-1)} \quad \textit{Factor.}$$
$$= \dfrac{x+1}{x} \quad \textit{Lowest terms}$$

23. $\dfrac{t^2 + 4t - 5}{t + 5} \div \dfrac{t - 1}{t^2 + 8t + 15}$
$$= \dfrac{t^2 + 4t - 5}{t + 5} \cdot \dfrac{t^2 + 8t + 15}{t - 1}$$

$\textit{Multiply by the reciprocal.}$

$$= \dfrac{(t+5)(t-1)}{t+5} \cdot \dfrac{(t+5)(t+3)}{t-1} \quad \textit{Factor.}$$
$$= (t+5)(t+3), \text{ or } t^2 + 8t + 15 \quad \textit{Lowest terms}$$

24. $\dfrac{y}{y^2 - 1} + \dfrac{y}{y + 1}$
$$= \dfrac{y}{(y+1)(y-1)} + \dfrac{y(y-1)}{(y+1)(y-1)}$$
$$= \dfrac{y + y(y-1)}{(y+1)(y-1)}$$
$$= \dfrac{y + y^2 - y}{(y+1)(y-1)} = \dfrac{y^2}{(y+1)(y-1)}$$

25. $\dfrac{2}{x+3} - \dfrac{4}{x-1}$
$$= \dfrac{2(x-1)}{(x+3)(x-1)} - \dfrac{4(x+3)}{(x-1)(x+3)}$$
$$= \dfrac{2(x-1) - 4(x+3)}{(x+3)(x-1)}$$
$$= \dfrac{2x - 2 - 4x - 12}{(x+3)(x-1)}$$
$$= \dfrac{-2x - 14}{(x+3)(x-1)}$$

26. $\dfrac{\frac{2}{3}+\frac{1}{2}}{\frac{1}{9}-\frac{1}{6}} = \dfrac{\frac{4}{6}+\frac{3}{6}}{\frac{2}{18}-\frac{3}{18}}$

$= \dfrac{\frac{7}{6}}{\frac{-1}{18}} = \dfrac{7}{6} \div \dfrac{-1}{18} = \dfrac{7}{6} \cdot \dfrac{18}{-1}$

$= 7 \cdot (-3) = -21$

27. $\qquad x^2 - 7x = -12$

$x^2 - 7x + 12 = 0$

$(x-3)(x-4) = 0$

$\qquad\qquad x = 3 \quad \text{or} \quad x = 4$

The solution set is $\{3, 4\}$.

28. $(x+4)(x-1) = -6$

$x^2 + 3x - 4 = -6$

$x^2 + 3x + 2 = 0$

$(x+2)(x+1) = 0$

$\qquad\qquad x = -2 \quad \text{or} \quad x = -1$

The solution set is $\{-2, -1\}$.

29. $\qquad \dfrac{x}{x+8} - \dfrac{3}{x-8} = \dfrac{128}{x^2 - 64}$

$\dfrac{x}{x+8} - \dfrac{3}{x-8} = \dfrac{128}{(x+8)(x-8)}$

Multiply by the LCD, $(x+8)(x-8)$.

$x(x-8) - 3(x+8) = 128$

$x^2 - 8x - 3x - 24 = 128$

$x^2 - 11x - 152 = 0$

$(x+8)(x-19) = 0$

$\qquad x = -8 \quad \text{or} \quad x = 19$

Check $x = 19$: $\quad \frac{19}{27} - \frac{3}{11} = \frac{128}{297}$ *True*

We cannot have -8 for a solution because that would result in division by zero. Thus, the solution set is $\{19\}$.

30. Solve $A = \dfrac{B+CD}{BC+D}$ for B.

$A(BC+D) = B + CD$

$ABC + AD = B + CD$

Get the B-terms on one side.

$ABC - B = CD - AD$

$B(AC - 1) = CD - AD$

$B = \dfrac{CD - AD}{AC - 1}$

31. $\qquad \sqrt{x} + 2 = x - 10$

$\sqrt{x} = x - 12$

$\left(\sqrt{x}\right)^2 = (x-12)^2$

$x = x^2 - 24x + 144$

$0 = x^2 - 25x + 144$

$0 = (x-16)(x-9)$

$\qquad x = 16 \quad \text{or} \quad x = 9$

Check $x = 16$: $\quad 6 = 6 \quad$ *True*
Check $x = 9$: $\quad 5 = -1 \quad$ *False*

The solution set is $\{16\}$.

32. $\sqrt{27} - 2\sqrt{12} + 6\sqrt{75}$

$= \sqrt{9} \cdot \sqrt{3} - 2\sqrt{4} \cdot \sqrt{3} + 6\sqrt{25} \cdot \sqrt{3}$

$= 3\sqrt{3} - 2\left(2\sqrt{3}\right) + 6\left(5\sqrt{3}\right)$

$= 3\sqrt{3} - 4\sqrt{3} + 30\sqrt{3} = 29\sqrt{3}$

33. $\dfrac{2}{\sqrt{3}+\sqrt{5}} = \dfrac{2\left(\sqrt{3}-\sqrt{5}\right)}{\left(\sqrt{3}+\sqrt{5}\right)\left(\sqrt{3}-\sqrt{5}\right)}$

$= \dfrac{2\left(\sqrt{3}-\sqrt{5}\right)}{3-5}$

$= \dfrac{2\left(\sqrt{3}-\sqrt{5}\right)}{-2}$

$= \dfrac{\sqrt{3}-\sqrt{5}}{-1} = -\sqrt{3} + \sqrt{5}$

34. $\left(3\sqrt{2}+1\right)\left(4\sqrt{2}-3\right)$

$= 3\sqrt{2}\left(4\sqrt{2}\right) - 3\sqrt{2}(3) + 1\left(4\sqrt{2}\right)$
$\quad + 1(-3)$

$= 12 \cdot 2 - 9\sqrt{2} + 4\sqrt{2} - 3$

$= 24 - 3 - 5\sqrt{2}$

$= 21 - 5\sqrt{2}$

35. $16^{5/4} = \left(\sqrt[4]{16}\right)^5 = 2^5 = 32$

CHAPTER 9   QUADRATIC EQUATIONS

9.1   Solving Quadratic Equations by the Square Root Property

9.1   Now Try Exercises

N1. **(a)** $x^2 - x - 20 = 0$

$(x - 5)(x + 4) = 0$ *Factor.*

$x - 5 = 0$ or $x + 4 = 0$ *Zero-factor property*
$x = 5$ or $x = -4$ *Solve each equation.*

The solution set is $\{-4, 5\}$.

(b) $x^2 = 36$

$x^2 - 36 = 0$ *Subtract 36.*
$(x + 6)(x - 6) = 0$ *Factor.*

$x + 6 = 0$ or $x - 6 = 0$ *Zero-factor property*
$x = -6$ or $x = 6$ *Solve each equation.*

The solution set is $\{-6, 6\}$.

N2. **(a)** $t^2 = 25$

Solve by using the square root property.

$t = \sqrt{25}$ or $t = -\sqrt{25}$
$t = 5$ or $t = -5$

The solution set is $\{-5, 5\}$, or $\{\pm 5\}$.

(b) $x^2 = 13$

Use the square root property.

$x = \sqrt{13}$ or $x = -\sqrt{13}$.

The solution set is $\{-\sqrt{13}, \sqrt{13}\}$, or $\{\pm \sqrt{13}\}$.

(c) $x^2 = -144$

Since -144 is a negative number and since the square of a real number cannot be negative, there is no real number solution of this equation. The solution set is $\emptyset$.

(d) $2x^2 - 5 = 35$

$2x^2 = 40$
$x^2 = 20$

Use the square root property.

$x = \sqrt{20}$ or $x = -\sqrt{20}$
$x = \sqrt{4} \cdot \sqrt{5}$ or $x = -\sqrt{4} \cdot \sqrt{5}$
$x = 2\sqrt{5}$ or $x = -2\sqrt{5}$

The solution set is $\{-2\sqrt{5}, 2\sqrt{5}\}$, or $\{\pm 2\sqrt{5}\}$.

N3. $(x - 2)^2 = 32$

Use the square root property.

$x - 2 = \sqrt{32}$ or $x - 2 = -\sqrt{32}$
$x = 2 + \sqrt{16} \cdot \sqrt{2}$ or $x = 2 - \sqrt{16} \cdot \sqrt{2}$
$x = 2 + 4\sqrt{2}$ or $x = 2 - 4\sqrt{2}$

Check $x = 2 \pm 4\sqrt{2}$:

$$\text{LS} = \left[(2 \pm 4\sqrt{2}) - 2 \right]^2$$
$$= \left(\pm 4\sqrt{2} \right)^2 = 16 \cdot 2 = 32 = \text{RS}$$

The solution set is $\left\{ 2 \pm 4\sqrt{2} \right\}$.

N4. $(2t - 4)^2 = 50$

$2t - 4 = \sqrt{50}$ or $2t - 4 = -\sqrt{50}$
$2t - 4 = \sqrt{25} \cdot \sqrt{2}$ or $2t - 4 = -\sqrt{25} \cdot \sqrt{2}$
$2t = 4 + 5\sqrt{2}$ or $2t = 4 - 5\sqrt{2}$
$t = \dfrac{4 + 5\sqrt{2}}{2}$ or $t = \dfrac{4 - 5\sqrt{2}}{2}$

The solution set is $\left\{ \frac{4 \pm 5\sqrt{2}}{2} \right\}$.

N5. $(2x + 1)^2 = -5$

Since the square root of -5 is not a real number, there is *no real number solution*. The solution set is $\emptyset$.

N6. $w = \dfrac{L^2 g}{1200}$ *Given formula*

$2.10 = \dfrac{L^2 \cdot 9}{1200}$ *Let w = 2.10, g = 9.*

$2520 = 9L^2$ *Multiply by 1200.*
$L^2 = 280$ *Divide by 9.*
$L \approx 16.73$ *Approximate L > 0.*

The length of the bass is approximately 16.73 in.

9.1   Section Exercises

1. $x^2 = 12$ has two irrational solutions, $\pm \sqrt{12} = \pm 2\sqrt{3}$. The correct choice is **C**.

3. $x^2 = \frac{25}{36}$ has two rational solutions that are not integers, $\pm \frac{5}{6}$. The correct choice is **D**.

5. $x^2 - x - 56 = 0$
$(x - 8)(x + 7) = 0$ *Factor.*

$x - 8 = 0$ or $x + 7 = 0$ *Zero-factor property*
$x = 8$ or $x = -7$ *Solve each equation.*

The solution set is $\{-7, 8\}$.

7.
$$x^2 = 121$$
$$x^2 - 121 = 0 \quad \text{Subtract 121.}$$
$$(x + 11)(x - 11) = 0 \quad \text{Factor.}$$

$$x + 11 = 0 \quad \text{or } x - 11 = 0 \quad \text{Zero-factor prop.}$$
$$x = -11 \text{ or } \qquad x = 11 \text{ Solve each eq.}$$

The solution set is $\{-11, 11\}$, or $\{\pm 11\}$.

9.
$$3x^2 - 13x = 30$$
$$3x^2 - 13x - 30 = 0$$
$$(x - 6)(3x + 5) = 0 \qquad \text{Factor.}$$

$$x - 6 = 0 \text{ or } 3x + 5 = 0 \quad \text{Zero-factor prop.}$$
$$x = 6 \text{ or } \qquad x = -\tfrac{5}{3} \text{ Solve each eq.}$$

The solution set is $\left\{-\tfrac{5}{3}, 6\right\}$.

11. $x^2 = 81$

Use the square root property to get
$$x = \sqrt{81} = 9 \quad \text{or} \quad x = -\sqrt{81} = -9.$$
The solution set is $\{\pm 9\}$.

13. $x^2 = 14$

Use the square root property to get
$$x = \sqrt{14} \quad \text{or} \quad x = -\sqrt{14}.$$
The solution set is $\left\{\pm \sqrt{14}\right\}$.

15. $t^2 = 48$
$$t = \sqrt{48} \quad \text{or} \quad t = -\sqrt{48}$$
Write $\sqrt{48}$ in simplest form.
$$\sqrt{48} = \sqrt{16} \cdot \sqrt{3} = 4\sqrt{3}$$
The solution set is $\left\{\pm 4\sqrt{3}\right\}$.

17. $x^2 = -100$

This equation has no real number solution because the square of a real number cannot be negative. The square root property cannot be used because it requires that k be positive. The solution set is $\emptyset$.

19. $x^2 = \tfrac{25}{4}$
$$x = \sqrt{\tfrac{25}{4}} \quad \text{or} \quad x = -\sqrt{\tfrac{25}{4}}$$
$$x = \tfrac{5}{2} \quad \text{or} \quad x = -\tfrac{5}{2}$$
The solution set is $\left\{\pm \tfrac{5}{2}\right\}$.

21. $x^2 = 2.25$
$$x = \sqrt{2.25} \quad \text{or} \quad x = -\sqrt{2.25}$$
$$x = 1.5 \quad \text{or} \quad x = -1.5$$
The solution set is $\{\pm 1.5\}$.

23. $r^2 - 3 = 0$
$$r^2 = 3$$
$$r = \sqrt{3} \quad \text{or} \quad r = -\sqrt{3}$$
The solution set is $\left\{\pm \sqrt{3}\right\}$.

25. $7x^2 = 4$
$$x^2 = \tfrac{4}{7} \quad \text{Divide by 7.}$$

$$x = \sqrt{\tfrac{4}{7}} \qquad \text{or} \quad x = -\sqrt{\tfrac{4}{7}}$$
$$= \frac{\sqrt{4}}{\sqrt{7}} \cdot \frac{\sqrt{7}}{\sqrt{7}} \qquad = -\frac{\sqrt{4}}{\sqrt{7}} \cdot \frac{\sqrt{7}}{\sqrt{7}}$$
$$= \frac{2\sqrt{7}}{7} \qquad = -\frac{2\sqrt{7}}{7}$$

The solution set is $\left\{\pm \tfrac{2\sqrt{7}}{7}\right\}$.

27. $3n^2 - 72 = 0$
$$3n^2 = 72$$
$$n^2 = 24$$
$$n = \pm \sqrt{24} = \pm \sqrt{4 \cdot 6} = \pm 2\sqrt{6}$$
The solution set is $\left\{\pm 2\sqrt{6}\right\}$.

29. $5x^2 + 4 = 8$
$$5x^2 = 4 \quad \text{Subtract 4.}$$
$$x^2 = \tfrac{4}{5} \quad \text{Divide by 5.}$$

$$x = \sqrt{\tfrac{4}{5}} \qquad \text{or} \quad x = -\sqrt{\tfrac{4}{5}}$$
$$= \frac{\sqrt{4}}{\sqrt{5}} \cdot \frac{\sqrt{5}}{\sqrt{5}} \qquad = -\frac{\sqrt{4}}{\sqrt{5}} \cdot \frac{\sqrt{5}}{\sqrt{5}}$$
$$= \frac{2\sqrt{5}}{5} \qquad = -\frac{2\sqrt{5}}{5}$$

The solution set is $\left\{\pm \tfrac{2\sqrt{5}}{5}\right\}$.

31. $2t^2 + 7 = 61$
$$2t^2 = 54$$
$$t^2 = 27$$
$$t = \pm \sqrt{27} = \pm \sqrt{9 \cdot 3} = \pm 3\sqrt{3}$$
The solution set is $\left\{\pm 3\sqrt{3}\right\}$.

33. $-8x^2 = -64$
$$x^2 = 8 \quad \text{Divide by } -8.$$
$$x = \pm \sqrt{8} = \pm \sqrt{4 \cdot 2} = \pm 2\sqrt{2}$$
The solution set is $\left\{\pm 2\sqrt{2}\right\}$.

35. It is not correct to say that the solution set of $x^2 = 81$ is $\{9\}$, because -9 also satisfies the equation.

When we solve an equation, we want to find *all* values of the variable that satisfy the equation. The completely correct answer is that the solution set of $x^2 = 81$ is $\{\pm 9\}$.

37. $(x-3)^2 = 25$

Use the square root property.

$$x - 3 = \sqrt{25} \quad \text{or} \quad x - 3 = -\sqrt{25}$$
$$x - 3 = 5 \quad \text{or} \quad x - 3 = -5$$
$$x = 8 \quad \text{or} \quad x = -2$$

The solution set is $\{-2, 8\}$.

39. $(z+5)^2 = -13$

The square root of -13 is not a real number, so there is no real number solution for this equation. The solution set is $\emptyset$.

41. $(x-8)^2 = 27$

Begin by using the square root property.

$$x - 8 = \sqrt{27} \quad \text{or} \quad x - 8 = -\sqrt{27}$$

Now simplify the radical.

$$\sqrt{27} = \sqrt{9} \cdot \sqrt{3} = 3\sqrt{3}$$

$$x - 8 = 3\sqrt{3} \quad \text{or} \quad x - 8 = -3\sqrt{3}$$
$$x = 8 + 3\sqrt{3} \quad \text{or} \quad x = 8 - 3\sqrt{3}$$

The solution set is $\left\{8 \pm 3\sqrt{3}\right\}$.

43. $(3x+2)^2 = 49$

$$3x + 2 = \sqrt{49} \quad \text{or} \quad 3x + 2 = -\sqrt{49}$$
$$3x + 2 = 7 \quad \text{or} \quad 3x + 2 = -7$$
$$3x = 5 \quad \text{or} \quad 3x = -9$$
$$x = \tfrac{5}{3} \quad \text{or} \quad x = -3$$

The solution set is $\left\{-3, \tfrac{5}{3}\right\}$.

45. $(4x-3)^2 = 9$

$$4x - 3 = \sqrt{9} \quad \text{or} \quad 4x - 3 = -\sqrt{9}$$
$$4x - 3 = 3 \quad \text{or} \quad 4x - 3 = -3$$
$$4x = 6 \quad \text{or} \quad 4x = 0$$
$$x = \tfrac{6}{4} = \tfrac{3}{2} \quad \text{or} \quad x = 0$$

The solution set is $\left\{0, \tfrac{3}{2}\right\}$.

47. $(5-2x)^2 = 30$

$$5 - 2x = \sqrt{30} \quad \text{or} \quad 5 - 2x = -\sqrt{30}$$
$$-2x = -5 + \sqrt{30} \quad \text{or} \quad -2x = -5 - \sqrt{30}$$
$$x = \frac{-5 + \sqrt{30}}{-2} \quad \text{or} \quad x = \frac{-5 - \sqrt{30}}{-2}$$
$$x = \frac{-5 + \sqrt{30}}{-2} \cdot \frac{-1}{-1} \quad \text{or} \quad x = \frac{-5 - \sqrt{30}}{-2} \cdot \frac{-1}{-1}$$
$$x = \frac{5 - \sqrt{30}}{2} \quad \text{or} \quad x = \frac{5 + \sqrt{30}}{2}$$

The solution set is $\left\{\frac{5 \pm \sqrt{30}}{2}\right\}$.

49. $(3k+1)^2 = 18$

$$3k + 1 = \sqrt{18} \quad \text{or} \quad 3k + 1 = -\sqrt{18}$$
$$3k = -1 + 3\sqrt{2} \quad \text{or} \quad 3k = -1 - 3\sqrt{2}$$

Note that $\sqrt{18} = \sqrt{9 \cdot 2} = 3\sqrt{2}$.

$$k = \frac{-1 + 3\sqrt{2}}{3} \quad \text{or} \quad k = \frac{-1 - 3\sqrt{2}}{3}$$

The solution set is $\left\{\frac{-1 \pm 3\sqrt{2}}{3}\right\}$.

51. $(\tfrac{1}{2}x + 5)^2 = 12$

$$\tfrac{1}{2}x + 5 = \sqrt{12} \quad \text{or} \quad \tfrac{1}{2}x + 5 = -\sqrt{12}$$
$$\tfrac{1}{2}x = -5 + 2\sqrt{3} \quad \text{or} \quad \tfrac{1}{2}x = -5 - 2\sqrt{3}$$

Note that $\sqrt{12} = \sqrt{4 \cdot 3} = 2\sqrt{3}$.

$$x = 2\left(-5 + 2\sqrt{3}\right) \quad \text{or} \quad x = 2\left(-5 - 2\sqrt{3}\right)$$
$$x = -10 + 4\sqrt{3} \quad \text{or} \quad x = -10 - 4\sqrt{3}$$

The solution set is $\left\{-10 \pm 4\sqrt{3}\right\}$.

53. $(4x-1)^2 - 48 = 0$

$$(4x - 1)^2 = 48$$

$$4x - 1 = \sqrt{48} \quad \text{or} \quad 4x - 1 = -\sqrt{48}$$
$$4x - 1 = 4\sqrt{3} \quad \text{or} \quad 4x - 1 = -4\sqrt{3}$$
$$4x = 1 + 4\sqrt{3} \quad \text{or} \quad 4x = 1 - 4\sqrt{3}$$
$$x = \frac{1 + 4\sqrt{3}}{4} \quad \text{or} \quad x = \frac{1 - 4\sqrt{3}}{4}$$

The solution set is $\left\{\frac{1 \pm 4\sqrt{3}}{4}\right\}$.

55. Jeff's first solution, $\frac{5 + \sqrt{30}}{2}$, is equivalent to Linda's second solution, $\frac{-5 - \sqrt{30}}{-2}$. This can be verified by multiplying $\frac{5 + \sqrt{30}}{2}$ by 1 in the form $\frac{-1}{-1}$. Similarly, Jeff's second solution is equivalent to Linda's first one.

57. $(k+2.14)^2 = 5.46$

$$k + 2.14 = \sqrt{5.46} \quad \text{or} \quad k + 2.14 = -\sqrt{5.46}$$
$$k = -2.14 + \sqrt{5.46} \quad \text{or} \quad k = -2.14 - \sqrt{5.46}$$
$$k \approx 0.20 \quad \text{or} \quad k \approx -4.48$$

To the nearest hundredth, the solution set is $\{-4.48, 0.20\}$.

59. $(2.11p + 3.42)^2 = 9.58$

$$2.11p + 3.42 = \pm\sqrt{9.58}$$

Remember that this represents two equations.

$$2.11p = -3.42 \pm \sqrt{9.58}$$
$$p = \frac{-3.42 \pm \sqrt{9.58}}{2.11}$$
$$\approx -3.09 \text{ or } -0.15$$

To the nearest hundredth, the solution set is $\{-3.09, -0.15\}$.

61. $d = 16t^2$

$4 = 16t^2$ *Let d = 4.*

$t^2 = \frac{4}{16} = \frac{1}{4}$

$t = \pm\sqrt{\frac{1}{4}} = \pm\frac{1}{2}$

Reject $-\frac{1}{2}$ as a solution, since negative time does not make sense. About $\frac{1}{2}$ second elapses between the dropping of the coin and the shot.

63. $A = \pi r^2$

$81\pi = \pi r^2$ *Let A = 81π.*

$81 = r^2$ *Divide by π.*

$r = 9$ or $r = -9$

Discard -9 since the radius cannot be negative. The radius is 9 inches.

65. Let $A = 104.04$ and $P = 100$.

$$A = P(1 + r)^2$$

$$104.04 = 100(1 + r)^2$$

$$(1 + r)^2 = \frac{104.04}{100} = 1.0404$$

$$1 + r = \pm\sqrt{1.0404}$$

$$1 + r = \pm 1.02$$

$$r = -1 \pm 1.02$$

So $r = -1 + 1.02 = 0.02$ or $r = -1 - 1.02 = -2.02$. Reject the solution -2.02. The rate is $r = 0.02$ or 2%.

67. $\frac{4}{5} + \sqrt{\frac{48}{25}} = \frac{4}{5} + \frac{\sqrt{48}}{\sqrt{25}} = \frac{4}{5} + \frac{\sqrt{16}\cdot\sqrt{3}}{5}$

$\qquad = \frac{4 + 4\sqrt{3}}{5}$

69. $\frac{6 + \sqrt{24}}{8} = \frac{6 + \sqrt{4}\cdot\sqrt{6}}{8} = \frac{6 + 2\sqrt{6}}{8}$

$\qquad = \frac{2\left(3 + \sqrt{6}\right)}{2\cdot 4} = \frac{3 + \sqrt{6}}{4}$

71. $x^2 - 10x + 25 = x^2 - 2\cdot 5\cdot x + 5^2$

$\qquad = (x - 5)^2$

9.2   Solving Quadratic Equations by Completing the Square

9.2   Now Try Exercises

N1. **(a)** $x^2 + 4x + \underline{}$

Here the middle term $4x$ must equal $2kx$.

$2kx = 4x$

$k = 2$ *Divide by 2x.*

Thus, $k = 2$ and $k^2 = 4$. The required trinomial is $x^2 + 4x + \underline{4}$, which factors as $(x + 2)^2$.

(b) $x^2 - 22x + \underline{}$

Here the middle term $-22x$ must equal $2kx$.

$2kx = -22x$

$k = -11$ *Divide by 2x.*

Thus, $k = -11$ and $k^2 = 121$. The required trinomial is $x^2 - 22x + \underline{121}$, which factors as $(x - 11)^2$.

N2. $x^2 + 10x + 8 = 0$

$x^2 + 10x = -8$

$x^2 + 10x + 25 = -8 + 25$

$\qquad$ *2kx = -10x, so k = -5 and k² = 25.*

$(x - 5)^2 = 17$

$x - 5 = \sqrt{17}$ or $x - 5 = -\sqrt{17}$

$\quad x = 5 + \sqrt{17}$ or $\quad x = 5 - \sqrt{17}$

The solution set is $\left\{5 \pm \sqrt{17}\right\}$.

N3. $x^2 - 6x = 9$

Take half of the coefficient of x and square the result.

$$\tfrac{1}{2}(-6) = -3, \text{ and } (-3)^2 = 9.$$

Add 9 to each side of the equation, and write the left side as a perfect square.

$$x^2 - 6x + 9 = 9 + 9$$

$$(x - 3)^2 = 18 \qquad \textit{Factor.}$$

Use the square root property to solve for x.

$x - 3 = \sqrt{18}$ or $x - 3 = -\sqrt{18}$

$x - 3 = \sqrt{9}\cdot\sqrt{2}$ or $x - 3 = -\sqrt{9}\cdot\sqrt{2}$

$\quad x = 3 + 3\sqrt{2}$ or $\quad x = 3 - 3\sqrt{2}$

The solution set is $\left\{3 \pm 3\sqrt{2}\right\}$.

N4. $4t^2 - 4t - 3 = 0$

$4t^2 - 4t = 3$ *Add 3.*

Divide each side by 4 to get a coefficient of 1 for the t^2-term.

$$t^2 - t = \tfrac{3}{4} \qquad \textit{Divide by 4.}$$

Take half the coefficient of t, or $\left(\tfrac{1}{2}\right)(-1) = -\tfrac{1}{2}$, and square the result: $\left(-\tfrac{1}{2}\right)^2 = \tfrac{1}{4}$. Then add $\tfrac{1}{4}$ to each side.

$t^2 - t + \tfrac{1}{4} = \tfrac{3}{4} + \tfrac{1}{4}$ *Add $\frac{1}{4}$.*

$\left(t - \tfrac{1}{2}\right)^2 = 1$ *Factor.*

Apply the square root property, and solve for t.

$t - \tfrac{1}{2} = \sqrt{1}$ or $t - \tfrac{1}{2} = -\sqrt{1}$

$t - \tfrac{1}{2} = 1$ or $t - \tfrac{1}{2} = -1$

$\quad t = \tfrac{1}{2} + 1$ or $\quad t = \tfrac{1}{2} - 1$

$\quad t = \tfrac{3}{2}$ or $\quad t = -\tfrac{1}{2}$

A check verifies that $-\frac{1}{2}$ and $\frac{3}{2}$ are solutions of the original equation. Using a calculator for your check is highly recommended.

The solution set is $\left\{-\frac{1}{2}, \frac{3}{2}\right\}$.

N5. $4x^2 + 9x - 9 = 0$

$$4x^2 + 9x = 9 \quad \text{Add 9.}$$

Divide each side by 4 to get a coefficient of 1 for the x^2-term.

$$x^2 + \frac{9}{4}x = \frac{9}{4} \quad \text{Divide by 4.}$$

Take half the coefficient of x, or $\left(\frac{1}{2}\right)\left(\frac{9}{4}\right) = \frac{9}{8}$, and square the result: $\left(\frac{9}{8}\right)^2 = \frac{81}{64}$. Then add $\frac{81}{64}$ to each side.

$$x^2 + \frac{9}{4}x + \frac{81}{64} = \frac{9}{4} + \frac{81}{64} \quad \text{Add } \frac{81}{64}.$$
$$x^2 + \frac{9}{4}x + \frac{81}{64} = \frac{225}{64} \qquad \frac{9}{4} = \frac{144}{64}$$
$$\left(x + \frac{9}{8}\right)^2 = \frac{225}{64} \qquad \text{Factor.}$$

Apply the square root property, and solve for x.

$$x + \frac{9}{8} = \sqrt{\frac{225}{64}} \quad \text{or} \quad x + \frac{9}{8} = -\sqrt{\frac{225}{64}}$$
$$x + \frac{9}{8} = \frac{15}{8} \quad \text{or} \quad x + \frac{9}{8} = -\frac{15}{8}$$
$$x = -\frac{9}{8} + \frac{15}{8} \quad \text{or} \quad x = -\frac{9}{8} - \frac{15}{8}$$
$$x = \frac{3}{4} \quad \text{or} \quad x = -3$$

A check verifies that -3 and $\frac{3}{4}$ are solutions of the original equation.

The solution set is $\left\{-3, \frac{3}{4}\right\}$.

N6. $3t^2 - 12t + 15 = 0$

$$t^2 - 4t + 5 = 0 \qquad \text{Divide by 3.}$$
$$t^2 - 4t = -5 \qquad \text{Subtract 5.}$$
$$t^2 - 4t + 4 = -5 + 4 \quad \begin{array}{l}\text{Add}\\ \left(\frac{1}{2} \cdot 4\right)^2 = 4.\end{array}$$
$$(t - 2)^2 = -1$$

The square root of -1 is not a real number, so the square root property does not apply. This equation has *no real number solution*. The solution set is $\emptyset$.

N7. $(x - 5)(x + 1) = 2$

$$x^2 - 4x - 5 = 2$$
$$x^2 - 4x = 7$$
$$x^2 - 4x + 4 = 7 + 4 \quad \begin{array}{l}\text{Add}\\ \left[\frac{1}{2}(-4)\right]^2 = 4.\end{array}$$
$$(x - 2)^2 = 11$$
$$x - 2 = \sqrt{11} \quad \text{or} \quad x - 2 = -\sqrt{11}$$
$$x = 2 + \sqrt{11} \quad \text{or} \quad x = 2 - \sqrt{11}$$

The solution set is $\left\{2 \pm \sqrt{11}\right\}$.

N8. $s = -16t^2 + 64t$

$$-16t^2 + 64t = 28 \qquad \text{Let } s = 28.$$
$$t^2 - 4t = -\frac{7}{4} \qquad \text{Divide by } -16.$$
$$t^2 - 4t + 4 = -\frac{7}{4} + 4 \quad \text{Add } \left[\frac{1}{2}(-4)\right]^2 = 4.$$
$$(t - 2)^2 = \frac{9}{4} \qquad \text{Factor.}$$
$$t - 2 = \sqrt{\frac{9}{4}} \quad \text{or} \quad t - 2 = -\sqrt{\frac{9}{4}}$$
$$t = 2 + \frac{3}{2} \quad \text{or} \qquad t = 2 - \frac{3}{2}$$
$$t = 3.5 \quad \text{or} \qquad t = 0.5$$

The ball will be 28 feet above the ground after 0.5 second and again after 3.5 seconds.

9.2 Section Exercises

1. $x^2 + 10x +$ ____

Here the middle term $10x$ must equal $2kx$.

$$2kx = 10x$$
$$k = 5 \qquad \text{Divide by 2x.}$$

Thus, $k = 5$ and $k^2 = 25$. The required trinomial is $x^2 + 10x + \underline{25}$, which factors as $(x + 5)^2$.

3. $z^2 - 20z +$ ____

Here the middle term $-20z$ must equal $2kz$.

$$2kz = -20z$$
$$k = -10 \qquad \text{Divide by 2z.}$$

Thus, $k = -10$ and $k^2 = 100$. The required trinomial is $z^2 - 20z + \underline{100}$, which factors as $(z - 10)^2$.

5. $x^2 + 2x +$ ____

Here the middle term $2x$ must equal $2kx$.

$$2kx = 2x$$
$$k = 1 \qquad \text{Divide by 2x.}$$

Thus, $k = 1$ and $k^2 = 1$. The required trinomial is $x^2 + 2x + \underline{1}$, which factors as $(x + 1)^2$.

7. $p^2 - 5p +$ ____

Here the middle term $-5p$ must equal $2kp$.

$$2kp = -5p$$
$$k = -\frac{5}{2} \qquad \text{Divide by 2p.}$$

Thus, $k = -\frac{5}{2}$ and $k^2 = \frac{25}{4}$. The required trinomial is $p^2 - 5p + \underline{\frac{25}{4}}$, which factors as $\left(p - \frac{5}{2}\right)^2$.

9. $2x^2 - 4x = 9$

Before completing the square, the coefficient of x^2 must be 1. Dividing each side of the equation by 2 is the correct way to begin solving the equation, and this corresponds to choice **D**.

11. $x^2 - 4x = -3$

Take half of the coefficient of x and square it. Half of -4 is -2, and $(-2)^2 = 4$. Add 4 to each side of the equation, and write the left side as a perfect square.

$$x^2 - 4x + 4 = -3 + 4$$
$$(x-2)^2 = 1$$

Use the square root property.

$$
\begin{array}{lll}
x - 2 = \sqrt{1} & \text{or} & x - 2 = -\sqrt{1} \\
x - 2 = 1 & \text{or} & x - 2 = -1 \\
x = 3 & \text{or} & x = 1
\end{array}
$$

A check verifies that the solution set is $\{1, 3\}$.

13. $x^2 + 2x - 5 = 0$

Add 5 to each side.

$$x^2 + 2x = 5$$

Take half the coefficient of x and square it.

$$\tfrac{1}{2}(2) = 1, \text{ and } 1^2 = 1.$$

Add 1 to each side of the equation, and write the left side as a perfect square.

$$x^2 + 2x + 1 = 5 + 1$$
$$(x+1)^2 = 6$$

Use the square root property.

$$
\begin{array}{lll}
x + 1 = \sqrt{6} & \text{or} & x + 1 = -\sqrt{6} \\
x = -1 + \sqrt{6} & \text{or} & x = -1 - \sqrt{6}
\end{array}
$$

A check verifies that the solution set is $\left\{-1 \pm \sqrt{6}\right\}$. Using a calculator for your check is highly recommended.

15.
$$x^2 - 8x = -4$$
$$x^2 - 8x + 16 = -4 + 16 \quad \left[\tfrac{1}{2}(-8)\right]^2 = 16$$
$$(x-4)^2 = 12 \qquad \textit{Factor.}$$

$$
\begin{array}{lll}
x - 4 = \sqrt{12} & \text{or} & x - 4 = -\sqrt{12} \\
x - 4 = \sqrt{4} \cdot \sqrt{3} & \text{or} & x - 4 = -\sqrt{4} \cdot \sqrt{3} \\
x = 4 + 2\sqrt{3} & \text{or} & x = 4 - 2\sqrt{3}
\end{array}
$$

The solution set is $\left\{4 \pm 2\sqrt{3}\right\}$.

17. $x^2 + 6x + 9 = 0$

The left-hand side of this equation is already a perfect square.

$$(x+3)^2 = 0$$
$$x + 3 = 0$$
$$x = -3$$

A check verifies that the solution set is $\{-3\}$.

19. $4x^2 + 4x = 3$

Divide each side by 4 so that the coefficient of x^2 is 1.

$$x^2 + x = \tfrac{3}{4}$$

The coefficient of x is 1. Take half of 1, square the result, and add this square to each side.

$$\tfrac{1}{2}(1) = \tfrac{1}{2} \text{ and } \left(\tfrac{1}{2}\right)^2 = \tfrac{1}{4}$$
$$x^2 + x + \tfrac{1}{4} = \tfrac{3}{4} + \tfrac{1}{4}$$

The left-hand side can then be written as a perfect square.

$$\left(x + \tfrac{1}{2}\right)^2 = 1$$

Use the square root property.

$$
\begin{array}{lll}
x + \tfrac{1}{2} = 1 & \text{or} & x + \tfrac{1}{2} = -1 \\
x = -\tfrac{1}{2} + 1 & \text{or} & x = -\tfrac{1}{2} - 1 \\
x = \tfrac{1}{2} & \text{or} & x = -\tfrac{3}{2}
\end{array}
$$

A check verifies that the solution set is $\left\{-\tfrac{3}{2}, \tfrac{1}{2}\right\}$.

21. $2p^2 - 2p + 3 = 0$

Divide each side by 2.

$$p^2 - p + \tfrac{3}{2} = 0$$

Subtract $\tfrac{3}{2}$ from both sides.

$$p^2 - p = -\tfrac{3}{2}$$

Take half the coefficient of p and square it.

$$\tfrac{1}{2}(-1) = -\tfrac{1}{2}, \text{ and } \left(-\tfrac{1}{2}\right)^2 = \tfrac{1}{4}.$$

Add $\tfrac{1}{4}$ to each side of the equation.

$$p^2 - p + \tfrac{1}{4} = -\tfrac{3}{2} + \tfrac{1}{4}$$

Factor on the left side and add on the right.

$$\left(p - \tfrac{1}{2}\right)^2 = -\tfrac{5}{4}$$

The square root of $-\tfrac{5}{4}$ is not a real number, so the solution set is $\emptyset$.

23. $3x^2 - 9x + 5 = 0$

Divide each side by 3.

$$x^2 - 3x + \tfrac{5}{3} = 0$$

Put constant terms on one side.

$$x^2 - 3x = -\tfrac{5}{3}$$

Take half of the coefficient of x and square it.

$$\tfrac{1}{2}(-3) = -\tfrac{3}{2} \text{ and } \left(-\tfrac{3}{2}\right)^2 = \tfrac{9}{4}.$$

Add $\tfrac{9}{4}$ to each side of the equation.

$$x^2 - 3x + \tfrac{9}{4} = -\tfrac{5}{3} + \tfrac{9}{4}$$
$$\left(x - \tfrac{3}{2}\right)^2 = \tfrac{7}{12}$$

Use the square root property.

$$x - \tfrac{3}{2} = \sqrt{\tfrac{7}{12}} \qquad \text{or} \qquad x - \tfrac{3}{2} = -\sqrt{\tfrac{7}{12}}$$
$$x - \tfrac{3}{2} = \tfrac{\sqrt{7}}{\sqrt{12}} \cdot \tfrac{\sqrt{3}}{\sqrt{3}} \qquad \text{or} \qquad x - \tfrac{3}{2} = -\tfrac{\sqrt{7}}{\sqrt{12}} \cdot \tfrac{\sqrt{3}}{\sqrt{3}}$$
$$x - \tfrac{9}{6} = \tfrac{\sqrt{21}}{6} \qquad \text{or} \qquad x - \tfrac{9}{6} = -\tfrac{\sqrt{21}}{6}$$
$$x = \tfrac{9}{6} + \tfrac{\sqrt{21}}{6} \qquad \text{or} \qquad x = \tfrac{9}{6} - \tfrac{\sqrt{21}}{6}$$
$$x = \tfrac{9 + \sqrt{21}}{6} \qquad \text{or} \qquad x = \tfrac{9 - \sqrt{21}}{6}$$

A check verifies that the solution set is $\left\{ \tfrac{9 \pm \sqrt{21}}{6} \right\}$.

25. $3x^2 + 7x = 4$

Divide each side by 3.

$$x^2 + \tfrac{7}{3}x = \tfrac{4}{3}$$

Take half of the coefficient of x and square it.

$$\tfrac{1}{2}\left(\tfrac{7}{3}\right) = \tfrac{7}{6} \quad \text{and} \quad \left(\tfrac{7}{6}\right)^2 = \tfrac{49}{36}.$$

Add $\tfrac{49}{36}$ to each side of the equation.

$$x^2 + \tfrac{7}{3}x + \tfrac{49}{36} = \tfrac{4}{3} + \tfrac{49}{36}$$
$$\left(x + \tfrac{7}{6}\right)^2 = \tfrac{97}{36}$$

Use the square root property.

$$x + \tfrac{7}{6} = \sqrt{\tfrac{97}{36}} \qquad \text{or} \qquad x + \tfrac{7}{6} = -\sqrt{\tfrac{97}{36}}$$
$$x + \tfrac{7}{6} = \tfrac{\sqrt{97}}{6} \qquad \text{or} \qquad x + \tfrac{7}{6} = -\tfrac{\sqrt{97}}{6}$$
$$x = -\tfrac{7}{6} + \tfrac{\sqrt{97}}{6} \qquad \text{or} \qquad x = -\tfrac{7}{6} - \tfrac{\sqrt{97}}{6}$$
$$x = \tfrac{-7 + \sqrt{97}}{6} \qquad \text{or} \qquad x = \tfrac{-7 - \sqrt{97}}{6}$$

A check verifies that the solution set is $\left\{ \tfrac{-7 \pm \sqrt{97}}{6} \right\}$.

27. $(x + 3)(x - 1) = 5$
$$x^2 + 2x - 3 = 5$$
$$x^2 + 2x = 8$$
$$x^2 + 2x + 1 = 8 + 1$$
$$(x + 1)^2 = 9$$
$$x + 1 = 3 \qquad \text{or} \qquad x + 1 = -3$$
$$x = 2 \qquad \text{or} \qquad x = -4$$

A check verifies that the solution set is $\{-4, 2\}$.

29. $(r - 3)(r - 5) = 2$
$$r^2 - 8r + 15 = 2$$
$$r^2 - 8r = -13$$
$$r^2 - 8r + 16 = -13 + 16$$
$$(r - 4)^2 = 3$$
$$r - 4 = \sqrt{3} \qquad \text{or} \qquad r - 4 = -\sqrt{3}$$
$$r = 4 + \sqrt{3} \qquad \text{or} \qquad r = 4 - \sqrt{3}$$

A check verifies that the solution set is $\left\{ 4 \pm \sqrt{3} \right\}$.

31. $-x^2 + 2x = -5$

Divide each side by -1.

$$x^2 - 2x = 5$$

Take half of the coefficient of x and square it. Half of -2 is -1, and $(-1)^2 = 1$. Add 1 to each side of the equation, and write the left side as a perfect square.

$$x^2 - 2x + 1 = 5 + 1$$
$$(x - 1)^2 = 6$$

Use the square root property.

$$x - 1 = \sqrt{6} \qquad \text{or} \qquad x - 1 = -\sqrt{6}$$
$$x = 1 + \sqrt{6} \qquad \text{or} \qquad x = 1 - \sqrt{6}$$

A check verifies that the solution set is $\left\{ 1 \pm \sqrt{6} \right\}$.

33.
$$3r^2 - 2 = 6r + 3$$
$$3r^2 - 6r = 5$$
$$r^2 - 2r = \tfrac{5}{3}$$
$$r^2 - 2r + 1 = \tfrac{5}{3} + 1$$
$$(r - 1)^2 = \tfrac{8}{3}$$
$$r - 1 = \pm \sqrt{\tfrac{8}{3}}$$

Simplify the radical.

$$\sqrt{\tfrac{8}{3}} = \tfrac{\sqrt{8}}{\sqrt{3}} = \tfrac{2\sqrt{2}}{\sqrt{3}} \cdot \tfrac{\sqrt{3}}{\sqrt{3}} = \tfrac{2\sqrt{6}}{3}$$
$$r = 1 \pm \tfrac{2\sqrt{6}}{3}$$
$$r = \tfrac{3}{3} \pm \tfrac{2\sqrt{6}}{3} = \tfrac{3 \pm 2\sqrt{6}}{3}$$

(a) The solution set with *exact* values is $\left\{ \tfrac{3 \pm 2\sqrt{6}}{3} \right\}$.

(b) $\tfrac{3 + 2\sqrt{6}}{3} \approx 2.633,\ \tfrac{3 - 2\sqrt{6}}{3} \approx -0.633$

The solution set with *approximate* values is $\{-0.633, 2.633\}$.

35. $(x + 1)(x + 3) = 2$
$$x^2 + 3x + x + 3 = 2$$
$$x^2 + 4x = -1$$
$$x^2 + 4x + 4 = -1 + 4$$
$$(x + 2)^2 = 3$$
$$x + 2 = \pm \sqrt{3}$$
$$x = -2 \pm \sqrt{3}$$

(a) The solution set with *exact* values is $\left\{ -2 \pm \sqrt{3} \right\}$.

(b) $-2 + \sqrt{3} \approx -0.268$
$$-2 - \sqrt{3} \approx -3.732$$

The solution set with *approximate* values is $\{-3.732, -0.268\}$.

37.
$$s = -13t^2 + 104t$$
$$195 = -13t^2 + 104t \quad \textit{Let s = 195.}$$
$$-15 = t^2 - 8t \qquad \textit{Divide by -13.}$$
$$t^2 - 8t + 16 = -15 + 16$$
$$\textit{Add } \left[\tfrac{1}{2}(-8)\right]^2 = 16.$$
$$(t - 4)^2 = 1$$
$$t - 4 = \pm\sqrt{1} = \pm 1$$
$$t = 4 \pm 1$$
$$= 3 \text{ or } 5$$

The object will be at a height of 195 feet at 3 seconds (on the way up) and 5 seconds (on the way down).

39. $s = -16t^2 + 96t$

Find the value of t when $s = 80$.

$$80 = -16t^2 + 96t \quad \textit{Let s = 80.}$$
$$-16t^2 + 96t = 80$$
$$t^2 - 6t = -5 \qquad \textit{Divide by -16.}$$
$$t^2 - 6t + 9 = -5 + 9 \qquad \textit{Add 9.}$$
$$(t - 3)^2 = 4 \qquad \textit{Factor; add.}$$
$$t - 3 = \pm\sqrt{4} = \pm 2$$
$$t = 3 + 2 = 5 \quad \text{or} \quad t = 3 - 2 = 1$$

The object will reach a height of 80 feet at 1 second (on the way up) and at 5 seconds (on the way down).

41. Let x = the width of the pen.
Then $175 - x$ = the length of the pen.

Use the formula for the area of a rectangle.

$$A = LW$$
$$7500 = (175 - x)x$$
$$7500 = 175x - x^2$$
$$x^2 - 175x + 7500 = 0$$

Solve this quadratic equation by completing the square.

$$x^2 - 175x = -7500$$
$$x^2 - 175x + \tfrac{30{,}625}{4} = -\tfrac{30{,}000}{4} + \tfrac{30{,}625}{4}$$
$$\textit{Add } \left(\tfrac{175}{2}\right)^2 = \tfrac{30{,}625}{4}.$$
$$\left(x - \tfrac{175}{2}\right)^2 = \tfrac{625}{4}$$
$$x - \tfrac{175}{2} = \pm\sqrt{\tfrac{625}{4}} = \pm\tfrac{25}{2}$$
$$x = \tfrac{175}{2} \pm \tfrac{25}{2}$$
$$x = \tfrac{175}{2} + \tfrac{25}{2} \quad \text{or} \quad x = \tfrac{175}{2} - \tfrac{25}{2}$$
$$x = \tfrac{200}{2} \qquad \text{or} \qquad x = \tfrac{150}{2}$$
$$x = 100 \qquad \text{or} \qquad x = 75$$

If $x = 100$, $175 - x = 175 - 100 = 75$.
If $x = 75$, $175 - x = 175 - 75 = 100$.
The dimensions of the pen are 75 feet by 100 feet.

43. Let x = the distance the slower car traveled.
Then $x + 7$ = the distance the faster car traveled.

Since the cars traveled at right angles, a right triangle is formed with hypotenuse of length 17. Use the Pythagorean theorem with $a = x$, $b = x + 7$, and $c = 17$.

$$a^2 + b^2 = c^2$$
$$x^2 + (x + 7)^2 = 17^2$$
$$x^2 + (x^2 + 14x + 49) = 289$$
$$2x^2 + 14x = 240$$
$$x^2 + 7x = 120$$
$$x^2 + 7x + \tfrac{49}{4} = 120 + \tfrac{49}{4}$$
$$\textit{Add } \left[\tfrac{1}{2}(7)\right]^2 = \tfrac{49}{4}.$$
$$\left(x + \tfrac{7}{2}\right)^2 = \tfrac{529}{4}$$
$$x + \tfrac{7}{2} = \pm\sqrt{\tfrac{529}{4}} = \pm\tfrac{23}{2}$$
$$x = -\tfrac{7}{2} \pm \tfrac{23}{2}$$
$$= 8 \text{ or } -15$$

Discard -15 since distance cannot be negative. The slower car traveled 8 miles.

45. The side has length x, so the area of the original square is $x \cdot x = x^2$.

46. The area of one rectangle is $x \cdot 1 = x$, so the area of 8 rectangles is $8x$, and the area of the figure is $x^2 + 8x$.

47. The area of a small square is $1 \cdot 1 = 1$, so the area of 16 small squares is 16, and the area of the figure is $x^2 + 8x + 16$.

48. It occurred when we added the 16 squares.

49. $\dfrac{8 - 6\sqrt{3}}{6} = \dfrac{2\left(4 - 3\sqrt{3}\right)}{2 \cdot 3} = \dfrac{4 - 3\sqrt{3}}{3}$

51. $\dfrac{6 - \sqrt{45}}{6} = \dfrac{6 - \sqrt{9} \cdot \sqrt{5}}{6} = \dfrac{3\left(2 - \sqrt{5}\right)}{3 \cdot 2}$

$$= \dfrac{2 - \sqrt{5}}{2}$$

53. $a = 1, b = 2, c = -4$
$$\sqrt{b^2 - 4ac} = \sqrt{2^2 - 4(1)(-4)} = \sqrt{4 + 16}$$
$$= \sqrt{20} = \sqrt{4} \cdot \sqrt{5} = 2\sqrt{5}$$

9.3 Solving Quadratic Equations by the Quadratic Formula

9.3 Now Try Exercises

N1. **(a)** $3x^2 - 7x + 4 = 0$ has the form of the standard quadratic equation $ax^2 + bx + c = 0$.

Thus, $a = 3$, $b = -7$, and $c = 4$.

(b) $x^2 - 3 = -2x$

Rewrite in $ax^2 + bx + c = 0$ form.

$$x^2 + 2x - 3 = 0$$

Then $a = 1$, $b = 2$, and $c = -3$.

(c) $2x^2 - 4x = 0$ has the form of the standard quadratic equation $ax^2 + bx + c = 0$.

Thus, $a = 2$, $b = -4$, and $c = 0$.

(d) $2(2x + 1)(x - 5) = -3$

$$\begin{aligned} 2(2x^2 - 9x - 5) &= -3 \quad \textit{FOIL} \\ 4x^2 - 18x - 10 &= -3 \quad \textit{Multiply by 2.} \\ 4x^2 - 18x - 7 &= 0 \quad \textit{Add 3.} \end{aligned}$$

The last equation is in standard form. The values of a, b, and c are 4, -18, and -7, respectively.

N2. $3x^2 + 5x - 2 = 0$

The quadratic equation is in standard form, so $a = 3$, $b = 5$, and $c = -2$. Substitute these values into the quadratic formula.

$$x = \frac{-b \pm \sqrt{b^2 - 4ac}}{2a}$$

$$x = \frac{-5 \pm \sqrt{5^2 - 4(3)(-2)}}{2(3)}$$

$$x = \frac{-5 \pm \sqrt{25 + 24}}{6}$$

$$x = \frac{-5 \pm \sqrt{49}}{6}$$

$$x = \frac{-5 \pm 7}{6}$$

$$x = \frac{-5 + 7}{6} \quad \text{or} \quad x = \frac{-5 - 7}{6}$$

$$x = \frac{2}{6} \quad \text{or} \quad x = \frac{-12}{6}$$

$$x = \frac{1}{3} \quad \text{or} \quad x = -2$$

A check verifies that the solution set is $\left\{-2, \frac{1}{3}\right\}$.

N3. $x^2 + 2 = 6x$

Write the equation in standard form.

$$x^2 - 6x + 2 = 0$$

Substitute $a = 1$, $b = -6$, and $c = 2$ into the quadratic formula.

$$x = \frac{-b \pm \sqrt{b^2 - 4ac}}{2a}$$

$$x = \frac{-(-6) \pm \sqrt{(-6)^2 - 4(1)(2)}}{2(1)}$$

$$x = \frac{6 \pm \sqrt{36 - 8}}{2} = \frac{6 \pm \sqrt{28}}{2}$$

$$= \frac{6 \pm \sqrt{4} \cdot \sqrt{7}}{2} = \frac{6 \pm 2\sqrt{7}}{2}$$

$$= \frac{2\left(3 \pm \sqrt{7}\right)}{2} = 3 \pm \sqrt{7}$$

The solution set is $\left\{3 \pm \sqrt{7}\right\}$.

N4.
$$16x^2 = 8x - 1$$
$$16x^2 - 8x + 1 = 0 \qquad \textit{Standard form}$$

Substitute $a = 16$, $b = -8$, and $c = 1$ into the quadratic formula.

$$x = \frac{-b \pm \sqrt{b^2 - 4ac}}{2a}$$

$$x = \frac{-(-8) \pm \sqrt{(-8)^2 - 4(16)(1)}}{2(16)}$$

$$x = \frac{8 \pm \sqrt{64 - 64}}{32}$$

$$x = \frac{8 \pm 0}{32} = \frac{8}{32} = \frac{8 \cdot 1}{8 \cdot 4} = \frac{1}{4}$$

Since there is just one solution, $\frac{1}{4}$, the trinomial $16x^2 - 8x + 1$ is a perfect square. The solution set is $\left\{\frac{1}{4}\right\}$.

N5.
$$\tfrac{1}{12}x^2 = \tfrac{1}{2}x - \tfrac{1}{3}$$
$$x^2 = 6x - 4 \quad \textit{Multiply by 12.}$$
$$x^2 - 6x + 4 = 0 \qquad \textit{Standard form}$$

Substitute $a = 1$, $b = -6$, and $c = 4$ into the quadratic formula.

$$x = \frac{-b \pm \sqrt{b^2 - 4ac}}{2a}$$

$$x = \frac{-(-6) \pm \sqrt{(-6)^2 - 4(1)(4)}}{2(1)}$$

$$x = \frac{6 \pm \sqrt{36 - 16}}{2} = \frac{6 \pm \sqrt{20}}{2}$$

$$= \frac{6 \pm \sqrt{4} \cdot \sqrt{5}}{2} = \frac{6 \pm 2\sqrt{5}}{2}$$

$$= \frac{2\left(3 \pm \sqrt{5}\right)}{2} = 3 \pm \sqrt{5}$$

The solution set is $\left\{3 \pm \sqrt{5}\right\}$.

9.3 Section Exercises

1. $3x^2 + 4x - 8 = 0$

Match the coefficients of the quadratic equation with the letters a, b, and c of the standard quadratic equation

$$ax^2 + bx + c = 0.$$

In this case, $a = 3$, $b = 4$, and $c = -8$.

3. $-8x^2 - 2x - 3 = 0$

Match the coefficients of the quadratic equation with the letters a, b, and c of the standard quadratic equation $ax^2 + bx + c = 0$.

In this case, $a = -8$, $b = -2$, and $c = -3$.

5. $3x^2 = 4x + 2$

First, write the equation in standard form,
$ax^2 + bx + c = 0$.

$$3x^2 - 4x - 2 = 0$$

Now, identify the values: $a = 3$, $b = -4$, and $c = -2$.

7. $3x^2 = -7x$

Write the equation in standard form.

$$3x^2 + 7x = 0$$

Now, identify the values: $a = 3$, $b = 7$, and $c = 0$.

9. $(x - 3)(x + 4) = 0$
   $x^2 + x - 12 = 0$ *FOIL*

The values are $a = 1$, $b = 1$, and $c = -12$.

11. $9(x - 1)(x + 2) = 8$
   $9(x^2 + x - 2) = 8$ *FOIL*
   $9x^2 + 9x - 18 = 8$ *Distributive property*
   $9x^2 + 9x - 26 = 0$ *Standard form*

The values are $a = 9$, $b = 9$, and $c = -26$.

13. $2a$ should be the denominator for $-b$ as well. The correct formula is

$$x = \frac{-b \pm \sqrt{b^2 - 4ac}}{2a}.$$

15. $k^2 + 12k - 13 = 0$

Use $a = 1$, $b = 12$, and $c = -13$.
Substitute these values into the quadratic formula.

$$k = \frac{-b \pm \sqrt{b^2 - 4ac}}{2a}$$

$$k = \frac{-12 \pm \sqrt{12^2 - 4(1)(-13)}}{2(1)}$$

$$= \frac{-12 \pm \sqrt{144 + 52}}{2}$$

$$= \frac{-12 \pm \sqrt{196}}{2} = \frac{-12 \pm 14}{2}$$

$$k = \frac{-12 + 14}{2} = \frac{2}{2} = 1$$

$$\text{or} \quad k = \frac{-12 - 14}{2} = \frac{-26}{2} = -13$$

The solution set is $\{-13, 1\}$.

17. $2x^2 + 12x = -5$

Write the equation in standard form.

$$2x^2 + 12x + 5 = 0$$

Substitute $a = 2$, $b = 12$, and $c = 5$ into the quadratic formula.

$$x = \frac{-b \pm \sqrt{b^2 - 4ac}}{2a}$$

$$x = \frac{-12 \pm \sqrt{12^2 - 4(2)(5)}}{2(2)}$$

$$= \frac{-12 \pm \sqrt{144 - 40}}{4} = \frac{-12 \pm \sqrt{104}}{4}$$

$$= \frac{-12 \pm \sqrt{4} \cdot \sqrt{26}}{4} = \frac{-12 \pm 2\sqrt{26}}{4}$$

$$= \frac{2\left(-6 \pm \sqrt{26}\right)}{2 \cdot 2} = \frac{-6 \pm \sqrt{26}}{2}$$

The solution set is $\left\{ \frac{-6 \pm \sqrt{26}}{2} \right\}$.

19. $p^2 - 4p + 4 = 0$

Substitute $a = 1$, $b = -4$, and $c = 4$ into the quadratic formula.

$$p = \frac{-b \pm \sqrt{b^2 - 4ac}}{2a}$$

$$p = \frac{-(-4) \pm \sqrt{(-4)^2 - 4(1)(4)}}{2(1)}$$

$$= \frac{4 \pm \sqrt{16 - 16}}{2}$$

$$= \frac{4 \pm 0}{2} = \frac{4}{2} = 2$$

The solution set is $\{2\}$. Note that the discriminant is 0.

21. $2x^2 = 5 + 3x$

Write the equation in standard form.

$$2x^2 - 3x - 5 = 0$$

Substitute $a = 2$, $b = -3$, and $c = -5$ into the quadratic formula.

$$x = \frac{-b \pm \sqrt{b^2 - 4ac}}{2a}$$

$$x = \frac{-(-3) \pm \sqrt{(-3)^2 - 4(2)(-5)}}{2(2)}$$

$$= \frac{3 \pm \sqrt{9 + 40}}{4}$$

$$= \frac{3 \pm \sqrt{49}}{4} = \frac{3 \pm 7}{4}$$

$$x = \frac{3 + 7}{4} = \frac{10}{4} = \frac{5}{2}$$

$$\text{or} \quad x = \frac{3 - 7}{4} = \frac{-4}{4} = -1$$

The solution set is $\left\{ -1, \frac{5}{2} \right\}$.

23. $6x^2 + 6x = 0$

Substitute $a = 6$, $b = 6$, and $c = 0$ into the quadratic formula.

$$x = \frac{-b \pm \sqrt{b^2 - 4ac}}{2a}$$

$$x = \frac{-6 \pm \sqrt{6^2 - 4(6)(0)}}{2(6)}$$

$$= \frac{-6 \pm \sqrt{36 - 0}}{12}$$

$$= \frac{-6 \pm 6}{12}$$

$$x = \frac{-6 + 6}{12} = \frac{0}{12} = 0$$

or $\quad x = \frac{-6 - 6}{12} = \frac{-12}{12} = -1$

The solution set is $\{-1, 0\}$.

25. $7x^2 = 12x$

Write the equation in standard form.

$$7x^2 - 12x = 0$$

Substitute $a = 7$, $b = -12$, and $c = 0$ into the quadratic formula.

$$x = \frac{-b \pm \sqrt{b^2 - 4ac}}{2a}$$

$$x = \frac{-(-12) \pm \sqrt{(-12)^2 - 4(7)(0)}}{2(7)}$$

$$= \frac{12 \pm \sqrt{144 - 0}}{14}$$

$$= \frac{12 \pm 12}{14}$$

$$x = \frac{12 + 12}{14} = \frac{24}{14} = \frac{12}{7}$$

or $\quad x = \frac{12 - 12}{14} = \frac{0}{14} = 0$

The solution set is $\left\{0, \frac{12}{7}\right\}$.

27. $x^2 - 24 = 0$

Substitute $a = 1$, $b = 0$, and $c = -24$ into the quadratic formula.

$$x = \frac{-b \pm \sqrt{b^2 - 4ac}}{2a}$$

$$x = \frac{-0 \pm \sqrt{0^2 - 4(1)(-24)}}{2(1)}$$

$$= \frac{\pm \sqrt{96}}{2} = \frac{\pm \sqrt{16} \cdot \sqrt{6}}{2}$$

$$= \frac{\pm 4\sqrt{6}}{2} = \pm 2\sqrt{6}$$

The solution set is $\left\{\pm 2\sqrt{6}\right\}$.

29. $25x^2 - 4 = 0$

Substitute $a = 25$, $b = 0$, and $c = -4$ into the quadratic formula.

$$x = \frac{-b \pm \sqrt{b^2 - 4ac}}{2a}$$

$$x = \frac{-0 \pm \sqrt{0^2 - 4(25)(-4)}}{2(25)}$$

$$= \frac{\pm \sqrt{400}}{50} = \frac{\pm 20}{50} = \pm \frac{2}{5}$$

The solution set is $\left\{\pm \frac{2}{5}\right\}$.

31. $3x^2 - 2x + 5 = 10x + 1$

Write the equation in standard form.

$$3x^2 - 12x + 4 = 0$$

Substitute $a = 3$, $b = -12$, and $c = 4$ into the quadratic formula.

$$x = \frac{-b \pm \sqrt{b^2 - 4ac}}{2a}$$

$$x = \frac{-(-12) \pm \sqrt{(-12)^2 - 4(3)(4)}}{2(3)}$$

$$= \frac{12 \pm \sqrt{144 - 48}}{6}$$

$$= \frac{12 \pm \sqrt{96}}{6} = \frac{12 \pm \sqrt{16} \cdot \sqrt{6}}{6}$$

$$= \frac{12 \pm 4\sqrt{6}}{6} = \frac{2\left(6 \pm 2\sqrt{6}\right)}{2 \cdot 3}$$

$$= \frac{6 \pm 2\sqrt{6}}{3}$$

The solution set is $\left\{\frac{6 \pm 2\sqrt{6}}{3}\right\}$.

33. $-2x^2 = -3x + 2$

Write the equation in standard form.

$$-2x^2 + 3x - 2 = 0$$

Substitute $a = -2$, $b = 3$, and $c = -2$ into the quadratic formula.

$$x = \frac{-b \pm \sqrt{b^2 - 4ac}}{2a}$$

$$x = \frac{-3 \pm \sqrt{3^2 - 4(-2)(-2)}}{2(-2)}$$

$$= \frac{-3 \pm \sqrt{9 - 16}}{-4}$$

$$= \frac{-3 \pm \sqrt{-7}}{-4}$$

Because $\sqrt{-7}$ does not represent a real number, the solution set is $\emptyset$.

35. $2x^2 + x + 5 = 0$

Substitute $a = 2$, $b = 1$, and $c = 5$ into the quadratic formula.

$$x = \frac{-1 \pm \sqrt{1^2 - 4(2)(5)}}{2(2)}$$

$$= \frac{-1 \pm \sqrt{1 - 40}}{4}$$

$$= \frac{-1 \pm \sqrt{-39}}{4}$$

Because $\sqrt{-39}$ does not represent a real number, the solution set is $\emptyset$.

37. $(x + 3)(x + 2) = 15$

$$x^2 + 5x + 6 = 15$$
$$x^2 + 5x - 9 = 0$$

Use $a = 1$, $b = 5$, and $c = -9$.

$$x = \frac{-5 \pm \sqrt{5^2 - 4(1)(-9)}}{2(1)}$$

$$= \frac{-5 \pm \sqrt{25 + 36}}{2}$$

$$= \frac{-5 \pm \sqrt{61}}{2}$$

The solution set is $\left\{ \frac{-5 \pm \sqrt{61}}{2} \right\}$.

39. $2x^2 = 5 - 2x$

$$2x^2 + 2x - 5 = 0$$

Use $a = 2$, $b = 2$, and $c = -5$.

$$x = \frac{-2 \pm \sqrt{2^2 - 4(2)(-5)}}{2(2)}$$

$$= \frac{-2 \pm \sqrt{4 + 40}}{4} = \frac{-2 \pm \sqrt{44}}{4}$$

$$= \frac{-2 \pm 2\sqrt{11}}{4} = \frac{2\left(-1 \pm \sqrt{11}\right)}{2 \cdot 2}$$

$$= \frac{-1 \pm \sqrt{11}}{2}$$

(a) The solution set with *exact* values is
$$\left\{ \frac{-1 \pm \sqrt{11}}{2} \right\}.$$

(b) The solution set with *approximate* values (to the nearest thousandth) is $\{-2.158, 1.158\}$.

41. $x^2 = 1 + x$

$$x^2 - x - 1 = 0$$

Use $a = 1$, $b = -1$, and $c = -1$.

$$x = \frac{-(-1) \pm \sqrt{(-1)^2 - 4(1)(-1)}}{2(1)}$$

$$= \frac{1 \pm \sqrt{1 + 4}}{2} = \frac{1 \pm \sqrt{5}}{2}$$

(a) The solution set with *exact* values is
$$\left\{ \frac{1 \pm \sqrt{5}}{2} \right\}.$$

(b) The solution set with *approximate* values (to the nearest thousandth) is $\{-0.618, 1.618\}$.

43. $\frac{3}{2}k^2 - k - \frac{4}{3} = 0$

Eliminate the denominators by multiplying each side by the least common denominator, 6.

$$9k^2 - 6k - 8 = 0$$

Substitute $a = 9$, $b = -6$, and $c = -8$ into the quadratic formula.

$$k = \frac{-(-6) \pm \sqrt{(-6)^2 - 4(9)(-8)}}{2(9)}$$

$$= \frac{6 \pm \sqrt{36 + 288}}{18} = \frac{6 \pm \sqrt{324}}{18}$$

$$= \frac{6 \pm 18}{18}$$

$$k = \frac{6 + 18}{18} = \frac{24}{18} = \frac{4}{3}$$

or $k = \frac{6 - 18}{18} = \frac{-12}{18} = -\frac{2}{3}$

The solution set is $\left\{ -\frac{2}{3}, \frac{4}{3} \right\}$.

45. $\frac{1}{2}x^2 + \frac{1}{6}x = 1$

Eliminate the denominators by multiplying each side by the least common denominator, 6.

$$3x^2 + x = 6$$
$$3x^2 + x - 6 = 0$$

Use $a = 3$, $b = 1$, and $c = -6$.

$$x = \frac{-1 \pm \sqrt{1^2 - 4(3)(-6)}}{2(3)}$$

$$= \frac{-1 \pm \sqrt{1 + 72}}{6}$$

$$= \frac{-1 \pm \sqrt{73}}{6}$$

The solution set is $\left\{ \frac{-1 \pm \sqrt{73}}{6} \right\}$.

47. $\frac{3}{8}x^2 - x + \frac{17}{24} = 0$

Multiply each side by the least common denominator, 24.

$$9x^2 - 24x + 17 = 0$$

Use the quadratic formula with $a = 9$, $b = -24$, and $c = 17$.

$$x = \frac{-(-24) \pm \sqrt{(-24)^2 - 4(9)(17)}}{2(9)}$$

$$= \frac{24 \pm \sqrt{576 - 612}}{18}$$

$$= \frac{24 \pm \sqrt{-36}}{18}$$

Because $\sqrt{-36}$ does not represent a real number, the solution set is $\emptyset$.

49. $0.5x^2 = x + 0.5$

To eliminate the decimals, multiply each side by 10.

$$5x^2 = 10x + 5$$
$$5x^2 - 10x - 5 = 0$$

Divide each side by 5 so that we can work with smaller coefficients in the quadratic formula.

$$x^2 - 2x - 1 = 0$$

Use the quadratic formula with $a = 1$, $b = -2$, and $c = -1$.

$$x = \frac{-(-2) \pm \sqrt{(-2)^2 - 4(1)(-1)}}{2(1)}$$

$$= \frac{2 \pm \sqrt{4 + 4}}{2} = \frac{2 \pm \sqrt{8}}{2}$$

$$= \frac{2 \pm \sqrt{4} \cdot \sqrt{2}}{2} = \frac{2 \pm 2\sqrt{2}}{2}$$

$$= \frac{2\left(1 \pm \sqrt{2}\right)}{2} = 1 \pm \sqrt{2}$$

The solution set is $\left\{1 \pm \sqrt{2}\right\}$.

51. $0.6x - 0.4x^2 = -1$

To eliminate the decimals, multiply each side by 10.

$$6x - 4x^2 = -10$$

Write this equation in standard form.

$$0 = 4x^2 - 6x - 10$$

Divide each side by 2 so that we can work with smaller coefficients in the quadratic formula.

$$0 = 2x^2 - 3x - 5$$

Use the quadratic formula with $a = 2$, $b = -3$, and $c = -5$.

$$x = \frac{-(-3) \pm \sqrt{(-3)^2 - 4(2)(-5)}}{2(2)}$$

$$x = \frac{3 \pm \sqrt{9 + 40}}{4} = \frac{3 \pm \sqrt{49}}{4}$$

$$= \frac{3 \pm 7}{4}$$

$$x = \frac{3 + 7}{4} = \frac{10}{4} = \frac{5}{2}$$

$$\text{or } x = \frac{3 - 7}{4} = \frac{-4}{4} = -1$$

The solution set is $\left\{-1, \frac{5}{2}\right\}$.

53. Solve $S = 2\pi rh + \pi r^2$ for r.

Write this equation in the standard form of a quadratic equation, treating r as the variable and S, π, and h as constants.

$$\pi r^2 + (2\pi h)r - S = 0$$

Use $a = \pi$, $b = 2\pi h$, and $c = -S$ in the quadratic formula.

$$r = \frac{-b \pm \sqrt{b^2 - 4ac}}{2a}$$

$$r = \frac{-2\pi h \pm \sqrt{(2\pi h)^2 - 4(\pi)(-S)}}{2(\pi)}$$

$$= \frac{-2\pi h \pm \sqrt{4\pi^2 h^2 + 4\pi S}}{2\pi}$$

$$= \frac{-2\pi h \pm \sqrt{4(\pi^2 h^2 + \pi S)}}{2\pi}$$

$$= \frac{-2\pi h \pm 2\sqrt{\pi^2 h^2 + \pi S}}{2\pi}$$

$$r = \frac{-\pi h \pm \sqrt{\pi^2 h^2 + \pi S}}{\pi}$$

55. $h = -0.5x^2 + 1.25x + 3$
$1.25 = -0.5x^2 + 1.25x + 3$ *Let h = 1.25.*
$0.5x^2 - 1.25x - 1.75 = 0$
$2x^2 - 5x - 7 = 0$ *Multiply by 4.*

Use $a = 2$, $b = -5$, and $c = -7$.

$$x = \frac{-(-5) \pm \sqrt{(-5)^2 - 4(2)(-7)}}{2(2)}$$

$$= \frac{5 \pm \sqrt{25 + 56}}{4}$$

$$= \frac{5 \pm \sqrt{81}}{4} = \frac{5 \pm 9}{4}$$

$$x = \frac{5 + 9}{4} = \frac{14}{4} = 3.5$$

$$\text{or } x = \frac{5 - 9}{4} = \frac{-4}{4} = -1$$

x must be positive, so the frog was 3.5 feet from the base of the stump when he was 1.25 feet above the ground.

57. $\left(\dfrac{d-4}{4}\right)^2 = 9$

$$\dfrac{d-4}{4} = \pm\sqrt{9} = \pm 3$$
$$d - 4 = \pm 12$$
$$d = 4 \pm 12$$
$$= 16 \text{ or } -8$$

The solution set for the equation is $\{-8, 16\}$. Only 16 feet is a reasonable answer.

59. $(4 + 6z) + (-9 + 2z) = 4 - 9 + 6z + 2z$
$$= -5 + 8z$$

61. $4 - (6 - 3k) = 4 - 6 + 3k$
$$= -2 + 3k$$

63. $(4 + 3r)(6 - 5r) = 24 - 20r + 18r - 15r^2$
$$= 24 - 2r - 15r^2$$

Summary Exercises on Quadratic Equations

1. $s^2 = 36$

Use the square root property.

$$s = \sqrt{36} \quad \text{or} \quad s = -\sqrt{36}$$
$$s = 6 \quad \text{or} \quad s = -6$$

The solution set is $\{\pm 6\}$.

3. $(x + 2)(x - 4) = 16$
$$x^2 - 2x - 8 = 16$$
$$x^2 - 2x - 24 = 0$$

Solve this equation by factoring.

$$(x + 4)(x - 6) = 0$$

$$x + 4 = 0 \quad \text{or} \quad x - 6 = 0$$
$$x = -4 \quad \text{or} \quad x = 6$$

The solution set is $\{-4, 6\}$.

5. $z^2 - 4z + 3 = 0$

Solve this equation by factoring.

$$(z - 3)(z - 1) = 0$$

$$z - 3 = 0 \quad \text{or} \quad z - 1 = 0$$
$$z = 3 \quad \text{or} \quad z = 1$$

The solution set is $\{1, 3\}$.

7. $z(z - 9) = -20$
$$z^2 - 9z = -20$$
$$z^2 - 9z + 20 = 0$$

Solve this equation by factoring.

$$(z - 4)(z - 5) = 0$$

$$z - 4 = 0 \quad \text{or} \quad z - 5 = 0$$
$$z = 4 \quad \text{or} \quad z = 5$$

The solution set is $\{4, 5\}$.

9. $(3x - 2)^2 = 9$

Use the square root property.

$$3x - 2 = \sqrt{9} \quad \text{or} \quad 3x - 2 = -\sqrt{9}$$
$$3x - 2 = 3 \quad \text{or} \quad 3x - 2 = -3$$
$$3x = 5 \quad \text{or} \quad 3x = -1$$
$$x = \tfrac{5}{3} \quad \text{or} \quad x = -\tfrac{1}{3}$$

The solution set is $\left\{-\tfrac{1}{3}, \tfrac{5}{3}\right\}$.

11. $(x + 6)^2 = 121$

Use the square root property.

$$x + 6 = \sqrt{121} \quad \text{or} \quad x + 6 = -\sqrt{121}$$
$$x + 6 = 11 \quad \text{or} \quad x + 6 = -11$$
$$x = 5 \quad \text{or} \quad x = -17$$

The solution set is $\{-17, 5\}$.

13. $(3r - 7)^2 = 24$

Use the square root property.

$$3r - 7 = \sqrt{24} \quad \text{or} \quad 3r - 7 = -\sqrt{24}$$

Now simplify the radical.

$$\sqrt{24} = \sqrt{4} \cdot \sqrt{6} = 2\sqrt{6}$$

$$3r - 7 = 2\sqrt{6} \quad \text{or} \quad 3r - 7 = -2\sqrt{6}$$
$$3r = 7 + 2\sqrt{6} \quad \text{or} \quad 3r = 7 - 2\sqrt{6}$$
$$r = \dfrac{7 + 2\sqrt{6}}{3} \quad \text{or} \quad r = \dfrac{7 - 2\sqrt{6}}{3}$$

The solution set is $\left\{\dfrac{7 \pm 2\sqrt{6}}{3}\right\}$.

15. $(5x - 8)^2 = -6$

The square root of -6 is not a real number, so the square root property does not apply. This equation has solution set $\emptyset$.

17. $-2x^2 = -3x - 2$
$$2x^2 - 3x - 2 = 0$$

Solve this equation by factoring.

$$(2x + 1)(x - 2) = 0$$

$$2x + 1 = 0 \quad \text{or} \quad x - 2 = 0$$
$$x = -\tfrac{1}{2} \quad \text{or} \quad x = 2$$

The solution set is $\left\{-\tfrac{1}{2}, 2\right\}$.

19. $8z^2 = 15 + 2z$
$$8z^2 - 2z - 15 = 0$$

Solve this equation by factoring.

$$(4z + 5)(2z - 3) = 0$$

$$4z + 5 = 0 \quad \text{or} \quad 2z - 3 = 0$$
$$z = -\tfrac{5}{4} \quad \text{or} \quad z = \tfrac{3}{2}$$

The solution set is $\left\{-\tfrac{5}{4}, \tfrac{3}{2}\right\}$.

21. $0.1x^2 - 0.2x = 0.1$
$$x^2 - 2x = 1$$
$$x^2 - 2x - 1 = 0$$

Use the quadratic formula with
$a = 1$, $b = -2$, and $c = -1$.

$$x = \frac{-b \pm \sqrt{b^2 - 4ac}}{2a}$$

$$x = \frac{-(-2) \pm \sqrt{(-2)^2 - 4(1)(-1)}}{2(1)}$$

$$= \frac{2 \pm \sqrt{4 + 4}}{2} = \frac{2 \pm \sqrt{8}}{2}$$

$$= \frac{2 \pm 2\sqrt{2}}{2} = \frac{2\left(1 \pm \sqrt{2}\right)}{2}$$

$$= 1 \pm \sqrt{2}$$

The solution set is $\left\{1 \pm \sqrt{2}\right\}$.

23. $5z^2 - 22z = -8$
$$5z^2 - 22z + 8 = 0$$

Solve this equation by factoring.

$$(5z - 2)(z - 4) = 0$$

$$5z - 2 = 0 \quad \text{or} \quad z - 4 = 0$$

$$z = \tfrac{2}{5} \quad \text{or} \quad z = 4$$

The solution set is $\left\{\tfrac{2}{5}, 4\right\}$.

25. $(x + 2)(x + 1) = 10$
$$x^2 + 3x + 2 = 10$$
$$x^2 + 3x - 8 = 0$$

Use the quadratic formula with
$a = 1$, $b = 3$, and $c = -8$.

$$x = \frac{-b \pm \sqrt{b^2 - 4ac}}{2a}$$

$$x = \frac{-3 \pm \sqrt{3^2 - 4(1)(-8)}}{2(1)}$$

$$= \frac{-3 \pm \sqrt{9 + 32}}{2}$$

$$= \frac{-3 \pm \sqrt{41}}{2}$$

The solution set is $\left\{\frac{-3 \pm \sqrt{41}}{2}\right\}$.

27. $4x^2 = -1 + 5x$
$$4x^2 - 5x + 1 = 0$$

Solve this equation by factoring.

$$(x - 1)(4x - 1) = 0$$

$$x - 1 = 0 \quad \text{or} \quad 4x - 1 = 0$$

$$x = 1 \quad \text{or} \quad x = \tfrac{1}{4}$$

The solution set is $\left\{\tfrac{1}{4}, 1\right\}$.

29. $3m(3m + 4) = 7$
$$9m^2 + 12m = 7$$
$$9m^2 + 12m - 7 = 0$$

Use the quadratic formula with
$a = 9$, $b = 12$, and $c = -7$.

$$m = \frac{-b \pm \sqrt{b^2 - 4ac}}{2a}$$

$$m = \frac{-12 \pm \sqrt{12^2 - 4(9)(-7)}}{2(9)}$$

$$= \frac{-12 \pm \sqrt{144 + 252}}{18}$$

$$= \frac{-12 \pm \sqrt{396}}{18} = \frac{-12 \pm \sqrt{36} \cdot \sqrt{11}}{18}$$

$$= \frac{-12 \pm 6\sqrt{11}}{18} = \frac{6\left(-2 \pm \sqrt{11}\right)}{6 \cdot 3}$$

$$= \frac{-2 \pm \sqrt{11}}{3}$$

The solution set is $\left\{\frac{-2 \pm \sqrt{11}}{3}\right\}$.

31. $\dfrac{r^2}{2} + \dfrac{7r}{4} + \dfrac{11}{8} = 0$

Multiply each side by the least common denominator, 8.

$$8\left(\frac{r^2}{2} + \frac{7r}{4} + \frac{11}{8}\right) = 8(0)$$

$$4r^2 + 14r + 11 = 0$$

Use the quadratic formula with
$a = 4$, $b = 14$, and $c = 11$.

$$r = \frac{-b \pm \sqrt{b^2 - 4ac}}{2a}$$

$$r = \frac{-14 \pm \sqrt{14^2 - 4(4)(11)}}{2(4)}$$

$$= \frac{-14 \pm \sqrt{196 - 176}}{8}$$

$$= \frac{-14 \pm \sqrt{20}}{8} = \frac{-14 \pm 2\sqrt{5}}{8}$$

$$= \frac{2\left(-7 \pm \sqrt{5}\right)}{2(4)} = \frac{-7 \pm \sqrt{5}}{4}$$

The solution set is $\left\{\frac{-7 \pm \sqrt{5}}{4}\right\}$.

33. $9x^2 = 16(3x + 4)$
$$9x^2 = 48x + 64$$
$$9x^2 - 48x - 64 = 0$$

Use the quadratic formula with
$a = 9$, $b = -48$, and $c = -64$.

$$x = \frac{-b \pm \sqrt{b^2 - 4ac}}{2a}$$

$$x = \frac{-(-48) \pm \sqrt{(-48)^2 - 4(9)(-64)}}{2(9)}$$

$$= \frac{48 \pm \sqrt{2304 + 2304}}{18} = \frac{48 \pm \sqrt{4608}}{18}$$

$$= \frac{48 \pm \sqrt{2304} \cdot \sqrt{2}}{18} = \frac{48 \pm 48\sqrt{2}}{18}$$

$$= \frac{6\left(8 \pm 8\sqrt{2}\right)}{6 \cdot 3} = \frac{8 \pm 8\sqrt{2}}{3}$$

The solution set is $\left\{ \frac{8 \pm 8\sqrt{2}}{3} \right\}$.

35. $x^2 - x + 3 = 0$

Use the quadratic formula with $a = 1$, $b = -1$, and $c = 3$.

$$x = \frac{-b \pm \sqrt{b^2 - 4ac}}{2a}$$

$$x = \frac{-(-1) \pm \sqrt{(-1)^2 - 4(1)(3)}}{2(1)}$$

$$= \frac{1 \pm \sqrt{1 - 12}}{2} = \frac{1 \pm \sqrt{-11}}{2}$$

Because $\sqrt{-11}$ does not represent a real number, the solution set is $\emptyset$.

37. $-3x^2 + 4x = -4$

$3x^2 - 4x - 4 = 0$

Solve this equation by factoring.

$$(3x + 2)(x - 2) = 0$$

$3x + 2 = 0$ or $x - 2 = 0$

$x = -\frac{2}{3}$ or $x = 2$

The solution set is $\left\{ -\frac{2}{3}, 2 \right\}$.

39. $5x^2 + 19x = 2x + 12$

$5x^2 + 17x - 12 = 0$

Solve this equation by factoring.

$$(5x - 3)(x + 4) = 0$$

$5x - 3 = 0$ or $x + 4 = 0$

$x = \frac{3}{5}$ or $x = -4$

The solution set is $\left\{ -4, \frac{3}{5} \right\}$.

41. $x^2 - \frac{4}{15} = -\frac{4}{15}x$

Multiply both sides by 15 to clear fractions.

$$15x^2 - 4 = -4x$$

Write this equation in standard form.

$$15x^2 + 4x - 4 = 0$$

Solve by factoring.

$$(3x + 2)(5x - 2) = 0$$

$3x + 2 = 0$ or $5x - 2 = 0$

$x = -\frac{2}{3}$ or $x = \frac{2}{5}$

The solution set is $\left\{ -\frac{2}{3}, \frac{2}{5} \right\}$.

9.4   Complex Numbers

9.4   Now Try Exercises

N1. $\sqrt{-12} = i\sqrt{12} = i \cdot \sqrt{4} \cdot \sqrt{3} = i \cdot 2 \cdot \sqrt{3}$, or $2i\sqrt{3}$

N2. **(a)** $(3 + 5i) + (-4 - 2i) = (3 - 4) + (5 - 2)i$
$$= -1 + 3i$$

(b) $(-2 - i) - (1 - 3i)$
$$= (-2 - i) + (-1 + 3i)$$
Use the definition of subtraction.
$$= (-2 - 1) + (-1 + 3)i$$
$$= -3 + 2i$$

N3. **(a)** $8i(1 - 3i)$
$$= 8i - 24i^2 \quad \textit{Distributive prop.}$$
$$= 8i - 24(-1) \quad i^2 = -1$$
$$= 8i + 24 \quad \textit{Multiply.}$$
$$= 24 + 8i \quad \textit{Standard form}$$

(b) $(2 - 4i)(3 + 2i)$
$$= 6 + 4i - 12i - 8i^2 \quad \textit{FOIL}$$
$$= 6 - 8i - 8(-1)$$
$$= 6 - 8i + 8$$
$$= 14 - 8i$$

(c) $(5 - 7i)(5 + 7i)$
$$= 5^2 - (7i)^2$$
$$= 25 - 49i^2$$
$$= 25 - 49(-1)$$
$$= 25 + 49 = 74$$

N4. **(a)** $\dfrac{3 - i}{2 + 3i}$ *Multiply numerator and denominator by the conjugate of denominator.*

$$= \frac{3 - i}{2 + 3i} \cdot \frac{2 - 3i}{2 - 3i}$$

$$= \frac{6 - 9i - 2i + 3i^2}{2^2 - 9i^2}$$

$$= \frac{6 - 9i - 2i + 3(-1)}{4 - 9(-1)} \quad i^2 = -1$$

$$= \frac{3 - 11i}{13}$$

$$= \frac{3}{13} - \frac{11}{13}i \quad \textit{Standard form}$$

(b) $\dfrac{5+i}{-i}$ *Multiply numerator and denominator by the conjugate of denominator.*

$= \dfrac{5+i}{-i} \cdot \dfrac{i}{i}$ *The conjugate of $0-i$ is $0+i$.*

$= \dfrac{5i+i^2}{-i^2}$

$= \dfrac{5i+(-1)}{-(-1)}$ $i^2=-1$

$= -1+5i$

N5. $(x-1)^2 = -49$ Use the square root property.

$x-1 = \sqrt{-49}$ or $x-1 = -\sqrt{-49}$
$x-1 = 7i$ or $x-1 = -7i$
$x = 1+7i$ or $x = 1-7i$

The solution set is $\{1 \pm 7i\}$.

N6. $3t^2 = 2t - 1$

$3t^2 - 2t + 1 = 0$

Use the quadratic formula with $a=3$, $b=-2$, and $c=1$.

$t = \dfrac{-b \pm \sqrt{b^2-4ac}}{2a}$

$t = \dfrac{-(-2) \pm \sqrt{(-2)^2-4(3)(1)}}{2(3)}$

$= \dfrac{2 \pm \sqrt{4-12}}{6} = \dfrac{2 \pm \sqrt{-8}}{6}$

$= \dfrac{2 \pm i\sqrt{8}}{6} = \dfrac{2 \pm 2i\sqrt{2}}{6}$

$= \dfrac{2(1 \pm i\sqrt{2})}{2(3)} = \dfrac{1}{3} \pm \dfrac{\sqrt{2}}{3}i$

The solution set is $\left\{\dfrac{1}{3} \pm \dfrac{\sqrt{2}}{3}i\right\}$.

9.4 Section Exercises

1. $\sqrt{-9} = i\sqrt{9} = i \cdot 3 = 3i$

3. $\sqrt{-20} = i\sqrt{20}$
$= i\sqrt{4 \cdot 5}$
$= i \cdot 2 \cdot \sqrt{5}$
$= 2i\sqrt{5}$

5. $\sqrt{-18} = i\sqrt{18}$
$= i\sqrt{9 \cdot 2}$
$= i \cdot 3 \cdot \sqrt{2}$
$= 3i\sqrt{2}$

7. $\sqrt{-125} = i\sqrt{125}$
$= i\sqrt{25 \cdot 5}$
$= i \cdot 5 \cdot \sqrt{5}$
$= 5i\sqrt{5}$

9. $(2+8i) + (3-5i)$
$= (2+3) + (8-5)i$
Add real parts; add imaginary parts.
$= 5+3i$

11. $(8-3i) - (2+6i)$
$= (8-3i) + (-2-6i)$
Use the definition of subtraction.
$= (8-2) + (-3-6)i$
$= 6-9i$

13. $4i + (-6-2i)$
$= 4i - 6 - 2i$
$= -6+2i$

15. $(-3+6i) - 5$
$= -3+6i-5$
$= -8+6i$

17. $(3-4i) + (6-i) - (3+2i)$
$= (9-5i) - (3+2i)$
Add the first two complex numbers.
$= (9-5i) + (-3-2i)$
Use the definition of subtraction.
$= (9-3) + (-5-2)i$
$= 6-7i$

19. $(2i)(i) = 2i^2$
$= 2(-1)$ $i^2=-1$
$= -2$

21. $(-6i)(-i) = 6i^2$
$= 6(-1)$ $i^2=-1$
$= -6$

23. $2i(4-3i) = 8i - 6i^2$
$= 8i - 6(-1)$ $i^2=-1$
$= 6+8i$

25. $(3+2i)(4-i)$ *Use FOIL.*
$= 3(4) + 3(-i) + (2i)(4) + (2i)(-i)$
$= 12 - 3i + 8i - 2i^2$
$= 12 + 5i - 2(-1)$ $i^2=-1$
$= 12 + 5i + 2$
$= 14+5i$

27. $(5-4i)(3-2i)$ *Use FOIL.*
$= 5(3) + 5(-2i) + (-4i)(3) + (-4i)(-2i)$
$= 15 - 10i - 12i + 8i^2$
$= 15 - 22i + 8(-1)$ $i^2=-1$
$= 15 - 22i - 8$
$= 7-22i$

29. $(3 + 6i)(3 - 6i)$ *Use FOIL.*

$= 3(3) + 3(-6i) + (6i)(3) + (6i)(-6i)$

$= 9 - 18i + 18i - 36i^2$

$= 9 + 0 - 36(-1)$ $i^2 = -1$

$= 9 + 36 = 45$

This product can also be found by using the formula for the product of the sum and difference of two terms, $(a + b)(a - b) = a^2 - b^2$.

$(3 + 6i)(3 - 6i)$

$= 3^2 - (6i)^2$

$= 9 - 36i^2$

$= 9 - 36(-1)$ $i^2 = -1$

$= 9 + 36 = 45$

The quickest way to find the product is to notice that $3 + 6i$ and $3 - 6i$ are conjugates and use the rule for the product of conjugates, $(a + bi)(a - bi) = a^2 + b^2$.

$(3 + 6i)(3 - 6i) = 3^2 + 6^2$ *Let a = 3, b = 6.*

$= 9 + 36 = 45$

31. $\dfrac{1}{1 - i}$ *Multiply numerator and denominator by the conjugate of denominator.*

$= \dfrac{1}{1 - i} \cdot \dfrac{1 + i}{1 + i}$

$= \dfrac{1 + i}{1^2 - i^2}$

$= \dfrac{1 + i}{1 - (-1)}$ $i^2 = -1$

$= \dfrac{1 + i}{2}$

$= \dfrac{1}{2} + \dfrac{1}{2}i$ *Standard form*

33. $\dfrac{40}{2 + 6i} = \dfrac{2(20)}{2(1 + 3i)} = \dfrac{20}{1 + 3i}$

Multiply numerator and denominator by the conjugate of denominator.

$= \dfrac{20}{1 + 3i} \cdot \dfrac{1 - 3i}{1 - 3i}$

$= \dfrac{20(1 - 3i)}{(1 + 3i)(1 - 3i)}$

$= \dfrac{20(1 - 3i)}{1^2 + 3^2}$ $(a + bi)(a - bi)$
$= a^2 + b^2$

$= \dfrac{20(1 - 3i)}{10}$

$= 2(1 - 3i)$

$= 2 - 6i$ *Standard form*

35. $\dfrac{i}{4 - 3i}$ *Multiply numerator and denominator by the conjugate of denominator.*

$= \dfrac{i}{4 - 3i} \cdot \dfrac{4 + 3i}{4 + 3i}$

$= \dfrac{i(4 + 3i)}{(4 - 3i)(4 + 3i)}$

$= \dfrac{4i + 3i^2}{4^2 + 3^2}$ $(a + bi)(a - bi)$
$= a^2 + b^2$

$= \dfrac{-3 + 4i}{25}$

$= -\dfrac{3}{25} + \dfrac{4}{25}i$ *Standard form*

37. $\dfrac{7 + 3i}{1 - i}$ *Multiply numerator and denominator by the conjugate of denominator.*

$= \dfrac{7 + 3i}{1 - i} \cdot \dfrac{1 + i}{1 + i}$

$= \dfrac{7 + 7i + 3i + 3i^2}{1^2 - i^2}$

$= \dfrac{7 + 10i + 3(-1)}{1 - (-1)}$ $i^2 = -1$

$= \dfrac{4 + 10i}{2}$

$= 2 + 5i$ *Standard form*

39. $\dfrac{17 + i}{5 + 2i}$ *Multiply numerator and denominator by the conjugate of denominator.*

$= \dfrac{17 + i}{5 + 2i} \cdot \dfrac{5 - 2i}{5 - 2i}$

$= \dfrac{85 - 34i + 5i - 2i^2}{5^2 - 4i^2}$

$= \dfrac{85 - 29i - 2(-1)}{25 - 4(-1)}$ $i^2 = -1$

$= \dfrac{87 - 29i}{29} = \dfrac{29(3 - i)}{29}$

$= 3 - i$ *Standard form*

41. $\dfrac{-5 + 10i}{6 + 12i} = \dfrac{5(-1 + 2i)}{6(1 + 2i)}$

Multiply numerator and denominator by the conjugate of denominator.

$= \dfrac{5(-1 + 2i)}{6(1 + 2i)} \cdot \dfrac{1 - 2i}{1 - 2i}$

$= \dfrac{5(-1 + 2i)(1 - 2i)}{6(1 + 2i)(1 - 2i)}$

$= \dfrac{5(-1 + 2i + 2i - 4i^2)}{6(1^2 - 4i^2)}$

$= \dfrac{5(-1 + 4i + 4)}{6(1 + 4)}$ $i^2 = -1$

$= \dfrac{5(3 + 4i)}{6(5)} = \dfrac{3 + 4i}{6}$

$= \dfrac{3}{6} + \dfrac{4}{6}i = \dfrac{1}{2} + \dfrac{2}{3}i$ *Standard form*

43. $(x + 1)^2 = -4$ Use the square root property.

$x + 1 = \sqrt{-4}$ or $x + 1 = -\sqrt{-4}$

$x + 1 = 2i$ or $x + 1 = -2i$

$x = -1 + 2i$ or $x = -1 - 2i$

The solution set is $\{-1 \pm 2i\}$.

45. $(x-3)^2 = -5$ Use the square root property.

$$x - 3 = \sqrt{-5} \qquad \text{or} \qquad x - 3 = -\sqrt{-5}$$
$$x - 3 = i\sqrt{5} \qquad \text{or} \qquad x - 3 = -i\sqrt{5}$$
$$x = 3 + i\sqrt{5} \qquad \text{or} \qquad x = 3 - i\sqrt{5}$$

The solution set is $\left\{ 3 \pm i\sqrt{5} \right\}$.

47. $(3x+2)^2 = -18$
$$3x + 2 = \pm\sqrt{-18}$$
$$3x + 2 = \pm i\sqrt{18}$$
$$3x + 2 = \pm 3i\sqrt{2}$$
$$3x = -2 \pm 3i\sqrt{2}$$
$$x = \frac{-2 \pm 3i\sqrt{2}}{3}$$
$$x = -\tfrac{2}{3} \pm i\sqrt{2} \quad \textit{Standard form}$$

The solution set is $\left\{ -\tfrac{2}{3} \pm i\sqrt{2} \right\}$.

49. $m^2 - 2m + 2 = 0$

Use $a = 1$, $b = -2$, and $c = 2$.

$$m = \frac{-b \pm \sqrt{b^2 - 4ac}}{2a}$$
$$m = \frac{-(-2) \pm \sqrt{(-2)^2 - 4(1)(2)}}{2(1)}$$
$$= \frac{2 \pm \sqrt{4 - 8}}{2} = \frac{2 \pm \sqrt{-4}}{2}$$
$$= \frac{2 \pm 2i}{2} = \frac{2(1 \pm i)}{2} = 1 \pm i$$

The solution set is $\{1 \pm i\}$.

51. $2r^2 + 3r + 5 = 0$

Use $a = 2$, $b = 3$, and $c = 5$.

$$r = \frac{-b \pm \sqrt{b^2 - 4ac}}{2a}$$
$$r = \frac{-3 \pm \sqrt{3^2 - 4(2)(5)}}{2(2)}$$
$$= \frac{-3 \pm \sqrt{9 - 40}}{4} = \frac{-3 \pm \sqrt{-31}}{4}$$
$$= \frac{-3 \pm i\sqrt{31}}{4} = -\frac{3}{4} \pm \frac{\sqrt{31}}{4}i$$

The solution set is $\left\{ -\frac{3}{4} \pm \frac{\sqrt{31}}{4}i \right\}$.

53. $p^2 - 3p + 4 = 0$

Use $a = 1$, $b = -3$, and $c = 4$.

$$p = \frac{-b \pm \sqrt{b^2 - 4ac}}{2a}$$
$$p = \frac{-(-3) \pm \sqrt{(-3)^2 - 4(1)(4)}}{2(1)}$$
$$p = \frac{3 \pm \sqrt{9 - 16}}{2} = \frac{3 \pm \sqrt{-7}}{2}$$
$$= \frac{3 \pm i\sqrt{7}}{2} = \frac{3}{2} \pm \frac{\sqrt{7}}{2}i$$

The solution set is $\left\{ \frac{3}{2} \pm \frac{\sqrt{7}}{2}i \right\}$.

55. $5x^2 + 3 = 2x$
$$5x^2 - 2x + 3 = 0 \quad \textit{Standard form}$$
Use $a = 5$, $b = -2$, and $c = 3$.

$$x = \frac{-b \pm \sqrt{b^2 - 4ac}}{2a}$$
$$x = \frac{-(-2) \pm \sqrt{(-2)^2 - 4(5)(3)}}{2(5)}$$
$$= \frac{2 \pm \sqrt{4 - 60}}{10} = \frac{2 \pm \sqrt{-56}}{10}$$
$$= \frac{2 \pm i\sqrt{4 \cdot 14}}{10} = \frac{2 \pm 2i\sqrt{14}}{10}$$
$$= \frac{2\left(1 \pm i\sqrt{14}\right)}{2 \cdot 5} = \frac{1 \pm i\sqrt{14}}{5}$$
$$= \frac{1}{5} \pm \frac{\sqrt{14}}{5}i \quad \textit{Standard form}$$

The solution set is $\left\{ \frac{1}{5} \pm \frac{\sqrt{14}}{5}i \right\}$.

57. $2m^2 + 7 = -2m$
$$2m^2 + 2m + 7 = 0 \quad \textit{Standard form}$$
Use $a = 2$, $b = 2$, and $c = 7$.

$$m = \frac{-b \pm \sqrt{b^2 - 4ac}}{2a}$$
$$m = \frac{-2 \pm \sqrt{2^2 - 4(2)(7)}}{2(2)}$$
$$= \frac{-2 \pm \sqrt{4 - 56}}{4} = \frac{-2 \pm \sqrt{-52}}{4}$$
$$= \frac{-2 \pm i\sqrt{52}}{4} = \frac{-2 \pm 2i\sqrt{13}}{4}$$
$$= \frac{2\left(-1 \pm i\sqrt{13}\right)}{4} = \frac{-1 \pm i\sqrt{13}}{2}$$
$$= -\frac{1}{2} \pm \frac{\sqrt{13}}{2}i \quad \textit{Standard form}$$

The solution set is $\left\{ -\frac{1}{2} \pm \frac{\sqrt{13}}{2}i \right\}$.

59. $x^2 - x + 3 = 0$

Use $a = 1, b = -1$, and $c = 3$.

$$x = \frac{-b \pm \sqrt{b^2 - 4ac}}{2a}$$

$$x = \frac{-(-1) \pm \sqrt{(-1)^2 - 4(1)(3)}}{2(1)}$$

$$= \frac{1 \pm \sqrt{1 - 12}}{2} = \frac{1 \pm \sqrt{-11}}{2}$$

$$= \frac{1 \pm i\sqrt{11}}{2} = \frac{1}{2} \pm \frac{\sqrt{11}}{2}i$$

The solution set is $\left\{ \frac{1}{2} \pm \frac{\sqrt{11}}{2}i \right\}$.

61. "Every real number is a complex number" is *true*. A real number m can be represented as $m + 0i$.

63. "Every complex number is a real number" is *false*. Examples of complex numbers which are not real numbers are $3i, -i\sqrt{2}, 4 + 6i$, and $\sqrt{3} - 2i$.

65. $2x - 3y = 6$
If $x = 0$, then $y = -2$.
If $y = 0$, then $x = 3$.
Graph the line through its intercepts, $(3, 0)$ and $(0, -2)$.

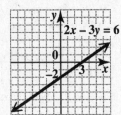

67. $2x^2 - x + 1$
$= 2 \cdot 3^2 - 3 + 1$ *Let x = 3.*
$= 18 - 3 + 1$
$= 15 + 1$
$= 16$

9.5 More on Graphing Quadratic Equations; Quadratic Functions

9.5 Now Try Exercises

N1. $y = x^2 - x - 2$

Find any x-intercepts by substituting 0 for y in the equation.

$y = x^2 - x - 2$
$0 = x^2 - x - 2$ *Let y = 0.*
$0 = (x + 1)(x - 2)$ *Factor.*

$x + 1 = 0$ or $x - 2 = 0$
$x = -1$ or $x = 2$

The x-intercepts are $(-1, 0)$ and $(2, 0)$.

Now find any y-intercepts by substituting 0 for x.

$y = x^2 - x - 2$
$y = 0^2 - 0 - 2$ *Let x = 0.*
$y = -2$

The y-intercept is $(0, -2)$. The x-value of the vertex is halfway between the x-intercepts, $(-1, 0)$ and $(2, 0)$.

$$x = \tfrac{1}{2}(-1 + 2) = \tfrac{1}{2}(1) = \tfrac{1}{2}$$

Find the y-value of the vertex by substituting $\frac{1}{2}$ for x in the given equation.

$y = x^2 - x - 2$
$y = (\tfrac{1}{2})^2 - \tfrac{1}{2} - 2$ *Let x = $\tfrac{1}{2}$.*
$y = \tfrac{1}{4} - \tfrac{1}{2} - 2$
$y = -\tfrac{9}{4}$

The vertex is at $\left(\tfrac{1}{2}, -\tfrac{9}{4} \right)$.

The axis of the parabola is the line $x = \tfrac{1}{2}$.

x	y
-2	4
-1	0
0	-2
$\tfrac{1}{2}$	$-\tfrac{9}{4}$
1	-2
2	0
3	4

Plot the intercepts, vertex, and additional points shown in the table of values and connect them with a smooth curve.

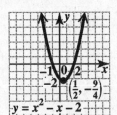

N2. $y = -x^2 + 4x + 2$

Use $a = -1, b = 4$, and $c = 2$.

The x-value of the vertex is

$$x = -\frac{b}{2a} = -\frac{4}{2(-1)} = 2.$$

The y-value of the vertex is

$$y = -2^2 + 4(2) + 2 = 6,$$

so the vertex is $(2, 6)$.

The axis of the parabola is the line $x = 2$.

Find the y-intercept by letting $x = 0$.

$$y = -0^2 + 4(0) + 2 = 2$$

The y-intercept is $(0, 2)$.

Now find the x-intercepts by using the quadratic formula.

$$x = \frac{-4 \pm \sqrt{4^2 - 4(-1)(2)}}{2(-1)}$$

$$x = \frac{-4 \pm \sqrt{24}}{-2}$$

Using a calculator, $x \approx 4.45$ and $x \approx -0.45$.

The x-intercepts are $(-0.45, 0)$ and $(4.45, 0)$.

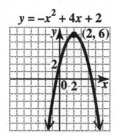

$$y = -x^2 + 4x + 2$$

N3. Since the graph of $f(x) = -x^2 + 4x - 4$ has *one* x-intercept, the corresponding equation, $-x^2 + 4x - 4 = 0$, has *one* real number solution. Since the x-intercept of the graph is $(2, 0)$, the solution set of the equation is $\{2\}$.

9.5 Section Exercises

1. $y = x^2 - 6$

If $x = 0$, $y = -6$, so the y-intercept is $(0, -6)$.

To find any x-intercepts, let $y = 0$.

$$0 = x^2 - 6$$
$$x^2 = 6$$
$$x = \pm\sqrt{6} \approx \pm 2.45$$

The x-intercepts are $\left(\pm\sqrt{6}, 0 \right)$.

The x-value of the vertex is

$$x = -\frac{b}{2a} = -\frac{0}{2(1)} = 0.$$

Thus, the vertex is the same as the y-intercept (since $x = 0$). The axis of the parabola is the vertical line $x = 0$.

Make a table of ordered pairs whose x-values are on either side of the vertex's x-value of $x = 0$.

x	y
0	-6
± 1	-5
± 2	-2
± 3	3

Plot these seven ordered pairs and connect them with a smooth curve.

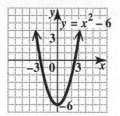

$$y = x^2 - 6$$

3. $y = (x + 3)^2 = x^2 + 6x + 9$

If $x = 0$, $y = 9$, so the y-intercept is $(0, 9)$.

To find any x-intercepts, let $y = 0$.

$$0 = (x + 3)^2$$
$$0 = x + 3$$
$$-3 = x$$

The x-intercept is $(-3, 0)$.

The x-value of the vertex is

$$x = -\frac{b}{2a} = -\frac{6}{2(1)} = -3.$$

Thus, the vertex is the same as the x-intercept. The axis of the parabola is the vertical line $x = -3$.

Make a table of ordered pairs whose x-values are on either side of the vertex's x-value of $x = -3$.

x	y
-6	9
-5	4
-4	1
-3	0
-2	1
-1	4
0	9

Plot these seven ordered pairs and connect them with a smooth curve.

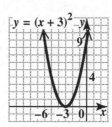

$$y = (x + 3)^2$$

5. $y = x^2 + 2x + 3$

If $x = 0$, $y = 3$, so the y-intercept is $(0, 3)$.

To find any x-intercepts, let $y = 0$.

$$0 = x^2 + 2x + 3$$

The trinomial on the right cannot be factored. Because the discriminant
$b^2 - 4ac = 2^2 - 4(1)(3) = -8$ is negative,

this equation has no real number solutions. Thus, the parabola has no x-intercepts.

The x-value of the vertex is

$$x = -\frac{b}{2a} = -\frac{2}{2(1)} = -1.$$

The y-value of the vertex is

$$y = (-1)^2 + 2(-1) + 3$$
$$= 1 - 2 + 3 = 2,$$

so the vertex is $(-1, 2)$. The axis of the parabola is the vertical line $x = -1$.

Make a table of ordered pairs whose x-values are on either side of the vertex's x-value of $x = -1$.

x	y
-4	11
-3	6
-2	3
-1	2
0	3
1	6
2	11

Plot these seven ordered pairs and connect them with a smooth curve.

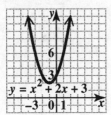

7. $y = x^2 - 8x + 16 = (x - 4)^2$

If $x = 0$, $y = 16$, so the y-intercept is $(0, 16)$.

Let $y = 0$ and solve for x.

$$0 = (x - 4)^2$$
$$0 = x - 4$$
$$4 = x$$

The only x-intercept is $(4, 0)$.

The x-value of the vertex is

$$x = -\frac{b}{2a} = -\frac{-8}{2(1)} = 4.$$

The y-value of the vertex has already been found. The vertex is $(4, 0)$, which is also the x-intercept. The axis of the parabola is the line $x = 4$.

Make a table of ordered pairs whose x-values are on either side of the vertex's x-value of $x = 4$.

x	y
1	9
2	4
3	1
4	0
5	1
6	4
7	9

Plot these seven ordered pairs and connect them with a smooth curve.

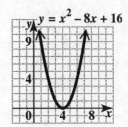

9. $y = -x^2 + 6x - 5$

If $x = 0$, $y = -5$, so the y-intercept is $(0, -5)$.

Let $y = 0$ and solve for x.

$$0 = -x^2 + 6x - 5$$
$$x^2 - 6x + 5 = 0$$
$$(x - 1)(x - 5) = 0$$

$$x - 1 = 0 \quad \text{or} \quad x - 5 = 0$$
$$x = 1 \quad \text{or} \quad x = 5$$

The x-intercepts are $(1, 0)$ and $(5, 0)$.

The x-value of the vertex is

$$x = -\frac{b}{2a} = -\frac{6}{2(-1)} = 3.$$

The y-value of the vertex is

$$y = -(3)^2 + 6(3) - 5$$
$$= -9 + 18 - 5 = 4,$$

so the vertex is $(3, 4)$.

The axis of the parabola is the line $x = 3$.

Make a table of ordered pairs whose x-values are on either side of the vertex's x-value of $x = 3$.

x	y
0	-5
1	0
2	3
3	4
4	3
5	0
6	-5

Plot these seven ordered pairs and connect them with a smooth curve.

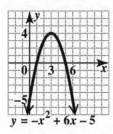

$$y = -x^2 + 6x - 5$$

11. $y = x^2 + 4x$

If $x = 0$, $y = 0$, so the y-intercept is $(0, 0)$.

Let $y = 0$ and solve for x.

$$0 = x^2 + 4x$$
$$0 = x(x + 4)$$

$$x = 0 \quad \text{or} \quad x = -4$$

The x-intercepts are $(0, 0)$ and $(-4, 0)$.

The x-value of the vertex is

$$x = -\frac{b}{2a} = -\frac{4}{2(1)} = -2.$$

The y-value of the vertex is

$$y = (-2)^2 + 4(-2)$$
$$= 4 - 8 = -4,$$

so the vertex is $(-2, -4)$.

The axis of the parabola is the line $x = -2$.

Make a table of ordered pairs whose x-values are on either side of the vertex's x-value of $x = -2$.

x	y
-5	5
-4	0
-3	-3
-2	-4
-1	-3
0	0
1	5

Plot these seven ordered pairs and connect them with a smooth curve.

$$y = x^2 + 4x$$

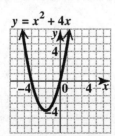

13. The corresponding equation has solutions when $y = 0$. Since the graph intersects the x-axis in only one point, the x-intercept $(2, 0)$, the corresponding equation has only one real number solution, 2. The solution set is $\{2\}$.

15. The graph crosses the x-axis in two places, the x-intercepts $(-2, 0)$ and $(2, 0)$, so the corresponding equation has two real number solutions, 2 and -2. The solution set is $\{\pm 2\}$.

17. Since the graph has no x-intercepts, the corresponding equation has no real number solution. The solution set is $\emptyset$.

19. If $a > 0$, the parabola $y = ax^2 + bx + c$ opens upward. If $a < 0$, the parabola $y = ax^2 + bx + c$ opens downward.

21. From the screens, we see that the solution set is $\{-2, 3\}$. We can verify this by factoring.

$$x^2 - x - 6 = 0$$
$$(x - 3)(x + 2) = 0$$

$$x = 3 \quad \text{or} \quad x = -2$$

23. Refer to the graph for Exercise 13. The value of x can be any real number, so the domain of the function is $(-\infty, \infty)$. The values of y are at least 0, so the range of the function is $[0, \infty)$.

25. Refer to the graph for Exercise 15. The value of x can be any real number, so the domain of the function is $(-\infty, \infty)$. The values of y are at most 4, so the range of the function is $(-\infty, 4]$.

27. Refer to the graph for Exercise 17. The value of x can be any real number, so the domain of the function is $(-\infty, \infty)$. The values of y are at least 1, so the range of the function is $[1, \infty)$.

29. $f(x) = 2x^2 - 5x + 3$
$f(0) = 2(0)^2 - 5(0) + 3 = 3$

31. $f(x) = 2x^2 - 5x + 3$
$f(-2) = 2(-2)^2 - 5(-2) + 3$
$= 2(4) + 10 + 3$
$= 8 + 10 + 3 = 21$

33. Let $x = $ one of the numbers and $80 - x = $ the other number.

The product P of the two numbers is given by

$$P = x(80 - x).$$

Writing this equation in standard form gives us

$$P = -x^2 + 80x.$$

Finding the maximum of the product is the same as finding the vertex of the graph of P. The x-value of the vertex is

$$x = -\frac{b}{2a} = -\frac{80}{2(-1)} = 40,$$

continued

which makes sense because 40 is halfway between 0 and 80 (the x-intercepts of $P = x(80 - x)$).

If x is 40, then $80 - x$ must also be 40. The two numbers are 40 and 40 and the product is $40 \cdot 40 = 1600$.

35. Because the vertex is at the origin, an equation of the parabola is of the form

$$y = ax^2.$$

As shown in the figure, one point on the graph has coordinates $(150, 44)$.

$$y = ax^2 \qquad \textit{General equation}$$
$$44 = a(150)^2 \quad \textit{Let x = 150, y = 44.}$$
$$44 = 22{,}500a$$
$$a = \frac{44}{22{,}500} = \frac{4 \cdot 11}{4 \cdot 5625} = \frac{11}{5625}$$

Thus, an equation of the parabola is

$$y = \tfrac{11}{5625}x^2.$$

38.

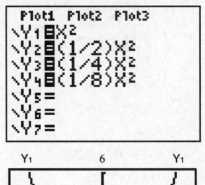

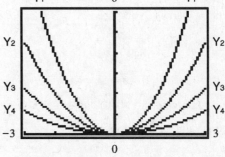

In each case, there is a vertical "shrink" of the parabola. It becomes wider as the coefficient gets smaller.

37.

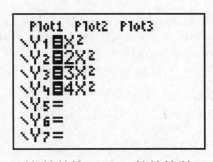

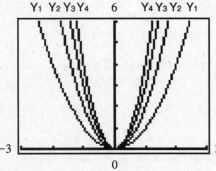

In each case, there is a vertical "stretch" of the parabola. It becomes narrower as the coefficient gets larger.

39.

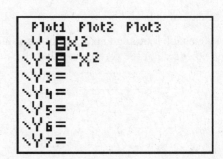

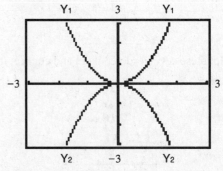

The graph of Y_2 is obtained from the graph of Y_1 by reflecting the graph of Y_1 across the x-axis.

40.

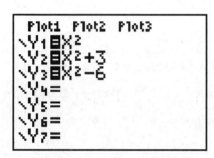

When the coefficient of x^2 is negative, the parabola opens downward.

41.

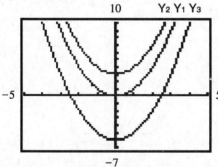

By adding a positive constant k, the graph is shifted k units upward. By subtracting a positive constant k, the graph is shifted k units downward.

42.

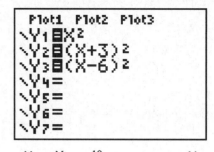

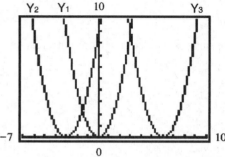

Adding a positive constant k before squaring moves the graph k units to the left.

Subtracting a positive constant k before squaring moves the graph k units to the right.

Chapter 9 Review Exercises

1. $z^2 = 144$

$$z = \sqrt{144} \quad \text{or} \quad z = -\sqrt{144}$$
$$z = 12 \quad \text{or} \quad z = -12$$

The solution set is $\{ \pm 12 \}$.

2. $x^2 = 37$

$$x = \sqrt{37} \quad \text{or} \quad x = -\sqrt{37}$$

The solution set is $\left\{ \pm \sqrt{37} \right\}$.

3. $m^2 = 128$

$$m = \pm \sqrt{128}$$
$$= \pm \sqrt{64 \cdot 2}$$
$$= \pm 8\sqrt{2}$$

The solution set is $\left\{ \pm 8\sqrt{2} \right\}$.

4. $(x + 2)^2 = 25$

$$x + 2 = \sqrt{25} \quad \text{or} \quad x + 2 = -\sqrt{25}$$
$$x + 2 = 5 \quad \text{or} \quad x + 2 = -5$$
$$x = 3 \quad \text{or} \quad x = -7$$

The solution set is $\{-7, 3\}$.

5. $(r - 3)^2 = 10$

$$r - 3 = \sqrt{10} \quad \text{or} \quad r - 3 = -\sqrt{10}$$
$$r = 3 + \sqrt{10} \quad \text{or} \quad r = 3 - \sqrt{10}$$

The solution set is $\left\{ 3 \pm \sqrt{10} \right\}$.

6. $(2p+1)^2 = 14$

$$2p+1 = \sqrt{14} \qquad \text{or} \qquad 2p+1 = -\sqrt{14}$$
$$2p = -1 + \sqrt{14} \qquad \text{or} \qquad 2p = -1 - \sqrt{14}$$
$$p = \frac{-1+\sqrt{14}}{2} \qquad \text{or} \qquad p = \frac{-1-\sqrt{14}}{2}$$

The solution set is $\left\{ \frac{-1\pm\sqrt{14}}{2} \right\}$.

7. $(3x+2)^2 = -3$

$$3x+2 = \sqrt{-3} \quad \text{or} \quad 3x+2 = -\sqrt{-3}$$

Because $\sqrt{-3}$ does not represent a real number, the solution set is $\emptyset$.

8. $(3-5x)^2 = 8$

$$3-5x = \sqrt{8} \qquad \text{or} \qquad 3-5x = -\sqrt{8}$$
$$-5x = -3+2\sqrt{2} \qquad \text{or} \qquad -5x = -3-2\sqrt{2}$$
$$x = \frac{-3+2\sqrt{2}}{-5} \qquad \text{or} \qquad x = \frac{-3-2\sqrt{2}}{-5}$$
$$x = \frac{3-2\sqrt{2}}{5} \qquad \text{or} \qquad x = \frac{3+2\sqrt{2}}{5}$$

The solution set is $\left\{ \frac{3\pm2\sqrt{2}}{5} \right\}$.

9. $m^2 + 6m + 5 = 0$

Rewrite the equation with the variable terms on one side and the constant on the other side.

$$m^2 + 6m = -5$$

Take half the coefficient of m and square it.

$$\tfrac{1}{2}(6) = 3, \quad \text{and} \quad (3)^2 = 9.$$

Add 9 to each side of the equation.

$$m^2 + 6m + 9 = -5 + 9$$
$$m^2 + 6m + 9 = 4$$
$$(m+3)^2 = 4 \quad \textit{Factor.}$$

$$m+3 = \sqrt{4} \quad \text{or} \quad m+3 = -\sqrt{4}$$
$$m+3 = 2 \quad \text{or} \quad m+3 = -2$$
$$m = -1 \quad \text{or} \quad m = -5$$

The solution set is $\{-5, -1\}$.

10. $p^2 + 4p = 7$

Take half the coefficient of p and square it.

$$\tfrac{1}{2}(4) = 2, \quad \text{and} \quad (2)^2 = 4.$$

Add 4 to each side of the equation.

$$p^2 + 4p + 4 = 7 + 4$$
$$(p+2)^2 = 11$$

$$p+2 = \sqrt{11} \qquad \text{or} \qquad p+2 = -\sqrt{11}$$
$$p = -2 + \sqrt{11} \qquad \text{or} \qquad p = -2 - \sqrt{11}$$

The solution set is $\left\{ -2 \pm \sqrt{11} \right\}$.

11. $-x^2 + 5 = 2x$

Divide each side of the equation by -1 to make the coefficient of the squared term equal to 1.

$$x^2 - 5 = -2x$$

Rewrite the equation with the variable terms on one side and the constant on the other side.

$$x^2 + 2x = 5$$

Take half the coefficient of x and square it.

$$\tfrac{1}{2}(2) = 1, \quad \text{and} \quad 1^2 = 1.$$

Add 1 to both sides of the equation.

$$x^2 + 2x + 1 = 5 + 1$$
$$(x+1)^2 = 6$$

$$x+1 = \sqrt{6} \qquad \text{or} \qquad x+1 = -\sqrt{6}$$
$$x = -1 + \sqrt{6} \qquad \text{or} \qquad x = -1 - \sqrt{6}$$

The solution set is $\left\{ -1 \pm \sqrt{6} \right\}$.

12. $2z^2 - 3 = -8z$

Divide both sides by 2 to get the z^2 coefficient equal to 1.

$$z^2 - \tfrac{3}{2} = -4z$$

Rewrite the equation with the variable terms on one side and the constant on the other side.

$$z^2 + 4z = \tfrac{3}{2}$$

Take half the coefficient of z and square it.

$$\tfrac{1}{2}(4) = 2, \quad \text{and} \quad 2^2 = 4.$$

Add 4 to both sides of the equation.

$$z^2 + 4z + 4 = \tfrac{3}{2} + 4$$
$$(z+2)^2 = \tfrac{11}{2}$$
$$z + 2 = \pm\sqrt{\frac{11}{2}}$$
$$z + 2 = \pm\frac{\sqrt{11}}{\sqrt{2}} \cdot \frac{\sqrt{2}}{\sqrt{2}}$$
$$z + 2 = \pm\frac{\sqrt{22}}{2}$$
$$z = -2 \pm \frac{\sqrt{22}}{2}$$
$$z = \frac{-4}{2} \pm \frac{\sqrt{22}}{2}$$
$$z = \frac{-4 \pm \sqrt{22}}{2}$$

The solution set is $\left\{ \frac{-4\pm\sqrt{22}}{2} \right\}$.

13. $5x^2 - 3x - 2 = 0$

Divide both sides by 5 to get the x^2 coefficient equal to 1.

$$x^2 - \tfrac{3}{5}x - \tfrac{2}{5} = 0$$

Rewrite the equation with the variable terms on one side and the constant on the other side.

$$x^2 - \tfrac{3}{5}x = \tfrac{2}{5}$$

Take half the coefficient of x and square it.

$$\tfrac{1}{2}\left(-\tfrac{3}{5}\right) = -\tfrac{3}{10}, \text{ and } \left(-\tfrac{3}{10}\right)^2 = \tfrac{9}{100}$$

Add $\tfrac{9}{100}$ to both sides of the equation.

$$x^2 - \tfrac{3}{5}x + \tfrac{9}{100} = \tfrac{2}{5} + \tfrac{9}{100}$$
$$\left(x - \tfrac{3}{10}\right)^2 = \tfrac{40}{100} + \tfrac{9}{100}$$
$$\left(x - \tfrac{3}{10}\right)^2 = \tfrac{49}{100}$$

$$x - \tfrac{3}{10} = \sqrt{\tfrac{49}{100}} \quad \text{or} \quad x - \tfrac{3}{10} = -\sqrt{\tfrac{49}{100}}$$
$$x - \tfrac{3}{10} = \tfrac{7}{10} \quad \text{or} \quad x - \tfrac{3}{10} = -\tfrac{7}{10}$$
$$x = \tfrac{10}{10} \quad \text{or} \quad x = -\tfrac{4}{10}$$
$$x = 1 \quad \text{or} \quad x = -\tfrac{2}{5}$$

The solution set is $\left\{-\tfrac{2}{5}, 1\right\}$.

14. $(4x + 1)(x - 1) = -7$

Multiply on the left side and then simplify. Get all variable terms on one side and the constant on the other side.

$$4x^2 - 4x + x - 1 = -7$$
$$4x^2 - 3x = -6$$

Divide both sides by 4 so that the coefficient of x^2 will be 1.

$$x^2 - \tfrac{3}{4}x = -\tfrac{6}{4} = -\tfrac{3}{2}$$

Square half the coefficient of x and add it to both sides.

$$x^2 - \tfrac{3}{4}x + \tfrac{9}{64} = -\tfrac{3}{2} + \tfrac{9}{64}$$
$$\left(x - \tfrac{3}{8}\right)^2 = -\tfrac{96}{64} + \tfrac{9}{64}$$
$$\left(x - \tfrac{3}{8}\right)^2 = -\tfrac{87}{64}$$

The square root of $-\tfrac{87}{64}$ is not a real number, so the solution set is $\emptyset$.

15. $h = -16t^2 + 32t + 50$

Let $h = 30$ and solve for t (which must have a positive value since it represents a number of seconds).

$$30 = -16t^2 + 32t + 50$$
$$16t^2 - 32t - 20 = 0$$

Divide both sides by 16.

$$t^2 - 2t - \tfrac{20}{16} = 0$$
$$t^2 - 2t = \tfrac{5}{4}$$

Half of -2 is -1, and $(-1)^2 = 1$.

Add 1 to both sides of the equation.

$$t^2 - 2t + 1 = \tfrac{5}{4} + 1$$
$$(t - 1)^2 = \tfrac{9}{4}$$

$$t - 1 = \sqrt{\tfrac{9}{4}} \quad \text{or} \quad t - 1 = -\sqrt{\tfrac{9}{4}}$$
$$t - 1 = \tfrac{3}{2} \quad \text{or} \quad t - 1 = -\tfrac{3}{2}$$
$$t = 1 + \tfrac{3}{2} \quad \text{or} \quad t = 1 - \tfrac{3}{2}$$
$$t = \tfrac{5}{2} = 2\tfrac{1}{2} \quad \text{or} \quad t = -\tfrac{1}{2}$$

Reject the negative value of t. The object will reach a height of 30 feet after $2\tfrac{1}{2}$ seconds.

16. Use the Pythagorean theorem with legs x and $x + 2$ and hypotenuse $x + 4$.

$$a^2 + b^2 = c^2$$
$$(x)^2 + (x + 2)^2 = (x + 4)^2$$
$$x^2 + x^2 + 4x + 4 = x^2 + 8x + 16$$
$$x^2 - 4x - 12 = 0$$
$$(x - 6)(x + 2) = 0$$

$$x - 6 = 0 \quad \text{or} \quad x + 2 = 0$$
$$x = 6 \quad \text{or} \quad x = -2$$

Reject the negative value because x represents a length. The value of x is 6. The lengths of the three sides are 6, 8, and 10.

17. Take half the coefficient of x and square the result.

$$\tfrac{1}{2} \cdot 3 = \tfrac{3}{2}$$
$$\left(\tfrac{3}{2}\right)^2 = \tfrac{9}{4}$$

Add $\left(\tfrac{3}{2}\right)^2$, or $\tfrac{9}{4}$, to $x^2 + 3x$ to get the perfect square $x^2 + 3x + \tfrac{9}{4}$.

18. $x^2 - 9 = 0$, or $1x^2 + 0x - 9 = 0$

(a) $(x + 3)(x - 3) = 0$

$$x + 3 = 0 \quad \text{or} \quad x - 3 = 0$$
$$x = -3 \quad \text{or} \quad x = 3$$

The solution set is $\{\pm 3\}$.

(b) $x^2 = 9$

$$x = \pm\sqrt{9} = \pm 3$$

The solution set is $\{\pm 3\}$.

(c) Use $a = 1$, $b = 0$, and $c = -9$.

$$x = \frac{-0 \pm \sqrt{0^2 - 4(1)(-9)}}{2(1)}$$
$$x = \frac{\pm\sqrt{36}}{2} = \frac{\pm 6}{2} = \pm 3$$

The solution set is $\{\pm 3\}$.

(d) Because there is only one solution set, we always get the same results, no matter which method of solution is used.

19. $x^2 - 2x - 4 = 0$

This equation is in standard form with $a = 1$, $b = -2$, and $c = -4$. Substitute these values into the quadratic formula.

$$x = \frac{-b \pm \sqrt{b^2 - 4ac}}{2a}$$

$$x = \frac{-(-2) \pm \sqrt{(-2)^2 - 4(1)(-4)}}{2(1)}$$

$$= \frac{2 \pm \sqrt{4 + 16}}{2} = \frac{2 \pm \sqrt{20}}{2}$$

$$= \frac{2 \pm 2\sqrt{5}}{2} = \frac{2\left(1 \pm \sqrt{5}\right)}{2}$$

$$= 1 \pm \sqrt{5}$$

The solution set is $\left\{1 \pm \sqrt{5}\right\}$.

20. $3k^2 + 2k = -3$

$3k^2 + 2k + 3 = 0$

Use $a = 3$, $b = 2$, and $c = 3$.

$$k = \frac{-2 \pm \sqrt{2^2 - 4(3)(3)}}{2(3)}$$

$$= \frac{-2 \pm \sqrt{4 - 36}}{6}$$

$$= \frac{-2 \pm \sqrt{-32}}{6}$$

Because $\sqrt{-32}$ does not represent a real number, the solution set is $\emptyset$.

21. $2p^2 + 8 = 4p + 11$

$2p^2 - 4p - 3 = 0$

Use the quadratic formula with $a = 2$, $b = -4$, and $c = -3$.

$$p = \frac{-(-4) \pm \sqrt{(-4)^2 - 4(2)(-3)}}{2(2)}$$

$$= \frac{4 \pm \sqrt{16 + 24}}{4} = \frac{4 \pm \sqrt{40}}{4}$$

$$= \frac{4 \pm \sqrt{4 \cdot 10}}{4} = \frac{4 \pm 2\sqrt{10}}{4}$$

$$p = \frac{2\left(2 \pm \sqrt{10}\right)}{2(2)} = \frac{2 \pm \sqrt{10}}{2}$$

The solution set is $\left\{\frac{2 \pm \sqrt{10}}{2}\right\}$.

22. $-4x^2 + 7 = 2x$

$0 = 4x^2 + 2x - 7$

Use $a = 4$, $b = 2$, and $c = -7$.

$$x = \frac{-2 \pm \sqrt{(2)^2 - 4(4)(-7)}}{2(4)}$$

$$= \frac{-2 \pm \sqrt{4 + 112}}{8} = \frac{-2 \pm \sqrt{116}}{8}$$

$$= \frac{-2 \pm 2\sqrt{29}}{8} = \frac{2\left(-1 \pm \sqrt{29}\right)}{2(4)}$$

$$= \frac{-1 \pm \sqrt{29}}{4}$$

The solution set is $\left\{\frac{-1 \pm \sqrt{29}}{4}\right\}$.

23. $\frac{1}{4}p^2 = 2 - \frac{3}{4}p$

$\frac{1}{4}p^2 + \frac{3}{4}p - 2 = 0$

Multiply both sides by the least common denominator, 4.

$$4\left(\tfrac{1}{4}p^2 + \tfrac{3}{4}p - 2\right) = 4(0)$$

$$p^2 + 3p - 8 = 0$$

Use the quadratic formula with $a = 1$, $b = 3$, and $c = -8$.

$$p = \frac{-3 \pm \sqrt{3^2 - 4(1)(-8)}}{2(1)}$$

$$= \frac{-3 \pm \sqrt{9 + 32}}{2}$$

$$= \frac{-3 \pm \sqrt{41}}{2}$$

The solution set is $\left\{\frac{-3 \pm \sqrt{41}}{2}\right\}$.

24. $3x^2 - x - 2 = 0$

Use the quadratic formula with $a = 3$, $b = -1$, and $c = -2$.

$$x = \frac{-(-1) \pm \sqrt{(-1)^2 - 4(3)(-2)}}{2(3)}$$

$$= \frac{1 \pm \sqrt{1 + 24}}{6}$$

$$= \frac{1 \pm \sqrt{25}}{6} = \frac{1 \pm 5}{6}$$

$$x = \frac{1 + 5}{6} = \frac{6}{6} = 1$$

or $\;x = \frac{1 - 5}{6} = \frac{-4}{6} = -\frac{2}{3}$

The solution set is $\left\{-\frac{2}{3}, 1\right\}$.

25. Since a negative radicand means that the radical $\sqrt{b^2 - 4ac}$ is not a real number, there are *no* real solutions for the equation.

26. $(3 + 5i) + (2 - 6i)$
$= (3 + 2) + (5 - 6)i$
$= 5 - i$

27. $(-2 - 8i) - (4 - 3i)$
$= (-2 - 8i) + (-4 + 3i)$
$= (-2 - 4) + (-8 + 3)i$
$= -6 - 5i$

28. $(6 - 2i)(3 + i)$
$= 18 + 6i - 6i - 2i^2 \quad FOIL$
$= 18 - 2(-1) \qquad\quad i^2 = -1$
$= 18 + 2 = 20$

29. $(2 + 3i)(2 - 3i)$
$= 2^2 + 3^2 \qquad Product\ of\ conjugates$
$= 4 + 9 = 13$

30. $\dfrac{1 + i}{1 - i} = \dfrac{1 + i}{1 - i} \cdot \dfrac{1 + i}{1 + i}$

Multiply numerator and denominator by the conjugate of the denominator.

$= \dfrac{1 + 2i + i^2}{1 - i^2}$

$= \dfrac{1 + 2i + (-1)}{1 - (-1)} \qquad i^2 = -1$

$= \dfrac{2i}{2} = i$

31. $\dfrac{5 + 6i}{2 + 3i} = \dfrac{5 + 6i}{2 + 3i} \cdot \dfrac{2 - 3i}{2 - 3i}$

$= \dfrac{10 - 15i + 12i - 18i^2}{4 - 9i^2}$

$= \dfrac{10 - 3i - 18(-1)}{4 - 9(-1)}$

$= \dfrac{10 + 18 - 3i}{4 + 9}$

$= \dfrac{28 - 3i}{13}$

$= \dfrac{28}{13} - \dfrac{3}{13}i \qquad Standard\ form$

32. The real number a can be written as $a + 0i$. The conjugate of a is $a - 0i$ or a. Thus, the conjugate of a real number is the real number itself.

33. No. As shown below, the product of a complex number and its conjugate is *always* a real number.
$(a + bi)(a - bi) = a^2 - b^2 i^2$
$= a^2 - b^2(-1)$
$= a^2 + b^2$

34. $(m + 2)^2 = -3$

Use the square root property.

$m + 2 = \pm\sqrt{-3}$
$m + 2 = \pm i\sqrt{3}$
$m = -2 \pm i\sqrt{3}$

The solution set is $\left\{-2 \pm i\sqrt{3}\right\}$.

35. $(3p - 2)^2 = -8$

Use the square root property.

$3p - 2 = \pm\sqrt{-8}$
$3p - 2 = \pm 2i\sqrt{2}$
$3p = 2 \pm 2i\sqrt{2}$
$p = \dfrac{2 \pm 2i\sqrt{2}}{3} = \dfrac{2}{3} \pm \dfrac{2\sqrt{2}}{3}i$

The solution set is $\left\{\dfrac{2}{3} \pm \dfrac{2\sqrt{2}}{3}i\right\}$.

36. $3x^2 = 2x - 1$

Rewrite the equation in standard form.

$$3x^2 - 2x + 1 = 0$$

Use the quadratic formula with
$a = 3$, $b = -2$, and $c = 1$.

$x = \dfrac{-b \pm \sqrt{b^2 - 4ac}}{2a}$

$x = \dfrac{-(-2) \pm \sqrt{(-2)^2 - 4(3)(1)}}{2(3)}$

$= \dfrac{2 \pm \sqrt{4 - 12}}{6}$

$= \dfrac{2 \pm \sqrt{-8}}{6} = \dfrac{2 \pm 2i\sqrt{2}}{2 \cdot 3}$

$= \dfrac{2\left(1 \pm i\sqrt{2}\right)}{2 \cdot 3} = \dfrac{1 \pm i\sqrt{2}}{3}$

$= \dfrac{1}{3} \pm \dfrac{\sqrt{2}}{3}i \qquad Standard\ form$

The solution set is $\left\{\dfrac{1}{3} \pm \dfrac{\sqrt{2}}{3}i\right\}$.

37. $x^2 + 3x = -8$

Rewrite the equation in standard form.

$$x^2 + 3x + 8 = 0$$

Use the quadratic formula with
$a = 1$, $b = 3$, and $c = 8$.

$x = \dfrac{-b \pm \sqrt{b^2 - 4ac}}{2a}$

$x = \dfrac{-3 \pm \sqrt{3^2 - 4(1)(8)}}{2(1)}$

$= \dfrac{-3 \pm \sqrt{9 - 32}}{2} = \dfrac{-3 \pm \sqrt{-23}}{2}$

$= -\dfrac{3}{2} \pm \dfrac{\sqrt{23}}{2}i \qquad Standard\ form$

The solution set is $\left\{-\dfrac{3}{2} \pm \dfrac{\sqrt{23}}{2}i\right\}$.

38. $4q^2 + 2 = 3q$

$4q^2 - 3q + 2 = 0$ *Standard form*

Use the quadratic formula with
$a = 4$, $b = -3$, and $c = 2$.

$$q = \frac{-(-3) \pm \sqrt{(-3)^2 - 4(4)(2)}}{2 \cdot 4}$$

$$= \frac{3 \pm \sqrt{9 - 32}}{8} = \frac{3 \pm \sqrt{-23}}{8}$$

$$= \frac{3 \pm i\sqrt{23}}{8} = \frac{3}{8} \pm \frac{\sqrt{23}}{8}i$$

The solution set is $\left\{ \frac{3}{8} \pm \frac{\sqrt{23}}{8}i \right\}$.

39. $9z^2 + 2z + 1 = 0$

Use the quadratic formula with
$a = 9$, $b = 2$, and $c = 1$.

$$z = \frac{-b \pm \sqrt{b^2 - 4ac}}{2a}$$

$$z = \frac{-2 \pm \sqrt{2^2 - 4(9)(1)}}{2(9)}$$

$$= \frac{-2 \pm \sqrt{4 - 36}}{18} = \frac{-2 \pm \sqrt{-32}}{18}$$

$$= \frac{-2 \pm i\sqrt{32}}{18} = \frac{-2 \pm i\sqrt{16 \cdot 2}}{18}$$

$$= \frac{-2 \pm 4i\sqrt{2}}{18} = \frac{2\left(-1 \pm 2i\sqrt{2}\right)}{2 \cdot 9}$$

$$= \frac{-1 \pm 2i\sqrt{2}}{9} = -\frac{1}{9} \pm \frac{2\sqrt{2}}{9}i$$

The solution set is $\left\{ -\frac{1}{9} \pm \frac{2\sqrt{2}}{9}i \right\}$.

40. $y = -3x^2$

If $x = 0$, $y = 0$, so the y- and x-intercepts are
$(0, 0)$.

The x-value of the vertex is

$$x = -\frac{b}{2a} = -\frac{0}{2(-3)} = 0.$$

Thus, the vertex is the same as the y-intercept
(since $x = 0$). The axis of the parabola is the
vertical line $x = 0$.

Make a table of ordered pairs whose x-values are
on either side of the vertex's x-value of $x = 0$.

x	y
0	0
± 1	-3
± 2	-12
± 3	-27

Plot these seven ordered pairs and connect them
with a smooth curve.

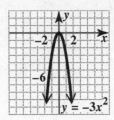

41. $y = -x^2 + 5$

If $x = 0$, $y = 5$, so the y-intercept is $(0, 5)$.

To find any x-intercepts, let $y = 0$.

$$0 = -x^2 + 5$$
$$x^2 = 5$$
$$x = \pm\sqrt{5} \approx \pm 2.24$$

The x-intercepts are $\left(\pm\sqrt{5}, 0 \right)$.

The x-value of the vertex is

$$x = -\frac{b}{2a} = -\frac{0}{2(-1)} = 0.$$

Thus, the vertex is the same as the y-intercept
(since $x = 0$). The axis of the parabola is the
vertical line $x = 0$.

Make a table of ordered pairs whose x-values are
on either side of the vertex's x-value of $x = 0$.

x	y
0	5
± 1	4
± 2	1
± 3	-4

Plot these seven ordered pairs and connect them
with a smooth curve.

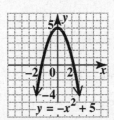

42. $y = (x + 4)^2 = x^2 + 8x + 16$

If $x = 0$, $y = 16$, so the y-intercept is $(0, 16)$.

To find any x-intercepts, let $y = 0$.

$$0 = (x + 4)^2$$
$$0 = x + 4$$
$$-4 = x$$

The x-intercept is $(-4, 0)$.

The x-value of the vertex is

$$x = -\frac{b}{2a} = -\frac{8}{2(1)} = -4.$$

Thus, the vertex is the same as the x-intercept. The axis of the parabola is the vertical line $x = -4$.

Make a table of ordered pairs whose x-values are on either side of the vertex's x-value of $x = -4$.

x	y
-7	9
-6	4
-5	1
-4	0
-3	1
-2	4
-1	9

Plot these seven ordered pairs and connect them with a smooth curve.

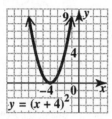

43. $y = x^2 - 2x + 1$

If $x = 0$, $y = 1$, so the y-intercept is $(0, 1)$.

Let $y = 0$ and solve for x.

$$0 = x^2 - 2x + 1$$
$$0 = (x - 1)^2$$
$$x - 1 = 0$$
$$x = 1$$

The x-intercept is $(1, 0)$.

The x-value of the vertex is

$$x = -\frac{b}{2a} = -\frac{-2}{2(1)} = 1.$$

The y-value of the vertex is

$$y = 1^2 - 2(1) + 1 = 0,$$

so the vertex is $(1, 0)$.

Make a table of ordered pairs whose x-values are on either side of the vertex's x-value of $x = 1$.

x	y
-1	4
0	1
1	0
2	1
3	4

Plot these five ordered pairs and connect them with a smooth curve.

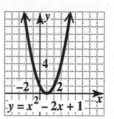

44. $y = -x^2 + 2x + 3$

If $x = 0$, $y = 3$, so the y-intercept is $(0, 3)$.

Let $y = 0$ and solve for x.

$$0 = -x^2 + 2x + 3$$
$$x^2 - 2x - 3 = 0$$
$$(x - 3)(x + 1) = 0$$

$$x - 3 = 0 \quad \text{or} \quad x + 1 = 0$$
$$x = 3 \quad \text{or} \quad x = -1$$

The x-intercepts are $(3, 0)$ and $(-1, 0)$.

The x-value of the vertex is

$$x = -\frac{b}{2a} = -\frac{2}{2(-1)} = 1.$$

The y-value of the vertex is

$$y = -1^2 + 2(1) + 3 = 4,$$

so the vertex is $(1, 4)$.

Make a table of ordered pairs whose x-values are on either side of the vertex's x-value of $x = 1$.

x	y
-1	0
0	3
1	4
2	3
3	0

Plot these five ordered pairs and connect them with a smooth curve.

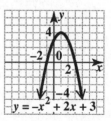

45. $y = x^2 + 4x + 2$

If $x = 0$, $y = 2$, so the y-intercept is $(0, 2)$.

Let $y = 0$ and solve for x.

$$x^2 + 4x + 2 = 0$$
$$x^2 + 4x = -2$$
$$x^2 + 4x + 4 = -2 + 4$$
$$(x + 2)^2 = 2$$

$$x + 2 = \sqrt{2} \quad \text{or} \quad x + 2 = -\sqrt{2}$$
$$x = -2 + \sqrt{2} \quad \text{or} \quad x = -2 - \sqrt{2}$$
$$x \approx -0.6 \quad \text{or} \quad x \approx -3.4$$

The x-intercepts are approximately $(-0.6, 0)$ and $(-3.4, 0)$.

The x-value of the vertex is

$$x = -\frac{b}{2a} = -\frac{4}{2(1)} = -2.$$

The y-value of the vertex is

$$y = (-2)^2 + 4(-2) + 2 = -2,$$

so the vertex is $(-2, -2)$.

Make a table of ordered pairs whose x-values are on either side of the vertex's x-value of $x = -2$.

x	y
-4	2
-3.4	0
-3	-1
-2	-2
-1	-1
-0.6	0
0	2

Plot these seven ordered pairs and connect them with a smooth curve.

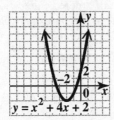

46. Since the graph has two x-intercepts, there are two real number solutions to the corresponding equation. Since the x-intercepts of the graph are $(-2, 0)$ and $(2, 0)$, the solution set of the equation is $\{\pm 2\}$.

The value of x can be any real number, so the domain of the function is $(-\infty, \infty)$. The values of y are at least -2, so the range of the function is $[-2, \infty)$.

47. The graph intersects the x-axis at its vertex, so it has one x-intercept. Thus, the corresponding equation has one real number solution. From the graph, the only x-intercept is $(2, 0)$, so the solution set of the equation is $\{2\}$.

The value of x can be any real number, so the domain of the function is $(-\infty, \infty)$. The values of y are at most 0, so the range of the function is $(-\infty, 0]$.

48. The graph has no x-intercepts, so the corresponding equation has no real number solution. The solution set is $\emptyset$.

The value of x can be any real number, so the domain of the function is $(-\infty, \infty)$. The values of y are at least 1, so the range of the function is $[1, \infty)$.

49. **[9.1]** $(2t - 1)(t + 1) = 54$

Write the equation in standard form.

$$2t^2 + t - 1 = 54$$
$$2t^2 + t - 55 = 0$$
$$(2t + 11)(t - 5) = 0 \quad \textit{Factor.}$$

$$2t + 11 = 0 \quad \text{or} \quad t - 5 = 0$$
$$t = -\frac{11}{2} \quad \text{or} \quad t = 5$$

The solution set is $\left\{ -\frac{11}{2}, 5 \right\}$.

50. **[9.1]** $(2p + 1)^2 = 100$

Use the square root property.

$$2p + 1 = \sqrt{100} \quad \text{or} \quad 2p + 1 = -\sqrt{100}$$
$$2p + 1 = 10 \quad \text{or} \quad 2p + 1 = -10$$
$$2p = 9 \quad \text{or} \quad 2p = -11$$
$$p = \frac{9}{2} \quad \text{or} \quad p = -\frac{11}{2}$$

The solution set is $\left\{ -\frac{11}{2}, \frac{9}{2} \right\}$.

51. **[9.3]** $(x + 2)(x - 1) = 3$

Write the equation in standard form.

$$x^2 + x - 2 = 3$$
$$x^2 + x - 5 = 0$$

The left side cannot be factored, so use the quadratic formula with $a = 1$, $b = 1$, and $c = -5$.

$$x = \frac{-b \pm \sqrt{b^2 - 4ac}}{2a}$$
$$x = \frac{-1 \pm \sqrt{1^2 - 4(1)(-5)}}{2(1)}$$
$$= \frac{-1 \pm \sqrt{1 + 20}}{2} = \frac{-1 \pm \sqrt{21}}{2}$$

The solution set is $\left\{ \frac{-1 \pm \sqrt{21}}{2} \right\}$.

52. **[9.1]** $6t^2 + 7t - 3 = 0$
$$(3t - 1)(2t + 3) = 0 \quad \textit{Factor.}$$

$$3t - 1 = 0 \quad \text{or} \quad 2t + 3 = 0$$
$$t = \frac{1}{3} \quad \text{or} \quad t = -\frac{3}{2}$$

The solution set is $\left\{ -\frac{3}{2}, \frac{1}{3} \right\}$.

53. **[9.3]** $2x^2 + 3x + 2 = x^2 - 2x$

Write the equation in standard form.

$$x^2 + 5x + 2 = 0$$

The left side cannot be factored, so use the quadratic formula with $a = 1$, $b = 5$, and $c = 2$.

$$x = \frac{-b \pm \sqrt{b^2 - 4ac}}{2a}$$

$$x = \frac{-5 \pm \sqrt{5^2 - 4(1)(2)}}{2(1)}$$

$$= \frac{-5 \pm \sqrt{25 - 8}}{2} = \frac{-5 \pm \sqrt{17}}{2}$$

The solution set is $\left\{ \frac{-5 \pm \sqrt{17}}{2} \right\}$.

54. **[9.3]** $x^2 + 2x + 5 = 7$

Write the equation in standard form.

$$x^2 + 2x - 2 = 0$$

The left side cannot be factored, so use the quadratic formula with $a = 1$, $b = 2$, and $c = -2$.

$$x = \frac{-2 \pm \sqrt{2^2 - 4(1)(-2)}}{2(1)}$$

$$= \frac{-2 \pm \sqrt{4 + 8}}{2} = \frac{-2 \pm \sqrt{12}}{2}$$

$$= \frac{-2 \pm 2\sqrt{3}}{2} = \frac{2\left(-1 \pm \sqrt{3}\right)}{2}$$

$$= -1 \pm \sqrt{3}$$

The solution set is $\left\{ -1 \pm \sqrt{3} \right\}$.

55. **[9.3]** $m^2 - 4m + 10 = 0$

Use the quadratic formula with $a = 1$, $b = -4$, and $c = 10$.

$$m = \frac{-(-4) \pm \sqrt{(-4)^2 - 4(1)(10)}}{2(1)}$$

$$= \frac{4 \pm \sqrt{16 - 40}}{2} = \frac{4 \pm \sqrt{-24}}{2}$$

Because $\sqrt{-24}$ does not represent a real number, the solution set is $\emptyset$.

56. **[9.3]** $k^2 - 9k + 10 = 0$

The left side cannot be factored, so use the quadratic formula with $a = 1$, $b = -9$, and $c = 10$.

$$k = \frac{-b \pm \sqrt{b^2 - 4ac}}{2a}$$

$$k = \frac{-(-9) \pm \sqrt{(-9)^2 - 4(1)(10)}}{2(1)}$$

$$= \frac{9 \pm \sqrt{81 - 40}}{2} = \frac{9 \pm \sqrt{41}}{2}$$

The solution set is $\left\{ \frac{9 \pm \sqrt{41}}{2} \right\}$.

57. **[9.1]** $(3x + 5)^2 = 0$

$$3x + 5 = 0$$

$$3x = -5$$

$$x = -\frac{5}{3}$$

The solution set is $\left\{ -\frac{5}{3} \right\}$.

58. **[9.3]** $0.5r^2 = 3.5 - r$

Multiply by 2 to clear fractions; then rewrite the result in standard form.

$$r^2 = 7 - 2r$$

$$r^2 + 2r - 7 = 0$$

The left side does not factor, so use the quadratic formula with $a = 1$, $b = 2$, and $c = -7$.

$$r = \frac{-2 \pm \sqrt{2^2 - 4(1)(-7)}}{2(1)}$$

$$= \frac{-2 \pm \sqrt{4 + 28}}{2} = \frac{-2 \pm \sqrt{32}}{2}$$

$$= \frac{-2 \pm 4\sqrt{2}}{2} = \frac{2\left(-1 \pm 2\sqrt{2}\right)}{2}$$

$$= -1 \pm 2\sqrt{2}$$

The solution set is $\left\{ -1 \pm 2\sqrt{2} \right\}$.

59. **[9.3]** $x^2 + 4x = 1$

$$x^2 + 4x - 1 = 0$$

The left side does not factor, so use the quadratic formula with $a = 1$, $b = 4$, and $c = -1$.

$$x = \frac{-4 \pm \sqrt{4^2 - 4(1)(-1)}}{2(1)}$$

$$= \frac{-4 \pm \sqrt{16 + 4}}{2} = \frac{-4 \pm \sqrt{20}}{2}$$

$$= \frac{-4 \pm 2\sqrt{5}}{2} = \frac{2\left(-2 \pm \sqrt{5}\right)}{2}$$

$$= -2 \pm \sqrt{5}$$

The solution set is $\left\{ -2 \pm \sqrt{5} \right\}$.

60. **[9.1]** $7x^2 - 8 = 5x^2 + 8$

$$2x^2 = 16$$

$$x^2 = 8$$

$$x = \pm \sqrt{8} = \pm 2\sqrt{2}$$

The solution set is $\left\{ \pm 2\sqrt{2} \right\}$.

61.   **[9.1]** $p = -(d-6)^2 + 10$

$\qquad\qquad 6 = -(d-6)^2 + 10 \quad Let\ p = 6.$

$\qquad\quad (d-6)^2 = 4$

$\qquad d-6 = \sqrt{4} \quad$ or $\quad d-6 = -\sqrt{4}$

$\qquad d-6 = 2 \quad$ or $\quad d-6 = -2$

$\qquad\quad d = 8 \quad$ or $\qquad\ d = 4$

Since the demand d is measured in hundreds, a demand of 400 or 800 cards produces a price of $6.

62.   **[9.5]** $p = -(d-6)^2 + 10$

$\qquad\qquad\ = -(d^2 - 12d + 36) + 10$

$\qquad\qquad\ = -d^2 + 12d - 36 + 10$

$\qquad\qquad\ = -d^2 + 12d - 26$

The d-value of the vertex is

$$d = -\frac{b}{2a} = -\frac{12}{2(-1)} = 6.$$

The p-value of the vertex is

$$p = -(6-6)^2 + 10 = 0 + 10 = 10.$$

So the vertex of the parabola is $(6, 10)$, which indicates that a demand of 600 cards ($d = 6$) produces a price of $10.

Chapter 9 Test

1. $x^2 = 39$

$$x = \sqrt{39} \quad \text{or} \quad x = -\sqrt{39}$$

The solution set is $\left\{ \pm\sqrt{39} \right\}$.

2. $(z+3)^2 = 64$

$\qquad z + 3 = \sqrt{64} \quad$ or $\quad z + 3 = -\sqrt{64}$

$\qquad z + 3 = 8 \quad$ or $\quad z + 3 = -8$

$\qquad\quad z = 5 \quad$ or $\qquad\ z = -11$

The solution set is $\{-11, 5\}$.

3. $(4x+3)^2 = 24$

$\qquad 4x + 3 = \sqrt{24} \quad$ or $\quad 4x + 3 = -\sqrt{24}$

Note that $\sqrt{24} = \sqrt{4 \cdot 6} = 2\sqrt{6}$.

$4x + 3 = 2\sqrt{6} \qquad$ or $\qquad 4x + 3 = -2\sqrt{6}$

$\quad 4x = -3 + 2\sqrt{6} \quad$ or $\qquad 4x = -3 - 2\sqrt{6}$

$\quad\ x = \dfrac{-3 + 2\sqrt{6}}{4} \quad$ or $\qquad x = \dfrac{-3 - 2\sqrt{6}}{4}$

The solution set is $\left\{ \frac{-3 \pm 2\sqrt{6}}{4} \right\}$.

4. $x^2 - 4x = 6$

$\quad x^2 - 4x + 4 = 6 + 4 \qquad Add\ \left[\frac{1}{2}(-4)\right]^2 = 4.$

$\qquad\quad (x-2)^2 = 10$

$\qquad x - 2 = \sqrt{10} \qquad$ or $\qquad x - 2 = -\sqrt{10}$

$\qquad\qquad x = 2 + \sqrt{10} \quad$ or $\qquad x = 2 - \sqrt{10}$

The solution set is $\left\{ 2 \pm \sqrt{10} \right\}$.

5. $2x^2 + 12x - 3 = 0$

$\qquad x^2 + 6x - \frac{3}{2} = 0$

$\qquad\quad x^2 + 6x = \frac{3}{2}$

$\quad x^2 + 6x + 9 = \frac{3}{2} + 9 \qquad Add\ \left[\frac{1}{2}(6)\right]^2 = 9.$

$\qquad\ (x+3)^2 = \frac{21}{2}$

$\quad x + 3 = \sqrt{\frac{21}{2}} \quad$ or $\quad x + 3 = -\sqrt{\frac{21}{2}}$

Note that

$$\sqrt{\frac{21}{2}} = \frac{\sqrt{21}}{\sqrt{2}} = \frac{\sqrt{21} \cdot \sqrt{2}}{\sqrt{2} \cdot \sqrt{2}} = \frac{\sqrt{42}}{2}.$$

$\quad x + 3 = \dfrac{\sqrt{42}}{2} \qquad$ or $\qquad x + 3 = -\dfrac{\sqrt{42}}{2}$

$\quad x = -3 + \dfrac{\sqrt{42}}{2} \quad$ or $\qquad x = -3 - \dfrac{\sqrt{42}}{2}$

$\quad x = \dfrac{-6 + \sqrt{42}}{2} \quad$ or $\qquad x = \dfrac{-6 - \sqrt{42}}{2}$

The solution set is $\left\{ \frac{-6 \pm \sqrt{42}}{2} \right\}$.

6. $5x^2 + 2x = 0$

Use $a = 5$, $b = 2$, and $c = 0$.

$$x = \frac{-b \pm \sqrt{b^2 - 4ac}}{2a}$$

$$x = \frac{-2 \pm \sqrt{2^2 - 4(5)(0)}}{2(5)}$$

$$= \frac{-2 \pm \sqrt{4}}{10} = \frac{-2 \pm 2}{10}$$

$$x = \frac{-2 + 2}{10} = \frac{0}{10} = 0$$

$$\text{or} \quad x = \frac{-2 - 2}{10} = \frac{-4}{10} = -\frac{2}{5}$$

The solution set is $\left\{ -\frac{2}{5}, 0 \right\}$.

7. $2x^2 + 5x - 3 = 0$

Use $a = 2$, $b = 5$, and $c = -3$.

$$x = \frac{-b \pm \sqrt{b^2 - 4ac}}{2a}$$

$$x = \frac{-5 \pm \sqrt{5^2 - 4(2)(-3)}}{2(2)}$$

$$= \frac{-5 \pm \sqrt{25 + 24}}{4}$$

$$= \frac{-5 \pm \sqrt{49}}{4} = \frac{-5 \pm 7}{4}$$

$$x = \frac{-5 + 7}{4} \quad \text{or} \quad x = \frac{-5 - 7}{4}$$

$$x = \frac{2}{4} = \frac{1}{2} \quad \text{or} \quad x = \frac{-12}{4} = -3$$

The solution set is $\left\{-3, \frac{1}{2}\right\}$.

8.
$$3w^2 + 2 = 6w$$
$$3w^2 - 6w + 2 = 0$$

Use $a = 3$, $b = -6$, and $c = 2$.

$$w = \frac{-(-6) \pm \sqrt{(-6)^2 - 4(3)(2)}}{2(3)}$$

$$= \frac{6 \pm \sqrt{36 - 24}}{6}$$

$$= \frac{6 \pm \sqrt{12}}{6} = \frac{6 \pm 2\sqrt{3}}{6}$$

$$= \frac{2\left(3 \pm \sqrt{3}\right)}{2(3)} = \frac{3 \pm \sqrt{3}}{3}$$

The solution set is $\left\{\frac{3 \pm \sqrt{3}}{3}\right\}$.

9. $4x^2 + 8x + 11 = 0$

Use $a = 4$, $b = 8$, and $c = 11$.

$$x = \frac{-8 \pm \sqrt{8^2 - 4(4)(11)}}{2(4)}$$

$$= \frac{-8 \pm \sqrt{64 - 176}}{8} = \frac{-8 \pm \sqrt{-112}}{8}$$

$$= \frac{-8 \pm i\sqrt{112}}{8} = \frac{-8 \pm i\sqrt{16 \cdot 7}}{8}$$

$$= \frac{-8 \pm 4i\sqrt{7}}{8} = \frac{4\left(-2 \pm i\sqrt{7}\right)}{4 \cdot 2}$$

$$= \frac{-2 \pm i\sqrt{7}}{2} = -1 \pm \frac{\sqrt{7}}{2}i$$

The solution set is $\left\{-1 \pm \frac{\sqrt{7}}{2}i\right\}$.

10.
$$t^2 - \frac{5}{3}t + \frac{1}{3} = 0$$
$$3\left(t^2 - \frac{5}{3}t + \frac{1}{3}\right) = 3(0)$$
$$3t^2 - 5t + 1 = 0$$

Use $a = 3$, $b = -5$, and $c = 1$.

$$t = \frac{-(-5) \pm \sqrt{(-5)^2 - 4(3)(1)}}{2(3)}$$

$$= \frac{5 \pm \sqrt{25 - 12}}{6} = \frac{5 \pm \sqrt{13}}{6}$$

The solution set is $\left\{\frac{5 \pm \sqrt{13}}{6}\right\}$.

11. $p^2 - 2p - 1 = 0$

Solve by completing the square.

$$p^2 - 2p = 1$$
$$p^2 - 2p + 1 = 1 + 1$$
$$(p - 1)^2 = 2$$

$$p - 1 = \sqrt{2} \quad \text{or} \quad p - 1 = -\sqrt{2}$$
$$p = 1 + \sqrt{2} \quad \text{or} \quad p = 1 - \sqrt{2}$$

The solution set is $\left\{1 \pm \sqrt{2}\right\}$.

12. $(2x + 1)^2 = 18$

Use the square root property.

$$2x + 1 = \pm\sqrt{18}$$
$$2x + 1 = \pm 3\sqrt{2}$$
$$2x = -1 \pm 3\sqrt{2}$$
$$x = \frac{-1 \pm 3\sqrt{2}}{2}$$

The solution set is $\left\{\frac{-1 \pm 3\sqrt{2}}{2}\right\}$.

13. $(x - 5)(2x - 1) = 1$
$$2x^2 - 11x + 5 = 1$$
$$2x^2 - 11x + 4 = 0$$

Use $a = 2$, $b = -11$, and $c = 4$.

$$x = \frac{-(-11) \pm \sqrt{(-11)^2 - 4(2)(4)}}{2(2)}$$

$$= \frac{11 \pm \sqrt{121 - 32}}{4} = \frac{11 \pm \sqrt{89}}{4}$$

The solution set is $\left\{\frac{11 \pm \sqrt{89}}{4}\right\}$.

14.
$$t^2 + 25 = 10t$$
$$t^2 - 10t + 25 = 0$$
$$(t - 5)^2 = 0$$
$$t - 5 = 0$$
$$t = 5$$

The solution set is $\{5\}$.

15. $s = -16t^2 + 64t$

$$-16t^2 + 64t = 64 \qquad \textit{Let } s = 64.$$
$$16t^2 - 64t + 64 = 0$$
$$t^2 - 4t + 4 = 0 \qquad \textit{Divide by 16.}$$
$$(t - 2)^2 = 0$$
$$t - 2 = 0$$
$$t = 2$$

The object will reach a height of 64 feet after 2 seconds.

16. Use the Pythagorean theorem.

$$c^2 = a^2 + b^2$$
$$(x+8)^2 = (x)^2 + (x+4)^2$$
$$x^2 + 16x + 64 = x^2 + x^2 + 8x + 16$$
$$0 = x^2 - 8x - 48$$
$$0 = (x-12)(x+4)$$

$$x - 12 = 0 \quad \text{or} \quad x + 4 = 0$$
$$x = 12 \quad \text{or} \qquad x = -4$$

Reject the negative length. The sides measure 12, $x + 4 = 12 + 4 = 16$, and $x + 8 = 12 + 8 = 20$.

17. $(3+i) + (-2+3i) - (6-i)$
$$= (3+i) + (-2+3i) + (-6+i)$$
$$= (3 - 2 - 6) + (1 + 3 + 1)i$$
$$= -5 + 5i$$

18. $(6+5i)(-2+i)$
$$= 6(-2) + 6(i) + 5i(-2) + 5i(i) \quad \textit{FOIL}$$
$$= -12 + 6i - 10i + 5i^2$$
$$= -12 - 4i + 5(-1)$$
$$= -12 - 4i - 5$$
$$= -17 - 4i$$

19. $(3-8i)(3+8i)$
$$= 3^2 - (8i)^2$$
$$= 9 - 64i^2$$
$$= 9 - 64(-1)$$
$$= 9 + 64 = 73$$

20. $\dfrac{15-5i}{7+i} = \dfrac{15-5i}{7+i} \cdot \dfrac{7-i}{7-i}$

$$= \frac{(15-5i)(7-i)}{(7+i)(7-i)}$$

$$= \frac{105 - 15i - 35i + 5i^2}{49 - i^2}$$

$$= \frac{105 - 50i + 5(-1)}{49 - (-1)}$$

$$= \frac{100 - 50i}{50}$$

$$= \frac{100}{50} - \frac{50}{50}i$$

$$= 2 - i \qquad \textit{Standard form}$$

21. $y = x^2 - 6x + 9 = (x-3)^2$

The x-value of the vertex is

$$x = -\frac{b}{2a} = -\frac{-6}{2(1)} = 3.$$

The y-value of the vertex is

$$y = (3-3)^2 = 0,$$

so the vertex is $(3, 0)$.

Make a table of ordered pairs whose x-values are on either side of the vertex's x-value of $x = 3$.

x	y
0	9
1	4
2	1
3	0
4	1
5	4
6	9

Plot these seven ordered pairs and connect them with a smooth curve.

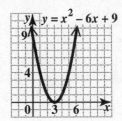

22. $y = -x^2 - 2x - 4$

The x-value of the vertex is

$$x = -\frac{b}{2a} = -\frac{-2}{2(-1)} = -1.$$

The y-value of the vertex is

$$y = -(-1)^2 - 2(-1) - 4 = -3,$$

so the vertex is $(-1, -3)$.

Make a table of ordered pairs whose x-values are on either side of the vertex's x-value of $x = -1$.

x	y
-3	-7
-2	-4
-1	-3
0	-4
1	-7

Plot these five ordered pairs and connect them with a smooth curve.

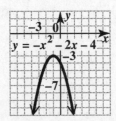

23. $f(x) = x^2 + 6x + 7$

The x-value of the vertex is

$$x = -\frac{b}{2a} = -\frac{6}{2(1)} = -3.$$

The y-value of the vertex is

$$y = (-3)^2 + 6(-3) + 7 = -2,$$

so the vertex is $(-3, -2)$.

Make a table of ordered pairs whose x-values are on either side of the vertex's x-value of $x = -3$.

x	y
-6	7
-5	2
-4	-1
-3	-2
-2	-1
-1	2
0	7

Plot these seven ordered pairs and connect them with a smooth curve.

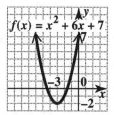

24. **(a)** Since the graph of

$$y = f(x) = x^2 + 6x + 7$$

has two x-intercepts, the equation $x^2 + 6x + 7 = 0$ has two real number solutions.

(b) $x^2 + 6x + 7 = 0$

Use the quadratic formula with $a = 1$, $b = 6$, and $c = 7$.

$$x = \frac{-6 \pm \sqrt{6^2 - 4(1)(7)}}{2(1)}$$

$$= \frac{-6 \pm \sqrt{36 - 28}}{2} = \frac{-6 \pm \sqrt{8}}{2}$$

$$= \frac{-6 \pm 2\sqrt{2}}{2} = \frac{2\left(-3 \pm \sqrt{2}\right)}{2}$$

$$= -3 \pm \sqrt{2}$$

The exact values of the solutions are $-3 + \sqrt{2}$ and $-3 - \sqrt{2}$, and the solution set is $\left\{-3 \pm \sqrt{2}\right\}$.

(c) $-3 - \sqrt{2} \approx -4.414$ and
$-3 + \sqrt{2} \approx -1.586$

25. Let $x =$ one of the numbers and $400 - x =$ the other number. The product P of the two numbers is given by

$$P = x(400 - x).$$

Writing this equation in standard form gives us

$$P = -x^2 + 400x.$$

Finding the maximum of the product is the same as finding the vertex of the graph of P. The x-value of the vertex is

$$x = -\frac{b}{2a} = -\frac{400}{2(-1)} = 200,$$

which makes sense because 200 is halfway between 0 and 400 (the x-intercepts of $P = x(400 - x)$).

If x is 200, then $400 - x$ must also be 200. The two numbers are 200 and 200 and the product is $200 \cdot 200 = 40{,}000$.

Cumulative Review Exercises (Chapters 1–9)

1. $\dfrac{-4 \cdot 3^2 + 2 \cdot 3}{2 - 4 \cdot 1} = \dfrac{-4 \cdot 9 + 6}{2 - 4}$

    $= \dfrac{-36 + 6}{-2} = \dfrac{-30}{-2} = 15$

2. $-9 - (-8)(2) + 6 - (6 + 2)$
    $= -9 - (-8)(2) + 6 - 8$
    $= -9 - (-16) + 6 - 8$
    $= -9 + 16 + 6 - 8$
    $= 7 + 6 - 8$
    $= 13 - 8 = 5$

3. $-4r + 14 + 3r - 7 = -r + 7$

4. $5(4m - 2) - (m + 7)$
    $= 5(4m - 2) - 1(m + 7)$
    $= 20m - 10 - m - 7$
    $= 19m - 17$

5. $x - 5 = 13$
    $x = 18$

    The solution set is $\{18\}$.

6. $3k - 9k - 8k + 6 = -64$
    $-14k + 6 = -64$
    $-14k = -70$
    $k = 5$

    The solution set is $\{5\}$.

7. $\frac{3}{5}t - \frac{1}{10} = \frac{3}{2}$
    Multiply each side by the LCD, 10.
    $10(\frac{3}{5}t - \frac{1}{10}) = 10(\frac{3}{2})$
    $6t - 1 = 15$
    $6t = 16$
    $t = \frac{16}{6} = \frac{8}{3}$

    The solution set is $\left\{\frac{8}{3}\right\}$.

8. $2(m - 1) - 6(3 - m) = -4$
    $2m - 2 - 18 + 6m = -4$
    $8m - 20 = -4$
    $8m = 16$
    $m = 2$

    The solution set is $\{2\}$.

9. Together, the two angles form a straight angle, so the sum of their measures is 180°.

$$(20x - 20) + (12x + 8) = 180$$
$$32x - 12 = 180$$
$$32x = 192$$
$$x = \frac{192}{32} = 6$$

If $x = 6$, $20x - 20 = 20(6) - 20$
$$= 120 - 20 = 100,$$

and $12x + 8 = 12(6) + 8$
$$= 72 + 8 = 80.$$

The measures of the angles are 100° and 80°.

10. Let $L =$ the length of the court.
Then $L - 44 =$ the width of the court.

Use the formula for the perimeter of a rectangle, $P = 2L + 2W$, with $P = 288$.

$$288 = 2L + 2(L - 44)$$
$$288 = 2L + 2L - 88$$
$$376 = 4L$$
$$94 = L$$

If $L = 94$, $L - 44 = 94 - 44 = 50$.
The length of the court is 94 feet and the width of the court is 50 feet.

11. Solve $P = 2L + 2W$ for L.

$$P - 2W = 2L$$
$$\frac{P - 2W}{2} = L, \quad \text{or} \quad L = \frac{P}{2} - W$$

12. $-8m < 16$

Divide each side by -8 and reverse the inequality symbol.

$$\frac{-8m}{-8} > \frac{16}{-8}$$
$$m > -2$$

The solution set is $(-2, \infty)$.

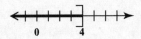

13. $-9p + 2(8 - p) - 6 \geq 4p - 50$
$$-9p + 16 - 2p - 6 \geq 4p - 50$$
$$-11p + 10 \geq 4p - 50$$
$$-15p \geq -60$$
$$\frac{-15p}{-15} \leq \frac{-60}{-15} \quad \textit{Divide by –15;}$$
$$\qquad\qquad\qquad \textit{reverse symbol.}$$
$$p \leq 4$$

The solution set is $(-\infty, 4]$.

14. $2x + 3y = 6$

Find the intercepts.

Let $x = 0$. $2(0) + 3y = 6$
$$3y = 6$$
$$y = 2$$

The y-intercept is $(0, 2)$.

Let $y = 0$. $2x + 3(0) = 6$
$$2x = 6$$
$$x = 3$$

The x-intercept is $(3, 0)$.

The graph is the line through the points $(0, 2)$ and $(3, 0)$.

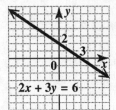

15. $y = 3$

For any value of x, the value of y will always be 3. Three ordered pairs are $(-2, 3)$, $(0, 3)$, and $(4, 3)$. Plot these points and draw a line through them. This will be a horizontal line.

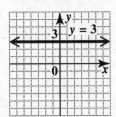

16. The slope m of the line passing through the points $(-1, 4)$ and $(5, 2)$ is

$$m = \frac{\text{change in } y}{\text{change in } x} = \frac{2 - 4}{5 - (-1)} = \frac{-2}{6} = -\frac{1}{3}.$$

17. Slope 2; y-intercept $(0, 3)$

Let $m = 2$ and $b = 3$ in slope-intercept form.

$$y = mx + b$$
$$y = 2x + 3$$

Now rewrite the equation in the form $Ax + By = C$.

$$-2x + y = 3$$
$$2x - y = -3 \quad \textit{Multiply by –1.}$$

18. $2x + y = -4$ (1)
     $-3x + 2y = 13$ (2)

Use the elimination method.

Multiply equation (1) by -2 and add the result to equation (2).

$$\begin{array}{rcrcr} -4x & - & 2y & = & 8 \\ -3x & + & 2y & = & 13 \\ \hline -7x & & & = & 21 \\ & & x & = & -3 \end{array}$$

To find y, substitute -3 for x in equation (1).

$$2x + y = -4$$
$$2(-3) + y = -4$$
$$-6 + y = -4$$
$$y = 2$$

The solution set is $\{(-3, 2)\}$.

19. $3x - 5y = 8$ (1)
     $-6x + 10y = 16$ (2)

Use the elimination method.

Multiply equation (1) by 2 and add the result to equation (2).

$$\begin{array}{rcrcr} 6x & - & 10y & = & 16 \\ -6x & + & 10y & = & 16 \\ \hline & & 0 & = & 32 \quad \textit{False} \end{array}$$

The false statement indicates that the solution set is $\emptyset$.

20.   Let x = the price of an AT&T phone and
      y = the price of a jWIN phone.

We have the system

$$3x + 2y = 84.95 \quad (1)$$
$$2x + 3y = 89.95 \quad (2)$$

To solve the system by the elimination method, we multiply equation (1) by 2 and equation (2) by -3, and then add the results.

$$\begin{array}{rcrcr} 6x & + & 4y & = & 169.90 \\ -6x & - & 9y & = & -269.85 \\ \hline & & -5y & = & -99.95 \\ & & y & = & 19.99 \end{array}$$

To find the value of x, substitute 19.99 for y in equation (1).

$$3x + 2(19.99) = 84.95$$
$$3x + 39.98 = 84.95$$
$$3x = 44.97$$
$$x = 14.99$$

The price of a single AT&T phone is \$14.99, and the price of a single jWIN phone is \$19.99.

21. $2x + y \le 4$ (1)
      $x - y > 2$ (2)

For inequality (1), draw a solid boundary line through $(2, 0)$ and $(0, 4)$, and shade the side that includes the origin (since substituting 0 for x and 0 for y results in a true statement). For inequality (2), draw a dashed boundary line through $(2, 0)$ and $(0, -2)$, and shade the side that *does not* include the origin.

The solution of the system of inequalities is the intersection of these two shaded half-planes. It includes the solid line and excludes the dashed line.

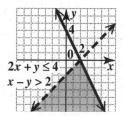

22. $\left(3^2 \cdot x^{-4}\right)^{-1} = \left(\dfrac{3^2}{x^4}\right)^{-1}$

$$= \left(\frac{x^4}{3^2}\right)^1 = \frac{x^4}{3^2}, \quad \text{or} \quad \frac{x^4}{9}$$

23. $\left(\dfrac{b^{-3}c^4}{b^5 c^3}\right)^{-2} = \left(b^{-3-5} c^{4-3}\right)^{-2}$

$$= \left(b^{-8} c^1\right)^{-2}$$
$$= \left(b^{-8}\right)^{-2} \left(c^1\right)^{-2}$$
$$= b^{16} c^{-2}$$
$$= b^{16} \cdot \frac{1}{c^2} = \frac{b^{16}}{c^2}$$

24. $\left(5x^5 - 9x^4 + 8x^2\right) - \left(9x^2 + 8x^4 - 3x^5\right)$

$$= 5x^5 - 9x^4 + 8x^2 - 9x^2 - 8x^4 + 3x^5$$
$$= 8x^5 - 17x^4 - x^2$$

25. $(2x - 5)(x^3 + 3x^2 - 2x - 4)$

Multiply vertically.

$$\begin{array}{rcrcrcrcr} & & x^3 & + & 3x^2 & - & 2x & - & 4 \\ & & & & & & 2x & - & 5 \\ \hline & & -5x^3 & - & 15x^2 & + & 10x & + & 20 \\ 2x^4 & + & 6x^3 & - & 4x^2 & - & 8x & & \\ \hline 2x^4 & + & x^3 & - & 19x^2 & + & 2x & + & 20 \end{array}$$

26. $\dfrac{3x^3 + 10x^2 - 7x + 4}{x + 4}$

$$
\begin{array}{r}
3x^2 - 2x + 1 \\
x + 4 \overline{\smash{\big)}\ 3x^3 + 10x^2 - 7x + 4} \\
\underline{3x^3 + 12x^2} \\
-2x^2 - 7x \\
\underline{-2x^2 - 8x} \\
x + 4 \\
\underline{x + 4} \\
0
\end{array}
$$

The remainder is 0, so the answer is the quotient, $3x^2 - 2x + 1$.

27. **(a)** $6{,}350{,}000{,}000 = 6.35 \times 10^9$

The decimal point was moved 9 places to the left.

(b) $2.3 \times 10^{-4} = 0.00023$

28. $16x^3 - 48x^2 y$
$= 16x^2(x - 3y)$ *GCF = 16x²*

29. $2a^2 - 5a - 3$

Use the grouping method. Look for two integers whose product is $2(-3) = -6$ and whose sum is -5. The integers are -6 and 1.

$$2a^2 - 5a - 3$$
$$= 2a^2 - 6a + 1a - 3$$
$$= 2a(a - 3) + 1(a - 3)$$
$$= (a - 3)(2a + 1)$$

30. $16x^4 - 1$
$= (4x^2 + 1)(4x^2 - 1)$ *Difference of squares*
$= (4x^2 + 1)(2x + 1)(2x - 1)$ *(again)*

31. $25m^2 - 20m + 4$

Since $25m^2 = (5m)^2$, $4 = 2^2$, and $20m = 2(5m)(2)$, $25m^2 - 20m + 4$ is a perfect square trinomial.

$$25m^2 - 20m + 4$$
$$= (5m)^2 - 2(5m)(2) + (2)^2$$
$$= (5m - 2)^2$$

32. $x^2 + 3x - 54 = 0$
$(x + 9)(x - 6) = 0$

$x + 9 = 0$ or $x - 6 = 0$
$x = -9$ or $x = 6$

The solution set is $\{-9, 6\}$.

33. Let x represent the width of the rectangle. Then $2.5x$ represents the length.

Use the formula for the area of a rectangle.

$$\mathcal{A} = LW$$
$$1000 = (2.5x)x \qquad \textit{Let } \mathcal{A} = 1000.$$
$$1000 = 2.5x^2$$
$$x^2 = \frac{1000}{2.5} = 400$$
$$x = \pm\sqrt{400} = \pm 20$$

Reject $x = -20$ since the width cannot be negative. The width is 20 meters, so the length is $2.5(20) = 50$ meters.

34. $\dfrac{2}{a - 3} \div \dfrac{5}{2a - 6}$

$= \dfrac{2}{a - 3} \cdot \dfrac{2a - 6}{5}$ *Multiply by reciprocal of divisor.*

$= \dfrac{2}{a - 3} \cdot \dfrac{2(a - 3)}{5}$ *Factor.*

$= \dfrac{4(a - 3)}{(a - 3)5}$ *Multiply.*

$= \dfrac{4}{5}$ *Lowest terms*

35. $\dfrac{1}{k} - \dfrac{2}{k - 1}$

$= \dfrac{1(k - 1)}{k(k - 1)} - \dfrac{2(k)}{(k - 1)k}$ *LCD = k(k − 1)*

$= \dfrac{(k - 1) - 2k}{k(k - 1)}$ *Subtract numerators.*

$= \dfrac{-k - 1}{k(k - 1)}$ *Combine terms.*

36. $\dfrac{2}{a^2 - 4} + \dfrac{3}{a^2 - 4a + 4}$

$= \dfrac{2}{(a + 2)(a - 2)} + \dfrac{3}{(a - 2)(a - 2)}$

Factor denominators.

$= \dfrac{2(a - 2)}{(a + 2)(a - 2)(a - 2)}$
$\quad + \dfrac{3(a + 2)}{(a - 2)(a - 2)(a + 2)}$

LCD = (a + 2)(a − 2)(a − 2)

$= \dfrac{2(a - 2) + 3(a + 2)}{(a + 2)(a - 2)(a - 2)}$

Add numerators.

$= \dfrac{2a - 4 + 3a + 6}{(a + 2)(a - 2)(a - 2)}$

Distributive property

$= \dfrac{5a + 2}{(a + 2)(a - 2)(a - 2)}$

Combine terms.

$= \dfrac{5a + 2}{(a + 2)(a - 2)^2}$

37. $\dfrac{\dfrac{1}{a}+\dfrac{1}{b}}{\dfrac{1}{a}-\dfrac{1}{b}}=\dfrac{ab\left(\dfrac{1}{a}+\dfrac{1}{b}\right)}{ab\left(\dfrac{1}{a}-\dfrac{1}{b}\right)}$

Multiply each term by the LCD, ab.

$$=\frac{ab\left(\frac{1}{a}\right)+ab\left(\frac{1}{b}\right)}{ab\left(\frac{1}{a}\right)-ab\left(\frac{1}{b}\right)}$$

$$=\frac{b+a}{b-a}$$

38. $\dfrac{1}{x+3}+\dfrac{1}{x}=\dfrac{7}{10}$

Multiply each side by the least common denominator, $10x(x+3)$.

$$10x(x+3)\left(\frac{1}{x+3}+\frac{1}{x}\right)=10x(x+3)\left(\frac{7}{10}\right)$$

$$10x(x+3)\left(\frac{1}{x+3}\right)+10x(x+3)\left(\frac{1}{x}\right)$$
$$=10x(x+3)\left(\tfrac{7}{10}\right)$$

$$10x+10(x+3)=7x(x+3)$$
$$10x+10x+30=7x^2+21x$$
$$20x+30=7x^2+21x$$
$$0=7x^2+x-30$$
$$0=(7x+15)(x-2)$$

$$7x+15=0 \quad \text{or} \quad x-2=0$$
$$x=-\tfrac{15}{7} \quad \text{or} \quad x=2$$

The solution set is $\left\{-\tfrac{15}{7},2\right\}$.

39. $\sqrt{100}=10$ since $10^2=100$ and $\sqrt{100}$ represents the positive square root.

40. $\dfrac{6\sqrt{6}}{\sqrt{5}}=\dfrac{6\sqrt{6}\cdot\sqrt{5}}{\sqrt{5}\cdot\sqrt{5}}=\dfrac{6\sqrt{30}}{5}$

41. $\sqrt[3]{\dfrac{7}{16}}=\dfrac{\sqrt[3]{7}}{\sqrt[3]{16}}$

Since $16=2^4$, we need to multiply by $\sqrt[3]{2^2}$ to get $\sqrt[3]{2^6}$ (6 is a multiple of 3).

$$\frac{\sqrt[3]{7}}{\sqrt[3]{16}}=\frac{\sqrt[3]{7}\cdot\sqrt[3]{4}}{\sqrt[3]{16}\cdot\sqrt[3]{4}}$$

$$=\frac{\sqrt[3]{28}}{\sqrt[3]{64}}=\frac{\sqrt[3]{28}}{4}$$

42. $3\sqrt{5}-2\sqrt{20}+\sqrt{125}$
$$=3\sqrt{5}-2\sqrt{4\cdot5}+\sqrt{25\cdot5}$$
$$=3\sqrt{5}-2\cdot2\sqrt{5}+5\sqrt{5}$$
$$=3\sqrt{5}-4\sqrt{5}+5\sqrt{5}$$
$$=(3-4+5)\sqrt{5}=4\sqrt{5}$$

43. $\sqrt{x+2}=x-4$
$$\left(\sqrt{x+2}\right)^2=(x-4)^2$$
$$x+2=x^2-8x+16$$
$$0=x^2-9x+14$$
$$0=(x-7)(x-2)$$

$$x-7=0 \quad \text{or} \quad x-2=0$$
$$x=7 \quad \text{or} \quad x=2$$

Check $x=7$: $\sqrt{9}=3$ *True*
Check $x=2$: $\sqrt{4}=-2$ *False*

The solution set is $\{7\}$.

44. **(a)** $8^{2/3}=\left(\sqrt[3]{8}\right)^2=(2)^2=4$

     (b) $-16^{1/4}=-\sqrt[4]{16}=-2$

45. $(3x+2)^2=12$
$$3x+2=\pm\sqrt{12}$$
$$3x+2=\pm2\sqrt{3}$$
$$3x=-2\pm2\sqrt{3}$$
$$x=\frac{-2\pm2\sqrt{3}}{3}$$

The solution set is $\left\{\frac{-2\pm2\sqrt{3}}{3}\right\}$.

46. $-x^2+5=2x$
$$x^2-5=-2x$$
$$x^2+2x=5$$
$$x^2+2x+1=5+1 \qquad \textit{Add } \left[\tfrac{1}{2}(2)\right]^2=1.$$
$$(x+1)^2=6$$
$$x+1=\pm\sqrt{6}$$
$$x=-1\pm\sqrt{6}$$

The solution set is $\left\{-1\pm\sqrt{6}\right\}$.

47. $2x(x-2)-3=0$
$$2x^2-4x-3=0$$

Use the quadratic formula with $a=2$, $b=-4$, and $c=-3$.

$$x=\frac{-(-4)\pm\sqrt{(-4)^2-4(2)(-3)}}{2(2)}$$

$$=\frac{4\pm\sqrt{16+24}}{4}=\frac{4\pm\sqrt{40}}{4}$$

$$=\frac{4\pm2\sqrt{10}}{4}=\frac{2\left(2\pm\sqrt{10}\right)}{2\cdot2}$$

$$=\frac{2\pm\sqrt{10}}{2}$$

The solution set is $\left\{\frac{2\pm\sqrt{10}}{2}\right\}$.

48. **(a)** $(-9 + 3i) + (4 + 2i) - (-5 - 3i)$

$\quad = (-9 + 3i) + (4 + 2i) + (5 + 3i)$

$\quad = (-9 + 4 + 5) + (3 + 2 + 3)i$

$\quad = 0 + 8i = 8i$

(b) $\dfrac{-17 - i}{-3 + i}$

$\quad = \dfrac{-17 - i}{-3 + i} \cdot \dfrac{-3 - i}{-3 - i}$

$\quad = \dfrac{(-17 - i)(-3 - i)}{(-3 + i)(-3 - i)}$

$\quad = \dfrac{51 + 17i + 3i + i^2}{9 - i^2}$

$\quad = \dfrac{51 + 20i + (-1)}{9 - (-1)}$

$\quad = \dfrac{50 + 20i}{10} = \dfrac{50}{10} + \dfrac{20}{10}i$

$\quad = 5 + 2i$ $\qquad$ *Standard form*

49. $\qquad 2x^2 + 2x = -9$

$\qquad 2x^2 + 2x + 9 = 0$

Use the quadratic formula with
$a = 2$, $b = 2$, and $c = 9$.

$x = \dfrac{-2 \pm \sqrt{2^2 - 4(2)(9)}}{2(2)}$

$\quad = \dfrac{-2 \pm \sqrt{4 - 72}}{4} = \dfrac{-2 \pm \sqrt{-68}}{4}$

$\quad = \dfrac{-2 \pm i\sqrt{68}}{4} = \dfrac{-2 \pm i \cdot 2\sqrt{17}}{4}$

$\quad = \dfrac{2\left(-1 \pm i\sqrt{17}\right)}{4} = \dfrac{-1 \pm i\sqrt{17}}{2}$

The solution set is $\left\{ -\frac{1}{2} \pm \frac{\sqrt{17}}{2}i \right\}$.

50. $f(x) = -x^2 - 2x + 1$

If $x = 0$, then $y = 1$, and the y-intercept is $(0, 1)$.

If $y = 0$, then $0 = -x^2 - 2x + 1$.

Use the quadratic formula with
$a = -1$, $b = -2$, and $c = 1$.

$x = \dfrac{-(-2) \pm \sqrt{(-2)^2 - 4(-1)(1)}}{2(-1)}$

$\quad = \dfrac{2 \pm \sqrt{8}}{-2} = \dfrac{2 \pm 2\sqrt{2}}{-2}$

$\quad = \dfrac{2\left(1 \pm \sqrt{2}\right)}{-2} = -1 \pm \sqrt{2}$

The x-intercepts are $(-1 + \sqrt{2}, 0)$ and $(-1 - \sqrt{2}, 0)$, which are approximately $(0.4, 0)$ and $(-2.4, 0)$.

The x-value of the vertex is

$x = -\dfrac{b}{2a} = -\dfrac{-2}{2(-1)} = -1$.

The y-value of the vertex is

$f(-1) = -(-1)^2 - 2(-1) + 1 = 2$,

so the vertex is $(-1, 2)$.

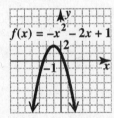

The value of x can be any real number, so the domain is $(-\infty, \infty)$. The value of y can be at most 2, so the range is $(-\infty, 2]$.

APPENDIX A   SETS

Appendix A Now Try Exercises

N1. The set of *odd* natural numbers less than 13 is

$$\{1, 3, 5, 7, 9, 11\}.$$

N2. (a) The set of negative integers
One way to list the elements is
$\{-1, -2, -3, -4, \ldots\}$. The set is *infinite*.

(b) The set of even natural numbers between 11 and 19 is $\{12, 14, 16, 18\}$. The set is *finite*.

N3. (a) $B \subseteq A$ is *true* since all the elements of $B = \{1, 5, 7, 9\}$ are also elements of $A = \{1, 3, 5, 7, 9, 11\}$.

(b) $C \subseteq B$ is *false* since $C = \{1, 9, 11\}$ has an element, 11, which is not in B.

(c) $C \nsubseteq A$ (C is *not* a subset of A) is *false* since all the elements of C are also elements of A.

N4. $U = \{2, 4, 6, 8, 10, 12, 14\}$

$M = \{2, 10, 12, 14\}$, so $M' = \{4, 6, 8\}$, which is the set of elements that are in U but not in M.

N5. $M = \{1, 3, 5, 7, 9\}$ and $N = \{0, 3, 6, 9\}$, so $M \cup N = \{0, 1, 3, 5, 6, 7, 9\}$, the set of elements that are in *either M or N*.

N6. $M = \{1, 3, 5, 7, 9\}$ and $N = \{0, 3, 6, 9\}$, so $M \cap N = \{3, 9\}$, the set of elements that are in *both M and N*.

N7. $U = \{1, 2, 4, 5, 7, 8, 9, 10\}$, $A = \{1, 4, 7, 9, 10\}$, $B = \{2, 5, 8\}$,   and $C = \{5\}$.

(a) $B \cup C = \{2, 5, 8\} = B$, the set of elements that are in either B or C.

(b) $A \cap B = \emptyset$ (the empty set), the set of elements that are in both A and B.

(c) $C' = \{1, 2, 4, 7, 8, 9, 10\}$, the set of elements that are in U but not in C.

Appendix A Section Exercises

1. The set of all natural numbers less than 8 is

$$\{1, 2, 3, 4, 5, 6, 7\}.$$

3. The set of seasons is

$$\{\text{winter, spring, summer, fall}\}.$$

The seasons may be written in any order within the braces.

5. To date, there have been no women presidents, so this set is the empty set, written $\emptyset$, or $\{\ \}$.

7. The set of letters of the alphabet between K and M is $\{L\}$.

9. The set of positive even integers is

$$\{2, 4, 6, 8, 10, \ldots\}.$$

11. The sets in Exercises 9 and 10 are infinite, since each contains an unlimited number of elements.

13. $5 \in \{1, 2, 5, 8\}$ ▪ 5 is an element of the set, so the statement is true.

15. $2 \in \{1, 3, 5, 7, 9\}$ ▪ 2 is not an element of the set, so the statement is false.

17. $7 \notin \{2, 4, 6, 8\}$ ▪ 7 is not an element of the set, so the statement is true.

19. $\{2, 4, 9, 12, 13\} = \{13, 12, 9, 4, 2\}$ ▪ The two sets have exactly the same elements, so they are equal. The statement is true. (The order in which the elements are written does not matter.)

21. $A \subseteq U$ ▪

Since all the elements of $A = \{1, 3, 4, 5, 7, 8\}$ are elements of $U = \{1, 2, 3, 4, 5, 6, 7, 8, 9, 10\}$, the statement $A \subseteq U$ is true.

23. $\emptyset \subseteq A$ ▪ Since the empty set contains no elements, the empty set is a subset of every set. The statement $\emptyset \subseteq A$ is true.

25. $C \subseteq A$ ▪

Since all the elements of $C = \{1, 3, 5, 7\}$ are elements of $A = \{1, 3, 4, 5, 7, 8\}$, the statement $C \subseteq A$ is true.

27. $D \subseteq B$ ▪

Since 1 and 3 are elements of $D = \{1, 2, 3\}$ but are not elements of $B = \{2, 4, 6, 8\}$, the statement $D \subseteq B$ is false.

29. $D \nsubseteq E$ ▪

Since 1 and 2 are elements of $D = \{1, 2, 3\}$ and are not elements of $E = \{3, 7\}$, D is not a subset of E, so the statement $D \nsubseteq E$ is true.

31. There are exactly 4 subsets of E. ▪

Since $E = \{3, 7\}$ has 2 elements, the number of subsets is $2^2 = 4$. The statement is true.

33. There are exactly 12 subsets of C. ▪

Since $C = \{1, 3, 5, 7\}$ has 4 elements, the number of subsets is $2^4 = 16$. The statement is false.

35. $\{4, 6, 8, 12\} \cap \{6, 8, 14, 17\} = \{6, 8\}$ ▪

The symbol $\cap$ means the intersection of the two sets, which is the set of elements that belong to both sets. Since 6 and 8 are the only elements belonging to both sets, the statement is true.

37. $\{3, 1, 0\} \cap \{0, 2, 4\} = \{0\}$ ▪ Only 0 belongs to both sets, so the statement is true.

39. $\{3, 9, 12\} \cap \emptyset = \{3, 9, 12\}$ ■

Since 3, 9, and 12 are not elements of the empty set, they are not in the intersection of the two sets. The intersection of any set with the empty set is the empty set. The statement is false.

41. $\{3, 5, 7, 9\} \cup \{4, 6, 8\} = \emptyset$ ■

The union of the two sets is the set of all elements that belong to either one of the sets or to both sets. Thus, $\{3, 5, 7, 9\} \cup \{4, 6, 8\} = \{3, 4, 5, 6, 7, 8, 9\} \neq \emptyset$. The statement is false.

43. $\{4, 9, 11, 7, 3\} \cup \{1, 2, 3, 4, 5\}$

$$= \{1, 2, 3, 4, 5, 7, 9, 11\}$$ ■

The union of the two sets is the set of all elements that belong to either one of the sets or to both sets. The statement is true.

45. A' ■ $U = \{a, b, c, d, e, f, g, h\}$
    $A = \{a, b, c, d, e, f\}$

A' contains all elements in U that are not in A, so

$$A' = \{g, h\}.$$

47. C' ■ $U = \{a, b, c, d, e, f, g, h\}$
    $C = \{a, f\}$

C' contains all elements in U that are not in C, so

$$C' = \{b, c, d, e, g, h\}.$$

49. $A \cap B$ ■ $A = \{a, b, c, d, e, f\}, B = \{a, c, e\}$

The intersection of A and B is the set of all elements belonging to both A and B, so

$$A \cap B = \{a, c, e\} = B.$$

51. $A \cap D$ ■ $A = \{a, b, c, d, e, f\}, D = \{d\}$

Since d is the only element in both A and D,

$$A \cap D = \{d\} = D.$$

53. $B \cap C$ ■ $B = \{a, c, e\}, C = \{a, f\}$

Since a is the only element that belongs to both sets, the intersection is the set with a as its only element, so

$$B \cap C = \{a\}.$$

55. $B \cup D$ ■ $B = \{a, c, e\}, D = \{d\}$

The union of B and D is the set of elements belonging to either B or D or both, so

$$B \cup D = \{a, c, d, e\}.$$

57. $C \cup B$ ■ $C = \{a, f\}, B = \{a, c, e\}$

The union of C and B is the set of elements belonging to either C or B or both, so

$$C \cup B = \{a, c, e, f\}.$$

59. $A \cap \emptyset$ ■ Since $\emptyset$ has no elements, there is no element that belongs to both A and $\emptyset$, so the intersection is the empty set. Thus,

$$A \cap \emptyset = \emptyset.$$

61. $A = \{a, b, c, d, e, f\}$ $C = \{a, f\}$
    $B = \{a, c, e\}$ $D = \{d\}$

Disjoint sets are sets which have no elements in common.

B and D are disjoint since they have no elements in common. Also, C and D are disjoint since they have no elements in common.